MW00814667

Elements of Fluid Mechanics

First Edition

by

David C. Wilcox

Elements of Fluid Mechanics

First Printing: August, 2005

DCW Industries, Inc.
5354 Palm Drive, La Cañada, California 91011
818/790-3844 (FAX) 818/952-1272
World Wide Web: http://www.dcwindustries.com

This book was prepared with LaTeX as implemented by Personal TeX, Inc. of Mill Valley, California. It was printed and bound in the United States of America by Birmingham Press, Inc., San Diego, California.

Library of Congress Cataloging in Publication Data

Wilcox, David C.
   Elements of Fluid Mechanics/David C. Wilcox—1st ed.
   Includes bibliography and index.
   1. Fluid Mechanics.
   2. Fluid Dynamics.
Catalog Card Number 2005904671

ISBN   1-928729-17-7

Dedicated to Dad
for teaching me about the glory of learning

Made in the USA

About the Author

Dr. David C. Wilcox, was born in Wilmington, Delaware. He did his undergraduate studies from 1963 to 1966 at the Massachusetts Institute of Technology, graduating with a Bachelor of Science degree in Aeronautics and Astronautics. From 1966 to 1967, he was employed by the McDonnell Douglas Aircraft Division in Long Beach, California, and began his professional career under the guidance of A. M. O. Smith. His experience with McDonnell Douglas focused on subsonic and transonic flow calculations. From 1967 to 1970, he attended the California Institute of Technology, graduating with a Ph.D. in Aeronautics. In 1970 he joined TRW Systems, Inc. in Redondo Beach, California, where he performed studies of both high- and low-speed fluid-mechanical and heat-transfer problems, such as turbulent hypersonic flow and thermal radiation from a flame. From 1972 to 1973, he was a staff scientist for Applied Theory, Inc., in Los Angeles, California. He participated in many research efforts involving numerical computation and analysis of fluid flows such as separated turbulent flow, transitional flow and hypersonic plume-body interaction.

In 1973, he founded DCW Industries, Inc., a La Cañada, California firm engaged in engineering research, software development and publishing, for which he is currently the President. He has taught several fluid mechanics and applied mathematics courses at the University of Southern California and at the University of California, Los Angeles.

Dr. Wilcox has numerous publications on turbulence modeling, computational fluid dynamics, boundary-layer separation, boundary-layer transition, thermal radiation, and rapidly rotating fluids. His publications include numerous college text books, namely *Basic Fluid Mechanics*, *Study Guide for Basic Fluid Mechanics*, *Turbulence Modeling for CFD* and *Perturbation Methods in the Computer Age*.

He is an Associate Fellow of the American Institute of Aeronautics and Astronautics (AIAA) and has served as an Associate Editor for the AIAA Journal.

In order to obtain an up-to-date list of all known typographical errors in this book, please visit DCW Industries' Home Page on the Worldwide Web at **http://www.dcwindustries.com**.

Contents

Index of Home Experiments   **xiii**

Notation   **xv**

Preface   **xxi**

1   Introduction   **1**
   1.1   Dimensions and Units 5
     1.1.1   Independent Dimensions 5
     1.1.2   Common Systems of Units 5
     1.1.3   Converting Dimensional Quantities 9
   1.2   Definition of a Fluid 12
     1.2.1   Simplistic Definition of a Fluid 12
     1.2.2   Rigorous Definition of a Fluid 12
   1.3   Continuum Approximation 13
   1.4   Microscopic and Macroscopic Views 15
     1.4.1   Density 15
     1.4.2   Temperature 17
     1.4.3   Pressure 17
     1.4.4   Vapor Pressure 18
   1.5   Thermodynamic Properties of Gases 20
   1.6   Compressibility 21
   1.7   Surface Tension 23
   1.8   Viscosity 27
   1.9   Examples of Viscosity-Dominated Flows 29
     1.9.1   Couette Flow 29
     1.9.2   Hagen-Poiseuille Flow 31
   1.10   Fluid-Flow Regimes 32
   1.11   Brief History of Fluid Mechanics 33
     Problems 35

2   Dimensional Analysis   **45**
   2.1   Basic Premises 46
   2.2   Buckingham Π Theorem 49
     2.2.1   Finding Dimensionless Groupings 49
     2.2.2   Interpretation of Results 54
     2.2.3   Nonuniqueness of Results 57
     2.2.4   E. S. Taylor's Method 58

	2.3	Common Dimensionless Groupings	60
	2.4	Dynamic Similitude	61
		2.4.1 Similitude for an Airplane	62
		2.4.2 Similitude for a Ship	64
		Problems	66

3 Pressure   **81**

	3.1	Pressure on an Infinitesimal Element	82
	3.2	The Hydrostatic Relation	84
	3.3	Atmospheric Pressure Variation	86
	3.4	Manometry	89
	3.5	Bernoulli's Equation	91
	3.6	Velocity-Measurement Techniques	94
		3.6.1 Stagnation Points	94
		3.6.2 Pitot Tube	94
		3.6.3 Pitot-Static Tube	96
	3.7	Hydrostatic Forces on Plane Surfaces	97
	3.8	Hydrostatic Forces on Curved Surfaces	106
	3.9	Equivalent Pressure Fields and Superposition	110
	3.10	Buoyancy	112
		Problems	114

4 Kinematics   **133**

	4.1	Eulerian Versus Lagrangian Description	133
	4.2	Steady and Unsteady Flows	139
	4.3	Vorticity and Circulation	140
	4.4	Streamlines, Streaklines and Pathlines	145
	4.5	Extensive and Intensive Properties	148
	4.6	Surface Fluxes	149
	4.7	Reynolds Transport Theorem	151
	4.8	Vector Calculus and Multiple Integrals	156
		Problems	158

5 Conservation of Mass and Momentum: Integral Form   **169**

	5.1	Control-Volume Method	169
	5.2	Conservation of Mass	171
	5.3	Conservation of Momentum	172
	5.4	Preliminaries	173
		5.4.1 Overview of the Method	174
		5.4.2 Useful Control-Volume Theorem	177
		5.4.3 Fluid Statics Revisited	178
	5.5	Mass Conservation Applications	179
		5.5.1 Steady Flow in a Pipe	180
		5.5.2 Sphere Falling in a Cylinder	182
		5.5.3 Deforming Control Volume	186
	5.6	Mass and Momentum Conservation Applications	188
		5.6.1 Channel Flow With Suction	188
		5.6.2 Indirect Force Computation—Rotational Flow	191
		5.6.3 Indirect Force Computation—Irrotational Flow	194

	5.6.4	Reaction Forces	196
	5.6.5	Nonuniform Cross-Sectional Velocity Profiles	198
5.7	Accelerating Control Volume		200
	5.7.1	Inertial and Noninertial Coordinate Frames	200
	5.7.2	Ascending Rocket	202
	Problems		206

6 Conservation of Energy: Integral Form   **233**

6.1	Thermodynamics		234
	6.1.1	Fundamental Concepts	234
	6.1.2	The First Law of Thermodynamics	237
	6.1.3	The Second Law of Thermodynamics	237
	6.1.4	Combined First and Second Laws	238
	6.1.5	The First Law for a Moving Fluid	240
6.2	Integral Form of the Energy Equation		240
6.3	Approximate Form of the Energy Equation		244
6.4	Flow in Pipes		248
	6.4.1	Friction and Head Loss in Straight Pipes	249
	6.4.2	Minor Losses and Non-Circular Cross Sections	254
	6.4.3	Multiple-Pipe Systems	262
6.5	Open-Channel Flow		265
	6.5.1	Uniform Flow	269
	6.5.2	The Equations of Chézy and Manning	271
	6.5.3	Surface Wave Speed and Flow Classification	273
	6.5.4	Specific Energy	278
	6.5.5	Hydraulic Jump	281
	Problems		283

7 Turbomachinery   **301**

7.1	Classification		301
7.2	Conservation of Angular Momentum		303
	7.2.1	Derivation	303
	7.2.2	Lawn Sprinkler	304
7.3	The Euler Turbomachine Equations		306
7.4	Application to a Centrifugal Pump		310
	7.4.1	Velocity Triangles	310
	7.4.2	Computing Pump Properties	314
	7.4.3	Efficiency	316
	7.4.4	Performance Curves	316
7.5	Dimensional Considerations		317
	7.5.1	Primary Dimensionless Groupings	317
	7.5.2	Specific Speed	320
7.6	Common Turbomachines		320
	7.6.1	Pumps and Compressors	320
	7.6.2	Reaction and Impulse Turbines	322
	Problems		326

8 One-Dimensional Compressible Flow   **333**
  8.1 Classification 334
  8.2 Thermodynamic Properties of Air 335
  8.3 Speed of Sound 336
  8.4 Subsonic Versus Supersonic Flow 339
  8.5 Analysis of a Streamtube 340
  8.6 Total Conditions 344
  8.7 Normal Shock Waves 346
  8.8 Directionality 348
  8.9 Laval Nozzle 350
  8.10 Fanno Flow 355
      8.10.1 Fanno-Flow Equations 356
      8.10.2 Solution of the Fanno-Flow Equations 358
      8.10.3 Subsonic Inlet 360
      8.10.4 Supersonic Inlet 361
      8.10.5 Mollier Diagram 362
  8.11 Rayleigh Flow 363
      8.11.1 Rayleigh-Flow Equations 364
      8.11.2 Solution of the Rayleigh-Flow Equations 364
      8.11.3 Subsonic Inlet With Heating 365
      8.11.4 Supersonic Inlet With Heating 366
      8.11.5 Mollier Diagram 367
      Problems 370

9 Conservation Laws: Differential Form   **383**
  9.1 Reynolds Transport Theorem Revisited 384
  9.2 Conservation of Mass 386
  9.3 Conservation of Momentum 387
  9.4 Conservation of Energy 388
  9.5 Summary of the Conservation Equations 389
  9.6 Mass Conservation for Incompressible Flows 390
  9.7 Euler's Equation 391
      9.7.1 Rotating Tank 393
      9.7.2 Galilean Invariance of Euler's equation 395
  9.8 Entropy Generation 397
  9.9 Derivation of Bernoulli's Equation 398
  9.10 Bernoulli's Equation and Energy Conservation 401
      9.10.1 Flow of a Liquid 401
      9.10.2 Flow of a Gas 401
      Problems 404

10 Potential Flow   **413**
  10.1 Differential Equations of Motion 414
  10.2 Mathematical Foundation 414
      10.2.1 Cylindrical Coordinates 415
      10.2.2 Velocity-Potential Representation 416
      10.2.3 Streamfunction Representation 417
  10.3 Streamlines and Equipotential Lines 418
  10.4 Bernoulli's Equation 420

10.5 Fundamental Solutions 421
    10.5.1 Uniform Flow 422
    10.5.2 Potential Vortex 423
    10.5.3 Source and Sink 424
    10.5.4 Doublet 426
    10.5.5 Comments on Use of Fundamental Solutions 428
10.6 Flow Past a Cylinder 430
    10.6.1 The Basic Solution 430
    10.6.2 Force on the Cylinder 433
    10.6.3 Adding a Vortex 435
    10.6.4 Force Computation Redone 437
10.7 Other Interesting Solutions 442
    10.7.1 Accelerating Cylinder 442
    10.7.2 Rankine Oval 444
    10.7.3 Wedge 446
    10.7.4 Stagnation Point 448
    10.7.5 The Method of Images 450
    Problems 454

11 Vorticity, Viscosity, Lift and Drag   **465**
11.1 The Vortex Force 466
11.2 Helmholtz's Theorem and d'Alembert's Paradox 467
11.3 Boundary Conditions at a Solid Boundary 469
11.4 Viscous Effects and Vorticity Generation 470
11.5 Navier-Stokes Equation and Diffusion of Vorticity 473
11.6 Prandtl's Boundary Layer 473
    11.6.1 High-Reynolds-Number Flow Near a Surface 474
    11.6.2 Boundary-Layer Equations 475
    11.6.3 Momentum-Integral Equation 481
    11.6.4 Blasius Solution 486
    11.6.5 Effects of Pressure Gradient 490
    11.6.6 Boundary-Layer Separation 493
11.7 Turbulence 495
    11.7.1 Laminar, Transitional and Turbulent Flow 496
11.8 Lift and Drag of Common Objects 500
    Problems 506

A Fluid Properties   **519**
A.1 Perfect-Gas Properties 519
A.2 Pressure 519
A.3 Density 520
A.4 Compressibility and Speed of Sound 521
A.5 Surface Tension 521
A.6 Viscosity 522
A.7 Vapor Pressure 523

B Compressible Flow Tables   **525**
   B.1   Isentropic-Flow and Normal-Shock Relations 525
   B.2   Fanno Flow 531
   B.3   Rayleigh Flow 534

C Vector Differential Operator ∇   **537**
   C.1   Definition of ∇ 537
   C.2   Gradient, Divergence and Curl 537
   C.3   Directional Derivative 538
   C.4   Operations and Identities 538

D Equations of Motion in Various Coordinate Systems   **539**
   D.1   Rectangular Cartesian Coordinates 540
   D.2   Cylindrical Polar Coordinates 541
   D.3   Spherical Coordinates 542
   D.4   Rotating Coordinate System 543
   D.5   Natural (Streamline) Coordinates 544

   Answers to Selected Problems   **547**

   Bibliography   **555**

   Index   **559**

Index of Home Experiments

The book contains several straightforward experiments that can be performed without special laboratory equipment. The experiments can be found on the following pages.

Chapter	Topic	Page
1	Surface Tension	26
3	Bernoulli's Equation	92
3	Buoyancy	113
10	Magnus Force	441
11	Vorticity	486
11	Turbulence	500

Notation

This section includes the most commonly used notation in this book. In order to avoid departing too much from conventions normally used in literature on general fluid mechanics, a few symbols denote more than one quantity.

English Symbols

Symbol	Definition
a	Speed of sound
$\mathbf{a}$	Acceleration vector
A	Area; constant in Sutherland's law
b	Open-channel width; impeller-blade width
$B(t)$	Extensive variable for a system
c	Celerity
c_f	Skin friction, $\tau_w/(\frac{1}{2}\rho u_e^2)$
c_p, c_v	Specific heat at constant pressure, volume
C_D	Drag coefficient, $D/(\frac{1}{2}\rho U^2 A)$
C_H	Head coefficient
C_L	Lift coefficient, $L/(\frac{1}{2}\rho U^2 A)$
C_p	Pressure coefficient
C_P	Power coefficient
C_Q	Capacity coefficient; suction coefficient
d, D	Diameter
dS	Differential surface area
dV	Differential volume
D, D_f, D_p	Drag, friction drag, pressure drag
D_h	Hydraulic diameter
$\mathcal{D}$	Doublet strength
e	Specific internal energy
E	Total energy; open-channel specific energy
E_v	Bulk modulus, $E_v = 1/\tau$ where τ is compressibility
$\mathcal{E}$	Specific total energy
f	Friction factor, $\tau_w/(\frac{1}{8}\rho\overline{u}^2)$; frequency
$\mathbf{f}$	Body force vector per unit mass
$\mathbf{f}_v$	Viscous force vector per unit area
F	Force vector magnitude, $|\mathbf{F}|$
F_{buoy}	Buoyancy force
F_D	Drag force

F_H, F_V	Horizontal force, vertical force
Fr_L	Froude number based on length L
F_x, F_y, F_z	Force components in x, y, z directions
$\mathbf{F}$	Force vector
$\mathbf{F}_b$, $\mathbf{F}_s$	Body-force vector, surface-force vector
$\mathbf{F}_D$	Drag force vector
g, $\mathbf{g}$	Gravitational acceleration magnitude, vector
g_c	Acceleration constant for EE units
h	Specific enthalpy; height
h_L	Head loss
h_p	Head supplied by a pump
h_t	Head given up to a turbine
H	Total enthalpy; channel height; shape factor, δ^*/θ
$\mathcal{H}$	Specific total enthalpy
$\mathbf{H}$	Angular-momentum vector
$\mathbf{i}$, $\mathbf{j}$, $\mathbf{k}$	Unit vectors in x, y, z directions
I	Moment of inertia
k	Boltzmann's constant; thermal conductivity
k_s	Surface roughness height
K	Loss coefficient
Kn	Knudsen number
ℓ_e	Entrance length in a pipe
ℓ_{mfp}	Mean free path
L	Lift force; characteristic length scale
L_e	Equivalent length
$\mathbf{L}$	Lift vector
m	Mass
$\dot{m}$	Mass-flow rate
M	Mach number; total mass; characteristic mass scale
$\mathbf{M}$	Moment vector
n	Number density; Manning roughness coefficient
$\mathbf{n}$	Outer unit normal vector
N	Number of molecules
N_s	Specific speed
p	Static pressure
p_a	Atmospheric pressure
p_v	Vapor pressure
P	Perimeter
Pr	Prandtl number
$\mathbf{P}$	Momentum vector
q	Specific heat transfer; heat added per unit mass
$\mathbf{q}$	Heat-flux vector
Q	Volume-flow rate; heat transfer; source strength
$\dot{Q}$	Heat-transfer rate
r	Recovery factor
r, θ, z	Cylindrical polar coordinates
$\mathbf{r}$	Position vector
$\mathbf{r}_{cp}$	Center of pressure vector
R	Perfect-gas constant; radius

R_h	Hydraulic radius
R, θ, ϕ	Spherical coordinates
$\mathcal{R}$	Radius of curvature
$\mathbf{R}$	Reaction force vector
Re_L, Re_x	Reynolds number based on length L, x
Re_θ	Reynolds number based on momentum thickness
s	Distance along a streamline; specific entropy
S	Surface area; constant in Sutherland's law
S_o	Bottom slope in an open channel
St	Strouhal number
SG	Specific gravity
SW	Specific weight, ρg
t	Time
$\mathbf{t}$	Unit vector tangent to a surface or curve
T	Temperature; thrust; time; characteristic time scale
u, v, w	Cartesian velocity components in x, y, z directions
u_r, u_θ, w	Cylindrical velocity components in r, θ, z directions
u_R, u_θ, u_ϕ	Spherical velocity components in R, θ, ϕ directions
u_n, u_t	Velocity components in normal, tangential directions
$\mathbf{u}$	Velocity vector
u_τ	Friction velocity, $\sqrt{\tau_w/\rho}$
U, V, W	Velocity components in x, y, z directions
U_j	Jet velocity
U_m	Maximum or centerline velocity
$\mathbf{U}_s$	Velocity vector of a moving surface
v_w	Surface injection velocity
V	Volume
$\mathbf{V}, \mathbf{w}$	Absolute, relative velocity
$\mathcal{V}$	Body-force potential
w_e	Exit velocity relative to control volume
W	Weight; work
We	Weber number
W_R	Rocket velocity
$\dot{W}$	Rate at which work is done (power)
$\dot{W}_m$	Mechanical power
$\dot{W}_p$	Hydraulic power (power supplied by a pump)
$\dot{W}_s$	Shaft work, $\dot{W}_t - \dot{W}_p$
$\dot{W}_t$	Power given up to a turbine
x, y, z	Rectangular Cartesian coordinates
x_{cp}, z_{cp}	Center of pressure components
y	Open-channel depth
y_c	Critical depth
$\bar{z}$	Centroid

Greek Symbols

Symbol	Definition
α	Lapse rate; kinetic-energy correction factor; angle
β	Intensive variable; Falkner-Skan parameter; shock angle; blade angle
γ	Specific-heat ratio, c_p/c_v
Γ	Circulation
Γ_a	Aerodynamic circulation, $-\Gamma$
δ	Boundary-layer thickness
δ^*	Displacement thickness, $\int_0^\delta \left(1 - \frac{u}{u_e}\right) dy$
$\Delta x, \Delta y, \Delta z$	Infinitesimal-element size in x, y, z direction
ϵ	Small parameter
η	Similarity variable
η_p, η_t	Pump, turbine efficiency
θ	Angle; momentum thickness, $\int_0^\delta \frac{u}{u_e} \left(1 - \frac{u}{u_e}\right) dy$
Θ	Characteristic temperature scale
μ	Molecular viscosity
$\mu(M)$	Mach angle, $\sin^{-1}(1/M)$
ν	Kinematic viscosity, μ/ρ
ρ	Mass density
σ	Surface tension
τ	Shear stress; compressibility; time
τ_w	Surface shear stress
$\boldsymbol{\tau}$	Torque vector
υ	Specific volume, $1/\rho$
ϕ	Velocity potential; angle
ψ	Streamfunction
ω	Vorticity magnitude
$\boldsymbol{\omega}$	Vorticity vector
Ω	Angular-velocity magnitude
$\boldsymbol{\Omega}$	Angular-velocity vector

Subscripts

Symbol	Definition
1	Ahead of shock
2	Behind shock
avg	Average
e	Boundary-layer edge; exit plane
inc	Incompressible
$irrev$	Irreversible
m	Model; maximum
max	Maximum
o	Centerline; initial value; total (stagnation, reservoir)
p	Prototype
rev	Reversible
t	Total (stagnation, reservoir)

w	Wall (surface)
∞	Freestream

Superscripts

Symbol	Definition
cv	Control volume
rel	Relative
$*$	Throat (sonic-point); best-efficiency point

Other

Symbol	Definition
$<f>$	Statistical average of f
$\overline{f}$	Statistical average of f
f'	Fluctuating part of f
Im	Imaginary part

Preface

This book has been developed from the author's lecture notes used in presenting fluid mechanics courses at the University of Southern California and the University of California, Los Angeles. The primary goal of this book is to provide a rigorous and understandable introduction to the fascinating field of fluid mechanics with a classical point of view. While maintaining a commitment to mathematical rigor throughout, the text continually emphasizes the physics of fluid motion. Mathematical results are repeatedly reinforced and verified by appealing to physical arguments.

Chapter 1 familiarizes the student with the nomenclature of fluid mechanics as well as some commonly observed features of flowing fluids. Chapter 2 covers dimensional analysis, a topic that immediately preceded the rapid development of theoretical fluid mechanics in the twentieth century and helped organize the empirical relationships, experimental data and formulas devised in the hydraulics era of the nineteenth century.

Chapter 3 discusses the pressure field in a fluid, including effects of gravity. The chapter introduces the famous equation attributed to Bernoulli and its limiting form for a motionless fluid, the hydrostatic relation. It also covers straightforward methods that permit computation of forces and moments on submerged stationary structures such as dams.

Chapter 4 derives the all-important Reynolds Transport Theorem for a one-dimensional geometry and heuristically generalizes it for three-dimensional flows. Chapters 5 and 6 develop the integral forms of the basic conservation laws for a control volume in a straightforward manner using the Reynolds Transport Theorem.

Chapters 5 through 8 focus upon the classical control-volume method, which provides a global view of fluid mechanics. Using the integral form of the conservation laws for mass, momentum and energy, these chapters discuss general fluid-flow phenomena, typical flowfields and common fluid-mechanical devices.

Applying the integral form of the conservation laws to a differential-sized control volume, Chapter 9 shifts the emphasis to a detailed view of fluid mechanics through the differential form of the conservation laws. Chapters 9 through 11 illustrate the detailed view for low-speed flows.

The text provides worked examples for addressing all of the key concepts presented in this book. To assist the student in developing the ability to apply the concepts, many of the most complex homework problems at the end of each chapter have multiple parts that lead the student through a logical sequence of steps to develop the solution. In this sense, some of the homework problems can be viewed as pseudo-examples. Numerous homework problems require only an algebraic result. More often than not, the answers are even rigged to involve only rational numbers to help simplify the algebra. This is a reflection of a desire to stress the importance of understanding the physics, rather than improving the student's calculator skills.

It also permits emphasizing the importance of good engineering practices such as checking mathematical results for dimensional consistency and examining limiting cases for which properties of the solution are evident. Of course, some feel for the magnitudes of quantities of interest in practical fluid flows is needed, and problems have been included throughout the text that require a numerical answer.

The material presented in this book is appropriate for a one-term, junior or senior level undergraduate course. Successful study of this material requires an understanding of basic calculus and elementary mechanics. Some exposure to thermodynamics is helpful but not required. Chapter 6 provides an overview of the most important concepts needed from the science of thermodynamics.

This text contains more than enough material for a one-quarter course at most universities. This permits the instructor to choose the topics most pertinent to the curriculum. All students will benefit from the material presented in Chapters 1 through 6, 8, 9 and 11. These chapters provide the foundation for learning the classical control-volume method, the invaluable tool of dimensional analysis and the differential form of the conservation laws. This exposes the student to the primary techniques that have been used so successfully in developing both design methods and general theoretical tools.

The student will find practical applications of the control-volume method in Chapter 6, viz., pipe flow and open-channel flow. With its focus on turbomachinery, Chapter 7 provides a quintessential example of how effective the combination of dimensional analysis and the control-volume method is in developing an understanding of complex fluid flows. For a course that includes some exposure to solving the differential equations of motion, Chapter 10 explores the fascinating topic of potential flow.

As usual, my all-time favorite student, Christopher Landry, reviewed the manuscript, taking me to task whenever I failed to explain things clearly.

I thank the hundreds of undergraduate students at USC and UCLA who have had their introduction to fluid mechanics in my classes, especially for their many thought-provoking questions that have helped me improve my own understanding of fluid mechanics. It is largely at the urging of those students that I have written this text.

Finally, I owe yet another debt to my wonderful wife Barbara for helping me survive the transition from strictly aerospace research activities to all these other things I do well.

David C. Wilcox

NOTE: We have taken great pains to assure the accuracy of this manuscript. However, if you find errors, please report them to DCW Industries' Home Page on the Worldwide Web at **http://www.dcwindustries.com**. As long as we maintain a WWW page, we will provide an updated list of known typographical errors.

Chapter 1

Introduction

There is little doubt that from the days when humans gazed up at the sky above the warm savannas and cold caves they marveled at the ability of birds to fly. And in the evenings they surely wondered about the lights that dotted the sky. It cannot be otherwise for, as the philosopher Aristotle said, "All men by nature desire to know."

The invention of the telescope provided the first great advance toward discovering the nature of the universe. But, it provided no clues about how mankind might travel beyond the confines of the Earth and its atmosphere.

The twentieth century ushered in the era of flight. The century had barely begun when, on December 17, 1903, Orville and Wilbur Wright piloted the first heavier-than-air craft to fly under its own power. The Wright Flyer I, shown in Figure 1.1, was the result of several achievements by the Wright brothers, including: extensive and painstaking wind-tunnel tests; a new lightweight engine design; and an efficient propeller. While their first flight lasted only about 12 seconds, it proved that powered flight was possible.

Figure 1.1: *With their historic flight at Kitty Hawk, NC on December 17, 1903, Orville and Wilbur Wright accomplished a long-sought goal of mankind—flight. [Photographs courtesy of and © www.arttoday.com]*

Orville was the pilot of the first flight, and he lay next to the motor on the lower wing. Wilbur steadied the vehicle at one of the wing tips. After running along the ground for about 40 feet, the Wright Flyer lifted off of a coastal sand dune near Kitty Hawk, North Carolina. In the 12 seconds before it returned to the ground, the airplane had flown 120 feet. They made several flights on that historic day, with Wilbur piloting the longest—852 feet in 59 seconds.

Since that historic day, aviation has brought us faster and more efficient vehicles that permit mankind to "soar with the eagles." Civil aviation developed rapidly during the next 40 years and by the end of World War II in 1945, estimates indicate that at least a million people had flown in an airplane.

Just 44 years after the Wright brothers' first powered flight, United States Air Force test pilot Chuck Yeager demonstrated that it is possible to fly at speeds in excess of the speed of sound, an issue that was very much in doubt at the time. He was selected from a group of volunteers to test the experimental X-1 aircraft. The goal of the X-1 was determine whether or not an airplane could fly faster than the speed of sound, and if a pilot could control the plane despite the effects of shock waves.

The flight took place on October 14, 1947 above Rogers Dry Lake in California. The X-1, depicted in Figure 1.2, was launched from a B-29 mother ship at an altitude of 25,000 feet. Yeager then flew the rocket-powered X-1 to an altitude of 43,000 feet. He succeeded in flying at a Mach number of 1.06 with the X-1 remaining intact, and thus became the first person to break the sound barrier. Yeager's rocket-powered flight is often cited as the beginning of the space age.

Figure 1.2: *Chuck Yeager broke through the sound barrier with his October 14, 1947 flight of the X-1—nicknamed the "Glamorous Glennis"—over Rogers Dry Lake in California. [Photographs courtesy of the U. S. Air Force]*

The development of rockets has taken us beyond the confines of Earth's gravitational field. Those of us old enough to be tuned in, *on color television broadcasting live from the Moon* (an achievement in its own right), still remember the chills up and down our spine on July 20, 1969, when Neil Armstrong (Figure 1.3) said, as he became the first man to set foot on the Moon, **"That's one small step for man, one giant leap for mankind."**

Today, we see routine flights of the Space Shuttle performing missions in Earth orbit and rocket launches from all over the globe. In particular, there have been numerous private-sector commercial achievements in space, most notably in the area of communications satellites.

Nevertheless, manned space travel has remained an unrealized revolution. In contrast to what occurred in the 40 years after the Wright brothers' flight, in a comparable time period since the first humans rocketed into space, fewer than 500 individuals have followed them.

At the dawn of the twenty-first century, another historic milestone has been reached that holds promise to advance mankind's desire for space travel and exploration. On October 4, 2004, the United States of America's private-enterprise system demonstrated its vast powers of creativity and can-do attitude in a most dramatic way. Scaled Composites, an entrepreneurial venture founded by aviation pioneer Burt Rutan, launched a vehicle named *SpaceShipOne* for

Figure 1.3: *On July 20, 1969, Neil Armstrong became the first human to set foot on the moon.* *[Photographs courtesy of NASA]*

the second time in two weeks. *SpaceShipOne*, shown in Figure 1.4, rocketed to an altitude of 69.6 miles above the Earth's surface and captured the ten-million dollar *Ansari X-Prize*.

As with Chuck Yeager's 1947 flight, *SpaceShipOne* was launched from a mother ship named the *White Knight*. The *White Knight* took off from the Mojave Airport and carried *SpaceShipOne* to an altitude of 47,000 feet. After detaching from the mother ship, *Space-ShipOne* then fired its rocket engine for 84 seconds, which propelled the vehicle to a speed in excess of Mach 3. At the maximum altitude, pilot Brian Binnie was weightless for three and a half minutes. To re-enter the Earth's atmosphere, *SpaceShipOne's* wings were folded into a self-stabilizing, high-drag configuration. After re-entry, its wings were folded back into a glider configuration for landing.

The saga of *SpaceShipOne* is punctuated with salutes to aviation history. For example, the private-sector *Ansari X-Prize* is modeled after the Orteg Prize that Charles Lindbergh won for his 1927 solo flight across the Atlantic Ocean. Brian Binnie also had the distinction of piloting *SpaceShipOne's* first powered, supersonic flight on November 17, 2003, the one-hundredth anniversary of the Wright brothers' flight.

Figure 1.4: *On October 4, 2004, entrepreneur Burt Rutan's SpaceShipOne rocketed into suborbital space, 69.6 miles above the Earth. [Photographs courtesy of Scaled Composites]*

A total of 26 teams competed for the honor of making the first private-enterprise space shot. Aerospace giants such as the Boeing Company and Lockheed Martin are now investigating the possibility of undertaking commercial space-travel ventures. Needless to say, the entrance of private enterprise into manned space travel puts an awesome new participant in quest of mankind's long-sought dream of traveling to the heavens.

All of these achievements have been accomplished partly because of the vast accumulation of knowledge regarding the motion of fluids. The lifting and propulsive forces generated by the Wright Flyer have been thoroughly understood, and dramatically improved designs have evolved for modern aircraft. Theoretical studies of flows at supersonic speeds were conducted by German researcher Ludwig Prandtl in the early part of the twentieth century, long before supersonic flight was even dreamed of. Those studies provided part of the basis for Chuck Yeager's historic flight. Intense theoretical and experimental analysis in the 1950's and 1960's provided accurate predictions of the heating that a space vehicle would have to be protected against to permit safe reentry into the Earth's atmosphere. The emergence of Computational Fluid Dynamics (CFD) has put the computer at our disposal to aid in the design of advanced vehicles such as the Space Shuttle and *SpaceShipOne*.

While evidence of its importance and relevance to real-world problems is most obvious in the achievements with aircraft and rockets, fluid mechanics is not limited solely to such problems. The aerodynamic drag of automobiles early in the twenty-first century is less than 40% of the drag of models designed in the early 1900's (cf. Figure 1.5). The flow of gases through the engine of an automobile and even its air-conditioning system are routinely computed by today's auto builders.

Figure 1.5: *Left: 1914 sedan with a drag coefficient of 0.80. Right: 2002 Corvette with a drag coefficient of 0.29. [Corvette photograph courtesy of William Watts.]*

A challenging area of fluid-mechanics research is the flow of blood through arteries and, ultimately, through the human heart. Developing a thorough understanding of blood flow through the body will surely help control and/or eliminate some of the most serious health hazards in today's world such as arteriosclerosis and heart disease.

Applications involving fluid motion extend to all branches of engineering. Aeronautical and mechanical engineers' interests range from basic studies of fluid motion, most notably turbulence, to everyday problems involving power generation, heating and ventilation, computer disk-drive design, etc. Civil engineers focus on interaction of aerodynamic forces with structures such as bridges and piers. Electrical engineers seek reduced costs in forming microchips, where acids must flow in a controlled manner to create desired patterns on silicon and other semiconductor materials. Chemical engineers must accurately determine reaction rates, a particularly acute problem when the velocity is high enough for the flow to be turbulent.

In this introductory chapter we will first discuss units and dimensions commonly used in fluid mechanics. Then, we will define a fluid, have a quick look at some of the properties of fluids, and conclude with a quick summary of how the science has evolved. By and large, all branches of engineering use the same mathematical tools and laws of physics, so that the primary distinguishing feature is nomenclature. Thus, one of the most important objectives of this chapter is to introduce fluid-mechanics nomenclature.

This book has been written with the object of introducing the reader to the exciting field of fluid mechanics. The book strives for mathematical rigor throughout, without concealing the exciting physical concepts involved. So, read on with the understanding that the intent is to challenge your intellect, and to encourage you to hone your skills in mathematical and physical reasoning. Aiming high is a good strategy to follow if you truly want to reach for the stars!

1.1 Dimensions and Units

Like all of the branches of physical science, the field of fluid mechanics has developed through a balance between theory and experiment. In order to have confidence in any theoretical development, we must demand that the quantitative predictions of the theory are consistent with and within the accuracy of measurements.

To describe physical properties in quantitative terms, we must select a system of units for expressing their dimensions. There are several systems of units used in general engineering practice, and this section describes four of the commonly used systems. Of these four, this text almost exclusively uses Standard International units and one of the subsets of the U. S. Customary System of units. In the text, we refer to these two sets of units as SI and USCS, respectively. Because of their inherent simplicity, SI units are becoming the standard in many branches of engineering. Several engineering journals, for example, require authors who use USCS units to parenthetically include corresponding values in SI units.

1.1.1 Independent Dimensions

The first thing we must decide upon is what dimensions are independent of all others, i.e., which are most basic in some sense. Just as the spatial coordinates x, y and z are independent variables in a three-dimensional rectangular Cartesian coordinate system, so we must choose a set of **independent dimensions**. Any other dimensions are then expressed in terms of the independent dimensions, and are called **secondary dimensions** or **dependent dimensions**.

For engineering work, we select either mass, length, time and temperature or force, length, time and temperature as the independent dimensions. In either case, Newton's second law of motion provides the defining relationship between mass and force, regardless of which is chosen as the independent dimension. We will return to the concepts of independent and secondary dimensions in Chapter 2.

1.1.2 Common Systems of Units

For the sake of uniform standards, various groups and countries have established systems of units and associated independent dimensions. Part of the objective in establishing these standards has been to promote use of **consistent units**. This means there are no special conversion factors required in a physically-based equation to make it dimensionally homogeneous, i.e., to make all terms in the equation have the same dimensions. Three of the four systems discussed in this section satisfy this constraint. As an interesting counter example, one of the systems, viz., the English Engineering system, violates this objective with regard to Newton's second law of motion.

The two most prevalent types of units are **Metric Units** and the **U. S. Customary System (USCS)**. Within each type of units, there are two sub-types. In the case of Metric Units, there is the **Standard International (SI)** system and the **Centimeter-Gram-Second (CGS)** system. Table 1.1 lists mass, length, time, temperature and force in the SI and CGS systems of units.

Table 1.1: *Metric Units*

Dimension	Standard International (SI)	Centimeter-Gram-Second (CGS)
Mass	kilogram (kg)	gram (g)
Length	meter (m)	centimeter (cm)
Time	second (sec)	second (sec)
Temperature	Kelvin (K)	Kelvin (K)
Force	Newton (N)	dyne (dyne)

For the USCS, the sub-types are known as the **British Gravitational (BG)** system and the **English Engineering (EE)** system. Table 1.2 lists mass, length, time, temperature and force in the BG and EE systems of units. This text exclusively uses the SI system and the BG subset of the U. S. Customary System. Because of its unusual nature, the text completely avoids the EE system.

Table 1.2: *U. S. Customary System of Units (USCS)*

Dimension	British Gravitational (BG)	English Engineering (EE)
Mass	slug (slug)	pound-mass (lbm)
Length	foot (ft)	foot (ft)
Time	second (sec)	second (sec)
Temperature	° Rankine (°R)	° Rankine (°R)
Force	pound (lb)	pound-force (lbf)

Standard International Units

The independent dimensions in the SI system[1] are as follows. The unit of mass is the kilogram (kg), the unit of length is the meter (m), the unit of time is the second (sec) and the unit of temperature is the Kelvin (K). The temperature expressed in Kelvins is absolute temperature, and is related to the commonly used Celsius, or centigrade, scale (°C) as follows.

$$K = {}^{\circ}C + 273.16 \tag{1.1}$$

The most important secondary dimension for applications using the laws of physics is force. The unit of force is called the Newton (N), which is defined as the force required to give a

---

[1]In older texts, this system is often referred to as the MKS system, which is an acronym for meter-kilogram-second. Its modern name is also Système Internationale.

mass of one kilogram an acceleration of one meter per second per second. In dimensional terms, we say

$$1 \text{ N} = (1 \text{ kg}) \cdot (1 \text{ m/sec}^2) = 1 \text{ kg} \cdot \text{m/sec}^2 \qquad (1.2)$$

Note that, since the acceleration of gravity is 9.807 m/sec^2, this means the weight of a one-kilogram mass is 9.807 Newtons. Note also that when—in common parlance—the "weight" of an object is expressed in kilograms, what is actually being quantified is the object's mass.

Two other important secondary dimensions are those of work and power. In the SI system, the units of work and power are the Joule (J) and the Watt (W), respectively. A Joule is the amount of work done when a one-Newton force produces a displacement of one meter in the direction of the force. A Watt is one Joule per second. Algebraically, we say

$$\left. \begin{aligned} 1 \text{ J} &= 1 \text{ N} \cdot \text{m} \\ 1 \text{ W} &= 1 \text{ J/sec} = 1 \text{ N} \cdot \text{m/sec} \end{aligned} \right\} \qquad (1.3)$$

Centimeter-Gram-Second Units

In the CGS system, we select the units of mass, length, time and temperature to be the gram (g), the centimeter (cm), the second (sec) and the Kelvin (K). The unit of force is called the dyne (dyne), which is defined as the force required to give a mass of one gram an acceleration of one centimeter per second per second. Hence,

$$1 \text{ dyne} = (1 \text{ g}) \cdot (1 \text{ cm/sec}^2) = 1 \text{ g} \cdot \text{cm/sec}^2 = 10^{-5} \text{ N} \qquad (1.4)$$

Because the acceleration of gravity is 980.7 cm/sec^2, this means the weight of a one-gram mass is 980.7 dynes.

In the CGS system, the unit of work is the erg (erg), while power is expressed in terms of ergs per second (erg/sec). An erg is defined as the amount of work done when a one-dyne force produces a displacement of one centimeter in the direction of the force. Table 1.3 lists the prefixes used with metric units.

Table 1.3: *Metric-Unit Prefixes*

Factor			Prefix	Symbol
1,000,000,000,000	=	10^{12}	tera	T
1,000,000,000	=	10^{9}	giga	G
1,000,000	=	10^{6}	mega	M
1,000	=	10^{3}	kilo	k
100	=	10^{2}	hecto	h
10	=	10^{1}	deka	da
0.1	=	10^{-1}	deci	d
0.01	=	10^{-2}	centi	c
0.001	=	10^{-3}	milli	m
0.000001	=	10^{-6}	micro	μ
0.000000001	=	10^{-9}	nano	n
0.000000000001	=	10^{-12}	pico	p

British Gravitational Units

In the BG system, the independent dimensions are force, length, time and temperature. The unit of force is the pound (lb), the unit of length is the foot (ft), the unit of time is the second (sec) and the unit of temperature is the degree Rankine (°R). The temperature expressed in degrees Rankine is absolute temperature, and is related to the Fahrenheit scale (°F) by

$$°R = °F + 459.67° \tag{1.5}$$

The unit of mass is called the slug (slug), which is defined as the mass upon which application of a one-pound force will cause an acceleration of one foot per second per second. Thus,

$$1 \text{ lb} = (1 \text{ slug}) \cdot (1 \text{ ft/sec}^2) = 1 \text{ slug} \cdot \text{ft/sec}^2 \tag{1.6}$$

The acceleration of gravity in BG units is 32.174 ft/sec^2, so that a mass of one slug weighs 32.174 pounds.

Finally, the units of work and power in the BG system are the British thermal unit (Btu) and the horsepower (hp), respectively. A Btu is the amount of heat required to raise the temperature of one pound of water (at its maximum density) by one degree Rankine. It is equal to 1,055 Joules. A horsepower is numerically equal to the rate of 33,000 foot-pounds of work per minute.[2] It is equal to 0.7457 kiloWatts.

$$\left.\begin{array}{l} 1 \text{ Btu} = 778 \text{ ft} \cdot \text{lb} \\ 1 \text{ hp} = 2545 \text{ Btu/hr} = 550 \text{ ft} \cdot \text{lb/sec} \end{array}\right\} \tag{1.7}$$

English Engineering Units

In the EE system, the units for mass and force are defined separately. The unit of force is the pound-force (lbf), while the unit of mass is the pound-mass (lbm). As with the BG system, the units of length, time and temperature are the foot, the second and the degree Rankine, respectively. Having separate units for mass and force adds the complication that Newton's second law of motion must be written as

$$\mathbf{F} = \frac{m\mathbf{a}}{g_c} \tag{1.8}$$

where $\mathbf{F}$ is force, m is mass, $\mathbf{a}$ is acceleration and g_c is a dimensional constant determined as follows. We define the pound-force and the pound-mass such that application of a one-lbf force to a one-lbm mass yields an acceleration equal to the acceleration of gravity on Earth, 32.174 ft/sec^2. Hence,

$$1 \text{ lbf} = \frac{1 \text{ lbm} \cdot 32.174 \text{ ft/sec}^2}{g_c} \quad \Longrightarrow \quad g_c = 32.174 \frac{\text{ft} \cdot \text{lbm}}{\text{lbf} \cdot \text{sec}^2} \tag{1.9}$$

Because of the dimensional parameter g_c appearing in a basic law of physics, EE units are inconvenient, and find use mainly when convention dictates. The primary fields in which EE units are found are thermodynamics and heat transfer.

---

[2]The term horsepower was originated by Boulton and Watt to state the power of their steam engines. In a practical test, it was found that the average horse could work constantly at the rate of 22,000 foot-pounds per minute. This was increased by one half in defining this arbitrary, and now universal, unit of power.

Example 1.1 *A nanotube researcher chooses to express length in Angstrom units (Å), mass in grams (gm) and time in microseconds (μsec). In this system of units, infer the logical choice of units for force.*

Solution. First, we note that $1 \overset{\circ}{A} = 10^{-10}$ m, 1 gm $= 10^{-3}$ kg and 1 μsec $= 10^{-6}$ sec. Since force has units of mass times length divided by time squared, the units are

$$\frac{10^{-3} \text{ kg} \cdot 10^{-10} \text{ m}}{(10^{-6} \text{ sec})^2} = 10^{-1} \frac{\text{kg} \cdot \text{m}}{\text{sec}^2} = 10^{-1} \text{ N} = 1 \text{ dN}$$

where we observe from Equation (1.2) that one Newton is $1 \text{ kg} \cdot \text{m/sec}^2$ and, from Table 1.3, "dN" denotes deciNewton. Thus, the logical choice of units for expressing force in this system would be the deciNewton.

1.1.3 Converting Dimensional Quantities

For the discussion to follow, it is convenient to identify the independent dimensions in algebraic terms. We will refer to mass as M, length as L, time as T, temperature as Θ, and force as F. As noted above, secondary dimensions are related by law or by definition to the independent dimensions. For instance, the dimensional representation of velocity, U, is

$$[U] = \frac{L}{T} \tag{1.10}$$

where $[U]$ means dimensions of U. Similarly, pressure has dimensions F/L^2 and acceleration is expressed dimensionally as L/T^2. We will focus further on the distinction between dimensions and units in Chapter 2.

Converting from one system of units to another is straightforward if we follow some simple algebraic rules. For example, Table 1.4, which lists conversion factors for properties commonly encountered in fluid mechanics, tells us that 1 pound is equal to 4.448 Newtons. Also, 1 foot is 0.3048 meters. Hence, to convert from pressure expressed in lb/ft² to N/m², we proceed as follows. We form the ratios of independent or secondary dimensions, and use them as dimensional multiplicative factors with a magnitude of unity. That is, we say

$$\left(\frac{4.448 \text{ N}}{1 \text{ lb}} \right) = 1 \quad \text{and} \quad \left(\frac{0.3048 \text{ m}}{1 \text{ ft}} \right) = 1 \tag{1.11}$$

We can now use these factors in a strict algebraic sense to convert from the USCS system to the SI system as follows.

$$1 \frac{\text{lb}}{\text{ft}^2} = \frac{(1 \text{ lb}) \cdot \left(4.448 \frac{\text{N}}{\text{lb}} \right)}{(1 \text{ ft}^2) \cdot \left(0.3048 \frac{\text{m}}{\text{ft}} \right)^2} = 47.88 \frac{\text{N}}{\text{m}^2} \tag{1.12}$$

Note the precise algebraic cancelation of "lb" and "ft²" in the two factors after the first equals sign. Organizing units computations in this manner greatly reduces the chance of arithmetic error in converting from one set of units to another.

Example 1.2 *The units of energy in the SI system are Newton·meter. What is the conversion factor from the SI system to foot·pounds in the USCS system?*

Solution. Using Equation (1.11), we can say

$$1 \text{ ft} \cdot \text{lb} = (1 \text{ ft}) \cdot \left(0.3048 \ \frac{\text{m}}{\text{ft}}\right) \cdot (1 \text{ lb}) \cdot \left(4.448 \ \frac{\text{N}}{\text{lb}}\right) = 1.3558 \text{ N} \cdot \text{m}$$

Both SI and USCS units are used in modern engineering applications, especially in the aerospace industry. While there is a trend toward the use of SI units, many important engineering journal articles and books are in common usage that provide information in terms of the USCS. Such articles and books continue to constitute important reference sources that reflect many decades of engineering research and experimentation. Hence, the conscientious student will make a point of being "bilingual" regarding these systems of units. In order to promote this end, approximately half of the homework problems in this text are cast in terms of each of the two predominant systems.

To avoid unnecessary error, be sure to work the problems in the units given. Converting to another system of units simply introduces potential arithmetic errors that have nothing to do with mastering the physical principles involved in the problem.

The homework problems throughout this book strive toward maximizing physical understanding while minimizing arithmetic error. Specifically, most of the problems require you to first derive an algebraic result. Formulas and equations, after all, are the things that a practicing engineer incorporates into a computer program in order to conduct a design study. Also, as any conscientious practicing engineer will tell you, the equations you derive will prove to be especially helpful in assuring that your answer is algebraically correct to the extent that it has the correct *dimensions* (Chapter 2 discusses this concept in great detail). To help familiarize you with typical engineering magnitudes for key flow properties, many of the problems provide numerical values that can be substituted into the algebraic formula.

Example 1.3 *If you are driving along a highway at a speed of 53 mph, what is your speed in Astronomical Units per Millennium (AUpM)?*

Solution. By definition, the Astronomical Unit (AU) is the average distance between the Earth and the sun, which is $93 \cdot 10^6$ miles. Also, a millennium (Mln) is 1000 years, each of which has 365.25 days that are 24 hours long. Hence, to convert from mph to AUpM, we first note that

$$1 \text{ Mln} = (1000 \text{ yr}) \left(365.25 \ \frac{\text{day}}{\text{yr}}\right) \left(24 \ \frac{\text{hr}}{\text{day}}\right) = 8.766 \cdot 10^6 \text{ hr}$$

Therefore, we have

$$1 \text{ AUpM} = \frac{\left(93 \cdot 10^6 \text{ mi/AU}\right) (1 \text{ AU})}{\left(8.766 \cdot 10^6 \text{ hr/Mln}\right) (1 \text{ Mln})} = 10.6 \text{ mph}$$

So, your speed of 53 mph expressed in Astronomical Units per millennium is

$$\text{Speed} = \frac{(53 \text{ mph})}{(10.6 \text{ mph/AUpM})} = 5 \text{ AUpM}$$

Note how the units cancel algebraically in the equations above.

Table 1.4: *Conversion Table*

Mass (M)			Velocity (LT^{-1})		
1 lbm	=	0.454 kg	1 mph	=	1.467 ft/sec
1 slug	=	14.594 kg	1 mph	=	0.447 m/sec
1 oz	=	28.35 g	1 knot	=	1.688 ft/sec
1 kg	=	2.205 lbm	1 knot	=	0.514 m/sec
1 kg	=	0.0685 slug	1 m/sec	=	3.281 ft/sec
Length (L)			**Work** (ML^2T^{-2})		
1 ft	=	0.3048 m	1 Btu	=	778 ft·lb
1 in	=	25.4 mm	1 ft·lb	=	1.3558 J
1 mi	=	5280 ft	1 Btu	=	1055 J
1 km	=	0.6214 mi	1 J	=	0.239 cal
1 m	=	3.2808 ft	1 J	=	0.7376 ft·lb
Temperature (Θ)			**Power** (ML^2T^{-3})		
°C	=	$\frac{5}{9}$(°F−32°)	1 hp	=	550 ft·lb/sec
K	=	°C + 273.16	1 hp	=	2545 Btu/hr
°R	=	$\frac{9}{5}$K	1 kW	=	1.341 hp
°R	=	°F + 459.67°	1 J/sec	=	1 W
Force (MLT^{-2})			**Volume** (L^3)		
1 lb	=	4.448 N	1 gal	=	0.0037854 m^3
1 kN	=	224.8 lb	1 gal	=	231 in^3
1 ton	=	2000 lb	1 gal	=	0.134 ft^3
1 ton	=	8.897 kN	1 L	=	0.001 m^3
Pressure $(ML^{-1}T^{-2})$			**Rotation** (T^{-1})		
1 psi	=	6.895 kPa	1 rpm	=	0.1047 sec^{-1}
1 psf	=	47.88 Pa	1 Hz	=	2π sec^{-1}
1 atm	=	101 kPa	1 cps	=	2π sec^{-1}
1 atm	=	2116.8 psf	1 rev	=	2π radians
1 atm	=	760 mmHg			
Gravity (LT^{-2})			**Area** (L^2)		
g	=	32.174 ft/sec^2	1 ft^2	=	0.0929 m^2
g	=	9.807 m/sec^2	1 in^2	=	6.4516 cm^2
			1 m^2	=	10.764 ft^2

1.2   Definition of a Fluid

To begin our study, we first note that the field we call **fluid mechanics** is the branch of physics concerned with fluid motion. Most substances found in nature can be thought of as either a solid or a fluid, where the two most common fluids are gases and liquids. The goals of this book are twofold. Our most important goal is to explain how to apply Newton's laws of motion to fluids. Our second goal is to present a brief overview of the general field of fluid mechanics. The first thing we need to do is define what we mean by a fluid.

1.2.1   Simplistic Definition of a Fluid

Anything that flows. Liquids and gases are the most obvious examples of substances that will flow. Traffic is a more subtle example that can be treated as a fluid, in the sense that its motion can be described with equations derived from fluid-mechanics principles. The flow of traffic has been modeled [Lighthill and Whitham (1955)] as a fluid whose density (number of automobiles per unit distance) varies with speed. A particularly interesting application of traffic-flow theory is to the timing of traffic lights. In the 1960's and 1970's, many cities posted signs indicating the speed required to avoid catching a red light on a given route. The speed and timing of the lights were determined from observations of normal traffic flow and a theory developed from the mass and momentum conservation laws of fluid mechanics.

Figure 1.6: *The flow of traffic can be modeled as a fluid whose density, defined as the number of automobiles per unit distance, varies with speed. [Photographs courtesy of California Department of Transportation]*

1.2.2   Rigorous Definition of a Fluid

A substance that cannot be in static equilibrium under the action of an oblique stress.[3] On the one hand, if only normal forces such as pressure act, a fluid will adjust to the applied pressure with a change in volume. After adjusting, it remains at rest as illustrated in Figure 1.7(a). On the other hand, if a tangential stress such as shear due to friction is present, the fluid deforms and continues moving. Such a flow has an oblique stress because the resultant force from the tangential shear and normal pressure acts at an oblique angle.

---

[3]A stress, by definition, is a force per unit area.

For example, consider flow in a channel as shown in Figure 1.7(b). If we apply different pressures $p_1 > p_2$ at the ends of the channel, frictional forces develop at the channel walls (and, in fact, throughout the channel) to balance the pressure difference. As a result, the initially rectangular section of fluid moves at different velocities across the channel, and thus undergoes significant distortion. By contrast, a solid will resist a shearing force until it yields.

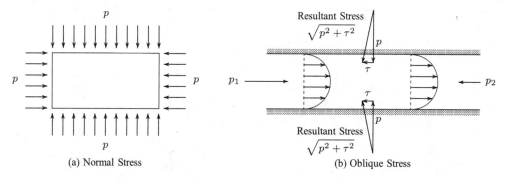

(a) Normal Stress

(b) Oblique Stress

Figure 1.7: *Fluid response to normal and oblique stresses. The forces are not drawn to scale in (b)—shear stresses are generally much smaller than pressure. Also, in (b), the solid lines indicate the boundaries of the moving fluid that was initially contained within the volume denoted by the dashed lines.*

1.3   Continuum Approximation

In reality, a fluid is made up of molecules. The molecules are widely spaced for a gas and more closely spaced for a liquid. In either case, the distance between molecules is very large compared to the diameter of a molecule. The molecules are not fixed in space. Rather, they are in constant motion, colliding with one another. In a given volume, we thus have a collection of molecules with empty space between them. This stands in distinct contrast to a continuous substance in which the fluid mass density (mass per unit volume) would vary smoothly from one point to another.

Since a fluid is not a continuous substance, in principle, we must apply the appropriate laws of motion to each molecule to describe the overall motion. This would be an extremely difficult task given the large number of molecules present for any practical flow. Table 1.5 quantifies just how large a number this would be for water and air. Note that, by definition, the number density, n, is the number of molecules per unit volume.

Table 1.5: *Molecular Number Density for Water and Air*

Fluid	Number of Molecules	Volume	Number density, n
Water	$3.34 \cdot 10^{19}$	1 mm^3	$3.34 \cdot 10^{19}$ mm^{-3}
Air	$2.69 \cdot 10^{19}$	1 cm^3	$2.69 \cdot 10^{19}$ cm^{-3}

The values quoted correspond to **Standard Temperature and Pressure (STP)**. By definition, STP is $0°$ C and 1 atm. One way of computing the motion of such a large system of particles is to use the mathematically complex theory of statistical mechanics.

To simplify our task, we introduce the **continuum approximation**, a concept that finds its origins with Swiss mathematician Leonhard Euler (1707-1783). The continuum approximation is used in many engineering disciplines such as solid mechanics, thermodynamics and electrodynamics. Using this concept means we regard a fluid as being a **continuous** substance as opposed to being made up of discrete molecules. That is, we regard the fluid as being an infinitely divisible substance with no empty space between particles. We pretend that the basic properties (e.g., pressure, density, viscosity) of the smallest subdivisions of a fluid approach a unique limiting value as we reduce the size of the subdivision. In other words, we assume a definite value of each fluid property exists at each point in space.

Our justification for making this approximation is as follows. We can average random thermal motion of molecules provided the number of molecules in a **fluid particle** is very large, a fluid particle being defined as a volume of fluid whose size is extremely small compared to the characteristic dimension of the flow under consideration.

Figure 1.8 illustrates the limiting process we use to define a fluid property such as mass density, ρ, where mass density is mass per unit volume at each point in a given fluid. In terms of a fluid particle, if its mass is Δm and its volume is ΔV, there is a limiting volume, ΔV_ℓ, below which an insufficient number of molecules can be found to define a meaningful statistical average. Also, as shown in the figure, for larger values of ΔV, say $\Delta V > \Delta V_u$, the ratio of Δm to ΔV will vary if the fluid has nontrivial spatial variations in its properties.

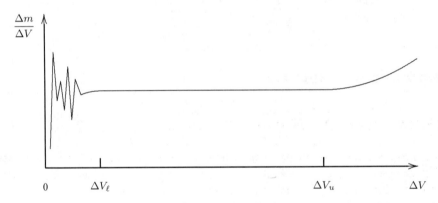

Figure 1.8: *Continuum definition of mass density at a point in a fluid.*

In the continuum limit, we postulate that a wide separation of scales exists, i.e., that

$$\Delta V_\ell \ll \Delta V_u \tag{1.13}$$

and we define the mass density by

$$\rho = \lim_{\Delta V \to \Delta V_\ell} \frac{\Delta m}{\Delta V} \tag{1.14}$$

When there is a wide separation of scales, the limiting process defined in Equation (1.14) is well defined, with $\Delta m/\Delta V$ approaching a constant limiting value at each point in the fluid. We have a continuum provided, of course, that we can define a fluid particle in which a sufficient number of molecules exists to define statistical averages of molecular motion. The vast majority of practical fluid-flow applications satisfy this constraint.

Example 1.4 *Consider flow past a sphere of 1 cm diameter in a standard wind tunnel. Determine the number of molecules in a volume the size of a dust particle (diameter 1 micron = 10^{-4} cm) for air at STP. If we regard this small a volume as a typical "fluid particle" for analyzing flow past the sphere, does the continuum approximation apply?*

Solution. Since the volume of the dust particle is $\frac{1}{6}\pi \cdot 10^{-12}$ cm^3 and Table 1.5 tells us there are $2.69 \cdot 10^{19}$ molecules in 1 cm^3 of air at STP, the number of molecules is

$$N = \frac{1}{6}\pi \cdot 10^{-12} \text{ cm}^3 \times 2.69 \cdot 10^{19} \frac{\text{molecules}}{\text{cm}^3} = 14 \cdot 10^6 \text{ molecules}$$

Clearly, we can define meaningful statistical averages if we have such a large number of molecules in our sample. Thus, the continuum approximation is valid for this flow.

Example 1.5 *Now let's move the sphere 200 miles above the Earth's surface. At this altitude, there are approximately 20,000 molecules in each cm^3. Again, compute the number of molecules in a volume the size of a dust particle (diameter 1 micron = 10^{-4} cm). Does the continuum limit apply in this case for analyzing flow past the sphere?*

Solution. Since the volume of the dust particle is $\frac{1}{6}\pi \cdot 10^{-12}$ cm^3 and there are $2 \cdot 10^4$ molecules in 1 cm^3, the number of molecules is

$$N = \frac{1}{6}\pi \cdot 10^{-12} \text{ cm}^3 \times 2 \cdot 10^4 \frac{\text{molecules}}{\text{cm}^3} = 1.0 \cdot 10^{-8} \text{ molecules}$$

At this altitude, our dust-particle sized volume contains 10^{-8} molecules, i.e., to find a single molecule we need a volume a hundred million times larger than that of a dust particle. The continuum limit is meaningless for this flow.

1.4 Microscopic and Macroscopic Views

With the continuum approximation, we view the fluid from the **macroscopic** level. By contrast, in viewing the fluid on a scale of the order of a molecule, we observe things on the **microscopic** level. Hence, we refer to continuum fluid properties like mass density as macroscopic properties. In this section, we will explore the manner in which density, pressure, temperature and other thermodynamic variables relate to what actually happens at the microscopic level.

1.4.1 Density

Mass density, usually called simply density, is denoted by ρ and has dimensions of mass per unit volume. Thus,

$$\rho = mn \qquad (1.15)$$

where m is mass of a molecule and n is the number density, i.e., the number of molecules per unit volume (see Table 1.5). Values of ρ for air and water at atmospheric pressure and a temperature of 20° C (68° F) are as follows.

$$\rho = \begin{cases} 1.20 & \text{kg/m}^3 \quad (0.00234 \text{ slug/ft}^3), \quad \text{Air} \\ 998 & \text{kg/m}^3 \quad (1.94 \text{ slug/ft}^3), \quad \text{Water} \end{cases} \qquad (1.16)$$

The density of a liquid is a weak function of pressure and depends mainly on temperature. Table A.2 in Appendix A includes densities of common liquids, while Table A.3 includes the variation of density with temperature for water.

The density of a gas varies with both pressure and temperature. The perfect-gas law, which is given below as Equation (1.25) and in Appendix A as Equation (A.1), can be used to compute the density of a given gas.

There are three additional quantities related to density that are used in fluid-flow applications, viz.,

- **Specific volume**, v: Used primarily in thermodynamics, this quantity is the volume per unit mass. Thus,

$$v = \frac{1}{\rho} \tag{1.17}$$

- **Specific weight**, SW: Used primarily for flows with constant density, this quantity is weight per unit volume. Thus, denoting gravitational acceleration by g,

$$SW = \rho g \tag{1.18}$$

- **Specific gravity**, SG: This is a dimensionless representation of density. For a liquid, it is the ratio of fluid density to the density of water at $4°\,\mathrm{C}$. For a gas, it is the ratio of gas density to the density of air at $20°\,\mathrm{C}$ and 1 atm. Hence, we have

$$SG = \frac{\rho}{\rho_{ref}} \quad \text{where} \quad \rho_{ref} = \left\{ \begin{array}{ll} 1000 & \mathrm{kg/m^3} \ \ (1.94 \ \mathrm{slug/ft^3}), \quad \text{Liquid} \\ 1.20 & \mathrm{kg/m^3} \ \ (0.00234 \ \mathrm{slug/ft^3}), \quad \text{Gas} \end{array} \right. \tag{1.19}$$

Example 1.6 *Using SI units, compute the specific volume, specific weight and specific gravity for glycerin.*

Solution. From Table A.2 the density of glycerin is 1260 $\mathrm{kg/m^3}$ and Table 1.4 tells us the gravitational acceleration, g, is 9.807 $\mathrm{m/sec^2}$. Thus, noting that 1 kN = 1000 kg·m/sec^2 in computing SW, we have

$$v = \frac{1}{1260 \ \mathrm{kg/m^3}} = 7.94 \cdot 10^{-4} \ \frac{\mathrm{m^3}}{\mathrm{kg}}$$

$$SW = 1260 \ \frac{\mathrm{kg}}{\mathrm{m^3}} \cdot 9.807 \ \frac{\mathrm{m}}{\mathrm{sec^2}} = 12.36 \ \frac{\mathrm{kN}}{\mathrm{m^3}}$$

$$SG = \frac{1260 \ \mathrm{kg/m^3}}{1000 \ \mathrm{kg/m^3}} = 1.26$$

Example 1.7 *Using USCS units, compute the specific volume, specific weight and specific gravity for a gas whose density is 0.00125 slug/ft^3.*

Solution. Table 1.4 tells us that $g = 32.174 \ \mathrm{ft/sec^2}$. Noting that 1 lb = 1 slug·ft/sec^2 in computing specific weight, there follows

$$v = \frac{1}{0.00125 \ \mathrm{slug/ft^3}} = 800 \ \frac{\mathrm{ft^3}}{\mathrm{slug}}$$

$$SW = 0.00125 \ \frac{\mathrm{slug}}{\mathrm{ft^3}} \cdot 32.174 \ \frac{\mathrm{ft}}{\mathrm{sec^2}} = 0.0402 \ \frac{\mathrm{lb}}{\mathrm{ft^3}}$$

$$SG = \frac{0.00125 \ \mathrm{slug/ft^3}}{0.00234 \ \mathrm{slug/ft^3}} = 0.534$$

1.4.2 Temperature

In describing the motion of a molecule, we separate its instantaneous velocity, $\mathbf{u}$, into two parts, i.e., $\mathbf{u} = \mathbf{U} + \mathbf{u}'$. The first is the bulk-motion or mean velocity, $\mathbf{U}$, and second is the fluctuating velocity, $\mathbf{u}'$. The latter contribution represents random molecular fluctuations. By definition,

$$\mathbf{U} = <\mathbf{u}> \tag{1.20}$$

where $<>$ denotes statistical average. The kinetic-theory definition of temperature, T, [cf. Jeans (1962)] is as follows.

$$\tfrac{3}{2}kT = <\tfrac{1}{2}m\mathbf{u}' \cdot \mathbf{u}'> \tag{1.21}$$

where k is Boltzmann's constant and m is molecular mass. That is, the temperature of a fluid is directly proportional to the average kinetic energy of the molecular motion. Standard temperature is the value at which water freezes, viz.,

$$T_{standard} = 0° \text{ Centigrade} = 32° \text{ Fahrenheit} \qquad [T(°\text{F}) = 1.8T(°\text{C}) + 32°] \tag{1.22}$$

In terms of **absolute temperature**, we have

$$T_{standard} = 273.16 \text{ Kelvins} = 491.67° \text{ Rankine} \qquad [T(°\text{R}) = 1.8T(\text{K})] \tag{1.23}$$

1.4.3 Pressure

For a fluid at rest we observe that the fluid exerts a pressure on the walls of its container. In fact, the only force exerted by the fluid is everywhere normal to the surface of the container. We also observe that if the fluid is moving, it exerts an oblique force on the walls of the container. We define the pressure, p, as the normal force per unit area exerted by a given fluid particle on its immediate neighbors, whether or not the fluid is moving.

Pressure is a scalar quantity and thus acts equally in all directions at a given point in the fluid. It can change from point to point in a given flow as required to balance local accelerations. The force attending pressure is formed as the product of the pressure, the area of the surface on which it acts and the unit normal to the surface. We refer to pressure as a normal stress. We will take up the notion of oblique stresses when we discuss effects of friction in Section 1.8.

At the microscopic level, fluid molecules are continuously moving in a random manner. Collisions with a container wall give rise to an instantaneous force (impulse) on the wall. As illustrated in Figure 1.9, if the number density is small, the resulting "pressure" appears as a series of spikes. The magnitude of the impulse increases as the angle of the molecule's trajectory approaches a right angle, which involves the maximum momentum change *normal to the wall*.

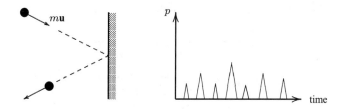

Figure 1.9: *Impulses from molecules striking a container wall in the limiting case of number density, $n \ll 1$.*

As depicted in Figure 1.10, for a very large number of collisions the effect of the collisions is similar to a continuous push or pressure on the wall. This is the continuum limit ($n \gg 1$), and we use a statistical average, $<p>$, to define the pressure. Pressure is expressed in many different units. For example, the standard value of pressure in the atmosphere, p_a, is

$$p_a = \begin{cases} 1 & \text{atm} & \text{(atmosphere)} \\ 1 & \text{torr} & \text{(Torr)} \\ 14.7 & \text{psi} & \text{(pound per square inch)} \\ 101 & \text{kPa} & \text{(kiloPascal)} \\ 760 & \text{mmHg} & \text{(millimeter of mercury)} \\ 2116.8 & \text{psf} & \text{(pound per square foot)} \end{cases} \qquad (1.24)$$

where one Pascal = one Newton per square meter. Pressure is often quoted in terms of **absolute** (psia) and **gage** (psig) values, with the latter relative to the atmospheric pressure. Thus, $p = 14.9$ psi is an absolute pressure of 14.9 psia and a gage pressure of 0.2 psig.

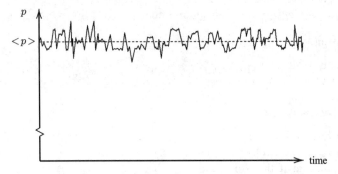

Figure 1.10: *Pressure from molecules striking a container wall in the limiting case of number density, $n \gg 1$. p is the instantaneous pressure and $<p>$ is the statistical average.*

1.4.4   Vapor Pressure

There is a special value of the pressure known as the vapor pressure, p_v, that manifests itself in the familiar phenomena of **evaporation** and **boiling**. It also affects a more subtle phenomenon known as **cavitation**.

To understand the concept of vapor pressure, consider the following. Imagine that we pour a liquid into a closed container and measure the pressure in the space above the liquid. As time passes, we will observe an increase in the pressure. At some point the pressure will reach a constant value. By definition, this is the vapor pressure for the liquid.

In our hypothetical closed container, some of the molecules within the liquid have sufficient kinetic energy to overcome the attractive forces between the molecules. As the number of molecules that have escaped increases, the number of collisions of the vapor molecules with the liquid surface increases and some of the vapor molecules rejoin the main body of liquid. Equilibrium is achieved when the number of molecules rejoining the liquid equals the number leaving to enter the vapor phase.

Clearly, vapor pressure is dependent upon the strength of the intermolecular attractive forces. These forces are very strong for liquids like mercury and glycerin. Consequently, a relatively small number of molecules enter the vapor phase and p_v is very small (see Table A.9 in Appendix A). By contrast, the intermolecular forces are much weaker for carbon tetrachloride and gasoline, wherefore p_v is much higher.

Since the molecular kinetic energy is greater at higher temperature, more molecules can escape, wherefore vapor pressure is correspondingly greater. And indeed, we observe that vapor pressure increases with temperature. Table A.9 in Appendix A tabulates values of p_v for water. Focusing on the pressure within a given liquid, p, we have three different possibilities that depend upon how it compares to p_v.

- **Evaporation** ($p > p_v$): If a liquid rests in an open container, $p > p_v$ and fluid will continually enter the vapor phase. Unlike the situation in a closed container, most of the vapor above the liquid escapes to the atmosphere. There is a one way transition from liquid to vapor phase, and the liquid evaporates.

- **Boiling** ($p \leq p_v$): When we increase the temperature of a liquid, vapor bubbles begin to form within the liquid. Since the pressure in an interior bubble is equal to the vapor pressure, such a bubble will rapidly collapse if the pressure in the fluid exceeds p_v. On the other hand, when the pressure in the fluid is such that $p \leq p_v$, the bubble will remain or expand and the liquid boils. The temperature at which p_v equals the pressure of the atmosphere is the **boiling point**. Water boils at 212° F at sea level where the atmospheric pressure is 14.7 psi. At altitudes far above sea level, air pressure drops appreciably with an attendant drop in boiling-point temperature. For example, at the top of Alaska's 17,000-foot Mount McKinley, atmospheric pressure is approximately 7.6 psi, and Table A.9 shows that water will boil at 180° F. Since cooking occurs at a lower temperature, it would take longer to cook a hard-boiled egg, or any food requiring boiling, at the top of Mount McKinley than it would in your kitchen.

- **Cavitation** ($p \leq p_v$): When a fluid is flowing, boiling can be caused *at any temperature* by regions of very low pressure. This occurs, for example, in valves, pumps and marine propellers. For flow through or past such devices, it is possible to have the local pressure drop below the vapor pressure. This can cause severe problems because, when the vapor bubbles are swept into higher-pressure regions in the flow, they will suddenly implode. If the implosion occurs close to a solid surface, it can—over a period of time—cause structural damage. Figure 1.11 shows cavitation emanating from the tips of a marine propeller. The bubbles are shed from the low-pressure regions near the tips of the propeller blades and form a helical pattern as the fluid moves from left to right.

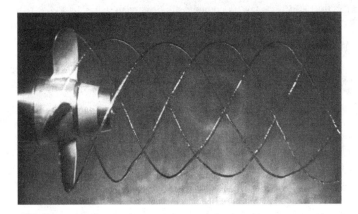

Figure 1.11: *Tip cavitation from a marine propeller. [Photograph courtesy of Garfield Thomas Water Tunnel, Pennsylvania State University]*

1.5   Thermodynamic Properties of Gases

When our fluid is a gas, there are some additional useful thermodynamic relations and properties of interest in the study of fluid mechanics. For most applications, we treat the gas as though it is a **perfect gas**. Also, when we are interested in heat transfer, there are several energy-related properties of interest to us.

If the intermolecular forces in a gas are negligible and intermolecular collisions are perfectly elastic, the pressure, density and temperature are related by the **perfect-gas equation of state**,

$$p = \rho R T \tag{1.25}$$

where R is the **perfect-gas constant**. For air, the value of R is

$$R_{air} = 287 \text{ J/(kg·K)} \quad [1716 \text{ ft·lb/(slug·°R)}] \tag{1.26}$$

Table A.1 in Appendix A includes values of R for common gases.

Alternatively, the perfect-gas law can be stated in terms of the universal gas constant, $\mathcal{R}$, according to

$$p = \frac{\rho \mathcal{R} T}{\mathcal{M}} \tag{1.27}$$

where $\mathcal{M}$ is the molecular weight of the gas. Note that the units of molecular weight are kg/(kg-mole) [slug/(slug-mole)]. The value of the universal gas constant is

$$\mathcal{R} = 8310 \ \frac{\text{J}}{\text{kg-mole} \cdot \text{K}} = 49700 \ \frac{\text{ft} \cdot \text{lb}}{\text{slug-mole} \cdot \text{°R}} \tag{1.28}$$

Example 1.8 *The pressure in a one cubic foot tank of compressed helium is 45 psi. The temperature inside the tank is 77°F. Compute the density of the helium. Also, compute the weight of the helium in the tank, expressing your answer in ounces.*

Solution. We use the perfect-gas law to compute the density so that

$$\rho = \frac{p}{RT}$$

Reference to Table A.1 tells us that $R = 12419$ ft·lb/(slug·°R). Also, converting to absolute temperature, we have $T = (77 + 459.67)$°R $= 536.67$°R. Then,

$$\rho = \frac{\left(45 \ \frac{\text{lb}}{\text{in}^2}\right)\left(144 \ \frac{\text{in}^2}{\text{ft}^2}\right)}{\left(12419 \ \frac{\text{ft} \cdot \text{lb}}{\text{slug·°R}}\right)(536.67\text{°R})} = 9.7225 \cdot 10^{-4} \ \frac{\text{slug}}{\text{ft}^3}$$

Letting V denote the tank volume, the weight, W of the helium within the tank is

$$W = \rho g V = \left(9.7225 \cdot 10^{-4} \ \frac{\text{slug}}{\text{ft}^3}\right)\left(32.174 \ \frac{\text{ft}}{\text{sec}^2}\right)\left(1 \ \text{ft}^3\right)\left(16 \ \frac{\text{oz}}{\text{lb}}\right) = 0.50 \ \text{oz}$$

We will make use of other thermodynamic quantities and notation. For example, the heat contained by a fluid is quantified in terms of the **internal energy**, e. In general, the basic postulates of thermodynamics tell us e is a function of any two independent thermodynamic state variables, a point we will address in Chapter 6. The internal energy is usually written as a function of temperature and specific volume, $v = 1/\rho$, so that, in general, $e = e(T, v)$.

There is a second thermodynamic quantity that characterizes the heat contained by a fluid. It is called **enthalpy** and is denoted by h. By definition,

$$h = e + p/\rho \qquad (1.29)$$

In contrast to internal energy, h is usually written as a function of temperature and pressure, i.e., $h = h(T, p)$. Enthalpy and internal energy have units of J/kg (ft·lb/slug).

There are two **specific-heat coefficients** denoted by c_v and c_p. By definition,

$$c_v = \left(\frac{\partial e}{\partial T}\right)_v \qquad \text{and} \qquad c_p = \left(\frac{\partial h}{\partial T}\right)_p \qquad (1.30)$$

where c_v is appropriate to a constant-volume process and c_p to a constant-pressure process. Both of the specific-heat coefficients have units of J/(kg·K) [ft·lb/(slug·°R)] and, in general, depend upon T.

Finally, the ratio of the specific heats is denoted by γ (the symbol introduced by Rankine) and referred to as the **specific-heat ratio**. Thus, we write

$$\gamma = \frac{c_p}{c_v} \qquad (1.31)$$

There are two limiting cases that are of interest regarding internal energy, enthalpy, the specific-heat coefficients and the specific-heat ratio.

- **Thermally-perfect gas**: For a wide range of practical conditions, we find that $e = e(T)$ and $h = h(T)$ for many gases. It is true, for example, for air at temperatures up to about 2000 K (3600° R). A gas is said to be thermally perfect when e and h are functions only of T.

- **Calorically-perfect gas**: Except at very high temperatures, c_v and c_p are constant for many gases. When c_v and c_p are constant, we say the gas is calorically perfect. It follows that, for a calorically-perfect gas, $e = c_v T$, $h = c_p T$ and $\gamma =$ constant.

Table A.1 in Appendix A includes values of γ and c_p for common gases.

1.6 Compressibility

If we hold temperature constant and apply a pressure to a container filled with gas, its density will change according to

$$\rho = \frac{p}{RT} \qquad (1.32)$$

so that density changes, $\Delta\rho$, are given by

$$\Delta\rho = \frac{\Delta p}{RT} \qquad \text{(constant } T) \qquad (1.33)$$

In so doing, we **compress** the gas. The formal definition of the **compressibility**, τ, is

$$\tau = \frac{1}{\rho} \frac{\partial \rho}{\partial p} \tag{1.34}$$

Some authors prefer to work with the **bulk modulus**, E_v, which is the reciprocal of the compressibility. For our isothermal perfect gas, we have $\partial \rho / \partial p = 1/(RT)$ so that

$$\tau = \frac{1}{\rho} \cdot \frac{1}{RT} = \frac{1}{p} \tag{1.35}$$

Compressibility is a property of the fluid. The compressibility for gases is much larger than corresponding values for liquids. At atmospheric pressure and constant temperature,

$$\tau = \begin{cases} 1.00 \cdot 10^{-5} & \text{m}^2/\text{N} \quad (4.79 \cdot 10^{-4} \text{ ft}^2/\text{lb}), \quad \text{Air} \\ 4.65 \cdot 10^{-10} & \text{m}^2/\text{N} \quad (2.23 \cdot 10^{-8} \text{ ft}^2/\text{lb}), \quad \text{Water} \end{cases} \tag{1.36}$$

Table A.4 in Appendix A lists τ for several common fluids.

Example 1.9 *Estimate how much greater a pressure must be applied to water, relative to that required for air, to obtain a small fractional change in ρ.*

Solution. Since Equation (1.34) implies $\Delta \rho / \rho = \tau \Delta p$, we have

$$\frac{(\Delta \rho / \rho)_{H_2O}}{(\Delta \rho / \rho)_{air}} = \frac{\tau_{H_2O}}{\tau_{air}} \frac{\Delta p_{H_2O}}{\Delta p_{air}} \quad \Longrightarrow \quad \frac{\Delta p_{H_2O}}{\Delta p_{air}} = \frac{\tau_{air}}{\tau_{H_2O}} \frac{(\Delta \rho / \rho)_{H_2O}}{(\Delta \rho / \rho)_{air}}$$

So, to achieve the same fractional change in density, $\Delta \rho / \rho$, the pressure change varies inversely with compressibility. Appealing to Equation (1.36), we find

$$\frac{\Delta p_{H_2O}}{\Delta p_{air}} = \frac{4.79 \cdot 10^{-4} \text{ ft}^2/\text{lb}}{2.23 \cdot 10^{-8} \text{ ft}^2/\text{lb}} = 2.15 \cdot 10^4$$

Fluids that require very large changes in pressure to cause even a minor change in density are termed **incompressible**. In a mathematical sense, we can say incompressible flow occurs when $\tau \to 0$, so that $\Delta \rho = \rho \tau \Delta p \to 0$. Hence, for our immediate purposes, we will consider an incompressible flow to be constant-density flow. As we will see in Chapter 8, this is not the conventional definition. It would imply that motion in a "stratified" medium such as the ocean whose density varies with depth should be classified "compressible." This is incorrect.

When we study compressibility effects we will find that a better definition of incompressible flow is **very low Mach number flow**, where Mach number is the ratio of flow speed to the speed of sound. This is a far better indicator of incompressibility, and is satisfactory for most practical applications.[4]

The Mach number appears in a more precise differentiation between incompressible and compressible flows because it tells us something about how the flow adjusts to changes. If there is a disturbance in the flow caused, for example, by the motion of an object, waves propagate through the flow. This mechanism permits the flow to adjust itself and reach a new steady equilibrium state corresponding to the altered flow conditions. For a compressible fluid, these waves travel at a finite speed. In the case that the disturbance is weak, the wave travels at the speed of sound.

---

[4]Even this definition has its limitations. For example, flow in an internal-combustion engine occurs at low Mach number, yet has large density changes caused by the very high pressure within the combustion chamber.

As we will show in Chapter 8, we can calculate the speed of sound, a, for any perfect gas from the following equation.

$$a = \sqrt{\gamma R T} \qquad \text{(Perfect gas)} \qquad (1.37)$$

where γ is specific heat ratio defined in Equation (1.31), R is the perfect-gas constant (see Table A.1) and T is absolute temperature. Now, using the perfect-gas law and Equation (1.35) for compressibility, τ, the speed of sound in a gas thus depends upon τ according to

$$a = \sqrt{\frac{\gamma}{\rho \tau}} \qquad (1.38)$$

So, for a gas, the speed of sound approaches ∞ as $\tau \to 0$. Although we must rely upon measurements, the same proves to be true for liquids. Table A.4 in Appendix A lists the speed of sound for several common fluids and shows, for example, that the speed of sound in water is more than four times the speed in air.

Thus, the speed of sound for an incompressible fluid is large so that acoustic waves travel at very high speed and the entire flow will approach its new equilibrium state very quickly.

Until we specifically address effects of compressibility in Chapter 8, most of our applications will be for incompressible flows. This greatly simplifies the equations we have to deal with and actually precludes the need to include the energy-conservation principle for most problems. This means that incompressible-flow problems can be solved using only mass and momentum conservation concepts. This by no means limits our study to a small class of fluid-flow problems. Incompressible flow covers the flow of liquids and "low-speed" gas flows, where low speed means the Mach number is less than about 0.3. Examples are the motion of a torpedo, an automobile and a Cessna 310 with typical speeds of 40 mph, 65 mph and 200 mph, respectively. Based on these speeds and the sound speeds listed in Table A.4, the Mach numbers would be 0.012, 0.085 and 0.262.

1.7 Surface Tension

There is an interesting effect that occurs near liquid-gas and immiscible-liquid interfaces, viz., surface tension. At such an interface, we find the surface to be like a stretched membrane. A spider can walk on water, a steel needle placed gently in a pan of water floats, water "beads up" on your freshly waxed car. These are all commonly observed examples of surface tension.

Surface tension is actually a very complex phenomenon, involving concepts from physical chemistry [cf. Adamson (1960)]. A simplified explanation that captures the dominant physics is as follows. On the one hand, molecules within the main body of the fluid are surrounded by molecules with attractive forces equal in all directions. On the other hand, molecules at the surface are attracted only by molecules from within the fluid and thus exhibit a stronger attraction for other surface molecules to achieve an overall balance of molecular forces. This modified "force field" results in the surface behaving like a stretched membrane.

We can thus describe surface tension mathematically by making an analogy to a membrane, in which a uniform tensile force, σ, acts tangent to the surface. However, there is no physical membrane present—only a distribution of molecular forces that acts like the tension in a membrane. *Note that, by definition, σ has dimensions force per unit length.*

For example, any curvilinear surface can be described in terms of two principal radii of curvature, $\mathcal{R}_1$ and $\mathcal{R}_2$ (Figure 1.12). In analogy to a stretched membrane, the pressure difference, Δp, that can be supported by a surface tension, σ, on a general surface is given by **Laplace's formula** [see Landau and Lifshitz (1966)]:

$$\frac{\sigma}{\mathcal{R}_1} + \frac{\sigma}{\mathcal{R}_2} = \Delta p \tag{1.39}$$

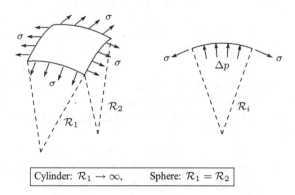

Cylinder: $\mathcal{R}_1 \to \infty$,   Sphere: $\mathcal{R}_1 = \mathcal{R}_2$

Figure 1.12: *Stretched-membrane analogy for surface tension.*

Alternatively, for a specific application, we can balance forces and arrive at the same result with some added insight into the physics involved. The pressure within a droplet, for example, is higher than the pressure outside. Surface tension supports the difference in pressure. Consider the bubble depicted in Figure 1.13, with the pressure forces and surface tension indicated. The net contribution from the atmosphere is p_a times the projection of the hemisphere onto a vertical plane. If this point is unclear, consider the limiting case $\Delta p = 0$ and $\sigma = 0$ in which the atmospheric pressure on the two surfaces must be in balance. Thus,

$$\sum F = (p_a + \Delta p)\,\pi R^2 - p_a \pi R^2 - 2\pi R\sigma = 0 \tag{1.40}$$

Simplifying, we find

$$\Delta p \pi R^2 = 2\pi R\sigma \quad \Longrightarrow \quad \Delta p = \frac{2\sigma}{R} \tag{1.41}$$

which is identical to what we obtained above using Laplace's formula (with $\mathcal{R}_1 = \mathcal{R}_2 = R$).

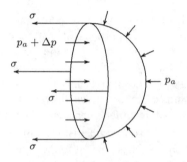

Figure 1.13: *Force diagram for a droplet.*

Example 1.10 *Compute the pressure within a 1 mm diameter spherical droplet of water relative to the atmospheric pressure outside.*

Solution. To apply Laplace's formula, we make an imaginary cut through the center of the droplet as shown in Figure 1.13. Since the principal radii of curvature are both equal to the sphere radius, $R = 5 \cdot 10^{-4}$ m, and the surface tension for water is 0.073 N/m, Equation (1.41) tells us

$$\Delta p = \frac{2\sigma}{R} = \frac{2(0.073 \text{ N/m})}{5 \cdot 10^{-4} \text{ m}} = 292 \ \frac{\text{N}}{\text{m}^2} = 0.003 \text{ atm}$$

As a second interesting example of the effect of surface tension, consider the phenomenon known as **capillary action**. A tube of diameter d is immersed in a container filled with a liquid of density ρ as shown in Figure 1.14. The fluid is drawn up into the tube a distance Δh. Surface tension, σ, acts around the perimeter of the tube and supports the weight of the column of fluid. The surface tension is tangent to the surface as shown, making an angle ϕ with the tube surface. The vertical component of the surface tension balances the weight of the column of fluid so that

$$\pi d\, \sigma \cos\phi = \rho g \Delta h(\pi d^2/4) \tag{1.42}$$

Thus, the average height to which the fluid is drawn up the tube is given by

$$\Delta h = \frac{4\sigma}{\rho g d} \cos\phi \tag{1.43}$$

The height is an average since our computation approximates the column of fluid as being exactly rectangular. In reality, its shape is curved as shown in Figure 1.14, corresponding to the commonly observed **meniscus**.

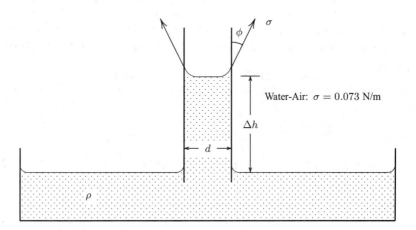

Figure 1.14: *Capillary action—cross-sectional view. Note that σ has dimensions of force per unit length. It acts along an arc, not on a surface.*

In general, surface tension depends upon the surface material as well as the nature of the two fluids involved, and it is a function of pressure and temperature. It is also considerably affected by dirt on the surface. Table A.5 in Appendix A lists surface tension for several common liquids. The angle ϕ is called the **wetting angle** or the **contact angle**. For water interfacing air in a glass tube, we find that $\phi \approx 0°$. By contrast, for mercury interfacing air in a glass tube, the surface tension acts downward with $\phi \approx 129°$. Similarly, for water, air and paraffin wax, the wetting angle is 105° [cf. Sabersky et al. (1989)]. As a consequence of having $\phi > 90°$, the column of fluid is drawn downward by capillary action in these two cases.

Example 1.11 *Consider a small capillary tube made of glass with diameter d = 1 mm. The liquid is water, and the tube is open to the atmosphere so that ρ = 998 kg/m³, σ = 0.073 N/m and ϕ = 0°. How many tube diameters will the water rise in the tube due to capillary action?*

Solution. Using Equation (1.43), the height of the column of water drawn up the tube is

$$\Delta h = \frac{(4)(0.073 \text{ N/m})(1)}{\left(998 \text{ kg/m}^3\right)\left(9.807 \text{ m/sec}^2\right)(0.001 \text{ m})} = 0.0298 \text{ m} \approx 30 \text{ mm}$$

Thus, due to capillary action, the water is drawn 30 tube diameters above the container surface.

Example 1.12 *Now, change the fluid to mercury leaving all other conditions the same as in Example 1.11. Using ρ = 13550 kg/m³, σ = 0.466 N/m and ϕ = 129° determine the change in column height, Δh, in terms of tube diameters.*

Solution. Again using Equation (1.43), we obtain

$$\Delta h = \frac{(4)(0.466 \text{ N/m})(-0.629)}{\left(13550 \text{ kg/m}^3\right)\left(9.807 \text{ m/sec}^2\right)(0.001 \text{ m})} = -0.0088 \text{ m} \approx -9 \text{ mm}$$

Thus, because the wetting angle exceeds 90°, capillary action draws the fluid 9 tube diameters below the container surface.

AN EXPERIMENT YOU CAN DO AT HOME
Contributed by Prof. Ron Blackwelder, University of Southern California

To observe the effect of the fluid and/or wetting angle on capillary action, assemble the following items: a transparent straw or small glass tube; a coffee cup; water; and some liquid soap. If all you can find is a standard drinking straw, use coffee or some other darkly colored fluid instead of water. Fill the cup with your working fluid and insert the straw. Dip it in and out a couple times to make sure the surface is uniformly wetted—this will maximize the height, Δh, to which the fluid rises. Measure Δh. Now, add a teaspoon full of liquid soap and mix thoroughly. Repeat the experiment. If you have done everything correctly, Δh is reduced.

The explanation for what happens is as follows. Adding the soap reduces the fluid's surface tension, and probably the wetting angle as well. Hence, the fluid rises to a reduced level. Note that you should not expect Equation (1.43) to provide an accurate estimate of Δh unless you very carefully sterilize the straw!

1.8 Viscosity

Fluids such as honey or molasses tend to flow very slowly down an inclined plane. Water goes much faster. The former fluids are more **viscous**. Recall that for a container filled with a fluid at rest we observe only normal stresses. If the fluid is moving, we also observe oblique stresses on the walls of the container. In fact, these oblique stresses act everywhere in the fluid if the fluid is moving.

To understand the origin of viscosity, we need to take another look at what goes on at the molecular level. Referring to Figure 1.15, consider an imaginary plane in a fluid at $y = 0$, which is shown as the shaded area. The figure shows one molecule above the plane ($y > 0$) that moves, on the average, with velocity U_1. A second molecule lies below the plane ($y < 0$) and it is moving with an average velocity U_2. For the purpose of this discussion, we assume $U_1 > U_2$.

As discussed in Section 1.4, in addition to their mean velocity, the molecules are moving with random velocity fluctuations in both magnitude and direction. Of particular interest, this includes fluctuations in the y direction that can carry molecules close to $y = 0$ across the plane. Molecules migrating across $y = 0$ are **typical of where they come from**. That is, molecules moving up bring a momentum deficit and vice versa. This incremental change in momentum gives rise to a stress on a surface whose unit normal lies in the y direction and that acts in the x direction. We define this type of stress as a shear stress.

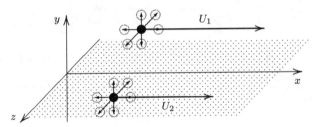

Figure 1.15: *Close-up view of molecular motion. Because of their random molecular motion, molecules can migrate across the (shaded) plane, $y = 0$. The momentum transfer gives rise to a shear stress.*

When you step off a conveyer belt (moving walkway), you experience this phenomenon. You are typical of where you came from, namely a region where you had nonzero forward momentum with respect to the stationary floor. As your feet touch the floor, you arrive in a region where you have an excess amount of momentum and your body will lurch forward until you regain your balance. Note that, assuming you step directly to your side to exit the belt, the "force" on your body is normal to the direction you have chosen to move.

In a viscous fluid, the velocity in the x direction, u, varies continuously in the y direction. For a majority of fluids encountered in engineering applications, we find that the shear stress, τ, is proportional to the slope of the velocity profile, du/dy. Thus, we can say that

$$\tau = \mu \frac{du}{dy} \tag{1.44}$$

where μ is the viscosity coefficient with units kg/(m·sec) [slug/(ft·sec)]. When τ is a linear function of du/dy, we have what is referred to as a **Newtonian fluid**. For some fluids like blood and long-chain polymers, we have a nonlinear relation between τ and du/dy. Such a fluid is called a **non-Newtonian fluid**.

Often in fluid mechanics, especially for incompressible flows, we find ourselves dealing with the ratio μ/ρ. This happens so frequently we choose to dignify the ratio with its own symbol and name. We define the **kinematic viscosity**, ν, as follows.

$$\nu = \frac{\mu}{\rho} \tag{1.45}$$

Kinematic viscosity has units of m^2/sec (ft^2/sec).

Measurements indicate that viscosity of both liquids and gases varies with temperature but shows virtually no dependence upon pressure. Usually $d\mu/dT$ and $d\nu/dT$ are positive for a gas while $d\mu/dT$ and $d\nu/dT$ are negative for a liquid. Table A.6 in Appendix A includes kinematic viscosities of common liquids and gases, while Table A.7 includes the variation of ν with temperature for water. For air and water at atmospheric pressure and 20° C (68° F), the kinematic viscosity is

$$\nu = \begin{cases} 1.51 \cdot 10^{-5} \text{ m}^2/\text{sec} \quad (1.62 \cdot 10^{-4} \text{ ft}^2/\text{sec}), & \text{Air} \\ 1.00 \cdot 10^{-6} \text{ m}^2/\text{sec} \quad (1.08 \cdot 10^{-5} \text{ ft}^2/\text{sec}), & \text{Water} \end{cases} \tag{1.46}$$

There are numerous empirical formulas that are used in general engineering applications. The following are three of the most commonly used equations. Note that, for all of these formulas, T is the absolute temperature, and is thus given either in Kelvins (K) or degrees Rankine (° R).

- **Sutherland's Law**: This is an empirical equation for the viscosity of gases that is quite accurate for a wide range of temperatures. The formula is

$$\mu = \frac{AT^{3/2}}{T + S} \tag{1.47}$$

where A and S are constants. The values of A and S for air in USCS units are

$$A = 2.27 \cdot 10^{-8} \frac{\text{slug}}{\text{ft·sec·}(°\text{R})^{1/2}}, \quad S = 198.6° \text{R} \qquad \text{(Air)} \tag{1.48}$$

The values for SI units are

$$A = 1.46 \cdot 10^{-6} \frac{\text{kg}}{\text{m·sec·K}^{1/2}}, \quad S = 110.3 \text{ K} \qquad \text{(Air)} \tag{1.49}$$

Appendix A.8 includes values of A and S for several common gases.

- **Power-law viscosity**: This is a common approximation for the viscosity of gases, which assumes the viscosity varies with temperature according to

$$\frac{\mu}{\mu_r} = \left(\frac{T}{T_r} \right)^{\omega} \tag{1.50}$$

where μ_r and T_r are reference viscosity and temperature, respectively. These constants and the exponent ω depend on the gas. Forsythe (1964) summarizes values for several gases.

- **Andrade's equation**: This is a useful empirical formula that is often used to approximate the viscosity of liquids. The viscosity varies with temperature according to

$$\frac{\mu}{\mu_r} = e^{T_r/T} \tag{1.51}$$

As with the power-law formula for gases, μ_r and T_r are reference viscosity and temperature, respectively. Using measured values of μ, the reference values can be determined.

1.9 Examples of Viscosity-Dominated Flows

There are some flows that are completely dominated by viscous effects. In this section, we consider two such flows, i.e., two-dimensional **Couette flow** and axisymmetric **Hagen-Poiseuille flow**. Our discussion here is, of necessity, heuristic and limited to examining properties of the solutions. A rigorous derivation requires developing the viscous-flow equations of motion, which is beyond the scope of this text. Wilcox (2000), for example, gives a complete development of the solutions for both of these flows.

1.9.1 Couette Flow

Couette flow occurs when we have a fluid between two infinite parallel planes as shown in Figure 1.16. The lower plane is at rest while the upper plane moves with constant velocity U, and the pressure is constant. This flow can be realized in a laboratory by having a conveyer belt close to a stationary surface. If the distance h is very small compared to the length of the belt, we approximate the idealized Couette-flow geometry provided we are not too close to the ends of the belt.

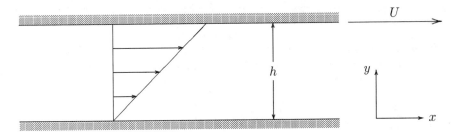

Figure 1.16: *Couette flow—the lower wall is at rest and the upper wall is moving.*

We expect the solution to depend only upon distance between the plates. If we look close to the solid boundaries, we observe that the fluid "sticks" to the boundaries. This is known as the **no-slip** surface boundary condition on the velocity. This boundary condition will be discussed in more detail in Section 11.3. Thus, our solution must satisfy $u(0) = 0$ and $u(h) = U$, where $u(y)$ is the velocity component in the x direction. Indeed, an exact solution to the fluid mechanics equations of motion exists in which flow properties are functions only of y. The velocity varies linearly with distance across the channel, viz.,

$$u(y) = U \frac{y}{h} \tag{1.52}$$

The corresponding shear stress is

$$\tau = \mu \frac{du}{dy} = \frac{\mu U}{h} \tag{1.53}$$

Hence, the shear stress is not confined to the walls of the channel. Rather, it is constant across the channel and acts throughout. We can rewrite Equation (1.53) in terms of nondimensional parameters as follows.

$$\frac{\tau}{\frac{1}{2}\rho U^2} = \frac{2}{(\rho U h / \mu)} \tag{1.54}$$

The dimensionless parameter on the left-hand side of Equation (1.54) is called the **skin-friction coefficient**, and is usually denoted by c_f. The dimensionless parameter in the denominator of the right-hand side of Equation (1.54) is called the **Reynolds number**, and is usually denoted by Re. The linear velocity profile, constant-shear-stress solution is observed experimentally provided

$$\frac{\rho U h}{\mu} < 1500 \tag{1.55}$$

This is known as the **laminar**-flow solution. At larger values of Reynolds number, this solution is unstable to small disturbances, and the flow becomes **turbulent**. When the flow is turbulent, the velocity and pressure vary rapidly with time and the effective shear stress is much larger than the laminar-flow value given above.

The Couette-flow solution is often used as a "building-block" solution. Any application that has flow through a small gap can be described in terms of the linear velocity variation in the gap. Practical applications involving flow in a small gap include flow between a computer hard-disk platter and read/write head, and the manufacturing process for a conventional recording tape. The Couette-flow solution (with the gap width, h, varying slowly in the streamwise direction) also plays a role in the classical *Lubrication Theory* developed by Osborne Reynolds in 1886.

Example 1.13 *In manufacturing recording tape, the tape is coated with a lubricant. This is done by pulling it through a narrow gap filled with the lubricant. Consider tape whose width and thickness are w and h, respectively. The maximum tensile force the tape can withstand is T. The tape is centered in a gap of height $3h$, the lubricant has viscosity μ, and the length of tape within the gap is ℓ. What is the maximum velocity at which the tape can be pulled through the gap?*

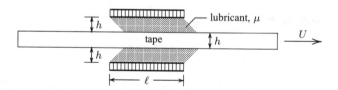

Solution. The forces on the tape are the tension force, T, and the friction force on each side of the tape. In steady state, these forces balance so that

$$T = 2\tau_w w\ell$$

where τ_w is the shear stress on the surface of the tape. Assuming the gap width is very narrow so that $h \ll \ell$, we can use the Couette-flow solution. Thus, the velocity in the gap is $u = Uy/h$, and the shear stress is

$$\tau_w = \mu\frac{du}{dy} = \frac{\mu U}{h}$$

So, the tension force is given by

$$T = 2\frac{\mu U w\ell}{h}$$

wherefore the maximum velocity at which the tape can be pulled through the gap is

$$U = \frac{1}{2}\frac{Th}{\mu w\ell}$$

1.9.2 Hagen-Poiseuille Flow

Hagen-Poiseuille Flow is a second simple example of a flow completely dominated by viscosity. We consider flow through an infinitely-long pipe with circular cross section (see Figure 1.17). As with Couette flow, this subsection discusses properties of what is actually an exact solution to the viscous-flow equations of motion.

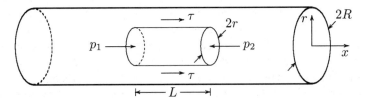

Figure 1.17: *Hagen-Poiseuille (pipe) flow.*

Consider the flow in a cylindrical section of the pipe of radius r and length L shown in Figure 1.17. Assume the pressure on the left side of the section is p_1 and the pressure on the right side of the section is p_2. We choose our sign convention so that positive shear stress on the cylindrical section is in the positive x direction as shown.

The pressure difference is balanced by the shear stress acting on the circumference of the cylindrical section. Hence, we have

$$p_1 \pi r^2 - p_2 \pi r^2 + 2\pi r L \tau = 0 \tag{1.56}$$

Thus, the shear stress is given by

$$\tau = -\frac{(p_1 - p_2)}{L} \frac{r}{2} \tag{1.57}$$

So, when $p_1 > p_2$, the shear stress is negative and therefore resists the motion. We can determine the velocity profile, $u(r)$, by integrating Equation (1.44) with r replacing y. The integration is done subject to the no-slip boundary condition that tells us $u(R) = 0$. The resulting expression for the velocity then becomes

$$u(r) = \frac{(p_1 - p_2)}{4\mu L} \left(R^2 - r^2 \right) = u_m \left(1 - \frac{r^2}{R^2} \right) \tag{1.58}$$

where u_m is the maximum velocity, which occurs on the centerline ($r = 0$) and is given by

$$u_m = u(0) = \frac{(p_1 - p_2)}{4\mu L} R^2 \tag{1.59}$$

We observe this (laminar-flow) solution experimentally provided the Reynolds number based on u_m is as follows.

$$\frac{\rho u_m R}{\mu} < 2300 \tag{1.60}$$

If great care is taken to prevent transition to turbulence by keeping disturbances to a minimum, laminar flow can be achieved for pipe flow at a Reynolds number as high as 40000 [Schlichting-Gersten (2000)]. We will study pipe flow in greater detail in Chapter 6.

Example 1.14 *The viscosity and density of human blood are* $\mu = 3.3 \cdot 10^{-3}$ *kg/(m·sec) and* $\rho = 1058$ *kg/m^3. The maximum speed of blood in the aorta of a large man is roughly* $u_m \approx 0.30$ *m/sec. If the diameter of the aorta is* $d \approx 3$ *cm, what is the Reynolds number,* $\rho u_m d / \mu$? *Will the flow be laminar or turbulent?*

Solution. For the given conditions, the Reynolds number based on diameter is

$$\frac{\rho u_m d}{\mu} = \frac{\left(1058 \ \frac{\text{kg}}{\text{m}^3}\right)\left(0.30 \ \frac{\text{m}}{\text{sec}}\right)(0.03 \ \text{m})}{3.3 \cdot 10^{-3} \ \frac{\text{kg}}{\text{m} \cdot \text{sec}}} = 2885$$

To have laminar flow, the Reynolds number based on *radius* must be less than 2300, which corresponds to Reynolds number based on *diameter* of 4600. Therefore, the flow is laminar.

1.10 Fluid-Flow Regimes

Figure 1.18 presents an outline of the major regimes of classical continuum fluid mechanics. It also serves as a "road map" for this text, indicating the chapters in which the various regimes are discussed.

The first major subdivision of the discipline is between the flow of inviscid and viscous fluids. An inviscid fluid is also referred to as an **ideal fluid**. We define an ideal fluid as one in which the stresses acting are everywhere normal to the element of surface on which they are measured, whether or not the fluid is moving. This is sometimes referred to as a **perfect fluid** or a **frictionless fluid**. This is a drastic simplification as compared to a viscous fluid, which is often referred to as a **real fluid**.

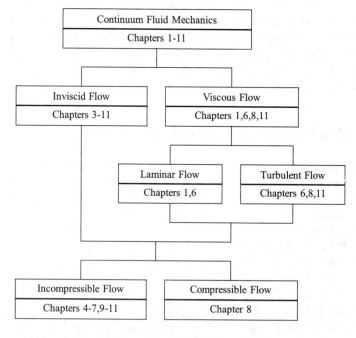

Figure 1.18: *Major regimes of fluid mechanics.*

For a moving fluid, the approximation of treating it as ideal cannot hold everywhere. For fluids of small viscosity such as water and air, we observe two important facts. On the one hand, since viscous stresses are proportional to velocity gradients [cf. Equation (1.44)], neglect of viscous stresses breaks down where velocity gradients are large. On the other hand, the flow around a streamlined body behaves almost as if the fluid were frictionless except in very thin layers next to the surface and in a thin wake downstream. These near-surface layers were discovered by Prandtl in 1904 who named them **boundary layers**. Velocity gradients are very large in boundary layers.

The concept of an Ideal Fluid is very useful, and much of our analysis in this text assumes effects of friction are unimportant. There are many interesting problems we can solve ignoring effects of friction, and the attendant mathematical simplification permits us to focus on the physical behavior of fluids without getting lost in the algebra.

However, there are problems for which an inviscid solution is completely unsatisfactory. Inviscid flow past a sphere or cylinder, for example, bears little similarity to the flow observed in nature. The inviscid solution predicts that there is no net force on the object which is contrary to physical observations. Even when we include viscosity, the flow past solid objects falls into two different categories.

On the one hand, at very low flow speeds, fluid motion is very smooth, and flow patterns remain unchanged from one instant to the next. This type of motion is termed **laminar flow**. We have seen two examples of exact laminar-flow solutions in Section 1.9. On the other hand, at higher flow speeds, laminar flow becomes unstable and undergoes transition to what is known as **turbulent flow**. Turbulence is a state in which all flow properties vary rapidly with time, with a wide range of excited frequencies. All laminar flows ultimately become turbulent, with 'higher flow speeds' corresponding to the mildest of motions such as a small puff of wind. Although extensive research efforts have focused on the topic, to paraphrase Prof. Richard Feynman, turbulence remains one of the most noteworthy unsolved scientific problems of modern times.

The final major subdivision addressed in this book is between compressible and incompressible flows. As discussed in Section 1.6, there are many incompressible flow applications in everyday life. It is equally true that compressible flow abounds in practical engineering design and application. Transmission of gas in pipelines at high pressure, flow past a commercial airliner, and flow beneath an Indy 500 racer all involve compressible flow. Chapter 8 focuses explicitly on compressible flow.

1.11 Brief History of Fluid Mechanics

This text focuses almost entirely on inviscid fluids. The concluding chapter touches briefly on the more complex analysis of viscous fluids. Approaching fluid mechanics in this manner, to some extent, reflects the chronological development of the subject. The quantitative approach to fluid mechanics began in earnest with Isaac Newton's seventeenth-century formulation of the laws of classical mechanics, including a postulate for how effects of friction should appear in a fluid. Eighteenth-century mathematicians, most notably Bernoulli, d'Alembert, Euler, Lagrange and Laplace, developed the theory of frictionless flow. One unsettling result of this early work was a famous corollary, known as d'Alembert's Paradox, proving that no forces can develop for inviscid flow past an object in an unbounded fluid. This is, of course, at variance with physical reality.

Eighteenth and nineteenth century attempts at applying Newton's law of friction to resolve d'Alembert's Paradox proved unsatisfactory, raising serious doubts about theoretical fluid mechanics. Largely rejecting theoretical methods, the field of hydraulics was formulated by researchers such as Chézy, Weber, Hagen, Poiseuille, Darcy and Weisbach. Their work was based almost entirely on empiricism and experimentation, with little relation to basic physics.

It wasn't until the end of the nineteenth century that a unified theory of fluid mechanics including both theoretical and experimental methods took hold. The first important developments that helped unify the theory (with key contributions by Froude, Rayleigh and Reynolds) were dimensional analysis and dynamic similitude. These techniques, although adding minimal physical understanding, helped correlate measurements and establish parallels between experimental observations and the differential equations of motion.

Prandtl's discovery of the boundary layer in 1904 resolved d'Alembert's Paradox, and showed that the early nineteenth century work of Navier and Stokes, which generalized Newton's law of friction, was indeed correct. The viscous theory of fluid motion advanced dramatically during the twentieth century with profound contributions from pioneers such as Kolmogorov, Prandtl, Taylor and von Kármán.

During the second half of the twentieth century, a third branch of the unified theory of fluid mechanics evolved, viz., Computational Fluid Dynamics, or CFD. Using digital computers, we can now solve very complex fluid-flow problems. Early researchers such as Richardson, Courant, Friedrichs, Lewy, Lax and von Neumann helped build the foundation of what has become an important tool for both basic and applied research activities. The list of active CFD researchers has expanded dramatically since the 1960's, and many of the most significant contributors in this rapidly developing field are currently expanding its horizons.

Figure 1.19: *The personal computer (PC) of the early twenty-first century has computing power comparable to that of a late 1980's Cray Y/MP supercomputer. Capable of performing in excess of 250 million floating-point operations per second, a PC with a 2.5 GHz Pentium IV processor brings complex CFD capability to the desktop of the engineer at very low cost. And, unlike the Cray Y/MP, it doesn't require liquid helium to cool any of its components!*

Problems

1.1 Using Tables 1.3 and 1.4, express 300 Btu, 275 hp and 14.6 knots in SI units.

1.2 Using Tables 1.3 and 1.4, express 500 cal, 30 L and 22 kW in USCS units.

1.3 A student weighs 110 lbs and stands 5 feet 4 inches tall. What is the student's mass in slugs? Determine the student's weight and height in kg and m.

1.4 In *Old Russian* units, a one-pound weight has a mass of 0.41 kg. What would the value of the gravitational acceleration, g_r, have to be for an object weighing 1 pound in *Old Russian* units to weigh 1 USCS pound with $g = 32.174$ ft/sec^2?

1.5 An astronaut weighs 2 *stones* on the Moon. Noting that a *stone* is 14 pounds, what is the astronaut's weight, in pounds, on Earth where the gravitational acceleration is 6 times that of the Moon?

1.6 In research studies involving highly explosive substances, a 19th-century scientist chose to express force in ounces (oz), time in milliseconds (msec) and length in inches (in). Beginning with the definition of the pound, 1 lb = 1 slug·ft/sec^2, infer the logical choice of units for mass.

1.7 In research studies involving highly explosive substances, a research scientist has chosen to express mass in kilograms (kg), time in milliseconds (msec) and force in kiloNewtons (kN). Beginning with the definition of the Newton, 1 N = 1 kg·m/sec^2, infer the logical choice of units for length.

1.8 In order to study atomic particles, a scientist chooses to express mass in atomic mass units (amu, 1 kg = 6.023·10^{26} amu), force in picoNewtons (pN) and length in nanometers (nm). Beginning with the definition of the Newton, 1 N = 1 kg·m/sec^2, infer the logical choice of units for time.

1.9 At an altitude of 100 km, the average distance between air molecules is approximately $3 \cdot 10^{-4}$ mm. Compute the number of molecules contained in a spherical volume of diameter 1 μm.

1.10 Experience shows that a gas fails to behave as a continuum when it has fewer than 10^{12} molecules per mm^3. If the temperature of air is 75° C, what is the lower bound on the pressure, in Pascals, for the continuum limit to be valid? **HINT:** Write the perfect gas law as $p = mnRT$, where m is mass of a molecule and n is number density.

1.11 A typical amoeba measures about 0.012 mm in diameter. You can assume its shape is a perfect sphere and that it is swimming in water.

(a) How many water molecules are there in an amoeba-sized volume?

(b) If an amoeba considers a fluid particle to be a spherical volume with 10^{-4} of its own diameter, how many molecules are there in such a fluid particle?

1.12 The average distance traveled by a molecule between collisions is called the *mean free path*, ℓ_{mfp}. The continuum limit is valid if the characteristic dimension of the flow is very large compared to ℓ_{mfp}. From the kinetic theory of gases, we know that $\ell_{mfp} = 1.5\nu/\sqrt{\gamma RT}$. Compute the ratio of the diameter of Imperial Standard 45 Gauge wire, $d = 0.0071$ cm, to the mean free path of helium at 15° C. Does flow of helium at this temperature past this type of wire fall within the continuum limit?

1.13 The average distance traveled by a molecule between collisions is called the *mean free path*, ℓ_{mfp}. The continuum limit is valid if the characteristic dimension of the flow is very large compared to ℓ_{mfp}. From the kinetic theory of gases, we know that $\ell_{mfp} = 1.5\nu/\sqrt{\gamma RT}$. Compute the ratio of the diameter of Imperial Standard 50 Gauge wire, $d = 0.001$ in, to the mean free path of carbon dioxide at 59° F. Does flow of carbon dioxide at this temperature past this type of wire fall within the continuum limit?

1.14 How long would it take to count the molecules in 1 mm^3 of air at STP with an extremely fast electronic counter that can count at a rate of ten billion molecules per minute?

1.15 How long would it take to count the molecules in a cubic micron of water (1 micron is 10^{-4} cm) with a very fast electronic counter that can count at a rate of five million molecules per second?

1.16 In a turbulent flow, the Kolmogorov length, η, is a measure of the smallest swirling motion—commonly referred to as an *eddy*. For flow near the windshield of an automobile moving at 65 mph, the Kolmogorov length is about 10^{-4} inch. Approximately how many air molecules are contained in a spherical volume with diameter η?

1.17 Given a temperature of $T = 40°\,C$ and a pressure of $p = 70$ kPa, determine the density for air, carbon dioxide, hydrogen, methane, nitrogen and oxygen.

1.18 If the pressure is 2000 psf and the temperature is $86°\,F$, what is the ratio of the density of water to the density of air, ρ_w/ρ_a?

1.19 Compute the weights, in Newtons, of 1 m^3 of carbon dioxide and of oxygen pressurized to 600 kPa and held at a temperature of $80°\,C$.

1.20 Compute the weights, in pounds, of 2 ft^3 of methane and of helium pressurized to 100 psi and held at a temperature of $-40°\,F$.

1.21 One kilogram of nitrogen is contained in a volume of 300 liters at a temperature of $-30°\,C$. What is the pressure?

1.22 Two pounds of methane are contained in a volume of 3 ft^3 at a pressure of 5 atm. What is the temperature?

1.23 What pressure (in atm) is required to compress air to a point where its density is 4% that of water if the temperature is $60°\,C$?

1.24 Find the effective molecular weight of air, where the units of molecular weight are slug/(slug-mole).

1.25 Referring to the periodic tables in your chemistry book, and noting that the units of molecular weight are kg/(kg-mole), determine the value of the perfect-gas constant, R, for methane, CH_4. How does this compare with the value of R in Table A.1 for methane?

1.26 Using the value of R in Table A.1 for carbon dioxide, what is its molecular weight in kg/(kg-mole)? Does this agree with what you learned in chemistry class?

1.27 Assuming water vapor can be treated as a perfect gas, what is its gas constant, R, if, for a pressure of 400 kPa and a temperature of $15°\,C$, the density is 3 kg/m^3?

1.28 Assuming water vapor can be treated as a perfect gas, what is its gas constant, R, if, for a pressure of 7200 psf and a temperature of $50°\,F$, the density is $5.1 \cdot 10^{-3}$ slug/ft^3?

1.29 Compute the specific gravity of ethyl alcohol, carbon tetrachloride, and mercury.

1.30 Verify that, at $20°\,C$ and 1 atm, if the gas constant is R, then the specific gravity of a gas is $SG = R_{air}/R$. Using this result, compute the specific gravity of CO_2, He, H_2, CH_4, N_2 and O_2.

1.31 Determine the specific weights of air and water, in lb/ft^3, at $68°\,F$ and 1 atm.

1.32 Determine the specific weights of air and water, in N/m^3, at $20°\,C$ and 1 atm.

1.33 For a *calorically-perfect* gas, the specific-heat coefficients, c_p and c_v, and hence the specific-heat ratio, $\gamma = c_p/c_v$, are all constant. For such a gas, derive expressions for c_p and c_v in terms of R and γ.

1.34 At high pressure and low temperature, the perfect-gas law is inaccurate. A better approximation is provided by the *van der Waal's* formula, which is

$$(p + a\rho^2)(1 - b\rho) = \rho RT$$

where $a = 27(RT_c/8)^2/p_c$ and $b = (RT_c/8)/p_c$. Determine the density, ρ_c, corresponding to p_c and T_c. **HINT:** Assume $p_c = N\rho_c RT_c$ and solve for N, noting that the resulting cubic equation for N has a triple root.

1.35 On a mountain-climbing expedition, climbers found that water boiled at $194°$ F. What was the atmospheric pressure? Express your answer in atm.

1.36 On a mountain-climbing expedition, climbers found that water boiled at $95°$ C at their first camp site. At the top of the mountain, they noted that water boiled at $85°$ C. How much lower was the atmospheric pressure at the top relative to its value at the first camp site? Express your answer in atm.

1.37 A closed tank contains liquid with open space above the liquid. All of the air has been removed.

(a) What will the pressure above the liquid be after several hours have passed if the liquid is carbon tetrachloride at a temperature of $20°$ C?

(b) If the liquid were water, at what temperature (to the nearest degree) would the same pressure be realized?

1.38 A closed tank contains liquid with open space above the liquid. All of the air has been removed.

(a) What will the pressure above the liquid be after several hours have passed if the liquid is gasoline at a temperature of $60°$ F?

(b) If the liquid were water, at what temperature (to the nearest degree) would the same pressure be realized?

1.39 The pressure near a $90°$ elbow of a piping system has been found by experiment to be well approximated by

$$p = p_a - \lambda p_a x/\mathcal{R}$$

where p_a is atmospheric pressure, x is distance from a point just upstream of the elbow, $\mathcal{R}$ is the radius of curvature and λ is a dimensionless constant. The working fluid is water.

(a) If the temperature is $68°$ F and cavitation occurs when $x = \mathcal{R}$, what is the value of λ?

(b) Using the results of Part (a) for λ, at what value of $x/\mathcal{R}$ will cavitation occur if the temperature increases to $86°$ F?

1.40 The pressure, p, near a pipe inlet can be estimated from a curve fit to measurements, which is

$$p = p_a e^{-s/s_o}$$

where p_a is atmospheric pressure, s is arclength and $s_o = 2$ m. The working fluid is water at $70°$ C.

(a) At what value of s will cavitation occur?

(b) To postpone the onset of cavitation to $s = 5$ m, what must the temperature be?

1.41 At a pressure of 500 psi, a sample of water occupies a volume of 1 ft^3. When the pressure increases to 3500 psi, the volume decreases by 1%. Estimate the value of the compressibility, τ, under these conditions, assuming the compressibility of water remains nearly constant.

1.42 What pressure change, Δp, starting from atmospheric pressure, is required to cause a 0.5% change in density for helium, mercury and water? Assume the process is isothermal, and express your answers in MPa (megaPascals).

1.43 What pressure change, Δp, starting from atmospheric pressure, is required to cause a 5% change in density for air, ether and glycerin? Assume $T \approx 60°$ F, and express your answers in atm.

1.44 When the pressure increases from 14.7 psi to 18 psi, what is the fractional change in density, $\Delta \rho / \rho$, for air? Assume constant temperature.

1.45 Assuming temperature is $15.6°$ C, compute the fractional change in density, $\Delta \rho / \rho$, for air and water caused by a change in pressure, $\Delta p = 0.213$ atm. The initial pressure is $p = 1$ atm.

1.46 If a gas is compressed through a frictionless process with no heat exchange, the pressure is related to density by the isentropic relation, $p = A \rho^\gamma$, where A is a constant and γ is the specific-heat ratio.

(a) Compute the compressibility, τ_s, for such a process. Express your answer in terms of γ and p.

(b) Compare your result with the isothermal τ for a gas when $p = 1$ atm.

1.47 At very high pressures, the compressibility of water is nearly constant. If the pressure on 1 liter of water increases from 50 atm to 500 atm, what is the change in volume?

1.48 An empirical formula relating pressure and density for seawater with temperature held constant is

$$p/p_a \approx (\alpha + 1)(\rho/\rho_a)^7 - \alpha$$

where p_a and ρ_a are conditions at the surface and the dimensionless constant α is approximately 3000. The pressure at the deepest part of the Pacific Ocean is about 1100 psi. What is the density of seawater at this depth in kg/m^3? What is the percentage difference from the value at the surface?

1.49 An empirical formula relating pressure and density for water with temperature held constant is

$$p/p_a \approx 3041(\rho/\rho_a)^n - 3040$$

where p_a and ρ_a are conditions at the surface and n is a dimensionless constant.

(a) According to this formula, what is the compressibility, τ, of water as a function of p, p_a and n?

(b) Appealing to Equation (1.36), determine the numerical value of the constant n.

1.50 Air surrounds a water droplet of diameter $D = 0.012$ inch. The pressure within the droplet is $\Delta p = 18$ psf above ambient. Determine the surface tension and the temperature.

1.51 Air surrounds a water droplet of diameter $D = 10$ mm. The pressure within the droplet is $\Delta p = 30$ Pa above ambient. Determine the surface tension and the temperature.

1.52 A small droplet of carbon tetrachloride is immersed in air. Compute the ratio of its diameter to the diameter it would have when immersed in water. You may assume the pressure within the droplet exceeds the ambient value by the same amount in both cases.

1.53 A small droplet of mercury is immersed in air. Compute the ratio of its diameter to the diameter it would have when immersed in water. You may assume the pressure within the droplet exceeds the ambient value by the same amount in both cases.

1.54 Compare the pressure difference between the interior and exterior of a water drop, Δp_{H_2O}, to that of a drop of mercury from a broken thermometer, Δp_{Hg}. The water drop is 1/3 cm in diameter and the drop of mercury is 1/2 cm in diameter. Assume the drops are spherical and that the air temperature is $20°$ C.

1.55 A spherical soap bubble has an inside radius, $R = 5$ mm, film thickness, $t \ll R$, and surface tension σ. Noting that a bubble has two liquid-air interfaces, find the difference between internal and ambient pressure, Δp, as a function of σ and R. Assuming $\sigma = 0.073$ N/m, what is Δp?

1.56 To what height will glycerin rise in a cylindrical capillary tube of diameter, $d = 0.01$ in? Assume the wetting angle is $\phi = 8°$.

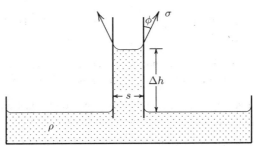

Problems 1.56, 1.57, 1.58, 1.59, 1.60

1.57 To what height will carbon tetrachloride rise in a cylindrical capillary tube of diameter, $d = 2$ mm? Assume the wetting angle is $\phi = 0°$.

1.58 To what height will mercury rise in a cylindrical capillary tube of diameter, $d = 1/6$ in? Assume the wetting angle is $\phi = 129°$.

1.59 Suppose a square tube of width s is immersed in a large container. If the fluid density is ρ, the surface tension is σ and the wetting angle is ϕ, to what height, Δh, will the fluid rise?

1.60 Suppose a rectangular tube s wide by $4s$ thick (out of the page) is immersed in a large container. If the fluid density is ρ, the surface tension is σ and the wetting angle is ϕ, to what height, Δh, will the fluid rise?

1.61 A round capillary tube of diameter d is inserted into a pan of water. A second tube, also of diameter d, is inserted into a pan of liquid whose density is $\frac{3}{2}\rho_w$, where ρ_w is the density of water. If the liquid rises to a height Δh in water and $0.2\Delta h$ in the unknown liquid, what is the surface tension of the unknown liquid? Assume the wetting angle is the same in both tubes and that $T = 68°$ F.

1.62 A small cylindrical capillary tube inserted in a pan of water has radius r. After the fluid rises in the tube to its equilibrium height, it is closed at the top and the pressure of the air trapped in the tube is increased to p_b. Assume that the wetting angle, ϕ, is $0°$, surface tension is σ and water density is ρ. Also, the gravitational acceleration is g and atmospheric pressure is p_a.

(a) Determine the value of Δh as a function of σ, ρ, g, r, p_a and p_b.

(b) Now, assuming $\sigma = 2rp_a$, what is the ratio of p_b to p_a required to make Δh half of what it would be if the capillary tube were open at the top?

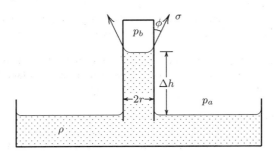

Problem 1.62

1.63 An open tube of diameter d is inserted into a pan of SAE 10W oil. A second tube of diameter $2d$ is inserted into a pan of seawater. Due to capillary action, the fluids rise to heights of Δh_1 and Δh_2 for the oil and seawater, respectively. Compute the ratio $\Delta h_2/\Delta h_1$. Assume the wetting angle is the same in both tubes and $T = 16°$ C.

1.64 Two tubes are used in a laboratory experiment to demonstrate capillary action. One tube has square cross section with side s. The other tube is cylindrical with radius r. Assume the wetting angle is the same for both tubes.

(a) Determine the value of s as a function of r required for the fluid to rise to the same level, h, in both tubes.

(b) Find the ratio of the volume of fluid in the square tube, V_{sq}, to the volume of fluid in the cylindrical tube, V_{cyl}. Explain why $V_{sq} > V_{cyl}$.

1.65 We want to approximate the viscosity of hydrogen with a power law, i.e., $\mu = \mu_r T^\omega$. Using *Sutherland's equation* (Table A.8) to determine μ when $T = 300$ K and 1000 K, compute μ_r and ω.

1.66 The viscosity of air can be approximated by a power law according to $\mu = AT^{0.7}$, where T is absolute temperature and A is a constant.

(a) Determine A by insisting that the value of μ at $T = 68°$ F matches the value given by *Sutherland's equation*.

(b) Make a graph of μ according to the power law developed in Part (a) and *Sutherland's equation* for $-250°$ F $\leq T \leq 2000°$ F.

1.67 The viscosity of water can be approximated by *Andrade's equation*, viz., $\mu = \mu_r e^{T_r/T}$, where T is absolute temperature. The quantities μ_r and T_r are empirical constants.

(a) Using values from Tables A.3 and A.7 for $T = 10°$ C and $T = 90°$ C, determine μ_r and T_r.

(b) Using the values of μ_r and T_r determined in Part (a), make a graph comparing the tabulated values of μ with *Andrade's equation* for $0°$ C $\leq T \leq 100°$ C.

1.68 The viscosity of water as a function of temperature can be represented by *Andrade's equation*., viz., $\mu = \mu_r e^{T_r/T}$. The empirical *Sutherland equation*, Equation (1.47), is often used to compute the viscosity of air. In both cases, the quantities A, S, T_r and μ_r are empirical coefficients that can be determined from measurements—all four are positive numbers. Using these formulas and assuming constant pressure, determine the sign of $d\nu/dT$ for water and for air.

1.69 Viscous flow close to a solid surface develops as shown in the figure. The region $0 \leq y \leq \delta$ is called the *boundary layer*, δ is the boundary-layer thickness and U_e is the boundary-layer edge velocity. Determine the surface shear stress (the value at $y = 0$) for the following velocity profiles.

(a) $u(y) = U_e \dfrac{y}{\delta}$.

(b) $u(y) = U_e \sin\left(\dfrac{\pi}{2}\dfrac{y}{\delta}\right)$.

(c) $u(y) = U_e \dfrac{y}{\delta}\left[\dfrac{3}{2} - \dfrac{1}{2}\left(\dfrac{y}{\delta}\right)^2\right]$.

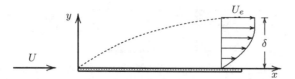

Problem 1.69

1.70 *Plane Poiseuille* flow corresponds to flow between two infinite parallel plates. The velocity distribution, $u(y)$, is given by

$$u(y) = u_m \left(\frac{y}{h} \right) \left(1 - \frac{y}{h} \right)$$

where u_m is constant, y is measured from the lower wall and the space between the walls is h. Plot the velocity profile as u/u_m versus y/h, and determine the shear stress at both walls as a function of μ, u_m and h.

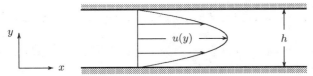

Problem 1.70

1.71 Heavy cream is a non-Newtonian fluid whose shear stress can be approximated by

$$\tau \approx 0.12 \left(\frac{du}{dy} \right)^{2/3}$$

where all quantities are expressed in SI units.

(a) Determine the effective viscosity, μ_e, such that $\tau = \mu_e du/dy$.

(b) Compute the ratio of μ_e to the viscosity of water, μ_w, at 20° C.

(c) Make a graph of μ_e/μ_w for values of du/dy between 0.01 sec^{-1} and 10 sec^{-1}.

1.72 A *rotating-cylinder viscometer* can be used to measure the viscosity of a liquid. This device consists of a cylinder of radius R_i and length L rotating at angular velocity Ω inside a concentric cylinder of radius R_o. The device is designed to have $R_o - R_i \ll R_i$ and $R_o - R_i \ll L$. Determine the torque, $\mathcal{T}$, on the rotating cylinder as a function of R_i, R_o, L, Ω and μ. Neglect any torque on the bottom of the inner cylinder.

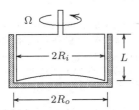

Problem 1.72

1.73 A plate moving with velocity U in its own plane lies between two stationary parallel plates as shown with $m = 2$. The viscosity of the fluid below the moving plate is μ, while the fluid above has viscosity $\mu_u = 2\mu$.

(a) Make a graph of the velocity, u/U, versus y/h for $0 \leq y/h \leq 3$.

(b) Compute the shear stress on the upper and lower walls.

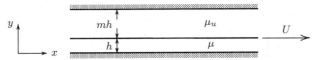

Problems 1.73, 1.74

1.74 A plate moving with velocity U in its own plane lies between two stationary parallel plates as shown. What must the viscosity of the fluid above the moving plate be, as a function of μ and m, in order for the magnitude of the shear stress above the plate to be one third the value below?

1.75 Compute the average velocity, $\overline{u}$, and the average specific kinetic energy, $\overline{\frac{1}{2}u^2}$, for Couette flow defined by

$$\overline{u} \equiv \frac{1}{h}\int_0^h u(y)dy \quad \text{and} \quad \overline{\frac{1}{2}u^2} \equiv \frac{1}{h}\int_0^h \frac{1}{2}u^2(y)dy$$

Problems 1.75, 1.76, 1.77

1.76 Measurements show that the shear stress for Couette flow is $\tau = 0.12$ lb/ft^2. The plate-separation distance is $h = 0.75$ in and the velocity is $U = 10$ ft/sec.

(a) What is the viscosity of the fluid between the plates?

(b) What is the maximum density for which the answer of Part (a) is valid?

1.77 Measurements show that the shear stress for Couette flow is $\tau = 18$ Pa. The plate-separation distance is $h = 0.4$ mm and the velocity is $U = 6$ m/sec.

(a) What is the viscosity of the fluid between the plates?

(b) What is the maximum density for which the answer of Part (a) is valid?

1.78 A solid circular cylinder of radius r and length ℓ slides inside a vertical smooth pipe having an inside radius R. A frictionless guide wire (not shown) maintains the cylinder's orientation. The small space between the cylinder and pipe is lubricated with a thin film that has viscosity μ. Derive a formula for the rate of descent, dU/dt, of the cylinder in the vertical pipe. Assume that the cylinder has weight W and is concentric with the pipe as it falls. Use the formula to find the terminal velocity (the velocity when $dU/dt \rightarrow 0$) of a 5.94 in diameter cylinder that slides inside a 6 in diameter pipe. The cylinder is 12 in long and weighs 3 lb.

(a) Determine the terminal velocity for water at 68° F, SAE 10W Oil at 100° F and glycerin at 68° F.

(b) For which of the three cases in Part (a) is the Couette-flow solution valid?

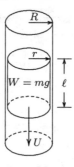

Problems 1.78, 1.79

1.79 A solid circular cylinder of radius r and length ℓ slides inside a vertical smooth pipe having an inside radius R. A frictionless guide wire (not shown) maintains the cylinder's orientation. The small space between the cylinder and pipe is lubricated with a thin film that has viscosity μ. Derive a formula for the rate of descent, dU/dt, of the cylinder in the vertical pipe. Assume that the cylinder has weight W and is concentric with the pipe as it falls. Use the formula to find the terminal velocity (the velocity when $dU/dt \rightarrow 0$) of a 99 mm diameter cylinder that slides inside a 101 mm diameter pipe. The cylinder is 200 mm long and weighs 20 N. The lubricant is glycerin at 20° C.

1.80 Insulating paint is applied to a wire by pulling it through a cylindrical orifice of radius R. The wire, of radius r, is centered in the orifice whose length is ℓ. The viscosity of the paint is μ. What force, F, is required to pull the wire at a constant velocity, U? You may assume $(R - r) \ll R$ and $(R - r) \ll \ell$.

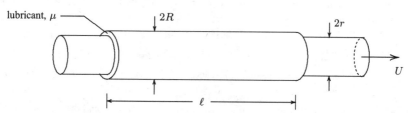

Problem 1.80

1.81 A cylinder of mass M—supported by a frictionless guide wire (not shown) to maintain its orientation—slides horizontally in a lubricated, but otherwise empty, pipe. The radial clearance between cylinder and pipe is d. The diameter and length of the cylinder are D and L, respectively. Assuming the cylinder's deceleration is $-A$ when the speed is U, derive a formula for the viscosity of the fluid. If the weight of the cylinder is 60 lb, $d = 0.012$ in, $A = 3$ ft/sec^2, $U = 2$ ft/sec, $D = 6$ in and $L = 10$ in, determine the value of μ in slug/(ft·sec).

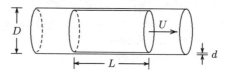

Problems 1.81, 1.82

1.82 A cylinder of mass M—supported by a frictionless guide wire (not shown) to maintain its orientation—slides horizontally in a lubricated, but otherwise empty, pipe. The radial clearance between cylinder and pipe is d. The diameter and length of the cylinder are D and L, respectively. Assuming the cylinder's deceleration is $-A$ when the speed is U, derive a formula for the viscosity of the fluid. If the weight of the cylinder is 100 N, $d = 0.025$ mm, $A = 1$ m/sec^2, $U = 7$ m/sec, $D = 15$ cm and $L = 8$ cm, determine the value of μ in kg/(m·sec).

1.83 The platter in a standard hard-disk drive rotates at $\Omega = 7200$ rpm. The platter diameter, D, is 9.5 cm and the read-write head flies $h = 12.5$ nanometers (nm) from the platter. Assuming the air in the drive is at 20° C, can flow between the head and the platter be approximated as Couette flow?

1.84 You were bored waiting for commercials to scan by while watching your favorite documentary on fluid mechanics. You decided to fix your VCR motor, increasing the scanning speed for a 6-hour tape. In Super Long Play (SLP) mode, 3 minutes of commercials now scan in 15 seconds instead of 50 seconds. Compute the tape speed, U, before and after the alteration. By what factor did you increase the tension in the tape as it passes by the head? You may assume the Couette-flow solution holds in the small gap between the tape and the head. A standard 6-hour tape is 807 feet long.

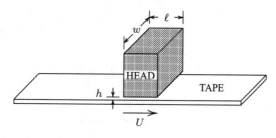

Problem 1.84

1.85 Using the pipe-flow solution, we can analyze laminar flow of blood in the aorta.

(a) Show that, for a given pressure difference, the volume-flow rate, Q, increases as R^4, where $Q \equiv 2\pi \int_0^R u(r)r dr$.

(b) Now, write your answer in terms of the maximum velocity, u_m, and the radius, R. The maximum velocity in the aorta is $u_m \approx 30$ cm/sec and the radius is $R \approx 1.5$ cm. How long does it take for 100 cm^3 of blood to flow through the aorta?

1.86 Using the pipe-flow solution, we can analyze laminar flow of blood in the aorta.

(a) Show that, for a given pressure difference, the volume-flow rate, Q, increases as R^4, where $Q \equiv 2\pi \int_0^R u(r)r dr$.

(b) Now, write your answer in terms of the maximum velocity, u_m, and the radius, R. The maximum velocity in the aorta is $u_m \approx 1$ ft/sec and the radius is $R \approx 0.6$ in. How long does it take for 10 in^3 of blood to flow through the aorta?

1.87 A defective pump moves water at $10°$ C through a small tube of diameter $D = 0.24$ in and length $L = 10$ in. It creates a suction-pressure difference of $p_1 - p_2 = 0.01$ psi.

(a) What is the flow speed at the center of the tube's cross section?

(b) What is the *Reynolds number*, $Re_R \equiv u_m R/\nu$?

(c) What is the total viscous force acting on the tube in pounds?

1.88 A defective pump moves water at $10°$ C through a small tube of diameter $D = 6$ mm and length $L = 20$ cm. It creates a suction-pressure difference of $p_1 - p_2 = 90$ Pa.

(a) What is the flow speed at the center of the tube's cross section?

(b) What is the *Reynolds number*, $Re_R \equiv u_m R/\nu$?

(c) What is the total viscous force acting on the tube in Newtons?

1.89 We wish to analyze flow in a pipe of radius R.

(a) Compute the average velocity for pipe flow, $\bar{u}$, defined by

$$\bar{u} \equiv \frac{1}{\pi R^2} \int_0^{2\pi} \int_0^R u(r)r dr d\theta$$

(b) Derive an equation for the *friction factor*, f, as a function of the *Reynolds number* based on pipe diameter, Re_D, according to the following definitions.

$$f \equiv \frac{-\tau(R)}{\frac{1}{8}\rho\bar{u}^2} \quad \text{and} \quad Re_D \equiv \frac{\rho\bar{u}D}{\mu}$$

1.90 Derive an equation for the *skin friction*, c_f, as a function of *Reynolds number* based on pipe radius, Re_R, for pipe flow, where R denotes pipe radius. By definition,

$$c_f \equiv \frac{-\tau(R)}{\frac{1}{2}\rho u_m^2} \quad \text{and} \quad Re_R \equiv \frac{\rho u_m R}{\mu}$$

1.91 Compute the average specific kinetic energy for pipe flow, $\overline{\frac{1}{2}u^2}$, defined by

$$\overline{\frac{1}{2}u^2} \equiv \frac{1}{\pi R^2} \int_0^{2\pi} \int_0^R \frac{1}{2}u^2(r)r dr d\theta$$

1.92 Water is flowing very slowly through a drain pipe of diameter $D = 5$ cm and length $L = 120$ m. The pressure difference between inlet and outlet is $(p_1 - p_2) = 0.04$ kPa, and the temperature is $20°$ C. Compute the maximum velocity and the Reynolds number of the water flowing in the pipe.

Chapter 2

Dimensional Analysis

Dimensional analysis is a particularly valuable tool whose applicability extends to all branches of physics. It is used extensively in fluid mechanics, especially in correlating experimental data. On the theoretical side, dimensional analysis is indispensable in checking consistency of algebraic relationships derived from basic principles. It is especially helpful in determining the relevant scales in a given problem, correlating experimental data and extrapolating measurements on small-scale models to large-scale objects. Some of the most important insight into the complicated phenomenon of turbulence has come from dimensional analysis [cf. Kolmogorov (1941), Bradshaw (1972), Tennekes and Lumley (1983), Landahl and Mollo-Christensen (1992) or Wilcox (1998)]. This chapter presents the methodology involved in dimensional analysis.

For many engineering students, these glowing remarks about dimensional concepts may appear surprising. Often, deriving an algebraic formula proves far less tedious than the frustrating exercise that follows in substituting actual numbers into the formula. The frustration is invariably associated with the units of dimensional quantities. For example, if a length is given in inches, it has to be divided by 12 to convert to feet in order to be consistent with a velocity given in feet per second. And that is a relatively minor issue compared to figuring out how many foot-pounds there are in a British thermal unit or foot-pounds per second in a horsepower! When the problem employs metric units, even an experienced engineer sometimes has to look up the definition of some of the more obscure units such as the Pascal, the Joule and the Watt.

Herein lies an important distinction that we touched on briefly in Section 1.1, namely, the difference between the separate concepts of **dimensions** and **units**. By convention, we refer to the dimensions of a physical quantity in a generic sense. For example, we say that the dimensions of velocity are length per unit time, without reference to any specific measure. When we use the term units, we make specific reference to a consistent set of measures that characterize the dimensions. Hence, we say the units of velocity are feet per second, meters per second or perhaps even (the legendary and rarely used) furlongs per fortnight. For most engineering students (and even those of us with years of experience), the problems encountered in computing numerical answers are usually associated with units.

With dimensional analysis, we deal with dimensions, so that our primary operations are unaffected by any particular choice of units. However, as we will discover, a solid grasp of the principles of dimensional analysis will even simplify the chore of cranking out a number from the elegant and glorious equations we work so hard to derive!

2.1   Basic Premises

Clearly, the laws of physics are independent of the system of units we choose. Recall from the discussion in Section 1.1 that, in fluid mechanics, the two most commonly used sets of units are Standard International (SI) units and the U. S. Customary System (USCS). In these two systems, we express mass, length, time and temperature as follows.

System	Mass, M	Length, L	Time, T	Temperature, Θ
SI	kilogram (kg)	meter (m)	second (sec)	Kelvin (K)
USCS	slug (slug)	foot (ft)	second (sec)	$^\circ$ Rankine ($^\circ$R)

In dimensional analysis, we introduce a symbolic representation of the basic units. Specifically, we let M, L, T and Θ denote mass, length, time and temperature, respectively. We have already found this to be useful in converting from one set of units to another (Subsection 1.1.3). As we will see in this chapter, this concept has even greater utility.

The dimensions of a given quantity are indicated by enclosing the quantity in square brackets. Thus, a typical dimensional-analysis equation for the viscosity of a fluid is

$$[\mu] = \frac{M}{LT} \tag{2.1}$$

In words, this equation says the dimensions of viscosity are mass per unit length per unit time. If we are using SI units, the viscosity thus has units kg/(m·sec). Similarly, in USCS units, its units are slug/(ft·sec).

Independent Dimensions. The four dimensions listed above are defined to be **independent dimensions**. All other dimensions of interest can be expressed in terms of the independent dimensions. For mechanical systems, M, L, T and Θ are sufficient to define all dimensional quantities. If electrical properties are included, we gain an additional independent dimension such as charge or resistance.

Example 2.1 *Express fluid pressure in terms of independent dimensions.*

Solution. Pressure is normally expressed as a force per unit area, such as Newtons per square meter or pounds per square foot. As a mnemonic device, we can take advantage of our knowledge of Newton's second law of motion to conclude that force is simply mass times acceleration. Hence,

$$[p] = \frac{\text{Force}}{L^2} = \frac{ML/T^2}{L^2} = \frac{M}{LT^2}$$

Thus, for the purposes of dimensional analysis, the units of pressure in SI and USCS units are kg/(m·sec^2) and slug/(ft·sec^2), respectively.

For the sake of maximum utility, any meaningful equation relating physical quantities should be valid regardless of the system of units employed. Not all formulas quoted in engineering literature satisfy this condition, and this limits their value. For example, most sailing enthusiasts are familiar with the formula for the **theoretical hull speed**[1] given by

$$U_{hull} = 1.34\sqrt{L_w} \tag{2.2}$$

---

[1]Beyond this speed, a single-hull boat encounters a rapid increase in the force resisting its forward motion known as "wave drag." This precludes any further increase in speed.

Figure 2.1: *A large sailing vessel from the past—its maximum speed was proportional to the square root of the length of its hull at the waterline. [Painting by William H. Partridge (1936)]*

where U_{hull} is the sailboat speed in knots (nautical miles per hour) and L_w is the length of the hull at the waterline in feet. This tells us that for $L_w = 20$ feet the theoretical hull speed is 6 knots. Our knowledge of dimensions also tells us that the coefficient 1.34 is not a pure number, but rather has the dimensions of $L^{1/2}/T$, which can be seen from the following dimensional equivalent of Equation (2.2).

$$\underbrace{[U_{hull}]}_{L/T} = \underbrace{[1.34]}_{L^{1/2}/T} \underbrace{[\sqrt{L_w}]}_{L^{1/2}} \tag{2.3}$$

Substituting a value for L_w in any units other than feet thus yields a meaningless result for the hull speed. The coefficient 1.34 would have to be replaced with an alternative value to permit specifying L_w in meters, for example.

For reasons that will become clear later in this chapter (when we define the *Froude number*), we should expect to see the acceleration due to gravity, g, under the square root in Equation (2.2). So let's revise Equation (2.2) in this manner, using the information that a 20 foot hull (at the waterline) yields a hull speed of 6 knots, which reference to Table 1.4 of Subsection 1.1 shows is 10.1 ft/sec. Hence, since the acceleration of gravity is $g = 32.174$ ft/sec^2, we find $U_{hull}/\sqrt{gL_w} = 0.40$. Therefore, we propose the following alternative to Equation (2.2).

$$U_{hull} = 0.40\sqrt{gL_w} \tag{2.4}$$

In this revised equation, the number 0.40 is dimensionless because $\sqrt{gL_w}$ has the same dimensions as velocity. Substituting a value for L_w of 20 feet yields a hull speed of 10.1 ft/sec, or 6 knots. However, unlike Equation (2.2), if we quote L_w in meters, we obtain the correct value of the hull speed in meters per second, viz., 3.09 m/sec. Reference to Table 1.4 shows that this is also 6 knots.

Dimensional Homogeneity. One of the most important facts that serves as the foundation of dimensional analysis is the following. All equations developed from the basic laws of physics are **dimensionally homogeneous**. That is, the dimensions of all terms appearing in such an equation are the same. Equation (2.4) is an example.

Despite the fact that we may be completely unfamiliar with a given equation, we can still assess its dimensional consistency with the information already at our disposal. This is an important point about dimensional analysis. Most of our focus is on the relationships amongst the dimensions of physical quantities, and not on the equations they happen to satisfy. This is especially useful when we begin our study of a new field, where we may be unfamiliar with the governing equations. Using dimensional analysis, we can quickly determine key parameters governing the basic physics of a given application, *independent of the detailed equations involved.*

Example 2.2 *The partial differential equation governing the one-dimensional diffusion of heat along a pipe filled with motionless water is*

$$\frac{\partial h}{\partial t} = \nu \frac{\partial^2 h}{\partial x^2}$$

where t is time, h is enthalpy (dimensions L^2/T^2), ν is kinematic viscosity (dimensions L^2/T) and x is distance along the pipe. Verify that this equation is dimensionally homogeneous.

Solution. First, summarize the dimensions of each quantity appearing in the equation.

$$[h] = L^2/T^2, \quad [t] = T, \quad [\nu] = L^2/T, \quad [x] = L$$

Differentiating with respect to time and space is equivalent to dividing by T and L, respectively. The dimensions of the left and right sides of the equation are as follows.

$$\left[\frac{\partial h}{\partial t}\right] = \frac{L^2/T^2}{T} = \frac{L^2}{T^3} \quad \text{and} \quad \left[\nu \frac{\partial^2 h}{\partial x^2}\right] = \frac{L^2}{T}\frac{L^2/T^2}{L^2} = \frac{L^2}{T^3}$$

Thus, the dimensions of both sides of the differential equation match so that, by definition, the equation is dimensionally homogeneous.

Obviously, one purpose dimensional analysis serves is to check for dimensional consistency of algebraic results. This is especially helpful when the result has been obtained from a lengthy algebraic exercise. If this were the only end served by dimensional analysis, it would hardly deserve an entire chapter of discussion. However, as we will see in the following sections, the method has even more important uses. There are nontrivial conclusions we can draw regarding a given problem where the only information we use is the dimensions of relevant physical quantities. Also, the method provides the proper scaling laws that permit extrapolation of measurements made on small-scale models to full-scale objects.

There are two key facts to keep in mind as we study dimensional analysis. On the one hand, as already mentioned, we have no need to solve any differential equations. In fact, *we don't even have to know what the governing equations are.* The method would appear to be very powerful from this point of view. The most noteworthy example is the pioneering work of Kolmogorov (1941), which laid the theoretical foundation for much of our present-day understanding of turbulence. Many important aspects of his research were based on dimensional analysis. On the other hand, dimensional analysis is no substitute for understanding of the physics, and its use as an analytical tool should not be overestimated. Of greatest importance, be aware that *we establish the relevant physical processes when we select the dimensional quantities in our analysis*. Thus, in Kolmogorov's case, he combined great physical insight into the nature of turbulence with dimensional analysis to discover some of the phenomenon's mysteries.

2.2 Buckingham Π Theorem

Buckingham (1915) proved that for any given physical problem, the number of independent **dimensionless groupings** needed to correlate is n less than the total number of dimensional quantities, where n is the number of independent dimensions. By definition, a dimensionless grouping is an algebraic combination of the dimensional quantities that has no dimensions. Equation (1.54), for example, involves the dimensionless groupings known as skin friction, $\tau/(\frac{1}{2}\rho U^2)$, and Reynolds number, $\rho Uh/\mu$. Also, our discussion of a sailboat's theoretical hull speed noted that $U_{hull}/\sqrt{gL_w}$ is a dimensionless grouping. This general result regarding the number of dimensionless groupings is known as the **Buckingham Π Theorem**.

There are exceptions, however, and the number of dimensionless groupings can be larger. As we will see in Example 2.4 below, it is obvious from the dimensional analysis when this somewhat pathological case arises.

If a problem involves only mass, length and time as independent dimensions, then n is 3. If the dimensional quantities also involve temperature, then n increases to 4, provided mass, length and time are still relevant. This section first introduces the classical **indicial method** developed by Buckingham. After illustrating how we interpret the results of a dimensional analysis, the section gives an alternative method for finding dimensionless groupings developed by E. S. Taylor.

2.2.1 Finding Dimensionless Groupings

To illustrate the ramifications of this theorem, consider viscous flow past a sphere as illustrated in Figure 2.2. Our object is to correlate the force that develops on the sphere, F, in terms of the dimensionless parameters in the problem. Assuming the sphere is not heated or cooled and that gravitational effects (or those of any other external force) are of no consequence, there are a total of five dimensional quantities in the problem, viz.,

F = force on the sphere (MLT^{-2}) U = velocity of the fluid (LT^{-1})
μ = viscosity of the fluid ($ML^{-1}T^{-1}$) D = diameter of the sphere (L)
ρ = density of the fluid (ML^{-3})

Note that in selecting our dimensional quantities, we have made two assumptions regarding the relevant physical processes. We ignore temperature of the fluid and the sphere as dimensional quantities because of our assumption about heating or cooling. We exclude gravitational

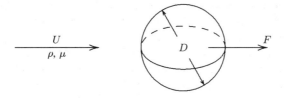

Figure 2.2: *Viscous flow past a sphere.*

acceleration, g, because of our assumption about external forces. This is the manner in which our physical understanding establishes the relevant dimensional quantities.

Appealing to the Buckingham Π theorem, we have a total of 5 dimensional quantities and 3 independent dimensions, i.e., M, L and T. Therefore, there are 2 dimensionless groupings. In order to find the two groupings, we postulate the **dimensional equation** relating the dimensions of F to the other four dimensional quantities, viz.,

$$[F] = [\mu]^{a_1}[\rho]^{a_2}[U]^{a_3}[D]^{a_4} \tag{2.5}$$

where a_1, a_2, a_3 and a_4 are unknown exponents. Our goal is to rearrange this equation so that it assumes a special form. In general, using dimensional analysis, we can show that the force on the sphere is the product of factors as follows.

$$[F] = [F_0] \prod_{i=1}^{N-1} [P_i]^{b_i} \tag{2.6}$$

where $N = 2$ is the number of dimensionless groupings, F_0 has dimensions of force, and the exponents b_i are functions of the a_i. The quantities P_i, known as **pi factors**, are dimensionless. To understand the motivation for the terminology "Π Theorem," note that by definition,

$$\prod_{i=1}^{N-1} [P_i]^{b_i} = [P_1]^{b_1}[P_2]^{b_2}[P_3]^{b_3} \cdots [P_{N-1}]^{b_{N-1}} \tag{2.7}$$

The operator $\prod$ is known as the **product pi**, and is analogous to the standard summation operator, $\sum$. In the present case, there will be one dimensionless grouping, P_1, on the right-hand side of our equation. The second dimensionless grouping predicted by the Buckingham Π Theorem, P_0, is the ratio of F to F_0. In other words, we are trying to arrive at a result of the form

$$\frac{F}{\text{Quantity with dimensions of force}} = \mathcal{F}(\text{dimensionless grouping}) \tag{2.8}$$

We use the following 4-step procedure, known as the **indicial method**, to arrive at the desired dimensionless groupings.

Step 1. To begin the process, we first identify the dimensions of all the dimensional quantities in the problem. While we have already done this above, we now state the results in algebraic form suitable for substitution into Equation (2.5). Thus, we have:

$$[F] = \frac{ML}{T^2}, \quad [\mu] = \frac{M}{LT}, \quad [\rho] = \frac{M}{L^3}, \quad [U] = \frac{L}{T}, \quad [D] = L \tag{2.9}$$

Step 2. Next, we substitute from Equation (2.9) into Equation (2.5). This yields an equation relating the independent dimensions, M, L and T, i.e.,

$$\begin{aligned}
MLT^{-2} &= M^{a_1}L^{-a_1}T^{-a_1}M^{a_2}L^{-3a_2}L^{a_3}T^{-a_3}L^{a_4} \\
&= M^{a_1+a_2}L^{-a_1-3a_2+a_3+a_4}T^{-a_1-a_3} \tag{2.10}
\end{aligned}$$

Since, from an algebraic point of view, M, L and T are independent variables, the only way this equation can be satisfied is to have identical exponents on both sides of the equation.

Thus, we arrive at the following three coupled, linear algebraic equations, known as the **indicial equations**.

$$\left.\begin{array}{r} a_1 + a_2 = 1 \\ -a_1 - 3a_2 + a_3 + a_4 = 1 \\ -a_1 - a_3 = -2 \end{array}\right\} \tag{2.11}$$

Observe that we have only 3 equations for 4 unknowns. In general, the number of unknowns will exceed the number of equations by one less than the number of dimensionless groupings predicted by the Buckingham Π Theorem.

While this will be the case in most applications, there is an exception. Specifically, the number of dimensionless groupings increases by one for each linearly-dependent equation present. An equation is **linearly dependent** if it can be derived as the sum of constant multiples of the other equations. For example, if the third of a set of three linear equations is the same as 2 times the first equation plus 3 times the second equation, then it is linearly dependent. None of Equations (2.11) can be derived as a linear combination of the others, and we thus describe them as being **linearly independent**. The indicial method almost always yields linearly-independent equations.

Step 3. We now solve the coupled set of equations for the exponents. This is usually straightforward since there is often weak coupling amongst the equations. In the present case, for example, the first and third of Equations (2.11) involve only two of the exponents. Thus, we find immediately that

$$a_1 = 2 - a_3 \tag{2.12}$$

while a_2 is given by

$$a_2 = 1 - a_1 = 1 - (2 - a_3) = a_3 - 1 \tag{2.13}$$

Thus, substituting these results into the second of Equations (2.11) permits solving for a_4 as a function of a_3, viz.,

$$-(2 - a_3) - 3(a_3 - 1) + a_3 + a_4 = 1 \qquad \Longrightarrow \qquad a_4 = a_3 \tag{2.14}$$

Step 4. We collect our solution for the exponents and substitute into Equation (2.5), wherefore

$$[F] = [\mu]^{2-a_3}[\rho]^{a_3-1}[U]^{a_3}[D]^{a_3} \tag{2.15}$$

Regrouping terms, we arrive at the final result, viz.,

$$[F] = \left[\frac{\mu^2}{\rho}\right]\left[\frac{\rho U D}{\mu}\right]^{a_3} \tag{2.16}$$

Therefore, the two dimensionless groupings are

$$P_0 = \frac{\rho F}{\mu^2} \quad \text{and} \quad P_1 = \frac{\rho U D}{\mu} \tag{2.17}$$

Finally, recalling Equation (2.8), we can also say that

$$F = \frac{\mu^2}{\rho}\mathcal{F}\left(\frac{\rho U D}{\mu}\right) \tag{2.18}$$

Example 2.3 *For flow of a viscous fluid close to a solid boundary rotating with angular velocity* Ω *(dimensions $1/T$), a region known as the Ekman layer develops. The thickness of the Ekman layer, δ, depends upon Ω, the freestream velocity, U, and the kinematic viscosity of the fluid, ν. Using dimensional analysis, develop an equation for δ as the product of a quantity independent of U with the same dimensions as δ and a function of all relevant dimensionless groupings.*

Solution. The dimensional quantities and their dimensions are

$$[\delta] = L, \qquad [\nu] = \frac{L^2}{T}, \qquad [\Omega] = \frac{1}{T}, \qquad [U] = \frac{L}{T}$$

There are 4 dimensional quantities and 2 independent dimensions (L, T), so that the number of dimensionless groupings is 2. The appropriate dimensional equation is

$$[\delta] = [\nu]^{a_1}[\Omega]^{a_2}[U]^{a_3}$$

Substituting the dimensions for each quantity yields

$$L = L^{2a_1}T^{-a_1}T^{-a_2}L^{a_3}T^{-a_3} = L^{2a_1+a_3}T^{-a_1-a_2-a_3}$$

Thus, equating exponents, we arrive at the following two equations:

$$
\begin{aligned}
1 &= 2a_1 + a_3 \\
0 &= -a_1 - a_2 - a_3
\end{aligned}
$$

We can solve immediately for a_1 from the first equation, viz.,

$$a_1 = \frac{1}{2} - \frac{1}{2}a_3$$

Substituting into the second equation yields

$$a_2 = -a_1 - a_3 = -\frac{1}{2} + \frac{1}{2}a_3 - a_3 \qquad \Longrightarrow \qquad a_2 = -\frac{1}{2} - \frac{1}{2}a_3$$

Substituting back into the dimensional equation, we have

$$[\delta] = [\nu]^{\frac{1}{2}-\frac{1}{2}a_3}[\Omega]^{-\frac{1}{2}-\frac{1}{2}a_3}[U]^{a_3} = \left[\sqrt{\frac{\nu}{\Omega}}\right]\left[\frac{U}{\sqrt{\nu\Omega}}\right]^{a_3}$$

Therefore, the Ekman layer thickness, δ is

$$\delta = \sqrt{\frac{\nu}{\Omega}} \, f\left(\frac{U}{\sqrt{\nu\Omega}}\right)$$

Before proceeding to interpretation of the results obtained in this subsection, it is worthwhile to pause and discuss three subtle points.

First, as an alternative to Equation (2.5), we could begin with the following.

$$1 = [F]^{a_0}[\mu]^{a_1}[\rho]^{a_2}[U]^{a_3}[D]^{a_4} \tag{2.19}$$

This equation corresponds to stating that products of all of the dimensional quantities raised to appropriate powers are dimensionless. Using the procedure above, we would arrive at an equation with two dimensionless groupings and two undetermined exponents, viz.,

$$1 = \left[\frac{\rho F}{\mu^2}\right]^{a_0}\left[\frac{\rho U D}{\mu}\right]^{a_3} \tag{2.20}$$

While the results are equivalent, the algebra is more tedious as we have 5 unknowns in place of 4. Thus, Equation (2.5) is the preferred starting point since it leads to fewer algebraic operations.

Second, suppose we had included a dimensional quantity such as the temperature of the freestream fluid, T_∞, and no other temperature. Upon completion of the dimensional-analysis algebra, we would find that

$$[F] = \left[\frac{\mu^2}{\rho}\right] \left[\frac{\rho U D}{\mu}\right]^{a_3} [T_\infty]^0 \tag{2.21}$$

i.e., the exponent of T_∞ is zero. This means either the temperature is irrelevant or we have inadvertently left out some other pertinent temperature. The converse is untrue, however. Had we included another irrelevant quantity such as surface tension, we would arrive at a perfectly valid dimensionless grouping, which would be of no importance to the application.

Third, there are rare occasions when the Buckingham Π Theorem underestimates the number of dimensionless groupings. This occurs when the indicial equations are linearly dependent. The following example illustrates such a case.

Example 2.4 *For a calorically-perfect gas, the specific internal energy, e, whose dimensions are L^2/T^2, is known to be a function of the perfect-gas constant, R, and temperature, T. Verify that the Buckingham Π Theorem implies there are no dimensionless groupings. Show that the indicial equations are linearly dependent, and verify that e = constant · RT.*

Solution. Noting that $[p] = [\rho R T]$, the dimensional quantities and their dimensions are:

$$[e] = \frac{L^2}{T^2}, \qquad [T] = \Theta, \qquad [R] = \frac{[p]}{[\rho][T]} = \frac{\text{Force} \cdot L^{-2}}{ML^{-3}\Theta} = \frac{LMLT^{-2}}{M\Theta} = \frac{L^2}{T^2\Theta}$$

There are 3 dimensional quantities and 3 independent dimensions (L, T, Θ), so that the number of dimensionless groupings is 0. The appropriate dimensional equation is

$$[e] = [R]^{a_1} [T]^{a_2}$$

Substituting the dimensions for each quantity yields

$$L^2 T^{-2} = L^{2a_1} T^{-2a_1} \Theta^{-a_1} \Theta^{a_2} = L^{2a_1} T^{-2a_1} \Theta^{-a_1 + a_2}$$

Thus, equating exponents, we arrive at the following three equations:

$$\begin{aligned} 2 &= 2a_1 \\ -2 &= -2a_1 \\ 0 &= -a_1 + a_2 \end{aligned}$$

By inspection, the second equation is -1 times the first equation. Thus, the equations are linearly dependent so that there must be 1 additional dimensionless grouping. In the present case, this means there is one such grouping. We can solve the first and third equations immediately for a_1 and a_2.

$$a_1 = 1 \quad \text{and} \quad a_2 = 1$$

Substituting the values of a_1 and a_2 into the dimensional equation, we have

$$[e] = [R]^1 [T]^1$$

Therefore, the dimensionless grouping is $e/(RT)$, so that we expect e to vary according to

$$e = \text{constant} \cdot RT$$

2.2.2   Interpretation of Results

It is important to understand that all we have shown is existence of the two dimensionless groupings in Equation (2.17). Because the number of unknown exponents exceeds the number of equations by one, the exponent a_3 is undetermined. Hence, while the formalism of the method led to Equation (2.16), we have not shown that $\rho F/\mu^2$ is proportional to some power of the dimensionless grouping known as **Reynolds number**, Re, defined by

$$Re = \frac{\rho U D}{\mu} \tag{2.22}$$

Rather, our analysis has shown that the dimensionless force, $\rho F/\mu^2$, is some function of Re. At this point, we have extracted as much analytical information as we can from dimensional analysis. However, this information is extremely useful for several purposes, including:

- Establishing the order of magnitude of flow properties;

- Correlating experimental data;

- Developing scaling laws to extrapolate measurements on small models to larger models.

For example, we can reasonably expect that μ^2/ρ provides the order of magnitude of the force on the sphere for some range of Reynolds numbers. As we will see below, this is indeed true at relatively small Reynolds numbers. However, as we will also find, flow past a sphere is a bit of a pathological case as the dimensionless force can assume a wide range of values as we vary Reynolds number over many orders of magnitude.

This reflects the fact that flow past a sphere undergoes dramatic changes as Reynolds number increases from very small to very large values. To appreciate the complexity of this flow, the curious reader might want to refer to the Van Dyke (1982) collection of fluid-motion photographs. The book provides a spectacular group of photographs depicting the many different Reynolds-number regimes for flow past a sphere.

One of the most powerful purposes of dimensional analysis is to aid in correlating measurements. If our object is to correlate the force on the sphere as a function of the relevant variables, without dimensional analysis we are dealing with a function of four variables, viz.,

$$F = \mathcal{G}(\rho, U, D, \mu) \tag{2.23}$$

By contrast, dimensional analysis shows that the dimensionless force is a function of a single variable, because our analysis tells us that

$$F = \frac{\mu^2}{\rho} \mathcal{F}(Re) \tag{2.24}$$

Obviously, Equation (2.24) is far easier to correlate than Equation (2.23). Using a single graph, all data from various experiments will fall on a single curve if we create the plot with the dimensionless groupings on the ordinate and abscissa.

Perhaps the most remarkable thing we have shown is that our measurements fall on the same curve even if one set of experiments is done in water and another in air. That is, dimensional analysis tells us that regardless of the fluid used, the (appropriately scaled) force depends only upon Reynolds number. This is a remarkable fact, and certainly not an intuitive one, whose discovery would require a large number of experiments varying ρ, U, D and μ independently. Nevertheless, we have discovered this fact with no knowledge of the complexities involved in viscous flow past a sphere, of which there are many.

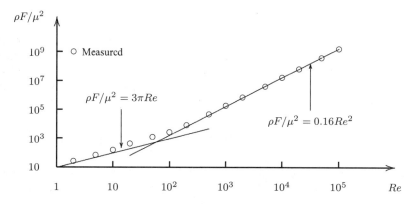

Figure 2.3: *Correlation of measured force on a sphere using $\rho F/\mu^2$.*

Figure 2.3 presents graphical results of hypothetical experiments for flow past a sphere. As shown, if we make a log-log plot of $\rho F/\mu^2$ versus Re for a wide range of Reynolds numbers, there are two clear limiting cases. On the one hand, in the limit $Re \to 0$, the measured values asymptotically approach linear variation of $\rho F/\mu^2$ with Re. If our measurements were very accurate, our graph would indicate that

$$\frac{\rho F}{\mu^2} \to 3\pi Re \quad \text{as} \quad Re \to 0 \tag{2.25}$$

This result can be shown to be valid for small Reynolds numbers by solving the exact equations of motion for flow past a sphere [see, for example, Panton (1996)]. This is the theoretical limiting case known as **Stokes flow**, or **creeping flow**.

On the other hand, for large Reynolds number, the measured values of $\rho F/\mu^2$ asymptotically approach quadratic variation with Re, viz.,

$$\frac{\rho F}{\mu^2} \to 0.16 Re^2 \quad \text{as} \quad Re \to \infty \tag{2.26}$$

Although this behavior, corresponding to turbulent fluid motion, has not been predicted from first principles, the coefficient 0.16 has been found experimentally by many researchers [cf. Schlichting-Gersten (2000)].

While we have made an important simplification in correlating the experimental data, we can extract additional useful information by manipulating the results of our dimensional analysis further. We might want to do this, for example, to focus on a specific range of Reynolds numbers where a discernible trend is evident.

Imagine that you are interested in low Reynolds number applications. Your experimental data clearly indicate a linear variation of $\rho F/\mu^2$ with Re. You can further verify this trend, including the theoretical coefficient 3π, by making measurements at even smaller Reynolds numbers. Furthermore, it would be advantageous to plot the ratio of $\rho F/\mu^2$ to Re versus Re, since the ratio asymptotes to a constant value as $Re \to 0$. This is permissible as the product of two dimensionless groupings remains dimensionless. Hence, the dimensionless force becomes

$$\frac{\rho F}{\mu^2} Re^{-1} = \frac{\rho F}{\mu^2} \frac{\mu}{\rho U D} = \frac{F}{\mu U D} \tag{2.27}$$

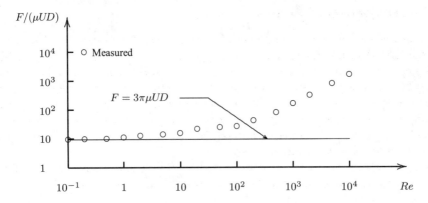

Figure 2.4: *Correlation of measured force on a sphere using $F/(\mu UD)$.*

Figure 2.4 shows the data plotted in terms of $F/(\mu UD)$. Clearly, the data tend toward the theoretical limit, thus providing confidence in the measured values. If you had no knowledge of the theoretical limiting value for $Re \to 0$ (which is normally the case for general applications), your measurements might provide a correlation that could be used for design purposes.

Finally, let's turn our attention to high Reynolds number applications, such as the flow past a tennis ball, a golf ball or a baseball. In this limit, your experimental data suggest that there is a quadratic variation of $\rho F/\mu^2$ with Re. Thus, plotting the ratio of $\rho F/\mu^2$ to Re^2 versus Re would appear to be appropriate. On a log-log plot, rescaling in this manner magnifies details of the shape of the curve in the limit $Re \to \infty$, just as the rescaling of Equation (2.27) magnifies details at low Reynolds number. Therefore, forming the desired ratio, the dimensionless force now becomes

$$\frac{\rho F}{\mu^2} Re^{-2} = \frac{\rho F}{\mu^2} \frac{\mu^2}{\rho^2 U^2 D^2} = \frac{F}{\rho U^2 D^2} \tag{2.28}$$

Figure 2.5 presents a log-log plot of the rescaled force, $F/(\rho U^2 D^2)$, as a function of Reynolds number. As shown in the figure, the force data tend toward a constant value of 0.16 as the Reynolds number increases.

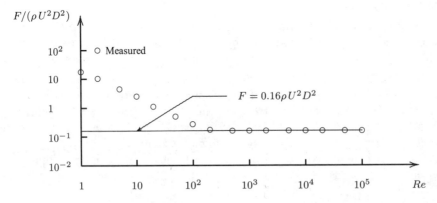

Figure 2.5: *Correlation of measured force on a sphere using $F/(\rho U^2 D^2)$.*

2.2.3 Nonuniqueness of Results

There is usually an element of nonuniqueness in dimensional analysis similar to what we have found in the last subsection for viscous flow past a sphere. In our analysis of the sphere, all of the proposed dimensionless forms of the force are valid. With a bit of reflection, this should not come as a surprise. Recall that we had more unknowns than equations when we formulated the problem, so that we can hardly expect to find a unique solution. Although there usually is an optimum choice of the dimensionless groupings, we don't know which form is best until we actually correlate our data. This is, of course, taking the view that we know little about the problem at the start.

To further underscore this point, suppose we had written the dimensional equation [Equation (2.5)] with viscosity appearing last rather than first on the right-hand side, viz.,

$$[F] = [\rho]^{a_1}[U]^{a_2}[D]^{a_3}[\mu]^{a_4} \tag{2.29}$$

After substituting the appropriate dimensions for each dimensional quantity and solving for the exponents, we would arrive at the following solution.

$$a_1 = 1 - a_4, \quad a_2 = 2 - a_4, \quad a_3 = 2 - a_4 \tag{2.30}$$

Then, combining Equations (2.29) and (2.30), we would conclude that

$$[F] = \rho U^2 D^2 \left[\frac{\mu}{\rho U D}\right]^{a_4} \tag{2.31}$$

Hence, the two dimensionless groupings are

$$\frac{F}{\rho U^2 D^2} \quad \text{and} \quad \frac{\mu}{\rho U D} \tag{2.32}$$

This is the form most appropriate for high Reynolds number applications that we discussed above. Note that the reciprocal of the Reynolds number falls naturally out of this computation. As you gain experience in fluid mechanics, you will immediately recognize the Reynolds number and select it as one of the dimensionless groupings rather than its reciprocal. In mathematical terms, you are simply appealing to the fact that if a number is dimensionless, then so is its reciprocal. As we saw in the sphere example, raising a dimensionless grouping to any power and multiplying two dimensionless groupings yields yet another dimensionless grouping. There are, in fact, an infinite number of solutions for the dimensionless groupings.

Example 2.5 *Dimensional analysis shows that for turbulent flow over a rough surface, the dimensionless groupings are*

$$P_0 = \frac{u}{u_\tau}, \qquad P_1 = \frac{u_\tau y}{\nu}, \qquad P_2 = \frac{u_\tau k_s}{\nu}$$

where u is velocity, y is distance normal to the surface, ν is kinematic viscosity, and k_s is the average height of surface roughness elements. The quantity $u_\tau = \sqrt{\tau_w/\rho}$ is the friction velocity, where τ_w is surface shear stress and ρ is density. Measurements indicate that when $u_\tau k_s/\nu > 100$ the dimensionless velocity, u/u_τ, is independent of kinematic viscosity. What would be the optimum choice of dimensionless groupings for correlating experimental data when $u_\tau k_s/\nu > 100$?

Solution. We can eliminate kinematic viscosity by combining the second and third dimensionless groupings as follows.

$$P_1 P_2^{-1} = \frac{u_\tau y}{\nu}\left(\frac{u_\tau k_s}{\nu}\right)^{-1} = \frac{y}{k_s}$$

Thus, the optimum choice would be to correlate u/u_τ versus y/k_s.

2.2.4   E. S. Taylor's Method

There is a useful alternative to the indicial method for determining dimensionless groupings developed by E. S. Taylor (1974). Briefly stated, it is a method of successive elimination of dimensions. It proceeds in several easy stages, with one important advantage over the indicial method. That is, an experienced person with knowledge of common dimensionless groupings can accelerate and guide the process. For example, an experienced fluid-flow engineer would expect Reynolds number, Re, to be one of the dimensionless groupings. As we saw in the preceding subsection, rearranging the order of dimensionless quantities in the dimensional equation led to $1/Re$. In fact, had we simply solved Equations (2.11) in a different order, we could arrive at $\sqrt{Re}$. This can be avoided completely with the Taylor method.

After using the Buckingham Π Theorem to determine the number of dimensionless groupings, the first thing we do is write down a matrix whose rows each represent the dimensions of one of the quantities. In our example for flow past a sphere, we have

	M	L	T
μ	1	-1	-1
ρ	1	-3	0
U	0	1	-1
D	0	1	0
F	1	1	-2

The columns are the coefficients of a_1, a_2, a_3 and a_4 and the right hand sides in Equations (2.11).

Now, we choose the simplest-looking column—in this problem the first—and divide all but one of the mass-containing quantities by that one. Division is accomplished by subtracting all elements of one row from those of another. Selecting ρ as the divisor, we rearrange our matrix by subtracting the second row from the first and the last. The rearranged matrix is

	M	L	T
μ/ρ	0	2	-1
ρ	1	-3	0
U	0	1	-1
D	0	1	0
F/ρ	0	4	-2

Note that ρ is now the only quantity that has mass as one of its dimensions, and therefore cannot appear in any dimensionless groups except as μ/ρ or F/ρ. Hence, we can delete it along with the resulting empty mass column to arrive at

	L	T
μ/ρ	2	-1
U	1	-1
D	1	0
F/ρ	4	-2

Inspection of the time column suggests dividing U by μ/ρ, which is clearly required as one step in forming the Reynolds number, $\rho U D/\mu$. Thus, we now eliminate the time dimension in all quantities except μ/ρ. To do this, we subtract the first row from the second row, and we subtract twice the first row from the last. The matrix simplifies to:

	L	T
μ/ρ	2	-1
$\rho U/\mu$	-1	0
D	1	0
$\rho F/\mu^2$	0	0

At this point, we eliminate μ/ρ from further consideration and delete the empty time column. All that remains at this point is a single column.

	L
$\rho U/\mu$	-1
D	1
$\rho F/\mu^2$	0

The 0 in the column next to $\rho F/\mu^2$ means this quantity is one of the dimensionless groupings. Also, further inspection of the nonzero entries in the length column shows that multiplying $\rho U/\mu$ by D yields the second dimensionless grouping. We do this by adding the first and second rows. Thus, we conclude that the two dimensionless groupings are

$$P_0 = \frac{\rho F}{\mu^2} \quad \text{and} \quad P_1 = \frac{\rho UD}{\mu} \tag{2.33}$$

This is identical to what we obtained [see Equation (2.17)] using the indicial method.

Example 2.6 *The load-carrying capacity, W, (a force) of a journal bearing is known to depend on three lengths, viz., bearing diameter, D, length, ℓ, and clearance, c. Additionally, W depends upon angular speed, ω, (dimensions 1/T) and viscosity of the lubricant, μ. How many dimensionless groupings are there? What are the dimensionless groupings?*

Solution. The dimensional quantities and their dimensions are

$$[W] = F = \frac{ML}{T^2}, \quad [D] = L, \quad [\ell] = L, \quad [c] = L, \quad [\omega] = \frac{1}{T}, \quad [\mu] = \frac{M}{LT}$$

There are 6 dimensional quantities and 3 independent dimensions (M, L, T), so that the number of dimensionless groupings is 3. Using the E. S. Taylor method, we proceed as follows.

	M	L	T
W	1	1	-2
D	0	1	0
ℓ	0	1	0
c	0	1	0
ω	0	0	-1
μ	1	-1	-1

$\rightarrow$

	M	L	T
W/μ	0	2	-1
D	0	1	0
ℓ	0	1	0
c	0	1	0
ω	0	0	-1
μ	1	-1	-1

$\rightarrow$

	L	T
W/μ	2	-1
D	1	0
ℓ	1	0
c	1	0
ω	0	-1

$\rightarrow$

	L	T
$W/(\mu D^2)$	0	-1
D	1	0
ℓ/D	0	0
c/D	0	0
ω	0	-1

$\rightarrow$

	T
$W/(\mu D^2)$	-1
ℓ/D	0
c/D	0
ω	-1

$\rightarrow$

	T
$W/(\omega\mu D^2)$	0
ℓ/D	0
c/D	0
ω	-1

$\rightarrow$

	T
$W/(\omega\mu D^2)$	0
ℓ/D	0
c/D	0

Therefore, the dimensionless groupings are:

$$\frac{W}{\omega\mu D^2}, \quad \frac{\ell}{D}, \quad \frac{c}{D}$$

2.3   Common Dimensionless Groupings

There are many dimensionless groupings that occur in fluid-mechanical applications. One of the customs that has evolved is to name these groupings after famous pioneers in the field. Some of the most frequently encountered follow.

- **Reynolds Number:** This grouping, named after Osborne Reynolds (1842-1912), provides a measure of the effects of viscous stresses on a flow. For most practical fluid flows, the Reynolds number is very large. It often appears with a subscript denoting the length that has been used. If fluid density is ρ, velocity is U, length is L and viscosity is μ, the Reynolds number is

$$Re_L = \frac{\rho U L}{\mu} \qquad (2.34)$$

- **Mach Number:** This famous parameter honors the important contributions of Ernst Mach (1838-1916), a pioneer in compressible fluid dynamics. The Mach number is the ratio of flow velocity, U, to the speed of sound, a, viz.,

$$M = \frac{U}{a} \qquad (2.35)$$

- **Froude Number:** Named after William Froude (1810-1879), this number provides a measure of the importance of gravity waves in a fluid. Normally used for liquids, it is analogous to the Mach number. As with the Reynolds number, it is customary to include the length used as a subscript. If flow velocity is U, gravitational acceleration is g and length is L, the Froude number is

$$Fr_L = \frac{U}{\sqrt{gL}} \qquad (2.36)$$

Inspection of Equation (2.4) shows that a sailboat's theoretical hull speed corresponds to a Froude number of 0.40.

- **Prandtl Number:** When heat transfer is important, the number named in honor of Ludwig Prandtl (1875-1953) appears. If the specific heat at constant pressure of the fluid is c_p, viscosity is μ and **thermal conductivity** is k, then the Prandtl number is

$$Pr = \frac{c_p \mu}{k} \qquad (2.37)$$

Unlike the other dimensionless parameters in our list, the Prandtl number is a property of the fluid alone, independent of the flow.

- **Strouhal Number:** This dimensionless grouping, named after Vincenz Strouhal (1850-1922), pertains to flows that vary with time. If the flow involves a periodic variation with frequency ω, for example, and if the flow velocity and length scale are U and L, respectively, the Strouhal number is

$$St = \frac{\omega L}{U} \qquad (2.38)$$

- **Weber Number:** This number honors the research of Moritz Weber (1871-1951), who investigated flows in which surface tension, σ, is important. If the fluid density is ρ, flow velocity is U, and characteristic length is L, the Weber number is defined by

$$We = \frac{\rho U^2 L}{\sigma} \tag{2.39}$$

Example 2.7 *The Stanton number, St, is a dimensionless grouping defined by*

$$St \equiv \frac{\dot{h}}{\rho U c_p}$$

where $\dot{h}$ is the heat-transfer coefficient, ρ is density, U is velocity and c_p is specific-heat coefficient. What are the dimensions of $\dot{h}$?

Solution. Noting that $[\dot{h}] = [c_p T]$, the dimensions of St, ρ, U and c_p are as follows:

$$[St] = 1, \quad [\rho] = \frac{M}{L^3}, \quad [U] = \frac{L}{T}, \quad [c_p] = \frac{[\dot{h}]}{[T]} = \frac{L^2 T^{-2}}{\Theta} = \frac{L^2}{T^2 \Theta}$$

Solving for $\dot{h}$ from the definition of the Stanton number, St, we have

$$[\dot{h}] = [\rho U c_p St] = \frac{M}{L^3} \cdot \frac{L}{T} \cdot \frac{L^2}{T^2 \Theta} \cdot 1 = \frac{M}{T^3 \Theta}$$

2.4 Dynamic Similitude

As mentioned in Subsection 2.2.2, dimensional analysis has a third important use that we have not yet discussed. We have seen how the method provides estimates of the appropriate scale and order of magnitude of flow properties. We have also demonstrated how powerful dimensional analysis is for correlating data. This section focuses on how we can use the results of dimensional analysis to extrapolate measurements made on small **models** to full-size objects, or **prototypes**.

To understand how we can accomplish such scaling, we must pause and consider how dimensional analysis relates to the physical equations governing a given problem. As an example, we use the differential equation for one-dimensional motion of a viscous fluid, viz., the *Navier-Stokes equation* [see Chapter 11]. The equation is

$$\rho \frac{\partial u}{\partial t} + \rho u \frac{\partial u}{\partial x} = -\frac{\partial p}{\partial x} + \mu \frac{\partial^2 u}{\partial x^2} \tag{2.40}$$

where we assume the temperature is constant so that μ is constant. Suppose a small one-dimensional sound source of size L moving at a constant velocity U emits a tone with frequency ω. For simplicity, we assume the motion can be represented by Equation (2.40). We can rewrite this equation in terms of dimensionless variables $\bar{u}$, $\bar{t}$, $\bar{x}$, $\bar{p}$ and $\bar{\rho}$, viz.,

$$\bar{u} = \frac{u}{U}, \quad \bar{t} = \omega t, \quad \bar{x} = \frac{x}{L}, \quad \bar{p} = \frac{p}{\rho_\infty U^2}, \quad \bar{\rho} = \frac{\rho}{\rho_\infty} \tag{2.41}$$

where ρ_∞ is the density of the fluid far from the sound source. Upon transforming, we find

$$St \bar{\rho} \frac{\partial \bar{u}}{\partial \bar{t}} + \bar{\rho} \bar{u} \frac{\partial \bar{u}}{\partial \bar{x}} = -\frac{\partial \bar{p}}{\partial \bar{x}} + \frac{1}{Re} \frac{\partial^2 \bar{u}}{\partial \bar{x}^2} \tag{2.42}$$

where $St = \omega L/U$ and $Re = \rho_\infty UL/\mu$ are the Strouhal and Reynolds numbers, respectively. The equation, in this dimensionless form, applies to any fluid and frequency. It is ideally suited for numerical solution on a computer, and the only problem-specific input required for a given geometry would be St and Re, regardless of the individual values of the dimensional quantities in the problem.

Although we could have deduced these two dimensionless groupings with no knowledge of Equation (2.40), understanding the connection with the governing equations provides additional insight that has a direct bearing on the concept known as **dynamic similitude**. This simple example shows that when we perform a dimensional analysis, we are finding the dimensionless groupings that appear when the governing equations are cast in nondimensional form. Since flow past both a small-scale model and the prototype is governed by the same physical laws and associated equations, matching the dimensionless groupings produces exactly the same solution in terms of dimensionless variables. To obtain the actual flowfield for both cases requires using equations such as Equation (2.41) to arrive at dimensional properties. When all relevant dimensionless groupings match for a model and a prototype, the flows are said to be **dynamically similar**, or to stand in **dynamic similitude**.

To see how we establish dynamic similitude between a model and a prototype, we focus on two general classes of applications, viz., airplanes and ships. For the airplane, which is an **aerodynamic** vehicle, we must match the Reynolds number, Re, and the Mach number, M. As we will see in the following example, we come pretty close to exact dynamic similitude in modern **wind tunnels**. By contrast, for the ship, which is a **hydrodynamic** vehicle, we must match the Reynolds number and, in place of the Mach number, we must match the Froude number, Fr. We'll see that simultaneously matching both Re and Fr cannot be done on Earth primarily because we cannot change g. Thus, we never achieve exact similitude for a ship.

2.4.1   Similitude for an Airplane

First consider an airplane. The relevant dimensional quantities are fluid density, ρ, viscosity, μ, speed of sound, a, velocity, U, and length, L. We would like to extrapolate measured lift or drag force, F, for the model to the prototype. Performing dimensional analysis shows that there are three dimensionless groupings, viz.,

$$\frac{F}{\rho U^2 L^2}, \quad M = \frac{U}{a}, \quad Re = \frac{\rho UL}{\mu}$$

Hence, we conclude that the force behaves according to

$$F = \rho U^2 L^2 \mathcal{F}(M, Re) \tag{2.43}$$

We thus achieve dynamic similitude by matching M and Re, and use Equation (2.43) to extrapolate the measured force.

Letting subscript m denote the model and subscript p the prototype, matching Mach and Reynolds numbers means we require

$$\frac{U_p}{a_p} = \frac{U_m}{a_m} \quad \text{and} \quad \frac{\rho_p U_p L_p}{\mu_p} = \frac{\rho_m U_m L_m}{\mu_m} \tag{2.44}$$

where L is the total length of the airplane (or any other convenient distance). Knowing the Mach and Reynolds numbers for the prototype, we select the flow properties and dimensions for our model in order to provide a match for both. When we have done this, the function $\mathcal{F}(M, Re)$ will have the same value for the model as for the prototype, wherefore

$$F_p = \rho_p U_p^2 L_p^2 \frac{F_m}{\rho_m U_m^2 L_m^2} \tag{2.45}$$

If you pause and think about this for a moment, what dimensional analysis provides us with is truly remarkable. For example, if we achieve dynamic similitude in a small wind tunnel, we can very accurately deduce the forces on a Boeing 747-400 whose wingspan is 64.5 m, by making measurements on a model with a wingspan of just 1 m! While there are wind tunnels such as the NASA Ames 80 ft × 120 ft facility (Figure 2.6) that are large enough to accommodate full-scale models, they are expensive to maintain and operate. Most wind tunnels require such scaling.

Figure 2.6: *The world's largest wind tunnel—the 80 ft × 120 ft facility at the NASA Ames Research Center, Moffett Field, California. The model shown is full scale. [Photograph courtesy of the NASA Ames Research Center]*

It is worthwhile to pause and investigate just how easy, or difficult, it might be to achieve dynamic similitude for aerodynamic applications. Although we can always use a different gas, we will assume our experiments are done in air. For air, it is a fact that over a wide range of temperatures, the viscosity and speed of sound vary with temperature according to[2]

$$\mu \propto T^{0.7} \quad \text{and} \quad a = \sqrt{\gamma R T} \propto T^{0.5} \tag{2.46}$$

So, matching Mach number means that the ratio of the wind-tunnel flow velocity to the prototype velocity is given by

$$\frac{U_m}{U_p} = \frac{a_m}{a_p} = \sqrt{\frac{T_m}{T_p}} \tag{2.47}$$

---

[2]The fact that $a = \sqrt{\gamma R T}$ is an exact theoretical result that we will derive in Chapter 8, while $\mu \propto T^{0.7}$ is purely empirical.

Also, matching Reynolds number means the model length to the prototype length ratio is

$$\frac{L_m}{L_p} = \frac{\rho_p}{\rho_m}\frac{U_p}{U_m}\frac{\mu_m}{\mu_p} = \frac{\rho_p}{\rho_m}\sqrt{\frac{T_p}{T_m}}\left(\frac{T_m}{T_p}\right)^{0.7} = \frac{\rho_p}{\rho_m}\left(\frac{T_m}{T_p}\right)^{0.2} \qquad (2.48)$$

Therefore, we achieve dynamic similitude provided:

$$\frac{U_m}{U_p} = \sqrt{\frac{T_m}{T_p}} \quad \text{and} \quad \frac{L_m}{L_p} = \frac{\rho_p}{\rho_m}\left(\frac{T_m}{T_p}\right)^{0.2} \qquad (2.49)$$

The weak dependence of L_m/L_p on temperature ratio means that we can almost independently control the model velocity and length by adjusting the air temperature and density, respectively. We have great flexibility with both quantities through heating or cooling the air supply and adjusting its pressure.

Because it is relatively easy to achieve dynamic similitude for an airplane, we can confidently use dimensional-analysis results to predict the forces and moments on prototypes based on measurements for a wind-tunnel model. This capability plays an important role in the fact that the airplane is one of the most efficient and reliable vehicles ever built.

Example 2.8 *A prototype airplane will fly at an altitude of 20 kilometers where the temperature is 218 Kelvins and the air density is 0.336 kg/m³. We wish to maintain atmospheric conditions in the wind tunnel. Determine how large a model we can use and the speed of the air in the tunnel. What is the ratio of prototype to model force, F_p/F_m?*

Solution. The air is at room temperature (293 Kelvins), atmospheric pressure ($1.01 \cdot 10^5$ N/m²) so that the density is 1.20 kg/m³. Thus, from Equation (2.49), we find

$$\frac{U_m}{U_p} = 1.16 \quad \text{and} \quad \frac{L_m}{L_p} = 1.06\frac{\rho_p}{\rho_m} = 0.30$$

Thus, maintaining atmospheric conditions in the tunnel, we can achieve exact dynamic similitude with a 30% scale model provided the air in the wind tunnel moves 16% faster than the prototype's velocity. In order to have a model one-tenth the size of the prototype, we can use a combination of pressurization and cooling of the tunnel's air supply to increase the density by about a factor of 3. Finally, combining Equations (2.43) and (2.49) shows that the force scales according to

$$\frac{F_p}{F_m} = \frac{\rho_p}{\rho_m}\frac{U_p^2}{U_m^2}\frac{L_p^2}{L_m^2} = \frac{\rho_m}{\rho_p}\left(\frac{T_p}{T_m}\right)^{1.4} = 2.36$$

2.4.2   Similitude for a Ship

For a ship, the situation is quite different. Consider a model to be tested in a **towing tank** (cf. Figure 2.7). Dimensional analysis tells us the wave drag, W, for example, scales according to

$$W = \rho g L^3 \mathcal{F}(Fr, Re) \qquad (2.50)$$

where Fr and Re are Froude and Reynolds numbers, respectively. Matching Froude numbers,

$$\frac{U_m}{\sqrt{gL_m}} = \frac{U_p}{\sqrt{gL_p}} \quad \Longrightarrow \quad \frac{U_m}{U_p} = \sqrt{\frac{L_m}{L_p}} \qquad (2.51)$$

so that the ratio of model to prototype velocity decreases as the ratio of model to prototype size decreases.

If we use water in the towing tank with no heating or cooling of the water in the tank (which would be very expensive), matching Reynolds numbers yields

$$\frac{\rho_m U_m L_m}{\mu_m} = \frac{\rho_p U_p L_p}{\mu_p} \quad \Longrightarrow \quad \frac{U_m}{U_p} = \frac{L_p}{L_m} \tag{2.52}$$

where we make use of the fact that the water density and viscosity are identical for the model and prototype. This tells us the ratio of model to prototype velocity increases as the ratio of model to prototype size decreases. Thus, the condition needed to match Reynolds number is the opposite of what is required to match Froude number. That is, matching both dimensionless groupings means

$$\sqrt{\frac{L_m}{L_p}} = \frac{L_p}{L_m} \quad \Longrightarrow \quad \left(\frac{L_p}{L_m}\right)^{3/2} = 1 \tag{2.53}$$

which is possible only if we have

$$L_p = L_m \tag{2.54}$$

Hence, using water at the same temperature as that of the prototype's environment, we can achieve dynamic similitude only with a full-scale model! We cannot practicably modify density or viscosity because inexpensive and safe liquids with a density greater than water are unavailable, and changing temperature of the water is expensive and difficult to control. If we could change g, we would have the same flexibility afforded in a wind tunnel where a different gas can be used. Unfortunately, the laws of gravitation prevent this. In practice, designers match Froude number only. This is not too bad if the model is large, say half size.

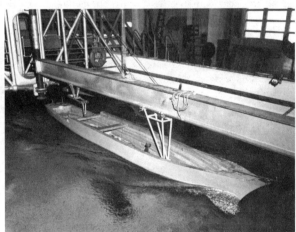

Figure 2.7: *A ship model being tested in a towing tank at the Naval Ship Research and Development Center (NSRDC) in Carderock, Maryland. [Photographs courtesy of NSRDC]*

Problems

2.1 The *Weber number*, We, is a dimensionless quantity that is important for flows with significant effects of surface tension. It is defined as

$$We \equiv \frac{\rho U^2 R}{\sigma}$$

where ρ is density, U is flow velocity, R is length and σ is surface tension. Verify that We is dimensionless.

2.2 The *Rossby number*, Ro, is a dimensionless quantity that characterizes the ratio of convective acceleration to Coriolis acceleration for a flow in a rotating coordinate system. It is defined as

$$Ro \equiv \frac{U}{\Omega R}$$

where U is flow velocity, Ω is angular velocity and R is length. Verify that Ro is dimensionless.

2.3 The gradient *Richardson number*, Ri, occurs in density-stratified flows. It is defined by

$$Ri \equiv \frac{-g d\rho/dz}{\rho \left(dU/dz\right)^2}$$

where g is gravitational acceleration, ρ is density, U is velocity and z is vertical distance. Verify that Ri is dimensionless.

2.4 The *Péclet number*, $Pé$, is a common dimensionless grouping for flows with heat transfer. It is defined by

$$Pé \equiv \frac{c_p \rho U s}{k}$$

The quantities c_p, ρ, U, s and k are specific heat for constant pressure, density, velocity, length and thermal conductivity, respectively. What are the dimensions of k?

2.5 The *Nusselt number*, Nu, is a dimensionless grouping defined by

$$Nu \equiv Pr \frac{s\dot{h}}{\mu c_p}$$

where Pr is the Prandtl number, $\dot{h}$ is the heat-transfer coefficient, s is arc length along a surface, μ is viscosity and c_p is specific-heat coefficient. What are the dimensions of $\dot{h}$?

2.6 The *Grashof number*, Gr, is a key dimensionless grouping for buoyancy-driven flows defined by

$$Gr \equiv \frac{\beta g \ell^3 \Delta T}{\nu^2}$$

The quantities β, g, ℓ, ΔT and ν are the thermal-expansion coefficient, gravitational acceleration, length, temperature difference and kinematic viscosity, respectively. What are the dimensions of β?

2.7 The *Ekman number*, Ek, is a dimensionless grouping that appears for viscous flows in a rotating coordinate system. By definition,

$$Ek \equiv \frac{\mu}{\rho \Omega R^2}$$

where μ is viscosity, ρ is density, Ω is angular velocity and R is length. Verify that Ek is dimensionless.

2.8 The *Eckert number*, Ec, is relevant for flows with heat transfer, and is defined by

$$Ec \equiv \frac{U^2}{c_p \Delta T}$$

where U is velocity, c_p is specific-heat coefficient and ΔT is temperature difference. Verify that Ec is dimensionless.

2.9 The *Dean number*, De, is pertinent for flow in curved pipes. It is defined by

$$De \equiv \frac{UR^{3/2}}{\nu \mathcal{R}^{1/2}}$$

where U is velocity, R is pipe radius, ν is kinematic viscosity and $\mathcal{R}$ is radius of curvature of the pipe. Verify that De is dimensionless.

2.10 What would the constant 1.34 in Equation (2.2) have to be replaced by to be valid for U_{hull} in mph and L_w in cubits?

2.11 What would the constant 1.34 in Equation (2.2) have to be replaced by to be valid for U_{hull} in knots and L_w in meters?

2.12 Turbulent channel flow can be accurately described using the following empirical differential equation for the mean velocity, u.

$$\nu \frac{d^2 u}{dy^2} + \frac{d}{dy}\left(\ell^2 \left|\frac{du}{dy}\right| \frac{du}{dy}\right) = \frac{1}{\rho}\frac{dp}{dx}$$

The quantities ν, ρ and p denote kinematic viscosity, density and pressure, respectively. Also, x and y are distance parallel to and normal to the channel walls. For this equation to be dimensionally homogeneous, what must the dimensions of ℓ be?

2.13 The following empirical differential equation determines a quantity ω known as the specific dissipation rate in a one-dimensional, steady flow.

$$u \frac{d\omega}{dx} = \alpha \left(\frac{du}{dx}\right)^2 - \beta \omega^2 + \frac{d}{dx}\left(\sigma \nu_t \frac{d\omega}{dx}\right)$$

The quantity u denotes velocity, x is spatial distance and ν_t is kinematic viscosity. Also, β and σ are dimensionless coefficients. For this equation to be dimensionally homogeneous, what must the dimensions of ω and α be?

2.14 An equation of motion for a rocket burning fuel is

$$\frac{dm}{dt} + \rho_e w_e A_e = 0$$

where m is the mass of the rocket (including fuel), t is time, ρ_e is fuel density at the rocket-nozzle exit plane, w_e is the exit velocity and A_e is the nozzle exit area. Assuming this equation is dimensionally homogeneous, verify that the velocity has dimensions length per unit time.

2.15 The shape of a hanging drop of liquid satisfies the empirical equation, $(\rho - \rho_a)gd^2/C = \sigma$. The quantities ρ and ρ_a are the density of the liquid and the surrounding air, respectively. Also, g is gravitational acceleration, d is the drop's diameter at its equator, σ is surface tension, and C is an empirical constant. What are the dimensions of C?

2.16 The classical *Stokes-Oseen equation* for the drag, F, in low Reynolds number flow past a sphere is

$$F = 3\pi \mu U D + \frac{9}{16}\pi \rho U^2 D^2$$

where ρ, μ, U and D are density, viscosity, velocity and sphere diameter, respectively.

(a) Is this equation dimensionally homogeneous?

(b) Rewrite this equation in terms of the Reynolds number, $Re = \rho U D/\mu$, and the drag coefficient, $C_D \equiv F/(\frac{1}{2}\rho U^2 A)$, where $A = \frac{\pi}{4}D^2$.

2.17 Formulas involving empirical coefficients determined from experiments with a single fluid are common in hydraulics literature of the 19th century. An example is the *Chézy-Manning* equation for average velocity, $\bar{u}$, in an open channel, as a function of hydraulic radius, R_h, and bottom slope, $S_o = \tan\beta$, viz., $\bar{u} = \chi S_o^{1/2} R_h^{2/3}/n$, where χ is an empirical coefficient and n is the dimensionless Manning roughness coefficient. What are the dimensions of the constant χ?

2.18 Formulas involving empirical coefficients determined from experiments with a single fluid are common in hydraulics literature of the 19th century. An example is the *Hazen-Williams* formula for volume-flow rate, Q, (dimensions L^3/T) in a pipe, as a function of pipe diameter, D, and pressure gradient, dp/dx, viz., $Q = 61.9 D^{2.63}(dp/dx)^{0.54}$.

(a) What are the dimensions of the constant 61.9?

(b) Suppose we rewrite this equation in terms of a revised empirical coefficient $C = 61.9\rho^{0.54}$, where ρ is the density of water. Show that C is only weakly dependent on dimensions, i.e., it is almost a dimensionless coefficient.

2.19 For turbulent flow close to a solid boundary, the fluctuating velocities scale with the *friction velocity*, u_τ. The friction velocity is a function of the density, ρ, and the surface shear stress, τ_w. Verify that the Buckingham Π Theorem implies there are no dimensionless groupings. Show that the indicial equations are linearly dependent, and (choosing the constant of proportionality to be 1) verify that $u_\tau = \sqrt{\tau_w/\rho}$.

2.20 For turbulent flow close to a solid boundary, the fluctuating velocities scale with the *friction velocity*, u_τ. The friction velocity is a function of the density, ρ, and the surface shear stress, τ_w. Verify that the Buckingham Π Theorem implies there are no dimensionless groupings. Using E. S. Taylor's method, explain how the linear dependence of the indicial equations manifests itself. Choosing the constant of proportionality to be 1, verify that $u_\tau = \sqrt{\tau_w/\rho}$.

2.21 For a perfect gas, the speed of sound, a, is known to be a function of pressure, p, and density, ρ. Verify that the Buckingham Π Theorem implies there are no dimensionless groupings. Show that the indicial equations are linearly dependent, and verify that $a = \text{constant} \cdot \sqrt{p/\rho}$.

2.22 For a perfect gas, the speed of sound, a, is known to be a function of pressure, p, and density, ρ. Verify that the Buckingham Π Theorem implies there are no dimensionless groupings. Using E. S. Taylor's method, explain how the linear dependence of the indicial equations manifests itself, and verify that $a = \text{constant} \cdot \sqrt{p/\rho}$.

2.23 The mean velocity, u_m, for flow through a horizontal tube is a function of the viscosity, μ, axial pressure gradient, dp/dx, and tube radius, R. Using dimensional analysis, deduce a formula for u_m as a function of μ, dp/dx and R.

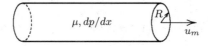

Problem 2.23

2.24 Once transients have settled out, the radial pressure gradient, dp/dr, in a rotating tank of liquid is a function of the density, ρ, angular velocity, Ω, and radial distance from the center of the tank, r. Deduce a formula for dp/dr as a function of ρ, Ω and r using dimensional analysis.

2.25 In analogy to laminar flow, a kinematic *eddy viscosity*, ν_t, can be used to describe many relatively simple turbulent flows. One empirical model for the eddy viscosity assumes it is a function of the specific turbulence kinetic energy, k, (dimensions L^2/T^2) and the rate of dissipation of energy, ϵ (dimensions L^2/T^3). Develop a formula for ν_t as a function of k and ϵ.

2.26 In analogy to laminar flow, a kinematic *eddy viscosity*, ν_t, can be used to describe many relatively simple turbulent flows. One empirical model for the eddy viscosity assumes it is a function of the specific turbulence kinetic energy, k, (dimensions L^2/T^2) and the specific rate of dissipation of energy, ω (dimensions $1/T$). Develop a formula for ν_t as a function of k and ω.

2.27 The sketch below illustrates how the centerline velocity, $u_m(x)$, develops in a turbulent *plane jet* of fluid blowing into a stationary fluid from a small orifice. Very far from the orifice, u_m is a function only of the specific momentum flux, J, (dimensions L^3/T^2) and distance from the origin, x. Using dimensional analysis, develop a formula for $u_m(x)$ as a function of J and x.

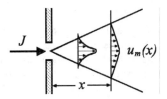

Problem 2.27

2.28 The energy, E, of an electron in orbit about the nucleus of an atom is a function of its rest mass, m_o, its charge, q, the permittivity of free space, ϵ_o, and Planck's constant, h. The dimensions of h and ϵ_o are $[h] = ML^2/T$ and $[\epsilon_o] = Q^2T^2/(ML^3)$, where Q denotes the independent dimension of electrical charge. Using dimensional analysis, deduce a formula for E as a function of m_o, q, ϵ_o and h.

2.29 A very strong spherical wave known as a *blast wave* develops after a large atmospheric explosion. If the energy released is E (dimensions ML^2/T^2) and the ambient density is ρ_a, develop an equation for the radius of the blast wave, $R(t)$, as a function of E, ρ_a and time, t.

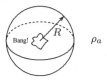

Problem 2.29

2.30 *Capillary waves* or *ripples* are waves of very short wavelength at the interface between a liquid and a gas. This type of wave is produced when a small object is dropped into a pool of water. The propagation speed, c, is a function of the liquid density, ρ, the surface tension, σ, and the wavelength, λ. Deduce a formula for c as a function of ρ, σ and λ.

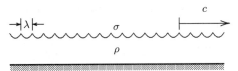

Problem 2.30

2.31 For the very smallest eddies in a turbulent flow, the kinetic energy of the turbulent fluctuations is dissipated through viscous processes. On the basis of dimensional analysis, Kolmogorov concluded that the velocity in the smallest eddies, υ, must be a function of dissipation rate, ϵ, (dimensions L^2/T^3) and the kinematic viscosity, ν. Deduce a formula for υ as a function of ν and ϵ.

2.32 For the very smallest eddies in a turbulent flow, the kinetic energy of the turbulent fluctuations is dissipated through viscous processes. On the basis of dimensional analysis, Kolmogorov concluded that the size of the smallest eddies, η, must be a function of dissipation rate, ϵ, (dimensions L^2/T^3) and the kinematic viscosity, ν. Deduce a formula for η as a function of ν and ϵ.

2.33 Kolmogorov argued that for high Reynolds number, in terms of Fourier analysis, the turbulence energy spectrum contains an *inertial subrange*. He further argued that in the inertial subrange, the turbulence-energy spectrum function, $E(\kappa)$, (dimensions L^3/T^2) depends only upon the dissipation rate, ϵ, (dimensions L^2/T^3) and the wave number, κ (dimensions $1/L$). Develop a formula for $E(\kappa)$ as a function of ϵ and κ.

2.34 Under certain conditions, liquid droplets in a moving gas will break up. The relevant dimensional quantities are gas density, ρ, velocity, U, droplet diameter, D, gas viscosity, μ, and surface tension, σ. How many dimensionless groupings are there? What are the dimensionless groupings?

2.35 Supported by surface tension, a small bug of mass $\mathcal{M}$ is standing on the surface of the water in a pond. The surface tension is σ, fluid density is ρ and the gravitational acceleration is g. Using dimensional analysis, determine how the radius of curvature of the surface of the water, $\mathcal{R}$, near the point where the bug is standing correlates with $\mathcal{M}$, ρ, σ and g. That is, propose an equation of the form $\mathcal{R} = P_0 f(P_1, ...)$, where P_0 has dimensions of length, $P_1, ...$ are dimensionless and f is a function that can be determined from measurements.

2.36 A small piece of steel of mass M is supported by surface tension in a container filled with water. The key dimensional quantities are the surface tension, σ, density of the water, ρ, gravitational acceleration, g, and the radius of curvature, $\mathcal{R}$, of the surface of the water near the piece of steel. How many dimensionless groupings are there? What are the dimensionless groupings?

2.37 Consider an air-hockey game, which uses a porous surface with air injected at a rate $\dot{m}$ (dimensions M/T). A puck of mass M moves at velocity V while floating at a distance h above the surface on a cushion of air. The kinematic viscosity of air is ν. How many independent dimensionless groupings are there? What are the dimensionless groupings? Manipulate your answer to show that one of the dimensionless groupings is the Reynolds number based on h.

2.38 The critical dimensional quantities in the rise of a liquid in a small tube inserted into a pan of water due to capillary action are the distance the liquid rises, Δh, the fluid density, ρ, the gravitational acceleration, g, the cross-sectional area of the tube, A, and the surface tension, σ. How many dimensionless groupings are there? What are the dimensionless groupings?

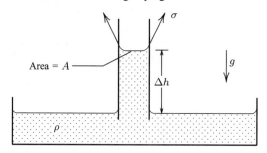

Problem 2.38

2.39 Classical *Couette flow* involves motion of a viscous fluid where one wall moves with a constant velocity, U, parallel to a stationary wall a distance h away. Consider the unsteady case where the upper wall starts moving from rest at time $t = 0$, with a constant velocity U. Assuming the velocity can be written as $u/U = f(y, \mu, t, h, \rho)$, determine the relevant dimensionless groupings.

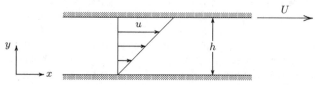

Problem 2.39

2.40 At very high temperatures, *quantum-mechanical* effects become important and the internal energy, e, of a gas becomes a function of the perfect-gas constant, R, temperature, T, Planck's constant, h, Boltzmann's constant, k, and the fundamental vibration frequency of the molecule, ν. The units of h are J·sec while those of k are J/K. How many independent dimensionless groupings are there? What are the dimensionless groupings?

2.41 The critical dimensional quantities for flow in a curved pipe of constant radius, R, are velocity, U, kinematic viscosity, ν, and radius of curvature, $\mathcal{R}$. How many independent dimensionless groupings are there? What are the dimensionless groupings? Manipulate your answer to show that one of the dimensionless groupings is the Reynolds number.

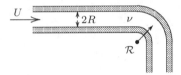

Problem 2.41

2.42 For natural convection above a mildly heated, infinite flat plate, the difference between the plate temperature and ambient temperature, ΔT, depends upon the specific-heat coefficient, c_p, thermal conductivity, k, viscosity, μ, fluid density, ρ, and gravitational acceleration, g. The units of c_p are ft²/(sec²·°R), while the units of k are slug·ft/(sec³·°R). How many independent dimensionless groupings are there? What are the dimensionless groupings?

2.43 A centrifugal pump operates with a volume-flow rate Q (units m³/sec), has an impeller diameter D and rotates with angular velocity Ω. The pump produces an increase in kinetic energy per unit mass of gh_p—note that we treat the product of g and h_p as a single quantity. How many independent dimensionless groupings are there? What are the dimensionless groupings?

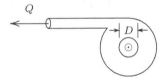

Problems 2.43, 2.44

2.44 For flow through a centrifugal pump, the pressure rise, Δp, is a function of pump diameter, D, angular speed, Ω, (dimensions $1/T$) volume-flow rate, Q, (dimensions L^3/T) and fluid density, ρ. Using dimensional analysis, develop an equation for Δp as the product of a quantity with the same dimensions as Δp and a function of all relevant dimensionless groupings.

2.45 Due to a phenomenon known as *vortex shedding*, a flagpole will oscillate at a frequency ω when the wind blows at velocity U. The diameter of the flagpole is D and the kinematic viscosity of air is ν. Using dimensional analysis, develop an equation for ω as the product of a quantity independent of ν with the same dimensions as ω and a function of all relevant dimensionless groupings.

2.46 In *magnetohydrodynamic (MHD)* flows (i.e., flows in which the fluid is electrically conducting), fluid-mechanics equations must be coupled with Maxwell's equations of electricity and magnetism. For sufficiently low frequencies, we introduce Ampere's law and Ohm's law in which the magnetic permeability of free space, μ_o, and electrical conductivity, σ, appear, where $[\mu_o] = \Omega T/L$ and $[\sigma] = 1/(\Omega L)$. The quantity Ω denotes the independent dimension of electrical resistance. For an incompressible, inviscid MHD flow with density ρ, use dimensional analysis to deduce the pertinent dimensionless parameters. Assume U and L are the appropriate velocity and length scales, and that the force on a body is F. How many independent dimensionless groupings are there? What are the dimensionless groupings?

2.47 One of the original problems discussed by Buckingham was the following. The pressure gradient, dp/dx, for flow through a smooth straight pipe depends on viscosity, μ, velocity, U, pipe diameter, D, and density, ρ. How many independent dimensionless groupings are there? What are the dimensionless groupings?

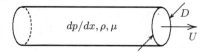

Problem 2.47

2.48 Suction is being used to reduce the drag of an airplane wing. The power that must be supplied by a pump, P, to reduce the drag by an amount ΔD is a function of the suction velocity, v_w, the freestream velocity, U, and the chord length, c. How many dimensionless groupings are there? What are the dimensionless groupings?

2.49 For an axial-flow turbine the torque, T, is a function of fluid density, ρ, rotor diameter, d, angular rotation rate, Ω, (dimensions $1/T$) and volume-flow rate, Q (dimensions L^3/T). Using dimensional analysis, develop an equation for T as the product of a quantity with the same dimensions as T and a function of all relevant dimensionless groupings.

2.50 The rotation torque, τ, (dimensions = force times distance) on a lawn sprinkler is a function of angular rotation rate, Ω (dimensions = 1/time), exit velocity, w_e, mass-flow rate, $\dot{m}$ and the length of the sprinkler arm, ℓ. How many dimensionless groupings are there? What are the dimensionless groupings?

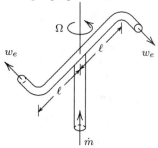

Problem 2.50

2.51 We wish to find a set of dimensionless parameters for organizing experimental data in flow past an oscillating body in a rotating fluid. The flow can be considered *inviscid*, and the only significant dimensional quantities are force, F, density, ρ, oscillation frequency, ω, coordinate system rotation rate, Ω (dimensions $1/T$), flow velocity, U, and length of the body, L. How many independent dimensionless groupings are there? What are the dimensionless groupings?

2.52 The pressure loading, Δp, of a journal bearing is known to depend on three lengths, viz., bearing diameter, D, length, ℓ, and clearance, c. Additionally, Δp depends upon angular speed, ω (dimensions $1/T$), and viscosity of the lubricant, μ. How many dimensionless groupings are there? What are the dimensionless groupings?

2.53 Experimental studies are planned for a rotational axisymmetric flowfield. The vorticity of the flow, ω, has dimensions $1/T$ and depends on initial circulation, Γ_o (dimensions L^2/T), radius, r, time, t, and kinematic viscosity of the fluid, ν. How many dimensionless groupings are there? What are the dimensionless groupings?

2.54 We are given that a force, F, is a function of velocity, U, fluid viscosity, μ, characteristic length, ℓ, fluid density, ρ, and gravitational acceleration, g. How many independent dimensionless groupings are there? What are the dimensionless groupings?

2.55 An object of length L translates with velocity U in a coordinate frame rotating with angular velocity Ω (dimensions $1/T$). The fluid has viscosity μ and density ρ, and the drag on the object is D. How many independent dimensionless groupings are there? What are the dimensionless groupings?

2.56 The figure depicts heat transfer to a fluid above a strongly heated plate. For natural convection, the difference between plate temperature and ambient temperature, ΔT, depends upon ambient temperature, T_∞, specific-heat coefficient, c_p, thermal conductivity, k, viscosity, μ, fluid density, ρ, and gravitational acceleration, g. The units of k are kg·m/(sec³·K). How many independent dimensionless groupings are there? What are the dimensionless groupings? Manipulate your answer so that one of the groupings is the Prandtl number and so that only the Prandtl number involves k.

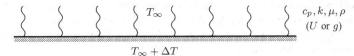

$$T_\infty + \Delta T$$

Problems 2.56, 2.57

2.57 The figure depicts heat transfer to a fluid above a heated plate. For parallel flow above the plate, the difference between plate temperature and ambient temperature, ΔT, depends upon ambient temperature, T_∞, specific-heat coefficient, c_p, thermal conductivity, k, viscosity, μ, fluid density, ρ, and freestream velocity, U. The units of k are slug·ft/(sec³·°R). How many independent dimensionless groupings are there? What are the dimensionless groupings? Manipulate your answer so that one of the groupings is the Prandtl number and so that only the Prandtl number involves k.

2.58 In an ink-jet printer, an ink droplet in the shape of an ellipsoid of revolution travels through air. The key dimensional quantities are the density of the air, ρ, velocity of the droplet, U, viscosity of the air, μ, surface tension, σ, and the diameters of the major and minor axes, $\mathcal{D}_1$ and $\mathcal{D}_2$. How many dimensionless groupings are there? What are the dimensionless groupings?

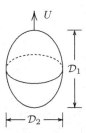

Problem 2.58

2.59 Study of flow between the read-write head of a floppy-disk drive and the disk surface indicates that the relevant dimensional quantities are the flow velocity, U, the kinematic viscosity of air, ν, the disk rotation rate, Ω, (dimensions $1/T$) the distance between the head and the disk, h, and the average height of disk-surface irregularities, k_s. What are the dimensionless parameters most appropriate for developing an empirical correlation for U?

2.60 In turbulent flow over a rough surface with mass injection, the velocity, U, is believed to be a function of the surface shear stress, τ_w, density, ρ, surface roughness height, k_s, surface-injection velocity, v_w, and viscosity, μ. How many independent dimensionless groupings are there? What are the dimensionless groupings? Express your answer in terms of $u_\tau \equiv (\tau_w/\rho)^{1/2}$.

2.61 For *inviscid* flow past a body in a rotating, stratified fluid, the significant dimensional quantities are force, F, density, ρ, gravitational acceleration, g, coordinate-system rotation rate, Ω, flow velocity, U, and body length, ℓ. How many independent dimensionless groupings are there? What are the dimensionless groupings?

2.62 The drag force, F, on a rough sphere of diameter D depends upon freestream velocity, U, density, ρ, viscosity, μ, and the height of a roughness element on the sphere, k_s. How many dimensionless groupings are there? What are the dimensionless groupings?

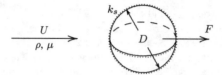

Problem 2.62

2.63 In order to enhance power and control, modern helicopters inject fluid from the tips of the rotor blades. The thrust force, T, of a helicopter blade with tip injection is a function of tip-injection velocity, V_t, the speed of sound, a, blade length, ℓ, rotation rate, Ω (dimensions $1/T$), translation speed, U, and fluid density, ρ. How many independent dimensionless groupings are there? What are the dimensionless groupings? Manipulate your answer to show that two of the dimensionless groupings are the Mach and Strouhal numbers.

2.64 In an experimental study of sedimentation, a thin layer of spherical particles rests on the bottom of a horizontal tube. An incompressible flow passes through the tube. If the experiment is conducted properly, we will observe that a critical velocity, U_c, exists at which the particles will rise and be transported along the tube. The critical velocity will be a function of fluid density, ρ, fluid viscosity, μ, pipe diameter, D, particle diameter, d, particle density, ρ_p, and the acceleration of gravity, g. How many independent dimensionless groupings are there? What are the dimensionless groupings?

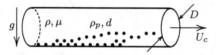

Problem 2.64

2.65 The power required to drive a propeller, P, is known to depend upon the freestream velocity, U, fluid density, ρ, speed of sound, a, angular velocity of the propeller, ω (dimensions $1/T$), diameter of the propeller, D, and the viscosity of the fluid, μ. How many independent dimensionless groupings are there? What are the dimensionless groupings? Manipulate your answer to show that three of the dimensionless groupings are the Mach, Strouhal and Reynolds numbers.

2.66 The figure below depicts flow in the immediate vicinity of a stagnation point. Ignoring viscous effects, the surface temperature, T_w, depends upon freestream temperature, T_∞, the specific-heat coefficients, c_p and c_v, freestream velocity, U, and speed of sound, a. Show that the indicial equations are linearly dependent and that the surface temperature can be written as $T_w = T_\infty F(\gamma, M, a/\sqrt{c_p T_\infty})$, where $\gamma = c_p/c_v$ and $M = U/a$.

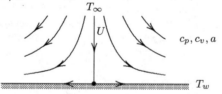

Problem 2.66

2.67 One of the original problems discussed by Buckingham was the following. The thrust force, T, of a screw propeller is known to be a function of fluid density, ρ, kinematic viscosity, ν, gravitational acceleration, g, screw diameter, D, rate of revolution, n (dimensions $1/T$), and speed of advance, U. How many independent dimensionless groupings are there? What are the dimensionless groupings? Manipulate your answer to show that three of the dimensionless groupings are the Froude, Reynolds and Strouhal numbers.

2.68 The head, Δh_d (dimensions L), developed by a turbomachine depends upon the rotor diameter, d, rotational speed, N (dimensions $1/T$), volume-flow rate, Q (dimensions L^3/T), kinematic viscosity, ν, and the acceleration due to gravity, g. How many independent dimensionless groupings are there? What are the dimensionless groupings?

2.69 We have a 1:20 scale model of an aircraft that will fly at very low Mach number, so low that we can consider the flow to be incompressible. For wind-tunnel testing, we cool the air to the point where its kinematic viscosity, ν, is one-fourth the value of normal atmospheric air at the intended altitude where the aircraft will fly. To guarantee dynamic similitude, what is the required ratio of the velocity in the wind tunnel to the velocity of the full-scale aircraft?

2.70 An airplane wing with a chord length of $c_p = 5$ ft, and a span of $S_p = 30$ ft has been designed for a cruise speed of $U_p = 180$ ft/sec. A 1:9 scale model will be tested in a water channel.

(a) If the model fluid and the prototype fluid both have a temperature of 68° F, what must the water-channel speed, U_m, be to match model and prototype Reynolds numbers?

(b) For the conditions of Part (a), what are the Mach numbers for the prototype and the model?

2.71 To analyze the dynamics of a golf ball, wind-tunnel tests are proposed for a full-scale ball in a tunnel whose maximum flow speed is $U_m = 165$ ft/sec. The temperature of the tunnel air supply can be adjusted to change the viscosity, μ_m. A professional golfer can hit a ball at $U_p = 220$ ft/sec with an angular spin rate of $\omega_p = 24$ sec^{-1}. Dimensional analysis shows that dynamic similitude can be achieved by matching Strouhal and Reynolds numbers, $St \equiv \omega D/U$ and $Re \equiv \rho U D/\mu$, where ρ is fluid density and $D = 1.67$ in is the diameter of a golf ball.

(a) If the air density for model and prototype is the same, what must ω_m and μ_m be?

(b) Assuming $\mu \propto T^{0.7}$ and $T_p = 75°$ F, what must the wind-tunnel temperature be?

2.72 Wind-tunnel tests show that the aerodynamic drag force on a 1:6 scale model of a minivan is 220 lb. The prototype will move at 30 mph in air at the same air conditions.

(a) What must the wind-tunnel velocity be to achieve dynamic similitude? Ignore Mach number in arriving at your answer.

(b) What will the aerodynamic drag force on the prototype be?

2.73 Water channel tests are planned for a 1:18 scale model to determine the drag of a blimp. The full-scale blimp will move at 25 ft/sec in 68° F air. The water in the channel will also be at 68° F.

(a) What must the water-channel velocity be to establish dynamic similitude? Ignore Mach and Froude numbers in arriving at your answer.

(b) If the measured drag is 487 lb, what is the drag (to the nearest lb) on the full-scale blimp?

(c) How much power, in hp, is required to move the blimp at this speed?

Problem 2.73

2.74 Using a pressurized wind tunnel with air of density $\rho_m = 0.00585$ slug/ft^3 and a temperature of 68° F, a 1:12 scale model of a tractor-trailer is being tested. The height of the model is $H_m = 1$ ft and the flow speed is $U_m = 240$ ft/sec. The measured aerodynamic drag is $F_m = 70$ lb.

(a) Assuming Mach number is unimportant and the prototype moves in air at 68° F and 1 atm, verify that to achieve dynamic similitude, the condition $F = \rho U^2 H^2 f(Re_H)$ implies $\rho_p F_p = \rho_m F_m$.

(b) Determine the prototype velocity and the prototype aerodynamic drag force.

2.75 A 1:16 scale model of a ship's screw propeller will be tested in water at the same temperature that will be encountered in practice. The flow speed past the model screw is $U_m = 2$ m/sec and its rotation rate is $N_m = 540$ rpm. Its diameter is $D_m = 18$ cm. Dimensional analysis indicates the Froude number, Fr, and Strouhal number, St, must be matched to achieve dynamic similitude, where

$$Fr \equiv U/\sqrt{gD} \quad \text{and} \quad St \equiv ND/U$$

(a) What are the corresponding forward velocity and rotation rate of the prototype screw?

(b) For the conditions of Part (a) and $T = 20°$ C, what are the Reynolds numbers based on D for the model and the prototype?

2.76 A researcher is conducting experiments in a wind tunnel to simulate operating conditions for a sonic buoy. The prototype buoy is $\ell_p = 6$ ft long and will travel at a speed of $U_p = 8$ ft/sec. The wind tunnel test section can accommodate a model buoy that is $\ell_m = 3$ ft long. Assume the kinematic viscosity of water for typical operating conditions is $\nu_p = 1.08 \cdot 10^{-5}$ ft²/sec, and wind tunnel conditions are such that the kinematic viscosity in the test section is $\nu_m = 1.62 \cdot 10^{-4}$ ft²/sec. Assume the flow is incompressible for prototype and model.

(a) At what speed must the air in the wind tunnel be moving to insure dynamic similitude?

(b) If the fluid temperature for prototype and model are $60°$ F, compare the Mach numbers. Comment on the suitability of the simulation, noting that a flow is considered to be incompressible if the Mach number is less than 0.3.

2.77 An airplane is designed to fly at a speed of 550 mph at an elevation of 30000 ft, where the temperature is $-56°$ F and the pressure is 4 psi. A 1:10 scale wind-tunnel model will be tested at a temperature of $70°$ F. The viscosity of air at 30000 ft is approximately $3.0 \cdot 10^{-7}$ slug/(ft·sec). It is $3.8 \cdot 10^{-7}$ slug/(ft·sec) in the wind tunnel. Noting that the speed of sound of air varies as $a \propto \sqrt{T}$, find the wind-tunnel velocity (in mph) and pressure (in atm) required to match both Mach and Reynolds numbers of the prototype.

2.78 A man named George is teaching his wife how to swing on a vine in the jungle. George's wife— who took a fluid mechanics course in college—has used dimensional analysis to show that the velocity required to topple a tree as a result of a collision is a function, $\mathcal{F}$, of m/M and ℓ/L, i.e.,

$$U = \sqrt{g\ell} \; \mathcal{F}\left(\frac{m}{M}, \frac{\ell}{L}\right)$$

where g is gravitational acceleration, ℓ is the height above the ground at which the swinger collides with the tree, m is the mass of the swinger, M is the mass of the tree and L is the height of the tree. George and his wife weigh 100 kg and 50 kg, respectively. After a heavy rainfall, George can topple a 1000 kg tree with a collision height of $\ell = 5$ m.

(a) How massive a tree can George's wife topple after the rainfall?

(b) Assuming $M \propto L^3$, at what height above the ground should her collision occur?

2.79 A 196:1 scale model of a harbor breakwater is constructed for testing. The object of the tests is to determine effects of tides and waves generated by a storm. Typical storm waves have an amplitude, h_p, of 1.5 m and travel with a velocity, U_p, of 9 m/sec. Dimensional analysis shows that dynamic similitude can be achieved by matching Froude number and h/H, where H is breakwater height.

(a) For identical model and prototype fluid properties, what must the amplitude and speed of the model waves be?

(b) To simulate effects of tides, we must match Strouhal number defined by $St \equiv H/(U\tau)$, where τ is the tidal period. If the time between tides is $\tau_p = 12$ hours, what must the model period, τ_m, be?

2.80 In order to help design a water-channel that will be constructed on the moon, an experiment is being conducted on Earth. The moon's gravitational acceleration is one-sixth that on Earth. The experiments match both Reynolds and Froude numbers using water at the prototype temperature. The prototype channel will be L_p = 2 m wide and will have a flow speed of U_p = 5 m/sec. Determine the width and flow speed of the model channel.

2.81 An inexperienced recent graduate has conducted a wind-tunnel experiment on a 1:20 scale model of a missile designed to fly at 400 mph. The wind-tunnel temperature and pressure were the same as those for the low-flying missile, and the tunnel flow speed was adjusted to match model and prototype Reynolds numbers.

(a) What was the wind-tunnel flow speed?

(b) The wind-tunnel temperature, T_m, was 60° F so that the speed of sound in the tunnel was a_m = 1119 ft/sec. Noting that the peak temperature on the model was given by

$$T_{max} = T_m \left[1 + \frac{\gamma - 1}{2} r M_m^2\right]$$

where T_m is absolute temperature, M_m is Mach number, $\gamma \approx 1.2$ and $r \approx 0.85$, explain why the aluminum model vaporized during the first test. **HINT:** Aluminum boils at 2057° C.

2.82 A graduate student performs wind-tunnel studies of a 1:25 scale submarine model. The goal of the experiments is to simulate conditions when the submarine moves at 10 knots in seawater at 4° C, for which the kinematic viscosity is ν_p = 1.54·10⁻⁶ m²/sec. The wind tunnel operates at 20° C and 1 atm.

(a) The graduate student's experiments match Reynolds numbers for model and prototype. What is the wind tunnel flow speed, U_m?

(b) Noting that the speed of sound in the wind tunnel is $a_m = \sqrt{\gamma R T_m}$ (T_m is absolute temperature, γ is specific-heat ratio and R is the perfect-gas constant), compute the Mach number in the wind tunnel. Can you explain why the student's thesis adviser is beside himself?

2.83 Using a highly pressurized wind tunnel with p = 20 atm and T = 60° C, a 1:16 scale model of a submarine is being tested. The prototype submarine will move at U_p = 6 m/sec in seawater, whose kinematic viscosity is ν_p = 1.04·10⁻⁶ m²/sec.

(a) Assuming dynamic similitude can be achieved by matching just the Reynolds number, what must the wind-tunnel flow speed be?

(b) Since the speed of sound of air at 60° C is 366 m/sec, what is the wind-tunnel Mach number?

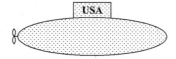

Problems 2.82, 2.83, 2.84

2.84 Experiments are planned for a small one-man submarine that uses surface heating to maintain laminar flow, and thus permit high enough velocity to outrun torpedoes due to its very low drag force. Pertinent flow parameters are the drag force, F, water density, ρ, velocity, U, submarine volume, V, freestream viscosity, μ_e, and viscosity corresponding to the surface temperature, μ_w.

(a) Using dimensional analysis, verify that the drag force can be written as

$$F = \rho U^2 V^{2/3} f(Re, \mu_w/\mu_e), \qquad Re \equiv \rho U V^{1/3}/\mu_e$$

(b) Experiments will be done on a model whose length is 1/3 that of the prototype with freestream and surface temperatures the same as will be encountered by the prototype. The prototype has a volume of V_p = 23 m³, and will move in fresh water at 28 knots. If the experiments are also done in fresh water, what must the model velocity, U_m, be?

2.85 Ignoring viscosity, the wave resistance on a ship, R, is known to be a function of density, ρ, gravitational acceleration, g, velocity, U, and length, L.

(a) Determine the appropriate dimensionless groupings.

(b) The wave resistance of a model of a ship at 1:25 scale is 6 lb at a model speed of 3 knots. What are the corresponding velocity and wave resistance of the prototype?

2.86 As part of an etching process, an acid flows over a solid surface at a velocity U. The flow includes vortices aligned with the flow direction that "dig" troughs of depth h, which depend upon the amount of time, τ, the acid flows over the surface. The kinematic viscosity of the acid is ν.

(a) Determine the dimensionless groupings.

(b) If it takes 1 μsec to dig a trough of depth $h = 10^{-3}$ mm with a Reynolds number $Uh/\nu = 100$ for the prototype material, how long will it take to dig a trough of 1 mm in a model experiment that is 1000 times larger than the prototype?

2.87 In manufacturing recording tape, the tape is coated with a lubricant. This is done by pulling it through a narrow gap filled with the lubricant. The tape width and thickness are w and h, respectively. The maximum tensile force the tape can withstand is T. The tape is centered in a gap of height $3h$, the lubricant has viscosity μ, and the length of tape within the gap is ℓ. The maximum velocity at which the tape can be pulled through the gap is U.

(a) Assuming the pertinent dimensional quantities are U, T, h, μ, w, ℓ, determine the appropriate dimensionless groupings required for dynamic similitude.

(b) If a model device is created using the same tape as in the prototype and a different lubricant, determine the ratio of model to prototype velocity, U_m/U_p.

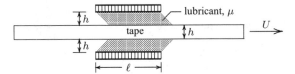

Problem 2.87

2.88 An oil-drilling platform is planned for a part of the ocean where average currents are $U = 0.12$ m/sec. Typical waves have a height of $h = 2.4$ m and a frequency of $\omega = 0.5$ sec^{-1}. Experiments will be conducted on a 1:16 scale model.

(a) If the width of the platform is L, verify that the pertinent dimensionless groupings are h/L, Strouhal number, $St = \omega h/U$, and Froude number, $Fr = U/\sqrt{gh}$.

(b) What current speed, wave height and wave frequency must be used in the experiments in order to establish dynamic similitude?

2.89 Consider the experimental determination of the force acting on a plate that is being dragged through a fluid. The force on the plate per unit width (into the page), F, is known to depend upon fluid density, ρ, fluid viscosity, μ, plate length, L, plate thickness, t, and the plate velocity, U.

(a) How many dimensionless groupings are there? Determine the dimensionless groupings. **NOTE:** The quantity F has dimensions of force/length.

(b) The prototype fluid has $\rho_p = 800$ kg/m^3 and $\mu_p = 0.03$ kg/(m·sec), while the model fluid has $\rho_m = 1000$ kg/m^3 and $\mu_m = 0.0015$ kg/(m·sec). We use the prototype plate in the model experiment. If the prototype velocity of interest is $U_p = 25$ m/sec, what should the model velocity, U_m, be to have similitude between the prototype and the model?

(c) The experiment is performed with the model fluid at velocity U_m determined in Part (b), and the force per unit width is found to be $F_m = 10$ N/m. What is the corresponding force per unit width, F_p, for the plate moving in the prototype fluid at the prototype velocity?

2.90 To simulate an accident in which a victim jumped from a building into a fireman's safety net that ripped upon impact, a lawyer dropped a dummy of twice the weight from one fourth the altitude of the victim's fall into an exact duplicate of the net. This guaranteed that the dummy's momentum in the experiment matched that of the victim in the accident. The net tore in the experiment, and left the net manufacturer in jeopardy of being liable for a huge financial award to the victim. The defense attorney hired an accident simulation expert who used dimensional analysis to demonstrate that the victim's lawyer's experiment failed to achieve dynamic similitude. The key dimensional quantities are the mass of the falling object, M, the object's velocity, U, the Young's modulus of the net material, E, the thickness of the net material, t, and the diameter of the net, d. Note that Young's modulus has dimensions of force per unit area.

   (a) Determine the dimensionless groupings.

   (b) Explain why the victim's lawyer's experiment is an invalid simulation.

2.91 Insulating paint is applied to a wire by pulling it through a cylindrical orifice of radius R. The wire, of radius r, is centered in the orifice whose length is ℓ. The viscosity of the paint is μ. A force F is required to pull the wire at a constant velocity U. To help visualize what goes on in the thin gap between the wire and the orifice, experiments are planned for a device three times larger than the prototype, using a lubricant with twice the viscosity of the paint.

   (a) Assuming the pertinent dimensional quantities are F, R, r, ℓ, μ and U, determine the appropriate dimensionless groupings required for dynamic similitude.

   (b) If the same force is applied to the model wire that is applied to the prototype wire, what is the ratio of the model velocity to the prototype velocity?

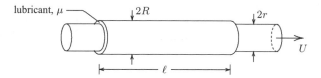

Problem 2.91

2.92 The maximum height that a dog can jump, H, is a function of the dog's mass, M, gravitational acceleration, g, and the energy the dog's leg muscles can deliver, E.

   (a) Determine the dimensionless groupings.

   (b) A 5-pound Chihuahua can jump 2.5 feet on Earth. How high could the same Chihuahua jump on the Moon where the gravitational acceleration is 1/6 that of the Earth?

Problem 2.92

2.93 To simulate flow through a centrifugal pump that will be used in a nuclear-reactor cooling system, a model is used that is 5 times larger than the prototype. The prototype coolant is liquid sodium. For the intended operating conditions, the kinematic viscosity of liquid sodium is $\nu_p = 3.41 \cdot 10^{-6}$ ft^2/sec. The prototype pump has an angular rotation speed of $N_p = 1800$ rpm and a volume-flow rate of $Q_p = 1.2$ ft^3/sec. The model fluid is water at 68° F.

   (a) If the pertinent dimensional quantities are Q, ν, N and impeller diameter, D, determine the appropriate dimensionless groupings required for dynamic similitude.

   (b) What are the model angular rotation speed, N_m and volume-flow rate, Q_m?

Chapter 3

Pressure

Our first topic as we begin our detailed exploration of fluid mechanics is the pressure that prevails whether or not a fluid is bounded by a container. We will quickly discover that fluid pressure, p, acts as a force potential, i.e., the gradient of the pressure, ∇p, is a force (per unit volume) that acts to accelerate and/or balance other forces acting on a fluid particle. Roughly speaking, pressure is analogous to voltage in electromagnetic theory. If it's constant, the fluid is at some base state. If it changes, then something happens.

We will develop two particularly interesting formulas in this chapter, viz., the **hydrostatic relation** for pressure and **Bernoulli's equation**. The hydrostatic relation shows why pressure increases with depth in the ocean and decreases with altitude in the atmosphere. It can also be used to measure the pressure at any point in a stationary fluid indirectly by simply measuring the height of a column of fluid. Bernoulli's equation, which gives an explicit relation between velocity and pressure for a moving fluid, is one of the most famous and widely used equations in fluid-flow theory. The section on Bernoulli's equation includes a simple experiment you can perform at home.

Even in the absence of fluid motion, the presence of a gravitational field gives rise to some interesting phenomena. On the one hand, we learned in Chapter 1 that pressure acts equally in all directions at a given point in a fluid. On the other hand, the gravitational field causes a **pressure gradient**, and we find that pressure increases with depth in a fluid. There is no contradiction as the former observation applies to a single point, while gravitational effects are felt over a finite distance.

After developing and applying the hydrostatic relation and Bernoulli's equation, the concluding sections of this chapter focus on fluid-pressure variation in a gravitational field for a motionless, or static fluid. For the obvious reason, general engineering literature describes this topic as **fluid statics**.

We will develop a method for computing forces and moments on both planar and curved surfaces. With a clever balance of forces, we will replace the problem of integrating a varying pressure over a general curvilinear surface with a simple set of algebraic operations. We will accomplish this by deriving formulas for a point force and moment arm that give the mechanical equivalent of the force and moment due to the distributed pressure. This method is useful for determining the loads on a structure such as a dam or a flow-control valve.

We conclude the chapter with a discussion of buoyancy and a famous result known as **Archimedes' Principle**. This principle says the buoyancy force on a floating or submerged body is equal to the weight of the displaced fluid. The section on buoyancy includes another home experiment.

3.1   Pressure on an Infinitesimal Element

Before proceeding, it is worthwhile to pause and derive a very useful mathematical relationship that we will make use of from time to time in developing the basic equations governing the behavior of fluids. Consider the differential element shown in Figure 3.1. The element is a rectangular parallelepiped and the pressure, p, acts on its six faces. We assume that the fluid element's center lies at the point (x, y, z) and that the lengths of the three sides are Δx, Δy and Δz in the x, y and z directions, respectively.

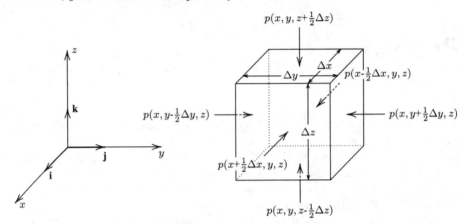

Figure 3.1: *Pressure acting on an infinitesimally small fluid element centered at the point* (x, y, z). **i**, **j** *and* **k** *are unit vectors in the* x, y *and* z *directions, respectively.*

We can calculate the net force on the fluid element by integrating the pressure over the surface bounding the fluid element. We indicate this symbolically as follows.

$$\mathbf{F} = -\oiint_S p\,\mathbf{n}\,dS \tag{3.1}$$

There are three key points that require discussion to explain the notation.

1. The 'O' joining the integrals denotes *closed-surface integral*, i.e., we must integrate over the entire surface that bounds the fluid element. For the infinitesimal element in Figure 3.1, this means

$$\oiint_S p\,\mathbf{n}\,dS = \sum_{m=1}^{6} \iint_{A_m} p\,\mathbf{n}\,dA_m$$

   where the six A_m are the areas of the six faces of the element.

2. The vector **n** is the *outer unit normal*, i.e., a unit vector normal to the bounding surface that points outward. In the present case we have

   - Face at $x+\frac{1}{2}\Delta x$: **n** = **i**, Face at $x-\frac{1}{2}\Delta x$: **n** = −**i**
   - Face at $y+\frac{1}{2}\Delta y$: **n** = **j**, Face at $y-\frac{1}{2}\Delta y$: **n** = −**j**
   - Face at $z+\frac{1}{2}\Delta z$: **n** = **k**, Face at $z-\frac{1}{2}\Delta z$: **n** = −**k**

   where **i**, **j** and **k** are unit vectors in the x, y and z directions, respectively.

3. We require a minus sign because the pressure pushes in the opposite direction of **n**.

So, for our infinitesimal element, we have

$$
\mathbf{F} = - \underbrace{\iint_A p\,\mathbf{i}\,dydz}_{\text{Face at } x+\frac{1}{2}\Delta x} - \underbrace{\iint_A p\,(-\mathbf{i})\,dydz}_{\text{Face at } x-\frac{1}{2}\Delta x} - \underbrace{\iint_A p\,\mathbf{j}\,dxdz}_{\text{Face at } y+\frac{1}{2}\Delta y}
$$

$$
- \underbrace{\iint_A p\,(-\mathbf{j})\,dxdz}_{\text{Face at } y-\frac{1}{2}\Delta y} - \underbrace{\iint_A p\,\mathbf{k}\,dxdy}_{\text{Face at } z+\frac{1}{2}\Delta z} - \underbrace{\iint_A p\,(-\mathbf{k})\,dxdy}_{\text{Face at } z-\frac{1}{2}\Delta z} \tag{3.2}
$$

Because we are dealing with infinitesimally small quantities, we can treat the pressure as being constant on each face. Our result will be exact in the limit as Δx, Δy and Δz approach zero. Thus,

$$
\begin{aligned}
\mathbf{F} \approx & -\left[p(x+\tfrac{1}{2}\Delta x, y, z) - p(x-\tfrac{1}{2}\Delta x, y, z)\right] \mathbf{i}\,\Delta y\Delta z \\
& -\left[p(x, y+\tfrac{1}{2}\Delta y, z) - p(x, y-\tfrac{1}{2}\Delta y, z)\right] \mathbf{j}\,\Delta x\Delta z \\
& -\left[p(x, y, z+\tfrac{1}{2}\Delta z) - p(x, y, z-\tfrac{1}{2}\Delta z)\right] \mathbf{k}\,\Delta x\Delta y
\end{aligned} \tag{3.3}
$$

Next, we rearrange terms to arrive at

$$
\begin{aligned}
\mathbf{F} \approx & -\frac{p(x+\tfrac{1}{2}\Delta x, y, z) - p(x-\tfrac{1}{2}\Delta x, y, z)}{\Delta x} \mathbf{i}\,\Delta x\Delta y\Delta z \\
& -\frac{p(x, y+\tfrac{1}{2}\Delta y, z) - p(x, y-\tfrac{1}{2}\Delta y, z)}{\Delta y} \mathbf{j}\,\Delta x\Delta y\Delta z \\
& -\frac{p(x, y, z+\tfrac{1}{2}\Delta z) - p(x, y, z-\tfrac{1}{2}\Delta z)}{\Delta z} \mathbf{k}\,\Delta x\Delta y\Delta z
\end{aligned} \tag{3.4}
$$

Finally, as Δx, Δy and Δz approach zero, each line in Equation (3.4) contains a partial derivative of the pressure. Defining the differential volume, ΔV, according to

$$
\Delta V = \Delta x \Delta y \Delta z \tag{3.5}
$$

we conclude that

$$
\mathbf{F} \approx -\left[\frac{\partial p}{\partial x}\mathbf{i} + \frac{\partial p}{\partial y}\mathbf{j} + \frac{\partial p}{\partial z}\mathbf{k}\right]\Delta V \tag{3.6}
$$

The quantity in square brackets is the familiar gradient of p, which we write symbolically as

$$
\nabla p = \frac{\partial p}{\partial x}\mathbf{i} + \frac{\partial p}{\partial y}\mathbf{j} + \frac{\partial p}{\partial z}\mathbf{k} \tag{3.7}
$$

Therefore, we have proven that, for an infinitesimally small fluid element,

$$
-\oiint_S p\,\mathbf{n}\,dS \approx -\nabla p\,\Delta V \quad \text{for} \quad \Delta V \to 0 \tag{3.8}
$$

In words, Equation (3.8) tells us that the net force on a differential fluid element due to the pressure is equal to the product of the element's volume and the pressure gradient within the element. We will use this result in the next section in order to derive a key result for fluids at rest.

3.2   The Hydrostatic Relation

For a fluid at rest in a gravitational field, the only forces acting are gravity and the pressure within the fluid. In the absence of gravity, the pressure in the fluid would obviously be the same at all points. Thus, the presence of a gravitational field must give rise to a variation in pressure. In this section, we will determine how the pressure varies.

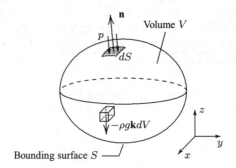

Figure 3.2: *Fluid volume in a gravitational field.*

Consider the arbitrary volume V bounded by a closed surface S shown in Figure 3.2. As noted in the preceding section, if $\mathbf{n}$ is the outer unit normal to the differential surface element dS, the net pressure force exerted *by the surroundings on the surface* is the closed-surface integral of $-p\,\mathbf{n}$ (the minus sign makes it an inward directed force). That is, the net pressure force acting on the volume is

$$\textbf{Pressure Force} = - \oiint_S p\,\mathbf{n}\,dS \qquad (3.9)$$

Now, the density of the fluid contained in the volume is ρ and the gravitational acceleration vector is given by $-g\mathbf{k}$ where $\mathbf{k}$ is a unit vector in the z direction. So, the weight of the fluid is the volume integral of $-\rho g\mathbf{k}$, i.e.,

$$\textbf{Weight} = - \iiint_V \rho g\mathbf{k}\,dV \qquad (3.10)$$

Since the volume is at rest, balancing forces yields

$$\oiint_S p\,\mathbf{n}\,dS + \iiint_V \rho g\mathbf{k}\,dV = \mathbf{0} \qquad (3.11)$$

Equation (3.11) applies to stationary volumes of arbitrary shape and size. We will make use of this integral relationship in Section 3.10 when we address buoyancy. Because of the generality of this equation, we can apply it to an infinitesimal fluid element such as the one depicted in Figure 3.1. We can use this fact to determine how the pressure behaves at an arbitrary point within the volume V. Clearly, as the volume size approaches zero, we have

$$\iiint_V \rho g\mathbf{k}\,dV \approx \rho g\mathbf{k}\,\Delta V \quad \text{as} \quad \Delta V \to 0 \qquad (3.12)$$

where ρ is an averaged value for the fluid within ΔV. In the limit $\Delta V \to 0$, the value of ρ approaches the value at the point we are interested in. Recalling Equation (3.8), the force balance simplifies to

$$\nabla p\,\Delta V + \rho g\mathbf{k}\,\Delta V = \mathbf{0} \quad \text{as} \quad \Delta V \to 0 \qquad (3.13)$$

Dividing through by ΔV yields

$$\nabla p = -\rho g \mathbf{k} \qquad (3.14)$$

Thus, in component form, we have shown that,

$$\frac{\partial p}{\partial x} = \frac{\partial p}{\partial y} = 0 \quad \text{and} \quad \frac{\partial p}{\partial z} = -\rho g \qquad (3.15)$$

In addition to demonstrating that a stationary fluid in a gravitational field has a pressure gradient parallel to the direction of gravitational acceleration, we have also shown that *pressure is a function only of depth*. This means $p = p(z)$ and we can replace Equation (3.15) with the following ordinary differential equation.

$$\frac{dp}{dz} = -\rho g \qquad (3.16)$$

We can integrate immediately for fluids such as liquids in which density is constant. The resulting pressure variation is

$$p + \rho g z = \text{constant} \qquad (3.17)$$

We call Equation (3.17) the **hydrostatic relation**. It tells us that pressure increases linearly with depth (remember that negative values of z correspond to increasing depth).

The physical meaning of the hydrostatic relation becomes obvious if we regard the pressure as a force potential. This is sensible since, by definition, a force potential is a function whose gradient is a force vector (the standard convention is to include a minus sign). In this sense pressure is analogous to voltage, whose gradient is proportional to electrostatic force. As demonstrated in the last section, the net force on an infinitesimal fluid element is $-\nabla p \Delta V$. Consequently, ∇p is force per unit volume and p is thus the pressure-force potential per unit volume.

Similarly, the term $\rho g z$ is potential energy per unit volume. Thus, Equation (3.17) tells us that the sum of the pressure-force potential per unit volume and the potential energy per unit volume is constant. **In other words, the hydrostatic relation is a mechanical-energy conservation principle for a motionless fluid.**

Example 3.1 *At what depth in a 20° C fresh-water lake is the pressure twice the pressure of the atmosphere?*

Solution. Since the pressure at the lake surface ($z = 0$) is atmospheric, the value of "constant" in Equation (3.17) is p_a. Thus, the depth at which $p = 2p_a$ is given by

$$2p_a + \rho g z = p_a$$

Solving for z yields

$$z = -\frac{p_a}{\rho g}$$

The density of water at 20° C is $\rho = 998$ kg/m³, $g = 9.807$ m/sec² and atmospheric pressure is $p_a = 1.01 \cdot 10^5$ Pa. Therefore,

$$z = -\frac{1.01 \cdot 10^5 \text{ kg}/(\text{m} \cdot \text{sec}^2)}{(998 \text{ kg/m}^3)(9.807 \text{ m/sec}^2)} = -10.32 \text{ m}$$

where we make use of the fact that 1 Pa = 1 kg/(m·sec²). Thus, the pressure is 2 atm at a depth of 10.32 m.

Figure 3.3: *Because of the increase in pressure with depth, which can cause the bends, sport diving is done no deeper than 130 feet. The hydrostatic relation tells us the pressure is nearly 5 atm at this depth. [Photograph courtesy of Barbara Wilcox]*

The hydrostatic relation tells us that the pressure at 130 feet in the ocean is 72.8 psi (note that the density of seawater is 2.0 slugs/ft^3). This high a pressure is sufficient to cause nitrogen to dissolve in a diver's bloodstream, leading to impaired judgment and a painful condition known as the **bends** if the diver returns to the surface too rapidly. Impaired judgment is very noticeable at a depth of 130 feet, which is the established maximum depth for sport diving. Using no breathing equipment, divers have exceeded this depth and risen to the surface rapidly with no ill effects. However, their pulse rates have gone so low at these great depths as to be in a state close to death—this practice is not recommended!

3.3   Atmospheric Pressure Variation

The pressure in the Earth's atmosphere varies in a more complicated manner than the simple linear relation in Equation (3.17). The reason for this is the following. Atmospheric air behaves like a perfect gas so that its density is given by

$$\rho = \frac{p}{RT} \tag{3.18}$$

Hence, Equation (3.16) assumes the following form.

$$\frac{dp}{dz} = -\frac{g}{RT}p \tag{3.19}$$

If we knew the variation of temperature T with altitude, integration of Equation (3.19) would be straightforward. There is a simple model for the atmosphere over the United States known as the U. S. Standard Atmosphere [U. S. Government Printing Office (1974)]. This model represents average conditions in the United States at 40° N latitude (e.g., New York City). In the U. S. Standard Atmosphere, the region from the Earth's surface ($z = 0$) up to $z = 11.0$ km (6.84 miles) is called the **troposphere** and the temperature decreases linearly with altitude according to

$$T = T_0 - \alpha z \quad \text{for} \quad 0 \le z \le 11.0 \text{ km} \tag{3.20}$$

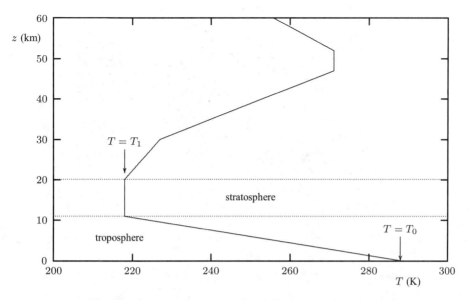

Figure 3.4: *Temperature in the U. S. Standard Atmosphere.*

where the coefficient α is the **lapse rate**, and T_0 is the average surface temperature. The region from $z = 11.0$ km (6.84 mi) to $z = 20.1$ km (12.5 miles) is called the **stratosphere**. The temperature in this idealized model is constant in the stratosphere, and denoted by T_1. Figure 3.4 shows the temperature variation in the U. S. Standard Atmosphere, with the troposphere and stratosphere clearly indicated. Above the stratosphere, temperature increases in a nontrivial manner. Table 3.1 lists values of α, T_0 and T_1 in SI and USCS units.

Table 3.1: *Properties of the U. S. Standard Atmosphere*

Property	SI Units	USCS Units
α	6.50 K/km	18.85° R/mi
T_0	288 K (15° C)	518.4° R (59° F)
T_1	218 K (−55° C)	392.4° R (−67° F)

Tropospheric Pressure Variation. Focusing first on the troposphere, combining Equations (3.19) and (3.20) yields

$$\frac{dp}{p} = -\frac{g\,dz}{R(T_0 - \alpha z)} \tag{3.21}$$

Integrating, we find that the pressure varies according to

$$p = p_0 \left[1 - \frac{\alpha z}{T_0}\right]^{g/(\alpha R)} \quad \text{for} \quad 0 \leq z \leq 11.0 \text{ km} \tag{3.22}$$

where $p_0 = 101$ kPa (14.7 psi) is the pressure at sea level. The exponent $g/(\alpha R)$ in Equation (3.22) is approximately 5.26. Note that the pressure falls to 22.5 kPa (3.28 psi) at the upper boundary of the troposphere, which is the approximate altitude at which modern airliners fly.

Stratospheric Pressure Variation. The integration is even easier in the stratosphere since temperature is constant. The pressure varies as follows.

$$p = p_1 \exp\left[-\frac{g(z - z_1)}{RT_1}\right] \qquad \text{for} \qquad 11.0 \text{ km} \le z \le 20.1 \text{ km} \qquad (3.23)$$

The pressure p_1 = 22.5 kPa (3.28 psi) follows from insisting that Equations (3.22) and (3.23) yield the same pressure at the interface between the troposphere and stratosphere. Thus, the pressure (and density) fall off exponentially in the stratosphere, and we sometimes refer to this as an **exponential atmosphere**.

Example 3.2 *The cabin pressure in a modern airliner at cruise altitude is typically about 12 psi. As the airplane descends to land, low pressure air is trapped inside your ears. This is what causes the "popping" sensation you experience when you yawn and allow the pressure to equilibrate. Estimate the altitude in the atmosphere at which this pressure prevails.*

Solution. We can use Equation (3.22) to compute the altitude in the U. S. Standard Atmosphere. Solving for z, we find

$$z = \frac{T_0}{\alpha}\left[1 - \left(\frac{p}{p_0}\right)^{\alpha R/g}\right] = \frac{518.4^\circ \text{ R}}{18.85^\circ \text{ R/mi}}\left[1 - \left(\frac{12.0 \text{ psi}}{14.7 \text{ psi}}\right)^{1/5.26}\right] = 1.04 \text{ mi}$$

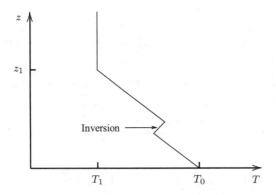

Figure 3.5: *Temperature variation over Los Angeles.*

The atmospheric temperature variation near Los Angeles (and many other large cities) deviates from the U. S. Standard Atmosphere in a significant manner. Specifically, there is a region in the troposphere, typically at an altitude of about 0.5 km (0.3 mi), known as the **inversion layer**, in which temperature increases with increasing altitude. This layer is present because the Los Angeles area is almost completely enclosed by high mountains. As air descends from the mountains, the sun heats it and creates a warm layer that rises above cooler air blowing in from the Pacific Ocean. This is the primary mechanism that creates the temperature inversion as in Figure 3.5, with the heavier cool air trapped near the surface. The inversion layer puts a "lid" on the area that traps surface emissions responsible for smog. The prevailing winds in the Los Angeles area are unable to relieve the pollution problem because of this lid. Rather, they merely move the smoggy air from one part of the region to another.

3.4 Manometry

The hydrostatic relation tells us that a change in depth is directly proportional to the corresponding change in pressure. This is the basis of the *mercury barometer* invented by Torricelli in 1644. As shown in Figure 3.6, a barometer consists of a glass tube closed at one end with the open end immersed in a container filled with mercury. The tube is initially filled with mercury and turned upside down when inserted in the container. Mercury vapor fills the open space above the column of mercury. However, because the vapor pressure of mercury is very small (Subsection 1.4.4 and Table A.9), there will be a near vacuum in the top of the tube.

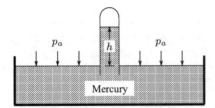

Figure 3.6: *Mercury barometer.*

Since pressure in the tube above the mercury, p, is negligible, the hydrostatic relation tells us

$$p + \rho g h_{Hg} = p_a \quad \Longrightarrow \quad h_{Hg} = \frac{p_a - p}{\rho g} \approx \frac{p_a}{\rho g} \tag{3.24}$$

Thus, for atmospheric pressure, $p_a = 1$ atm $= 101$ kPa, the height, h_{Hg}, is

$$h_{Hg} = \frac{101000 \ \text{N/m}^2}{\left(13550 \ \text{kg/m}^3\right)\left(9.807 \ \text{m/sec}^2\right)} = 760 \ \text{mm} \tag{3.25}$$

The hydrostatic relation is also the basis of a pressure-measurement device known as the **U-Tube Manometer**, an example of which is shown in Figure 3.7. To measure pressure in a channel filled with water (at Point 4), we attach a U-shaped tube containing mercury. The tube is open to the atmosphere at Point 1. The hydrostatic relation tells us that the pressure at Point 2 equals the pressure at Point 1 plus the weight of the column of mercury above. Hence,

$$p_1 + \rho_{Hg} g \Delta h = p_2 \tag{3.26}$$

Similarly, the pressure at Point 3 is equal to the pressure at Point 4 plus the weight of the column of water above, i.e.,

$$p_4 + \rho_{H_2O} g L = p_3 \tag{3.27}$$

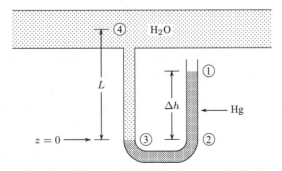

Figure 3.7: *U-Tube manometer.*

Finally, since $z_3 = z_2$, the hydrostatic relation also tells us that $p_3 = p_2$. Consequently, we can equate the left-hand sides of Equations (3.26) and (3.27), which yields the following.

$$p_4 - p_1 = (\rho_{H_g}\Delta h - \rho_{H_2O}L)g \qquad (3.28)$$

The implication of Equation (3.28) is clear. Since the U-Tube is open to the atmosphere, then p_1 is the pressure of the atmosphere. The density of mercury and water are known quantities. Thus, by simply measuring L and the height of the mercury column, Δh, we can infer the pressure in the water channel.

When a manometry configuration involves layers or columns of different fluids, pressure can be computed by repeated use of the hydrostatic relation, with one important word of caution. Since Equation (3.17) follows from Equation (3.15) only when density is constant, *we must stay within the same fluid when applying Equation (3.17)*. The following example illustrates how this is done when several fluids are present.

Example 3.3 *Consider the arrangement shown in the figure below, which is designed to monitor the pressure at the interface of the water and oil in the tank. To make the problem a little more interesting, we assume some air has been trapped in the attached manometer. The measured heights are $L_1 = 100$ mm, $L_2 = 450$ mm and $L_3 = 400$ mm. Determine the pressure difference $p_1 - p_0$, and how much of a difference ignoring the contribution from the air makes.*

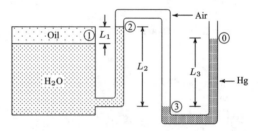

Solution. Working first in the column of mercury, we have

$$p_3 = p_0 + \rho_{Hg}gL_3$$

Next, applying the hydrostatic relation in the column of air yields

$$p_3 = p_2 + \rho_{Air}gL_2$$

Finally, for the column of water, we have

$$p_1 = p_2 + \rho_{H_2O}gL_1$$

We thus have three equations for the three unknown pressures, p_1, p_2 and p_3. A little straightforward algebra shows that the pressure at the oil-water interface is as follows.

$$p_1 = p_0 + (\rho_{Hg}L_3 + \rho_{H_2O}L_1 - \rho_{Air}L_2)g$$

Using densities corresponding to a temperature of $20°$ C, we find that the pressure difference, $p_1 - p_0$, is 54.127 kPa. Since L_1, L_2 and L_3 are all of the same order of magnitude, we can neglect the contribution from the trapped air because $\rho_{Air} \ll \rho_{Hg}, \rho_{H_2O}$. Therefore,

$$p_1 \approx p_0 + (\rho_{Hg}L_3 + \rho_{H_2O}L_1)g$$

This equation indicates that $p_1 - p_0$ is 54.133 kPa—a difference of 0.01% from the value above.

3.5 Bernoulli's Equation

While there are many important applications in which a fluid is motionless, the "real fun" begins when a fluid moves. That is, fluids are in motion for any vehicle that moves in the atmosphere, the ocean or the laboratory. The Earth's atmosphere is in continuous motion, often bringing cool breezes and sometimes bringing violent storms. Blood flows throughout your body on a continuous basis. The complicated motions involved are rich with interesting phenomena that have fascinated mankind for many centuries.

We can reasonably ask if the hydrostatic relation [Equation (3.17)], which is an energy conservation principle, has any relevance for a fluid that is moving. While the answer is essentially no (there are simple applications where it does, most notably for constant velocity), our knowledge of basic physics tells us that total energy is conserved. By simply replacing the hydrostatic relation with an equation that includes the fluid's kinetic energy, we arrive at one of the best known equations in fluid-flow theory.

Recall the discussion immediately following Equation (3.17) where we noted that the hydrostatic relation represents conservation of mechanical energy **per unit volume**. Thus, to include kinetic energy, we must determine the appropriate way to represent kinetic energy per unit volume. If ρ and $\mathbf{u}$ are the fluid density and velocity, respectively, at a given point, then the kinetic energy of a fluid particle with volume ΔV is $\frac{1}{2}\rho\Delta V\mathbf{u}\cdot\mathbf{u}$ and the corresponding kinetic energy per unit volume is $\frac{1}{2}\rho\mathbf{u}\cdot\mathbf{u}$ (this quantity is also referred to as the **dynamic pressure** or **dynamic head**). Therefore, in the presence of a gravitational field, we replace the hydrostatic relation by

$$p + \frac{1}{2}\rho\mathbf{u}\cdot\mathbf{u} + \rho g z = \text{constant} \tag{3.29}$$

This equation is one of the most famous in fluid mechanics literature and we call it **Bernoulli's equation**. It tells us that the sum of the pressure, p, kinetic energy per unit volume, $\frac{1}{2}\rho\mathbf{u}\cdot\mathbf{u}$, and potential energy per unit volume, $\rho g z$, is constant. Clearly, the hydrostatic relation is the limiting form of Bernoulli's equation for zero velocity.

Although we have arrived at this equation heuristically, it can be rigorously derived. However, this requires first developing the differential equations governing fluid motion, which we will do in Chapter 9. When we derive Bernoulli's equation we will find that several conditions must be satisfied in order for the equation to be valid. There are five conditions and they are as follows.

1. Viscous effects must be negligible;

2. The flow must not be changing with time, i.e., it must have $(\partial\mathbf{u}/\partial t = \mathbf{0})$;

3. The flow must be incompressible (ρ = constant);

4. The fluid must be subject only to a conservative forces, i.e., only forces that can be defined in terms of a force potential ($\mathbf{f} = -\nabla\mathcal{V}$ where $\mathcal{V}$ is a force potential);

5. The flow must be irrotational, i.e., the fluid velocity must have zero curl ($\nabla\times\mathbf{u} = \mathbf{0}$).

The first four conditions are straightforward and require no further elaboration. The last condition is more subtle and deserves a bit more discussion. With some analysis, we can show that the curl of a fluid particle's velocity is proportional to the particle's angular-rotation rate about its center of gravity. In Chapter 9, we will see explicitly why Bernoulli's equation holds only in a restricted way when a fluid particle rotates and tumbles as it moves.

AN EXPERIMENT YOU CAN DO AT HOME
Contributed by Prof. Peter Bradshaw, Stanford University

To observe one of the implications of Bernoulli's equation, you will need a paper napkin, a thin sheet of paper such as toilet paper, or even a sheet of notebook paper. Grasp the paper in each hand and hold it up to your mouth. Blow gently, being careful that you blow only on the top side. It may take a couple practice tries, but when done correctly, the paper will rise to a horizontal orientation. You might even want to try all of the types of paper noted above to observe that you must blow harder for the more massive sheets.

From Bernoulli's equation, it should be clear that by creating a moving stream of air over the top of the paper, the pressure decreases. Thus, you create a suction on the top surface that causes the paper to rise.

Example 3.4 *Consider a large tank with a small hole a distance h below the surface. A jet of fluid issues from the hole with velocity U. You may assume that, as is generally true for thin jets, the surrounding air impresses atmospheric pressure, p_a, throughout the jet. Assuming the tank diameter, D, is very large compared to the diameter of the small hole, d, determine U.*

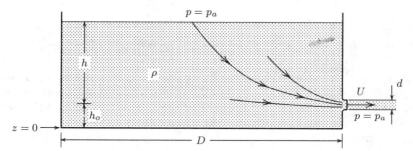

Solution. Bernoulli's equation tells us that

$$p + \frac{1}{2}\rho\, \mathbf{u} \cdot \mathbf{u} + \rho g z = \text{constant}$$

The tank is open to the atmosphere so that the pressure at the top of the tank is also p_a. Since the tank diameter, D, is very large compared to the diameter of the small hole, d, we can ignore the velocity of the fluid at the top of the tank. Hence, we can use Bernoulli's equation to relate a point at the top of the tank to a point in the jet to show that

$$p_a + \rho g(h + h_o) = p_a + \frac{1}{2}\rho U^2 + \rho g h_o$$

Note that we can set the origin, $z = 0$, anywhere we wish. This is true since the potential energy appears on both sides of this equation, so that only the difference in potential energy between the surface and the jet, $\rho g h$, matters. Thus, the velocity in the jet is given by

$$U = \sqrt{2gh}$$

In Example 3.4, we used Bernoulli's equation to relate conditions at two specified points in the flow. Often it is helpful to determine the "constant" in Bernoulli's equation by evaluating each term at a point in the flow where all terms are known. Typically, we seek a point that lies very far from solid boundaries where the flow is uniform. The most important point about using Bernoulli's equation in this way is that, because we have selected one universal reference point, the resulting equation applies at *every point in the flow*.

The following example illustrates how to implement Bernoulli's equation in this manner. It involves flow with what we call a **free surface**, i.e., a flow with an interface between a liquid and a gas. As in the preceding example, we make use of the fact that the surrounding air impresses atmospheric pressure, p_a, throughout the jet of fluid issuing from the tube.

Example 3.5 *Consider water flowing with uniform velocity, U_1, as shown in the figure. The water enters a uniform-diameter tube at some point below the surface. Determine the velocity of the water leaving the tube at a height z_2 above the surface. Also, determine the pressure in the primary stream of fluid.*

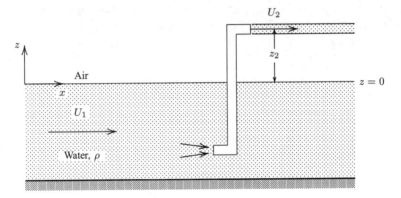

Solution. Clearly, one point where we know everything about the flow is at the free surface very far upstream. Because of the interface with the air, the pressure must be atmospheric so that $p = p_a$. Letting the free surface lie at $z = 0$, the potential energy is zero. Since the flow is uniform far upstream, we also know that the velocity is U_1. Thus, we have

$$p + \frac{1}{2}\rho \mathbf{u} \cdot \mathbf{u} + \rho g z = p_a + \frac{1}{2}\rho U_1^2$$

Hence, since $p = p_a$ in the jet of fluid issuing from the tube, we find

$$p_a + \frac{1}{2}\rho U_2^2 + \rho g z_2 = p_a + \frac{1}{2}\rho U_1^2$$

wherefore

$$U_2 = \sqrt{U_1^2 - 2g z_2}$$

Provided we are not too close to either the tube or the bottom, we expect the flow to be uniform with velocity $\mathbf{u} = U_1 \mathbf{i}$. Then, Bernoulli's equation becomes

$$p + \frac{1}{2}\rho U_1^2 + \rho g z = p_a + \frac{1}{2}\rho U_1^2 \qquad \Longrightarrow \qquad p = p_a - \rho g z$$

Thus, in the moving stream, the pressure satisfies the hydrostatic relation. Close to the tube, the velocity will deviate from the freestream value and, correspondingly, the pressure will depart from the hydrostatic relation. In a real fluid, viscous effects would result in nonuniform velocity near the bottom, which would also cause the pressure to differ from the hydrostatic value.

3.6   Velocity-Measurement Techniques

We can use Bernoulli's equation to infer velocity from a pressure measurement. This is useful because pressure is fundamentally easier to measure than velocity. To understand how this is done, we must first introduce the notion of a **stagnation point**. Then, we discuss two measurement devices known as the **Pitot tube** and the **Pitot-static tube**.

3.6.1   Stagnation Points

Figure 3.8 illustrates ideal (i.e., frictionless) two-dimensional flow past a cylinder. The figure includes several contours that we call streamlines (we will discuss streamlines in detail in Section 4.4). These are contours that are everywhere parallel to the flow velocity, which means fluid particles move along the streamlines. Since the contours are parallel to the velocity there is no flow across (normal to) streamlines. Because there is no flow across a solid boundary, the cylinder surface is a streamline. We call the streamline coincident with the x axis upstream and downstream of the cylinder the **dividing streamline**. This streamline splits at the front of the cylinder so that half of the fluid moves over the cylinder and half moves below. The streamlines rejoin at the back of the cylinder.

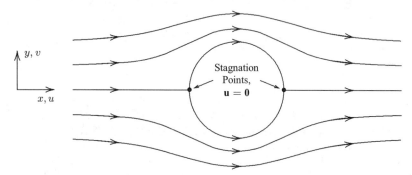

Figure 3.8: *Ideal flow past a cylinder.*

We have a very special situation at these two points. Specifically, we have the intersection of two perpendicular streamlines, namely, the dividing streamline and the cylinder surface. Now, since streamlines are parallel to the velocity, necessarily the vertical velocity component, v, on the dividing streamline must be zero approaching the cylinder. Also, very close to the front of the cylinder we must have zero horizontal velocity, u, on the surface (see Figure 3.8) in order to have flow tangent to the surface. Thus, at the point where the streamlines collide,

$$\mathbf{u} = \mathbf{0} \qquad \text{(Stagnation Point)} \qquad (3.30)$$

So, we have two points on the cylinder where the velocity vanishes, and we refer to these points as **stagnation points**.

3.6.2   Pitot Tube

The **Pitot tube** is one of the simplest devices based on Bernoulli's equation that provides an indirect measurement of velocity. Figure 3.9 illustrates flow in the immediate vicinity of a Pitot tube placed a distance d below the surface in a flowing stream of water. For precise measurements, the tube must have a very small diameter such as that characteristic of a hypodermic needle. As shown, the water fills the Pitot tube up to a point a distance h above the free surface. Because the fluid in the tube cannot move, there must be a stagnation point at the tube's entrance below the surface. As we will now show, this device permits determining the velocity by a simple measurement of the distance the fluid rises above the surface. It is thus the analog, for moving fluids, of the U-Tube manometer discussed in Section 3.4.

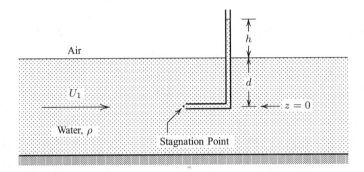

Figure 3.9: *Pitot tube.*

Now, Bernoulli's equation tells us that for this flow we have

$$p + \frac{1}{2}\rho\, \mathbf{u} \cdot \mathbf{u} + \rho g z = \text{constant} \tag{3.31}$$

We can evaluate the constant by selecting a point in the flowfield where the values of pressure, velocity and z are all known. As with the vertical-tube example considered earlier (see Example 3.3), we select a point far upstream of the Pitot tube at the free surface. At this point, the pressure is equal to the atmospheric pressure, p_a, the velocity assumes its freestream value, U_1, and $z = d$ (note that we are choosing the origin to be coincident with the lower portion of the Pitot tube). Hence,

$$\text{constant} = p_a + \frac{1}{2}\rho\, U_1^2 + \rho g d \tag{3.32}$$

Now, at the top of the tube, which is open to the atmosphere, we know the pressure is p_a, the velocity is zero and $z = d + h$. Hence, applying Bernoulli's equation, we have

$$p_a + \rho g(d + h) = p_a + \frac{1}{2}\rho\, U_1^2 + \rho g d \tag{3.33}$$

Simplifying, we arrive at the following straightforward relation between the flow velocity and the height of the column of fluid in the Pitot tube.

$$U_1 = \sqrt{2gh} \tag{3.34}$$

While the Pitot tube permits a correlation between the height of a column of fluid and the fluid velocity, it is clearly limited to flows with uniform velocity, and requires a point where velocity and pressure are both known. The device has no provision for flows in which the velocity varies with z, e.g., near the bottom of the channel shown in Figure 3.9 where viscous effects are important.

3.6.3 Pitot-Static Tube

The **Pitot-static tube** is a measuring device based on Bernoulli's equation that can be used for more general velocity distributions. This device makes two separate pressure measurements. The first measurement is done with a standard Pitot tube, which measures the pressure at the tip of the probe. Because this is a stagnation point, the Pitot tube measures the **stagnation pressure**, p_{stag}. The second measurement is at a point downstream of the probe tip sufficiently distant (typically 8 tube diameters) that the flow has returned to its freestream value, U. The pressure at this point is the freestream pressure, also referred to as the **static pressure**, p_{static}. Figure 3.10 schematically depicts a Pitot-static tube.

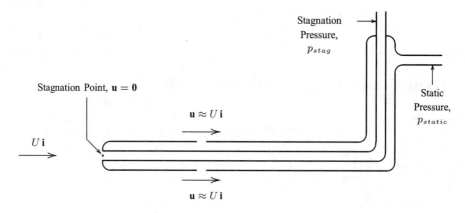

Figure 3.10: *Pitot-static tube.*

Assuming the tube is extremely thin (as it must be to avoid changing the flow), we can ignore the difference in depth of the stagnation point and the static pressure tap. Hence, from Bernoulli's equation, we have

$$p_{stag} = p_{static} + \frac{1}{2}\rho U^2 \tag{3.35}$$

That is, the stagnation pressure is the sum of the static pressure and the **dynamic pressure**, $\frac{1}{2}\rho U^2$. Therefore the local velocity is given by

$$U = \sqrt{\frac{2\left(p_{stag} - p_{static}\right)}{\rho}} \tag{3.36}$$

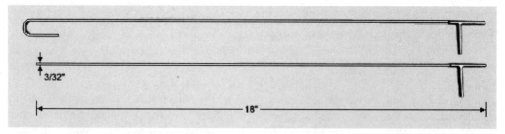

Figure 3.11: *Hand-held Pitot-static tubes used in the automobile racing engine industry. [Photograph courtesy of and © Audie Technology, Inc.]*

Clearly, the Pitot-static tube is not limited to uniform velocity distributions. Furthermore, the device is essentially self calibrating in the sense that no reference pressure or velocity is needed. Although somewhat sensitive to misalignment with flow direction, it is one of the most useful tools in experimental fluid mechanics.

Example 3.6 *A Pitot-static tube is placed in a flow of helium with $\rho = 0.16$ kg/m^3. The stagnation- and static-pressure taps read 103 kPa and 101 kPa, respectively. What is the velocity of the helium?*

Solution. Using Equation (3.36), the flow velocity is

$$U = \sqrt{\frac{2(103000 - 101000) \text{ N/m}^2}{0.16 \text{ kg/m}^3}} = 158 \frac{\text{m}}{\text{sec}}$$

3.7 Hydrostatic Forces on Plane Surfaces

The hydrostatic relation, Equation (3.17), tells us that pressure in a non-moving fluid varies only with depth. Thus, the force on a submerged horizontal surface of area A is simply $F = pA$, where p is the pressure at the specified depth. When the surface is inclined to the horizontal we must work a bit harder to evaluate the force. Specifically, as illustrated in Figure 3.12, we must integrate the varying pressure over the area of interest.

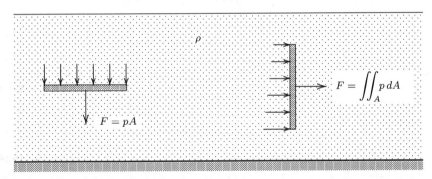

Figure 3.12: *Submerged surfaces and pressure distributions for horizontal and vertical orientations.*

We can illustrate the problem more explicitly by introducing the hydrostatic relation. Focusing first on the horizontal planar object to the left in Figure 3.12, if the plane is a distance h below the surface (so that $z = -h$) and the area of the surface is A, then the force on the upper surface of the object, F, is

$$F = (p_a + \rho g h)A \tag{3.37}$$

Turning now to the vertical surface to the right, we have the following equation for the force acting on the left side of the object.

$$F = \iint_A (p_a - \rho g z)\, dA \tag{3.38}$$

This integral depends, of course, on the detailed shape of the object.

As we will see in this section, because p varies *linearly* with depth, we can compute its integral over a surface inclined to the horizontal in terms of simple geometrical properties. Specifically, if we know the area and the centroid of the submerged surface, we can compute not only the force, but also the moment, on the surface. To see how this is done, we consider a plane surface inclined to the horizontal at an angle α as shown in Figure 3.13. It is submerged in a liquid of density ρ that has an air-liquid interface, often referred to as a **free surface**, at $z = 0$. Note that the coordinates are chosen so that z is positive in the direction of the gravitational acceleration vector,[1] x is parallel to the free surface, and y is out of the page. The figure includes a second set of Cartesian coordinates $(\xi\eta\zeta)$ aligned with the submerged surface that we will make use of below.

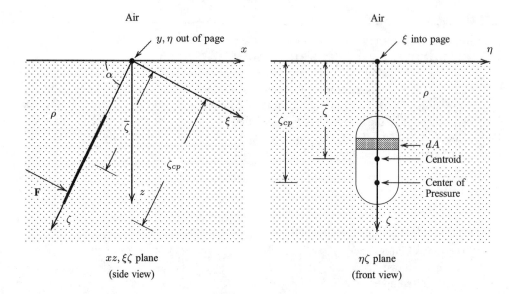

Figure 3.13: *Submerged surface inclined to the horizontal.*

It is important to pause at this point and stress that we are dealing with a *three-dimensional geometry*. Figure 3.13 shows a *side view* of the inclined surface, i.e., the xz plane. Although its thickness is exaggerated in the figure, the surface appears as a straight line from this view. The force vector, **F**, is normal to the surface and lies in the xz plane. As shown, the angle α determines the direction of the force. The figure also shows a *front view* of the surface in the $\eta\zeta$ plane, which follows from rotating the side view 90° about the ζ axis. The geometry of the surface determines the magnitude of the force vector.

The hydrostatic relation tells us the absolute pressure, p_{abs}, increases linearly with depth. That is, changing the sign of z in Equation (3.16) yields

$$\frac{dp_{\text{abs}}}{dz} = \rho g \tag{3.39}$$

If the atmospheric pressure is p_a, the absolute pressure is given by

$$p_{\text{abs}} = p_a + \rho g z \tag{3.40}$$

---

[1]In considering free-surface problems, we often define positive z downward for convenience. This obviates carrying minus signs throughout the analysis to follow.

To simplify our analysis, we will work with the gage pressure, p, i.e.,

$$p = p_{\text{abs}} - p_a \tag{3.41}$$

and note that $z = \zeta \sin \alpha$. Hence, we have

$$p = \rho g \zeta \sin \alpha \tag{3.42}$$

Now, on a surface element dA, the differential force exerted by the fluid to the left of the surface is given by $dF = p\,dA$ so that the force on the entire surface is given by the following integral.

$$F = \iint_A p\,dA = \rho g \sin \alpha \iint_A \zeta\,d\eta d\zeta \tag{3.43}$$

By definition, the centroid, $\overline{\zeta}$, of the planar surface A is

$$\overline{\zeta} \equiv \frac{1}{A} \iint_A \zeta\,d\eta d\zeta \tag{3.44}$$

Hence, the magnitude of the force on the inclined surface, $F = |\mathbf{F}|$, can be expressed in terms of the centroid of the planar surface as follows.

$$F = \rho g \overline{\zeta} A \sin \alpha \tag{3.45}$$

Since the force is normal to the submerged surface, in vector form we have

$$\mathbf{F} = \rho g \overline{\zeta} A \sin \alpha \left(\mathbf{i} \sin \alpha + \mathbf{k} \cos \alpha \right) \tag{3.46}$$

What we have computed is a *single point force* equivalent to the distributed pressure acting on the surface. If we know the depth of submergence of the planar surface's centroid and its area, we can compute the force from Equation (3.45), thus eliminating the need to do an integral.

Example 3.7 *What is the force F on the vertically-aligned planar object in Figure 3.12 if it is a square of side h and its centroid lies $2h$ below the surface?*

Solution. Referring to Equation (3.45), the angle is $\alpha = 90°$, the centroid is given to be $\overline{\zeta} = 2h$ and the area is $A = h^2$. Thus,

$$F = \rho g (2h) h^2 \sin(90°) = 2\rho g h^3$$

When forces are applied to an object, they cause the object's "center of gravity" to accelerate. The center of gravity and centroid are coincident for a constant density object. Because pressure increases with depth, it should be obvious that the equivalent force, $\mathbf{F}$, must act through a point a bit deeper than the centroid. This will give rise to a moment about the centroid.

We will now determine an effective lever arm to go with $\mathbf{F}$ that yields the same moment as the distributed pressure. The point corresponding to this lever arm is called the **center of pressure**. We denote the center of pressure by ζ_{cp} (see Figure 3.13). Taking moments about the η axis, we have

$$\mathbf{M} = \mathbf{r}_{cp} \times \mathbf{F} = \zeta_{cp} F \,\mathbf{j} \tag{3.47}$$

Hence, the center of pressure must be such that

$$\zeta_{cp} F = \int \zeta \, dF \tag{3.48}$$

Then, since $dF = p \, dA$, we have

$$\zeta_{cp} F = \iint_A \zeta p \, dA = \rho g \sin \alpha \iint_A \zeta^2 \, d\eta d\zeta \tag{3.49}$$

The integral of ζ^2 appearing in Equation (3.49) is the familiar **moment of inertia** of the planar surface A relative to the η axis. We denote the moment of inertia by I_o, so that[2]

$$I_o = \iint_A \zeta^2 \, d\eta d\zeta \tag{3.50}$$

For our purposes, it is more convenient to reference the moment of inertia to the centroid. We accomplish this by shifting the coordinate axes from the free surface to the centroid. That is, we simply replace ζ by $\zeta - \bar{\zeta}$. Thus, the moment of inertia relative to the centroid, I, is given by

$$
\begin{aligned}
I &\equiv \iint_A (\zeta - \bar{\zeta})^2 dA \\
&= \iint_A (\zeta^2 - 2\bar{\zeta}\zeta + \bar{\zeta}^2) dA \\
&= I_o - 2\bar{\zeta} \iint_A \zeta \, dA + \bar{\zeta}^2 \iint_A dA \\
&= I_o - 2\bar{\zeta}^2 A + \bar{\zeta}^2 A \\
&= I_o - \bar{\zeta}^2 A
\end{aligned}
\tag{3.51}
$$

Therefore, we can express the moment of inertia relative to the surface, I_o, as a function of I, A and $\bar{\zeta}$, viz.,

$$I_o = I + \bar{\zeta}^2 A \tag{3.52}$$

Equation (3.52) is known as the **parallel-axis theorem**. Substituting Equations (3.50) and (3.52) into Equation (3.49) yields

$$\zeta_{cp} F = \rho g \sin \alpha \left(I + \bar{\zeta}^2 A \right) \tag{3.53}$$

Then, using Equation (3.45), we can eliminate F and solve for ζ_{cp}. That is, substituting for F, we have

$$\zeta_{cp} \rho g A \bar{\zeta} \sin \alpha = \rho g \sin \alpha \left(I + \bar{\zeta}^2 A \right) \tag{3.54}$$

Dividing through by $\rho g A \bar{\zeta} \sin \alpha$, the final result is as follows.

$$\zeta_{cp} = \bar{\zeta} + \frac{I}{\bar{\zeta} A} \tag{3.55}$$

---

[2]This moment of inertia is usually denoted by I_{zz} in classical mechanics, and it is one of 9 components of a 3×3 tensor (matrix). Since we are using only one component of the inertia tensor, we omit the customary subscripts.

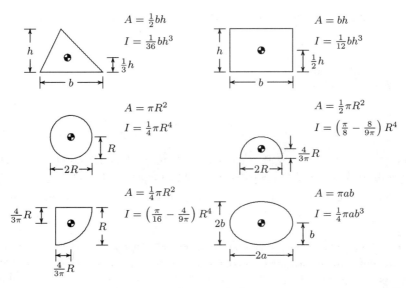

Figure 3.14: *Areas, moments of inertia and centroids for common geometries; in all cases, the moment of inertia is relative to a horizontal axis passing through the centroid.*

Equation (3.55) is especially useful because I and A depend only upon the shape of the planar surface in the $\eta\zeta$ plane. It is completely independent of the fluid, and the only way in which depth of submergence affects ζ_{cp} is through the dependence upon $\overline{\zeta}$. The same is true of the force, which is proportional to the area, A, and $\overline{\zeta}$. Since the moment of inertia, area and centroid are all positive, necessarily ζ_{cp} lies below the centroid, and $\zeta_{cp} \rightarrow \overline{\zeta}$ as $\overline{\zeta} \rightarrow \infty$.

Because our formulas for the force and center of pressure can be easily determined once A, I and $\overline{\zeta}$ are known, it is worthwhile to tabulate their values for common geometries. Figure 3.14 includes area, moment of inertia and centroid for a triangle, a rectangle, a circle, a semicircle a quarter circle and an ellipse.

Example 3.8 *Use the parallel-axis theorem to determine the moment of inertia for the double triangular geometry shown below.*

Solution. As shown, the centroid of each triangle is $\frac{2}{3}h$ from the common vertex. By symmetry, the centroid of the double triangle lies at the common vertex. Also, the moment of inertia is twice the moment of inertia of a single triangle, shifted by $\frac{2}{3}h$ according to the parallel-axis theorem. So, first note that for a single triangle, the area and moment of inertia are $A = \frac{1}{2}bh$ and $I = \frac{1}{36}bh^3$. Hence, using the parallel-axis theorem,

$$I_o = I + z^2 A = \frac{1}{36}bh^3 + \left(\frac{2}{3}h\right)^2 \frac{1}{2}bh = \frac{1}{4}bh^3$$

Thus, doubling this result to account for both triangles, the moment of inertia is

$$I = \frac{1}{2}bh^3$$

To illustrate how we can use Equations (3.45) and (3.55) to analyze fluid-statics problems, consider the forces acting on the triangular gate in Figure 3.15. From the left, the fluid develops a force of magnitude F that is balanced by the sum of the reaction force at the hinge, R_H, and the reaction force at the top of the gate, R_T, i.e.,

$$F = R_T + R_H \tag{3.56}$$

There is actually a fourth force acting, viz., the atmospheric pressure on the right-hand side of the gate. However, we implicitly account for this force when we use Equation (3.45). That is, since we reference all pressures to the atmospheric level [see Equation (3.41)], F is the difference between the integral of the absolute fluid pressure from the left, $p_{\text{abs}}(z)$, and the atmospheric pressure from the right, p_a. Equivalently, we can regard the adjusted pressure, $p = p_{\text{abs}} - p_a$, as being zero in the atmosphere, so that the net force is zero. We will return to this point in greater detail below.

The essence of the problem is as follows. There are three unknown forces to be determined, i.e., F, R_H and R_T. We thus need three equations to determine the solution. Equation (3.56) is the first of the three, while Equation (3.45) provides a second. To arrive at a third equation, we must balance moments about a suitable axis. As we will see below, computing the lever arm for the hydrostatic force involves the center of pressure.

Turning first to the force F, it is given by Equation (3.45). Hence, we must determine the area, A, and the location of the centroid, $\bar{\zeta}$. As indicated in the $\eta\zeta$-plane view of Figure 3.15, the base of the triangular gate is w and its altitude is $h/\sin\alpha$. Thus, the area of the gate is

$$A = \frac{1}{2}\frac{wh}{\sin\alpha} \tag{3.57}$$

Measuring along the ζ axis, which is aligned with the gate, the top of the gate is located a distance $h/\sin\alpha$ below the surface. The centroid is an additional $\frac{2}{3}h/\sin\alpha$ below the top of the gate. Adding these two contributions, we arrive at the following.

$$\bar{\zeta} = \frac{h}{\sin\alpha} + \frac{2}{3}\frac{h}{\sin\alpha} = \frac{5}{3}\frac{h}{\sin\alpha} \tag{3.58}$$

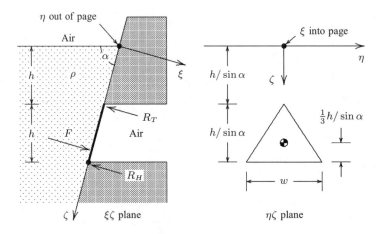

Figure 3.15: *Forces acting on a triangular gate.*

Therefore, substituting Equations (3.57) and (3.58) into Equation (3.45) yields the hydrostatic force normal to the gate, viz.,

$$F = \rho g \left(\frac{1}{2} \frac{wh}{\sin \alpha} \right) \left(\frac{5}{3} \frac{h}{\sin \alpha} \right) \sin \alpha = \frac{5}{6} \frac{\rho g w h^2}{\sin \alpha} \tag{3.59}$$

At this point, we have two of the three required equations for determining the complete solution. As noted earlier, this is as much information as we can extract from balancing forces. The third equation required to close our system of equations follows from balancing moments on the gate.

Hence, turning to the moment calculation, it is convenient to take moments about the axis perpendicular to the page passing through the hinge. As shown in Figure 3.16, the lever arms for the three forces are

$$\text{Lever Arm} = \begin{cases} 0, & R_H \\ \dfrac{h}{\sin \alpha}, & R_T \\ 2\dfrac{h}{\sin \alpha} - \zeta_{cp}, & F \end{cases} \tag{3.60}$$

So, balancing moments yields:

$$F \left(2\frac{h}{\sin \alpha} - \zeta_{cp} \right) = R_T \frac{h}{\sin \alpha} + R_H \cdot 0 \quad \Longrightarrow \quad \frac{R_T}{F} = 2 - \frac{\zeta_{cp}}{h} \sin \alpha \tag{3.61}$$

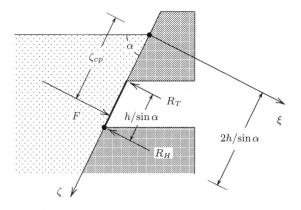

Figure 3.16: *Lever arms for the forces.*

Now, the center of pressure is given by Equation (3.55). For the triangular gate, reference to Figure 3.14 shows that the moment of inertia is

$$I = \frac{1}{36} w \left(\frac{h}{\sin \alpha} \right)^3 \tag{3.62}$$

Thus, the center of pressure is located at

$$\zeta_{cp} = \bar{\zeta} + \frac{I}{\bar{\zeta} A} = \frac{5}{3} \frac{h}{\sin \alpha} + \frac{\dfrac{1}{36} w \left(\dfrac{h}{\sin \alpha} \right)^3}{\left(\dfrac{5}{3} \dfrac{h}{\sin \alpha} \right) \left(\dfrac{1}{2} \dfrac{wh}{\sin \alpha} \right)} = \frac{17}{10} \frac{h}{\sin \alpha} \tag{3.63}$$

Hence, we have

$$\frac{R_T}{F} = 2 - \frac{17}{10} = \frac{3}{10} \qquad \Longrightarrow \qquad \frac{R_H}{F} = \frac{7}{10} \tag{3.64}$$

Therefore, the reaction forces acting on the gate are

$$R_H = \frac{7}{12}\frac{\rho g w h^2}{\sin\alpha} \qquad \text{and} \qquad R_T = \frac{1}{4}\frac{\rho g w h^2}{\sin\alpha} \tag{3.65}$$

Example 3.9 *Consider the L-shaped gate, ABC, shown below. The gate has width H out of the page and is free to pivot about the hinge at B. You may neglect the weight of the gate. At what depth h will the gate automatically open?*

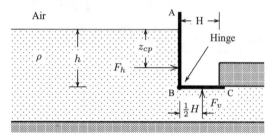

Solution. For the rectangular vertical part of the gate, the centroid, area and moment of inertia are given by

$$\bar{z} = \frac{1}{2}h, \qquad A = hH, \qquad I = \frac{1}{12}h^3 H$$

Hence, the horizontal force on the vertical face of the gate, F_h, is

$$F_h = \rho g \bar{z} A = \rho g\left(\frac{1}{2}h\right)(hH) = \frac{1}{2}\rho g h^2 H$$

Since the bottom part of the gate is horizontal, the pressure is constant and given by $p - p_a = \rho g h$. So, the vertical force on the horizontal surface points upwards as shown in the figure, and has magnitude, F_v, given by

$$F_v = \rho g h H^2$$

The gate will open when there is a net clockwise moment about the hinge. Thus, taking moments about the hinge, the gate just begins to open when

$$F_h\left(h - z_{cp}\right) = F_v\left(\tfrac{1}{2}H\right)$$

Note that the lever arm for F_v is $\frac{1}{2}H$ because it acts uniformly over the horizontal part of the gate. The center of pressure for the vertical part of the gate is

$$z_{cp} = \bar{z} + \frac{I}{\bar{z}A} = \frac{1}{2}h + \frac{\frac{1}{12}h^3 H}{\left(\frac{1}{2}h\right)(hH)} = \frac{1}{2}h + \frac{1}{6}h = \frac{2}{3}h$$

So, substituting the forces and center of pressure into the moment equation, we find

$$\frac{1}{2}\rho g h^2 H\left(h - \frac{2}{3}h\right) = \rho g h H^2\left(\frac{1}{2}H\right) \qquad \Longrightarrow \qquad h^2 = 3H^2$$

Therefore, the gate will open when

$$h = \sqrt{3}\,H$$

As a concluding comment on the calculation of forces and moments on submerged surfaces, consider the atmospheric pressure, p_a. Recall from Equation (3.41) that we have been dealing with what is referred to as **gage pressure**, $p = p_{abs} - p_a$. That is, by definition, p is the difference between the **absolute pressure**, p_{abs}, and the **atmospheric pressure**, p_a. This is sensible because pressure is similar to potential energy in the sense that changes in pressure give rise to forces and motion. Hence, a reference value can be chosen arbitrarily. Atmospheric pressure is a natural reference since it is the value that prevails at the free surface, i.e., at $z = 0$.

For many applications, the net force and moment on an object follow from integrands proportional to $(p_{abs} - p_a)$. This will be true, for example, when we have a liquid on one side of a barrier and air on the other side. Figure 3.17(a) shows such a configuration, including details of the pressure forces acting on both sides of the barrier. If the area and centroid of the submerged part of the barrier are A and $\bar{z}$, respectively (note that since the barrier is normal to the surface, we have $\bar{\zeta} = \bar{z}$), and the total area of the barrier is A_{tot}, then the net force, F, is given by

$$F = \underbrace{\rho g \bar{z} A + p_a A_{tot}}_{\substack{Force\ from \\ the\ left}} - \underbrace{p_a A_{tot}}_{\substack{Force\ from \\ the\ right}} = \rho g \bar{z} A \qquad (3.66)$$

Thus, the contribution from the atmospheric pressure cancels exactly, and all that remains is the hydrostatic contribution to the force. The same is true for the net moment.

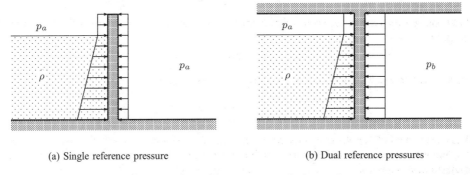

(a) Single reference pressure   (b) Dual reference pressures

Figure 3.17: *Pressure forces on a barrier including reference pressures.*

By contrast, if different reference pressures occur in a problem, we must account for them in computing forces and moments. For example, consider Figure 3.17(b), which corresponds to a closed chamber pressurized on one side. In the left chamber, the air is at atmospheric pressure, p_a. In the right chamber, the pressure is $p_b > p_a$. The net force on the barrier in this case is

$$F = \underbrace{\rho g \bar{z} A + p_a A_{tot}}_{\substack{Force\ from \\ the\ left}} - \underbrace{p_b A_{tot}}_{\substack{Force\ from \\ the\ right}} = \rho g \bar{z} A - (p_b - p_a) A_{tot} \qquad (3.67)$$

Because there is a difference in reference pressure, we see that the net force is not equal to the hydrostatic force. The additional contribution, $(p_b - p_a)A_{tot}$, is exactly equal to the difference between the reference pressure forces acting on each side of the barrier. The moment would also have a non-hydrostatic contribution proportional to the product of $(p_b - p_a)A_{tot}$ and an appropriate lever arm.

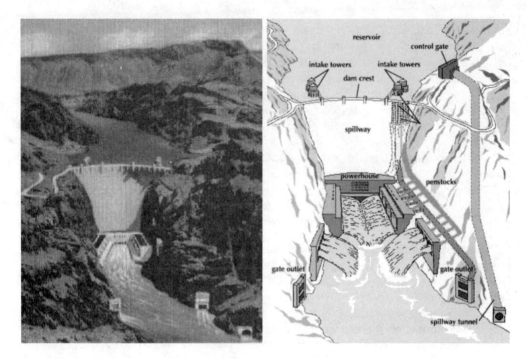

Figure 3.18: *Photo and sketch of Hoover Dam on the Colorado River at the southern tip of Nevada. The curved-surface methods of this chapter can be used to accurately predict the forces and moments on such a structure. [Photographs courtesy of United States Department of the Interior, Bureau of Reclamation - Lower Colorado Region]*

3.8   Hydrostatic Forces on Curved Surfaces

In principle, computing the hydrostatic force on a curved surface is more difficult than on a plane surface. We must now do an integral on a potentially complex contour, which can prove to be a nontrivial task. However, with a little cleverness, we can reduce this more-complex problem to one we have already solved. To understand how we can accomplish this end, consider the arc AB shown in Figure 3.19. We proceed as follows.

1. Resolve the force vector into its x and z components, i.e., $\mathbf{F} = F_x \mathbf{i} + F_z \mathbf{k}$, where $\mathbf{i}$ and $\mathbf{k}$ are unit vectors in the x and z directions, respectively.

2. Compute F_H, the force on surface OA, using plane surface methods, i.e., $F_H = \rho g \bar{z} A_{OA}$.

3. Compute F_V, the weight of the column of fluid above AB from the geometry, i.e., $F_V = \rho g V_{ABCD}$.

4. Balance forces in the x and z directions to determine F_x and F_z in terms of F_H and F_V. Hence, in the x direction,

$$F_x + F_{CB} = F_H + F_{OD} \quad \Longrightarrow \quad F_x = F_H$$

where $F_{CB} = F_{OD}$ because CB and OD have the same pressure but opposite unit normals. Similarly, balancing forces in the z direction shows that

$$F_z = F_V$$

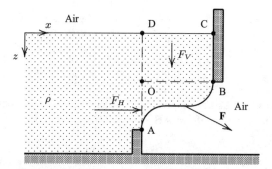

Figure 3.19: *Force on a curved surface. OA is the projection of the arc AB on a vertical plane.*

To see how this procedure works, consider the rectangular gate of unit width out of the page aligned with arc AB in Figure 3.20 separating a fluid of density ρ and air. We will use this problem as a first illustration of the curved-arc methodology. Because the arc is a straight line segment oriented at 45° to the horizontal, and since **F** is normal to the arc, necessarily $F_x = F_z$. We can use this fact as a check on our computations.

The x component of **F** is equal to the hydrostatic force on the projection of the gate surface onto the vertical plane. The projection is a surface of unit width out of the page aligned with arc OA. Because the projected surface is a rectangle, its centroid is $\frac{1}{2}H$ below point O, which is $\frac{3}{2}H$ below the surface. Therefore, the centroid location is $\bar{z} = \frac{3}{2}H$ and the area of the surface is $A = H \cdot 1 = H$ (the 1 represents unit width). Thus, we can use Equation (3.45) with $\alpha = 90°$ wherefore

$$F_x = \frac{3}{2}\rho g H^2 \qquad (3.68)$$

The weight of the column of fluid above AB is the sum of the weight of the fluid in OBCD and the fluid in OAB. The volumes of these columns are $V_{OBCD} = H^2 \cdot 1 = H^2$ and $V_{OAB} = \frac{1}{2}H^2 \cdot 1 = \frac{1}{2}H^2$. Thus, the z component of **F** is as follows.

$$F_z = \rho g H^2 + \frac{1}{2}\rho g H^2 = \frac{3}{2}\rho g H^2 \qquad (3.69)$$

Hence, as expected, we have shown that $F_x = F_z$.

We can compute the center of pressure using Equation (3.55) where $\bar{z}$, A and I are for the projection of AB onto the vertical plane. For this gate geometry, i.e., a rectangle of

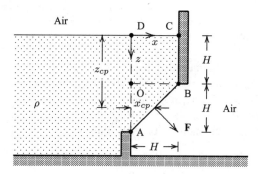

Figure 3.20: *Fluid-statics problem—the gate has unit width out of the page.*

height $h = H$ and width $b = 1$, reference to Figure 3.14 shows that $I = \frac{1}{12}H^3$. Thus, the z component of the center of pressure is given by

$$z_{cp} = \frac{3}{2}H + \frac{\frac{1}{12}H^3}{\frac{3}{2}H \cdot H} = \frac{14}{9}H \tag{3.70}$$

For planar surfaces, the center of pressure lies on the surface. Hence, for this problem, relative to point D, the equation of arc AB is $x = 2H - z$ so that

$$x_{cp} = 2H - \frac{14}{9}H = \frac{4}{9}H \tag{3.71}$$

We cannot assume that $\mathbf{r}_{cp}$ lies on the surface for curvilinear arcs. After all, it is an equivalent lever arm for an equivalent force. To find the x component of the center of pressure, x_{cp}, for general curvilinear arcs, we again take moments about the y axis. Assuming, for simplicity, that our surface is symmetric about $y = 0$, the moment is

$$\mathbf{M} = \mathbf{r}_{cp} \times \mathbf{F} = (z_{cp}F_x - x_{cp}F_z)\mathbf{j} \tag{3.72}$$

In general, to determine z_{cp}, we simply use Equation (3.55) on OA, the projection of AB on the vertical plane. To determine x_{cp}, we must satisfy

$$x_{cp}F_z = \int x \, dF \tag{3.73}$$

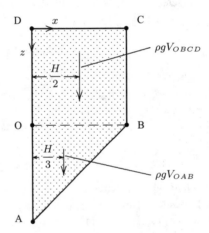

Figure 3.21: *Moment arms for computing* x_{cp}.

As an example of a straightforward way to evaluate the integral in Equation (3.73), consider the sample problem. The contribution to the moment from the combined arc ABC relative to the z axis is the sum of the moments from the weight of OBCD and the weight of OAB. As shown in Figure 3.21, the lever arm for OBCD (a square) is $\frac{1}{2}H$ relative to the z axis, while the lever arm for OAB (a triangle) is $\frac{1}{3}H$. These are distances of the centroids of these volumes from the z axis in the xz plane. (Note that we have previously considered the surface geometry in the yz plane.) The moment on ABC must exactly balance the moments on OBCD and OAB. Thus,

$$x_{cp}F_z = \int x \, dF = \frac{1}{2}H \cdot \rho g \underbrace{V_{OBCD}}_{H^2} + \frac{1}{3}H \cdot \rho g \underbrace{V_{OAB}}_{\frac{1}{2}H^2} = \frac{2}{3}\rho g H^3 \tag{3.74}$$

Combining Equations (3.69), (3.73) and (3.74) yields

$$\frac{3}{2}\rho g H^2 x_{cp} = \frac{2}{3}\rho g H^3 \quad \Longrightarrow \quad x_{cp} = \frac{4}{9}H \tag{3.75}$$

This matches the result quoted in Equation (3.71).

Example 3.10 *A dam has a parabolic shape defined by $z = 2h(1 - x/h)^2$ for $z > 0$. The dam is 6h wide (out of the page). Determine the force, **F**, on the dam and the z coordinate of the center of pressure, z_{cp}.*

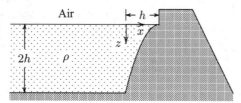

Solution. The projection of the dam onto a vertical plane is a rectangle of height $2h$ and width $6h$. So, the centroid, area and moment of inertia are

$$\bar{z} = h, \quad A = 12h^2, \quad I = \frac{1}{12}(6h)(2h)^3 = 4h^4$$

Thus, the horizontal component of the force on the dam is

$$F_x = \rho g \bar{z} A = \rho g h \left(12h^2\right) = 12\rho g h^3$$

Turning to the vertical component, the cross-sectional area of the fluid above the dam in the xz plane is

$$
\begin{aligned}
A_{cs} &= \int_0^h \int_0^{2h(1-x/h)^2} dz\,dx = \int_0^h 2h(1 - x/h)^2\,dx \\
&= 2h^2 \int_0^1 (1 - \xi)^2 d\xi = 2h^2 \int_0^1 \left(1 - 2\xi + \xi^2\right) d\xi \qquad (\xi \equiv x/h) \\
&= 2h^2 \left[\xi - \xi^2 + \frac{1}{3}\xi^3\right]_{\xi=0}^{\xi=1} = \frac{2}{3}h^2
\end{aligned}
$$

So, the volume of the fluid above the dam is

$$V = 6hA_{cs} = 4h^3$$

Therefore, the vertical force on the dam is

$$F_z = \rho g V = 4\rho g h^3$$

Combining the results for the two force components, F_x and F_z, into vector form, the net force on the dam is

$$\mathbf{F} = 4\rho g h^3 \left(3\mathbf{i} + \mathbf{k}\right)$$

Finally, the vertical component of the center of pressure, z_{cp}, is

$$z_{cp} = \bar{z} + \frac{I}{\bar{z}A} = h + \frac{4h^4}{h\left(12h^2\right)} = h + \frac{1}{3}h = \frac{4}{3}h$$

3.9   Equivalent Pressure Fields and Superposition

Because pressure varies only with depth, the magnitudes of the forces acting on the arcs shown in Figure 3.22 are identical. Hence, if we compute the force vector, F_1, with fluid of density ρ to the left of the arc, clearly the force is $-F_1$ when the same fluid lies to the right of the arc. Mathematically, we can say

$$F_2 = -F_1 \tag{3.76}$$

This is a useful fact since the method described in Section 3.8 is ideally suited for problems with fluid above the arc, but requires modification when the fluid is below the arc.

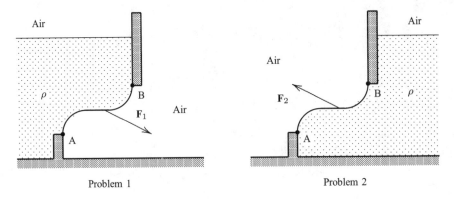

Problem 1　　　　　　　　　　　　　　　　　　　Problem 2

Figure 3.22: *Equivalent static pressure fields.*

Another useful technique for solving fluid-statics problems is to replace a given problem with a series of simpler problems. We do this by using the principle of superposition. Again, we seek replacement problems whose sum or difference has an equivalent pressure field.

As an example, consider an arc separating two fluids of different densities as shown in Figure 3.23. To solve this problem, we use superposition. That is, we first replace the problem in Figure 3.23 by the sum of Problem 1 of Figure 3.22 with $\rho = \rho_1$ and Problem 2 with $\rho = \rho_2$. The force from Problem 1 points downward while the force from Problem 2 points upward. Hence, the total force will be

$$F = F_1 + F_2 \tag{3.77}$$

Another way of viewing this problem is to say the solution is the difference between Problem 1 with $\rho = \rho_1$ and another Problem 1 with $\rho = \rho_2$. But, this is identical to solving Problem 1

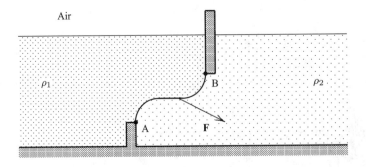

Figure 3.23: *Superposition example.*

with $\rho = \rho_1 - \rho_2$. The reason this works is because the pressure depends only upon depth, and pressure is proportional to the density. Consequently, at a given depth z, the pressure at any point on the upper part of the arc is $\rho_1 gz$ pushing to the right while the contribution from the lower part of the arc is $\rho_2 gz$ pushing in the opposite direction. This means the net pressure acting on the arc at a depth z is $(\rho_1 - \rho_2)gz$, which is exactly the pressure exerted by a single fluid above the arc of density $(\rho_1 - \rho_2)$.

Example 3.11 *Consider a container of water (density $= \rho$) with the shape shown below. The container has width h in the direction normal to the page. Determine the hydrostatic force, $\mathbf{F}$, and center of pressure, z_{cp}, on the circular arc AB.*

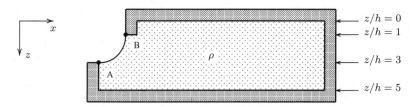

Solution. To begin, use superposition as illustrated in the figure below.

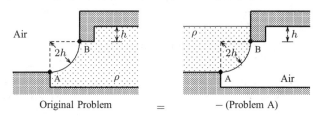

Original Problem $=$ $-$ (Problem A)

For Problem A, the gate's projection on a vertical plane is a rectangle of height $2h$ and width h. So, the centroid, area and moment of inertia are

$$\bar{z} = 2h, \qquad A = 2h^2, \qquad I = \frac{1}{12}h(2h)^3 = \frac{2}{3}h^4$$

The force components for Problem A are

$$F_{x_A} = \rho g \bar{z} A = \rho g(2h)\left(2h^2\right) = 4\rho g h^3$$

and

$$F_{z_A} = \rho g V = \rho g\left(2h^3\right) + \rho g\left[\frac{\pi}{4}(2h)^2\right] h = (2+\pi)\rho g h^3$$

Therefore, in vector form, the force for Problem A is

$$\mathbf{F}_A = \rho g h^3\left[4\mathbf{i} + (2+\pi)\mathbf{k}\right]$$

Then, reversing signs, the force on the gate for the original problem is

$$\mathbf{F} = -\rho g h^3\left[4\mathbf{i} + (2+\pi)\mathbf{k}\right]$$

Finally, the vertical component of the center of pressure is

$$z_{cp} = \bar{z} + \frac{I}{\bar{z}A} = 2h + \frac{\frac{2}{3}h^4}{(2h)\left(2h^2\right)} = 2h + \frac{1}{6}h = \frac{13}{6}h$$

3.10   Buoyancy

By definition, the **buoyancy force** on an object, $\mathbf{F}_{buoy}$, is the net pressure force exerted by the fluid on a given object submerged in a non-moving fluid. If the object is more dense than the surrounding fluid, its weight exceeds the buoyancy force and it sinks. If the object is less dense than the fluid, it will float. We express the buoyancy force mathematically as

$$\mathbf{F}_{buoy} = - \oiint_S p\,\mathbf{n}\,dS \qquad (3.78)$$

where we integrate about the surface bounding the object.

Now, we know that the pressure in a stationary fluid is a function only of depth. So, it makes no difference if the surface bounds a solid object or the fluid itself. That is, the pressure on the bounding surface S would be the same for a solid object as it would be if the object was replaced by fluid of density ρ. Since a solid object would, in fact, replace fluid of density ρ, the integration actually extends over the volume of fluid displaced by the object. Hence, we can apply Equation (3.11) to conclude that the buoyancy force is

$$\mathbf{F}_{buoy} = \iiint_V \rho g\,\mathbf{k}\,dV \qquad (3.79)$$

where g is gravitational acceleration and $\mathbf{k}$ is a unit normal vector directed upwards. Thus, since g and $\mathbf{k}$ are constant, we conclude finally that

$$\mathbf{F}_{buoy} = g\,\mathbf{k} \iiint_V \rho\,dV = Mg\,\mathbf{k} \qquad (3.80)$$

where M is the **mass of the displaced fluid**. Note that $\mathbf{F}_{buoy}$ acts vertically upward. This is known as **Archimedes' Principle**. It is valid for both floating and submerged bodies.

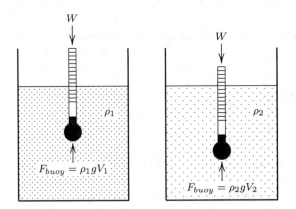

Figure 3.24: *Mass Hydrometer;* $\rho_1 > \rho_2$.

Figure 3.24 illustrates a practical application of Archimedes' Principle. The figure shows a **hydrometer**, which is a bottle of known weight, W, with a graded scale to indicate depth of submergence. The hydrometer sinks to a greater depth in the lighter fluid because more fluid must be displaced in order to generate a buoyancy force that balances W. In the figure, $\rho_1 > \rho_2$. If the submerged volumes are V_1 and V_2, then we have

$$\rho_1 g V_1 = W = \rho_2 g V_2 \qquad (3.81)$$

Consequently, the densities stand in the following ratio.

$$\rho_1/\rho_2 = V_2/V_1 \tag{3.82}$$

We can thus calibrate our hydrometer so that by reading a simple graded scale we know the density of a given fluid relative to a known reference density of a fluid such as water.

Example 3.12 *Consider a coal-carrying barge loaded with 100 tons of coal making its way down the Mississippi River. The empty barge weighs 25 tons. If the barge is 15 ft wide, 40 ft long and 8 ft high, what is its depth below the surface of the water?*

Solution. The buoyancy force on the barge, F_{buoy}, is

$$F_{buoy} = \rho g V = \rho g \ell w d$$

where ρ is the density of water. Also, ℓ, w and d are the length, width and submerged depth of the barge, respectively. The buoyancy force is balanced by the weight of the barge, wherefore

$$W = \rho g \ell w d \quad \Longrightarrow \quad d = \frac{W}{\rho g \ell w}$$

We are given that the combined weight of the barge and the coal is $W = 125$ tons $= 250{,}000$ lb. Also, $\ell = 40$ ft and $w = 15$ ft. Since the Mississippi River contains fresh water, assume the water density is $\rho = 1.94$ slug/ft^3. Thus,

$$d = \frac{250000 \text{ lb}}{\left(1.94 \text{ slug/ft}^3\right)\left(32.174 \text{ ft/sec}^2\right)(40 \text{ ft})(15 \text{ ft})} = 6.68 \text{ ft}$$

AN EXPERIMENT YOU CAN DO AT HOME

To discover for yourself a subtle effect of buoyancy in the comfort of your home, try the following simple experiment. Obtain a soda bottle, a wine bottle, or any container with a top small enough to be covered by your thumb. Fill it nearly to the top with water or any other readily available liquid. Insert a piece of cork in the bottle or container so that it floats. Now cover the top completely with your thumb. Press down firmly and, if you have done everything correctly, the cork sinks. Remove the pressure, and it returns to the surface.

The explanation for what happens is as follows. When you press down, you increase the pressure we have called p_a. This, in turn, increases the absolute pressure, $p_{\text{abs}} = p_a + \rho g z$ [see Equation (3.40)]. The cork, being a porous material, experiences two changes. First it is compressed, which reduces its volume and hence the buoyancy force. Second, it absorbs some of the water because of this increased pressure, and becomes more massive. Pressing hard enough causes the cork to undergo compression and to absorb a sufficient amount of water that its density becomes greater than that of water. As a result, it sinks. Removing the increased pressure causes the cork to expand, release the absorbed water and return to its original density.

Problems

3.1 James Bond has just thrown Blofeld into a vat of mercury. If Blofeld, chained to a piece of Plutonium, sinks to a depth of 2.5 m, will he have to worry about the bends as he floats back to the surface?

3.2 When we measure the pressure at one point in a container filled with a liquid of unknown density we find that it is 2134 psf. At another point 10 inches deeper in the container we measure a pressure of 2200 psf. What is the density of the liquid?

3.3 When we measure the pressure at one point in a container filled with a liquid of unknown density we find that it is 102 kPa. At another point 57 mm deeper in the container we measure a pressure of 102.7 kPa. What is the density of the liquid?

3.4 A tank contains three layers of unmixed fluids. The bottom layer is mercury, the central layer is a fluid of unknown density, ρ_u, and the top layer is SAE 10W Oil. The temperature is $38°\,C$ and the pressure at the bottom of the tank is 200 kPa. Also, the tank is open to the atmosphere at the top. You can assume the density of Mercury is independent of temperature. What is ρ_u if $h = 60$ cm?

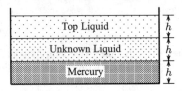

Problems 3.4, 3.5, 3.6, 3.7

3.5 A tank contains three layers of unmixed fluids. The bottom layer is mercury, the central layer is a fluid of unknown density, ρ_u, and the top layer is kerosene. The temperature is $68°\,F$ and the pressure at the bottom of the tank is 2350 psf. Also, the tank is open to the atmosphere at the top. What is the density of the unknown fluid, ρ_u, if $h = 3$ inches?

3.6 A tank contains three layers of unmixed fluids. The bottom layer is mercury, the central layer is a fluid of density $\rho_u = 1780$ kg/m^3 and the top layer is sea water. The temperature is $16°\,C$ and the tank is open to the atmosphere at the top. You can assume the density of Mercury is independent of temperature. What is the pressure 50 cm from the bottom of the tank if $h = 1$ meter?

3.7 A tank contains three layers of unmixed fluids. The bottom layer is mercury, the central layer is a fluid of density $\rho_u = 1600$ kg/m^3 and the top layer is ethyl alcohol. The temperature is $20°\,C$ and the pressure at the bottom of the tank is 258 kPa. Also, the tank is open to the atmosphere at the top. What is the layer depth, h?

3.8 An empirical formula relating pressure and density for seawater with temperature held constant is $p/p_a \approx (\alpha + 1)(\rho/\rho_a)^7 - \alpha$, where p_a and ρ_a are conditions at the surface and α is a dimensionless constant. Using this formula in the hydrostatic relation, determine the pressure as a function of depth.

3.9 Consider a density-stratified fluid in which the density varies with depth, z, as $\rho = \rho_a(1 + 2z/z_o)$, where z is measured downward from the surface, ρ_a is the density at the surface and z_o is a constant. Determine the pressure as a function of depth.

3.10 Consider the two layers of fluid shown. What is the pressure a distance h from the bottom?

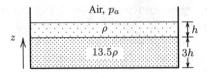

Problem 3.10

3.11 Consider the Torricelli barometer. Recall that it consists of a glass tube closed at one end with the open end immersed in a container filled with mercury. The tube is initially filled with mercury and is turned upside down when inserted in the container, creating a near vacuum in the top of the tube. When the temperature is $20°\,C$ and the atmospheric pressure, p_a, is 1 atm, mercury rises to $h = 760$ mm. If the barometer used glycerin instead of mercury, what would h be? Express your answer in meters.

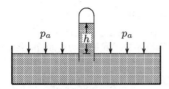

Problem 3.11

3.12 Assuming atmospheric conditions are accurately represented by the U. S. Standard Atmosphere, determine the pressure and temperature outside a small aircraft cruising at an altitude of 4/5 mile.

3.13 Assuming atmospheric conditions are accurately represented by the U. S. Standard Atmosphere, determine the pressure and temperature outside an airliner cruising at an altitude of 9 km.

3.14 Assuming atmospheric conditions are accurately represented by the U. S. Standard Atmosphere, determine the pressure and temperature outside a high-altitude aircraft cruising at an altitude of 8 miles.

3.15 On a day when the surface temperature is $38°\,C$, the pressure at the top of the spire on the Petronas Towers in Kuala Lumpur, Malaysia is 96.1 kPa. Assuming atmospheric pressure at the surface, estimate the height of this building using the U. S. Standard Atmosphere to describe conditions in Kuala Lumpur.

3.16 The world's tallest structure is a television transmitting station in North Dakota. On a day when the surface temperature is $50°\,F$, the temperature at the top of the TV tower is $42.6°\,F$. Estimate the height of the tower using the U. S. Standard Atmosphere to describe conditions in North Dakota.

3.17 The Washington Monument in the District of Columbia is 550 feet high. Assuming atmospheric conditions are as given by the U. S. Standard Atmosphere, determine the decrease in pressure and temperature—relative to surface values—at the top of the Monument.

3.18 Chuck Yeager flew the X-1 at a peak altitude of 43,000 ft and a peak velocity of 700 mph, corresponding to a Mach number of 1.06. Using the fact that the speed of sound is $a = \sqrt{\gamma R T}$, where $\gamma = 1.4$ is the specific-heat ratio and $R = 1716$ ft^2/(sec^2·$°$R) is the perfect-gas constant, estimate the ground temperature on the day of the flight. Assume the U. S. Standard Atmosphere applies.

3.19 The *White Knight* mother ship carried *SpaceShipOne* to an altitude of 14.3 km. Assuming the U. S. Standard Atmosphere applies, estimate the atmospheric pressure in atm and temperature in $°$C at this altitude.

3.20 If a Cadillac Eldorado delivers 300 horsepower at sea level, what is the corresponding power delivered in Denver, Colorado, which is approximately 1 mile above sea level? Use the fact that the power output of an automobile engine is proportional to the mass-flow rate of the air supplied to the carburetor. The mass-flow rate, in turn, is proportional to the atmospheric air density. Assume the U. S. Standard Atmosphere accurately describes conditions in Denver and at sea level.

3.21 In an *adiabatic atmosphere*, the pressure varies with density according to

$$p = A\rho^{\gamma}$$

where γ is the specific-heat ratio and A is a constant. Determine the adiabatic lapse rate, α, in such an atmosphere assuming the temperature at the surface is T_a. Express your answer in terms of γ, g, and the perfect-gas constant, R.

3.22 Consider an atmosphere with an inversion layer for which the temperature varies according to:

$$T = \begin{cases} T_a - \alpha z, & 0 \leq z \leq \frac{1}{4}T_a/\alpha \\ \frac{3}{4}T_a + \alpha(z - \frac{1}{4}T_a/\alpha), & \frac{1}{4}T_a/\alpha \leq z \leq \frac{3}{10}T_a/\alpha \\ \frac{4}{5}T_a, & z \geq \frac{3}{10}T_a/\alpha \end{cases}$$

(a) Make a graph of the temperature profile.

(b) Assuming the atmosphere is a perfect gas, determine the pressure in the inversion layer as a function of altitude. Express your answer in terms of the pressure at sea level, p_a, the perfect-gas constant, R, gravitational acceleration, g, as well as T_a, α and z.

3.23 A *hydraulic jack* uses a fluid of density ρ and a lever of length h to lift an object of weight W. If the piston areas are A and $5A$, what force, F, must be applied to support an object of weight W?

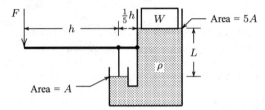

Problem 3.23

3.24 An object is resting on top of a piston in a cylindrical tank that is filled with a liquid of density $\rho = 1.72$ slug/ft³. The object weighs 2 tons and the diameter of the tank, D, is 4 ft. If the attached pressure gage indicates a pressure, p_g, of 263 psf, how far above the piston, Δz, is the gage located?

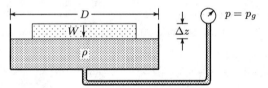

Problems 3.24, 3.25, 3.26

3.25 An object is resting on top of a piston in a cylindrical tank that is filled with a liquid of density $\rho = 1.75$ slug/ft³. The attached pressure gage indicates a pressure, p_g, of 1000 psf. How much does the object weigh if the gage is $\Delta z = 2$ ft above the piston as shown? The tank diameter, D, is 3 ft.

3.26 An object with a weight, W, of 100 kN rests on top of a piston in a cylindrical tank filled with a liquid of density ρ. The diameter of the tank, D, is 2 m. If the attached pressure gage, located a distance $\Delta z = 3$ m above the piston, indicates a pressure, p_g, of 6 kPa, what is the liquid density, ρ?

3.27 For the setup shown, what is the pressure at Point A if $\tilde{\rho} = \frac{1}{4}\rho$?

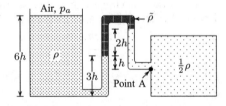

Problems 3.27, 3.28

3.28 For the setup shown, what is the density $\tilde{\rho}$ if the pressure at Point A is $p_a + 4.2\rho gh$, where p_a is atmospheric pressure?

3.29 An open-ended mercury manometer is attached to an ethyl-alcohol tank at $20°$ C. Water has entered the tank and lies at the bottom in a layer of thickness h. The air trapped in the tank is also pressurized at a pressure, p_b. Compute the differential difference, Δz, in the manometer's mercury column, assuming $N = 4$, $h = 1$ m and $p_b = 2p_a$, where $p_a = 101$ kPa is atmospheric pressure.

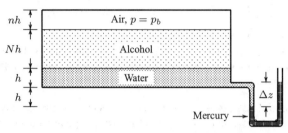

Problems 3.29, 3.30

3.30 An open-ended mercury manometer is attached to an ethyl-alcohol tank at $68°$ F. Water has entered the tank and lies at the bottom in a layer of thickness h. Some of the air trapped in the tank has been evacuated so that $p_b < p_a$, where $p_a = 2116.8$ psf is atmospheric pressure. Find p_b if the differential difference, Δz, in the manometer's mercury column is 1 ft, assuming $N = 10$ and $h = 2$ ft.

3.31 A *differential manometer* is a device connected between two tanks as shown. It is designed to measure the pressure difference $p_1 - p_2$.

(a) Determine $p_1 - p_2$ between the two tanks as a function of ρ_1, ρ_2, g, h_1, h_2, λ and ρ_{Hg}.

(b) Suppose $h_1 = h_2 = 1$ m, $\lambda = 10$ cm, fluid 1 is kerosene, fluid 2 is glycerin and the temperature is $20°$ C. What is $p_1 - p_2$?

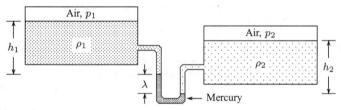

Problems 3.31, 3.32

3.32 A *differential manometer* is a device connected between two tanks as shown. It is designed to measure the pressure difference $p_1 - p_2$.

(a) Determine the distance λ as a function of ρ_1, ρ_2, g, h_1, h_2, p_1, p_2 and ρ_{Hg}.

(b) Suppose $h_1 = 1$ ft, $h_2 = 0.75$ ft, $p_1 = p_2$, fluid 1 is ethyl alcohol, fluid 2 is carbon tetrachloride and the temperature is $68°$ F. What is λ?

3.33 Compute the pressure at Point A in terms of atmospheric pressure, p_a, density of water, ρ, gravitational acceleration, g, and the distance h. Use the fact that $\rho_{H_2O} = \rho$, $\rho_{Oil} = 0.88\rho$ and $\rho_{Hg} = 13.55\rho$.

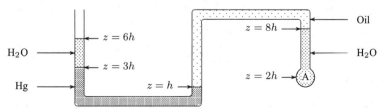

Problem 3.33

3.34 A *micromanometer* can be used to measure very small pressure differences. The device uses two fluids of slightly different densities, ρ and $\rho + \Delta\rho$, where $\Delta\rho$ is small compared to ρ. Assume the fluid levels in the cylindrical reservoirs are equal when $p_1 = p_2$. Develop a formula for h as a function of p_1, p_2, ρ, $\Delta\rho$, D, d and g.

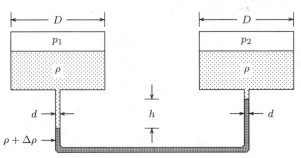

Problems 3.34, 3.35

3.35 A *micromanometer* can be used to measure very small pressure differences. The device uses two fluids of slightly different densities, ρ and $\rho + \Delta\rho$, where $\Delta\rho$ is small compared to ρ. The cylindrical reservoirs are very large ($D \gg d$) so that the reservoir levels are nearly equal when fluid moves from the reservoir to the tube and vice versa.

    (a) Develop a formula for h as a function of p_1, p_2, ρ, $\Delta\rho$ and g.

    (b) Assume the pressure difference is $p_1 - p_2 = 12$ Pa and that the lighter fluid is ethyl alcohol. Contrast the height, h, when the heavier fluid is kerosene with the value when the heavier fluid is SAE 10W oil. Assume the temperature is $20°$ C, and express your answers in millimeters.

3.36 An *inclined manometer* is a large spherical container with an inclined tube of small diameter attached. The tube is inclined at an angle α to the horizontal and has a scale with markings separated by 1 in. We would like each marking to correspond to a change in pressure, Δp_b, of 0.01 psi. We are considering several fluids including ethyl alcohol, water, glycerin and mercury. Assuming the temperature is $68°$ F, compute the angle α required for all 4 of the candidate fluids.

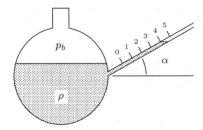

Problems 3.36, 3.37

3.37 An *inclined manometer* is a large spherical container with an inclined tube of small diameter attached. The tube is inclined at an angle α to the horizontal and has a scale with markings separated by 1 cm. If the liquid in the manometer is carbon tetrachloride, what must the angle α be if each marking corresponds to a change in pressure, Δp_b, of 100 Pa? Assume the temperature is $20°$ C.

3.38 We wish to determine the maximum pressure on your hand when you hold it out the window of your automobile on a day when the ambient pressure is 1 atm and the temperature is $68°$ F. Assuming the conditions required for Bernoulli's equation to hold are satisfied, compute the maximum pressure (in atm) when you are in the following two situations.

    (a) Cruising along a highway at 70 mph.

    (b) Driving your Indy 500 racer at 200 mph.

3.39 The velocity in the outlet pipe from a large reservoir of depth $h = 10$ m is $U = 11$ m/sec. Due to the rounded entrance to the pipe, the flow can be assumed to be irrotational. Also, the reservoir is so large that the flow is essentially steady. With these conditions, what is the pressure at point A as a function of the fluid density, ρ, gravitational acceleration, g, atmospheric pressure, p_a, as well as U and h? Determine the value of $p - p_a$ in kPa for water whose density is $\rho = 1000$ kg/m^3.

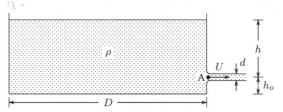

Problems 3.39, 3.40

3.40 The velocity in the outlet pipe from a large reservoir of depth $h = 25$ ft is $U = 30$ ft/sec. Due to the rounded entrance to the pipe, the flow can be assumed to be irrotational. Also, the reservoir is so large that the flow is essentially steady. With these conditions, what is the pressure at point A as a function of the fluid density, ρ, gravitational acceleration, g, atmospheric pressure, p_a, as well as U and h? Determine the value of $p - p_a$ in psi for water whose density is $\rho = 1.94$ slug/ft^3.

3.41 The streamlines for inviscid flow past an airfoil are as shown. If the freestream velocity, U_0, is 90 m/sec, what is the pressure difference, $p_2 - p_1$, at points where the velocities are $U_1 = 100$ m/sec and $U_2 = 80$ m/sec? Assume that the fluid density, ρ, is 1.20 kg/m^3 and that Bernoulli's equation applies. Express your answer in kPa.

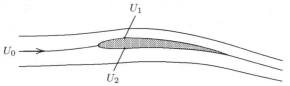

Problems 3.41, 3.42

3.42 The streamlines for inviscid flow past an airfoil are as shown. If the freestream velocity, U_0, is 120 ft/sec, what is the pressure difference, $p_2 - p_1$, at points where the velocities are $U_1 = 180$ ft/sec and $U_2 = 110$ ft/sec? Assume that the fluid density, ρ, is 0.00234 slug/ft^3 and that Bernoulli's equation applies. Express your answer in psi.

3.43 The figure depicts incompressible flow through a pipe and nozzle that emits a vertical jet. The flow is steady, irrotational, has density ρ and the only body force is gravity. What is the velocity of the jet at the nozzle exit, U_j? To what height, z_{max}, will the jet of fluid rise? Express your answers in terms of ρ, g, h, U_i and $p_i - p_a$. Determine the numerical values of U_j and z_{max} if $p_i - p_a = 60$ kPa, $h = 1$ m, $U_i = 4$ m/sec and $\rho = 1000$ kg/m^3.

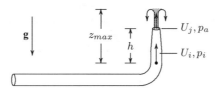

Problems 3.43, 3.44

3.44 The figure depicts incompressible flow through a pipe and nozzle that emits a vertical jet. The flow is steady, irrotational, has density ρ and the only body force is gravity. If the jet of fluid rises to a height $z_{max} = 2h$, what is the value of $p_i - p_a$ as a function of ρ, g, h and U_i? Determine the numerical value of $p_i - p_a$ in psf if $\rho = 1.98$ slug/ft^3, $h = 4$ ft and $U_i = 12$ ft/sec.

3.45 A siphon tube of constant diameter d is attached to a large tank as shown. You can assume the flow is quasi-steady, incompressible, irrotational and that gravity is the only body force.

(a) Find the outlet velocity, U, and the minimum pressure in the siphon tube, p_{min}, as functions of gravitational acceleration, g, fluid density, ρ, atmospheric pressure, p_a, and the distances ℓ, h_1 and h_2. **HINT:** Use the fact that the mass flux through the tube, $\dot{m} = \frac{\pi}{4}\rho|\mathbf{u}|d^2$, is constant.

(b) Calculate U and p_{min} for $h_1 = 4$ m, $h_2 = 4$ m, $\ell = 3$ m, $\rho = 1000$ kg/m^3 and $p_a = 1$ atm.

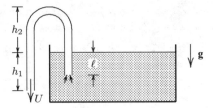

Problems 3.45, 3.46

3.46 A siphon tube of constant diameter d is attached to a large tank as shown. You can assume the flow is quasi-steady, incompressible, irrotational and that gravity is the only body force.

(a) Find the outlet velocity, U, and the minimum pressure in the siphon tube, p_{min}, as functions of gravitational acceleration, g, fluid density, ρ, atmospheric pressure, p_a, and the distances ℓ, h_1 and h_2. **HINT:** Use the fact that the mass flux through the tube, $\dot{m} = \frac{\pi}{4}\rho|\mathbf{u}|d^2$, is constant.

(b) Calculate U and p_{min} for $h_1 = 5$ ft, $h_2 = 3$ ft, $\ell = 3$ ft, $\rho = 1.94$ slug/ft^3 and $p_a = 1$ atm.

3.47 An eager student performs the home experiment described in this chapter. The student blows a stream of air over one side of a 21 cm by 29.7 cm sheet of paper. The density of air is $\rho = 1.20$ kg/m^3. Assuming the stream blows over the entire surface at a velocity $U = 1.20$ m/sec and the paper is in a horizontal position, what is the weight of the paper in Newtons? What is the pressure difference between the upper and lower surfaces of the paper?

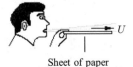

Problems 3.47, 3.48   Sheet of paper

3.48 An eager student performs the home experiment described in this chapter. The student blows a stream of air over one side of an $8\frac{1}{2}$ inch by 11 inch sheet of paper weighing $W = 0.01$ lb. The density of air is $\rho = 0.0024$ slug/ft^3. Assuming the stream blows over the entire surface, what velocity, U, is required to support the weight of the sheet of paper in a horizontal position? What is the pressure difference between the upper and lower surfaces of the paper?

3.49 The figure shows a downward-facing round nozzle that is attached to a hose through which an incompressible fluid of density ρ flows. The hose diameter is D and the (constant) mass flux is $\dot{m} = \frac{\pi}{4}\rho|w|d^2$, where d is nozzle diameter and w is vertical velocity. Treating the flow as one-dimensional, determine how the nozzle diameter must vary with z in order to have atmospheric pressure, p_a, throughout the nozzle. Assume the flow is steady, irrotational and the only body force acting is gravity.

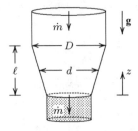

Problems 3.49, 3.50

3.50 The figure shows a downward-facing round nozzle that is attached to a hose through which an incompressible fluid of density ρ flows. The hose diameter is D and the (constant) mass flux is $\dot{m} = \frac{\pi}{4}\rho|w|d^2$, where d is nozzle diameter and w is vertical velocity. Treating the flow as one-dimensional, determine how the nozzle diameter must vary with z in order to have $dp/dz = -\frac{1}{4}\rho g$ throughout the nozzle. The quantity g is gravitational acceleration. Assume the flow is quasi-steady, irrotational and the only body force acting is gravity.

3.51 A large closed tank is pressurized as shown. A jet of fluid issues from a small hole. Assume the flow is incompressible, irrotational, quasi-steady and the only body force acting is gravity.

(a) Determine the pressure, p_b, required to increase the jet velocity by 50% relative to the value realized for atmospheric pressure, p_a, in the upper chamber.

(b) Compute the value of p_b in atm when $h = 10$ m and $\rho = 998$ kg/m^3. What is the jet velocity, U, for this pressure?

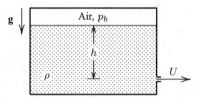

Problems 3.51, 3.52

3.52 A large closed tank is pressurized as shown. A jet of fluid issues from a small hole. Assume the flow is incompressible, irrotational, quasi-steady and the only body force acting is gravity.

(a) Determine the pressure, p_b, required to triple the jet velocity relative to the value realized for atmospheric pressure, p_a, in the upper chamber.

(b) Compute the value of p_b in atm when $h = 18$ ft and $\rho = 1.99$ slug/ft^3. What is the jet velocity, U, for this pressure?

3.53 Use the parallel-axis theorem to verify that the moments for a semicircle and a circle quoted in Figure 3.14 are consistent.

3.54 Use the parallel-axis theorem to determine the moment of inertia about the point at which the square and circle touch.

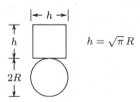

Problem 3.54

3.55 Use the parallel-axis theorem to determine the moment of inertia about the centroid of the double-circle geometry shown.

Problem 3.55

3.56 Use the parallel-axis theorem to compute the centroid location, $\bar{z}$, area, A, and moment of inertia, I, for the diamond-shaped geometry shown.

Problem 3.56

3.57 Determine the pivot location, h_p, for a triangular gate of base $4h$ (out of the page) such that it will open when $H = h$. The base lies below the pivot.

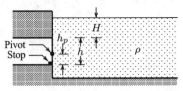

Problems 3.57, 3.58, 3.59, 3.60

3.58 Determine the pivot location, h_p, for an elliptical gate of width $2h$ (out of the page) such that it will open when $H = h$.

3.59 Consider a square gate of side h that can rotate about a pivot as shown. Determine the depth of the gate's top, H, at which the gate will open if $h_p = \frac{2}{5}h$.

3.60 The circular gate of diameter h shown can rotate about a pivot as shown. If the gate opens when $h_p = \frac{15}{32}h$, compute the depth of the gate's top, H.

3.61 The force on the gate's hinge, F_h, is 3/8 times the hydrodynamic force acting on the gate, F. The gate is rectangular and has width w out of the page. Determine the depth of the hinge, h.

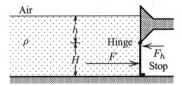

Problems 3.61, 3.62

3.62 The force on the gate's hinge, F_h, is 7/12 times the hydrodynamic force acting on the gate, F. The gate has the shape of an inverted triangle with base $12H$ at the hinge. Determine the hinge depth, h.

3.63 Compute the magnitude of the force, F, to the nearest ton, and the center of pressure on the dam, h_{cp}, if the 60° F seawater is $h = 5$ ft deep, the dam is $8h$ wide and $\alpha = 53°$.

Problems 3.63, 3.64

3.64 Find the magnitude of the force, F, and the center of pressure on the dam, h_{cp}, if the 20° C fresh water is $h = 9$ m deep, the dam is $4h$ wide and $\alpha = 50°$. Round F to the nearest MegaNewton (MN).

3.65 Determine the force due to hydrostatic pressure acting on the hinge of the gate shown. The fluid density is ρ and the width of the elliptical gate (out of the page) is $4H$.

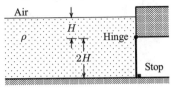

Problem 3.65

3.66 A circular gate of radius R pivots about a horizontal axis through its centroid, and is held in place by a stop. Regarding the center of pressure as a function of the depth of the top of the gate, i.e., $z_{cp} = z_{cp}(H)$, compute the ratio of z_{cp} for $H = R$ to its value for $H = 0$.

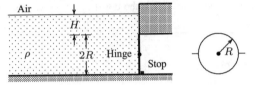

Problem 3.66

3.67 For the rectangular gate shown, the stop can resist a force of up to 5,000 kN. The gate is $4h$ wide (out of the page). For what depth H will the stop fail if $h = 4$ m and the liquid is $16°$ C seawater?

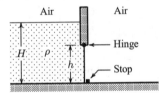

Problem 3.67

3.68 Develop a general expression for the force on the gate hinge, valid for arbitrary gate shape. Apply your result to:

   (a) an elliptical gate of width $6L$.

   (b) a triangular gate with base width $4L$ (the base is at the stop).

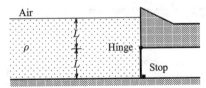

Problem 3.68

3.69 The rectangular gate shown is hinged at point A and its width (normal to the page) is $2H$. Compute the force at point B required to hold the gate closed. You may ignore the gate's weight.

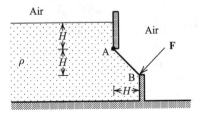

Problem 3.69

3.70 A circular gate of radius R pivots about a horizontal axis through its centroid, and is held in place by a stop. What force must be applied to the bottom of the gate by the stop to hold it closed?

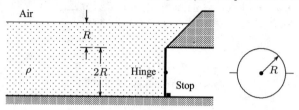

Problem 3.70

3.71 A triangular access port, hinged as shown below, is provided in the side of a form containing liquid concrete. Determine the hydrodynamic force exerted on the access port by the concrete, which has density ρ. Also, compute the force needed at the upper tip of the access port to prevent it from opening.

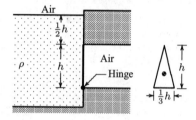

Problem 3.71

3.72 Consider the hinged gate with a weight connected through a pulley arrangement as shown. The gate is square with side H. The weight is a cube with side s and density 5ρ. You may ignore the gate's weight and any friction that might develop in the pulley. For an angle $\alpha = 36°$, at what depth, h, will the fluid cause the gate to rotate in the clockwise direction?

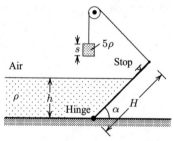

Problem 3.72

3.73 The width of the rectangular gate (normal to the page) is $3L$, where L is the distance (measured parallel to the gate) from the bottom to the fulcrum. You may neglect the gate's weight.

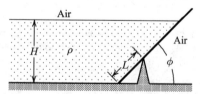

Problem 3.73

(a) Compute the height H to which a fluid of density ρ must rise in order to tip over the gate. Express your answer in terms of L and ϕ.

(b) What is the force normal to the gate when the gate is just about to tip over? Express your answer in terms of ρ, g, L and ϕ.

3.74 A rectangular gate of width $3H$ (out of the page) is designed to open and release fresh water when the tide goes out. The density of fresh water is ρ and the density of ocean water is $(1+\epsilon)\rho$. Neglecting the weight of the gate, verify that the ocean level, h, at which the gate will open can be determined from a cubic equation for h/H.

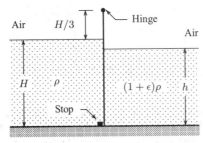

Problem 3.74

3.75 Determine the force, **F**, on the dam shown below and the z coordinate of the center of pressure, z_{cp}. The dam has a parabolic shape and is $6h$ wide (out of the page). The cross-sectional area of the region between the dam and the surface is $\frac{2}{3}h^2$.

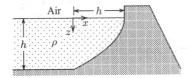

Problem 3.75

3.76 The dam shown below is $8h$ wide (out of the page). It is made up of three linear sections with $\alpha = \tan^{-1}(1/2)$. Find the force, **F**, on the dam and the z coordinate of the center of pressure, z_{cp}.

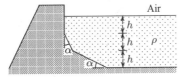

Problem 3.76

3.77 Ignoring the weight of the gate, compute the applied horizontal force, F, required to hold the gate closed. The gate side view is a quarter circle of radius H, and it is $4H$ wide (out of the page).

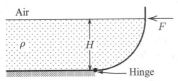

Problem 3.77

3.78 A swimming pool has a set of n steps at one end. Each step has a horizontal and vertical length of h/n, where h is the total depth. The width of the steps (out of the page) is $6h$. Compute the hydrostatic force on the set of steps.

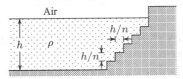

HINT: Use the fact that

$$\sum_{i=1}^{n} i = \frac{n(n+1)}{2}$$

Problem 3.78

3.79 Determine the force, **F**, on the dam below and the z coordinate of the center of pressure, z_{cp}. The dam has a quarter-circle shape and is $5h$ wide (out of the page).

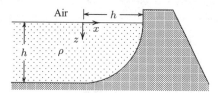

Problem 3.79

3.80 Determine the depth h as a function of H such that the net force in the x direction vanishes on arc AB separating two liquids of density ρ_1 and ρ_2 as shown. For this value of h, what is the vertical force on the arc assuming its weight is negligible? The area of OAB is $\frac{5}{8}H^2$, the arc has width $8H$ (out of the page), $\rho_1 = \rho$ and $\rho_2 = 2\rho$.

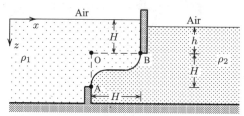

Problems 3.80, 3.81, 3.82, 3.83

3.81 A gate AB of constant width, H, (out of the page) separates two liquids of density ρ_1 and ρ_2. The net vertical force on the gate, F_z, is zero. Assume the area of OAB is $\frac{1}{3}H^2$ and $h = \frac{1}{3}H$.

(a) Determine the density ratio, ρ_2/ρ_1

(b) For the density ratio determined in (a), compute the horizontal force, F_x.

3.82 Compute the net force on the arc AB, assuming that the area of OAB is $\frac{2}{5}H^2$, $h = \frac{1}{5}h$, $\rho_1 = \rho$ and $\rho_2 = \frac{5}{2}\rho$. Also, assume unit width (out of the page).

3.83 Compute the net force on the arc AB, assuming that the area of OAB is $\frac{1}{2}H^2$. The gate has width H (out of the page), $h = \frac{3}{4}H$, $\rho_1 = \rho$ and $\rho_2 = 4\rho$.

3.84 The width of the rectangular gate (normal to the page) is w, and L is the distance (measured parallel to the gate) from the bottom to the fulcrum. You may neglect the gate's weight. A murky liquid of density $\frac{3}{2}\rho$ lies beneath a clean liquid of density ρ, with the interface between the liquids at the same depth as the top of the fulcrum. Verify that the height to which the clean fluid must rise in order to tip over the gate, $H/\sqrt{2}$, satisfies the following cubic equation.

$$\left(\frac{L}{H}\right)^3 + 3\left(\frac{L}{H}\right) - 1 = 0$$

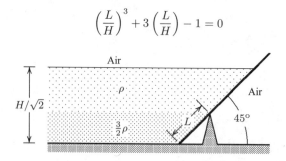

Problem 3.84

3.85 Determine the force on the gate AB if the upper layer of fluid has density $\rho_1 = \rho$ and the lower layer has density $\rho_2 = 4\rho$. The gate is rectangular and has width H (out of the page).

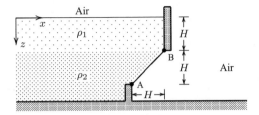

Problem 3.85, 3.86

3.86 Determine the force on the gate AB if the upper layer of fluid has density $\rho_1 = \rho$ and the lower layer has density $\rho_2 = 2\rho$. The gate is rectangular and has width $3H$ (out of the page).

3.87 The width of the circular-arc gate shown is 4ℓ (out of the page). Express your answers in terms of ρ, g and ℓ.

(a) Compute the hydrostatic force components, F_x and F_z, acting on the arc AB.

(b) Compute the vertical location of the center of pressure, z_{cp}.

(c) Compute the horizontal location of the center of pressure, x_{cp}.

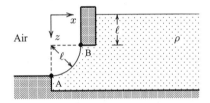

Problem 3.87

3.88 Arc AB separates fluids of density ρ_1 and ρ_2. The arc has width H normal to the page. The area of OAB is $\frac{2}{3}H^2$, and the centroid of OAB lies at $x = \frac{1}{4}H$.

(a) Compute the force components, F_x and F_z, acting on the arc AB. Is your answer sensible for the limiting case $\rho_1 = \rho_2$?

(b) Compute the vertical location of the center of pressure, z_{cp}.

(c) Compute the horizontal location of the center of pressure, x_{cp}. (Take moments relative to OA.)

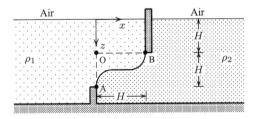

Problem 3.88

3.89 One fourth of an object of density ρ_a and volume V is submerged in a fluid of density ρ. What is the magnitude of the buoyancy force on the object?

3.90 A cube of balsa wood (density, $\rho_b = 124$ kg/m^3) measures 15 cm on its sides. What force is required to hold it in place completely submerged in seawater at $16°$ C?

3.91 In air, an irregularly-shaped object made of copper weighs 5 tons. When it is submerged in a tank filled with water at 50° F, a force of 4.5 tons is required to lift it from the bottom of the tank. What is the volume of the object? What is its density?

3.92 An irregularly-shaped cement slab weighs 51 N in air. When it is submerged in a tank filled with water at 20° C, a force of 33 N is required to lift it from the bottom of the tank. What is the volume of the slab? What is its density?

3.93 King Hiro ordered a new crown to be made of pure gold. When he received the crown, he suspected that other metals had been used in its construction. Archimedes found that, when immersed in water (density ρ), the crown's weight was 91% of its actual weight. Assuming the crown was made of a mixture of steel (density 7.8ρ) and gold (density 19.3ρ), what percentage, by weight, was gold?

3.94 Consider the object of density 2ρ immersed in a two-layer fluid as shown. If the object has constant cross-sectional area, A (out of the page), how much of the object's volume lies in the lower layer?

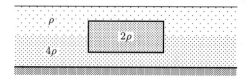

Problem 3.94

3.95 The density of the block shown below is $\rho_b = \frac{3}{5}\rho$ and it has width $5L$ out of the page. At what depth, ℓ, will it float in the two-liquid reservoir?

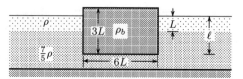

Problem 3.95

3.96 Compute the depth, ℓ, at which the cube shown below will float in the two-liquid reservoir. The density of the cube is $\rho_c = 1.45\rho$.

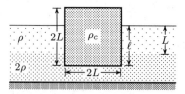

Problem 3.96

3.97 A cube-shaped "balloon" filled with helium (density ρ_{He}) rises through the stratosphere for which the density is $\rho = \rho_a \exp(-gz/RT)$. The "balloon" is designed so that ρ_{He} and h remain constant for all z. The lower face is located at z and the upper face at $(z + h)$. Compute the buoyancy force on the balloon. Now, assuming $gh/RT \ll 1$, to what altitude will the balloon rise?

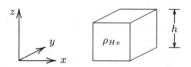

Problem 3.97

3.98 Compute the tension in the mooring line on the spherical balloon. The balloon has diameter 9 m, is filled with helium and is pressurized to 124 kPa. The surrounding air is at 1 atm and 25° C. Ignore the weight of the material from which the balloon is constructed.

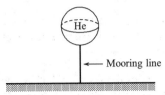

Problem 3.98

3.99 The fully enclosed tank shown below is divided into two independent chambers by a thin wall of negligible thickness. One chamber contains a fluid of density ρ while the other contains a fluid of density 3ρ. A prism of height H out of the page is attached to the wall as shown. Assuming the density of the prism is 2ρ, compute the force required to hold the prism in place.

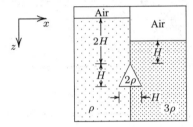

Problem 3.99

3.100 A cylindrical barrier holds water (density ρ) back as shown. If the length of the cylinder is L (out of the page) and its radius is R, determine its weight and the force exerted against the wall.

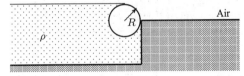

Problem 3.100

3.101 The figure below depicts a cylindrical *weir*, or low dam, of diameter D. The weir separates two reservoirs containing a fluid of density ρ as shown, and has a width w (out of the page). Determine the hydrostatic force acting on the weir.

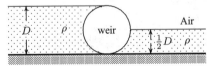

Problem 3.101

3.102 A hemispherical dome is located one dome radius, R, below the surface of a liquid of density, ρ. Determine the force components needed to hold the dome in place, and the vertical coordinate of the center of pressure, z_{cp}.

Problem 3.102

3.103 A liquid of density $N\rho$ lies below a liquid of density ρ. A rectangular piston of width H out of the page is attached to a totally submerged object of density $\frac{1}{3}\rho$ and volume $4\sqrt{2}H^3$, with a pulley attachment as shown. The pulley cable makes a right angle with the piston, which is always at a $45°$ angle to the horizontal, and the pressure below the piston is atmospheric. If there is zero net force on the piston, what is the value of N?

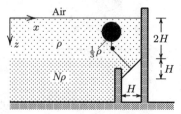

Problem 3.103

3.104 A compressor is used to regulate the air pressure in an enclosed storage tank that is designed to take water in from a reservoir when the water becomes too deep. To do this, it uses a gate that opens. The gate is hinged at the top, is $6L$ high and $6L$ wide (out of the page). Find the minimum pressure differential, $p_b - p_a$, required to keep the gate closed when water with depth $2L$ is in the tank.

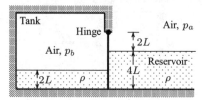

Problem 3.104

3.105 A wooden block with square cross section is in equilibrium when immersed in a fluid of density ρ as shown. The width of the block is H out of the page, and the side of the cross section is L. Ignoring any effects of friction in the hinge, determine the density of the block, ρ_b.

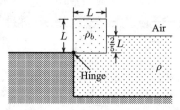

Problem 3.105

3.106 A rectangular gate of width w out of the page has a pivot at its center as shown. The gate separates fluids of density 2ρ and ρ. A cube of density 8ρ and side $\frac{1}{2}H$ is connected to the gate through the point indicated with a pulley arrangement so that the cable tension exerts a force normal to the gate. What must the width of the gate be in order to have zero net moment about the pivot?

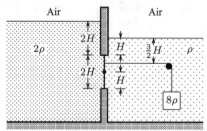

Problem 3.106

3.107 A rectangular gate of width $2L$ out of the page is hinged at point A and holds back a fluid of density ρ as shown. A cube of density ρ_c and side $\frac{1}{2}L$ is connected to the gate through its centroid with a pulley arrangement so that the weight exerts a force normal to the gate. The minimum force, P, required to hold the gate closed is $P = 12.25\rho g L^3$. Neglecting the weight of the gate, determine ρ_c.

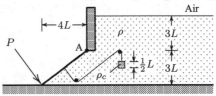

Problems 3.107, 3.108

3.108 A rectangular gate of width $6L$ out of the page is hinged at point A and holds back a fluid of density ρ as shown. A cube of density $\rho_c = 5\rho$ and side $\frac{1}{2}L$ is connected to the gate through its centroid with a pulley arrangement so that the weight exerts a force normal to the gate. Neglecting the weight of the gate, determine the minimum force, P, required to hold the gate closed.

3.109 The rectangular gate of width $3h$ (out of the page) shown below is designed to open when the water level, H, drops below $2h$. A cube-shaped weight of density ρ_b is connected to the gate to regulate its motion. Determine the threshold value for ρ_b.

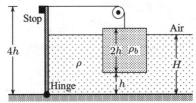

Problem 3.109

3.110 A large hollow cube-shaped flotation device holds a gate closed through a cable and pulley arrangement connecting the cube to the bottom of the gate. The effective density of the flotation device is ρ_o. The density of the fluid is ρ and the gate has width $2h$ out of the page. The gate is on the brink of opening when the depth of submergence of the flotation device is $h/2$. Determine ρ_o/ρ and the corresponding value of the force borne by the hinge when the gate is just about to open.

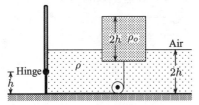

Problem 3.110

3.111 Determine the minimum volume of steel-reinforced concrete needed to keep the rectangular gate shown in a closed position. The width of the gate out of the page is 3ℓ, the density of water is ρ and the density of the concrete block is 3ρ.

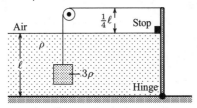

Problem 3.111

3.112 Compute the force, F, and center of pressure, z_{cp}, on the vertical wall below, assuming width $2h$ (out of the page). **NOTE:** The formulas developed for constant density do not apply to this problem.

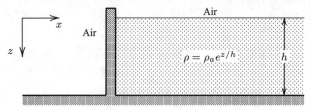

Problem 3.112

3.113 Consider the gate AB for a fluid in which the density is given by $\rho(z) = \rho_a(1 + \alpha z/H)$, where α is a constant. The projection of gate on OA is rectangular and has width H (out of the page). Also, the weight of the fluid in OAB is $\frac{1}{4}\rho_a gH^3(1+\alpha/2)$. Compute the force on the gate. Does your answer make sense in the limit $\alpha \to 0$?

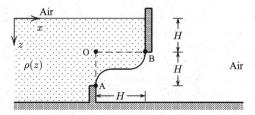

Problem 3.113

3.114 An object of cross-sectional area, A, height, h, and density, ρ_m, is floating in a fluid whose density varies with depth as $\rho = \rho_a(1+8z/h)$. If $\rho_m = 3\rho_a$, what fraction of the object, h_s/h, is submerged? What is the minimum ρ_m/ρ_a for which the object is totally submerged?

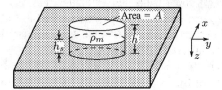

Problem 3.114

Chapter 4

Kinematics

Before we can embark on a study of fluids in motion, we must develop the equations governing conservation of mass, momentum and energy. While this point is obvious, there is a more subtle point we must address. Specifically, a central issue in developing the equations is the way we choose to describe spatial coordinates and velocities. The choice is not arbitrary, and the one we make for most fluid-mechanical applications differs from the choice made for elementary applications of Newton's laws of motion, i.e., for classical mechanics.

In this chapter, we address this fundamental difference. It is far more convenient in fluid mechanics to focus upon a fixed point or volume rather than to track individual fluid particles. The mathematical description used in fluid mechanics is known as the **Eulerian** description as opposed to the **Lagrangian** description used in classical mechanics. This is a key element of the topic known as **kinematics**. We take the term kinematics to mean basic relationships amongst coordinates and velocities required to describe a given problem, as opposed to developing and solving the dynamical equations of motion.

We begin by showing the difference between the Eulerian and Lagrangian descriptions. Then, we introduce two interrelated flow properties known as **vorticity** and **circulation**. Although our discussion is necessarily brief at this point, the central importance of vorticity in understanding the physics of fluid motion warrants special attention at all levels of study. The chapter also discusses **streamlines**, **streaklines** and **pathlines**. All are useful for visualizing details of a given flowfield, especially in experimental investigations.

After introducing these basic kinematical concepts, we derive a key integral relationship expressing the rate of change of properties in a fixed volume that contains different fluid particles at each instant. This is the famous continuum-mechanics theorem known as the **Reynolds Transport Theorem**. We will make use of this theorem in Chapters 5, 6, 7 and 9 in deriving the basic conservation laws for a fluid.

4.1 Eulerian Versus Lagrangian Description

Regardless of which kinematical description we choose, we must express position and velocity as vectors. Throughout this text, we will use the same notation, so it is worthwhile to identify our notation at this point. Most of our work will be done in rectangular Cartesian coordinates for which the position vector, $\mathbf{r}$, and velocity vector, $\mathbf{u}$, are

$$\mathbf{r} = x\,\mathbf{i} + y\,\mathbf{j} + z\,\mathbf{k} \qquad \text{and} \qquad \mathbf{u} = u\,\mathbf{i} + v\,\mathbf{j} + w\,\mathbf{k} \qquad (4.1)$$

133

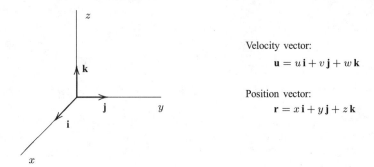

Velocity vector:
$$\mathbf{u} = u\,\mathbf{i} + v\,\mathbf{j} + w\,\mathbf{k}$$

Position vector:
$$\mathbf{r} = x\,\mathbf{i} + y\,\mathbf{j} + z\,\mathbf{k}$$

Figure 4.1: *Rectangular Cartesian coordinates.*

As shown in Figure 4.1, the quantities (x, y, z) are the three spatial coordinates, while (u, v, w) and $(\mathbf{i}, \mathbf{j}, \mathbf{k})$ are the velocity components and unit vectors, respectively, in the (x, y, z) directions.

In applying Newton's second law of motion, we compute the motion of an object by setting its mass times its acceleration equal to the sum of all forces acting. Inherent in the approach taken in classical mechanics is the way we describe the object's location, velocity and acceleration. Specifically, classical mechanics uses the **Lagrangian** description in which we track a specified object for all time so that its position vector, $\mathbf{r}$, is given by

$$\mathbf{r} = \mathbf{r}(\mathbf{r}_o, t) \quad \text{where} \quad \mathbf{r} = \mathbf{r}_o \quad \text{at} \quad t = t_o \tag{4.2}$$

Equation (4.2) states the essence of the Lagrangian description. The position of a given fluid particle at a time t is a function of time and its initial position. The velocity vector, $\mathbf{u}$, is the rate of change of position vector, $\mathbf{r}$, with respect to time:

$$\mathbf{u} = \left(\frac{\partial \mathbf{r}}{\partial t}\right)_{\mathbf{r}_o} \tag{4.3}$$

where the subscript $\mathbf{r}_o$ indicates that the particle's initial position is held constant in computing the derivative. This is a consequence of the fact that we are following the same particle around as it moves through space. If the sum of all external forces acting on an object of mass m is $\mathbf{F}$, its motion according to the laws of classical mechanics is governed by the *linear* partial differential equation:

$$\mathbf{F} = m\mathbf{a} = m\left(\frac{\partial^2 \mathbf{r}}{\partial t^2}\right)_{\mathbf{r}_o} \tag{4.4}$$

where $\mathbf{a}$ denotes acceleration.

The primary advantage of using the Lagrangian description is the linearity of the momentum equation, i.e., Equation (4.4). Nevertheless, the Lagrangian description is inconvenient in fluid mechanics for two key reasons. First, to implement this description we would have to keep track of a very large number of fluid particles to adequately resolve our continuum field. Second, the relative positions of fluid particles change by nontrivial amounts, so that viscous stresses are very complicated functions in the Lagrangian description. Its primary applications are in rarefied (very low density) gasdynamics, free-surface (water-air interface) flows and some combustion applications.

The vast majority of fluid-mechanics problems are solved using the **Eulerian** description. In this alternative kinematical formulation, we focus upon a fixed point in space and observe fluid particles as they pass by. The velocity is expressed as a function of time and of the position in space at which we make our observations, viz.,

$$\mathbf{u} = \mathbf{u}(\mathbf{r}, t) = \mathbf{u}(x, y, z, t) \tag{4.5}$$

Clearly, we cannot compute the acceleration of a fluid particle by taking the partial derivative with respect to time. If we did this, we would, in effect, be taking the difference between the velocities of two different fluid particles in forming the derivative. Instead, we must follow the fluid particle for a differential time increment to determine the differential change in its velocity. Using this information we can compute the acceleration. In other words, we still must differentiate the velocity with respect to time, holding the fluid particle's initial Lagrangian coordinates constant.

To illustrate why $\partial \mathbf{u}/\partial t$ is not the true acceleration in the Eulerian description, consider the following example (see Figure 4.2). Suppose you are focusing on a fixed point on a highway. At time $t = 0$, Granny B passes the point you are observing, and she is moving at a speed of 60 mph = 88 ft/sec. Because a Highway Patrol car is just ahead, all of the traffic is also moving at 60 mph, so that Granny B maintains a constant speed. Four seconds later, Leadfoot D passes your observation point moving at 90 mph = 132 ft/sec. Having noticed the "slow-moving" traffic, Leadfoot D has been rapidly decelerating to avoid colliding with Granny B and/or receiving a speeding ticket from the Highway Patrolman. Using your observations at the fixed point on the highway, you would estimate that traffic is *accelerating* at your observation point according to

$$a \approx \frac{\Delta u}{\Delta t} = \frac{132 \text{ ft/sec} - 88 \text{ ft/sec}}{4 \text{ sec}} = 11 \text{ ft/sec}^2 \tag{4.6}$$

In reality, Granny B has zero acceleration and Leadfoot D is *decelerating*, so that your computation is obviously incorrect.

Although the required process for computing the acceleration is conceptually straightforward, the mathematical details are not. The problem we face is the following. On the one hand, we need information about the same fluid particle's velocity at two different times in

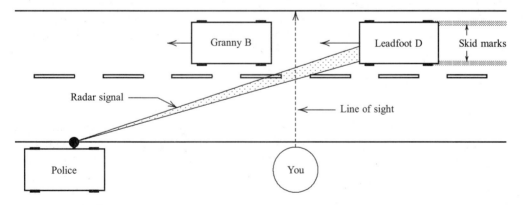

Figure 4.2: *An illustration of why $\partial \mathbf{u}/\partial t$ is not the true acceleration in the Eulerian description. Granny B is traveling at constant speed while Leadfoot D is decelerating. But, $\partial \mathbf{u}/\partial t > 0$, which would imply traffic is accelerating. Do you think the highway patrolman is likely to make this mistake?*

order to compute its acceleration. In terms of the Lagrangian description, all we have to do is form the following limit:

$$\mathbf{a} = \left(\frac{\partial \mathbf{u}}{\partial t}\right)_{\mathbf{r}_o} = \lim_{\Delta t \to 0} \frac{\mathbf{u}\left(\mathbf{r}_o, t + \Delta t\right) - \mathbf{u}\left(\mathbf{r}_o, t\right)}{\Delta t} \tag{4.7}$$

The operation is simple because $\mathbf{r}_o$ and t are independent of each other, and following the same fluid particle means holding $\mathbf{r}_o$ constant.

On the other hand, our choice of the Eulerian description means we have information about the velocity throughout the flowfield as a function of time and position relative to a fixed set of coordinates, (x, y, z). However, the Eulerian description includes no explicit information about where an individual fluid particle has been in the past nor where it will be at a future time. To obtain the information we need, we must regard the position vector as being an implicit function of time, i.e., $\mathbf{r} = \mathbf{r}(t)$, so that

$$\mathbf{a} = \lim_{\Delta t \to 0} \frac{\mathbf{u}\left(\mathbf{r} + \Delta \mathbf{r}, t + \Delta t\right) - \mathbf{u}\left(\mathbf{r}, t\right)}{\Delta t} \tag{4.8}$$

If we consider an infinitesimally small time increment, Δt, in computing distance traveled we can approximate the velocity as being $\mathbf{u}(\mathbf{r}, t)$ for the entire time it moves.[1] Hence, the distance traveled by the fluid particle is approximately $\mathbf{u}\Delta t$. Thus, as illustrated in Figure 4.3, a fluid particle located at position $\mathbf{r}(t)$ at time t, will move to

$$\mathbf{r} + \Delta \mathbf{r} \approx \mathbf{r}(t) + \mathbf{u}(\mathbf{r}, t)\Delta t \tag{4.9}$$

Then, in terms of Cartesian coordinates ($\mathbf{r} = x\mathbf{i} + y\mathbf{j} + z\mathbf{k}$ and $\mathbf{u} = u\mathbf{i} + v\mathbf{j} + w\mathbf{k}$), we can say that

$$\Delta \mathbf{r} = \mathbf{u}\Delta t \quad \Longrightarrow \quad \Delta x = u\Delta t, \quad \Delta y = v\Delta t, \quad \Delta z = w\Delta t \tag{4.10}$$

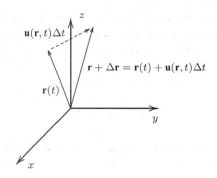

Figure 4.3: *Trajectory of a fluid particle.*

Now, expanding the velocity at time $t + \Delta t$ in Taylor series, and retaining only terms linear in Δt, we find

$$\mathbf{u}(\mathbf{r} + \Delta \mathbf{r}, t + \Delta t) = \mathbf{u}(x + \Delta x, y + \Delta y, z + \Delta z, t + \Delta t)$$

$$\approx \mathbf{u}(x, y, z, t) + \frac{\partial \mathbf{u}}{\partial x}\Delta x + \frac{\partial \mathbf{u}}{\partial y}\Delta y + \frac{\partial \mathbf{u}}{\partial z}\Delta z + \frac{\partial \mathbf{u}}{\partial t}\Delta t \tag{4.11}$$

---

[1]The error in $\Delta \mathbf{r} \approx \mathbf{u}\Delta t$ attending this approximation goes to zero as $(\Delta t)^2$ because the velocity approaches its value at time t in the limit $\Delta t \to 0$.

Therefore, combining Equations (4.10) and (4.11) and returning to the vector shorthand $\mathbf{u}(\mathbf{r}, t)$ for $\mathbf{u}(x, y, z, t)$, the change in velocity of the fluid particle between times $t + \Delta t$ and t is

$$\mathbf{u}(\mathbf{r} + \Delta\mathbf{r}, t + \Delta t) - \mathbf{u}(\mathbf{r}, t) \approx u\frac{\partial \mathbf{u}}{\partial x}\Delta t + v\frac{\partial \mathbf{u}}{\partial y}\Delta t + w\frac{\partial \mathbf{u}}{\partial z}\Delta t + \frac{\partial \mathbf{u}}{\partial t}\Delta t \qquad (4.12)$$

With a little rearrangement of terms, we can rewrite Equation (4.12) as

$$\lim_{\Delta t \to 0} \frac{\mathbf{u}(\mathbf{r} + \Delta\mathbf{r}, t + \Delta t) - \mathbf{u}(\mathbf{r}, t)}{\Delta t} = \frac{\partial \mathbf{u}}{\partial t} + u\frac{\partial \mathbf{u}}{\partial x} + v\frac{\partial \mathbf{u}}{\partial y} + w\frac{\partial \mathbf{u}}{\partial z} \qquad (4.13)$$

Consequently, denoting the acceleration by $d\mathbf{u}/dt$, Equation (4.13) tells us

$$\frac{d\mathbf{u}}{dt} = \frac{\partial \mathbf{u}}{\partial t} + u\frac{\partial \mathbf{u}}{\partial x} + v\frac{\partial \mathbf{u}}{\partial y} + w\frac{\partial \mathbf{u}}{\partial z} \qquad (4.14)$$

Finally, we can rewrite Equation (4.14) in more compact vector notation by introducing the ∇ operator. Hence,

$$\frac{d\mathbf{u}}{dt} = \frac{\partial \mathbf{u}}{\partial t} + \left(u\frac{\partial}{\partial x} + v\frac{\partial}{\partial y} + w\frac{\partial}{\partial z}\right)\mathbf{u} = \frac{\partial \mathbf{u}}{\partial t} + \mathbf{u} \cdot \nabla \mathbf{u} \qquad (4.15)$$

where

$$\nabla = \mathbf{i}\frac{\partial}{\partial x} + \mathbf{j}\frac{\partial}{\partial y} + \mathbf{k}\frac{\partial}{\partial z} \qquad (4.16)$$

Equation (4.15) is known as the **Eulerian derivative** of the velocity. In general, we regard the differential operator

$$\frac{d}{dt} = \frac{\partial}{\partial t} + \mathbf{u} \cdot \nabla \qquad (4.17)$$

as the *rate of change following a fluid particle*. Note that in Equation (4.17), we call the partial derivative with respect to time the **unsteady** contribution, while the second term on the right-hand side is known as **convection** (some authors call it **advection**).

Example 4.1 *For constant A and B, compute the acceleration for a flow with the following velocity vector.*

$$\mathbf{u} = Axt\,\mathbf{i} + By\,\mathbf{j}$$

Solution. Since the individual velocity components are $u = Axt$ and $v = By$, the acceleration corresponding to this velocity is

$$\frac{d\mathbf{u}}{dt} = \underbrace{\frac{\partial}{\partial t}(Axt\,\mathbf{i} + By\,\mathbf{j})}_{\mathbf{u}} + \underbrace{Axt\underbrace{\frac{\partial}{\partial x}(Axt\,\mathbf{i} + By\,\mathbf{j})}_{\mathbf{u}}}_{u\partial/\partial x} + \underbrace{By\underbrace{\frac{\partial}{\partial y}(Axt\,\mathbf{i} + By\,\mathbf{j})}_{\mathbf{u}}}_{v\partial/\partial y}$$

$$= Ax\,\mathbf{i} + Axt(At\,\mathbf{i}) + By(B\,\mathbf{j}) = Ax\left(1 + At^2\right)\mathbf{i} + B^2 y\,\mathbf{j}$$

The Eulerian derivative effectively bridges the gap between the Lagrangian and Eulerian descriptions. To help illustrate why this is true, consider the convective part of the Eulerian derivative. We can form a unit vector, $\mathbf{n}$, parallel to the velocity vector by writing

$$\mathbf{n} = \frac{\mathbf{u}}{|\mathbf{u}|} \qquad (4.18)$$

Then, the convective part of the Eulerian derivative can be rewritten as

$$\mathbf{u} \cdot \nabla = |\mathbf{u}|\, \mathbf{n} \cdot \nabla = |\mathbf{u}|\frac{\partial}{\partial n} \qquad (4.19)$$

where $\partial/\partial n$ is the derivative in the direction of $\mathbf{n}$. Hence, the convective part of the Eulerian derivative represents the rate of change in the direction of motion, i.e., *following the fluid particle.*

Example 4.2 *In a one-dimensional flow, each fluid particle begins motion in the x direction at time $t = 0$ with velocity $u = Kx_o$ from point $x = x_o$, and continues at that speed. Develop the Lagrangian and Eulerian descriptions of this motion.*

Solution. Since each particle moves at constant velocity, its position, x, after a time t has elapsed is given by

$$x = x_o + ut = x_o + Kx_ot$$

The velocity and acceleration follow by taking the first and second partial derivatives, respectively, with respect to time. Thus, we find that

$$u = \frac{\partial x}{\partial t} = Kx_o \quad \text{and} \quad a = \frac{\partial u}{\partial t} = 0$$

Because the description of the motion is stated in Lagrangian terms, i.e., in terms of the motion of individual particles, we are able to write the expressions above almost by inspection. Summarizing, the Lagrangian description of this motion is as follows.

$$x(x_o, t) = x_o(1 + Kt), \quad u(x_o, t) = Kx_o, \quad a(x_o, t) = 0 \qquad \text{(Lagrangian)}$$

In the Eulerian description, we focus upon a fixed value of x. Since each fluid particle travels with constant velocity $u = Kx_o$, after a time t it has traveled a distance Kx_ot. By definition, the distance traveled is also $x - x_o$. So, for this motion we know that

$$Kx_ot = x - x_o \qquad \Longrightarrow \qquad x_o = \frac{x}{1 + Kt}$$

That is, a particle with velocity Kx_o came from this point. Now, since $u = Kx_o$, we can express the velocity as a function of x and t as

$$u = Kx_o = \frac{Kx}{1 + Kt}$$

Finally, we compute the acceleration by taking the Eulerian derivative of the velocity. There follows

$$a = \frac{du}{dt} = \underbrace{\frac{\partial u}{\partial t}}_{\partial u/\partial t} + \underbrace{u}_{u}\underbrace{\frac{\partial u}{\partial x}}_{\partial u/\partial x} = -\frac{K^2 x}{(1 + Kt)^2} + \frac{Kx}{(1 + Kt)}\frac{K}{(1 + Kt)} = 0$$

Hence, we again find that the acceleration is zero, as it must be. Summarizing, the velocity and acceleration in the Eulerian description are:

$$u(x, t) = \frac{Kx}{1 + Kt}, \quad a(x, t) = 0 \qquad \text{(Eulerian)}$$

Note that in the Eulerian description we observe the motion at a specified point, and where a fluid particle came from is of no interest. Hence, we specify only velocity and acceleration. This requires us to readjust our thinking in order to understand what is happening. In Example 4.2, the Eulerian velocity is $u(x,t) = Kx/(1 + Kt)$, indicating that u decreases monotonically with time at a given point x. This reflects the fact that as time goes by, we see a fluid particle that has traveled from an increasingly distant point, i.e., closer to $x = 0$. But, the closer the particle's initial location is to the origin, the smaller its velocity. Thus, the velocity must indeed decrease as time passes.

The advantages of using the Eulerian description are that we have no need to track a large number of fluid particles, and stresses are very simple to express. However, there is one important disadvantage to using the Eulerian description that we can regard as an example of the applied mathematician's **conservation of difficulty** principle. Specifically, we find that the equations of motion are no longer linear. Rather, because of the presence of products of velocities and their derivatives in terms such as $u\partial u/\partial x$, the momentum equation is **quasi-linear**.[2] We will discuss this point further in Chapter 9. As a final comment, note that some authors refer to the Eulerian derivative[3] as the **material derivative**, the **substantial derivative** and some denote it as D/Dt.

4.2 Steady and Unsteady Flows

To further illustrate ramifications of the Eulerian description, consider flow past an airplane. If you are seated on the airplane and observe a point above one of the wings, you see the velocity relative to the vehicle. If the plane is cruising at constant velocity, and if you imagine a vector aligned with the fluid velocity, the vector has unchanging magnitude and direction. As illustrated in Figure 4.4, you observe a different velocity vector if you focus on another point above the wing. However, provided you are not looking at a point too close to the surface, the velocity will be independent of time at all points. If the local velocity is **u**, we state what you see in this situation as follows.

$$\frac{\partial \mathbf{u}}{\partial t} = \mathbf{0} \tag{4.20}$$

Any flow satisfying Equation (4.20) is called a **steady flow**. By contrast, when $\partial \mathbf{u}/\partial t$ is nonzero, the flow is said to be **unsteady**.

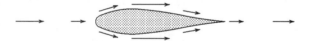

Figure 4.4: *Velocity vectors as seen by a wing-fixed observer.*

Steady flows are clearly easier to analyze both experimentally and analytically because the number of independent variables is reduced by one (time). Usually, a flow is steady if the flow geometry is constant in time, although this is not a guarantee, i.e., it is a necessary but not a sufficient condition. Many practical fluid-flow problems can be treated as steady, and many computational methods have been developed specifically for steady flows.

---

[2]A differential equation is termed **nonlinear** only if the highest derivative appears nonlinearly.

[3]It has even been called the **Lagrangian derivative**, which is true from a strict mathematical point of view. However, because it is ultimately expressed in terms of Eulerian variables, this terminology can be a bit confusing.

   There are many important unsteady flows such as flow past an advancing helicopter blade, a rapidly maneuvering airplane, flow in a reciprocating engine, etc. Even flow past a stationary cylinder can be unsteady, depending upon flow conditions. While many of these examples involve **periodic flow** (e.g., varying sinusoidally in time), there are also flows that involve **random motions** (e.g., varying in a non-periodic manner with time). For example, if we move our observation point very close to the airplane wing discussed above, we will see high-frequency variations in both magnitude and direction of the velocity vector. This apparently random motion is known as **turbulence**, and is present in most practical flows. However, we often approximate turbulent motion as being steady in the mean, so that the problem can still be treated by extensions of steady-flow methods.

   As discussed above, for steady flow the properties at a given point in space are constant. However, unless the flow is **uniform**, i.e., the same at all spatial locations, the properties of a fluid particle will change with time as it makes its way from one point in the flow to another. As shown in Figure 4.4, a fluid particle decelerates as it approaches the leading edge of a wing, accelerates to a maximum value larger than the freestream velocity at some point over the wing, and eventually decelerates back toward the freestream value. At each point along the wing we see steady velocities, although the velocity varies with position. There is no contradiction here as we simply observe different fluid particles at each instant in the Eulerian description.

Example 4.3 *Compute the acceleration vector for the following velocity vector, where U and H are constant.*

$$\mathbf{u} = \frac{U}{H} \left(x\,\mathbf{i} + y\,\mathbf{j} - 2z\,\mathbf{k} \right)$$

Solution. Because the velocity does not depend upon time, we can say

$$\mathbf{a} = \frac{\partial \mathbf{u}}{\partial t} + \mathbf{u} \cdot \nabla \mathbf{u} = \mathbf{u} \cdot \nabla \mathbf{u}$$

The differential operator $\mathbf{u} \cdot \nabla$ is

$$\mathbf{u} \cdot \nabla = \frac{U}{H} \left(x \frac{\partial}{\partial x} + y \frac{\partial}{\partial y} - 2z \frac{\partial}{\partial z} \right)$$

Therefore, the acceleration is

$$\mathbf{a} = \frac{U}{H} \left(x \frac{\partial}{\partial x} + y \frac{\partial}{\partial y} - 2z \frac{\partial}{\partial z} \right) \frac{U}{H} \left(x\,\mathbf{i} + y\,\mathbf{j} - 2z\,\mathbf{k} \right) = \frac{U^2}{H^2} \left(x\,\mathbf{i} + y\,\mathbf{j} + 4z\,\mathbf{k} \right)$$

4.3 Vorticity and Circulation

One of the most important quantities in fluid mechanics is known as **vorticity**. We denote it by ω, and define it to be the curl of the velocity.

$$\omega = \nabla \times \mathbf{u} \tag{4.21}$$

As a detailed kinematic analysis shows [cf. Wilcox (2000)], the vorticity is twice the local rate of rotation of a fluid particle as it moves through the flow. It is a bit like angular velocity, and gives rise to gyroscopic-like effects in a fluid flow. Flows that have zero vorticity are said to be **irrotational**. When the vorticity is nonzero, the flow is **rotational**.

Most importantly, as we will see in Chapter 11, vorticity plays a central role in the development of forces on an object such as the lifting force that develops on an airplane wing. Vorticity is so important that entire books have been written about it [e.g. Saffman (1993)].

While vorticity plays a critical role in theoretical and computational fluid dynamics, it has one important drawback for the experimentalist. It cannot be measured directly, and has to be obtained from derivatives of measured velocity data. Unless measurements are made at closely-spaced locations (preferably with non-intrusive techniques), differentiating measured quantities is a notoriously inaccurate process, so that reliable experimental data for vorticity are very difficult to obtain.

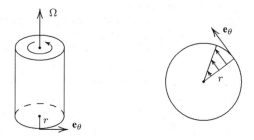

Figure 4.5: *Rigid-body rotation.*

For an example of a flow with vorticity, consider the flow inside a rotating cylinder (see Figure 4.5). If the cylinder has been rotating for a long time, all of the fluid inside will be rotating as though the container and its contents were a rigid body. We thus refer to this situation as **rigid-body rotation**. The velocity vector is given by

$$\mathbf{u} = \Omega r \mathbf{e}_\theta \tag{4.22}$$

where Ω is the angular velocity of the container, r is radial distance from the center of the cylinder and $\mathbf{e}_\theta$ is a unit vector in the circumferential direction. Taking the curl of this velocity, we find that the vorticity vector is (see Section D.2 in Appendix D for the curl in cylindrical coordinates):

$$\boldsymbol{\omega} = \frac{1}{r}\frac{\partial}{\partial r}(r u_\theta)\mathbf{k} = \frac{1}{r}\frac{d}{dr}(\Omega r^2)\mathbf{k} = 2\Omega\,\mathbf{k} \tag{4.23}$$

Thus, we see that rigid-body rotation is a rotational flow and its vorticity is twice the angular velocity.

Now consider the fluid motion associated with a hurricane. Figure 4.6 illustrates the approximate variation of a hurricane's circumferential velocity. Near the center, usually referred to as the eye, the air is essentially in rigid-body rotation and its velocity is given by Equation (4.22). Hence, very close to the center, velocities are low, which is the reason the

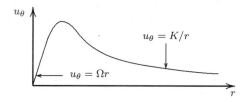

Figure 4.6: *Idealized variation of velocity in a hurricane.*

eye is found to be relatively calm. Moving outward from the center, the velocity achieves a maximum and then decreases with radius. Beyond the point where it reaches a maximum, the velocity can be approximated by

$$\mathbf{u} = \frac{K}{r}\mathbf{e}_\theta \tag{4.24}$$

where K is a constant. Taking the curl, we find

$$\omega = \frac{1}{r}\frac{\partial}{\partial r}\left(ru_\theta\right)\mathbf{k} = \frac{1}{r}\frac{dK}{dr}\mathbf{k} = \mathbf{0} \tag{4.25}$$

Hence, the flow in the outer part of a hurricane such as the one shown in Figure 4.7 is an example of irrotational flow. Interestingly, the flow defined by the velocity in Equation (4.24) is known as the **potential vortex**.

Figure 4.7: *An aerial view of a hurricane. [Photograph courtesy of the U. S. Air Force]*

At first glance, it might appear odd to call a flow that has zero vorticity a "potential vortex." To understand the nomenclature, we must introduce an important vorticity-related quantity known as **circulation**, Γ. Circulation is defined by

$$\Gamma = \oint_C \mathbf{u} \cdot d\mathbf{s} \tag{4.26}$$

where C is a closed contour and $d\mathbf{s}$ is a differential vector tangent to the contour. Equation (4.26) assumes integration on contour C proceeds in a counterclockwise direction, which is the classical definition of a contour integral. In aerodynamic applications, the convention is to integrate in the clockwise direction, and this yields the same magnitude but the opposite sign. We distinguish this difference by defining the **aerodynamic circulation**, Γ_a, as follows.

$$\Gamma_a = -\Gamma = -\oint_C \mathbf{u} \cdot d\mathbf{s} \tag{4.27}$$

To see the explicit connection between circulation and vorticity, consider the differential-sized rectangular contour shown in Figure 4.8. In general, to evaluate a line integral, we treat

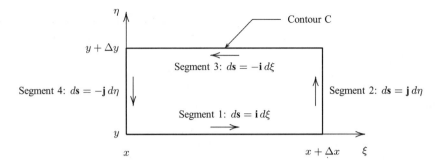

Figure 4.8: *Differential-sized contour for computing circulation.*

the integral as the sum of conventional integrals on each segment of the contour, with the sign determined by the differential distance vector, ds. Hence, referring to the figure, we see that

$$\Gamma = \underbrace{\int_x^{x+\Delta x} \mathbf{u}(\xi, y) \cdot (\mathbf{i}\, d\xi)}_{\text{Contour Segment 1}} + \underbrace{\int_y^{y+\Delta y} \mathbf{u}(x + \Delta x, \eta) \cdot (\mathbf{j}\, d\eta)}_{\text{Contour Segment 2}}$$

$$+ \underbrace{\int_x^{x+\Delta x} \mathbf{u}(\xi, y + \Delta y) \cdot (-\mathbf{i}\, d\xi)}_{\text{Contour Segment 3}} + \underbrace{\int_y^{y+\Delta y} \mathbf{u}(x, \eta) \cdot (-\mathbf{j}\, d\eta)}_{\text{Contour Segment 4}}$$

$$= \int_x^{x+\Delta x} [u(\xi, y) - u(\xi, y + \Delta y)]\, d\xi + \int_y^{y+\Delta y} [v(x + \Delta x, \eta) - v(x, \eta)]\, d\eta \quad (4.28)$$

Since the integration range is infinitesimal, we can replace the final line in Equation (4.28) by

$$\Gamma \approx [u(x, y) - u(x, y + \Delta y)]\, \Delta x + [v(x + \Delta x, y) - v(x, y)]\, \Delta y \quad (4.29)$$

This result will be more and more nearly exact as Δx and Δy become smaller and smaller. Rearranging terms yields

$$\Gamma \approx -\frac{u(x, y + \Delta y) - u(x, y)}{\Delta y} \Delta x \Delta y + \frac{v(x + \Delta x, y) - v(x, y)}{\Delta x} \Delta x \Delta y \quad (4.30)$$

Finally, as Δx and Δy approach zero, the right-hand side of Equation (4.30) contains partial derivatives of the velocity. Defining the differential area, ΔA according to

$$\Delta A = \Delta x \Delta y \quad (4.31)$$

we conclude that the circulation about this differential contour is

$$\Gamma = \left(\frac{\partial v}{\partial x} - \frac{\partial u}{\partial y} \right) \Delta A \quad (4.32)$$

The quantity in parentheses is the z component of the vorticity vector. To the order of approximation we have used in our derivation, multiplication by ΔA is the same as integration over the area bounded by the differential contour. Consequently, we have shown that

$$\Gamma = \iint_A \boldsymbol{\omega} \cdot \mathbf{n}\, dA \quad (4.33)$$

Thus, the circulation provides a measure of the strength of the vorticity contained within the bounding contour. Although our derivation is strictly valid only for the differential-sized contour of Figure 4.8, the result quoted in Equation (4.33) holds for arbitrary contours.

Now consider a potential vortex. Using Equation (4.26) with a circular contour of radius r, the differential vector tangent to the contour is $r\,d\theta\,\mathbf{e}_\theta$. Hence, since velocity is given by Equation (4.24), the circulation for a potential vortex is

$$\Gamma = \int_0^{2\pi} \left(\frac{K}{r}\right) r\,d\theta = 2\pi K \tag{4.34}$$

This result can be shown to hold on any contour that bounds $r = 0$. Hence, the potential vortex is a flow with constant circulation. This would appear to be inconsistent with Equation (4.33) as the vorticity is zero. However, the velocity is singular at $r = 0$, so that its derivative with respect to r is indeterminate. Since the circulation is the same regardless of how small a circular contour we use (such as the differential-sized contour in Figure 4.8), we conclude that this flow is irrotational everywhere except at the origin. That is, a potential vortex has all of its vorticity concentrated at a single point.

As a concluding remark, although very useful in inviscid flow theory, the potential vortex is a mathematical idealization and cannot be realized in nature all the way to the origin. Physical flowfields have no singular points, and vorticity is always distributed over a finite region. It is nevertheless a useful approximation in high Reynolds number flows, as the size of the finite region over which vorticity is typically distributed is often very small on the overall scale of the flow.

Example 4.4 *Consider the flow whose velocity is given by*

$$\mathbf{u} = U\frac{y}{H}\,\mathbf{i} - U\frac{x}{H}\,\mathbf{j}$$

where U and H are constants. Compute the circulation, $\Gamma = \oint_C \mathbf{u} \cdot d\mathbf{s}$, on rectangular contour C.

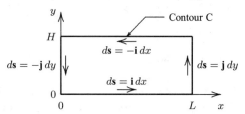

Solution. As noted in the main text, to evaluate a line integral, we treat the integral as the sum of conventional integrals on each segment of the contour, with the sign determined by the differential distance vector, $d\mathbf{s}$. Hence, referring to the figure, we see that

$$\begin{aligned}
\Gamma &= \int_0^L \mathbf{u}(x,0) \cdot (\mathbf{i}\,dx) + \int_0^H \mathbf{u}(L,y) \cdot (\mathbf{j}\,dy) + \int_0^L \mathbf{u}(x,H) \cdot (-\mathbf{i}\,dx) + \int_0^H \mathbf{u}(0,y) \cdot (-\mathbf{j}\,dy) \\
&= \int_0^L [u(x,0) - u(x,H)]\,dx + \int_0^H [v(L,y) - v(0,y)]\,dy
\end{aligned}$$

Thus, for the given velocity vector,

$$\Gamma = \int_0^L [0 - U]\,dx + \int_0^H \left[-U\frac{L}{H} - 0\right] dy = -UL - U\frac{L}{H} \cdot H = -2UL$$

4.4 Streamlines, Streaklines and Pathlines

One of the most commonly used concepts for describing a given flowfield is the **streamline**. By definition, a streamline is a curve in the flowfield to which the local velocity vector is tangent. We can derive a differential equation for a streamline by noting that the cross product of two parallel vectors is zero. So, if $d\mathbf{r} = dx\mathbf{i} + dy\mathbf{j} + dz\mathbf{k}$ is a differential element on the streamline, then

$$\mathbf{u} \times d\mathbf{r} = \mathbf{i}(v\,dz - w\,dy) + \mathbf{j}(w\,dx - u\,dz) + \mathbf{k}(u\,dy - v\,dx) = \mathbf{0} \qquad (4.35)$$

Hence, the differential equation of a streamline in Cartesian coordinates is:

$$\frac{dx}{u} = \frac{dy}{v} = \frac{dz}{w} \qquad (4.36)$$

In two dimensions, we can rewrite this in a more illuminating form,

$$\frac{dy}{dx} = \frac{v}{u} \qquad (4.37)$$

Equation (4.37) says the slope of the streamline, dy/dx, is equal to the angle the velocity vector makes with the horizontal, i.e., the velocity is tangent to the streamline. Figure 4.9 shows the streamlines for a **converging flow** and for the potential vortex. In converging flow the streamlines are straight lines, while the streamlines for vortex flow are concentric circles. Note that, by definition, there is no flow across (normal to) a streamline. Hence, solid boundaries, across which there can be no flow, are streamlines in any flow.

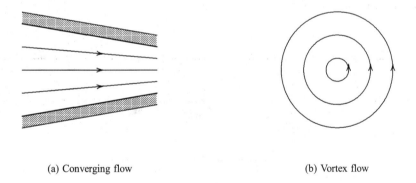

(a) Converging flow   (b) Vortex flow

Figure 4.9: *Examples of streamlines.*

For steady flows, streamlines can be visualized experimentally by injecting dye, smoke or some other tracer substance. For unsteady flows, however, there is no practical way to visualize streamlines. When we inject tracer material continuously into an unsteady flow, we observe the path traveled by all particles that have passed through the injection point. There is no reason for this path to be a streamline. We call such a path a **streakline**. Streaklines are primarily an experimenter's tool, and are rarely used in analytical work.

Experimenters have another flow-visualization method that is based on the Lagrangian description. Specifically, by taking a time exposure photograph of a marked fluid particle, we can trace the fluid-particle's **pathline**. The pathline follows the Lagrangian coordinates of a single fluid particle.

To illustrate these concepts, consider the following idealized unsteady flow. For time $t \leq \tau$, the velocity is uniform with magnitude U in the x direction. At $t = \tau$, the velocity is instantaneously rotated through an angle ϕ, and held constant thereafter. That is, the velocity vector is

$$\mathbf{u} = \begin{cases} U\mathbf{i}, & t \leq \tau \\ U(\mathbf{i}\cos\phi + \mathbf{j}\sin\phi), & t > \tau \end{cases} \tag{4.38}$$

Streamlines: For $t \leq \tau$, we have

$$\frac{dy}{dx} = \frac{v}{u} = 0 \qquad \Longrightarrow \qquad y = \text{constant} \tag{4.39}$$

while, for $t > \tau$,

$$\frac{dy}{dx} = \frac{v}{u} = \tan\phi \qquad \Longrightarrow \qquad y = x\tan\phi + \text{constant} \tag{4.40}$$

Thus, the streamlines for this flow are given by families of straight lines, i.e.,

$$y = \begin{cases} \text{constant}, & t \leq \tau \\ x\tan\phi + \text{constant}, & t > \tau \end{cases} \tag{4.41}$$

Pathlines: If we inject dye at the origin, the pathline is defined by the solution to the following initial-value problem.

$$\frac{d\mathbf{r}}{dt} = \mathbf{u}, \qquad \mathbf{r}(0) = \mathbf{0} \tag{4.42}$$

where $\mathbf{r} = x\mathbf{i} + y\mathbf{j}$ is the (two-dimensional) position vector. In component form, we have the following initial value problem for $t \leq \tau$.

$$\frac{dx}{dt} = U, \quad \frac{dy}{dt} = 0, \quad x(0) = 0, \quad y(0) = 0 \tag{4.43}$$

Integrating, we find

$$x(t) = Ut \quad \text{and} \quad y(t) = 0, \qquad t \leq \tau \tag{4.44}$$

To solve for $t > \tau$, we must follow the same fluid particle. The solution just obtained tells us that at $t = \tau$, the particle's position is $x = U\tau$ and $y = 0$. Thus, we must solve the following problem for $t > \tau$.

$$\frac{dx}{dt} = U\cos\phi, \quad \frac{dy}{dt} = U\sin\phi, \quad x(\tau) = U\tau, \quad y(\tau) = 0 \tag{4.45}$$

The solution is

$$x(t) = U\tau + U(t - \tau)\cos\phi \quad \text{and} \quad y(t) = U(t - \tau)\sin\phi \tag{4.46}$$

Hence, summarizing, the complete solution is

$$x\mathbf{i} + y\mathbf{j} = \begin{cases} Ut\mathbf{i}, & t \leq \tau \\ U\tau\mathbf{i} + U(t - \tau)(\mathbf{i}\cos\phi + \mathbf{j}\sin\phi), & t > \tau \end{cases} \tag{4.47}$$

so that the pathline is given by

$$y = \begin{cases} 0, & t \leq \tau \\ (x - U\tau)\tan\phi, & t > \tau \end{cases} \tag{4.48}$$

Streaklines: Finally, we can determine the streaklines by a simple graphical construction. First, we note that injected dye travels along the instantaneous pathlines. Hence, for $t \leq \tau$, the streakline is defined by $y = 0$. At $t = \tau$, the streakline is the line segment between $x = 0$ and $x = U\tau$. For $t > \tau$, the dye injected at the origin travels along the line $y = x \tan \phi$. The line of dye along the x axis at $t = \tau$ propagates up and to the right at an angle ϕ. Figure 4.10 illustrates the streamlines, a pathline and a streakline for this idealized flow. As shown, all three are different. However, until the velocity changes at $t = \tau$, the streamlines, streaklines and pathlines are all coincident. This is true in general for steady flow.

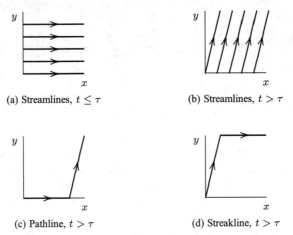

(a) Streamlines, $t \leq \tau$ (b) Streamlines, $t > \tau$

(c) Pathline, $t > \tau$ (d) Streakline, $t > \tau$

Figure 4.10: *Streamlines, a pathline and a streakline for an idealized unsteady flow.*

Normally, the streamlines and pathlines are different for unsteady flows. However, if $\mathbf{u}$ is parallel to $\partial \mathbf{u}/\partial t$, so that their cross product vanishes, they are coincident [Eringen (1980)]. Thus, we can say

$$\mathbf{u} \times \frac{\partial \mathbf{u}}{\partial t} = \mathbf{0} \qquad \Longrightarrow \qquad \text{Streamlines} \parallel \text{Pathlines} \qquad (4.49)$$

A straightforward example of such a flow is one for which the velocity varies sinusoidally with time such as $\mathbf{u} = \mathbf{u}_o(x, y, z) \sin(\omega t)$.

Example 4.5 *For flow through a variable-area channel, the velocity is given by*

$$\mathbf{u} = U\frac{A_o}{A}\left[\mathbf{i} + \frac{y}{A}\frac{dA}{dx}\mathbf{j}\right]$$

where $A(x)$ is the channel cross-sectional area, U is a constant reference velocity and A_o is the area at $x = 0$. Determine the equation describing the streamline that passes through $y = y_o$ when $x = 0$ for this flow.

Solution. The differential equation for the streamlines is

$$\frac{dy}{dx} = \frac{v}{u} = \frac{UA_o}{A^2}\frac{dA}{dx}y \cdot \frac{A}{UA_o} = \frac{1}{A}\frac{dA}{dx}y \qquad \Longrightarrow \qquad \frac{dy}{y} = \frac{1}{A}\frac{dA}{dx}dx = \frac{dA}{A}$$

Integrating yields

$$\ell ny = \ell nA(x) + \text{constant} \qquad \Longrightarrow \qquad y = y_o\frac{A(x)}{A_o}$$

where we evaluate the constant of integration by using the fact that $y = y_o$ when $A = A_o$.

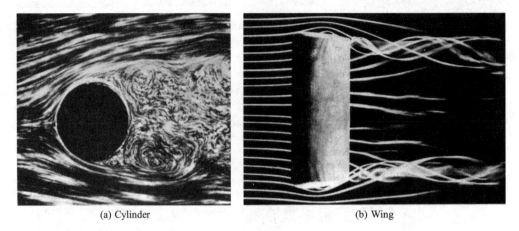

(a) Cylinder   (b) Wing

Figure 4.11: *Streamlines for high-Reynolds number flow past a cylinder and for flow over a wing. In both figures, flow is from left to right.*

Various experimental techniques have been devised to permit visualization of the streamlines in a flow. These include injection of dye, small particles and bubbles. With suitable illumination, the injectant makes the streamlines visible. This is especially helpful in revealing subtle and interesting features of flow about objects of engineering interest. Plotting streamlines from a computer simulation is also of great value, particularly in complex flows.

For example, Figure 4.11(a) shows the streamlines for flow past a cylinder. On the one hand, the streamlines are smooth and follow the shape of the cylinder as the flow approaches and moves over the front face. On the other hand, beyond about 90° measured from the leading edge (the forward-most point on the cylinder), the flow does not follow the shape of the cylinder. Rather, it *separates* from the body and a wake forms. Flow in the wake of the cylinder (to the right in the figure) is extremely complicated. This is called *turbulence*, which is an irregular eddying motion.

Figure 4.11(b) displays streamlines for flow over a wing viewed from above. As shown, the streamlines are intertwined at the wingtips. These streamlines correspond to *wingtip vortices* that form as part of the process through which lift develops on a wing. While flow over most of the wing is two dimensional, it is inherently three dimensional near the tips because of these vortices.

4.5   Extensive and Intensive Properties

We turn our attention now to the derivation of an important kinematical concept known as the **Reynolds Transport Theorem**. We will use this theorem in Chapters 5 and 6 to formulate the basic laws for conservation of mass, momentum and energy for a fluid. However, before proceeding to the formal derivation in Section 4.7, we must first introduce some preliminary concepts. That is the purpose of this and the next section.

In deriving the Reynolds Transport Theorem and the basic conservation laws, we must define three important concepts. Specifically, we must introduce the **system**, **extensive** properties and **intensive** properties. They are defined as follows.

- A **system** is the same collection of fluid particles for all time. A system thus has constant mass.

- An **extensive property** is dependent on the amount of fluid. Examples are mass (M), momentum (**P**), total energy (E), potential energy (Mgz) and kinetic energy ($\frac{1}{2}M\mathbf{u}\cdot\mathbf{u}$).

- An **intensive property** is one that is independent of the amount of fluid. Examples are velocity (**u**), internal energy (e), **specific** potential energy (gz) and specific kinetic energy ($\frac{1}{2}\mathbf{u}\cdot\mathbf{u}$). Note that each intensive-property example cited here corresponds to an extensive variable per unit mass, which is usually referred to by adding the leading word "specific."

Extensive and intensive properties are related through a volume integral. In general, if the extensive variable is B and the corresponding intensive variable is β, we have

$$B = \iiint_V \rho\,\beta\,dV \tag{4.50}$$

For example, the mass ($B = M$, $\beta = 1$) and momentum ($B = \mathbf{P}$, $\beta = \mathbf{u}$) of a system are given by

$$M = \iiint_V \rho\,dV \quad \text{and} \quad \mathbf{P} = \iiint_V \rho\,\mathbf{u}\,dV \tag{4.51}$$

4.6   Surface Fluxes

Now we consider the rate at which fluid flows across a plane of area A. Imagine that the density, ρ, and the velocity in the x direction, u, are constant. As indicated in Figure 4.12, the volume of fluid crossing the plane in a time Δt is $\Delta V = Au\Delta t$. Hence, the mass crossing the plane is

$$\Delta m = \rho u A\Delta t \tag{4.52}$$

Thus, in the limit $\Delta t \to 0$, the instantaneous mass flux, $\dot{m}$, is given by

$$\dot{m} \equiv \frac{dm}{dt} = \rho u A \tag{4.53}$$

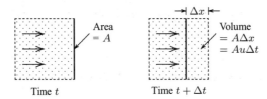

Figure 4.12: *Mass flow across a plane.*

We can easily generalize this result for the case where ρ and u vary over the plane. That is, we apply the analysis above to a differential area dA to conclude that $d\dot{m} = \rho u\,dA$. Then, integrating over the plane, we obtain

$$\dot{m} = \iint_A \rho u\,dA \tag{4.54}$$

We can also generalize the mass-flux integral to non-planar surfaces that are at an oblique angle to the velocity vector. In terms of a general velocity vector, **u**, only the normal component

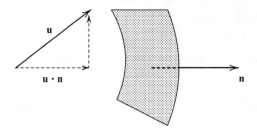

Figure 4.13: *Mass flow across a curvilinear surface.*

carries fluid across the surface (Figure 4.13) so that the differential mass flux is given by $d\dot{m} = \rho\,\mathbf{u}\cdot\mathbf{n}\,dA$, where $\mathbf{n}$ is a unit normal to the surface. Thus, the most general result for the mass flux across a surface is given by

$$\dot{m} = \iint_A \rho\,\mathbf{u}\cdot\mathbf{n}\,dA \tag{4.55}$$

Finally, the arguments leading to Equation (4.55) focus specifically upon mass flux. Clearly, if we are interested in the flux of any property such as momentum or energy, all that is needed is to replace ρ by the appropriate property per unit volume. In other words, the differential flux of extensive property B is given by $dB = \rho\beta\,\mathbf{u}\cdot\mathbf{n}\,dA$, where β is the corresponding intensive property. Hence, the flux of extensive property B across a general curvilinear surface is given by

$$\dot{B} = \iint_A \rho\beta\,\mathbf{u}\cdot\mathbf{n}\,dA \tag{4.56}$$

Example 4.6 *Consider a uniform flow with* $\mathbf{u} = U\mathbf{i}$. *Compute the mass-flow rate for constant width* h *(out of the page) through a* $30°$ *circular arc of radius* R.

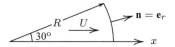

Solution. First, note that the unit normal in terms of Cartesian coordinates is

$$\mathbf{e}_r = \mathbf{i}\cos\theta + \mathbf{j}\sin\theta$$

Thus, the normal velocity component is

$$\mathbf{u}\cdot\mathbf{n} = U\cos\theta$$

Now, a differential surface element along the circular arc is

$$dS = hR\,d\theta$$

So, the mass flux across the arc is

$$\dot{m} = \rho h\int_0^{\pi/6} U\cos\theta\,R\,d\theta = \rho URh\sin\theta\,\Big|_{\theta=0}^{\theta=\pi/6} = \frac{1}{2}\rho URh$$

4.7 Reynolds Transport Theorem

We now have sufficient background information to derive a famous theorem of continuum mechanics known as the **Reynolds Transport Theorem**, whose origin traces from the pioneering work of Reynolds (1895). Modern use of the theorem in fluid mechanics began in the 1950's. It establishes a mathematical link between a Lagrangian or **system-oriented** approach in which we track the same collection of fluid particles, and an Eulerian or **control-volume** approach in which we focus on a fixed volume that contains different fluid particles at different times. This linkage is necessary for developing conservation laws in fluid mechanics where the latter approach is the most natural. By contrast, Newton's second law and the laws of thermodynamics, upon which conservation of momentum and energy are based, are formulated from a system-oriented point of view.

Once we establish the theorem, it becomes a simple task to derive the conservation principles for fluid motion. To understand why we need this theorem, consider what we have discussed in this chapter thus far. We have seen that using the Eulerian description permits us to focus on a fixed point in space and to observe the flow as time passes. Because we see different fluid particles at each instant, we have learned that special care must be taken in computing the rate of change of fluid properties. This led to the Eulerian derivative defined in Equation (4.17). We have also learned that one of the primary reasons we use the Eulerian description rather than the Lagrangian description in fluid mechanics is to obviate the need to follow a large number of fluid particles through the flowfield. Thus far, we have only partially accomplished this end. That is, we know how to focus our attention on a fixed point in space. However, space consists of an infinite number of points in a continuum. It would be far more convenient if we could expand our focus to a finite volume, or at least to a differential volume element. We refer to such a region as a **control volume**.

Hence, we pose the following problem. Given a fixed control volume, how do we compute the rate of change of its mass, momentum and energy? Just as with observing the flow at a single point, we face the problem that our control volume contains different fluid particles at each instant. This stands in distinct contrast to a **system**, which always contains the same fluid particles. Our strategy is the same as that applied in establishing the Eulerian derivative, which we aptly describe as the *rate of change following a fluid particle*. By analogy, we derive the Reynolds Transport Theorem by computing the *rate of change following a system* that is coincident with our control volume at a specified time, t. This is depicted in a one-dimensional geometry in Figure 4.14.

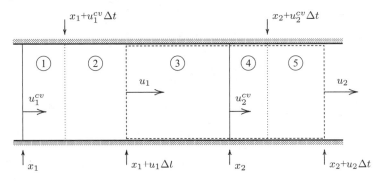

Figure 4.14: *A control volume and a system coincident at time t. At time t, the control volume and the system extend over Regions 1, 2 and 3. At time $t + \Delta t$, the control volume includes Regions 2, 3 and 4 while the system consists of the fluid in Regions 3, 4 and 5.*

Our goal in the following derivation is to determine the rate of change of a given extensive property, B, for the control volume. For the sake of generality, we will assume the control-volume boundaries are also moving. At time t, the control volume is the region between $x = x_1$ and $x = x_2$. In an infinitesimally small time increment, Δt, the fluid in the control volume at the initial time, t, moves as indicated in Figure 4.14. Because Δt is infinitesimal, we can approximate the velocity at each end of the system coincident with the control volume at time t as being constant during the motion. Any error introduced by this approximation will vanish when we take the limit $\Delta t \to 0$. So, our geometry evolves as follows.

- At time $t + \Delta t$, the *moving* system lies between $x_1 + u_1 \Delta t$ and $x_2 + u_2 \Delta t$, where u_1 and u_2 denote the velocities of the left and right boundaries of the system, respectively.

- At time $t + \Delta t$, the *moving* control volume lies between $x_1 + u_1^{cv} \Delta t$ and $x_2 + u_2^{cv} \Delta t$, where u_1^{cv} and u_2^{cv} denote the velocities of the left and right boundaries of the control volume, respectively.

Now, at time t, the value of $B(t)$ is given by

$$B(t) = \iiint_V \rho\beta \, dV = A \int_{x_1}^{x_2} \rho(x,t)\beta(x,t)dx \tag{4.57}$$

where β is the intensive variable corresponding to B, A is the (constant) cross-sectional area and the volume, V, is just $(x_2 - x_1)A$. Because this is a one-dimensional geometry, A can assume any value we wish. At time $t + \Delta t$,

$$B(t + \Delta t) = A \int_{x_1 + u_1 \Delta t}^{x_2 + u_2 \Delta t} \rho(x, t + \Delta t)\beta(x, t + \Delta t)dx \tag{4.58}$$

Thus, we find that

$$\frac{B(t+\Delta t) - B(t)}{\Delta t} = \underbrace{A \int_{x_1}^{x_2} \frac{\rho(x,t+\Delta t)\beta(x,t+\Delta t) - \rho(x,t)\beta(x,t)}{\Delta t} dx}_{\text{Regions 1, 2 and 3}}$$

$$+ \underbrace{\frac{A}{\Delta t} \int_{x_2}^{x_2 + u_2^{cv}\Delta t} \rho(x,t+\Delta t)\beta(x,t+\Delta t)\, dx}_{\text{Region 4}}$$

$$+ \underbrace{\frac{A}{\Delta t} \int_{x_2 + u_2^{cv}\Delta t}^{x_2 + u_2\Delta t} \rho(x,t+\Delta t)\beta(x,t+\Delta t)\, dx}_{\text{Region 5}}$$

$$- \underbrace{\frac{A}{\Delta t} \int_{x_1}^{x_1 + u_1^{cv}\Delta t} \rho(x,t+\Delta t)\beta(x,t+\Delta t)\, dx}_{\text{Region 1}}$$

$$- \underbrace{\frac{A}{\Delta t} \int_{x_1 + u_1^{cv}\Delta t}^{x_1 + u_1\Delta t} \rho(x,t+\Delta t)\beta(x,t+\Delta t)\, dx}_{\text{Region 2}} \tag{4.59}$$

Before we can take the limit $\Delta t \to 0$, we must pay special attention to the last four integrals in Equation (4.59). To determine their limiting values, we observe that because the integration limits differ by an infinitesimally small amount in the limit $\Delta t \to 0$, they are simply the product of the integrand (at any point within the integration range) and the difference between the integration limits. For the first of the four integrals (Region 4) we have

$$\int_{x_2}^{x_2+u_2^{cv}\Delta t} \rho(x,t+\Delta t)\beta(x,t+\Delta t)dx \approx \rho(x_2,t)\beta(x_2,t)\,u_2^{cv}\Delta t$$

$$\approx \rho(x_2,t)\beta(x_2,t)\frac{dx_2}{dt}\Delta t \qquad (4.60)$$

where we note that, by definition, u_2^{cv} is the rate of change of the control-volume boundary position at $x = x_2$. In an entirely similar way, the second of the four integrals (Region 5) is,

$$\int_{x_2+u_2^{cv}\Delta t}^{x_2+u_2\Delta t} \rho(x,t+\Delta t)\beta(x,t+\Delta t)dx \approx \rho(x_2,t)\beta(x_2,t)\,(u_2-u_2^{cv})\,\Delta t \qquad (4.61)$$

Consequently, as Δt approaches zero,

$$\lim_{\Delta t \to 0}\frac{A}{\Delta t}\int_{x_2}^{x_2+u_2^{cv}\Delta t} \rho(x,t+\Delta t)\beta(x,t+\Delta t)dx = A(\rho\beta)_2\frac{dx_2}{dt} \qquad (4.62)$$

$$\lim_{\Delta t \to 0}\frac{A}{\Delta t}\int_{x_2+u_2^{cv}\Delta t}^{x_2+u_2^{cv}\Delta t} \rho(x,t+\Delta t)\beta(x,t+\Delta t)dx = A\left[\rho(u-u^{cv})\beta\right]_2 \qquad (4.63)$$

with subscript 2 denoting values at $x = x_2$ and time t. A similar result holds for the two integrals over Regions 1 and 2 in Equation (4.59). Thus, in the limit $\Delta t \to 0$, we arrive at

$$\frac{dB}{dt} = A\int_{x_1}^{x_2}\frac{\partial}{\partial t}(\rho\beta)dx + \underbrace{A(\rho\beta)_2\frac{dx_2}{dt}}_{\text{Region 4}} - \underbrace{A(\rho\beta)_1\frac{dx_1}{dt}}_{\text{Region 1}} + \underbrace{A(\rho u^{rel}\beta)_2}_{\text{Region 5}} - \underbrace{A(\rho u^{rel}\beta)_1}_{\text{Region 2}} \qquad (4.64)$$

Note that the quantity u^{rel} denotes the fluid velocity relative to the control-volume bounding surface, i.e.,

$$u^{rel} \equiv u - u^{cv} \qquad (4.65)$$

We can simplify this equation by using **Leibnitz's Theorem**, which is quoted in most elementary calculus books. It tells us that differentiating an integral with variable limits consists of two contributions. The first contribution accounts for the time variation of the integrand. The second accounts for the fact that the limits of the integral are changing. Hence,

$$\frac{d}{dt}\int_{x_1}^{x_2}\rho(x,t)\beta(x,t)dx = \int_{x_1}^{x_2}\frac{\partial}{\partial t}(\rho\beta)dx + (\rho\beta)_2\frac{dx_2}{dt} - (\rho\beta)_1\frac{dx_1}{dt} \qquad (4.66)$$

Using Equation (4.66) and noting that the cross-sectional area of the control volume is constant, we can rewrite Equation (4.64) as follows.

$$\frac{dB}{dt} = \frac{d}{dt}\iiint_V \rho\beta\,dV + \underbrace{\iint_A (\rho u^{rel}\beta)_2 dA}_{\dot{B}_{\text{out}}} - \underbrace{\iint_A (\rho u^{rel}\beta)_1 dA}_{\dot{B}_{\text{in}}} \qquad (4.67)$$

For this simple one-dimensional geometry, we know that the **outer unit normal** at the left boundary is $\mathbf{n} = -\mathbf{i}$ and the velocity vector is $\mathbf{u}^{rel} = u_1^{rel}\,\mathbf{i}$, wherefore $\mathbf{u}^{rel}\cdot\mathbf{n} = -u_1^{rel}$. At the right boundary, we have $\mathbf{n} = \mathbf{i}$ and $\mathbf{u}^{rel} = u_2^{rel}\,\mathbf{i}$ so that $\mathbf{u}^{rel}\cdot\mathbf{n} = u_2^{rel}$. Hence, we can rewrite Equation (4.67) as

$$\frac{dB}{dt} = \frac{d}{dt}\iiint_V \rho\beta\,dV + \iint_A (\rho\beta\,\mathbf{u}^{rel}\cdot\mathbf{n})_2\,dA + \iint_A (\rho\beta\,\mathbf{u}^{rel}\cdot\mathbf{n})_1\,dA \qquad (4.68)$$

Finally, for this admittedly simple geometry, the surface bounding the control volume consists of the left and right cross sections. All other boundaries are of no consequence because the motion is strictly one dimensional. Hence, for this simple geometry, the sum of the two area integrals in Equation (4.68) can be replaced by a single area integral over the "closed" surface bounding the control volume, viz.,

$$\frac{dB}{dt} = \frac{d}{dt}\iiint_V \rho\beta\,dV + \oiint_S \rho\beta\,\mathbf{u}^{rel}\cdot\mathbf{n}\,dS \qquad (4.69)$$

In words, Equation (4.69) says the rate of change of a specified variable B for the system coincident with the control volume at time t is equal to the sum of the instantaneous rate of change of the property within the control volume and the net flux of the property out of the control volume.

Equation (4.69) is the Reynolds Transport Theorem in its most basic form, and it applies to general three-dimensional geometries such as the one shown in Figure 4.15. Although we have not proven so here, the theorem is valid for general three-dimensional motion. A three-dimensional derivation requires more advanced mathematical concepts, and will not be presented here.

As we will see in Chapters 5 and 6, using the Reynolds Transport Theorem permits expressing the classical laws for conservation of mass, momentum and energy in integral form. The most important point to note about Equation (4.69) is the following. Both integrals pertain to the control volume rather than to the system. This is important because the physical principles we will apply to deduce conservation laws are most naturally stated for a moving system. As discussed at the beginning of this section, the Reynolds Transport Theorem bridges the gap between the Eulerian-description-oriented control volume and the Lagrangian-description-oriented system.

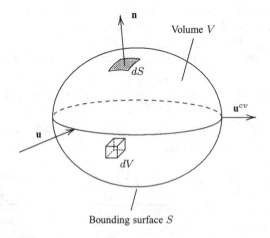

Figure 4.15: *A general three-dimensional geometry.*

Example 4.7 *Ants, frightened by the presence of an anteater, are vacating their underground nest. The number of ants per unit volume near the exit is n, the exit area is A and the ants are moving at velocity U as they exit. The anteater is consuming ants within the nest at a rate of $\dot{N}_{zap}$. Using the Reynolds Transport Theorem (with number density replacing mass density), determine the rate of change of the number of live ants in the nest.*

Solution. We can apply the Reynolds Transport Theorem to this problem if we consider the ants to be a system and the nest a control volume. Since the nest is stationary, $\mathbf{u}^{rel} = \mathbf{u}$, so that the Reynolds Transport Theorem tells us

$$-\dot{N}_{zap} = \frac{d}{dt} \iiint_V \rho \, dV + \oiint_S \rho \, (\mathbf{u} \cdot \mathbf{n}) \, dS$$

where $\dot{N}_{zap}$ is the rate at which the anteater is consuming ants and ρ is the number density of ants, i.e., number of ants per unit volume. Also, by definition, the number of live ants in the nest, N, is

$$N = \iiint_V \rho \, dV$$

So, the Reynolds Transport Theorem simplifies to

$$-\dot{N}_{zap} = \frac{dN}{dt} + \oiint_S \rho \, (\mathbf{u} \cdot \mathbf{n}) \, dS = \frac{dN}{dt} + nUA$$

Therefore, the rate of change of live ants in the nest is

$$\frac{dN}{dt} = -\dot{N}_{zap} - nUA$$

There is an alternate form of the Reynolds Transport Theorem that will prove to be useful when we develop the differential equations governing the motion of a fluid in Chapter 9. To arrive at this alternate form, we must revisit Equation (4.59). Recalling the fact that $dx_i/dt = u_i^{cv}$ for $i = 1, 2$ and Equation (4.65), which tells us $u^{rel} = u - u^{cv}$, we can say

$$\underbrace{A(\rho\beta)_2 dx_2/dt}_{\text{Region 4}} + \underbrace{A(\rho u^{rel}\beta)_2}_{\text{Region 5}} = A(\rho u\beta)_2 = \iint_A (\rho\beta \, \mathbf{u} \cdot \mathbf{n})_2 \, dA \qquad (4.70)$$

$$\underbrace{A(\rho\beta)_1 dx_1/dt}_{\text{Region 1}} + \underbrace{A(\rho u^{rel}\beta)_1}_{\text{Region 2}} = A(\rho u\beta)_1 = -\iint_A (\rho\beta \, \mathbf{u} \cdot \mathbf{n})_1 \, dA \qquad (4.71)$$

Hence, Equation (4.68) assumes the following alternate form.

$$\frac{dB}{dt} = \iiint_V \frac{\partial(\rho\beta)}{\partial t} \, dV + \iint_A (\rho\beta \, \mathbf{u} \cdot \mathbf{n})_2 \, dA + \iint_A (\rho\beta \, \mathbf{u} \cdot \mathbf{n})_1 \, dA \qquad (4.72)$$

Finally, combining the surface-flux integrals into a closed-surface integral as before, we conclude that

$$\frac{dB}{dt} = \iiint_V \frac{\partial(\rho\beta)}{\partial t} \, dV + \oiint_S \rho\beta \, \mathbf{u} \cdot \mathbf{n} \, dS \qquad (4.73)$$

As noted above, this form of the Reynolds Transport Theorem will be especially helpful when we derive the differential form of the conservation laws in Chapter 9.

4.8   Vector Calculus and Multiple Integrals

This book makes extensive use of vector calculus and multiple (volume and surface) integrals. At first glance, this might appear a bit formidable for the beginning engineering student. So why use these concepts?

The answer is simple. Force is a *vector*. Velocity is a *vector*. The direction of the force due to pressure acting on a surface is in the direction of the *vector* normal to the surface on which the pressure acts. Everywhere we look, we find a vector! The only kind of motion that can be represented strictly with scalar quantities is one dimensional. There are few real applications involving any kind of motion that can be described as one dimensional. Hence, of necessity, we must deal with vectors in order to tackle any but the most trivial problems.

Although we could still avoid vectors by working with individual components of forces and velocities, for example, the compactness of vector notation saves a great deal of writing (and reading). To fully appreciate this aspect of vector notation, simply consider the curl of the velocity, i.e., the vorticity vector. Calling the vorticity ω, in vector notation we say

$$\omega = \nabla \times \mathbf{u} \qquad (4.74)$$

To avoid vector notation, we must deal with its three components, ω_x, ω_y and ω_z:

$$\omega_x = \frac{\partial w}{\partial y} - \frac{\partial v}{\partial z}, \qquad \omega_y = \frac{\partial u}{\partial z} - \frac{\partial w}{\partial x}, \qquad \omega_z = \frac{\partial v}{\partial x} - \frac{\partial u}{\partial y} \qquad (4.75)$$

The compactness of vector notation is an attractive aspect motivating its use.

Concerning the use of multiple integrals, most of the integrands we will deal with in this text will vary in, at worst, just one direction. Often, we will even assume the integrands are constant so that the integrals can be replaced by a summation involving products of integrands and areas or volumes. So, their use is more symbolic than anything else for the highly idealized (for the sake of algebraic simplicity) classroom examples. However, in practical applications, the integrands will not be so simple, and evaluating the integrals will be important. There is no point in developing bad habits at this fundamental a level, so we retain the full integrals.

The greatest advantage we will have in working with volume and bounding-surface integrals will occur when we derive basic laws of fluid motion. All of our derivations will be valid for arbitrary geometries, as opposed to the standard approach of using a simple geometry with special symmetries that permit easy evaluation of integrals. Without using the integral approach, we generally confine our derivations to rectangular Cartesian coordinates, which is an unnecessary limitation.

Perhaps the most important thing that needs clarification is the meaning of a closed-surface integral and how it must be evaluated. If the surface bounding a control volume consists of N distinct segments, then for a given integrand Φ,

$$\oiint_S \Phi \, dS = \sum_{n=1}^{N} \iint_{A_n} \Phi \, dA_n \qquad (4.76)$$

where integration is always done in the positive sense over each of the N surface segments. Also, if the unit normal, $\mathbf{n}$, appears in the integrand, it is always the *outer unit normal*. That is, the unit normal points outward rather than toward the control-volume interior.

The reason, then, for using vectors (including the ∇ differential operator) along with volume and surface integrals is simple. We can avoid the tedium of developing differential, and other, relationships for contrived geometries, with no guarantee that the results are valid for any but the special geometry chosen. We minimize the amount of tedious math required to derive basic principles, *thus facilitating a greater focus on the physics of fluid motion.* This, above all, is the primary rationale for using vector calculus and multiple integrals.

Now that you understand the need for these mathematical tools, take the time to review the material in Appendix C if you have not already done so. It includes some properties of the ∇ operator that you will find useful.

Problems

4.1 Compute the acceleration vector for the following velocity vectors, where U, δ and a are constant.

   (a) $\mathbf{u} = U\cos(x + at)\mathbf{i}$

   (b) $\mathbf{u} = U\left(1 - e^{-y/\delta}\right)\mathbf{i}$

   (c) $\mathbf{u} = f(y)\mathbf{i} + g(x)\mathbf{j}$

4.2 Compute the acceleration vector for the following velocity vectors, where U and H are constant.

   (a) $\mathbf{u} = \dfrac{U}{H^2}\left(x^2\mathbf{i} + 2xy\mathbf{j} - 4xz\mathbf{k}\right)$

   (b) $\mathbf{u} = \dfrac{U}{H}\left(2x\mathbf{i} - 3y\mathbf{j} + z\mathbf{k}\right)$

4.3 Compute the acceleration vector for the following velocity vectors.

   (a) $\mathbf{u} = A(t)x\mathbf{i} + A(t)y\mathbf{j}$

   (b) $\mathbf{u} = A(t)x^2\mathbf{i} - A(t)y^2\mathbf{j}$

4.4 The velocity components for a steady, two-dimensional, incompressible flow are

$$u = v = \frac{U}{L}(y - x)$$

where U and L are constant velocity and length scales, respectively. Compute the acceleration vector, **a**.

4.5 The axial velocity in a conical nozzle is $\mathbf{u} = U_o\left(1 - x/x_o\right)^{-2}\mathbf{i}$ where U_o and x_o are constant velocity and length scales, respectively.

   (a) If we treat the flow as being steady and one dimensional, what is the acceleration, **a**?

   (b) If the magnitude of the acceleration at $x = x_2$ is 2/5 of its value at $x = x_1$, what is the ratio of the velocity at $x = x_2$ to its value at $x = x_1$?

4.6 The velocity for steady, incompressible flow in a converging nozzle can be approximated by $\mathbf{u} = U(1 + x/L)\mathbf{i} - Uy/L\mathbf{j}$ where U and L are characteristic velocity and length scales, respectively.

   (a) Compute the acceleration vector, **a**.

   (b) For $y = 0$, at what value of x is the acceleration double its value at $x = 0$?

4.7 The velocity for steady *Couette flow* is $\mathbf{u} = Uy\mathbf{i}/h$, where U is the velocity of the moving plate and h is the distance between the plates. What is the acceleration, **a**?

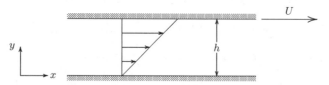

   Problem 4.7

4.8 Consider a one-dimensional flow whose velocity is $\mathbf{u} = (U_o - x/\tau)e^{-t/\tau}\mathbf{i}$, where U_o and τ are constant velocity and time scales. Compute the unsteady, convective and total accelerations.

4.9 Consider the one-dimensional flow whose velocity is given by $\mathbf{u} = (x/t)\mathbf{i}$.

   (a) Determine the unsteady, convective and total accelerations.

   (b) Repeat your computations for $u = -x/t$.

4.10 For flow near the *stagnation point* of an accelerating cylinder, the velocity is

$$\mathbf{u} = \frac{2U(t)}{R}(x\,\mathbf{i} - y\,\mathbf{j})$$

where R is the cylinder's radius and $U(t)$ is the cylinder's speed.

(a) Compute the unsteady, convective and total accelerations.

(b) Determine $U(t)$ and a_x if $a_y = 0$. Assume $U(0) = U_o$.

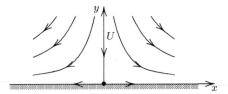

Problem 4.10

4.11 Consider the steady, two-dimensional, incompressible flow whose velocity components are given by $u = Ux/H$, $v = -Uy/H$, where U and H are characteristic velocity and length scales, respectively. Determine the acceleration vector, $\mathbf{a}$.

4.12 For cylindrical coordinates, the acceleration in two-dimensional flow is

$$a_r = \frac{\partial u_r}{\partial t} + u_r \frac{\partial u_r}{\partial r} + \frac{u_\theta}{r}\frac{\partial u_r}{\partial \theta} - \frac{u_\theta^2}{r}$$

$$a_\theta = \frac{\partial u_\theta}{\partial t} + u_r \frac{\partial u_\theta}{\partial r} + \frac{u_\theta}{r}\frac{\partial u_\theta}{\partial \theta} + \frac{u_r u_\theta}{r}$$

Find the acceleration vector for the following flows.

(a) Flow toward a sink, $\mathbf{u} = -\dfrac{Q}{2\pi r}\,\mathbf{e}_r$ (Q = constant)

(b) Rigid-body rotation, $\mathbf{u} = \Omega r\,\mathbf{e}_\theta$ (Ω = constant)

4.13 For cylindrical coordinates, the acceleration in two-dimensional flow is

$$a_r = \frac{\partial u_r}{\partial t} + u_r \frac{\partial u_r}{\partial r} + \frac{u_\theta}{r}\frac{\partial u_r}{\partial \theta} - \frac{u_\theta^2}{r}$$

$$a_\theta = \frac{\partial u_\theta}{\partial t} + u_r \frac{\partial u_\theta}{\partial r} + \frac{u_\theta}{r}\frac{\partial u_\theta}{\partial \theta} + \frac{u_r u_\theta}{r}$$

Find the acceleration vector for the following flows.

(a) Potential vortex, $\mathbf{u} = \dfrac{\Gamma}{2\pi r}\,\mathbf{e}_\theta$ (Γ = constant)

(b) Doublet flow, $\mathbf{u} = -\dfrac{D}{2\pi r^2}(\mathbf{e}_r \cos\theta + \mathbf{e}_\theta \sin\theta)$ (D = constant)

4.14 For low Mach number flow through a converging nozzle, the velocity vector, $\mathbf{u}$, can be approximated by $\mathbf{u} = U(1+x/L)\,\mathbf{i} - Uy/L\,\mathbf{j}$, where U and L are characteristic velocity and length scales, respectively. Determine the Lagrangian description of the fluid-particle coordinates, x and y, in terms of U, L, t and the initial values of the coordinates, x_o and y_o.

4.15 Consider a flow for which the velocity vector is $\mathbf{u} = U[(y/\ell)\,\mathbf{i} + \mathbf{j}]$, where U and ℓ are characteristic velocity and length scales, respectively. Determine the Lagrangian description of the fluid-particle position vector, $\mathbf{r} = x\,\mathbf{i} + y\,\mathbf{j}$, in terms of the constants U and ℓ, time t and the initial values of the coordinates, x_o and y_o.

4.16 Consider a flow for which the velocity vector is $\mathbf{u} = U\mathbf{i} + x/(t+\tau)\mathbf{j}$ where U and τ are characteristic velocity and time scales, respectively. Determine the Lagrangian description of the fluid-particle position vector, $\mathbf{r} = x\mathbf{i} + y\mathbf{j}$, in terms of the constants U and τ, time t and the initial values of the coordinates, x_o and y_o.

4.17 For flow near the *stagnation point* of a cylinder, the velocity is $\mathbf{u} = (2U/R)(x\mathbf{i} - y\mathbf{j})$, where R is the cylinder's radius and U is the speed of the incident flow. Determine the Lagrangian description of the fluid-particle position vector, $\mathbf{r} = x\mathbf{i} + y\mathbf{j}$, in terms of U, R, t and initial values of the coordinates, x_o and y_o.

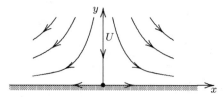

Problem 4.17

4.18 The velocity for incompressible flow above an infinite flat wall at $y = 0$ oscillating with velocity $u(0,t) = U\cos(\Omega t)$ (*Stokes' Second Problem*) is given by

$$u(y,t) = Ue^{-ky}\cos(\Omega t - ky), \quad v(y,t) = w(y,t) = 0$$

where t is time, Ω is frequency, y is distance normal to the surface and k is a constant of dimensions 1/length. Determine the Lagrangian description of the fluid-particle coordinates, x and y, in terms of U, Ω, k, t and initial values of the coordinates, x_o and y_o.

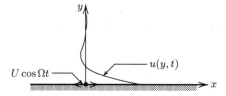

Problem 4.18

4.19 In cylindrical coordinates, the rate of change of the position vector, $\mathbf{r}$, is given by

$$\frac{d\mathbf{r}}{dt} = \frac{d}{dt}(r\mathbf{e}_r) = \frac{dr}{dt}\mathbf{e}_r + r\frac{d\theta}{dt}\mathbf{e}_\theta$$

Determine the Lagrangian description of the fluid-particle coordinates, r and θ, for *rigid-body rotation* in terms of Ω, t and initial values of the coordinates, r_o and θ_o.

4.20 In cylindrical coordinates, the rate of change of the position vector, $\mathbf{r}$, is given by

$$\frac{d\mathbf{r}}{dt} = \frac{d}{dt}(r\mathbf{e}_r) = \frac{dr}{dt}\mathbf{e}_r + r\frac{d\theta}{dt}\mathbf{e}_\theta$$

Determine the Lagrangian description of the fluid-particle coordinates, r and θ, for a *potential vortex* in terms of K, t and initial values of the coordinates, r_o and θ_o.

4.21 In cylindrical coordinates, the rate of change of the position vector, $\mathbf{r}$, is given by

$$\frac{d\mathbf{r}}{dt} = \frac{d}{dt}(r\mathbf{e}_r) = \frac{dr}{dt}\mathbf{e}_r + r\frac{d\theta}{dt}\mathbf{e}_\theta$$

Expressing your answer in terms of Q, t and initial values of the coordinates, r_o and θ_o, determine the Lagrangian description of the fluid-particle coordinates, r and θ, for a *point source*, whose velocity vector is

$$\mathbf{u} = \frac{Q}{2\pi r}\mathbf{e}_r$$

4.22 Assuming U and L are constant, compute the vorticity for $\mathbf{u} = U\left[(x/L)\mathbf{i} - 2(y/L)\mathbf{j} + (z/L)\mathbf{k}\right]$.

4.23 Assuming U and δ are constant, compute the vorticity for the following velocity vectors.

(a) $\mathbf{u} = U\left[(x/\delta)\mathbf{i} - (y/\delta)\mathbf{j}\right]$

(b) $\mathbf{u} = f_1(y)\mathbf{i} + f_2(x)\mathbf{j}$

(c) $\mathbf{u} = U\left(1 - e^{-y/\delta}\right)\mathbf{i}$

4.24 Assuming U, A and δ are constant, compute the vorticity for the following velocity vectors.

(a) $\mathbf{u} = A(x\mathbf{i} + y\mathbf{j} - 2z\,\mathbf{k})$

(b) $\mathbf{u} = A(y\mathbf{i} + x\mathbf{j})$

(c) $\mathbf{u} = U\left(1 - e^{-z/\delta}\right)(\mathbf{i} + \mathbf{j})$

4.25 For cylindrical coordinates, the vorticity is given by

$$\boldsymbol{\omega} = \begin{vmatrix} \dfrac{1}{r}\mathbf{e}_r & \mathbf{e}_\theta & \dfrac{1}{r}\mathbf{k} \\[2mm] \dfrac{\partial}{\partial r} & \dfrac{\partial}{\partial \theta} & \dfrac{\partial}{\partial z} \\[2mm] u_r & r u_\theta & w \end{vmatrix}$$

Compute the vorticity of the following flowfield for which all quantities are dimensionless.

$$\mathbf{u} = \frac{2\cos\theta}{r^2}\mathbf{e}_r + \frac{\sin\theta}{r^2}\mathbf{e}_\theta + \frac{z\cos\theta}{r^3}\mathbf{k}$$

4.26 In cylindrical coordinates for two-dimensional flow, the vorticity is given by

$$\boldsymbol{\omega} = \frac{1}{r}\left[\frac{\partial}{\partial r}(r u_\theta) - \frac{\partial u_r}{\partial \theta}\right]\mathbf{k}$$

Compute the vorticity for the following velocity vectors, where U and R are constant.

(a) $\mathbf{u} = U(r/R)^2 \sin 2\theta\,\mathbf{e}_r + U\left[R\theta/r + 2(r/R)^2\cos 2\theta\right]\mathbf{e}_\theta$

(b) $\mathbf{u} = (UR/r)(\mathbf{e}_r + 3\mathbf{e}_\theta)$

(c) $\mathbf{u} = U(r/R)^2\mathbf{e}_r - 2U(r/R)^2\theta\,\mathbf{e}_\theta$

4.27 In cylindrical coordinates for two-dimensional flow, the vorticity is given by

$$\boldsymbol{\omega} = \frac{1}{r}\left[\frac{\partial}{\partial r}(r u_\theta) - \frac{\partial u_r}{\partial \theta}\right]\mathbf{k}$$

Compute the vorticity for the following velocity vectors, where U and R are constant.

(a) $\mathbf{u} = 4U\dfrac{r}{R}\cosh\theta\,\mathbf{e}_r + 2U\dfrac{r}{R}\sinh\theta\,\mathbf{e}_\theta$

(b) $\mathbf{u} = U\left[1 - (R/r)^2\right]\cos\theta\,\mathbf{e}_r - U\left[1 + (R/r)^2\right]\sin\theta\,\mathbf{e}_\theta$

(c) $\mathbf{u} = 2U\dfrac{r}{R}\cos\theta\,\mathbf{e}_r - U\dfrac{r}{R}\sin\theta\,\mathbf{e}_\theta$

4.28 A velocity field is given in terms of dimensionless quantities by $\mathbf{u} = Ay^3 t\mathbf{i} + 3xy^2 t\mathbf{j}$. Is there any value of the constant A for which the flow is irrotational?

4.29 A velocity field is given in terms of dimensionless quantities by $\mathbf{u} = Ax\mathbf{i} + z\mathbf{j} + (y - z)\mathbf{k}$. Is there any value of the constant A for which the flow is irrotational?

4.30 Consider the following velocity vector.

$$\mathbf{u} = \left(3Ay^2z^2 - 5y^2z^2\right)\mathbf{i} - 4Bxyz^2\mathbf{j} + 5xy^2z\,\mathbf{k}$$

All quantities, including the unknown constants A and B, are dimensionless. Determine the values of A and B that yield an irrotational flow.

4.31 A flowfield has the following velocity vector

$$\mathbf{u} = y^2z^2\mathbf{i} + 3xyz^2\mathbf{j} - 3xz^3\,\mathbf{k}$$

where all quantities are dimensionless.

(a) Determine whether or not this flow is irrotational.

(b) If we drop the z component of the velocity, is the flow irrotational?

4.32 The velocity for incompressible flow above an infinite flat wall at $y = 0$ oscillating with velocity $u(0, t) = U\cos(\Omega t)$ (*Stokes' Second Problem*) is given by

$$u(y, t) = Ue^{-ky}\cos(\Omega t - ky), \quad v(y, t) = w(y, t) = 0$$

where t is time, Ω is frequency, y is distance normal to the surface and k is a constant. Compute the acceleration vector and the vorticity vector.

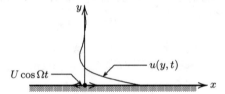

Problem 4.32

4.33 The velocity for incompressible flow above an infinite flat wall at $y = 0$ impulsively set in motion with constant velocity U at $t = 0$ (*Stokes' First Problem*) is given by

$$u(y, t) = U\left[1 - \mathrm{erf}\left(\frac{y}{2\sqrt{\nu t}}\right)\right], \quad v(y, t) = w(y, t) = 0$$

where t is time, y is distance normal to the surface, ν is kinematic viscosity and $\mathrm{erf}(\eta)$ is the *error function* defined by

$$\mathrm{erf}(\eta) = \frac{2}{\sqrt{\pi}}\int_0^{\eta} e^{-\hat{\eta}^2}\,d\hat{\eta}$$

Compute the acceleration vector and the vorticity vector.

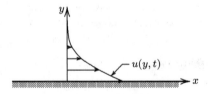

Problem 4.33

4.34 Verify that for steady, irrotational, two-dimensional flow

$$\frac{d\mathbf{u}}{dt} = \frac{1}{2}\nabla\left(u^2 + v^2\right)$$

4.35 Verify that the circulation, $\Gamma = \oint_C \mathbf{u} \cdot d\mathbf{s}$, of a *potential vortex*, for which $\mathbf{u} = (K/r)\,\mathbf{e}_\theta$, is $2\pi K$ for the rectangular contour shown. Note that $\tan^{-1}(H/L) + \tan^{-1}(L/H) = \pi/2$.

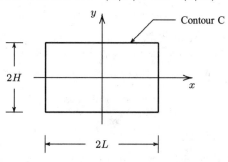

Problems 4.35, 4.36, 4.37

4.36 An incompressible flow has velocity given by $\mathbf{u} = U\,\mathbf{i} + \epsilon U(x/H)^2\,\mathbf{j}$, where ϵ, U and H are constants. Compute the circulation, $\Gamma = \oint_C \mathbf{u} \cdot d\mathbf{s}$, on the rectangular contour shown. Verify that your result is consistent with Equation (4.33).

4.37 Verify that the circulation, $\Gamma = \oint_C \mathbf{u} \cdot d\mathbf{s}$, of a *rigid-body rotation*, for which $\mathbf{u} = \Omega r\,\mathbf{e}_\theta$, is zero for the rectangular contour shown.

4.38 A low-speed flow has velocity vector $\mathbf{u} = 2U(xy/H^2)\,\mathbf{i} - U(y^2/H^2)\,\mathbf{j}$, where U and H are constants. Compute the circulation, $\Gamma = \oint_C \mathbf{u} \cdot d\mathbf{s}$, on the rectangular contour shown. Verify that your result is consistent with Equation (4.33).

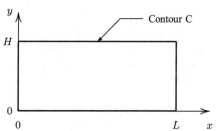

Problems 4.38, 4.39, 4.40

4.39 For *stagnation-point flow*, the velocity is given by $\mathbf{u} = Ax\,\mathbf{i} - Ay\,\mathbf{j}$, where A is a constant. Compute the circulation, $\Gamma = \oint_C \mathbf{u} \cdot d\mathbf{s}$, on the rectangular contour shown. Verify that your result is consistent with Equation (4.33).

4.40 Consider the two-dimensional velocity field given by

$$\mathbf{u} = U\frac{xy^2}{H^3}\mathbf{i} - U\frac{x^2y}{H^3}\mathbf{j}$$

where U and H are constants of dimensions length/time and length, respectively. Compute the circulation, $\Gamma = \oint_C \mathbf{u} \cdot d\mathbf{s}$, on the rectangular contour shown. Verify that your result is consistent with Equation (4.33).

4.41 Consider the flowfield whose velocity is given by

$$\mathbf{u} = \frac{U}{L^2}\left(\frac{1}{2}x^2\,\mathbf{i} - xy\,\mathbf{j}\right)$$

where U and L are constant velocity and length scales, respectively. Derive an equation defining the streamlines for this flow. Sketch a few streamlines for the right half plane, i.e., for $x > 0$. Indicate flow direction on the streamlines.

4.42 Suppose the velocity components for a two-dimensional, incompressible flowfield are given by

$$u = f(y) \quad \text{and} \quad v = g(x)$$

where $f(y)$ and $g(x)$ are arbitrary functions.

(a) What must these two functions be in order to have irrotational flow?

(b) Determine an equation for the streamlines, using your result of Part (a).

4.43 The velocity vector for an incompressible flowfield is

$$\mathbf{u} = \frac{Cx}{x^2 + y^2}\,\mathbf{i} + \frac{Cy}{x^2 + y^2}\,\mathbf{j}$$

where C is a constant of dimensions L^2/T. Derive an equation defining the streamlines for this flow. Sketch a few streamlines for the entire xy plane. Indicate flow direction on the streamlines for $C > 0$.

4.44 The velocity vector for an incompressible flowfield is

$$\mathbf{u} = \frac{Cy}{x^2 + y^2}\,\mathbf{i} - \frac{Cx}{x^2 + y^2}\,\mathbf{j}$$

where C is a constant of dimensions L^2/T. Derive an equation defining the streamlines for this flow. Sketch a few streamlines for the entire xy plane. Indicate flow direction on the streamlines for $C > 0$.

4.45 Consider the flowfield whose velocity is given by $\mathbf{u} = Cy\,\mathbf{i} + Cx\,\mathbf{j}$, where C is a constant of dimensions $1/T$. Derive an equation defining the streamlines for this flow. Sketch a few streamlines for the region $y > |x|$. Indicate flow direction on the streamlines for $C > 0$.

4.46 Consider the flowfield whose velocity is given by $\mathbf{u} = Cx\,\mathbf{i} - Cy\,\mathbf{j}$, where C is a constant of dimensions $1/T$. Derive an equation defining the streamlines for this flow. Sketch a few streamlines for the upper half plane, i.e., for $y > 0$. Indicate flow direction on the streamlines for $C > 0$.

4.47 The velocity for ideal flow past a cylinder is

$$\mathbf{u} = U\left[1 - \frac{R^2}{r^2}\right]\cos\theta\,\mathbf{e}_r - U\left[1 + \frac{R^2}{r^2}\right]\sin\theta\,\mathbf{e}_\theta$$

where U is freestream velocity and R is cylinder radius. The differential equation (in two dimensions) for streamlines in cylindrical coordinates is

$$\frac{dr}{u_r} = \frac{r\,d\theta}{u_\theta}$$

Derive an equation defining the streamlines for this flow. **HINT:** Use the fact that expanding in partial fractions,

$$\frac{1}{(r^2 - R^2)\,r} = \frac{1}{2R^2(r - R)} + \frac{1}{2R^2(r + R)} - \frac{1}{R^2 r}$$

4.48 Consider the flowfield whose velocity is given by

$$\mathbf{u} = \frac{\mathcal{D}\cos\theta}{r^2}\,\mathbf{e}_r + \frac{\mathcal{D}\sin\theta}{r^2}\,\mathbf{e}_\theta$$

where $\mathcal{D}$ is a constant. The differential equation (in two dimensions) for streamlines in cylindrical coordinates is

$$\frac{dr}{u_r} = \frac{r\,d\theta}{u_\theta}$$

Derive an equation defining the streamlines for this flow. Sketch a few streamlines. Indicate flow direction on the streamlines for $\mathcal{D} > 0$. **HINT:** After developing the equation for the streamlines, convert to Cartesian coordinates.

4.49 Derive an equation defining the streamlines for the incompressible flow whose velocity field is

$$\mathbf{u} = \frac{2K \cos \theta \sin \theta}{r^2} \, \mathbf{e}_r - \frac{K \cos^2 \theta}{r^2} \, \mathbf{e}_\theta$$

All quantities are dimensionless and K is a constant.

4.50 Consider the unsteady flow with velocity given by $\mathbf{u} = U\,\mathbf{i} + U \sin ft\,\mathbf{j}$, where U and f are constant velocity and frequency, respectively.

  (a) Derive an equation for the instantaneous streamlines. Sketch a few streamlines for $ft = 0$, $\pi/2$, π and $3\pi/2$.

  (b) Solve for the pathline of a fluid particle passing through $x = y = 0$ at $t = 0$. Eliminate t from your answer and sketch the pathline.

4.51 Consider the unsteady flow with velocity given by $\mathbf{u} = U\,\mathbf{i} + U \cos[k(x - Ut)]\,\mathbf{j}$, where U and k are constant velocity and wave number, respectively.

  (a) Derive an equation for the instantaneous streamlines. Sketch a few streamlines for a fixed time.

  (b) Solve for the pathline of a fluid particle passing through $x = x_o$ and $y = 0$ at $t = 0$. Eliminate t from your answer and sketch the pathline.

4.52 For flow near the *stagnation point* of an accelerating body, the flow is given by $\mathbf{u} = \dot{F}(t)\,[x\,\mathbf{i} - y\,\mathbf{j}]$, where $\dot{F}(t)$ denotes dF/dt.

  (a) Derive an equation for the instantaneous streamlines.

  (b) Solve for the pathline of a fluid particle passing through $x = x_o$ and $y = y_o$ at $t = 0$. Assume that $F(0) = 0$ and eliminate $F(t)$ from your answer.

  (c) Compute the cross product of $\mathbf{u}$ and $\partial \mathbf{u}/\partial t$. Are your answers of Parts (a) and (b) consistent with this result?

4.53 For *Couette flow*, the velocity is $u(y) = Uy/h$, where U is the velocity of the upper wall and h is channel height. Compute the mass flux, $\dot{m}$, momentum flux, $\dot{\mathbf{P}}$, and kinetic energy flux, $\dot{K}$, through a channel cross section. Assume the fluid density is ρ and the channel has width $20h$ (out of the page).

4.54 A good approximation for the turbulent-flow velocity distribution in a horizontal *open channel* of height h is $u(y) = U_{max}(y/h)^{1/n}$, where U_{max} is maximum velocity, y is normal distance and $6 \le n \le 8$.

  (a) Assuming constant width $8h$ (out of the page) and fluid density ρ, compute the mass-flow rate as a function of U_{max}, h and n.

  (b) Compute the ratio of mass flux for $n = 7$ to that for Couette flow, which is the limiting case $n = 1$.

4.55 The velocity vector for *stagnation point* flow is $\mathbf{u} = Ar \cos 2\theta\,\mathbf{e}_r - Ar \sin 2\theta\,\mathbf{e}_\theta$, where A is a positive constant. For a fluid of density ρ, compute the mass flux, $\dot{m}$, (per unit width out of the page) across the quarter-circle arc of radius R in the first quadrant.

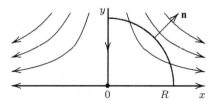

Problem 4.55

4.56 The velocity vector for *stagnation point* flow is $\mathbf{u} = Ax\,\mathbf{i} - Ay\,\mathbf{j}$, where A is a positive constant. For fluid density ρ, compute the mass flux, $\dot{m}$, and momentum flux, $\dot{\mathbf{P}}$, per unit width (out of the page) across Plane 1 and Plane 2.

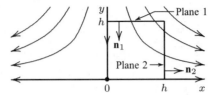

Problem 4.56

4.57 Consider a flow for which the velocity vector is $\mathbf{u} = (U/L)(x\,\mathbf{i} + y\,\mathbf{j} - 2z\,\mathbf{k})$, where U and L are constant velocity and length scales, respectively. Compute the mass-flow rate across a plane of area A that is perpendicular to the y axis and lies at $y = 2L$. The fluid density is ρ.

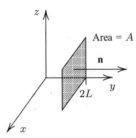

Problem 4.57

4.58 An incompressible flow has a velocity vector given by $\mathbf{u} = Uy^2 z/h^3\,\mathbf{j} - Uyz^2/h^3\,\mathbf{k}$, where U and h are constant velocity and length scales, respectively. Compute the mass flux, $\dot{m}$, and momentum flux, $\dot{\mathbf{P}}$, across the square planar area of side h shown, which is parallel to the xz plane at $y = h$.

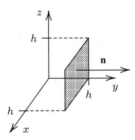

Problems 4.58, 4.59, 4.60

4.59 An incompressible flow has a velocity vector given by $\mathbf{u} = Uxz/h^2\,\mathbf{i} - Uyz/h^2\,\mathbf{j} + Uxy/h^2\,\mathbf{k}$, where U and h are constant reference velocity and length scales, respectively. For a fluid of density ρ, compute the z-momentum flux, $\dot{P}_z$, across the square planar area of side h shown, which is parallel to the xz plane at $y = h$.

4.60 An incompressible flow has a velocity vector given by $\mathbf{u} = U(1 + x/h)\,\mathbf{i} - Uy/h\,\mathbf{j} - U\,\mathbf{k}$, where U and h are constant reference velocity and length scales, respectively. For a fluid of density ρ, compute the x-momentum flux, $\dot{P}_x$, across the square planar area of side h shown, which is parallel to the xz plane at $y = h$.

4.61 A group of people is leaving a room in single file through a 0.8 m by 2.0 m doorway. Allowing for space between people, each person occupies 1 m³ as he or she approaches the doorway. The line of people is moving at a rate of 6.25 cm/sec. Using the Reynolds Transport Theorem (with number density replacing mass density), determine the rate of change of the number of people in the room, dN_R/dt.

4.62 Students are running out of the classroom after their fluid mechanics exam at a rate of 4 students/sec. Including space between students, each occupies 40 ft^3. Using the Reynolds Transport Theorem (with number density replacing mass density), determine the speed of the students as they leave the room. The door is $7\frac{1}{2}$ ft by $3\frac{1}{2}$ ft.

4.63 Helium-filled balloons are escaping through a hole of area A in the top of a circus tent. The number of balloons per unit volume just below the hole is n, and their velocity as they escape is U. Because of ragged edges at the hole, some of the balloons burst and fall back into the tent at a rate of $\dot{N}_{pop}$. Using the Reynolds Transport Theorem (with number density replacing mass density), determine the rate of change of the number of inflated balloons in the circus tent, dN/dt.

4.64 Spectre thugs, pursued by Agent 007, are vacating Spectre headquarters. The number of thugs per unit volume near the exit is n, the exit area is A and the thugs are moving at a velocity v as they exit. Bond is terminating thugs within Spectre headquarters at a rate of $\dot{N}_{007}$. Also, CIA Agent Felix Leiter is arresting half of those who escape from 007. Using the Reynolds Transport Theorem (with number density replacing mass density), determine the rate of change of the number of live thugs in Spectre headquarters, dN/dt.

4.65 You have just arrived in Las Vegas at your favorite hotel. Once inside, you enter the general check-in line. The archway area leading to the front desk is A_g, and each person occupies a volume V, including space between people. The line moves at an average velocity U_g. While waiting in line, you observe that a new line opens up for those with casino cards. As they learn of the new line, people with casino cards migrate from the general check-in line to the casino-card line. You also observe that someone has just won a huge jackpot so that no new people are entering the check-in line (they're all watching the gleeful winner). The archway area leading to this line is A_c, and the line moves at an average velocity U_c. Using the Reynolds Transport Theorem (with number density replacing mass density), derive an equation for the rate of change of people in the two lines. Use two separate control volumes, one for each line. The numbers of people in the general and casino-card lines are N_g and N_c, respectively.

4.66 A truck is moving down the highway at a constant speed U. Several kegs of nails near the back of the truck are damaged because the driver backed into a guard rail on the loading dock. As a result, nails are dropping onto the highway. The density of the nails is ρ (nails per ft^3), the combined area of the holes in the kegs through which nails are falling is A and they are ejected onto the highway at speed $\frac{1}{5}U$ relative to the truck. Use the Reynolds Transport Theorem to determine the rate of change of the number of nails remaining in the truck, dN_{truck}/dt.

4.67 For any scalar function ϕ, a useful vector identity is $\nabla \times \nabla\phi = \mathbf{0}$. By direct substitution, verify this identity in Cartesian coordinates. Do your work in three dimensions.

4.68 Verify, in three dimensions, that the divergence of the vorticity vector is zero.

4.69 The two-dimensional *Laplacian* of the velocity vector, $\mathbf{u}$, is defined by

$$\nabla^2\mathbf{u} = \frac{\partial^2\mathbf{u}}{\partial x^2} + \frac{\partial^2\mathbf{u}}{\partial y^2}$$

By direct substitution, verify the following identity in Cartesian coordinates for two-dimensional flows.

$$\nabla^2\mathbf{u} = \nabla(\nabla \cdot \mathbf{u}) - \nabla \times (\nabla \times \mathbf{u})$$

4.70 A useful vector identity for theoretical fluid mechanics is

$$\mathbf{u} \cdot \nabla\mathbf{u} = \nabla\left(\frac{1}{2}\mathbf{u} \cdot \mathbf{u}\right) - \mathbf{u} \times (\nabla \times \mathbf{u})$$

By direct substitution, verify this identity in Cartesian coordinates for two-dimensional flows.

Chapter 5

Conservation of Mass and Momentum: Integral Form

In this chapter, we will derive and apply the integral form of the basic conservation of mass and momentum laws for a fluid. Because of their simplicity and common occurrence in nature, our primary focus will be on incompressible flows. Mass- and momentum-conservation laws provide a sufficient number of equations to solve incompressible-flow problems for which temperature and heat transfer are of no interest. If thermal considerations are required or if viscous losses are important, we require an additional equation based on conservation of energy. We defer development of the energy-conservation law to Chapter 6.

We develop the mass- and momentum-conservation laws for a control volume by applying what we learned in Chapter 4, most notably the Reynolds Transport Theorem. Application of the conservation laws in their integral form constitutes a useful tool for analysis of complex fluid-flow problems that warrants detailed analysis and discussion. We present complete details on application of the integral forms by developing a solution procedure known as the *control-volume method*. The control-volume method permits us to implement the integral conservation laws for a carefully chosen **control volume** on which we can indirectly compute forces arising from complex pressure fields within.

5.1 Control-Volume Method

Before launching on our study of this powerful and useful method, it is worthwhile to pause and discuss the method's generality. As we will see, one way of using the **control-volume method** is to provide a *global view* as opposed to a *detailed view* of the fluid motion through a finite-sized volume. When used in this way, the method does not discern precise details of how fluid velocity, pressure, etc. vary throughout a flow. Rather, it provides integrated values such as the total mass flux across a given surface, the net pressure force on a solid boundary or the total force acting on a specified volume.

The control-volume method can thus be used as a consistency check of measured or computed properties. For example, if we measure or compute density and velocity at several points across the inlets to and outlets from a specified volume, we can compute the mass flux at the inlets and outlets by numerically integrating the experimental values. These integrated values can then be compared to results of a control-volume analysis.

Although it is a relatively new concept, the control-volume method has accelerated the development of fluid mechanics as a science and as a practical engineering discipline. As observed by Kline (1999):

> "A history of control-volume theory is given by Vincenti (1990). Without control volume theory fluid mechanics would be nearly impotent and our analysis of rocket engines would not get our astronauts into space and back. The differential forms of the theory trace back to Leonhard Euler in the mid-18th century, but it is still little known by chemists and physicists. The integral theory for finite volumes was only completed in the early 1950s by J. C. Hunsaker, Brandon Rightmire, Ascher Shapiro and J. H. Keenan at MIT."

In this chapter, for the sake of simplicity, the problems we will apply the control-volume method to will involve finite volumes. That is, we will confine our attention to a "global view." However, the method is not limited to such applications and, in fact, is used in modern fluid mechanics analysis to actually achieve a "detailed view." That is, we can observe more flow details by subdividing a finite control volume into a number of smaller volumes. In the limit of infinitesimal subdivisions, our solution will approach the continuum solution of the differential equations that govern the fluid motion. Consistency between the integral (global) and differential (detailed) forms will become even more evident when we derive the differential conservation laws in Chapter 9. We will deduce the differential equations from the very integral-conservation laws upon which the control-volume method is based.

The primary field in which we implement the control-volume method for a collection of small sub-volumes is called **Computational Fluid Dynamics (CFD)**. In order to replace the differential equations governing motion of a fluid by algebraic equations suitable for computer analysis, the standard starting point is to first establish a computational grid [cf. Thompson et al. (1998) or Knupp and Steinberg (1993)]. This involves dividing physical space into a collection of contiguous cells as illustrated in Figure 5.1. The cells are close enough to be considered infinitesimal on the overall scale of the problem. Flow properties are then defined at either the vertices or the centroid of each cell. To derive the algebraic equations that will be solved by a computer, the integral conservation equations are applied to each cell [cf. Anderson (1995)]. This is done to insure that properties are conserved on both a global and

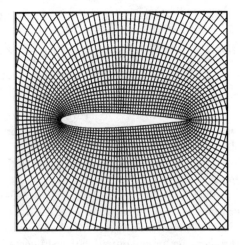

Figure 5.1: *Portion of a computational grid for an airfoil.*

a detailed level. This is so important for compressible flows that older numerical procedures that are not based on the exact conservation principles have been almost universally rejected by CFD researchers.

We begin this chapter by first deriving the conservation laws for mass and momentum. Then, we explore numerous examples that elucidate proper implementation of the control-volume method. Because of their relative simplicity, applications involving the mass-conservation principle are considered first. More complex problems that require combined use of mass and momentum conservation follow, including applications for which forces are computed indirectly in terms of mass and momentum flowing through the control volume. In Section 3.5, we learned that Bernoulli's equation can also be used to compute pressure forces, but only when the flow is inviscid, steady, incompressible, irrotational and subject to conservative body forces. By contrast, the momentum-conservation integral always applies.

5.2 Conservation of Mass

We derived the Reynolds Transport Theorem in Chapter 4 to bridge the gap between the Lagrangian and Eulerian descriptions. It is now possible to apply the familiar conservation laws of classical physics to a volume that contains different fluid particles at each instant. To derive an equation expressing conservation of mass, we appeal to the definition of a system.

Consider a general control volume, V, bounded by a closed surface S. Figure 5.2 illustrates such a volume, including a differential volume element, dV, a differential surface element, dS, and an **outer unit normal** vector, **n**. For the sake of generality, we assume the control volume is moving and denote the bounding-surface velocity by $\mathbf{u}^{cv}$. The fluid velocity is **u**. Since the total mass in the volume is given by

$$M = \iiint_V \rho \, dV \tag{5.1}$$

the appropriate extensive/intensive variables are $B = M$ and $\beta = 1$ [see Equation (4.50)]. By definition, a *system* always contains the same collection of fluid particles. Consequently, its

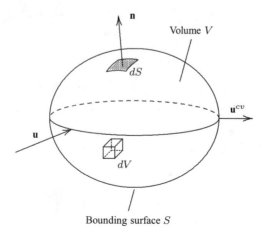

Figure 5.2: *A general control volume for mass conservation.*

mass is constant for all time. Thus, the rate of change of the system's mass is zero, which means

$$\frac{dM}{dt} = 0 \qquad (5.2)$$

Therefore, invoking the Reynolds Transport Theorem, we arrive at the integral form of the mass-conservation principle for a *control volume*.

$$\frac{d}{dt} \iiint_V \rho \, dV + \oiint_S \rho \, \mathbf{u}^{rel} \cdot \mathbf{n} \, dS = 0 \qquad (5.3)$$

where $\mathbf{u}^{rel}$ is the flow velocity relative to the control volume velocity on the bounding surface, viz.,

$$\mathbf{u}^{rel} \equiv \mathbf{u} - \mathbf{u}^{cv} \qquad (5.4)$$

The first term on the left-hand side of Equation (5.3) represents the *instantaneous rate of change of mass in the control volume*. The second term represents the *net flux of mass out of the control volume*.

5.3   Conservation of Momentum

For simplicity, we consider an inviscid, or perfect, fluid so that only pressure acts on any surface. We will address viscous effects briefly in Chapter 11. Letting $\mathbf{P}$ denote the momentum vector, the momentum of the control volume shown in Figure 5.3 is

$$\mathbf{P} = \iiint_V \rho \mathbf{u} \, dV \qquad (5.5)$$

so that our extensive variable is $\mathbf{P}$, while the intensive variable is $\mathbf{u}$. Now, we know from Newton's second law of motion applied to the system coincident with the control volume at an instant in time that

$$\frac{d\mathbf{P}}{dt} = \mathbf{F}_s + \mathbf{F}_b \qquad (5.6)$$

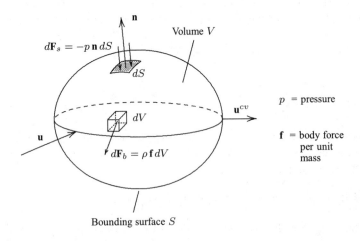

Figure 5.3: *A general control volume for momentum conservation.*

where $\mathbf{F}_s$ and $\mathbf{F}_b$ denote **surface force** and **body force** exerted by the surroundings on the system, respectively. We define a surface force as one that is transmitted across the surface bounding the system. By contrast, a body force acts at a distance, the most common examples being gravitational, electrical and magnetic forces.

Because we have confined our focus to a perfect fluid, the only surface force acting is the fluid pressure. Since $\mathbf{n}$ is an outer unit normal, the pressure force exerted by the surroundings on a differential surface element dS is $-p\,\mathbf{n}\,dS$. Hence, the net surface force due to pressure imposed by the surroundings on the system is

$$\mathbf{F}_s = - \oiint_S p\,\mathbf{n}\,dS \qquad (5.7)$$

As discussed in Chapter 3, this would be the buoyancy force acting on an object submerged or floating in a liquid. Turning to the body force, we introduce the **specific body force vector**, $\mathbf{f}$, whose dimensions are force per unit mass. The net body force on the system is given by the following volume integral.

$$\mathbf{F}_b = \iiint_V \rho\,\mathbf{f}\,dV \qquad (5.8)$$

As an example, for gravity we would say that $\mathbf{f} = \mathbf{g} = -g\mathbf{k}$, where $g = 32.174$ ft/sec^2 (9.807 m/sec^2) is the acceleration due to gravity. In this case, the force $\mathbf{F}_b$ would be the control volume's weight.

Thus, invoking the Reynolds Transport Theorem, we arrive at the conservation of momentum principle for a control volume, viz.,

$$\frac{d}{dt} \iiint_V \rho\,\mathbf{u}\,dV + \oiint_S \rho\,\mathbf{u}\,(\mathbf{u}^{rel} \cdot \mathbf{n})\,dS = - \oiint_S p\,\mathbf{n}\,dS + \iiint_V \rho\,\mathbf{f}\,dV \qquad (5.9)$$

The first term on the left-hand side of Equation (5.9) represents the *instantaneous rate of change of momentum in the control volume.* The second term represents the *net flux of momentum out of the control volume.* The two terms on the right-hand side are the *net pressure force* and *net body force* exerted by the surroundings on the control volume.

Note that, consistent with Newton's laws of motion, Equation (5.9) is a conservation law for **absolute momentum**, $\rho\,\mathbf{u}$, not for momentum relative to the control volume. Hence, $\mathbf{u}^{rel}$ appears only in the surface-flux integral (the second term on the left-hand side of Equation (5.9). We will discuss the issue of absolute momentum in greater detail when we focus on accelerating control volumes in Section 5.7.

5.4 Preliminaries

Before proceeding to applications of the control-volume method, it is worthwhile at this point to provide a framework and some useful preliminary information. To do so, this section discusses the following three key issues:

- Guidelines for selecting control-volume boundaries;

- A useful theorem that simplifies handling the pressure integral;

- A quick test of the method for a problem we solved earlier, viz., the fluid-statics problem.

5.4.1 Overview of the Method

The basic objective of the control-volume method is to use the conservation laws in their integral form to analyze global properties of a given fluid-flow problem. Because the integral laws involve volume integrals and integrals over the surface bounding the volume, our first requirement is to select a suitable volume for evaluating the integrals. There are two primary principles that influence selection of the control volume.

- Portions of the control surface across which fluid flows should have their unit normal as nearly as possible parallel to the fluid velocity vector, $\mathbf{u}^{rel}$.

- To the greatest extent possible, the pressure should be either known or superfluous on portions of the control surface that are not coincident with a solid boundary.

Although these principles are helpful as a starting point in establishing control-volume boundaries, they can be mutually exclusive. Hence, a bit of thought should always precede selection of the control volume.

Figure 5.4 illustrates application of these principles for simple geometries. Flow A has fluid leaving at an angle to the horizontal. Using a rectangular control volume would introduce a sinusoidal function in evaluating the flow rate, which is proportional to ($\mathbf{u}^{rel} \cdot \mathbf{n}$). However, selecting a control volume identical to the shape of the object obviates the need to introduce trigonometric functions in the flow rate. Such functions will be needed to properly address momentum conservation, but the overall computations are simplified by using the non-rectangular control volume. In this case, there is no obvious reason to consider the pressure, so only the first principle noted above affects the choice of the control-volume boundaries.

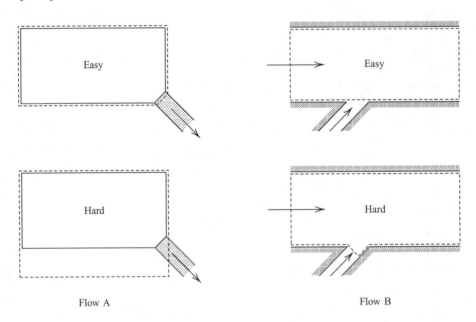

Figure 5.4: *Selecting a control volume (denoted by dashed lines).*

By contrast, Flow B is an example in which the two principles cannot both be satisfied, and we are forced to choose one over the other. The channel shown has fluid injected into the main channel from an inlet below in which the pressure is unknown. If we are interested only in mass and horizontal-momentum conservation, a rectangular control volume is most appropriate. That is, the unit normal at the inlet is vertical so that the horizontal component of the pressure integral vanishes, regardless of the pressure in the inlet. By contrast, the alternative control volume shown, which is consistent with having the boundaries everywhere perpendicular to $\mathbf{u}^{rel}$, would require a rather detailed computation to show that the pressure integral is zero (see Example 5.1).

There are several additional steps that should be taken to properly implement the control-volume method. In general, once the control volume is selected, the sequence of computations should be as follows.

1. **Apply Conservation of Mass:** This is the simplest of the conservation principles as it is a scalar equation and involves, at most, one volume integral and one closed-surface integral. In addition to establishing one equation relating flow properties, this step yields $(\mathbf{u}^{rel} \cdot \mathbf{n})$ at all points on the control-volume boundary, including sign. This factor appears in all of the conservation principles and can be re-used in subsequent steps.

2. **Apply Conservation of Momentum:** Use as many components of this vector equation as are appropriate for the problem at hand. Most importantly, when you evaluate the momentum-flux integral, viz.,

$$\oiint_S \rho\, \mathbf{u}\, (\mathbf{u}^{rel} \cdot \mathbf{n})\, dS$$

   treat the integrand as the product of two separate factors. The first is the *quantity being carried across the control surface*, i.e., the momentum per unit volume, $\rho\,\mathbf{u}$. The second is the *rate at which it is being carried across the surface*, $(\mathbf{u}^{rel} \cdot \mathbf{n})\, dS$. If you make this distinction, you will minimize the number of sign errors you make.

3. **Count Equations and Unknowns:** Check to see if the number of unknowns and the number of equations you have at this point are equal to determine if you have a **closed** system of equations. By definition, a system of equations is closed if the number of equations matches the number of unknowns.

4. **Appeal to Energy Conservation or Bernoulli's Equation if Necessary:** If there are more unknowns than equations, in the most general case, you must use the energy-conservation principle (see Chapter 6) to determine an additional equation. However, if the flow is inviscid, steady, incompressible, irrotational and involves conservative body forces, then Bernoulli's equation can be used—it is an expression of the conservation of mechanical energy.

5. **Check the Physics:** Throughout the process, pause and use physical reasoning to determine the validity of signs and limiting cases. Also check for dimensional consistency of your results.

To reinforce the importance of the final step, note that the control-volume method provides an excellent bookkeeping tool for analyzing fluid-flow problems. It provides a methodology for keeping track of mass and momentum fluxes and forces on a control volume. However, it is a means to an end, not the end itself. That is, it is a mathematical tool whose purpose is to aid in understanding the physics of the problem at hand. *You should not lose sight of the fact that understanding the physics through application of the conservation principles to the problem at hand is the desired end of the control-volume method.*

Example 5.1 *Consider the control volume indicated by the dashed lines in the inclined channel whose width (out of the page) is $3H$. Compute the area, A, and outer unit normal, $\mathbf{n}$, for all three faces shown. Verify from your results that, if the pressure, p, is constant throughout the control volume, then the net pressure force on the control volume is zero, viz.,*

$$\oiint_S p\,\mathbf{n}\,dS = p\,\mathbf{n}_1 A_1 + p\,\mathbf{n}_2 A_2 + p\,\mathbf{n}_3 A_3 = \mathbf{0}$$

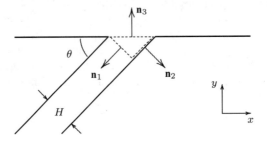

Solution. The areas of the three surfaces of interest are as follows.

$$
\begin{aligned}
A_1 &= H \cdot 3H &&= 3H^2 \\
A_2 &= H \cot\theta \cdot 3H &&= 3H^2 \cot\theta \\
A_3 &= H \csc\theta \cdot 3H &&= 3H^2 \csc\theta
\end{aligned}
$$

The outer unit normals are

$$
\begin{aligned}
\mathbf{n}_1 &= -\mathbf{i}\cos\theta - \mathbf{j}\sin\theta \\
\mathbf{n}_2 &= \mathbf{i}\sin\theta - \mathbf{j}\cos\theta \\
\mathbf{n}_3 &= \mathbf{j}
\end{aligned}
$$

Hence, the net pressure force is

$$
\begin{aligned}
\sum_{i=1}^{3} p\,\mathbf{n}_i A_i &= p\left[-3H^2\left(\mathbf{i}\cos\theta + \mathbf{j}\sin\theta\right) + 3H^2\cot\theta\left(\mathbf{i}\sin\theta - \mathbf{j}\cos\theta\right) + 3H^2\csc\theta\,\mathbf{j}\right] \\
&= -3H^2 p\left[\cos\theta\,\mathbf{i} + \sin\theta\,\mathbf{j} - \cos\theta\,\mathbf{i} + \cos^2\theta\csc\theta\,\mathbf{j} - \csc\theta\,\mathbf{j}\right] \\
&= -3H^2 p\left[\left(\cos\theta - \cos\theta\right)\mathbf{i} + \left(\sin\theta + \cos^2\theta\csc\theta - \csc\theta\right)\mathbf{j}\right] \\
&= -3H^2 p\,\mathbf{j}\left[\sin\theta + \cos^2\theta\csc\theta - \csc\theta\right] \\
&= -3H^2 p\,\mathbf{j}\left[\sin\theta + \left(\cos^2\theta - 1\right)\csc\theta\right] \\
&= -3H^2 p\,\mathbf{j}\left[\sin\theta - \sin^2\theta\csc\theta\right] \\
&= \mathbf{0}
\end{aligned}
$$

5.4.2 Useful Control-Volume Theorem

In evaluating the force on a control volume caused by the pressure field, we can replace p by $(p - p_a)$ where p_a is a constant. In applications, we will often select p_a as atmospheric pressure.

The proof is simple and requires use of the result we derived in Section 3.1, viz., for an infinitesimal volume element of volume ΔV [see Equation (3.8)]

$$\oiint_S p\,\mathbf{n}\,dS \approx \nabla p\,\Delta V \quad \text{for} \quad \Delta V \to 0 \tag{5.10}$$

Clearly, we can geometrically construct an arbitrary, finite-sized volume from a collection of infinitesimal sub-volumes. Now, the pressure on two adjacent sub-volume faces will cancel exactly since the pressure is the same while the outer unit normals are in opposite directions (see Figure 5.5). Thus, the overall pressure force on any collection of sub-volumes will add up to the pressure on the *bounding surface* of the finite-sized volume that they comprise. Consequently, summing the contributions from all of the infinitesimal cells, we conclude that Equation (5.10) holds for an arbitrary volume and we can write it in its general form as follows.

$$\oiint_S p\,\mathbf{n}\,dS = \sum_i (\nabla p)_i \,\Delta V_i \to \iiint_V \nabla p\,dV \tag{5.11}$$

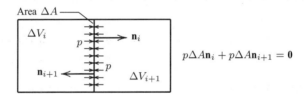

Figure 5.5: *Internal pressure force on adjacent faces of infinitesimal sub-volume cells cancel exactly because the pressure is equal while the unit normals are in opposite directions.*

To prove that we can replace p by $(p - p_a)$, we observe that integration is a linear operation, wherefore

$$\oiint_S (p - p_a)\,\mathbf{n}\,dS = \oiint_S p\,\mathbf{n}\,dS - \oiint_S p_a\,\mathbf{n}\,dS \tag{5.12}$$

For the integral of p_a, we have

$$\oiint_S p_a\,\mathbf{n}\,dS = \iiint_V \nabla p_a\,dV \tag{5.13}$$

But, since p_a is a constant, necessarily $\nabla p_a = \mathbf{0}$ so that this integral vanishes. Therefore,

$$\oiint_S (p - p_a)\,\mathbf{n}\,dS = \oiint_S p\,\mathbf{n}\,dS \tag{5.14}$$

This should come as no great surprise as it is pressure differences that cause fluid motion. Simply pumping up the ambient value has no effect on a closed volume. Nevertheless, failing to reference pressure to the ambient value, p_a, is a common source of error in the control-volume method, and is thus worthy of mention.

5.4.3   Fluid Statics Revisited

In this section we will employ a useful engineering technique to begin learning how to use the conservation principles in their integral form. Specifically, we will apply the conservation forms in a special limiting case for which we already know the answer. In particular, we will use the integral forms to determine the forces in a non-moving, or static, fluid. We can compare our results to those obtained in Chapter 3.

Consider the arc AB illustrated in Figure 5.6. We determined the force on the arc, $\mathbf{F}$, in Section 3.8. Our computations showed that the vertical force is equal to the weight of the column of fluid above the arc, while the horizontal force is equal to the force on arc AB's projection on a vertical plane. The task at hand is to show that the control-volume method yields the same result for the components of the force.

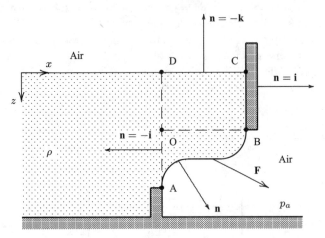

Figure 5.6: *Force on a curved surface for a stationary fluid. OA is the projection of the arc AB on a vertical plane.*

To obtain the force on arc AB, we appeal to the momentum-conservation principle with gravity as the body force ($\mathbf{f} = \mathbf{g} = g\mathbf{k}$):

$$\frac{d}{dt}\iiint_V \rho\mathbf{u}\,dV + \oiint_S \rho\mathbf{u}(\mathbf{u}^{rel}\cdot\mathbf{n})\,dS = -\oiint_S p\,\mathbf{n}\,dS + \iiint_V \rho\mathbf{g}\,dV \qquad (5.15)$$

For fluid statics, the velocity vector vanishes. Substituting $\mathbf{u} = \mathbf{u}^{rel} = \mathbf{0}$ into Equation (5.15) yields a balance between the net pressure force on the control volume and its weight, i.e.,

$$\oiint_S p\,\mathbf{n}\,dS = \iiint_V \rho\mathbf{g}\,dV \qquad (5.16)$$

Also, with zero velocity, Bernoulli's equation simplifies to

$$p - \rho g z = \text{constant} = p_a \qquad (5.17)$$

which, with a sign change to account for z being positive downward, is simply the **hydrostatic relation**, Equation (3.17), in which z is positive upward.

Now, as shown in the preceding section, we can replace p by $(p - p_a)$ in the pressure integral appearing in Equation (5.16). Defining our control volume as the region bounded by the closed contour OABCD, the closed-surface integral is the sum of integrals on arcs

AB, BC, CD, DO and OA. Since our geometry is actually three dimensional, we also have integrals on the end planes whose unit normals point into and out of the page. For the sake of brevity, we omit these contributions in the following equation. We will explain why they sum to zero below. Thus, proceeding in the counterclockwise direction about the surface bounding the chosen control volume, Equation (5.16) becomes:

$$\underbrace{\iint_{S_{AB}} (p - p_a)\,\mathbf{n}\,dS}_{\text{Arc AB}} \quad + \quad \underbrace{\iint_{S_{BC}} (p - p_a)\,\mathbf{i}\,dS}_{\text{Wall above Arc AB}}$$

$$+ \quad \underbrace{\iint_{S_{CD}} (p - p_a)(-\mathbf{k})\,dS}_{\text{Free Surface}} \quad + \quad \underbrace{\iint_{S_{DO}} (p - p_a)(-\mathbf{i})\,dS}_{\text{Projection of Wall on Vertical Plane}}$$

$$+ \quad \underbrace{\iint_{S_{OA}} (p - p_a)(-\mathbf{i})\,dS}_{\text{Projection of Arc AB on Vertical Plane}} \quad = \quad \underbrace{\iiint_{V_{OABCD}} \rho g\,\mathbf{k}\,dV}_{\text{Weight of Fluid above Arc AB}} \tag{5.18}$$

Note that when we evaluate a closed-surface integral, it is a sum of integrals with integration always in the positive direction. Thus, for example, the z integration on surfaces BC and DO begins at the free surface so that the lower limit of integration in both integrals is 0. *The sign of the integrals is determined solely by the unit normal.*

Hence, the sum of the integrals on arcs BC and DO cancels. This is true because p depends only upon z, and both arcs include the same range of values of z. Thus, the forces on both arcs have the same magnitude, but act in opposite directions. The sum of the forces on the end planes vanishes for precisely the same reason, so that their omission in Equation (5.18) is justified.

Also, because $p = p_a$ on the free surface (arc CD), the integral is zero. Finally, the force on the arc is the difference between the integral along arc AB of the hydrostatic pressure, p, (from above) and the integral along arc AB of the atmospheric pressure, p_a, (from below) i.e.,

$$\mathbf{F} = \iint_{S_{AB}} (p - p_a)\,\mathbf{n}\,dS \tag{5.19}$$

Therefore, since the integrals on S_{BC}, S_{CD} and S_{DO} sum to zero, Equation (5.18) simplifies to:

$$\mathbf{F} = \mathbf{i} \iint_{S_{OA}} (p - p_a)\,dS + \mathbf{k} \iiint_{V_{OABCD}} \rho g\,dV \tag{5.20}$$

In words, we have shown that, consistent with our analysis of fluid statics in Chapter 3, the vertical force on arc AB is equal to the weight of the column of fluid above the arc, while the horizontal force is equal to the force on arc AB's projection on a vertical plane.

5.5 Mass Conservation Applications

Now we turn to problems in which fluid is moving. The first of our applications to moving fluids will be for relatively simple problems that can be analyzed using only the mass-conservation integral. In this section, we discuss three such problems. Each brings out important aspects of the control-volume method and the integral law for mass conservation.

5.5.1   Steady Flow in a Pipe

Consider steady flow in a pipe of varying cross section as shown in Figure 5.7. If the pipe is curved as illustrated in the figure, the flow within is very complicated. In addition to the primary motion along the axis of the pipe, cross-sectional regions of swirling flow appear. The swirling patterns are known as **secondary motions**, and were first observed in curved pipe flow (with constant cross-sectional area) by Prandtl.

The precise details aren't really important for our present purposes. We simply mention secondary motions to stress that an innocent looking geometry can have a very complicated flow—and we can tackle it fearlessly with the control-volume method! Theoretical computations that predict details of secondary motions are very difficult to perform and are often very inaccurate. Nevertheless, the control-volume method is very easy to apply to this flow and results obtained are exact, albeit lacking subtle details.

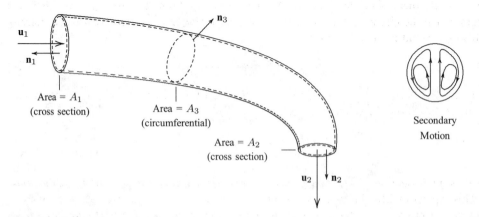

Figure 5.7: *Flow in a pipe; the secondary-motion offset shows cross-sectional streamlines in flow through a curved pipe first observed by Prandtl.*

We would like to develop a relation between flow properties at the inlet (subscript 1) and at the outlet (subscript 2). To begin our analysis, we must first select a suitable control volume. The boundaries of the control volume are indicated with dashed lines in Figure 5.7. The control volume is coincident with the pipe and is therefore motionless. In general, the mass conservation law is

$$\frac{d}{dt} \iiint_V \rho\, dV + \oiint_S \rho\, \mathbf{u}^{rel} \cdot \mathbf{n}\, dS = 0 \tag{5.21}$$

Because the flow is steady, the first integral vanishes. Also, because our control volume is stationary, $\mathbf{u}^{rel} = \mathbf{u}$. Thus, all that remains of Equation (5.21) is

$$\oiint_S \rho\, \mathbf{u} \cdot \mathbf{n}\, dS = 0 \tag{5.22}$$

In words, this equation says that *the net flux of fluid out of the control volume is zero*. It is a *net flux* because we are evaluating a closed-surface integral. It is a flux *out of the control volume* because we are using an *outer unit normal*.

To evaluate the closed-surface integral, we note that the control volume is bounded by three distinct surfaces as follows.

1. A cross section at the inlet of area A_1 and unit normal $\mathbf{n}_1$

2. A cross section at the outlet of area A_2 and unit normal $\mathbf{n}_2$

3. The circumferential area coincident with the pipe between inlet and outlet of area A_3 and unit normal $\mathbf{n}_3$

Thus, the closed-surface integral expands to the following.

$$\iint_{A_1} \rho_1 \mathbf{u}_1 \cdot \mathbf{n}_1 \, dA + \iint_{A_2} \rho_2 \mathbf{u}_2 \cdot \mathbf{n}_2 \, dA + \iint_{A_3} \rho_3 \mathbf{u}_3 \cdot \mathbf{n}_3 \, dA = 0 \qquad (5.23)$$

Now, because there is no flow through the pipe walls, we know that $\mathbf{u}_3 \cdot \mathbf{n}_3 = 0$. Thus, the integral along the circumference (i.e., on A_3) is exactly zero. Also, because we are working with an outer unit normal, necessarily

$$\mathbf{u}_1 \cdot \mathbf{n}_1 = -u_1 \quad \text{and} \quad \mathbf{u}_2 \cdot \mathbf{n}_2 = u_2 \qquad (5.24)$$

where u_1 and u_2 denote the magnitudes of the velocity vectors $\mathbf{u}_1$ and $\mathbf{u}_2$, respectively. Therefore, Equation (5.23) simplifies to:

$$\iint_{A_1} \rho_1 u_1 \, dA = \iint_{A_2} \rho_2 u_2 \, dA \qquad (5.25)$$

In words, this equation says *the mass flux into the control volume equals the mass flux out*.

This expression illuminates several points. We can use it, for example, to verify a numerical solution that provides the actual values of density and velocity at the inlet and outlet (and, presumably, on all cross sections). By verifying that Equation (5.25) is satisfied, we check that mass is conserved in a global sense. If not, the numerical solution is defective. Similarly, we can determine the accuracy and error bounds for measured density and velocity profiles.

In many of the applications to follow, we will simplify the analysis by treating flow properties as though they are uniform on cross sections. For a real fluid this is never true as the no-slip boundary condition (cf. Section 1.9) always gives rise to non-uniform velocity profiles. What we will be doing is working with averaged flow properties. For example, the average mass flux per unit area, $\overline{\rho u}$, on a given cross section is

$$\overline{\rho u} \equiv \frac{1}{A} \iint_A \rho u \, dA \qquad (5.26)$$

Thus, in terms of averaged mass flux per unit area, Equation (5.25) simplifies to

$$\overline{\rho_1 u_1} A_1 = \overline{\rho_2 u_2} A_2 \quad \Longrightarrow \quad \overline{\rho u} A = \text{constant} \qquad (5.27)$$

Since A_2 can be any cross section in the pipe, we conclude that the mass-flow rate, $\dot{m} = \overline{\rho u} A$, is constant through all cross sections.

While this example is extremely simple, it provides a clear glimpse into a key aspect of the control-volume method. Evaluating the net mass-flux integral, i.e., the closed-surface integral appearing in Equation (5.21), requires us to account for *all parts* of the surface bounding the control volume we have selected. To enunciate the importance of accounting for the entire bounding surface—i.e., being careful to do our "bookkeeping" correctly—the development above includes the circumferential area coincident with the pipe walls. We find that the integral on this surface is exactly zero, of course, because there can be no flow through the pipe walls. The following example illustrates how the method is impacted when fluid can pass through the pipe walls.

Example 5.2 *A wind tunnel has a porous wall in its test section to help reduce viscous effects. The diameter of the cylindrical test section is D, its length is L and the inlet velocity is U_i. The test-section surface has N holes per unit area of diameter d, and the suction velocity is v_w. Assuming the flow is steady and incompressible with density ρ, determine the outlet velocity, U_o.*

Solution. For a stationary control volume coincident with the wind-tunnel test-section boundaries, the steady-flow mass-conservation principle simplifies to

$$\oiint_S \rho\, \mathbf{u} \cdot \mathbf{n}\, dS = 0$$

where the fact that the control volume isn't moving means $\mathbf{u}^{rel} = \mathbf{u}$. Since there is flow through the tunnel walls as well as at the inlet and outlet, we have

$$\oiint_S \rho\, \mathbf{u} \cdot \mathbf{n}\, dS \;=\; \underbrace{\rho\left(-U_i \frac{\pi}{4} D^2\right)}_{\text{Inlet}} + \underbrace{\rho\left(U_o \frac{\pi}{4} D^2\right)}_{\text{Outlet}} + \underbrace{\rho\left(v_w \pi D L N \frac{\pi}{4} d^2\right)}_{\text{Suction holes}}$$

$$=\; \frac{\pi}{4}\rho D^2 \left[-U_i + U_o + \pi D L N v_w \left(\frac{d}{D}\right)^2\right] = 0$$

Therefore, the outlet velocity is

$$U_o = U_i - \pi D L N v_w \left(\frac{d}{D}\right)^2$$

5.5.2 Sphere Falling in a Cylinder

Another problem that can be solved using only the mass-conservation law is a sphere of radius R that is falling in a closed cylinder filled with an incompressible fluid of density, ρ [Figure 5.8(a)]. We assume that the sphere falls axially, i.e., it remains centered within the cylinder, whose radius is, say, twice that of the sphere, $2R$. Finally, we observe that the sphere falls at velocity U. We would like to determine the average velocity of the surrounding fluid at the midsection of the sphere.

This problem differs from the application of the preceding section in an important way. Specifically, the fluid motion in this problem is unsteady, while the pipe-flow problem was postulated to be steady. We can use either of the following strategies for solving this problem.

1. If the sphere falls at a constant velocity, we can use a Galilean transformation to recast the problem as a steady-flow problem. This permits using a stationary control volume.

2. We can select a control volume that moves with the sphere. This method is more general as it applies even if the sphere's velocity varies with time.

Considering the constant-velocity case first, we use Strategy 1 and change coordinate frames with a Galilean transformation. We know from elementary physics that the mass- and momentum-conservation laws for distinct particles are invariant under a Galilean transformation, and there is no reason for this invariance to fail for a fluid. By definition, we perform a Galilean transformation by simply subtracting a constant velocity from the velocity at all points in the flow. For the case at hand, we subtract $-U\mathbf{k}$ from the flow velocity.

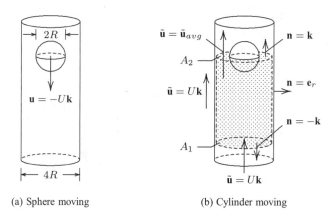

(a) Sphere moving   (b) Cylinder moving

Figure 5.8: *Sphere falling in a cylinder; the control volume is the shaded region.*

In the transformed frame, the sphere is at rest and the cylinder moves upward with velocity vector $\tilde{\mathbf{u}} = U\mathbf{k}$ as indicated in Figure 5.8(b). The flow far below the sphere approaches at a constant velocity $\tilde{\mathbf{u}} = U\mathbf{k}$ also. Because the cylinder wall is always vertical, the geometry for an *infinitely-long cylinder* is unchanged as time passes under this transformation. That is, unless we paint stripes on the cylinder walls, for example, an observer riding on the sphere cannot see any difference in the geometry as the cylinder passes. Even with such stripes, the fluid motion would experience no change, of course, so we clearly replace the original unsteady problem with an equivalent steady problem. For the transformed flowfield, the control volume is fixed in space (with the cylinder wall slipping by) so that $\mathbf{u}^{rel} = \tilde{\mathbf{u}}$.

The control volume is the shaded region bounded by the dashed lines in Figure 5.8(b). The upper boundary is the horizontal surface between the cylinder and the sphere and the hemispherical surface below the midsection. The lower boundary is a horizontal surface far below the sphere. The boundary is presumed to be sufficiently distant that the incoming flow is uniform, i.e., it is unaffected by the presence of the sphere. To close the surface bounding the control volume, we select a cylindrical surface coincident with the portion of the cylinder between the upper and lower control-volume boundaries. The outer unit normal, $\mathbf{n}$, is shown on various parts of the bounding surface. Note that, although this is not shown, in order to be an outer unit normal on the surface of the sphere, $\mathbf{n}$ must point *into* the sphere.

Because the flow is steady and the control volume is stationary in this reference frame, conservation of mass tells us that the net flux of fluid out of the control volume is zero, viz.,

$$\oiint_S \rho\, \tilde{\mathbf{u}} \cdot \mathbf{n}\, dS = 0 \tag{5.28}$$

where $\tilde{\mathbf{u}}$ denotes the Galilean-transformed velocity defined by

$$\tilde{\mathbf{u}} = \mathbf{u} - \mathbf{u}_{sphere} = \mathbf{u} + U\mathbf{k} \tag{5.29}$$

Noting that $\tilde{\mathbf{u}} \cdot \mathbf{n} = 0$ on the sphere and the cylinder, our mass-conservation integral assumes the following form:

$$\iint_{A_1} \rho\, U\mathbf{k} \cdot (-\mathbf{k})\, dA + \iint_{A_2} \rho\, \tilde{\mathbf{u}} \cdot \mathbf{k}\, dA = 0 \tag{5.30}$$

where A_1 is the area of the lower boundary, which is simply the area of the cylinder, i.e., $4\pi R^2$. The area A_2 is the part of the control-volume boundary between the sphere and the cylinder. It is the difference between the cylinder area and a circle of radius R, which is $3\pi R^2$. Thus, treating $\tilde{\mathbf{u}}$ as a constant on the upper control-volume boundary (which, as we saw in the preceding section, is the equivalent of using the average velocity), we can perform the indicated integrals. Thus, we have

$$-4\pi R^2 \rho\, U + 3\pi R^2 \rho\, \tilde{u}_{avg} = 0 \quad \Longrightarrow \quad \tilde{u}_{avg} = \frac{4}{3} U \tag{5.31}$$

We can pause at this point to check for physical consistency. That is, note that the first term on the left-hand side of Equation (5.31) has a minus sign and the second term has a plus sign. The first term corresponds to fluid entering the control volume, which is a negative flux out. Conversely, the second term corresponds to fluid leaving the control volume, which is a positive flux out. Thus, the signs make sense from a physical point of view.

Finally, in terms of the original coordinate frame in which the cylinder is stationary and the sphere is moving, the average velocity at the midsection of the sphere, u_{avg}, is

$$u_{avg} = \frac{1}{3} U \tag{5.32}$$

In general, because the Galilean transformation formally applies only to a coordinate frame translating at a constant velocity, we cannot use the approach just described if the velocity varies with time. Rather, we must implement Strategy 2 and use a moving control volume. For general (time-dependent) fluid motion, we know that

$$\frac{d}{dt} \iiint_V \rho\, dV + \oiint_S \rho\, \mathbf{u}^{rel} \cdot \mathbf{n}\, dS = 0 \tag{5.33}$$

We again select a control volume extending from the midsection of the sphere to a distance far below the sphere. In contrast to the formulation above, the control volume now moves downward with a velocity $\mathbf{u}^{cv} = -U\mathbf{k}$, wherefore the relative velocity, $\mathbf{u}^{rel}$, is given by

$$\mathbf{u}^{rel} = \mathbf{u} - \mathbf{u}^{cv} = \mathbf{u} + U\mathbf{k} \tag{5.34}$$

The control volume maintains the same shape as it moves. So, since the flow is incompressible, we regard the density as a constant. Thus, the volume integral of ρ, i.e., the mass of fluid in the control volume, is constant and its rate of change is zero. Because the lower surface of the sphere forms part of the control-volume boundary, clearly there is no mass flux across this part of the boundary (i.e., $\mathbf{u} - \mathbf{u}^{cv} = \mathbf{0}$ on the sphere). Since the unit normal lies at a right angle to the control-volume velocity vector on its cylindrical boundary, the mass flux vanishes there also. Hence, conservation of mass tells us that

$$\iint_{A_1} \rho\, (\mathbf{0} + U\mathbf{k}) \cdot (-\mathbf{k})\, dA + \iint_{A_2} \rho\, (u_{avg}\mathbf{k} + U\mathbf{k}) \cdot \mathbf{k}\, dA = 0 \tag{5.35}$$

Simplifying, we arrive at a result identical to Equation (5.32), viz.,

$$-4\pi R^2 \rho\, U + 3\pi R^2 \rho\, (u_{avg} + U) = 0 \quad \Longrightarrow \quad u_{avg} = \frac{1}{3} U \tag{5.36}$$

Example 5.3 *An automobile passes through a tunnel of cross-sectional area A_o at a speed U. The area between the top of the automobile and the top of the tunnel is A, and the area beneath the automobile can be neglected. Determine the average velocity, u, between the automobile and the upper tunnel wall as a function of U and A_o/A.*

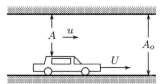

Solution. On the one hand, we can make a Galilean transformation so that the automobile is at rest as shown in Part (a) of the figure below. The fluid and tunnel walls move to the left with velocity U. The flow is steady in this coordinate frame and the control volume is stationary.

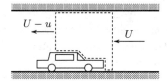
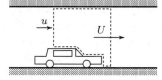

(a) Automobile and control volume at rest   (b) Automobile and control volume moving

For the indicated control volume, the steady-flow mass-conservation principle simplifies to

$$\oiint_S \rho \, \tilde{\mathbf{u}} \cdot \mathbf{n} \, dS = 0$$

where $\tilde{\mathbf{u}}$ is the velocity in the transformed coordinate system. Since there is no flow through the tunnel walls, we have

$$\oiint_S \rho \, \tilde{\mathbf{u}} \cdot \mathbf{n} \, dS = \rho \left[-U A_o \right] + \rho \left[(U - u) A \right] = 0$$

Hence,

$$\frac{U - u}{U} = \frac{A_o}{A} \qquad \Longrightarrow \qquad 1 - \frac{u}{U} = \frac{A_o}{A}$$

Therefore, the average velocity between the automobile and the upper tunnel wall is

$$u = -U \left[\frac{A_o}{A} - 1 \right]$$

On the other hand, we can use a control volume that moves to the right with velocity $\mathbf{u}^{cv} = U\mathbf{i}$ as shown in Part (b) of the figure. Since the flow is incompressible and the control-volume size and shape are constant, the mass-conservation equation simplifies to

$$\oiint_S \rho \left(\mathbf{u} - \mathbf{u}^{cv} \right) \cdot \mathbf{n} \, dS = 0 \qquad \Longrightarrow \qquad \rho(u\mathbf{i} - U\mathbf{i}) \cdot (-\mathbf{i}) A + \rho (\mathbf{0} - U\mathbf{i}) \cdot \mathbf{i} A_o = 0$$

So, we conclude that

$$\rho(U - u)A - \rho U A_o = 0 \qquad \Longrightarrow \qquad \frac{U - u}{U} = \frac{A_o}{A}$$

Therefore, as above, we find

$$u = -U \left[\frac{A_o}{A} - 1 \right]$$

As a final comment, it may seem to be more trouble than it's worth to use a Galilean transformation for control-volume problems that require only the mass-conservation principle. So why introduce the concept? The answer is simple. Implementing the Galilean transformation at this early stage increases our familiarity with the method and paves the way for using it when we really need it.

Specifically, some of the problems we will tackle in the following sections also use the momentum-conservation law. Sometimes we will find that we haven't enough equations to solve. For incompressible flows, we often appeal to Bernoulli's equation to provide the additional equations we need. However, Bernoulli's equation applies only to steady flows, so that a Galilean transformation provides a convenient avenue to achieving a closed system.

5.5.3   Deforming Control Volume

Consider a cylindrical plunger that moves downward into a conical receptacle filled with an incompressible fluid of density ρ as shown in Figure 5.9. The plunger moves downward with velocity $\mathbf{u} = -W\mathbf{k}$, and forces fluid to move upward through the area between the plunger and cone walls. Because the area changes and the fluid is incompressible, the average upward velocity of the fluid escaping across the plane the plunger is crossing must vary as the plunger moves deeper into the cone. By average, we mean the average value over the cross section. The chosen geometry causes complicated motion that is clearly not one dimensional. To make the problem specific, we want to determine the point at which the average upward flow speed is 1/3 that of the plunger. For present purposes, it will suffice to determine the cone diameter at which this condition holds.

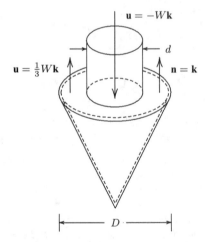

Figure 5.9: *Cylindrical plunger moving into a conical receptacle.*

Clearly, because the geometry of the plunger and the cone are dissimilar, we cannot use a Galilean transformation to recast this problem as a steady flow. This problem is an obvious case for which we must use a deforming control volume, i.e., a control volume for which parts of the bounding surface are moving.

We select our control volume so that its upper boundary moves with the plunger. In general, the mass-conservation principle is

$$\frac{d}{dt} \iiint_V \rho \, dV + \oiint_S \rho \mathbf{u}^{rel} \cdot \mathbf{n} \, dS = 0 \tag{5.37}$$

Since the flow is incompressible, the first term on the left-hand side of Equation (5.37) can be written as

$$\frac{d}{dt} \iiint_V \rho \, dV = \rho \frac{d}{dt} \iiint_V dV = \rho \frac{dV}{dt} \tag{5.38}$$

The total mass of fluid in the control volume decreases in direct proportion to the rate at which the control volume advances into the fluid. As illustrated in Figure 5.10, in time Δt, the control volume decreases in size by an amount

$$\Delta V = -\frac{\pi}{4} W D^2 \Delta t \qquad \Longrightarrow \qquad \frac{dV}{dt} = -\frac{\pi}{4} W D^2 \tag{5.39}$$

where D is the current diameter of the cone (see Figure 5.10). Therefore,

$$\frac{d}{dt} \iiint_V \rho \, dV = -\frac{\pi}{4} \rho W D^2 \tag{5.40}$$

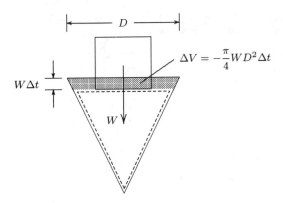

Figure 5.10: *Volume change as the plunger moves into the cone—moving control volume; the shaded region is the incremental volume change.*

Turning to the mass-flux integral, the only mass flowing across the control-volume surface is on the plane between the plunger and the cone. Therefore, we can evaluate the net mass-flux (surface) integral as follows.

$$\begin{aligned}
\oiint_S \rho \mathbf{u}^{rel} \cdot \mathbf{n} \, dS &= \rho \left[\frac{1}{3} W \mathbf{k} - (-W \mathbf{k}) \right] \cdot \mathbf{k} \frac{\pi}{4} \left(D^2 - d^2 \right) \\
&= \frac{\pi}{3} \rho W \left(D^2 - d^2 \right) \tag{5.41}
\end{aligned}$$

So, using Equations (5.40) and (5.41), we can evaluate the terms in the mass-conservation Equation (5.37) to arrive at

$$-\frac{\pi}{4} \rho W D^2 + \frac{\pi}{3} \rho W \left(D^2 - d^2 \right) = 0 \tag{5.42}$$

Solving for D is straightforward, and the final answer is

$$D = 2d \tag{5.43}$$

Note that, as with the pipe-flow and falling-sphere problems of the two preceding subsections, the detailed flow for this problem is quite complicated and certainly is not one dimensional.

Nevertheless, the control-volume method yields an exact answer for the average vertical velocity. It is exact because the method is based on the integral form of the mass-conservation principle, which holds regardless of how complicated the flowfield happens to be. While the method may conceal the complexity of the detailed flowfield, it still provides an accurate global view.

Example 5.4 *A piston advances into a tube of diameter D with a speed u. The piston forces an incompressible fluid of density ρ through a small hole of diameter d. Determine the speed U at which the fluid passes through the hole.*

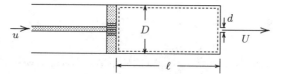

Solution. This problem is most conveniently solved using the deforming control volume indicated by the dashed boundary. Conservation of mass tells us that

$$\frac{d}{dt}\iiint_V \rho \, dV + \oiint_S \rho \mathbf{u}^{rel} \cdot \mathbf{n} \, dS = 0$$

For this cylindrical geometry, we have

$$\frac{d}{dt}\iiint_V \rho \, dV = \frac{d}{dt}\left(\rho\frac{\pi}{4}D^2\ell\right) = \frac{\pi}{4}\rho D^2 \frac{d\ell}{dt} = -\frac{\pi}{4}\rho D^2 u$$

where we observe that the rate of change of ℓ is given by $d\ell/dt = -u$. Since the only boundary across which fluid passes is the small hole, we have

$$\oiint_S \rho \mathbf{u}^{rel} \cdot \mathbf{n} \, dS = \frac{\pi}{4}\rho d^2 U$$

Hence, there follows

$$-\frac{\pi}{4}\rho D^2 u + \frac{\pi}{4}\rho d^2 U = 0 \quad \Longrightarrow \quad U = \frac{D^2}{d^2}u$$

5.6   Mass and Momentum Conservation Applications

There is only a limited class of problems that can be solved using the mass-conservation principle alone. In order to analyze more general problems, we must appeal to momentum conservation, including as many components as the geometry or other considerations dictate. For full three-dimensional applications, the momentum-conservation principle introduces three additional equations. This section focuses on problems that require combined use of the mass- and momentum-conservation principles.

5.6.1   Channel Flow With Suction

Consider two-dimensional flow between two parallel surfaces as shown in Figure 5.11. We refer to such a region as a channel, and the flow within is known as **channel flow**. We also have mass removal, or **suction**, at the upper wall, with the vertical velocity given by

$$v(x, H) = C_Q u, \qquad C_Q = C_Q(x) \tag{5.44}$$

The upper wall is made of a porous material, and the dimensionless coefficient, C_Q, is known as the **suction coefficient**. In the analysis to follow, we will neglect body forces and friction. Also, we assume the flow is incompressible and steady. Finally, for simplicity, we assume p and u are uniform across the channel, so that we say

$$p = p(x) \quad \text{and} \quad u = u(x) \tag{5.45}$$

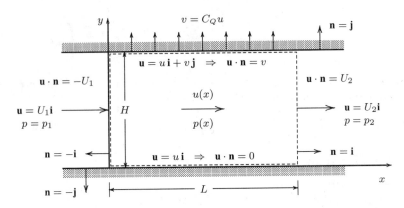

Figure 5.11: *Flow in a channel with mass removal from the porous upper wall.*

The goal of our analysis in this section is to illustrate how the combined mass- and momentum-conservation principles can be used to derive simplified one-dimensional equations governing the averaged velocity and pressure profiles in the channel. First, we consider mass conservation. The dashed lines delineate the control volume, which is stationary. Because the flow is steady and the size of the control volume is constant, we know that

$$\oiint_S \rho \mathbf{u} \cdot \mathbf{n} \, dS = 0 \tag{5.46}$$

where we use the fact that $\mathbf{u}^{rel} = \mathbf{u}$ for a motionless control volume. Since there is no flow through the bottom of the channel, the closed-surface integral is the sum of three integrals. Specifically, it is the sum of the integrals over the inlet, the outlet and the upper surface. Expanding the integral here is trivial since we assume constant velocities at the inlet and outlet so that

$$-\rho U_1 H + \rho U_2 H + \int_0^L \rho v \, dx = 0 \tag{5.47}$$

or, substituting from Equation (5.44) for v on the upper surface,

$$\int_0^L \rho C_Q u \, dx = \rho (U_1 - U_2) H \tag{5.48}$$

In words, this equation tells us the total mass removed from the upper wall [the integral on the left-hand side of Equation (5.48)] equals the difference between the mass that flows into the channel at $x = 0$ and the mass that flows out at $x = L$.

Turning now to momentum conservation, present purposes allow us to consider only the x direction. Thus, noting that taking a dot product with $\mathbf{i}$ yields the x component of a vector,

$$\oiint_S \rho u (\mathbf{u} \cdot \mathbf{n})\, dS = -\mathbf{i} \cdot \oiint_S p \mathbf{n}\, dS \tag{5.49}$$

As discussed in Subsection 5.4.1, we regard the integrand in the momentum-flux integral, i.e., the integral on the left-hand side of Equation (5.49), as consisting of the product of two quantities. The first is the property being carried across the surface, ρu. The second is the rate at which it is being carried across the boundary, $(\mathbf{u} \cdot \mathbf{n})\, dS$. Hence, keeping track of unit-normal orientation on all bounding surfaces, we expand the integral as follows.

$$\begin{aligned}
\oiint_S \rho u\, (\mathbf{u} \cdot \mathbf{n})\, dS &= (\rho U_1)(-U_1 H) + (\rho U_2)(U_2 H) + \int_0^L (\rho u) v\, dx \\
&= -\rho U_1^2 H + \rho U_2^2 H + \int_0^L \rho C_Q u^2\, dx
\end{aligned} \tag{5.50}$$

The integral on the right-hand side of Equation (5.50) arises because x momentum is carried out of the control volume at $y = H$ by the flux in the y direction. The pressure integral is

$$\oiint_S p \mathbf{n}\, dS = p_1(-\mathbf{i})H + p_2(\mathbf{i})H + \int_0^L p(x)(\mathbf{j})dx + \int_0^L p(x)(-\mathbf{j})dx \tag{5.51}$$

Now, we know that the unit vectors satisfy the following relations.

$$\mathbf{i} \cdot \mathbf{i} = 1 \quad \text{and} \quad \mathbf{i} \cdot \mathbf{j} = 0 \quad \Longrightarrow \quad -\mathbf{i} \cdot \oiint_S p \mathbf{n}\, dS = (p_1 - p_2)H \tag{5.52}$$

Observe that the unknown pressure on the channel walls has no effect on momentum conservation in the x direction because it exerts a force only in the y direction. As a check on the physics, note that the pressure integral, including the minus sign, is the net force on the control volume. If the pressure at the inlet exceeds that at the outlet, i.e., if $p_1 > p_2$, the net force must be positive. Inspection of Equation (5.52) shows that this is true. Hence, x-momentum conservation for the channel simplifies to

$$\rho \left(U_2^2 - U_1^2 \right) H + \int_0^L \rho C_Q u^2\, dx = (p_1 - p_2) H \tag{5.53}$$

It is helpful to pause and discuss what we have accomplished in our analysis of the channel-flow problem considered in this section. The most important point is the appearance of terms involving C_Q in the mass-conservation Equation (5.48) and in the momentum Equation (5.53). While it is obvious that mass removal should affect mass conservation, it is a bit more subtle why it appears in the momentum equation. The origin of the term with C_Q in the momentum equation is the momentum-flux integral on the left-hand side of Equation (5.49), and our assumption[1] that u is a function only of x. Its presence could be easily overlooked if we are not careful in separating what is being carried across the control-volume surface, ρu, and the rate at which it is carried across, $v dx = (\mathbf{u} \cdot \mathbf{n})\, dx$. Thus, the most important lesson to be gleaned from this application is the importance of separating the terms in the momentum-flux integral, and how helpful the control-volume method is in delineating the proper separation.

---

[1] In a viscous flow, $u = 0$ at the wall so that no horizontal momentum would be carried out of the control volume.

5.6.2   Indirect Force Computation—Rotational Flow

An interesting use of the control-volume method is for the indirect computation of forces. When we have flow about or through a solid object, the fluid exerts a force on the object at its surface. If we know the pressure (and viscous stresses for a viscous fluid), we can integrate over the surface of the object to find the net force. We do this in analyzing fluid-statics problems where we know the pressure from the hydrostatic relation.

Alternatively, if we know something about the flow such as velocities at the surface bounding a specified volume, we can compute the net force on the object indirectly. We have already shown this for the non-moving fluid case (Subsection 5.4.3). The following subsections show how it can be done in the more complex case where the fluid is moving.

Consider steady flow through the channel shown in Figure 5.12 with two branches exhausting to the atmosphere. When the lower branch develops a leak, we find that the velocity in the upper branch is 4/5 that of the lower branch and a jet of velocity U_j is emitted. Additionally, there is no net vertical force on the channel. Because the channel has rectangular cross sections, the flow is very complicated internally and is strongly rotational. Our objectives are to compute V and U_j and to determine the horizontal force, F_x, on the channel, ignoring body forces.

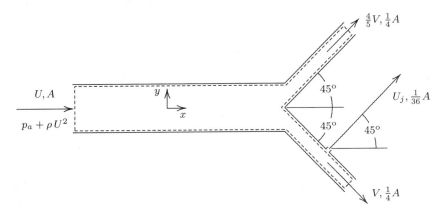

Figure 5.12: *Flow in a channel with a leak; the flow yields zero normal force ($F_y = 0$) on the channel. The dashed line denotes the control volume's bounding surface.*

As indicated in the figure, we use a stationary control volume entirely inside the channel. Since the flow is steady and $\mathbf{u}^{rel} = \mathbf{u}$, mass conservation tells us that

$$\oiint_S \rho\,\mathbf{u}\cdot\mathbf{n}\,dS = 0 \tag{5.54}$$

Hence, integrating about the closed surface, we find

$$\rho(-UA) + \rho\left(\frac{4}{5}V\frac{A}{4}\right) + \rho\left(V\frac{A}{4}\right) + \rho\left(U_j\frac{A}{36}\right) = 0 \tag{5.55}$$

Therefore,

$$\frac{9}{20}V + \frac{1}{36}U_j = U \tag{5.56}$$

Because the vertical force is known to be zero, it is most convenient in this problem to consider y momentum before considering x momentum. Replacing p by $(p - p_a)$ where p_a is atmospheric pressure and ignoring body forces, the exact equation is

$$\oiint_S \rho v(\mathbf{u} \cdot \mathbf{n}) dS = -\mathbf{j} \cdot \oiint_S (p - p_a)\mathbf{n}\, dS \tag{5.57}$$

where we take the dot product of the pressure integral and $\mathbf{j}$ to extract the y component. Remembering to segregate the integrand into the product of what's being carried across the surface and the rate at which it's being carried across, conservation of y momentum simplifies to the following.

$$\left(\frac{4}{5}\rho V \sin 45°\right)\left(\frac{4}{5}V\frac{A}{4}\right) + (\rho U_j \sin 45°)\left(U_j\frac{A}{36}\right) + (-\rho V \sin 45°)\left(V\frac{A}{4}\right) = 0 \tag{5.58}$$

Note that we take advantage of the fact that $F_y = 0$. Since the factor $\rho A \sin 45°$ appears in all terms, the equation simplifies immediately to

$$\frac{4}{25}V^2 + \frac{1}{36}U_j^2 - \frac{1}{4}V^2 = 0 \quad \implies \quad U_j = \frac{9}{5}V \tag{5.59}$$

At this point, we can solve directly for V and U_j by substituting into Equation (5.56), viz.,

$$\frac{9}{20}V + \frac{1}{36}\cdot\frac{9}{5}V = U \quad \implies \quad \frac{1}{2}V = U \tag{5.60}$$

Therefore, the velocities are

$$V = 2U \quad \text{and} \quad U_j = \frac{18}{5}U \tag{5.61}$$

In a similar way, the equation for conservation of x momentum is

$$\oiint_S \rho u(\mathbf{u} \cdot \mathbf{n}) dS = -\mathbf{i} \cdot \oiint_S (p - p_a)\mathbf{n}\, dS \tag{5.62}$$

Again being careful to segregate the integrand's two factors, we find

$$(\rho U)(-UA) + \left(\frac{4}{5}\rho V \cos 45°\right)\left(\frac{4}{5}V\frac{A}{4}\right) + (\rho U_j \cos 45°)\left(U_j\frac{A}{36}\right)$$
$$+ (\rho V \cos 45°)\left(V\frac{A}{4}\right) = -F_x + \rho U^2 A \tag{5.63}$$

Note that in evaluating the pressure integral, the x component of the force *exerted by the fluid on the pipe*, F_x, is the integral of $(p - p_a)\,\mathbf{n}$ over all solid parts of the channel. This is true because $\mathbf{n}$ points outward, so that we are computing the force exerted by the control volume on the surroundings. The complete integration domain also includes the three outlets and the inlet. Since the outlets all exhaust to the atmosphere, $p = p_a$ so that they contribute nothing to the integral. By contrast, $p = p_a + \rho U^2$ at the inlet, which yields a contribution of $\rho U^2 A$. Simplifying, there follows:

$$-\rho U^2 + \frac{4}{25}\rho V^2 \cos 45° + \frac{1}{36}\rho U_j^2 \cos 45° + \frac{1}{4}\rho V^2 \cos 45° = -\frac{F_x}{A} + \rho U^2 \tag{5.64}$$

Finally, solving for the force we have

$$F_x = 2(1 - \cos 45°)\rho U^2 A = 0.586\rho U^2 A \tag{5.65}$$

Example 5.5 *Incompressible fluid of density ρ flows through a container. We wish to design the container appendages so that for a given velocity U, the horizontal force on the container vanishes. Pressure is atmospheric at all three openings, the flow is steady and body forces can be neglected. What is the angle ϕ? Use the control volume indicated by the dashed boundary.*

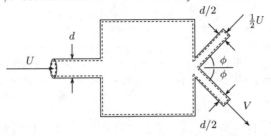

Solution. Since the flow is steady and the control volume is stationary (so that the mass contained in the control volume is constant and $\mathbf{u}^{rel} = \mathbf{u}$), the mass-conservation principle is

$$\iint_S \rho \mathbf{u} \cdot \mathbf{n} \, dS = 0$$

Expanding the closed-surface integral gives

$$\underbrace{\rho\left(-U\frac{\pi}{4}d^2\right)}_{\text{Left inlet}} + \underbrace{\rho\left(-\frac{1}{2}U\frac{\pi}{4}\frac{d^2}{4}\right)}_{\text{Right inlet}} + \underbrace{\rho\left(V\frac{\pi}{4}\frac{d^2}{4}\right)}_{\text{Outlet}} = 0$$

Simplifying, we have

$$-U - \frac{1}{8}U + \frac{1}{4}V = 0 \quad \Longrightarrow \quad V = \frac{9}{2}U$$

Turning to x-momentum conservation, we take the dot product of the pressure integral with $\mathbf{i}$ to extract the x component.

$$\iint_S \rho u \left(\mathbf{u} \cdot \mathbf{n}\right) dS = -\mathbf{i} \cdot \iint_S \left(p - p_a\right) \mathbf{n} \, dS$$

Therefore, noting that $p - p_a = 0$ on the chosen control volume's inlets and outlet, the net pressure force is exactly equal to the force on the container, F_x. So, we have

$$\underbrace{(\rho U)\left(-U\frac{\pi}{4}d^2\right)}_{\text{Left inlet}} + \underbrace{\left(-\frac{1}{2}\rho U \cos\phi\right)\left(-\frac{1}{2}U\frac{\pi}{4}\frac{d^2}{4}\right)}_{\text{Right inlet}} + \underbrace{(\rho V \cos\phi)\left(V\frac{\pi}{4}\frac{d^2}{4}\right)}_{\text{Outlet}} = F_x$$

which can be rearranged to read

$$\frac{\pi}{4}\rho U^2 d^2\left[-1 + \frac{1}{16}\cos\phi + \frac{1}{4}\left(\frac{V}{U}\right)^2\cos\phi\right] = F_x$$

Thus, substituting for V from above,

$$F_x = \frac{\pi}{4}\rho U^2 d^2\left[\frac{41}{8}\cos\phi - 1\right]$$

If the force F_x vanishes, necessarily

$$\phi = \cos^{-1}\left(\frac{8}{41}\right) = 78.7°$$

5.6.3   Indirect Force Computation—Irrotational Flow

Sometimes mass and momentum conservation fail to yield enough equations for the unknown properties in the problem. When this happens, we must in general appeal to conservation of energy to close the set of equations. For steady, inviscid, irrotational, incompressible flow problems with conservative body forces, we can appeal to Bernoulli's equation to obtain the required additional equations.

As an example of such a problem, consider the "fluid mechanical device" shown in Figure 5.13. An incompressible fluid flows *steadily* and *irrotationally* through the device and exhausts to the atmosphere. We would like to determine the angle ϕ that causes the vertical component of the force on the device to vanish. Also, we would like to know the horizontal component of the force for this angle. We assume properties are constant across inlets to and the outlet from the device and that body forces can be ignored.

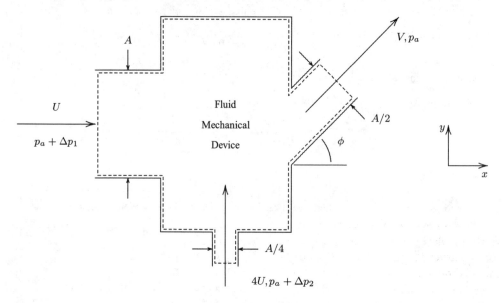

Figure 5.13: *Irrotational flow through a fluid mechanical device; A denotes area and the dashed line denotes the control volume's bounding surface.*

As in the example of the preceding section, we select a control volume that lies entirely within the device. Since the flow is assumed steady and $\mathbf{u}^{rel} = \mathbf{u}$, the mass-conservation equation simplifies to

$$\oiint_S \rho \mathbf{u} \cdot \mathbf{n} \, dS = 0 \tag{5.66}$$

Using the assumption that velocities are constant across inlets and the outlet, the closed-surface integral can be trivially evaluated, i.e.,

$$\rho(-UA) + \rho\left(-4U\frac{A}{4}\right) + \rho\left(V\frac{A}{2}\right) = 0 \tag{5.67}$$

We can solve for V immediately, wherefore

$$V = 4U \tag{5.68}$$

Since the flow is steady and no body forces are present, the x component of the momentum-conservation equation is

$$\oiint_S \rho\, u(\mathbf{u} \cdot \mathbf{n})\, dS = -\mathbf{i} \cdot \oiint_S (p - p_a)\, \mathbf{n}\, dS \qquad (5.69)$$

where we again replace the pressure p by $(p - p_a)$. Evaluating in the usual manner yields:

$$(\rho U)(-UA) + (\rho V \cos \phi)\left(V\frac{A}{2}\right) = \Delta p_1 A - F_x \qquad (5.70)$$

where we denote the x component of the force *exerted by the fluid on the device* by F_x. Taking advantage of Equation (5.68) to eliminate V, the force becomes

$$F_x = \Delta p_1 A - (8\cos \phi - 1)\rho U^2 A \qquad (5.71)$$

Similarly, the y component of the momentum-conservation equation is

$$\oiint_S \rho\, v(\mathbf{u} \cdot \mathbf{n})\, dS = -\mathbf{j} \cdot \oiint_S (p - p_a)\, \mathbf{n}\, dS \qquad (5.72)$$

Performing the integration around the closed surface yields:

$$(4\rho U)\left(-4U\frac{A}{4}\right) + (\rho V \sin \phi)\left(V\frac{A}{2}\right) = \Delta p_2 \frac{A}{4} - F_y \qquad (5.73)$$

so that

$$F_y = \frac{1}{4}\Delta p_2 A - (8\sin \phi - 4)\rho U^2 A \qquad (5.74)$$

At this point, we have insufficient information to solve for the stated unknown properties in the problem. That is, there are *six unknowns* in the problem, viz., V, ϕ, F_x, F_y, Δp_1 and Δp_2. However, we have only *four equations*, three of which are Equations (5.68), (5.71), (5.74). The fourth equation is the given fact that F_y vanishes, i.e.,

$$F_y = 0 \qquad (5.75)$$

the latter being part of the statement of the problem. Because the problem is cast as one involving conditions under which Bernoulli's equation applies, we thus use it to obtain the missing equations.

Since Bernoulli's equation holds throughout the flow, we can use it as many times as we need to arrive at a closed system of equations. First, we apply it between Inlet 1 and the outlet. Hence,

$$p_a + \Delta p_1 + \frac{1}{2}\rho U^2 = p_a + \frac{1}{2}\rho V^2 = p_a + 8\rho U^2 \qquad (5.76)$$

where we again make use of Equation (5.68). Therefore, solving for the pressure differential,

$$\Delta p_1 = \frac{15}{2}\rho U^2 \qquad (5.77)$$

Now, applying Bernoulli's equation between Inlet 2 and the outlet, we find:

$$p_a + \Delta p_2 + \frac{1}{2}\rho(4U)^2 = p_a + 8\rho U^2 \qquad (5.78)$$

Solving for Δp_2,

$$\Delta p_2 = 0 \tag{5.79}$$

Hence, in order to have vanishing vertical force, substituting Equation (5.79) into Equation (5.74) shows that

$$(8 \sin \phi - 4)\rho U^2 = 0 \tag{5.80}$$

Because ρ and U are nonzero, we conclude that

$$\sin \phi = \frac{1}{2} \quad \Longrightarrow \quad \phi = 30° \tag{5.81}$$

Finally, combining Equations (5.71) and (5.77) yields the horizontal force, i.e.,

$$F_x = \frac{15}{2}\rho U^2 A - (8 \cos \phi - 1)\rho U^2 A = \left(\frac{17}{2} - 8 \cos \phi\right)\rho U^2 A \tag{5.82}$$

Using the fact that ϕ is 30°, the x component of the force is

$$F_x = 1.57 \rho U^2 A \tag{5.83}$$

5.6.4   Reaction Forces

In all of the examples presented thus far, we have selected a control volume that contains only fluid. In order to compute the force on an object, we have chosen a control volume for which part of the bounding surface is coincident with the object's surface. Then, the force is given by the integral of pressure (and viscous stress if present) over the surface of the object. By contrast, for some applications, it is more straightforward to select a control volume that contains the object under consideration. The question then arises about how we might determine *the force exerted by the fluid on the object*. The answer is, we introduce a *reaction force*, which is *the force required to hold the control volume in place*. As we will see in this section, the reaction force is the negative of the force on the object.

To illustrate these concepts, consider a jet of fluid impinging at a right angle on a disk as shown in Figure 5.14. We select the stationary control volume indicated by the dashed

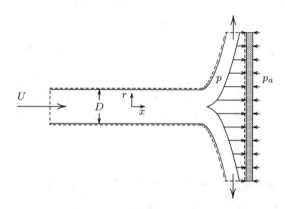

Figure 5.14: *Jet impinging on a disk with the disk outside the control volume.*

contour. The diameter of the jet is D and its average velocity is U. Since $\mathbf{u}^{rel} = \mathbf{u}$ and the flow is steady, momentum conservation tells us that

$$\oiint_S \rho\,\mathbf{u}(\mathbf{u}\cdot\mathbf{n})\,dS = -\oiint_S (p - p_a)\,\mathbf{n}\,dS \tag{5.84}$$

By symmetry, there can be no net force or momentum flux in the radial direction so that the momentum flux and the force must lie in the x direction. The only surface across which x momentum moves is the jet cross section upstream of the disk. Hence, the momentum-flux integral is

$$\oiint_S \rho\,\mathbf{u}(\mathbf{u}\cdot\mathbf{n})\,dS = -\frac{\pi}{4}\rho U^2 D^2\,\mathbf{i} \tag{5.85}$$

The force on the disk, $\mathbf{F}_{disk}$, is the difference between (a) the integral of the pressure on the disk surface exposed to the jet, and (b) the force exerted by the atmosphere on the opposite side of the disk. That is, the force on the disk is

$$\mathbf{F}_{disk} = \mathbf{i}\iint_{disk} (p - p_a)\,dS \tag{5.86}$$

Finally, provided the disk diameter is large compared to the jet diameter, the pressure is approximately p_a on the disk far from its center. Also, the surrounding fluid impresses atmospheric pressure both on the edge of and throughout the jet. Thus, the integral of $(p - p_a)$ on the control volume's bounding surface is zero except on the disk. Hence,

$$-\oiint_S (p - p_a)\,\mathbf{n}\,dS = -\mathbf{i}\iint_{disk} (p - p_a)\,dS = -\mathbf{F}_{disk} \tag{5.87}$$

Therefore, conservation of momentum tells us that

$$\mathbf{F}_{disk} = \frac{\pi}{4}\rho U^2 D^2\,\mathbf{i} \tag{5.88}$$

Now, consider the control volume shown in Figure 5.15. The disk is included within the control volume. An immediate advantage of this control volume is the fact that $p = p_a$ at all points on the bounding surface. Hence, the pressure integral vanishes. We must revise the momentum-conservation equation for this control volume by adding a **reaction force**, $\mathbf{R}$, viz.,

$$\oiint_S \rho\,\mathbf{u}(\mathbf{u}\cdot\mathbf{n})\,dS = -\oiint_S (p - p_a)\,\mathbf{n}\,dS + \mathbf{R} \tag{5.89}$$

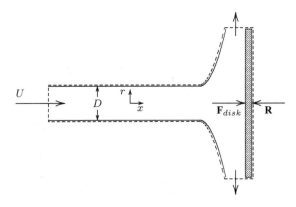

Figure 5.15: *Jet impinging on a disk with the disk inside the control volume.*

The physical meaning of the reaction force is as stated above—it is the force required to hold the control volume in place.

Now, the momentum flux is still given by Equation (5.85). Since the pressure integral is zero, we conclude immediately that the reaction force is given by

$$\mathbf{R} = -\frac{\pi}{4}\rho U^2 D^2\,\mathbf{i} \qquad (5.90)$$

Clearly, the force required to hold the disk in place is just the negative of the force exerted by the fluid on the disk, i.e.,

$$\mathbf{R} + \mathbf{F}_{disk} = \mathbf{0} \qquad (5.91)$$

wherefore the force is again given by Equation (5.88). Since both types of control volumes yield the same results, either can be used successfully. It is a matter of convenience as to which type you should use for a given problem.

As a final comment, we can check the reasonableness of the sign of our computed force for this problem. Clearly the force must be to the right, else we would be predicting that a jet will draw a disk toward its source! Equation (5.88) tells us the force is in the correct direction.

5.6.5 Nonuniform Cross-Sectional Velocity Profiles

In every control-volume example of the preceding sections, we have assumed the velocity is constant on cross sections. This is untrue for the flow of a real fluid, for example, because viscous effects give rise to nonuniform velocity profiles. Since all fluids in nature are viscous, and since other common effects such as body forces can lead to nonuniform velocity profiles, we must assess the importance of assuming uniform cross-sectional velocity. We saw in Subsection 5.5.1 that using uniform profiles is equivalent to working with average properties on a given cross section. Although this is convenient and exact for mass conservation, it is an approximation for momentum conservation. This is because the velocity appears linearly in the mass-conservation equation, while it appears quadratically in the momentum-conservation equation.

To illustrate why this is a problem, consider Couette flow (see Section 1.9) for which the plates are separated by a distance h. We know that the velocity varies linearly with distance from the lower plate, i.e., the horizontal velocity is

$$u(y) = U\frac{y}{h} \qquad (5.92)$$

Then, the average velocity is given by

$$\overline{U} = \frac{1}{h}\int_0^h U\frac{y}{h}\,dy = \frac{1}{2}U \qquad (5.93)$$

Using this average velocity, if the channel width is w, we would then conclude that the cross-sectional mass flux is

$$\iint_A \rho u\,dA = \rho\overline{U}hw = \frac{1}{2}\rho U hw \qquad (5.94)$$

Similarly, using the average velocity, the momentum flux through the same cross section would be

$$\iint_A \rho u^2\,dA \approx \rho\overline{U}^2 hw = \frac{1}{4}\rho U^2 hw \qquad (5.95)$$

Although the mass flux is correct, the momentum flux is not. If we substitute the exact velocity into the integrand, we find

$$\iint_A \rho u^2 \, dA = \rho U^2 w \int_0^h \left(\frac{y}{h}\right)^2 dy = \frac{1}{3}\rho U^2 hw \qquad (5.96)$$

Thus, using the average velocity yields a 25% error in the momentum flux! Mathematically speaking, our error results from the fact that the square of the mean is not equal to the mean of the square, i.e.,

$$\overline{U}^2 \neq \overline{U^2} \qquad (5.97)$$

Couette flow is an extreme case in this respect. In most practical viscous-flow applications, the error in the momentum flux will be less than 10%. This is true because most flows of engineering interest are turbulent, and the attending velocity profiles are not too far from uniform in pipes and channels, for example. We will explore this in more detail when we analyze pipe flow in Chapter 6. Nevertheless, for the sake of rigor and completeness, we should know how to account for nonuniform velocity profiles.

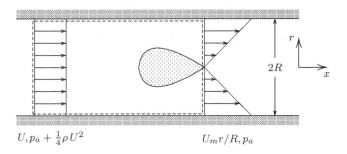

$$U, p_a + \tfrac{1}{4}\rho U^2 \qquad\qquad\qquad U_m r/R, p_a$$

Figure 5.16: *Cross-sectional view of low-speed viscous flow past a pear-shaped axisymmetric body in a cylindrical wind tunnel.*

To illustrate how the control-volume method is implemented when nonuniform velocity profiles are accounted for, consider flow past a pear-shaped axisymmetric body in a cylindrical wind tunnel of radius R. For algebraic simplicity, we will assume the body produces a linear, axisymmetric velocity profile as indicated in Figure 5.16. The measured pressures upstream and downstream of the body are as indicated and are uniform over the cross section, which is usually an excellent approximation for this type of flow. Neglecting the viscous force exerted by the tunnel wall on the air flow, we want to compute the drag force on the body. The flow is assumed to be steady and incompressible, with no body forces acting.

We select a control volume that extends from the uniform velocity profile upstream of the body to the linear profile downstream. Since the flow is steady and incompressible and the control volume is stationary, the integral form of mass conservation simplifies to

$$\oiint_S \rho\, \mathbf{u} \cdot \mathbf{n}\, dS = 0 \qquad (5.98)$$

Substituting the appropriate velocity profiles into the integral yields

$$-\rho U \pi R^2 + \int_0^{2\pi}\int_0^R \underbrace{(\rho)}_{\rho}\ \underbrace{(U_m r/R)}_{(\mathbf{u}\cdot\mathbf{n})}\ \underbrace{(r\,dr\,d\theta)}_{dS} = 0 \qquad (5.99)$$

Performing the indicated integrations, a little straightforward algebra shows that the maximum velocity, U_m, is

$$U_m = \frac{3}{2}U \qquad (5.100)$$

Focusing now on momentum conservation, since the body lies within the control volume, we must add a reaction force. Noting that the reaction force is minus the drag force on the body, F_x, the x component of the momentum-conservation equation is

$$\oiint_S \rho u(\mathbf{u} \cdot \mathbf{n})\, dS = -\mathbf{i} \cdot \oiint_S (p - p_a)\, \mathbf{n}\, dS - F_x \qquad (5.101)$$

The sign chosen corresponds to a positive value for F_x. That is, the body will tend to resist the flow, and should thus tend to reduce the net momentum flux, i.e., it will decelerate the flow. We can use this observation to check our result after we complete the computation. Substituting the various profiles into the integrals yields the following equation.

$$(\rho U)\left(-\pi R^2 U\right) + \int_0^{2\pi} \int_0^R \underbrace{(\rho U_m r/R)}_{\rho u}\, \underbrace{(U_m r/R)}_{(\mathbf{u} \cdot \mathbf{n})}\, \underbrace{(r dr d\theta)}_{dS} = \frac{1}{4}\rho U^2 \pi R^2 - F_x \qquad (5.102)$$

Solving for the force, we have

$$\begin{aligned}
F_x &= \frac{1}{4}\pi \rho U^2 R^2 + \pi \rho U^2 R^2 - 2\pi \rho U_m^2 R^{-2} \int_0^R r^3\, dr \\
&= \frac{5}{4}\pi \rho U^2 R^2 - \frac{1}{2}\pi \rho U_m^2 R^2
\end{aligned} \qquad (5.103)$$

Finally, observing that U_m is given in terms of U by Equation (5.100), a bit more algebra shows that the drag force on the body is:

$$F_x = \frac{1}{8}\pi \rho U^2 R^2 \qquad (5.104)$$

As expected, F_x is positive so that the computed force is in the correct direction.

5.7   Accelerating Control Volume

In Subsection 5.5.2 we saw that using a moving control volume is advantageous for some problems. However, the falling-sphere problem we analyzed in that section required using only the mass-conservation principle. In deriving the momentum-conservation equation in Section 5.3, we briefly mentioned the fact that Newton's law states that **absolute momentum** is conserved. In this section, we will first elaborate on the basic physical concepts involved in the momentum-conservation principle for a moving control volume. Then, we will use it to describe the motion of an accelerating rocket.

5.7.1   Inertial and Noninertial Coordinate Frames

Before we can proceed, we must discuss the concept of an **inertial frame**. Newton's laws of motion are valid in what we call an inertial frame of reference. The ultimate inertial frame lies at the "center of the universe," which is completely at rest. Any coordinate system that moves at constant velocity relative to the center of the universe is also an inertial frame, i.e., all inertial frames are related by a **Galilean transformation.** Since the laws of motion ultimately involve *changes* in mass, momentum and energy, we have no need to know where either the center of the universe is or what the velocity of our coordinate system is relative to it. This is true for momentum, for example, because the constant velocity of our coordinate frame cancels when we take differences between velocities observed in our moving frame.

However, we don't always work in an inertial frame. For example, because of its motion about the sun, or more generally about the center of the universe, a coordinate system fixed on the Earth is not an inertial system. To be precise in our calculations, we would have to account for the centrifugal and Coriolis accelerations attending the Earth's motion. Nevertheless, it would be exceedingly pedantic to account for such effects in typical Earth-bound applications. Engineering fluid-flow problems occur on such a small scale relative to galactic distances that the Earth's motion relative to the center of the universe can be ignored. We are all familiar with such a contrast between observations on vastly differing scales. Consider, for example, the flat appearance of the horizon as observed from a beach or a low-flying aircraft. Hence, an Earth-fixed coordinate frame may be regarded as an inertial frame for many engineering applications.

In the problems we analyze in general engineering practice, it is sometimes convenient to work in a **noninertial frame** relative to the Earth.[2] When we do this, we must account for the fact that *momentum relative to the inertial frame is conserved.* With this understanding, we have written the equations for mass and momentum conservation as follows.

$$\frac{d}{dt} \iiint_V \rho \, dV + \oiint_S \rho \mathbf{u}^{rel} \cdot \mathbf{n} \, dS = 0 \tag{5.105}$$

$$\frac{d}{dt} \iiint_V \rho \mathbf{u} \, dV + \oiint_S \rho \mathbf{u} \left(\mathbf{u}^{rel} \cdot \mathbf{n} \right) dS = \mathbf{F} \tag{5.106}$$

where $\mathbf{u}$ is the absolute velocity relative to the inertial frame, $\mathbf{u}^{rel}$ is the velocity relative to the control volume, and $\mathbf{F}$ is the sum of all external forces acting on the control volume. As discussed in Subsection 5.5.2, the rate at which fluid crosses the control-volume surface is proportional to the velocity relative to the moving control volume, regardless of the control-volume velocity. Also, the absolute velocity, $\mathbf{u}$, is related to the control-volume velocity, $\mathbf{u}^{cv}$, as follows.

$$\mathbf{u} = \mathbf{u}^{cv} + \mathbf{u}^{rel} \tag{5.107}$$

Before tackling a non-trivial problem, let's check our generalized conservation principles on a problem for which we know the answer. Specifically, we can examine the limiting case of a control volume translating at constant velocity. If Equations (5.105) and (5.106) are correct, then we should be able to replace the absolute velocity by the relative velocity in both equations. Since Equation (5.105) involves only $\mathbf{u}^{rel}$, all we actually have to show is that the momentum equation reduces to the proper form. So, we assume that

$$\mathbf{u}^{cv} = \mathbf{constant} \tag{5.108}$$

Then, we can rewrite Equation (5.106) as follows.

$$\frac{d}{dt} \iiint_V \rho \mathbf{u}^{rel} dV + \oiint_S \rho \mathbf{u}^{rel} \left(\mathbf{u}^{rel} \cdot \mathbf{n} \right) dS + \mathbf{u}^{cv} \left[\frac{d}{dt} \iiint_V \rho \, dV + \oiint_S \rho \mathbf{u}^{rel} \cdot \mathbf{n} \, dS \right] = \mathbf{F} \tag{5.109}$$

---

[2]Hurricanes are an example. Because of their large extent (hundreds of miles), they are strongly affected by the Earth's rotation.

But, the sum of the terms in brackets multiplied by the constant control-volume velocity vanishes by virtue of the mass-conservation principle. This shows that, under a Galilean transformation, the equations of motion for a moving control volume can be rewritten in terms of the velocity observed in the moving frame. Hence, our generalized control-volume conservation principles are consistent with Galilean invariance, as they must be.

5.7.2   Ascending Rocket

To illustrate how we use the conservation principles for an accelerating control volume, consider a rocket as it ascends through the atmosphere. Referring to Figure 5.17, we select a control volume whose bounding surface is coincident with the outer surface of the rocket. The speed of the rocket relative to the Earth is W_R. Hence, the velocity of the control volume for this application is

$$\mathbf{u}^{cv} = W_R\,\mathbf{k} \qquad (5.110)$$

We assume the rocket's orientation remains vertical as it rises so that the velocity of the exhausting rocket gases is given by

$$\mathbf{u}^{rel} = -w_e\,\mathbf{k} \qquad \text{at the nozzle exit plane} \qquad (5.111)$$

where $\mathbf{u}^{rel}$ is the velocity relative to the rocket. The area of the nozzle exit plane is A_e, and the density of the rocket exhaust gases is ρ_e. We will assume the flow through the rocket engine as seen by an observer riding on the rocket is steady.

Finally, the forces acting on the rocket are aerodynamic drag, D, and its weight, Mg, where M is the instantaneous mass and g is acceleration due to gravity. Our goal is to derive equations of motion for the rocket in terms of its instantaneous mass and velocity.

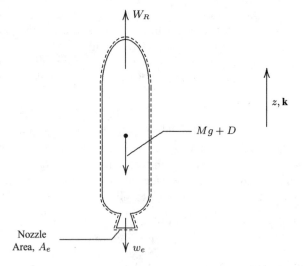

Figure 5.17: *Motion of an ascending rocket. The rocket's velocity relative to the Earth is W_R and the rocket-exhaust velocity is w_e relative to the accelerating rocket.*

To begin our analysis, we consider mass conservation. The mass within the control volume includes the rocket and the fuel contained within the rocket. Because exhaust products are being ejected from the rocket nozzle, the mass of the control volume changes with time. Hence, we must retain the instantaneous rate of change of mass integral, and mass conservation tells us that

$$\frac{d}{dt} \iiint_V \rho \, dV + \oiint_S \rho \mathbf{u}^{rel} \cdot \mathbf{n} \, dS = 0 \tag{5.112}$$

Now, by definition, the mass of the control volume is the volume integral of the density of all material within the rocket, i.e.,

$$M = \iiint_V \rho \, dV \tag{5.113}$$

This integral includes the rocket hardware as well as any fluids within the rocket. Also, the only part of the control volume across which mass passes is at the nozzle exit plane. Since the unit normal at the exit plane is $\mathbf{n} = -\mathbf{k}$, we can do the net-mass-flux integral to obtain

$$\frac{dM}{dt} + \rho_e w_e A_e = 0 \tag{5.114}$$

It's worthwhile to pause and check the sign of the predicted rate of change of the rocket's mass. Because the rocket is constantly ejecting fuel-exhaust products, the combined mass of the rocket and its fuel must decrease. Since ρ_e, w_e and A_e are all positive, clearly Equation (5.114) implies that $dM/dt < 0$. Thus, we have the correct sign.

Turning now to momentum conservation, assuming the rocket is axially symmetric, we can focus specifically on the z component of the conservation equation. Letting w denote the z component of $\mathbf{u}$, the conservation principle becomes

$$\frac{d}{dt} \iiint_V \rho \left(w + W_R \right) dV + \oiint_S \rho \left(w + W_R \right) \left(\mathbf{u}^{rel} \cdot \mathbf{n} \right) dS$$

$$= -\mathbf{k} \cdot \oiint_S p \mathbf{n} \, dS - \iiint_V \rho g \, dV - D_f \tag{5.115}$$

First, let's consider the forces acting on the rocket. As indicated, there are three types of forces, viz.,

- The net pressure force: $D_p = -\mathbf{k} \cdot \oiint_S p \mathbf{n} \, dS$

- The weight of the rocket: $\iiint_V \rho g \, dV$

- The friction drag force: D_f

We include the **friction drag**, D_f, which is one part of the aerodynamic drag. The other contribution that we regard as part of the total aerodynamic drag is contained within the pressure integral, and we call it **pressure drag**, D_p. By definition, the total aerodynamic drag, D_{tot}, is the sum of the friction and pressure drag, i.e.,

$$D_{tot} = D_f + D_p \tag{5.116}$$

The drag for the rocket can be obtained, for example, from wind-tunnel measurements. However, because of the complications attending simulation of the rocket exhaust, typical tests would determine D_f and D_p in the absence of the rocket exhaust. To account for this difference, we make a subtle distinction. Specifically, the primary difference between measured and actual pressures occurs at the nozzle exhaust plane, where the actual pressure, p_e, is much greater than the exhaust-free value. Hence, the total pressure drag is the value measured under no-exhaust conditions, plus the contribution due to the pressure increase caused by the exhaust. Denoting the pressure on the rocket without the exhausting fuel by p_b, we can write the net pressure force as

$$-\mathbf{k} \cdot \oiint_S p\,\mathbf{n}\,dS = -\mathbf{k} \cdot \oiint_S p_b\,\mathbf{n}\,dS + (p_e - p_b)\,A_e \equiv -D_{pa} + (p_e - p_b)\,A_e \qquad (5.117)$$

where D_{pa} is the pressure drag in the absence of the rocket exhaust. Finally, the weight of the rocket and fuel contained within the control volume is given by

$$\iiint_V \rho g\,dV = Mg \qquad (5.118)$$

Substituting these results into Equation (5.115), the equation for the rocket's motion simplifies to

$$\frac{d}{dt} \iiint_V \rho\,(w + W_R)\,dV + \oiint_S \rho\,(w + W_R)\,(\mathbf{u}^{rel} \cdot \mathbf{n})\,dS$$
$$= -D - Mg + (p_e - p_b)\,A_e \qquad (5.119)$$

where D is given by

$$D \equiv D_f + D_{pa} \qquad (5.120)$$

Turning now to the unsteady term, we are assuming that the flow within the rocket engine, as observed from the engine room, is steady. This assumption means the instantaneous rate of change of relative momentum is zero, i.e.,

$$\frac{d}{dt} \iiint_V \rho\,w^{rel}\,dV = 0 \qquad (5.121)$$

We can simplify the remaining term in the instantaneous rate of change of momentum integral as follows.

$$\frac{d}{dt} \iiint_V \rho W_R\,dV = \frac{d}{dt}\left(W_R \iiint_V \rho\,dV\right) = \frac{d}{dt}\,(MW_R) \qquad (5.122)$$

Because the nozzle exit plane is the only part of the surface with fluid crossing, evaluation of the net momentum-flux integral is especially simple, viz.,

$$\oiint_S \rho\,(w + W_R)\,(\mathbf{u}^{rel} \cdot \mathbf{n})\,dS = \iint_{A_e} \rho_e\,(-w_e + W_R)\,w_e\,dA$$
$$= \rho_e\,(-w_e + W_R)\,w_e\,A_e \qquad (5.123)$$

Substituting Equations (5.122) and (5.123) into Equation (5.119), the momentum equation becomes

$$\frac{d}{dt}\,(MW_R) + \rho_e w_e\,(W_R - w_e)\,A_e = -D - Mg + (p_e - p_b)\,A_e \qquad (5.124)$$

We can make an additional simplification by expanding the first term and regrouping as follows.

$$M\frac{dW_R}{dt} + W_R\left[\frac{dM}{dt} + \rho_e w_e A_e\right] - \rho_e w_e^2 A_e = -D - Mg + (p_e - p_b)\,A_e \qquad (5.125)$$

But, the term in brackets multiplied by W_R vanishes by virtue of mass conservation [Equation (5.114)]. Therefore, the final equation for the rocket's motion is

$$M\frac{dW_R}{dt} = T - D - Mg, \qquad T \equiv \left[(p_e - p_b) + \rho_e w_e^2\right] A_e \qquad (5.126)$$

The quantity T is known as the **thrust**, and is used to rate the performance of a rocket engine. Note that in words, Equation (5.126) says

$$(\text{Mass}) \cdot (\text{Acceleration}) = (\text{Thrust}) - (\text{Drag}) - (\text{Weight})$$

which is consistent with what our knowledge of basic mechanics would lead us to expect. Hence, application of the control-volume method to an accelerating vehicle produces an equation that is completely consistent with our experience in elementary physics.

Example 5.6 *A balloon moves vertically upward. In the initial moments of its ascent, air is ejected at a constant speed, w, relative to the balloon. You can neglect aerodynamic drag and you can assume the air exits at a pressure equal to the ambient pressure. The initial mass of the balloon and air within is M_o and the instantaneous mass is M. Determine the vertical speed of the balloon, W, as a function of w, M, M_o, gravitational acceleration, g, and time, t.*

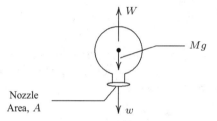

Solution. Dropping subscripts, we can use Equations (5.114) and (5.126) to solve this problem. Letting ρ denote air density and A the area of the balloon "nozzle," we have

$$\frac{dM}{dt} + \rho w A = 0 \quad \text{and} \quad M\frac{dW}{dt} = T - D - Mg$$

where

$$T \approx \rho w^2 A \quad \text{and} \quad D \approx 0$$

Taking advantage of the mass-conservation equation, we can rewrite the thrust as

$$T \approx -w\frac{dM}{dt}$$

So, the momentum equation becomes

$$M\frac{dW}{dt} = -w\frac{dM}{dt} - Mg \quad \Longrightarrow \quad \frac{dW}{dt} = -\frac{w}{M}\frac{dM}{dt} - g$$

Integrating over time yields
$$W = -w\ell n M - gt + \text{constant}$$

Finally, since the balloon has zero velocity at $t = 0$, necessarily constant $= w\ell n M_o$, so that the instantaneous velocity is

$$W = w\ell n\left(\frac{M_o}{M}\right) - gt$$

Problems

5.1 The velocity vector at the outlet from a tank is $\mathbf{u} = U\left[\mathbf{i}\cos\phi - \mathbf{j}\sin\phi\right]$. The jet cross-sectional area is A and the flow is incompressible. Using each of the two control volumes shown, for the part of the control-volume surface passing through the jet, determine $\mathbf{n}$, $(\mathbf{u}\cdot\mathbf{n})$, $\iint \rho u(\mathbf{u}\cdot\mathbf{n})dA$ and $\iint \rho v(\mathbf{u}\cdot\mathbf{n})dA$.

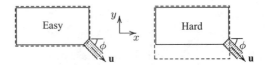

Problem 5.1

5.2 The pipe cross-sectional area and velocity at the inlet and the outlet are the same and equal to A and U, respectively for incompressible flow into a $180°$ bend. Assume the velocity is constant on all cross sections. At the inlet and outlet, determine $\mathbf{n}$, $(\mathbf{u}\cdot\mathbf{n})$, and $\iint \rho u(\mathbf{u}\cdot\mathbf{n})dA$.

Problem 5.2

5.3 For the duct section shown, you can assume the velocity is constant on all cross sections and that the flow is incompressible. Also, the duct width (out of the page) is $9H$. At the inlet and outlets, determine $\mathbf{n}$, $(\mathbf{u}\cdot\mathbf{n})$, $\iint \rho u(\mathbf{u}\cdot\mathbf{n})dA$ and $\iint \rho v(\mathbf{u}\cdot\mathbf{n})dA$.

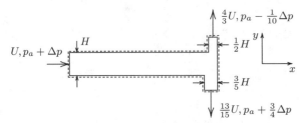

Problems 5.3, 5.4

5.4 Determine the net pressure force exerted by the surroundings on the duct section shown. The duct width (out of the page) is w. The control volume (indicated by the dashed contour) lies entirely outside of the duct, where the pressure is atmospheric ($p = p_a$) everywhere except on duct cross sections. Assume pressure is constant on all duct cross sections.

5.5 The figure depicts an inclined channel whose width (out of the page) is $2h$. The velocity is constant throughout the control volume. Using the control volume indicated by the dashed lines, compute the area, A, outer unit normal, $\mathbf{n}$, normal velocity, $\mathbf{u}\cdot\mathbf{n}$, and volume flux, $\mathbf{u}\cdot\mathbf{n}A$, at the inlet and the outlet.

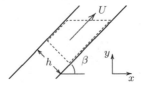

Problem 5.5

5.6 Compute the net pressure force exerted by the surroundings on the channel shown. The control volume (indicated by the dashed contour) lies entirely outside of the channel where the pressure is atmospheric ($p = p_a$) everywhere except at the inlets. Assume pressure is constant on all cross sections.

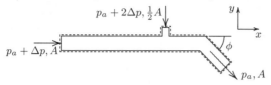

Problem 5.6

5.7 For flow in a channel of height H, the pressure decreases linearly from p_1 to p_2 over a distance L. For the control volume indicated by the dashed lines, determine the pressure force per unit width (out of the page) on the control volume for each of the four faces. What is the net pressure force?

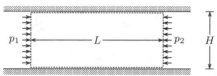

Problem 5.7

5.8 Compute the net pressure force exerted by the surroundings on the pipe shown. The control volume (indicated by the dashed contour) lies entirely outside of the pipe where the pressure is equal to its atmospheric value, p_a, everywhere except at the inlet where it is $\frac{3}{2}p_a$. Assume pressure is constant on all pipe cross sections.

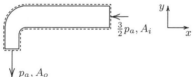

Problem 5.8

5.9 The velocity is constant throughout the pipe segment shown, which has one face slanted to the horizontal at angle α. Compute the area, A, outer unit normal, $\mathbf{n}$, normal velocity, $\mathbf{u} \cdot \mathbf{n}$, and volume flux, $\mathbf{u} \cdot \mathbf{n} A$, at the inlet and the outlet. **HINT:** The area of an ellipse with semimajor axis a and semiminor axis b is πab.

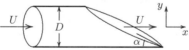

Problem 5.9

5.10 Consider the branching pipe of circular cross section shown. Using the control volume indicated by the dashed contour, compute the unit normals, $\mathbf{n}_1$, $\mathbf{n}_2$ and $\mathbf{n}_3$, the normal velocities, $\mathbf{u}_1 \cdot \mathbf{n}_1$, $\mathbf{u}_2 \cdot \mathbf{n}_2$ and $\mathbf{u}_3 \cdot \mathbf{n}_3$, the volume fluxes, $\mathbf{u}_1 \cdot \mathbf{n}_1 A_1$, $\mathbf{u}_2 \cdot \mathbf{n}_2 A_2$ and $\mathbf{u}_3 \cdot \mathbf{n}_3 A_3$ and the net pressure force exerted by the surroundings on the control volume.

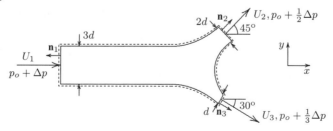

Problem 5.10

5.11 Water at 68° F flows steadily with a mass flow rate $\dot{m} = 2.12$ slug/sec through the nozzle shown. What are the average velocities U and u if the diameters are $d = 2$ in and $D = 8$ in?

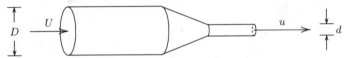

Problems 5.11, 5.12

5.12 Water at 10° C flows steadily with a mass flow rate $\dot{m} = 31.4$ kg/sec through the nozzle shown. What are the average velocities U and u if the diameters are $d = 6$ cm and $D = 20$ cm?

5.13 A cylindrical tank of diameter D is supplied with an incompressible fluid of density ρ by a pipe of diameter $\frac{1}{10}D$ and velocity U. Fluid leaves the tank through another horizontal pipe of diameter $\frac{1}{10}D$ and a vertical pipe of diameter $\frac{1}{5}D$. If the water level does not change with time and the velocity in the horizontal pipe is $U_h = \frac{1}{4}U$, what is the velocity in the vertical pipe, U_v?

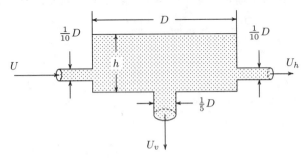

Problems 5.13, 5.14

5.14 A cylindrical tank of diameter D is supplied with an incompressible fluid of density ρ by a pipe of diameter $\frac{1}{10}D$ and velocity U. Fluid leaves the tank through another horizontal pipe of diameter $\frac{1}{10}D$ and a vertical pipe of diameter $\frac{1}{5}D$. If the velocity in the horizontal pipe is $U_h = \frac{1}{4}U$ and the velocity in the vertical pipe is $U_v = \frac{1}{8}U$, at what rate, dh/dt, is the level changing in the tank?

5.15 The constant α is 1 for the cylindrical tank with attached cylindrical pipes shown. Determine the value of the constant β for which the water level in the tank is constant.

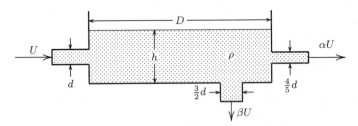

Problems 5.15, 5.16, 5.17

5.16 The amount of water in the cylindrical tank with attached cylindrical pipes shown does not change with time. Determine α as a function of β. In the special case $\beta = 3/5$, what is happening at the horizontal outlet, i.e., the outlet to the right where $d = \frac{4}{5}d$?

5.17 For the cylindrical tank with attached cylindrical pipes shown, compute the rate at which the water level is changing for $\alpha = \frac{1}{4}$ and $\beta = \frac{2}{5}$. Indicate whether the tank is filling or emptying. Express your answer for dh/dt as a function of U, d and D.

5.18 Compute the rate of change of the mass of the (incompressible) fluid contained in the tank shown. Express your answer as a function of fluid density, ρ, inlet flow velocity, U, and pipe diameter, d. Is the tank filling or emptying?

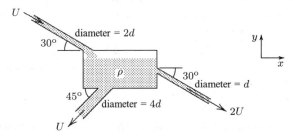

Problem 5.18

5.19 A liquid of density ρ flows steadily through the piping system shown. The pipe cross sections are circular, and you can assume flow properties are uniform on all cross sections. Determine the velocity V as a function of U.

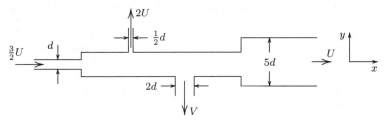

Problem 5.19

5.20 The figure illustrates a *jet pump*. At Section 1, a high-speed jet of fluid is injected into a uniform flow of velocity U_1 in a duct of area A. The fluid mixes and, at Section 2, returns to nominally uniform flow with velocity $U_2 = \frac{5}{2}U_1$. If the jet area is $A_j = \frac{1}{5}A$, what is the jet velocity, U_j? Assume the flow is steady and incompressible with density ρ.

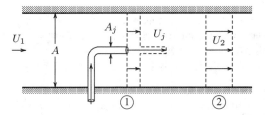

Problems 5.20, 5.21

5.21 The figure illustrates a *jet pump*. At Section 1, a high-speed jet of fluid is injected into a uniform flow of velocity U_1 in a duct of area A. The fluid mixes and, at Section 2, returns to nominally uniform flow with velocity U_2. If the jet velocity is $U_j = 15U_1$ and the jet area is $A_j = \frac{1}{9}A$, what is the velocity at Section 2? Assume the flow is steady and incompressible with density ρ.

5.22 Consider a sphere of radius R that falls axially through an infinitely long cylinder of radius $\frac{7}{5}R$ at a speed U. What is the mean upward velocity of the surrounding (incompressible) fluid at the midsection of the sphere? Use a moving control volume to solve.

5.23 Consider a sphere of radius R that falls axially through an infinitely long cylinder of radius $\frac{9}{5}R$ at a speed U. What is the mean upward velocity of the surrounding (incompressible) fluid at the midsection of the sphere? Use a Galilean transformation and a stationary control volume to solve.

5.24 In order to destroy Blofeld's headquarters, James Bond wants to launch a spherical object filled with explosives into an air-conditioning duct of square cross section. Q has designed the sphere to move at the same speed as the air, U, but in the opposite direction. The maximum flow speed, U_{max}, in the duct will occur on the cross section at the center plane of the sphere. In order for the sphere to move undetected due to the noise associated with its motion, the maximum flow speed must not exceed the undisturbed duct speed by more than 8%. Determine the maximum diameter, D, of the sphere as a function of the duct width, H. Use a moving control volume to solve.

Problems 5.24, 5.25

5.25 In order to destroy Blofeld's headquarters, James Bond wants to launch a spherical object filled with explosives into an air-conditioning duct of square cross section. Q has designed the sphere to move at the same speed as the air, U, but in the opposite direction. The maximum flow speed, U_{max}, in the duct will occur on the cross section at the center plane of the sphere. In order for the sphere to move undetected due to the noise associated with its motion, the maximum flow speed must not exceed the undisturbed duct speed by more than 12%. Determine the maximum diameter, D, of the sphere as a function of the duct width, H. Use a Galilean transformation and a stationary control volume to solve.

5.26 A subway train has a cross-sectional area, $A_{cs} = \frac{1}{8}A$, where A is the area of the tunnel through which it moves. The train is traveling at a constant velocity U. What is the average velocity, u, between the train and the tunnel walls in the indicated direction? Use a moving control volume to solve.

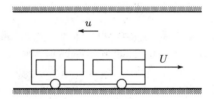

Problems 5.26, 5.27

5.27 A subway train moves at a constant velocity U through a tunnel of area A. If the average velocity between the train and the tunnel walls is $u = \frac{1}{7}U$ in the direction shown, what is the cross-sectional area of the train, A_{cs}? Use a Galilean transformation and a stationary control volume to solve.

5.28 The velocity in the outlet pipe from the reservoir is U. Assuming the diameter of the reservoir, D, is 10 times the diameter, d, at point A, what is the rate at which the surface recedes, dh/dt? Express your result in terms of U.

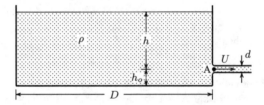

Problems 5.28, 5.29

5.29 The velocity in the outlet pipe from the reservoir is U. Assuming the rate at which the surface recedes, $dh/dt = -U/25$, what is the diameter, d, at point A? Express your result in terms of the diameter of the reservoir, D.

5.30 A plunger moves downward in a conical receptacle as shown. The receptacle is filled with an incompressible fluid of density ρ. For $N = 80$, at what level (z as a function of d) above the bottom of the receptacle will the mean upward velocity of the fluid between the bottom tip of the plunger and the receptacle wall be $\lambda = 2/3$ of the downward velocity of the plunger?

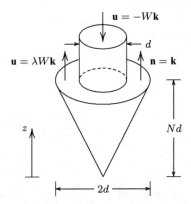

Problems 5.30, 5.31

5.31 A plunger moves downward in a conical receptacle as shown. The receptacle is filled with an incompressible fluid of density ρ. For $N = 100$, at what level (z as a function of d) above the bottom of the receptacle will the mean upward velocity of the fluid between the bottom tip of the plunger and the receptacle wall be $\lambda = 3/4$ of the downward velocity of the plunger?

5.32 The mass flow rate through a nozzle is $\dot{m} = 0.65 p_c A_t / \sqrt{RT_c}$, where p_c and T_c are pressure and temperature in the rocket chamber and R is the gas constant of the gases in the chamber. The propellant burning rate (surface-regression rate) can be expressed as $\dot{r} = a p_c^n$, where a and n are constants. As a result, gas of density ρ_p that is initially motionless enters the chamber.

(a) Denoting the propellant surface burning area by A_p, show that

$$p_c = (a\rho_p/0.65)^{1/(1-n)} (A_p/A_t)^{1/(1-n)} (RT_c)^{1/[2(1-n)]}$$

(b) If the operating chamber pressure of a rocket motor is 68000 psf and the exponent $n = 0.33$, how much will the chamber pressure increase if a crack develops in the grain, increasing the burning area by 12%? You may assume all variables except p_c and A_p remain constant.

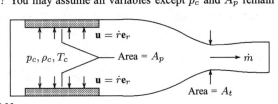

Problems 5.32, 5.33

5.33 The mass flow rate through a nozzle is $\dot{m} = 0.65 p_c A_t / \sqrt{RT_c}$, where p_c and T_c are pressure and temperature in the rocket chamber and R is the gas constant of the gases in the chamber. The propellant burning rate (surface-regression rate) can be expressed as $\dot{r} = a p_c^n$, where a and n are constants. As a result, gas of density ρ_p that is initially motionless enters the chamber.

(a) Denoting the propellant surface burning area by A_p, show that

$$p_c = (a\rho_p/0.65)^{1/(1-n)} (A_p/A_t)^{1/(1-n)} (RT_c)^{1/[2(1-n)]}$$

(b) If the operating chamber pressure of a rocket motor is 4 megaPascals (MPa) and the exponent $n = 0.29$, how much will the chamber pressure increase if a crack develops in the grain, increasing the burning area by 18%? You may assume all variables except p_c and A_p remain constant.

5.34 A spherical ball of diameter d is dropped into a tank with square cross section of width h as shown below. The tank is filled with an incompressible fluid of density ρ. Determine the ball's diameter if the mean upward velocity of the fluid between the ball and the tank walls is one tenth the downward velocity of the ball.

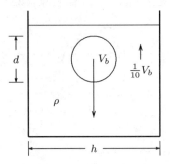

Problem 5.34

5.35 Two parallel disks of diameter D are moving away from one another, each having a speed V as shown. The fluid between the disks is incompressible with density ρ. You may assume uniform velocity across any cylindrical section, i.e., that radial velocity, u, depends only upon radial distance from the origin.

  (a) Using a cylindrical control volume of radius $r < D/2$, determine u as a function of r, h and V.

  (b) Show that the unsteady acceleration at $r = D/2$ is $A_{\text{unsteady}} = V^2 D/h^2$ and the convective acceleration is $A_{\text{convective}} = \frac{1}{2} V^2 D/h^2$.

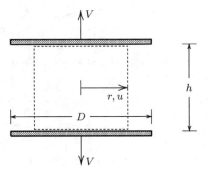

Problem 5.35

5.36 A disk of diameter d is moving at speed U and is approaching a stationary disk, also of diameter d as shown. The fluid between the disks is incompressible and has density ρ. Using a deforming cylindrical control volume between the disks (indicated by the dashed lines), determine the radial velocity, U_r, of the fluid. Express your answer for U_r as a function of U, radial distance, r, and instantaneous distance between the disks, ℓ.

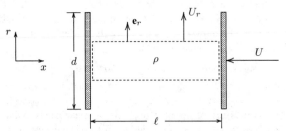

Problem 5.36

5.37 The flow in a channel inlet develops from uniform flow, $u_1(y) = U_o$, to a parabolic distribution downstream given by $u_2(y) = Ay(h - y)$, where A is a constant and h is distance between the plates. If the flow is incompressible and steady, what is the maximum velocity at the downstream location?

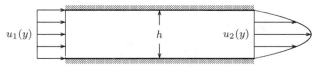

Problems 5.37, 5.38

5.38 The flow in a channel inlet develops from uniform flow, $u_1(y) = U_o$, to a parabolic distribution downstream given by $u_2(y) = Ay(h - y)$, where A is a constant and h is distance between the plates. If the flow is incompressible and steady, what is the average velocity at the downstream location?

5.39 The velocity profile in a channel changes from $u_1(y) = Ay(h - y)$ to $u_2(y) = By(2h - y)$ for $0 \leq y \leq h$, and the profiles are symmetric about the centerline at $y = h$. The quantities A and B are constants of dimensions $1/(LT)$. Assuming the flow is steady and incompressible, solve for the constant B as a function of A.

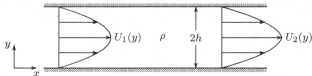

Problems 5.39, 5.40

5.40 The velocity profile in a channel changes from $u_1(y)$ to $u_2(y)$, where

$$u_1(y) = 2U\left[y/h - \tfrac{1}{2}(y/h)^2\right] \quad \text{and} \quad u_2(y) = K\left[y/h + (y/h)^2 - (y/h)^3\right]$$

for $0 \leq y \leq h$, and the profiles are symmetric about the centerline at $y = h$. Assuming the flow is steady and incompressible, solve for the constant K.

5.41 The rotating vessel shown has height h normal to the page. There is no flow into the vessel, but a chemical reaction that occurs inside generates gas that leaves through the four openings, each $(n - 1)r_o$ by h in cross-sectional area as shown, where $n = 3/2$. The velocity relative to the vessel, V, varies with radius as $V = 2V_o(2 - r/r_o)$. The flow is incompressible with density ρ_a. Find the rate of change of mass in the vessel, dM/dt. Express your answer in terms of ρ_a, V_o, r_o and h.

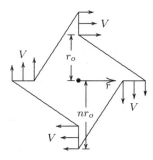

Problems 5.41, 5.42

5.42 The rotating vessel shown has height h normal to the page. There is no flow into the vessel, but a chemical reaction that occurs inside generates gas that leaves through the four openings, each $(n - 1)r_o$ by h in cross-sectional area as shown, where $n = 7/5$. The velocity relative to the vessel, V, and the density of the gas, ρ, leaving vary with radius as $V = V_o(r_o/r)$ and $\rho = \rho_a(1 + r/r_o)$. Find the rate of change of mass in the vessel, dM/dt. Express your answer in terms of ρ_a, V_o, r_o and h.

5.43 A two-dimensional channel of width H has a slot of width h as shown. Fluid is injected through the slot at an angle ϕ to the horizontal. The fluid is incompressible with density ρ and body forces can be neglected. Assume velocity and pressure are constant across the channel.

(a) Derive an equation stating conservation of mass for the channel.

(b) Derive an equation stating conservation of x momentum for the channel.

(c) Now, assume $\phi = 60°$ so that $\cos\phi = 1/2$. Combine your results of Parts (a) and (b) with Bernoulli's equation to determine U_2 as a function of U_1, H and h.

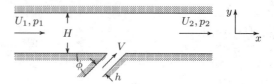

Problem 5.43

5.44 An incompressible fluid of density ρ flows in a duct with velocity U and area A as shown. Fluid of the same density is injected with velocity $2U$ from the upper duct wall. Assume the flow is steady, and neglect effects of gravity and viscosity. If all flow properties are constant on cross sections, compute the pressure difference between points 1 and 2. **NOTE:** U_2 is not given, you must solve for it.

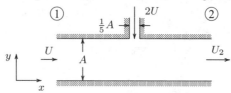

Problem 5.44

5.45 A two-dimensional channel of width $H = 9h$ has two slots of width h as shown. Fluid is injected at the indicated velocities through the lower slot at an angle ϕ to the horizontal and normal to the flow direction from the upper slot. The fluid is incompressible with density ρ and body forces can be neglected. Velocity and pressure can be assumed constant across the channel. The value of λ is 2.

(a) Derive an equation stating conservation of mass for the channel.

(b) Derive an equation stating conservation of x momentum for the channel.

(c) Now, using Bernoulli's equation, eliminate p_1 and p_2 from the momentum equation.

(d) If $U_2 = 6U_1$, what is the angle ϕ?

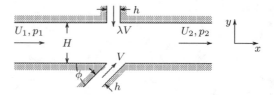

Problems 5.45, 5.46

5.46 A two-dimensional channel of width H has two slots of width h as shown. Fluid is injected at the indicated velocities through the lower slot at an angle ϕ to the horizontal and normal to the flow direction from the upper slot. The fluid is incompressible with density ρ and body forces can be neglected. Velocity and pressure can be assumed constant across the channel. The value of λ is 1.

(a) Derive an equation stating conservation of mass for the channel.

(b) Derive an equation stating conservation of x momentum for the channel.

(c) Now, assume $\phi = 60°$ so that $\cos\phi = 1/2$. Combine your results of Parts (a) and (b) with Bernoulli's equation to determine U_2 as a function of U_1, H and h.

5.47 The figure depicts two incompressible plane jets of the same velocity, V, and density, ρ. The jets meet head on and the flow shown results. In the absence of body forces, how does $\cos\alpha$ depend upon h_1 and h_2? Discuss the limiting case $\alpha = 90°$.

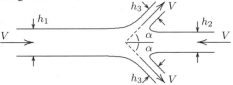

Problem 5.47

5.48 An incompressible fluid flows in a pipe with branches as shown below. The flow is steady, inviscid and no body forces are acting. Solve for U_1 and the pressure difference $p_4 - p_1$ as a function of p_2, ρ, U and atmospheric pressure, p_a. Assume velocity and pressure are constant on all cross sections, all of which are circular. Also assume the fluid exerts no net force on the pipe in the x direction.

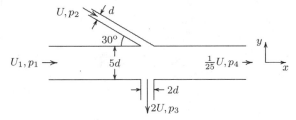

Problem 5.48

5.49 A pipe carries water to three jets that discharge into the atmosphere as shown. Because the pipe is horizontal, you can neglect body forces. The velocity of all three jets is V, and there is no net force on the pipe. What is the angle ϕ and what is the overpressure, Δp at the upstream section in the pipe? Express your answer for Δp in terms of water density, ρ, and inlet velocity, U. **NOTE:** V is not given, you must solve for it.

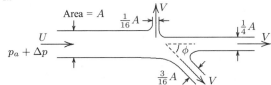

Problem 5.49

5.50 A water jet of velocity $U_j = 10$ ft/sec impinges on a small cart, causing it to move at a constant velocity $U = 4$ ft/sec. If the jet diameter is $d = 2$ inches and the water density is $\rho = 1.94$ slug/ft^3, determine the force on the cart. Use a moving control volume to solve. Explain why $d/dt \int_V \rho\mathbf{u}\,dV = \mathbf{0}$.

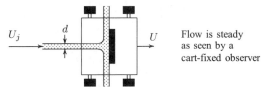

Flow is steady
as seen by a
cart-fixed observer

Problems 5.50, 5.51

5.51 A water jet of cross-sectional area A with velocity U_j and density ρ causes a cart to move at a constant velocity $U = \frac{1}{4}U_j$. Use a Galilean transformation and a stationary control volume to solve.

(a) What is the rolling resistance of the cart?

(b) The jet velocity is changed to $\hat{U}_j$ and we observe that the cart velocity doubles, i.e., $U = \frac{1}{2}U_j$, where U_j is the original jet velocity. What must the incident jet velocity be? **NOTE:** The rolling resistance is independent of cart velocity.

5.52 A hemisphere of diameter $2d$ advances to the left at a speed U into a tube of diameter $3d$. An incompressible fluid of density ρ flows to the right at a speed $2U$. The flow speed and pressure across a plane coincident with the base of the hemisphere are U_2 and p_2, where $p_2 = p_\infty - 16\rho U^2$. Assuming the pressure on the base of the hemisphere is constant and equal to p_2, compute the flow speed U_2 and the net force, F, on the hemisphere. The flow is steady as seen by a hemisphere-fixed observer. Use a Galilean transformation and a stationary control volume to solve.

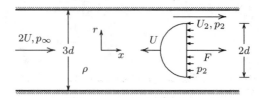

Problem 5.52

5.53 An incompressible jet of density ρ and velocity U is deflected by a cone of half angle $\theta = 60°$. The cone moves toward the jet at speed $U_c = U$. Determine the thickness of the radially spreading jet as it leaves the cone, τ, and the force required to move the cone. You may assume the flow is irrotational, that body forces can be neglected and the flow is steady as seen by a cone-fixed observer. Use a Galilean transformation and a stationary control volume with the boundaries indicated in the figure to solve.

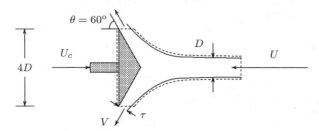

Problem 5.53

5.54 A vane deflects a jet of incompressible fluid with density ρ through $180°$. Because of a specially designed circular groove on the vane, both the incident and reflected jets are circular with diameter d. The incident absolute jet velocity is U_i and the vane is moving with absolute velocity $U_v = \frac{2}{5}U_i$. The flow is steady as seen by a vane-fixed observer. Ignore gravity and the vane's weight. Find, as a function of ρ, U_i and d, the net force on the vane. Use a Galilean transformation and a stationary control volume to solve.

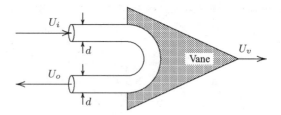

Problems 5.54, 5.55

5.55 A vane deflects a jet of incompressible fluid with density ρ through $180°$. Because of a specially designed circular groove on the vane, both the incident and reflected jets are circular with diameter d. The incident absolute jet velocity is U_i and the vane is moving with absolute velocity $U_v = \frac{1}{5}U_i$. The flow is steady as seen by a vane-fixed observer. Ignore gravity and the vane's weight. Find, as a function of ρ, U_i and d, the net force on the vane. Use a moving control volume to solve. Explain why $d/dt \int_V \rho \mathbf{u}\, dV = \mathbf{0}$.

5.56 Consider the triple-branched channel shown. There is no net force in the x direction. The flow is steady, incompressible and is not subject to any body forces. Because of the sharp corners, however, the flow is *strongly rotational*. The inlet velocity, U_i, channel area, A, and the angle $\phi = 60°$ are given. Also, the pressure at the inlet and central outlet branch are $p_i = p_a + \Delta p$ and $p_o = p_a + \frac{11}{2}\Delta p$, where p_a is atmospheric pressure and the overpressure is $\Delta p = \frac{1}{8}\rho U_i^2$. The upper and lower branches exhaust to the atmosphere at velocity V. Determine the velocities V and U_o, and the vertical force, F_y, required to hold the channel in place. **NOTE:** There are two possible solutions for this problem, and you should find both.

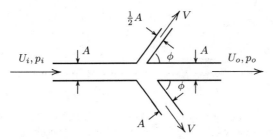

Problem 5.56

5.57 Consider incompressible flow of a fluid of density ρ in a pipe. Gravitational effects can be ignored, the flow is *rotational* and $\Delta p = \frac{1}{5}\rho U^2$. What are the velocities, V_1 and V_2, and what must the angle ϕ be in order to have no net force on the pipe?

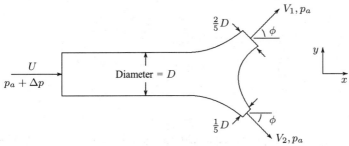

Problem 5.57

5.58 Consider steady, incompressible flow through the device shown. Body forces can be ignored and the flow is *strongly rotational*. The flow enters with a known flow speed U, and the areas of the inlets and outlets are as shown. The flow exhausts to the atmosphere at the outlets where the flow speed is V. The force required to hold the device in place is $\mathbf{R} = R_x\mathbf{i} + R_y\mathbf{j}$. Assuming $R_y = 0$, solve for V, the pressure differential, Δp, and R_x. Also, assume flow properties are constant on all cross sections.

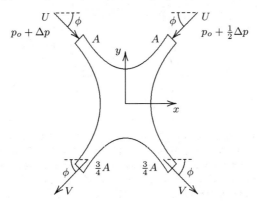

Problem 5.58

5.59 The flow in the pipe shown below is asymmetric because of the sharp bend that causes flow separation at the lower corner. The flow is steady, incompressible and *strongly rotational*. We observe that the force required to hold the pipe in place is in the direction shown. What is the overpressure, Δp, at the inlet? **NOTE:** V is not given, you must solve for it. You may neglect body forces and assume all properties are constant across inlet and outlets.

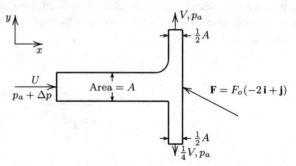

Problem 5.59

5.60 The flow in the device shown below is steady, incompressible and *strongly rotational*. We observe that the force required to hold the pipe in place is in the direction shown. What is the overpressure, Δp, at the inlet? **NOTE:** V is not given, you must solve for it. You may neglect body forces and assume all properties are constant across inlet and outlets.

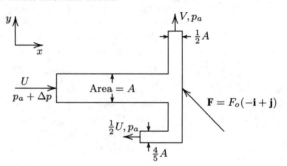

Problem 5.60

5.61 The figure illustrates a *jet pump*. Compute the pressure coefficient, $C_p = (p_2 - p_1)/\rho U_o^2$, assuming the flow is steady, incompressible, body forces can be neglected and that velocity and pressure are uniform across Sections 0, 1, and 2. In the jet, the flow is *rotational*. In computing the velocity at Station 1, be sure to account for the blockage due to the pump, whose area is $A_j = A_o/100$. Also, $U_j = 10U_o$.

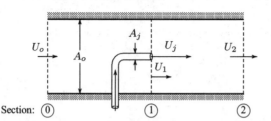

HINT: Use one control volume between Sections 0 and 1 to find U_1, and another between Sections 1 and 2 to complete the solution.

Problems 5.61, 5.62

5.62 The figure illustrates a *jet pump*. We want to design the pump so that the pressure coefficient is $C_p = (p_2 - p_1)/\rho U_o^2 = 9/11$. Assume the flow is steady, incompressible, body forces can be neglected and that velocity and pressure are uniform across Sections 0, 1, and 2. In the jet, the flow is *rotational*. In computing the velocity at Station 1, be sure to account for the blockage due to the pump, whose area is $A_j = A_o/12$. What is the jet velocity, U_j?

5.63 Wind tunnel tests are being conducted on a blunt-based body mounted on a *sting*, which is a hollow tube that holds the body in place. The flow is incompressible, steady and *strongly rotational*. Fluid supplied through the sting is injected from the base of the model with velocity U_b. The base pressure, p_b, is constant and given by $C_p = (p_b - p_\infty)/(\frac{1}{2}\rho U^2) = -11/45$, where p_∞ is the pressure far upstream of the model. Assume that the pressure on the plane extending above and below the model is also p_b. The thrust caused by the base injection exactly counters the model's drag so that there is no net force, T, on the model. In computing the velocity at Station 1, be sure to account for the blockage due to the model. You may neglect the cross-sectional area of the sting. Determine the base-injection velocity, U_b.

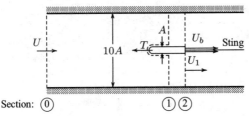

HINT: Use one control volume between Sections 0 and 1 to find U_1, and another between Sections 0 and 2 to complete the solution.

Section: ⓪ ①②

Problems 5.63, 5.64

5.64 Wind tunnel tests are being conducted on a blunt-based body mounted on a *sting*, which is a hollow tube that holds the body in place. The flow is incompressible, steady and *strongly rotational*. Fluid supplied through the sting is injected from the base of the model with velocity $U_b = 2U$. The thrust caused by the base injection exceeds the model's drag such that the net force on the body, T, is to the left and given by $T = 6\rho U^2 A$. Assume that the pressure on the plane extending above and below the model equals the base pressure, p_b. In computing the velocity at Station 1, be sure to account for the blockage due to the model. You may neglect the cross-sectional area of the sting. Determine p_b as a function of ρ, U and p_∞, where p_∞ is the pressure far upstream of the model.

5.65 Consider the incompressible, steady, strongly-rotational flow from a pipe of diameter D into a spherical container as shown. Fluid exits the sphere through two jets as shown. Both jets have velocity U_j and diameter $\frac{1}{2}D$. The pressure in the pipe is $p_i = p_a + \frac{1}{8}\rho W^2$, where ρ is the fluid density, p_a is atmospheric pressure and W is the inlet velocity. The weight of the sphere, including the fluid inside is $Mg = \frac{\pi}{32}\rho W^2 D^2$. If the vertical component of the force required to hold the sphere in place, F_z, is triple the horizontal component, F_x, what is the angle ϕ?

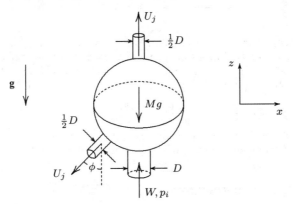

Problems 5.65, 5.66

5.66 Consider the incompressible, steady, strongly-rotational flow from a pipe of diameter D into a spherical container as shown. The mass of the sphere, including the fluid inside is M. Fluid exits the sphere through two jets as shown. Both jets have velocity U_j and diameter $\frac{1}{2}D$. The pressure in the pipe is $p_i = p_a + \frac{1}{5}\rho W^2$, where ρ is the fluid density, p_a is atmospheric pressure and W is the inlet velocity. If the vertical component of the force required to hold the sphere in place is zero, what is the horizontal component, F_x? Express your answer in terms of M, g and the angle ϕ.

5.67 Consider steady, incompressible flow through the device shown. Body forces can be ignored and the flow is strongly rotational. The flow enters through two inlet pipes with a known flow speed U_i, and the diameters of the inlet and outlet pipes are as shown. The flow exhausts to the atmosphere at the outlet where the flow speed is U_o. Determine U_o as a function of U_i and the ratio, R_y/R_x, as a function of the angle ϕ, where R_x and R_y are the x and y components of the rection force. Assume flow properties are constant on all cross sections.

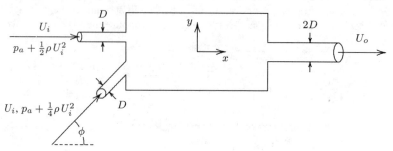

Problem 5.67

5.68 A *splitter plate* inserted part way into a two-dimensional incompressible stream produces the split stream shown. You may ignore viscous effects and body forces, and assume the flow is steady. The fluid density is ρ.

(a) Determine the width of the upper stream in terms of h and λ.

(b) Determine $\sin \phi$ as a function of λ.

(c) Compute the horizontal force, F_x, exerted by the fluid on the splitter plate as a function of ρ, U, h and λ. Explain why your answer for F_x makes sense for the limiting case $\lambda \to 0$.

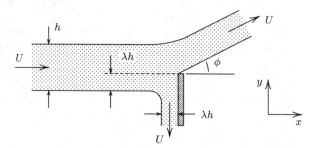

Problem 5.68

5.69 A steady stream of incompressible, inviscid fluid flowing in the xy plane is deflected by two vanes as shown below. The deflection angle for each vane is α, and the inlet and outlet streams are a distance Δy apart. Gravity acts normal to the xy plane. The mass flow rate is $\dot{m}$, and the speed at all points in the stream is U. Note that this stream is a free stream, i.e., it is not under pressure in a pipe. Compute the force exerted by the fluid on the top vane, and the total force exerted by the fluid on both vanes.

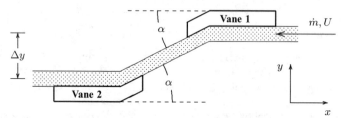

Problem 5.69

5.70 An incompressible jet of density ρ encounters a fixed vane that turns the jet through an angle β. The flow is steady, body forces can be ignored and velocity can be assumed constant on cross sections. Determine the force required to hold the vane in place as a function of U_1, U_2, β and the mass flux in the jet, $\dot{m}$.

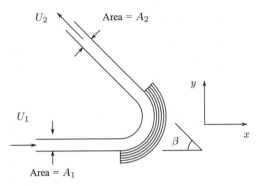

Problem 5.70

5.71 Sand of density ρ is being loaded onto a barge as shown. The velocity of the sand is U, the cross-sectional area of the pipe is A, and the mass-flow rate through the pipe is $\dot{m} = \rho U A$.

(a) If the mass of the barge is M_o, what is the total mass of sand and barge a time t after the pipe starts emitting sand?

(b) Ignoring the instantaneous rate of change of momentum in your control volume, determine the tension, T, in the mooring line as a function of $\dot{m}$, U and α.

(c) Find the buoyancy force on the barge as a function of M_o, g, $\dot{m}$, t, U and α.

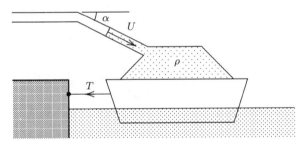

Problem 5.71

5.72 A really cheap faucet made of tin springs a leak so that an extra horizontal spray makes its use practical only when the user is wearing a raincoat. Assuming the flow to be *rotational*, what is the vertical force required to hold the faucet in place if (thanks to the leak) the net horizontal force is zero? Express your answer in terms of water density, ρ, inlet velocity, U_i, and the constant cross-sectional area, A. Neglect body forces and note that neither U_f nor U_j are given, you must solve for them.

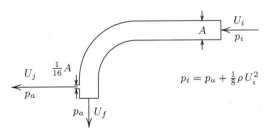

Problem 5.72

5.73 The device shown draws in water as indicated. The flow is steady. The upper inlet has a blockage, which generates *substantial vorticity* throughout the channel. Because of the asymmetry caused by the blockage, a side force develops, which is 1/7 of the horizontal force required to hold the device in place. What is the underpressure, Δp, at the outlet? **NOTE:** V is not given, you must solve for it. You may neglect body forces and assume all properties are constant across inlets and outlet as shown.

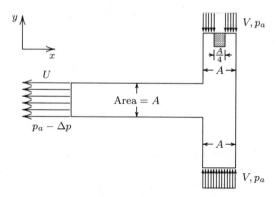

Problem 5.73

5.74 A pipe has a 90° bend and exhausts to the atmosphere. Ahead of the bend, the pipe diameter is d, and it decreases to $\frac{1}{2}d$ at the outlet. In terms of Cartesian coordinates, the inlet and outlet velocities are $\mathbf{u}_1 = -U\mathbf{i}$ and $\mathbf{u}_2 = (V/\sqrt{2})(\mathbf{j} + \mathbf{k})$. The flow is steady, incompressible, *strongly rotational*, and the pressure at the inlet is $p_1 = p_a + 3\rho U^2$, where ρ is the fluid density and p_a is atmospheric pressure. Ignoring body forces, solve for V and determine the force required to hold the pipe in place.

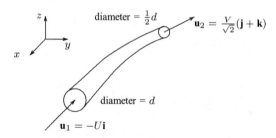

Problem 5.74

5.75 An incompressible fluid of density ρ flows through the multiple nozzle shown and discharges to the atmosphere at all three outlets. You may assume the flow is steady, *strongly rotational* and that properties are uniform on all cross sections. Also, the nozzle lies in a horizontal plane so that effects of gravity can be neglected. We observe that the components of the net force required to hold the nozzle in place are such that $R_y = 4R_x$. Determine the pressure differential at the inlet, Δp.

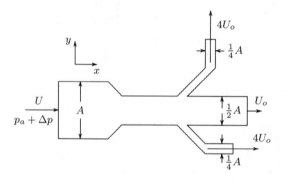

Problem 5.75

5.76 An incompressible fluid of density ρ flows steadily through the device shown. The inlet velocity is U. The fluid at both exits exhausts to the atmosphere. Ignoring body forces, determine the inlet pressure, p_i, as a function of ρ, U and atmospheric pressure, p_a, if the force exerted by the fluid on the device is $\mathbf{F} = 1.05\pi\rho U^2 d^2\,\mathbf{i}$.

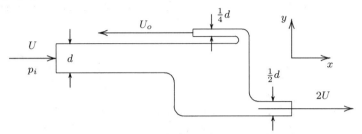

Problems 5.76, 5.77

5.77 An incompressible fluid of density ρ flows steadily through the device shown. The inlet velocity is U. The fluid at both exits exhausts to the atmosphere. Ignoring body forces, determine the force exerted by the fluid on the device as a function of ρ, U and diameter d if the inlet pressure is $p_i = p_a + \frac{2}{5}\rho U^2$.

5.78 The device shown has fluid of density ρ and velocity U entering at an angle ϕ. The flow is steady, incompressible and no body forces are acting. The fluid turns after entering the chamber and exits at the right to the atmosphere. Also, there is a small bleed valve in the upper left corner to regulate the flow. The bleed valve exit pressure is the same as that of the atmosphere, p_a. Determine the bleed-valve exit velocity, V, and the force required to hold the device in place, $\mathbf{F} = F_x\,\mathbf{i} + F_y\,\mathbf{j}$, as a function of ρ, U, A and ϕ.

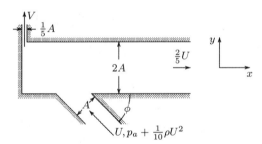

Problem 5.78

5.79 The nozzle shown below discharges water into the atmosphere through two outlets. You may assume the flow is steady, irrotational and you may neglect effects of gravity. The jet velocities, V, (for which you must solve) are equal. Determine the force through the mounting bolts required to hold the nozzle in place as a function of fluid density, ρ, inlet velocity, U, cross-sectional area, A, and the angle ϕ.

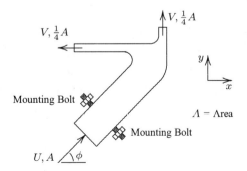

Problem 5.79

5.80 An incompressible fluid of density ρ flows through the device shown below. The flow is steady and irrotational. Fluid exits to the atmosphere at all three outlets. Determine the inlet overpressure, Δp, the outlet velocity, U_o, and the force required to hold the device in place. A denotes area and no body forces are acting. Also, assume flow properties are constant on all cross sections.

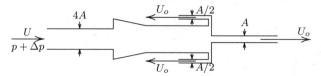

Problem 5.80

5.81 An incompressible fluid flows steadily and irrotationally through the two-dimensional channel as shown and exhausts to the atmosphere. What must the angle ϕ be in order for the vertical component of the force on the channel walls to vanish, and what is the horizontal component of the force for this angle? Assume properties are constant across the channel and body forces can be ignored. Express your answers in terms of ρ, U, and A. **NOTE:** V, Δp_1 and Δp_2 are not given, you must solve for them.

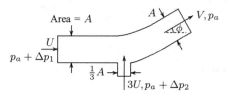

Problem 5.81

5.82 Consider steady, incompressible flow into a $180°$ bend. The pipe diameters at Sections 1 and 2 are $D_1 = D$ and $D_2 = 1.10D$, while the pressure at both Sections 1 and 2 is $p_a + \Delta p$, where p_a is atmospheric pressure. Assume the velocity and pressure are constant on all cross sections and that body forces can be ignored. If the horizontal force exerted by the fluid on the portion of the pipe between Sections 1 and 2 is $F_x = \frac{1}{2}\pi\rho U_1^2 D^2$, what is the overpressure, Δp? Do conditions at Sections 1 and 2 satisfy Bernoulli's equation?

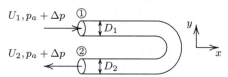

Problems 5.82, 5.83

5.83 Consider steady, incompressible flow into a $180°$ bend. The pipe diameter is constant so that $D_1 = D_2 = D$. The pressure at both Sections 1 and 2 is $p_a + \Delta p$, where p_a is atmospheric pressure. Assume the velocity and pressure are constant on all cross sections and that body forces can be ignored. What is the force exerted by the fluid on the portion of the pipe between Sections 1 and 2? Do conditions at Sections 1 and 2 satisfy Bernoulli's equation?

5.84 Consider steady, incompressible flow into a $180°$ bend that springs a leak. Body forces can be ignored, and the flow is irrotational. The cross-sectional areas at the inlet and the outlet are the same and equal to A, while the area of the jet of fluid emanating from the leak is $\frac{1}{25}A$. Assume flow properties are constant on all cross sections, and that p_a is atmospheric pressure. Solve for the jet velocity, U_j, the pressure differentials, Δp_1 and Δp_2, and the force exerted by the fluid on the bend.

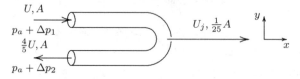

Problem 5.84

5.85 A pipe carries water to two jets that discharge into the atmosphere from the pipe as shown. The exit velocity for both jets is V and the jet diameters are half the inlet pipe diameter. What force is required to hold the pipe in place? Express your answer in terms of fluid density, ρ, jet velocity, V, and pipe diameter, D. You may assume the flow to be steady, irrotational and that body forces can be ignored.

Problem 5.85

5.86 Consider a nozzle through which an incompressible fluid of density ρ flows and discharges to the atmosphere. You may assume the flow is irrotational and that all flow properties are uniform across the nozzle. Find the force, as a function of ρ, U and D, required to hold the nozzle in place.

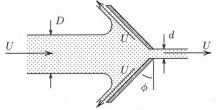

Problem 5.86

5.87 A round jet of diameter D, density ρ, and speed U flows toward a cone concentrically and exits through a hole of diameter d at its center. What is the force required to hold the cone in place if the jet issuing from the hole also has speed U? You may assume the flow is incompressible, steady and that body forces can be neglected.

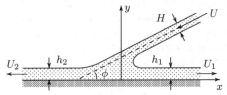

Problem 5.87

5.88 A two-dimensional jet of velocity U and width H impinges on a wall at an angle ϕ. The fluid is incompressible and has density ρ. Far from the impingement point, the velocities of the *wall jets* are equal so that $U_1 = U_2 = U$. Neglect body forces and assume the flow is steady, inviscid, and that velocities are uniform across the jets. Determine h_1 and h_2 as functions of H and ϕ.

Problems 5.88, 5.89

5.89 A two-dimensional jet of velocity U and width H impinges on a wall at an angle ϕ. The fluid is incompressible and has density ρ. Far from the impingement point, the widths of the *wall jets* are equal so that $h_1 = h_2 = h$. Neglect body forces and assume the flow is steady, inviscid, and that velocities are uniform across the jets.

(a) What is the force on the wall? *Be sure to indicate direction.*

(b) What are the velocities U_1 and U_2 as functions of H, h, U and ϕ?

(c) Check the limiting case $\phi = 90°$ and explain why your answer to Part (b) is sensible.

5.90 A water jet of density ρ with initial velocity W and diameter d_o issues vertically from a wall and supports a hemispherical shell at height h as shown. You can assume the flow is steady, irrotational and that velocities are constant across all cross sections. Also, assume the volume of the fluid contained in the indicated control volume is $V = \frac{\pi}{2}d_o^2 h$. **NOTE:** Neither the downflow velocity, w, nor the area of the downflowing fluid are given, you must solve for them.

(a) Using the indicated control volume, determine the weight of the hemispherical shell, Mg.

(b) Compute Mg for $\rho = 1.94$ slug/ft^3, $W = 20$ ft/sec, $d_o = 2$ in and $h = 9$ in.

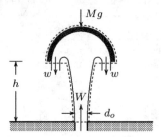

Problems 5.90, 5.91

5.91 A water jet of density ρ with initial velocity W and diameter d_o issues vertically from a wall and supports a hemispherical shell at height h as shown. You can assume the flow is steady, irrotational and that velocities are constant across all cross sections. Also, you can ignore the weight of the fluid in the control volume. **NOTE:** Neither the downflow velocity, w, nor the area of the downflowing fluid are given, you must solve for them.

(a) Using the indicated control volume, determine the weight of the hemispherical shell, Mg.

(b) Compute Mg for $\rho = 1000$ kg/m^3, $W = 25$ m/sec, $d_o = 8$ cm and $h = 2$ m.

5.92 A water jet of density ρ with initial velocity W and diameter d_o issues vertically from a wall and supports a conical object that rises to height h as shown. You can assume the flow is steady, irrotational and that velocities are constant across all cross sections. Also, you can ignore the weight of the fluid in the control volume. **NOTE:** Neither the upflow velocity, w, nor the area of the upflowing fluid at the base of the cone are given, you must solve for them.

(a) Using the indicated control volume, determine the weight of the cone, Mg.

(b) Compute Mg for $\rho = 1.94$ slug/ft^3, $W = 20$ ft/sec, $d_o = 2$ in, $h = 1$ ft, $\phi = 30°$.

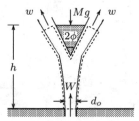

Problems 5.92, 5.93

5.93 A water jet of density ρ with initial velocity W and diameter d_o issues vertically from a wall and supports a conical object that rises to height h as shown. You can assume the flow is steady, irrotational and that velocities are constant across all cross sections. Also, assume the volume of the fluid contained in the indicated control volume is $V = \frac{\pi}{3}d_o^2 h$. **NOTE:** Neither the upflow velocity, w, nor the area of the upflowing fluid at the base of the cone are given, you must solve for them.

(a) Using the indicated control volume, determine the weight of the cone, Mg.

(b) Compute Mg for $\rho = 998$ kg/m^3, $W = 20$ m/sec, $d_o = 10$ cm, $h = 1$ m, $\phi = 33°$.

5.94 An incompressible jet of fluid of density ρ, velocity W and diameter d_o issues vertically from a wall and supports a disk of mass M and diameter D. We want to determine the height h to which the disk will rise. Solve this problem using two control volumes, CV_1 and CV_2, as shown. You can assume the flow is steady, the flow in CV_1 is irrotational, and velocities are constant across all cross sections.

(a) Derive the basic equations required to solve the problem for the three unknown quantities w, d and h. You can regard ρ, W, d_o, M and τ as known quantities.

(b) Now, simplify the equations derived in Part (a) assuming that $\tau \ll h$ and that $\rho D^2 \tau \ll M$.

(c) Solve for the height h. Express your answer in terms of the dimensionless grouping $2gh/W^2$.

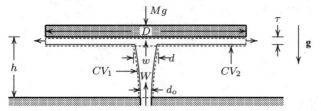

Problem 5.94

5.95 Water flows steadily and irrotationally up a vertical pipe of diameter D as shown, and flows radially outwards between two circular, horizontal end plates. The diameter of the end plates is $2D$.

(a) Assuming the distance between the plates is $t = D/16$, determine the radial velocity, v, as a function of the inlet velocity, U.

(b) Assuming $t \ll h$ and the pressure at the inlet is $p_a + 3\rho g h$ (p_a is atmospheric pressure), determine the inlet velocity, U, as a function of g and h.

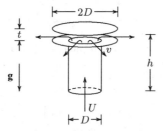

Problem 5.95

5.96 A large reservoir contains a liquid of density ρ. At depth h there is a small hole of diameter d through which a jet of velocity V issues. The jet strikes a cube with side $8d$ made of a material whose density is 1.75ρ. The coefficient of sliding friction between cube and surface is 0.75. Assume the flow is steady, incompressible, irrotational, and neglect effects of gravity *in the jet*.

(a) What is the jet velocity?

(b) What is the force on the cube from the jet? **NOTE:** Ignore any horizontal momentum in the fluid as it moves beyond the edges of the cube.

(c) What is the maximum value of h for which the cube will not move?

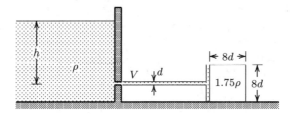

Problem 5.96

5.97 An incompressible fluid of density ρ flows steadily through the nozzle shown and discharges to the atmosphere. The pressure at the inlet is $p = p_a + \frac{1}{4}\rho U^2$. Assume the pressure is uniform on all cross sections, body forces are unimportant and that velocity is uniform on the inlet plane.

(a) Determine the force required to hold the nozzle in place assuming u is uniform $[u(r) = U_e]$ on the exit plane.

(b) Determine the force if $u(r) = U_e(1 - 4r/D)^{1/7}$ on the exit plane, where r is radial distance from the centerline and D is the initial nozzle diameter. How does the force differ from the value determined in Part (a)? **HINT:** To make evaluation of the integrals on the exit plane easier, use the change of integration variable defined by $\eta = 1 - 4r/D$.

Problems 5.97, 5.98

5.98 An incompressible fluid of density ρ flows steadily through the nozzle shown and discharges to the atmosphere. The pressure at the inlet is $p = p_a + \frac{1}{8}\rho U^2$. Assume the pressure is uniform on all cross sections, body forces are unimportant and that velocity is uniform on the inlet plane.

(a) Determine the force required to hold the nozzle in place assuming u is uniform $[u(r) = U_e]$ on the exit plane.

(b) Determine the force if $u(r) = U_e(1 - 16r^2/D^2)$ on the exit plane, where r is radial distance from the centerline and D is the initial nozzle diameter. How does the force differ from the value determined in Part (a)?

5.99 Because of an imposed pressure difference, the velocity profile in a very wide channel changes from $U_1(y) = 2U[y/h - \frac{1}{2}(y/h)^2]$ to $U_2 = K[y/h + (y/h)^2 - (y/h)^3]$. The profiles are valid for $0 \le y \le h$, and the flow is symmetric about the centerline at $y = h$. Assuming the flow is steady and incompressible, solve for the constant K.

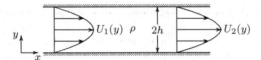

Problem 5.99

5.100 Consider steady, incompressible flow of a fluid with density ρ in a pipe of constant radius R with two outlets as shown. Although the configuration is symmetric in every other respect, the inner surface of the upper outlet is rough while the lower outlet is smooth. As a result, the velocities in the upper and lower outlets are $V_u(r) = 1.173V(1 - r/R)^{1/9}$ and $V_\ell(r) = 1.224V(1 - r/R)^{1/7}$, respectively. Body forces are unimportant and both outlets exhaust to the atmosphere.

(a) Verify that the average velocity at both outlets is V. Ignoring the variation of the velocities on cross sections, what is the vertical force required to hold the pipe in place?

(b) Taking account of the variable velocities, determine the force as a function of ρ, V and R. **HINT:** To make evaluation of the integrals on the exit plane easier, use the change of integration variable defined by $\eta = 1 - 4r/D$.

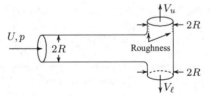

Problem 5.100

5.101 At the inlet of a two-dimensional *reducing bend*, the velocity varies linearly with distance across the channel and the pressure is unknown. The flow is uniform at both outlets, incompressible with density ρ and exhausts to the atmosphere. The flow is steady and body forces can be neglected. Determine V as a function of U, and compute the vertical force exerted by the fluid on the bend as a function of ρ, U, h and α.

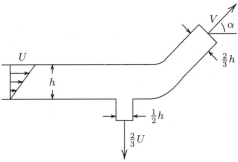

Problem 5.101

5.102 Near the entrance to a pipe, for laminar viscous flow the velocity distribution changes from uniform ($u_1 = U$) to parabolic as shown. At the fully-developed outlet section, the velocity varies according to $u_2(r) = U_{\max}[1 - 4(r/D)^2]$. Derive a formula for the net resisting frictional force, F_τ, on the pipe section as a function of U, ρ, p_1, p_2 and D.

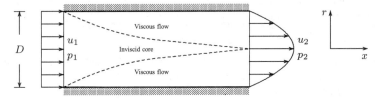

Problem 5.102

5.103 A circular cylinder of diameter D and length L (out of page) is mounted in a *rectangular* duct. Far upstream, the flow is uniform with velocity U_∞ and pressure p_1. Far downstream, the velocity varies with y while the pressure is p_2. Using the stationary control volume indicated, the symmetry of the geometry, the fact that the drag coefficient is $C_D = F_D/(\rho U_\infty^2 DL) = 2$, and that the outlet velocity is $u_2 = U(y/3D)^2$ where U is a constant to be determined, determine $p_1 - p_2$. Ignore effects of friction on the duct walls and assume pressure is independent of y at the duct inlet and outlet.

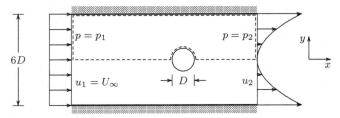

Problems 5.103, 5.104

5.104 A circular cylinder of diameter D and length L (out of page) is mounted in a *rectangular* duct. Far upstream, the flow is uniform with velocity U_∞ and pressure p_1. Far downstream, the velocity varies with y while the pressure is p_2. Using the stationary control volume indicated, the symmetry of the geometry, the fact that the drag coefficient is $C_D = F_D/(\rho U_\infty^2 DL) = 1$, and that the outlet velocity is $u_2 = U(y/3D)^4$ where U is a constant to be determined, determine $p_1 - p_2$. Ignore effects of friction on the duct walls and assume pressure is independent of y at the duct inlet and outlet.

5.105 An incompressible fluid of density ρ enters a square duct of constant height h with uniform velocity U. The duct makes a 90° bend that distorts the flow to produce the linear velocity profile shown at the exit, with $v_{max} = 2v_{min}$. You may assume the flow is steady and that body forces are negligible. Determine v_{min} as a function of U. Also, assuming Bernoulli's equation holds between the inlet and the low-speed side of the exit, determine the inlet pressure coefficient, $C_p = (p - p_a)/(\frac{1}{2}\rho U^2)$.

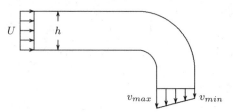

Problem 5.105

5.106 The figure depicts steady, incompressible, viscous flow above a flat plate. The velocity above is $u(x, y) = U_\infty f(y/\delta)$, where U_∞ is the constant freestream velocity and $\delta(x)$ is the thickness of the viscous region. Pressure is constant and equal to p_∞. Assume $u = U_\infty$ for $y \geq \delta$.

(a) Show that, with ρ denoting fluid density, the drag force per unit width (out of the page) is

$$F_D = \int_0^\delta \rho(U_\infty - u)\, u\, dy$$

(b) For turbulent flow, a good approximation for very high Reynolds numbers is $u = U_\infty (y/\delta)^{1/7}$. Compute F_D as a function of ρ, U_∞ and δ.

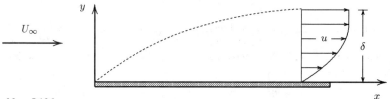

Problem 5.106

5.107 Consider steady, incompressible, viscous flow above a flat plate of length L with surface mass removal, which is referred to as *suction*. The velocity above the plate is given by $u(x, y) = U_\infty f(y/\delta)$, where U_∞ is the constant freestream velocity and $\delta(x)$ is the thickness of the viscous region. Pressure is constant and equal to p_∞, and you can assume that $u = U_\infty$ for $y \geq \delta$. The surface velocity is given by $\mathbf{u} = -C_Q U_\infty \mathbf{j}$, where C_Q is the constant suction coefficient. Show that, with ρ denoting fluid density and $\dot{m} = \rho U_\infty C_Q L$, the drag force per unit width (out of the page) is

$$F_D = \dot{m} U_\infty + \int_0^\delta \rho(U_\infty - u)\, u\, dy$$

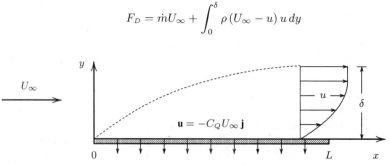

Problem 5.107

5.108 A small cart receives water from a jet as shown. The speed and cross-sectional area of the jet are U_j and A, respectively. At time $t = 0$, the cart is at rest and its mass is m_o. Assume the motion is such that aerodynamic forces on the cart are negligible, i.e., ignore viscosity and assume $p = p_a$ at the cart surface, where p_a is atmospheric pressure. The only force acting on the cart is rolling friction, μmg, where μ is the coefficient of rolling friction, m is the instantaneous mass of the cart and g is gravitational acceleration. Assume the motion of the fluid inside the cart is steady as seen by an observer riding on the cart.

(a) Using the translating (and accelerating) control volume indicated by the dashed contour, develop differential equations for m and U that describe the motion of the cart.

(b) Assuming the surface is frictionless so that $\mu = 0$, combine the equations developed in Part (a) into a differential equation involving dU/dm.

(c) Solve the equation derived in Part (b) and determine the velocity when $m(t) = 5m_o$.

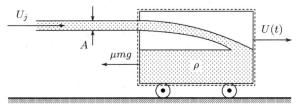

Problem 5.108

5.109 Captain Picard is returning to the Enterprise in a shuttle when Klingons turn on their *tractor beam*. To counter the beam's effect, i.e., to continue moving on the same course at a constant velocity, V, Picard fires a rocket as shown. The tractor-beam force is

$$\mathbf{F} = -T^2 M^2 \mathbf{i}$$

where T is a constant, and M is the total mass of the shuttle (fuel plus vehicle). Assuming fuel density, ρ_e, velocity, u_e, and exit area, A_e, are constant at the rocket exit plane, and that pressure at the rocket exit plane is negligibly small, determine the total mass of Picard's shuttle as a function of time. The initial mass (at time $t = 0$) is M_o.

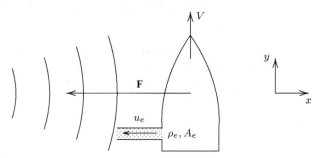

Problems 5.109, 5.110

5.110 Captain Kirk is returning to the Enterprise in a shuttle when Klingons turn on their *tractor beam*. To counter the beam's effect, i.e., to continue moving on the same course at a constant velocity, V, Kirk fires a rocket as shown. The tractor-beam force is

$$\mathbf{F} = -B^2 M^4 \mathbf{i}$$

where B is a constant, and M is the total mass of the shuttle (fuel plus vehicle). Assuming fuel density, ρ_e, velocity, u_e, and exit area, A_e, are constant at the rocket exit plane, and that pressure at the rocket exit plane is negligibly small, determine the total mass of Kirk's shuttle as a function of time. The initial mass (at time $t = 0$) is M_o.

5.111 A small boat is powered by an air jet of diameter d, density ρ_e and velocity u_e relative to the boat. The density of water is ρ and the hydrodynamic drag on the boat's hull is $F_D = \rho U^2 A C_D$, where U is the speed of the boat, A is an effective cross-sectional area and C_D is the drag coefficient. You can neglect any other form of drag on the boat including the net pressure force.

(a) Derive appropriate equations for conservation of mass and momentum.

(b) Show that the instantaneous velocity, $U(t)$, is

$$U(t) = U_f \left[\frac{1 - (M/M_o)^{2u_e/U_f}}{1 + (M/M_o)^{2u_e/U_f}} \right]$$

where $M(t)$ is the instantaneous mass of the boat and compressed air, $M_o = M(0)$ and U_f is a velocity to be determined as part of the solution.

(c) Compute the steady-state speed, U_f, in knots for $\rho A C_D = 20$ kg/m, $d = 8$ cm, $\rho_e = 1.2$ kg/m^3 and $u_e = 100$ m/sec.

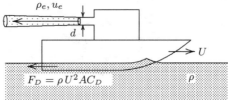

Problems 5.111, 5.112

5.112 A small boat is powered by an air jet of diameter d, density ρ_e and velocity u_e relative to the boat. The density of water is ρ and the hydrodynamic drag on the boat's hull is $F_D = \rho U^2 A C_D$, where U is the speed of the boat, A is an effective cross-sectional area and C_D is the drag coefficient. You can neglect any other form of drag on the boat including the net pressure force.

(a) Derive appropriate equations for conservation of mass and momentum.

(b) Solve for the steady-state velocity, U, in terms of u_e, ρ_e, ρ, d, A and C_D.

(c) Compute the steady-state speed in knots for $\rho A C_D = 0.60$ slug/ft, $d = 2$ in, $\rho_e = 0.00234$ slug/ft^3 and $u_e = 400$ ft/sec.

5.113 A tank of water sits on a sled. High pressure in the tank is maintained by a compressor so that the water leaving the tank through the orifice does so at a constant speed, u_e, relative to the tank. The orifice area is A. The instantaneous mass of the sled, water, tank and compressor is M, the water density is ρ and the coefficient of sliding friction between sled and ice is μ_s. The sled starts from rest with an initial mass, M_o. Ignore the aerodynamic drag on the sled.

(a) Derive appropriate equations for conservation of mass and momentum.

(b) Solve for the instantaneous velocity, $U(t)$, in terms of u_e, M, M_o, μ_s, g and t (g = gravitational acceleration, t = time).

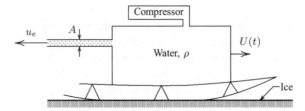

Problem 5.113

Chapter 6

Conservation of Energy: Integral Form

In Chapter 5, we found that some fluid-flow problems arise in which we need an additional equation to have a complete description. We have been able to use Bernoulli's equation as an energy-conservation principle, but only when several limiting conditions hold.

We can understand the origin of the problem by simply counting unknowns and equations. We obtain four equations from the conservation laws, viz., one from mass conservation (a scalar equation) and three from momentum conservation (a vector equation). A fifth equation comes from the thermodynamic equation of state, which is

$$\rho = \begin{cases} \dfrac{p}{RT}, & \text{gases} \\[2mm] \text{constant}, & \text{liquids} \end{cases} \qquad (6.1)$$

Considering liquids first, the unknowns are density, ρ, pressure, p, and the three velocity components, (u, v, w). Thus, we have a total of five unknowns. Our mathematical system is said to be closed as we have a sufficient number of equations to solve for the unknowns.

Turning to gases, note that the equation of state introduces temperature as an additional unknown. We thus have six unknowns for a gas. However, conservation of mass, momentum and the state equation still account for only five equations. Our system is not closed as we lack a sufficient number of equations to solve for all of the unknowns.

Actually, we don't have enough equations for a liquid either if temperature is required, as it would be for a flow with heat transfer. There are many practical flows for which thermal energy cannot be ignored in the overall energy balance. Common examples are compressible flows and incompressible flows with heat transfer. Thus, we must consider conservation of energy, explicitly including thermal energy, to complete our analysis.

This chapter first provides a brief overview of thermodynamics. We then derive the integral conservation law for energy in a control volume. We also develop a simplified form of the energy equation that can be used in place of Bernoulli's equation when heat transfer and frictional losses are important. The chapter concludes with a section on pipe flow and a section on open-channel flow. These topics are ideally suited for control-volume analysis and illustrate the type of approach to fluid-mechanics problems characteristic of the nineteenth-century era of hydraulics.

6.1   Thermodynamics

Before we can formulate an equation for conservation of energy, we require some familiarity with a few basic concepts of **thermodynamics**. The science of thermodynamics deals with relations between heat and work. Its foundation consists of two general laws of nature, i.e., the first and second laws of thermodynamics. This section presents a brief overview of thermodynamics, focusing on those concepts that will be needed in the following sections. It is not intended as a thorough exposition of the topic. For a more complete introduction to thermodynamics, see an introductory text such as Lee and Sears (1963), Wark (1966), Reynolds and Perkins (1977), Van Wylen and Sonntag (1986), or Cengel and Boles (1994).

6.1.1   Fundamental Concepts

There are four fundamental concepts from thermodynamics central to formulation of an energy-conservation principle. The first concerns basic definitions, most of which we have already discussed in Section 4.5 and Chapter 5. The second is the concept of thermodynamic equilibrium. The third is the type of process, while the fourth is that of the equation of state and state variables.

- **Basic Definitions.** We have already used some of the nomenclature of thermodynamics in formulating the control-volume method. The concepts of a system, a control volume, extensive properties and intensive properties that we introduced earlier all find their origins in classical thermodynamics.

  The primary focus in thermodynamics is usually on a *system*, which refers to "a definite quantity of matter bounded by some closed surface which is impervious to the flow of matter" [cf. Lee and Sears (1963)]. The surface can be either solid or a streamsurface, i.e., a surface parallel to streamlines. This is entirely consistent with the definition presented in Section 4.5, which, in the context of a fluid, is *the same collection of fluid particles for all time*.

  Just as in fluid mechanics, thermodynamicists find it convenient to introduce the concept of the control volume, which is sometimes referred to as an *open system*. The control volume is bounded by a control surface, across which matter may flow. There is no difference between thermodynamic control volumes and those analyzed and discussed in Chapter 5.

  Everything outside the boundary of a system is referred to as the *surroundings*. As we will see, distinguishing between a system (or control volume) and the surroundings is important in determining the transfer direction of quantities such as heat.

  Extensive and intensive properties are usually defined for a thermodynamic system as follows. We imagine that the system is divided into several parts. A property of the system whose value, for the entire system, is equal to the sum of its values for the separate parts of the system, is referred to as an extensive property. Extensive properties are a function of the quantity of matter (fluid) present, while intensive properties are not. Volume, mass and total energy are all examples of extensive properties. Properties that are independent of the size of the system (supposing them to be the same in all parts of the system) are known as intensive properties. Examples are temperature, pressure and density.

  As noted in Section 4.5, a property expressed per unit mass is called a *specific* property. Thus, if M is the total mass of a volume of fluid moving with constant speed U, then

$\frac{1}{2}MU^2$ is the total kinetic energy, an extensive property. By contrast, $\frac{1}{2}U^2$ is the specific kinetic energy.

- **Thermodynamic Equilibrium.** In general, when a system with nonuniform properties is isolated from interaction with its surroundings, those properties will generally change as time passes. If the pressure varies, for example, parts of the system may move, perhaps expanding or contracting. Ultimately, such motion stops, and when this occurs, we say the system has reached a state of *mechanical equilibrium.*

  In thermodynamics, we consider more properties than those associated with mechanical issues. If the temperature varies through the system, for example, we observe that it varies in a manner that may or may not be directly coupled with the mechanical motion. In fact, the temperature can vary on a time scale quite different from that of the mechanical time scale, and can even occur in the absence of all motion. When all changes in temperature cease, the system is in *thermal equilibrium.*

  Finally, the most general system can contain substances that undergo chemical reactions. The reactions will occur on a third time scale dictated by the nature of the reactants. The chemical time scale is often much shorter than the mechanical and thermal time scales, but in any case is distinct from the former two. When all chemical-reaction activity ceases, a state of *chemical equilibrium* is reached.

  A system that is simultaneously in mechanical, thermal and chemical equilibrium is said to be in *thermodynamic equilibrium.*

- **Processes.** The strict definition of thermodynamic equilibrium requires a totally static end state for a system. A natural question to pose is, of what value is the concept of thermodynamic equilibrium in the study of a fluid in motion? The answer is contained in the classical approach to thermodynamics, which addresses systems changing in such a manner that the departure from equilibrium is infinitesimally small. That is, a system that changes in time is viewed as taking a succession of small steps from one equilibrium state to another. This methodology is often described as a *quasi-static* or a *quasi-steady* approach. We used the quasi-steady approach to simplify the analysis of the leaky tank in Section 3.5.

  Many processes are characterized by having some property of the system held constant during the process. We generally append the prefix "iso" to name the process. For example, a constant-volume process is called *isovolumetric*, while a constant-energy process is called *isoenergetic*. Somewhat more subtle variations sometimes occur in naming such processes. That is, a constant-temperature process is called *isothermal*, while a process with constant pressure is referred to as *isobaric* (from the Greek word *baros*, which means heavy—an isobar on a weather map is a line of constant pressure).

  We call a process *reversible* if, upon completion of the process, the initial state of *both the surroundings and the system* can be restored. This will be possible only if the process remains in thermodynamic equilibrium at all times, and has no losses due, for example, to friction. If the initial states of the system and its surroundings cannot be simultaneously returned to their initial state, the process is *irreversible*. We will discuss reversible and irreversible processes in more detail below in the context of entropy and the second law of thermodynamics.

  Finally, when a process involves no transfer of heat between a system and its surroundings, we call the process *adiabatic.*

- **Equation of State and State Variables.** Experimentation has shown that, for a pure substance, there is a lower bound on the number of properties of the substance that can be given arbitrary values. That is, there is a distinct number of independent variables (properties) sufficient to define the thermodynamic state of a system. The functional form of the relationship amongst the properties is called the *equation of state*.

Even more significantly, when the properties of a system are expressed exclusively in terms of intensive properties, the equation of state of a substance can be expressed in terms of *just three* such properties. The perfect-gas law is an example—pressure, p, density, ρ, and temperature, T, are related according to $p = \rho R T$, where R is the perfect-gas constant for the gas under consideration. In general, we can write

$$F(p, \rho, T) = 0 \tag{6.2}$$

where F is the appropriate functional relationship for the substance. The interrelated properties appearing in the state equation are referred to as *state variables*, and are understood to correspond to a given equilibrium state. The practical implication of having such a relationship is the following: Equation (6.2) provides an implicit relationship defining any one of the state variables as a function of the other two, i.e., any two state variables define the state uniquely.

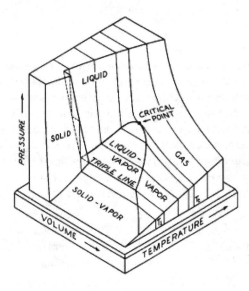

Figure 6.1: *p-v-T diagram for water.*

Of course, any three state variables can appear as variables in the equation of state—it is not restricted to p, ρ and T. For example, Figure 6.1 graphically depicts the equation of state for water as a three-dimensional surface with pressure, volume and temperature (p, v, T) as the state variables. We can make two interesting observations based on the shape of this surface. First, note that volume increases at temperature T_1 as water passes from liquid to solid state (ice). This corresponds to the fact that water expands when it freezes. Second, *vapor* is defined as a gas in equilibrium with its liquid. At the *critical point*, the specific volumes of liquid and vapor are equal.

Example 6.1 *A rigid tank contains 200 ft³ of air at a pressure of 125 psi and a temperature of 230°F. If the temperature decreases to 70°F, what is the new pressure?*

Solution. We use the perfect-gas equation of state, viz.,

$$p = \rho R T$$

Since the volume and mass of air remain constant, necessarily the density is unchanged. Hence, denoting initial and final conditions with subscripts i and f, respectively, we have

$$\frac{p_f}{p_i} = \frac{T_f}{T_i} \qquad \Longrightarrow \qquad p_f = p_i \frac{T_f}{T_i}$$

Since temperature must be expressed in degrees Rankine, there follows

$$p_f = (125 \text{ psi}) \frac{(70 + 459.67)°\text{R}}{(230 + 459.67)°\text{R}} = 96 \text{ psi}$$

6.1.2 The First Law of Thermodynamics

The first law of thermodynamics is the basic energy-conservation principle relating all modes of energy transfer. In words, the first law says that the change in a system's energy, dE, equals the heat added to the system, δQ, minus the work done by the system on its surroundings, δW. Thus, we can say

$$dE = \delta Q - \delta W \tag{6.3}$$

In writing this equation, we make a distinction between **exact** and **inexact differentials**. Because the energy is a state variable, it depends only on the initial and final states. Thus, a differential change is independent of the process, and we represent this as a perfect differential, dE. By contrast, both the heat added and the work done are very much dependent on the process. For example, the heat added might occur as a result of radiation or thermal conduction, which are quite different processes. Similarly, work might be done under isothermal conditions or perhaps at constant pressure, again quite different processes. Simply specifying the end states in the limits of an integral is insufficient as the integration path affects the final result. We distinguish this difference by denoting the differential changes as inexact differentials, δQ and δW.

A useful result that holds for a reversible process is as follows. If we let V denote volume, the work done in a reversible process (sometimes called **piston work**) is $\delta W = p dV$. Thus, we can rewrite the first law as

$$dE = \delta Q - p dV \qquad \text{(Reversible process)} \tag{6.4}$$

As a corollary result, we can also write the first law of thermodynamics in terms of specific variables. When we do this, we introduce the specific internal energy, e, heat added per unit mass, q, and specific volume, $v = 1/\rho$. The first law becomes:

$$de = \delta q - p d(1/\rho) \qquad \text{(First Law per unit mass)} \tag{6.5}$$

6.1.3 The Second Law of Thermodynamics

The second law of thermodynamics makes a statement about the directionality of thermodynamic processes. We need it because the first law is insufficient as it stands. To understand why, consider a familiar situation. Suppose we place an ice cube in contact with a hot object. Experience and common sense tells us the ice cube will heat up (and melt) while the object will have its temperature reduced. However, the first law does not prevent the reverse from occurring, so long as energy is conserved.

The second law of thermodynamics says, in words, that the **entropy** of a closed system and its surroundings must always increase. The entropy, s, is a state variable defined by

$$ds = \frac{\delta q_{rev}}{T} \tag{6.6}$$

where δq_{rev} denotes an incremental amount of heat added reversibly to the system, and T is the system temperature. Note that δq_{rev} is not necessarily the actual heat added to the system. It is the effective heat that would have to be added reversibly to achieve the differential change in entropy from one equilibrium state to another according to Equation (6.6). If the actual heat added to the system is δq, we can rewrite this equation as

$$ds = \frac{\delta q}{T} + ds_{irrev} \tag{6.7}$$

where ds_{irrev} is the change in entropy due to irreversible processes such as viscous dissipation, heat conduction and mass diffusion within the system. Nature dictates that these processes always cause the entropy to increase, i.e., that

$$ds_{irrev} \geq 0 \tag{6.8}$$

We have equality in the special case of a reversible process. Since δq is positive by definition, the second law states that the **Clausius inequality** holds, viz.,

$$ds \geq \frac{\delta q}{T} \tag{6.9}$$

6.1.4   Combined First and Second Laws

One of the most useful equations of thermodynamics is known as **Gibbs' equation**. It is the result of combining the first and second laws, and permits a quantitative measure of the entropy of a system. To derive this equation, consider a reversible process. Then, the first law of thermodynamics is given by Equation (6.5) so that, per unit mass,

$$de = \delta q - pd(1/\rho) \tag{6.10}$$

Now, since the process is reversible, necessarily we have $ds_{irrev} = 0$ in Equation (6.7), wherefore

$$\delta q = T\,ds \tag{6.11}$$

Thus, combining Equations (6.10) and (6.11), we can rewrite the first law as

$$T ds = de + pd(1/\rho) \tag{6.12}$$

Equation (6.12) is the famous **Gibbs' equation**. Because it involves only state variables (and hence perfect differentials), it relates conditions between different equilibrium states, and is thus independent of the process. That is, Gibbs' equation holds for both reversible and irreversible processes.

As an example of how Gibbs' relation permits developing a quantitative expression for entropy, consider a perfect gas that is also calorically perfect. Because we have a perfect gas, the equation of state is $p = \rho R T$. By definition (see Section 1.5), a calorically-perfect gas has constant specific-heat coefficients, c_p and c_v, so that the internal energy and enthalpy are given by $e = c_v T$ and $h = c_p T$. Thus, from the definition of enthalpy,

$$h = e + \frac{p}{\rho} = c_v T + R T = (c_v + R)T \tag{6.13}$$

Now, since $h = c_p T$,

$$c_p T = (c_v + R)T \quad \Longrightarrow \quad c_p = c_v + R \tag{6.14}$$

Additionally, if we make use of the fact that the specific-heat ratio is defined by $\gamma = c_p/c_v$, then

$$\gamma = \frac{c_p}{c_v} = \frac{c_v + R}{c_v} \quad \Longrightarrow \quad c_v = \frac{R}{\gamma - 1}, \quad c_p = \frac{\gamma R}{\gamma - 1} \tag{6.15}$$

So, for a calorically-perfect gas, $de = c_v dT$, wherefore

$$T ds = c_v dT + d\left(\frac{1}{\rho}\right) = c_v dT - \frac{p}{\rho^2} d\rho \tag{6.16}$$

The differential change in entropy is

$$ds = c_v \frac{dT}{T} - \frac{p}{\rho T} \frac{d\rho}{\rho} \tag{6.17}$$

Then, using the perfect-gas law and noting from Equation (6.15) that the perfect-gas constant can be written as $R = (\gamma - 1)c_v$, we arrive at:

$$ds = c_v \frac{dT}{T} - R \frac{d\rho}{\rho} = c_v \frac{dT}{T} - (\gamma - 1)c_v \frac{d\rho}{\rho} \tag{6.18}$$

Hence, integration yields

$$s = c_v \, \ell n \left(\frac{T}{\rho^{\gamma - 1}}\right) + \text{constant} \tag{6.19}$$

But, we can rewrite the argument of the natural logarithm as follows.

$$\frac{T}{\rho^{\gamma - 1}} = \frac{\rho T}{\rho^{\gamma}} = \frac{p}{R \rho^{\gamma}} \tag{6.20}$$

Thus, absorbing R in the constant, the entropy for a perfect gas is given by

$$s = c_v \, \ell n \left(\frac{p}{\rho^{\gamma}}\right) + \text{constant} \tag{6.21}$$

Of particular interest is the special case of **isentropic** flow, i.e., flow with constant entropy. Clearly, the argument of the logarithm is constant so that

$$p = A\rho^{\gamma}, \quad A = \text{constant} \tag{6.22}$$

Example 6.2 *Determine the temperature as a function of pressure for isentropic flow of a perfect gas.*

Solution. We use the perfect-gas equation of state and Equation (6.22), viz.,

$$p = \rho R T \quad \text{and} \quad p = A\rho^{\gamma}, \quad A = \text{constant}$$

So, the temperature is

$$T = \frac{p}{\rho R} = \frac{p}{(p/A)^{1/\gamma} R} = Bp^{(\gamma-1)/\gamma}$$

where $B = A^{1/\gamma}/R$ is a constant.

6.1.5   The First Law for a Moving Fluid

The basis of energy conservation for a fluid is the first law of thermodynamics. As noted above, the first law in its most elementary form says that, for a system:

$$dE = \delta Q - \delta W \tag{6.23}$$

where

$$
\begin{aligned}
E &= \text{Total energy of the system} \\
Q &= \text{Heat transferred to the system} \\
W &= \text{Work done by the system on the surroundings}
\end{aligned}
$$

Equivalently, assuming the flow remains close to thermodynamic equilibrium, we can regard the flow as being quasi-steady (in the thermodynamic sense) and rewrite Equation (6.23) as a differential equation, viz,

$$\frac{dE}{dt} = \dot{Q} - \dot{W} \tag{6.24}$$

where $\dot{Q} \equiv \delta Q/dt$ and where $\dot{W} \equiv \delta W/dt$. In the following sections, we will evaluate the terms in Equation (6.24) as functions of properties in a control volume. This will provide the final equation needed to analyze general fluid-flow problems.

6.2   Integral Form of the Energy Equation

Consider a system in which the only body forces acting are conservative, i.e., represented by $\mathbf{f} = -\nabla \mathcal{V}$, where $\mathcal{V}$ is the force potential. We also assume the force potential is independent of time. For general fluid motion, the total energy consists of the sum of internal energy, kinetic energy, and potential energy. Then, the appropriate extensive/intensive variable pair is

$$B = E \quad \text{and} \quad \beta = e + \frac{1}{2}\mathbf{u}\cdot\mathbf{u} + \mathcal{V} \tag{6.25}$$

Alternatively, we could exclude potential energy and include the body force in the work term—we will discuss this alternative below. The three terms in the intensive variable, β, are internal, kinetic, and potential energy per unit mass, respectively. Thus, the Reynolds Transport Theorem tells us that

$$\frac{dE}{dt} = \frac{d}{dt}\iiint_V \left[\rho\left(e + \frac{1}{2}\mathbf{u}\cdot\mathbf{u} + \mathcal{V}\right)\right] dV + \oiint_S \rho\left(e + \frac{1}{2}\mathbf{u}\cdot\mathbf{u} + \mathcal{V}\right)\left(\mathbf{u}^{rel}\cdot\mathbf{n}\right) dS \tag{6.26}$$

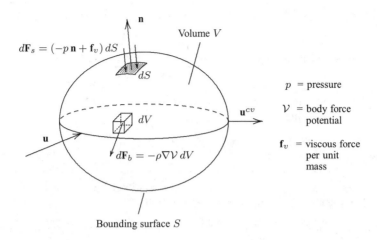

Figure 6.2: *A general control volume for energy conservation with a conservative body force.*

In general, the work done by a force, **F**, in moving a distance $d\mathbf{r}$ is defined by the dot product of **F** and $d\mathbf{r}$, i.e.,

$$dW = \mathbf{F} \cdot d\mathbf{r} = \mathbf{F} \cdot \frac{d\mathbf{r}}{dt}\, dt = \mathbf{F} \cdot \mathbf{u}\, dt \qquad (6.27)$$

The system does work on its surroundings through the surface force, $\mathbf{F}_s$. This force will do work on the surroundings only if the surface bounding the control volume is in motion relative to the surroundings, which means $\mathbf{u}^{rel}$ is the appropriate velocity. Thus,

$$\dot{W} = \oiint_S \mathbf{F}_s \cdot \mathbf{u}^{rel}\, dS \qquad (6.28)$$

The surface force exerted by a system on its surroundings is minus the force exerted by the surroundings on the system. With no loss of generality, we can say for a viscous fluid that the surface force acting on a system is $-p\,\mathbf{n} + \mathbf{f}_v$, where the vector $\mathbf{f}_v$ denotes the surface force due to viscosity (see Figure 6.3). Although a mathematical representation of $\mathbf{f}_v$ requires detailed analysis [cf. Wilcox (2000)], we can proceed with development of the energy equation using this symbolic representation of the friction force. So, the surface force exerted by the system on its surroundings is

$$\mathbf{F}_s = p\,\mathbf{n} - \mathbf{f}_v \qquad (6.29)$$

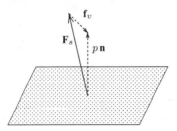

Figure 6.3: *Pressure and viscous forces on a surface element; the surface force, $\mathbf{F}_s$, can have both normal and tangential components.*

Thus, the rate at which the system does work on its surroundings becomes

$$\dot{W} = \underbrace{\oiint_S p\left(\mathbf{u}^{rel} \cdot \mathbf{n}\right) dS}_{Pressure\ Work} - \underbrace{\oiint_S \mathbf{f}_v \cdot \mathbf{u}^{rel} dS}_{Friction\ Work} \tag{6.30}$$

Substituting Equations (6.26) and (6.30) into the first law of thermodynamics [Equation (6.24)], we arrive at the following.

$$\frac{d}{dt} \iiint_V \left[\rho \left(e + \frac{1}{2}\mathbf{u}\cdot\mathbf{u} + \mathcal{V} \right) \right] dV + \oiint_S \rho \left(e + \frac{1}{2}\mathbf{u}\cdot\mathbf{u} + \mathcal{V} \right) \left(\mathbf{u}^{rel} \cdot \mathbf{n} \right) dS$$
$$= \dot{Q} - \oiint_S p\left(\mathbf{u}^{rel}\cdot\mathbf{n}\right) dS + \oiint_S \mathbf{f}_v \cdot \mathbf{u}^{rel} dS \tag{6.31}$$

Note that since we have not specified the type of heat transfer, we represent the net rate of heat transfer by the original symbol, $\dot{Q}$.

We can move the pressure-work integral from the right-hand to the left-hand side of this equation by first observing that

$$\oiint_S p\left(\mathbf{u}^{rel}\cdot\mathbf{n}\right) dS = \oiint_S \rho\left(\frac{p}{\rho}\right)\left(\mathbf{u}^{rel}\cdot\mathbf{n}\right) dS \tag{6.32}$$

We then combine the pressure-work integral with the energy-flux integral [i.e., the closed-surface integral on the left-hand side of Equation (6.31)]. To help accomplish this end, recall that the enthalpy, h, is defined by

$$h = e + p/\rho \tag{6.33}$$

This yields a slightly more compact energy-conservation equation, viz.,

$$\frac{d}{dt} \iiint_V \left[\rho \left(e + \frac{1}{2}\mathbf{u}\cdot\mathbf{u} + \mathcal{V} \right) \right] dV + \oiint_S \rho \left(h + \frac{1}{2}\mathbf{u}\cdot\mathbf{u} + \mathcal{V} \right) \left(\mathbf{u}^{rel} \cdot \mathbf{n} \right) dS$$
$$= \dot{Q} + \oiint_S \mathbf{f}_v \cdot \mathbf{u}^{rel} dS \tag{6.34}$$

In words, this equation says that *the instantaneous rate of change of total energy in the control volume plus the net flux of total enthalpy out of the control volume equals the sum of the heat transfer rate to the control volume and the friction work done by the control volume on the surroundings.* For purposes of using the integral form, all of the applications in this chapter involve the energy equation in the following form.

$$\frac{d}{dt} \iiint_V \left[\rho \left(e + \frac{1}{2}\mathbf{u}\cdot\mathbf{u} + \mathcal{V} \right) \right] dV + \oiint_S \rho \left(h + \frac{1}{2}\mathbf{u}\cdot\mathbf{u} + \mathcal{V} \right) \left(\mathbf{u}^{rel} \cdot \mathbf{n} \right) dS$$
$$= \dot{Q} - \dot{W}_s \qquad (\mathbf{f} = -\nabla\mathcal{V}) \tag{6.35}$$

where we define $\dot{W}_s$ as

$$\dot{W}_s = - \oiint_S \mathbf{f}_v \cdot \mathbf{u}^{rel} dS \tag{6.36}$$

Either the quantities $\dot{Q}$ and $\dot{W}_s$ are given or sufficient information is provided to permit evaluation of the two integrals, with $\dot{Q}$ and $\dot{W}_s$ being part of the solution. The negative of the friction work, $\dot{W}_s$, is generally called the **shaft work**. It represents the useful work that is done by the fluid-mechanical device contained in the control volume. The name, shaft work, originates from devices, such as a turbine or a pump, that operate with a shaft. Strictly speaking, shaft work is the integral over the area enclosing the shaft. All other contributions to the integral are sometimes called shear work, which is non-zero only if the surface velocity is non-zero. Shear work is rarely significant compared to shaft work, so that the entire closed-surface integral is very nearly equal to the shaft work.

Example 6.3 *Consider a simple turbine as illustrated in the figure below. Assume the flow is steady, and that the mass flux through the turbine is $\dot{m}$. The heat transfer, $\dot{Q}$, is given, while properties at the inlet are denoted by subscript 1 and those at the outlet by subscript 2. Both are assumed uniform over the cross section. Ignoring effects of body forces, compute the shaft work, including the heat transfer rate from the surroundings to the turbine, $\dot{Q}$. Use the control volume whose surface (indicated by the dashed lines) coincides with the turbine walls as shown.*

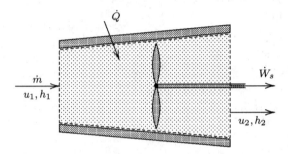

Solution. We select a stationary control volume indicated by the dashed lines. Turning first to mass conservation, the fact that the flow is steady tells us the net mass flux out of the control volume is zero, i.e.,

$$\oiint_S \rho\, \mathbf{u} \cdot \mathbf{n}\, dS = 0$$

where the fact that our control volume is stationary tells us that $\mathbf{u}^{rel} = \mathbf{u}$. Thus, denoting inlet and outlet areas by A_1 and A_2, respectively, we have

$$-\rho_1 u_1 A_1 + \rho_2 u_2 A_2 = 0 \quad \Longrightarrow \quad \dot{m} = \rho_1 u_1 A_1 = \rho_2 u_2 A_2$$

We learn nothing from momentum conservation because, for each component introduced, we introduce an unknown reaction force. Thus, we turn to the energy equation. Because the flow is steady and we are ignoring effects of body forces for simplicity, we have

$$\oiint_S \rho \left(h + \frac{1}{2} \mathbf{u} \cdot \mathbf{u} \right) (\mathbf{u} \cdot \mathbf{n})\, dS = \dot{Q} - \dot{W}_s$$

wherefore the shaft work is given by

$$
\begin{aligned}
\dot{W}_s &= \dot{Q} - \oiint_S \rho \left(h + \frac{1}{2} \mathbf{u} \cdot \mathbf{u} \right) (\mathbf{u} \cdot \mathbf{n})\, dS \\
&= \dot{Q} - \left[\rho_1 \left(h_1 + \frac{1}{2} u_1^2 \right) (-u_1 A_1) + \rho_2 \left(h_2 + \frac{1}{2} u_2^2 \right) (u_2 A_2) \right] \\
&= \dot{Q} + \rho_1 u_1 A_1 \left(h_1 + \frac{1}{2} u_1^2 \right) - \rho_2 u_2 A_2 \left(h_2 + \frac{1}{2} u_2^2 \right)
\end{aligned}
$$

Finally, using the result obtained from mass conservation, we can rewrite the expression for $\dot{W}_s$ in terms of $\dot{m}$, viz.,

$$\dot{W}_s = \dot{Q} + \dot{m} \left[(h_1 - h_2) + \frac{1}{2} \left(u_1^2 - u_2^2 \right) \right]$$

As a final comment, if a nonconservative body force is acting, we cannot define a force potential, $\mathcal{V}$. To develop the energy-conservation principle in this case, we must exclude potential energy from B and β in Equation (6.25). That is, we replace Equation (6.25) by

$$B = E \quad \text{and} \quad \beta = e + \frac{1}{2}\mathbf{u} \cdot \mathbf{u} \qquad (\mathbf{f} \neq -\nabla\mathcal{V}) \qquad (6.37)$$

Then, we must develop an expression for the work done by the body force in a manner similar to the way we handled the surface forces. The work done, per unit volume, by the body force on the control volume is $\rho\,\mathbf{f} \cdot \mathbf{u}$. Note that the appropriate velocity for a body force, which acts within the interior, is $\mathbf{u}$. In contrast to what occurs on the bounding surface, the work done within the control volume is unaffected the motion of the control volume. The work done by the body force on the surroundings, per unit volume, is $-\rho\,\mathbf{f} \cdot \mathbf{u}$. Hence, the rate at which the body force effectively does work on the surroundings is $dW = -\rho\,\mathbf{f} \cdot \mathbf{u}\,dV$. Consequently, the work term defined in Equation (6.28) must be replaced by

$$\dot{W} = \oiint_S \mathbf{F}_s \cdot \mathbf{u}^{rel}\,dS - \iiint_V \rho\,\mathbf{f} \cdot \mathbf{u}\,dV \qquad (\mathbf{f} \neq -\nabla\mathcal{V}) \qquad (6.38)$$

Aside from these two adjustments, all other steps in the derivation are the same as above. Thus, the resulting energy-conservation principle for a general body force is

$$\frac{d}{dt}\iiint_V \left[\rho\left(e + \frac{1}{2}\mathbf{u} \cdot \mathbf{u}\right)\right]dV + \oiint_S \rho\left(h + \frac{1}{2}\mathbf{u} \cdot \mathbf{u}\right)(\mathbf{u}^{rel} \cdot \mathbf{n})\,dS$$
$$= \dot{Q} - \dot{W}_s + \iiint_V \rho\,\mathbf{f} \cdot \mathbf{u}\,dV \qquad (\mathbf{f} \neq -\nabla\mathcal{V}) \qquad (6.39)$$

6.3 Approximate Form of the Energy Equation

One important application that requires use of the energy-conservation principle, including heat-transfer and viscous-work terms, is flow in pipes. Delivery of gas and oil from an oil well to a refinery, for example, requires pipes that can be hundreds of miles long. Viscous losses are nontrivial for flow through such pipe systems, especially at tees or corners and at junctions between pipes of differing areas. In this section, we develop an approximate energy-conservation equation that accounts for viscous and thermal losses and that can be used as a replacement for Bernoulli's equation in such flows.

We confine our attention to steady, incompressible flow and assume that the body force is gravity so that $\mathcal{V} = gz$, where g is the acceleration due to gravity and z is vertical distance. For steady flow through a motionless control volume, the energy-conservation principle simplifies to

$$\oiint_S \rho\left(h + \frac{1}{2}\mathbf{u} \cdot \mathbf{u} + gz\right)(\mathbf{u} \cdot \mathbf{n})\,dS = \dot{Q} - \dot{W}_s \qquad (6.40)$$

To apply this principle to flow through a pipe, we consider a control volume that is coincident with the pipe walls. Because the pipe walls are solid surfaces, the only boundaries of the control volume across which fluid passes are the inlet and outlet. Thus, referring to Figure 6.4, the net flux of total enthalpy is

$$\oiint_S \rho\left(h + \frac{1}{2}\mathbf{u} \cdot \mathbf{u} + gz\right)(\mathbf{u} \cdot \mathbf{n})\,dS = -\iint_{A_1} \rho\left(h_1 + \frac{1}{2}u_1^2 + gz_1\right)u_1\,dA$$
$$+ \iint_{A_2} \rho\left(h_2 + \frac{1}{2}u_2^2 + gz_2\right)u_2\,dA \qquad (6.41)$$

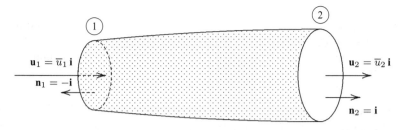

Figure 6.4: *Incompressible pipe flow.*

Note that u_1 and u_2 are the speeds at the inlet and outlet, respectively. In general, they vary over the cross sections. At this point, we have made no approximation to the exact conservation principle. To make further headway, we assume that, on each cross section of the pipe, enthalpy and elevation are nearly constant across the section so that

$$
\iint_{A_i} \rho\,(h_i + gz_i)\,u_i\,dA \;\approx\; \rho\,(h_i + gz_i) \iint_{A_i} u_i\,dA
$$
$$
= \; (h_i + gz_i)\,\rho \bar{u}_i A_i = (h_i + gz_i)\,\dot{m} \qquad (6.42)
$$

where $\bar{u}_i$ is the average velocity on cross section i defined by

$$
\bar{u}_i = \frac{1}{A_i} \iint_{A_i} u_i\,dA \qquad (6.43)
$$

We exercise a bit of caution in computing the kinetic-energy flux. That is, as discussed in Subsection 5.6.5, we must take account of the fact that the average of u_i^3 is not equal to the cube of the average of u_i (the former is always larger for positive u_i). To account for the difference, we rearrange the kinetic-energy flux integral, i.e.,

$$
\iint_{A_i} \rho \left(\frac{1}{2} u_i^2 \right) u_i\,dA = \frac{1}{2}\rho \iint_{A_i} u_i^3\,dA = \frac{1}{2}\rho \bar{u}_i^3 \iint_{A_i} \left(\frac{u_i}{\bar{u}_i} \right)^3 dA \qquad (6.44)
$$

where $\bar{u}_i$ is again the average value of u_i. We now define the **kinetic-energy correction factor**, α_i, as

$$
\alpha_i \equiv \frac{1}{A_i} \iint_{A_i} \left(\frac{u_i}{\bar{u}_i} \right)^3 dA \qquad (6.45)
$$

Inspection of Equation (6.45) shows that it is simply the ratio of the average value of u_i^3, i.e., $\overline{u_i^3}$, to the cube of the average velocity, $\bar{u}_i^3$. Therefore, the kinetic-energy flux integral can be written in a more compact form as follows.

$$
\iint_{A_i} \rho \left(\frac{1}{2} u_i^2 \right) u_i\,dA = \frac{1}{2}\rho \alpha_i \bar{u}_i^3 A_i = \frac{1}{2}\alpha_i \bar{u}_i^2 \dot{m} \qquad (6.46)
$$

Substituting these averaged values into the exact energy-conservation equation yields a straightforward algebraic equation that relates appropriately averaged flow properties in our pipe.

$$
\dot{Q} - \dot{W}_s = -\left(h_1 + \frac{1}{2}\alpha_1 \bar{u}_1^2 + gz_1 \right)\dot{m} + \left(h_2 + \frac{1}{2}\alpha_2 \bar{u}_2^2 + gz_2 \right)\dot{m} \qquad (6.47)
$$

As is customary in classical pipe-flow analysis, we divide through by $\dot{m}g$ in order to recast the equation with all terms having the dimension of length, viz.,

$$\frac{\dot{Q} - \dot{W}_s}{\dot{m}g} + \frac{h_1}{g} + \alpha_1 \frac{\overline{u}_1^2}{2g} + z_1 = \frac{h_2}{g} + \alpha_2 \frac{\overline{u}_2^2}{2g} + z_2 \qquad (6.48)$$

Next, we rewrite the shaft work, $\dot{W}_s$, as the difference between the power given up to a device such as a turbine, $\dot{W}_t$, and the power supplied by a device such as a pump, $\dot{W}_p$. In mathematical terms, we say

$$\dot{W}_s = \dot{W}_t - \dot{W}_p \qquad (6.49)$$

The idea here is that we will apply this approximate equation [cf. Equation (6.42) for one of the approximations] to pipe systems including components that add and/or remove energy. For example, a pump adds energy to the system, while a turbine extracts energy from the system. As we will see below, there are viscous and heat-transfer effects inherent to real fluids that extract energy. Rewriting the enthalpy as $h = e + p/\rho$, the equation for energy conservation becomes

$$\frac{\dot{W}_p}{\dot{m}g} + \frac{p_1}{\rho g} + \alpha_1 \frac{\overline{u}_1^2}{2g} + z_1 = \frac{\dot{W}_t}{\dot{m}g} + \frac{p_2}{\rho g} + \alpha_2 \frac{\overline{u}_2^2}{2g} + z_2 + \left[\frac{(e_2 - e_1)}{g} - \frac{\dot{Q}}{\dot{m}g} \right] \qquad (6.50)$$

Finally, we define the **head supplied by a pump**, h_p, the **head given up to a turbine**, h_t, and the **head loss**, h_L, by the following equations.

$$h_p \equiv \frac{\dot{W}_p}{\dot{m}g}, \quad h_t \equiv \frac{\dot{W}_t}{\dot{m}g}, \quad h_L \equiv \left[\frac{(e_2 - e_1)}{g} - \frac{\dot{Q}}{\dot{m}g} \right] \qquad (6.51)$$

Clearly, all terms in Equations (6.50) and (6.51) have dimensions of length. Regarding h_L, there is always a head loss in a viscous fluid. Such losses occur because mechanical energy is converted to thermal energy through heat transfer and dissipative processes. This appears as a loss in the overall energy balance. In terms of these parameters, the final form of our approximate equation for incompressible pipe flow is:

$$\frac{p_1}{\rho g} + \alpha_1 \frac{\overline{u}_1^2}{2g} + z_1 + h_p = \frac{p_2}{\rho g} + \alpha_2 \frac{\overline{u}_2^2}{2g} + z_2 + h_t + h_L \qquad (6.52)$$

In contrast to the classical equation developed by Bernoulli, Equation (6.52) tells us that total energy available, i.e., the sum of the pressure potential, kinetic energy and potential energy, does not remain constant. In a pipe system, a pump will increase the total energy available while devices such as turbines extract energy. Also, heat transfer and dissipative losses reduce total energy available.

Equation (6.52) can be used as a replacement for Bernoulli's equation when viscous and heat-transfer effects are important. Keep in mind that this relation is approximate, and is intended mainly for application to steady, incompressible flow through pipe systems. The equation can be quite helpful in estimating the power output and/or needs in hydroelectric plants, water-irrigation systems and many other engineering applications.

It is worthwhile at this point to pause and discuss the kinetic-energy correction factor, α, including an example of how it can be computed. The first point is that, in practical pipe-flow applications, its value is a bit larger than unity—typically 1.05. Thus, you might be tempted to ignore it and simply approximate $\alpha = 1$. Some of the homework problems include this approximation for the sake of algebraic simplicity. However, you should be aware that this is inconsistent with the spirit of Equation (6.52), and should not be done for practical applications. For example, the head loss, h_L, is often a small fraction of the kinetic energy— sometimes less than 10%. Thus, if you approximate $\alpha \approx 1$, then all head losses should be ignored as well. This would, of course, defeat much of the purpose of using Equation (6.52) as a replacement for Bernoulli's equation. The following example is a typical case where head loss is quite small compared to the contributions attending $\alpha \neq 1$.

Example 6.4 *Consider a pipe system that pumps water from one elevation to another as shown in the figure. Determine how powerful a pump is needed to have an outlet velocity that is twice the inlet velocity. The pressure is the same at inlet and outlet so that $p_1 = p_2$, while the kinetic-energy correction factors are equal and given by $\alpha_1 = \alpha_2 = 1.06$. The flow rate is $\dot{m} = 0.5$ kg/sec, the inlet velocity is $U = 30$ m/sec, and the change in elevation is $\Delta z = 100$ m. Neglect head loss.*

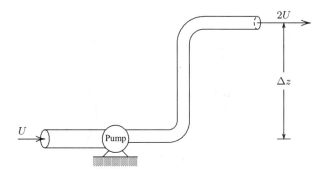

Solution. We first determine the head supplied by the pump, h_p, by using Equation (6.52). Ignoring any head loss in the pipe system, we thus have

$$\frac{p_1}{\rho g} + \alpha_1 \frac{\overline{u}_1^2}{2g} + z_1 + h_p = \frac{p_2}{\rho g} + \alpha_2 \frac{\overline{u}_2^2}{2g} + z_2$$

where we omit h_t because we are not driving a turbine or any other energy-extracting device. Rearranging terms yields

$$0.53 \frac{U^2}{g} + h_p = 0.53 \frac{(2U)^2}{g} + \Delta z$$

Therefore, the head supplied by the pump must be

$$h_p = 1.59 \frac{U^2}{g} + \Delta z$$

To compute the power supplied by the pump, we appeal to the definition of head [Equation (6.51)], to conclude that

$$\dot{W}_p = \dot{m} g h_p = 1.59 \dot{m} U^2 + \dot{m} g \Delta z$$

For the given conditions, the power required of the pump is

$$\begin{aligned} \dot{W}_p &= (1.59)(0.5)(900) + (0.5)(9.807)(100) \text{ kg} \cdot \text{m}^2/\text{sec}^3 \\ &= 1206 \text{ Joules/sec} = 1.206 \text{ kWatts} \end{aligned}$$

The following example illustrates the procedure for computing α from a known velocity profile. The velocity specified is, in fact, the exact velocity variation for laminar flow through a pipe. Although laminar pipe flow is uncommon in engineering applications, this computation underscores the caution we exercised in evaluating the kinetic-energy flux integral. In this special case, assuming $\overline{u}^3 = \overline{u^3}$ would introduce a factor-of-two error!

Example 6.5 *Consider laminar pipe flow, which we discussed in Subsection 1.9.2. The velocity for this type of flow is*

$$u(r) = u_m \left(1 - r^2/R^2\right)$$

where u_m is maximum velocity, r is distance from the centerline and R is pipe radius. Assuming the pipe has circular cross section, determine the kinetic-energy correction factor, α.

Solution. First, we must compute the average velocity, $\overline{u}$. Because the pipe has circular cross section, it is most convenient to work in cylindrical coordinates. Thus, since the cross-sectional area is $A = \pi R^2$ and the differential area is $dA = r dr d\theta$, we have

$$
\begin{aligned}
\overline{u} &= \frac{1}{\pi R^2} \int_0^{2\pi} \int_0^R u(r)\, r dr d\theta \\
&= \frac{2\pi u_m}{\pi R^2} \int_0^R \left(1 - \frac{r^2}{R^2}\right) r dr \\
&= 2 u_m \int_0^1 \left(1 - \frac{r^2}{R^2}\right) \frac{r}{R} d\left(\frac{r}{R}\right) \\
&= 2 u_m \int_0^1 \left(1 - \xi^2\right) \xi d\xi = 2 u_m \int_0^1 \left(\xi - \xi^3\right) d\xi \qquad (\xi \equiv r/R) \\
&= 2 u_m \left(\frac{1}{2}\xi^2 - \frac{1}{4}\xi^4\right)\Big|_{\xi=0}^{\xi=1} = \frac{1}{2} u_m
\end{aligned}
$$

Then, from the definition of α given in Equation (6.45),

$$
\begin{aligned}
\alpha &= \frac{1}{\pi R^2} \int_0^{2\pi} \int_0^R \left[\frac{u_m \left(1 - r^2/R^2\right)}{u_m/2}\right]^3 r dr d\theta \\
&= \frac{8 \cdot (2\pi)}{\pi R^2} \int_0^R \left(1 - \frac{r^2}{R^2}\right)^3 r dr \\
&= 16 \int_0^1 \left(1 - \frac{r^2}{R^2}\right)^3 \frac{r}{R} d\left(\frac{r}{R}\right) \\
&= 16 \int_0^1 \left(1 - \xi^2\right)^3 \xi d\xi = 16 \int_0^1 \left(1 - 3\xi^2 + 3\xi^4 - \xi^6\right) \xi d\xi \\
&= 16 \left(\frac{1}{2}\xi^2 - \frac{3}{4}\xi^4 + \frac{1}{2}\xi^6 - \frac{1}{8}\xi^8\right)\Big|_{\xi=0}^{\xi=1} = 2
\end{aligned}
$$

6.4   Flow in Pipes

Recall that in Subsection 1.9.2, we analyzed incompressible flow in an infinitely-long pipe. We found that while the velocity varies across the pipe cross section, it is independent of distance along the pipe, x. This independence of x is called **fully-developed** flow. The

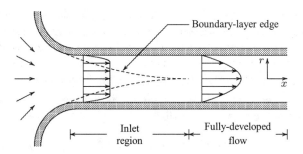

Figure 6.5: *Fully-developed pipe flow with the radial scale magnified. The length of the inlet region depends upon Reynolds number.*

solution we obtained is observed in finite-length pipes provided the pipe's overall length, L, is very large compared to its diameter, D, and provided we make our observations at a sufficient distance from the ends of the pipe. Figure 6.5 schematically depicts the way in which pipe flow develops from a rounded inlet to the fully-developed region.

As shown, near the inlet viscous effects are confined to the region close to the pipe wall, known as the boundary layer (Chapter 11 discusses boundary layers). Eventually, these boundary layers grow in thickness until they merge on the centerline. At a distance, ℓ_e, from the inlet, measurements show that the flow becomes fully developed, where

$$\frac{\ell_e}{D} = \begin{cases} 0.06 Re_D, & \text{Laminar} \\ 4.4 Re_D^{1/6}, & \text{Turbulent} \end{cases} \tag{6.53}$$

Thus, for example, the entrance length for laminar pipe flow with $Re_D = 1000$ is 60 pipe diameters. While we have not developed the viscous-flow equations needed to analyze pipe flow, we can nevertheless use analytical and empirical results regarding head loss for pipes and pipe systems. Formal justification for some of the relationships discussed in the following subsections can be found in most books on viscous flow [cf. Wilcox (2000) or White (1994)].

6.4.1 Friction and Head Loss in Straight Pipes

As discussed in Subsection 1.9.2, laminar flow is possible only at relatively small Reynolds numbers. Unless great care is exercised to minimize flow disturbances (and to thus permit larger Reynolds numbers), the laminar velocity profile, $u(r) = u_m \left(1 - r^2/R^2\right)$ is observed only if the Reynolds number, Re_D, satisfies[1]

$$Re_D = \frac{\overline{u}D}{\nu} < 2300 \qquad \text{(Laminar flow)} \tag{6.54}$$

At greater values of the Reynolds number, the flow undergoes transition to turbulence, characterized by time-varying fluctuations in all flow properties. Again, in the absence of any attempts to minimize flow disturbances, the flow in a pipe will be completely turbulent when

$$Re_D = \frac{\overline{u}D}{\nu} > 3000 \qquad \text{(Turbulent flow)} \tag{6.55}$$

For intermediate Reynolds numbers, the flow is described as being **transitional**, i.e., in a state of transition from laminar to turbulent flow.

---

[1]This criterion is consistent with Equation (1.60) because $\overline{u} = u_m/2$ and $D = 2R$.

The head loss in a pipe, h_L, is a function of the friction at the pipe walls, the average kinetic energy of the flow and the length to diameter ratio. It is expressed by the **Darcy-Weisbach equation** as follows.

$$h_L = f \frac{\overline{u}^2}{2g} \frac{L}{D} \tag{6.56}$$

In Equation (6.56), L is the length of the pipe and f is the dimensionless **Darcy friction factor** defined in terms of the shear stress at the pipe wall, τ_w, by

$$f \equiv \frac{\tau_w}{\frac{1}{8}\rho \overline{u}^2} \tag{6.57}$$

It is worthwhile to pause and discuss how the friction factor is actually determined for pipe flow. While instruments are available to measure it directly, the special nature of pipe flow permits an indirect measurement. In Subsection 1.9.2, we showed that the shear stress varies linearly with distance across a pipe [Equation (1.57)]. This result holds for both laminar and turbulent flow. The shear stress at the wall is[2]

$$\tau_w = \frac{R}{2}\left(\frac{p_1 - p_2}{L}\right) \quad \Longrightarrow \quad f = \frac{D}{\frac{1}{2}\rho \overline{u}^2}\left(\frac{p_1 - p_2}{L}\right) \tag{6.58}$$

Thus, the friction factor can be inferred from the imposed pressure gradient and the measured average velocity.

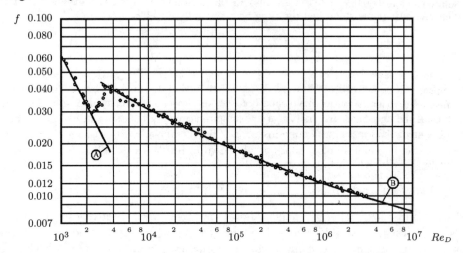

Figure 6.6: *Fiction factor for a perfectly-smooth pipe:* ◦-*measured, A-laminar, B-turbulent.*

For **perfectly-smooth pipes**, the friction factor depends upon whether or not the flow is turbulent (see Figure 6.6). Its value is

$$\left.\begin{array}{ll} f = \dfrac{64}{Re_D}, & \text{Laminar} \\[4mm] 1/\sqrt{f} = -2\log_{10}\left(\dfrac{2.51}{Re_D\sqrt{f}}\right), & \text{Turbulent} \end{array}\right\} \tag{6.59}$$

The laminar value is the exact solution, while the turbulent-flow formula has been justified by asymptotic analysis [cf. Wilcox (1995)] and confirmed by correlation of measurements.

---

[2]For reasons that require analysis beyond the scope of this text, positive shear at the pipe surface corresponds to a change of sign in Equation (1.57).

Example 6.6 *Develop an equation for the rate at which energy is lost in flow through a pipe,* $\dot{m}gh_L$*, expressing your answer as a function of the shear stress at the pipe wall,* τ_w*, average flow velocity,* $\bar{u}$*, and circumferential area of the pipe,* $A = \pi DL$*.*

Solution. Substituting Equation (6.57) into Equation (6.56), the head loss is

$$h_L = \frac{\tau_w}{\frac{1}{8}\rho\bar{u}^2} \frac{\bar{u}^2}{2g} \frac{L}{D} = \frac{4\tau_w L}{\rho g D}$$

Hence, since the mass-flow rate is $\dot{m} = \frac{\pi}{4}\rho\bar{u}D^2$, the rate at which energy is being lost is

$$\dot{m}gh_L = \frac{\pi}{4}\rho\bar{u}D^2 g \frac{4\tau_w L}{\rho g D} = \tau_w\bar{u}\,(\pi DL) = \tau_w\bar{u}A$$

Example 6.7 *With great care, a graduate student has achieved laminar flow in a pipe at a Reynolds number of* $Re_D = 10^4$*. Compare the friction factors and head loss for laminar and turbulent flow when* $Re_D = 10^4$*, assuming pipe diameter, D, length, L, and kinematic viscosity,* ν*, are the same. If* Re_D*, D and* ν *remain the same, how much longer a pipe can we have with laminar flow that sustains the same overall head loss as in a turbulent flow?*

Solution. From Equation (6.59), the laminar and turbulent friction factors are

$$f_{lam} = 0.0064, \quad \text{Laminar}$$

$$f_{turb} = 0.0309, \quad \text{Turbulent}$$

Note that for turbulent-flow, Equation (6.59) is an implicit equation for f. The fastest way to compute f is to first determine an approximate value from Figure 6.6. This value can be used as a first guess in an iterative solution or, by simple trial and error. For the value quoted above, the left-hand and right-hand sides of the equation for f differ by 0.04%. This difference in the value of f is much smaller than measurement errors (cf. Figure 6.6).

Turning to the head loss, since Re_D, D, L and ν are all the same, so is the flow velocity, $\bar{u}$. Thus, Equation (6.56) tells us that the ratio of the turbulent-flow head loss, $(h_L)_{turb}$, to the laminar-flow head loss, $(h_L)_{lam}$, is

$$\frac{(h_L)_{turb}}{(h_L)_{lam}} = \frac{f_{turb}}{f_{lam}} = \frac{0.0309}{0.0064} = 4.828$$

Thus, the loss in turbulent pipe flow is nearly 5 times that of laminar pipe flow.

Now, if we extend the length of the pipe from the laminar case value of L_{lam}, while maintaining the same values for Re_D, D, and ν, Equation (6.56) tells us that the turbulent-flow length, L_{turb}, satisfies the following equation.

$$\frac{(h_L)_{turb}}{(h_L)_{lam}} = \frac{f_{turb}}{f_{lam}} \frac{L_{turb}}{L_{lam}}$$

Therefore, if the head loss is the same, necessarily

$$\frac{L_{lam}}{L_{turb}} = \frac{f_{turb}}{f_{lam}} = 4.828$$

So, if we could maintain laminar flow at this Reynolds number, we could maintain flow through a pipe nearly 5 times longer with the same change in head.

For **rough pipes**, if we have uniformly distributed, or **sand-grain**, roughness elements of average height k_s, the Colebrook (1939) friction-factor formula provides a value accurate to within about 15% over a wide range of Reynolds numbers. The **Colebrook formula** is as follows.

$$1/\sqrt{f} = -2\log_{10}\left(\frac{k_s/D}{3.7} + \frac{2.51}{Re_D\sqrt{f}}\right) \tag{6.60}$$

Alternatively, we can use Figure 6.7, which is known as the **Moody diagram** [Moody (1944)], to find the friction factor for laminar and turbulent flow, including effects of roughness. The figure consists of two families of curves.

The first family is friction factor, f, as a function of Reynolds number, Re_D, for several values of the dimensionless roughness height, k_s/D. The curves are graphical representations of Equation (6.60). The family consists of the nearly horizontal curves whose axes are at the left and bottom of the plot. As shown in the figure, for rough surfaces the friction factor tends toward a constant value at high Reynolds number. We can use the friction factor curves to determine head loss when the velocity (or flow rate) and pipe dimensions are known.

Example 6.8 *A 100 foot length of steel pipe has a diameter of 1 inch. The fluid in the pipe is water at 68° F flowing at an average speed, $\overline{u}$, of 10 ft/sec. Determine the friction factor and the head loss.*

Solution. First, we compute the Reynolds number. Reference to Table A.7 shows that the kinematic viscosity, ν, is $1.08 \cdot 10^{-5}$ ft²/sec. Thus,

$$Re_D = \frac{\overline{u}D}{\nu} = \frac{(10 \text{ ft/sec}) \cdot (1/12 \text{ ft})}{1.08 \cdot 10^{-5} \text{ ft}^2/\text{sec}} = 7.716 \cdot 10^4$$

Next, from the insert in Figure 6.7, the roughness height is $k_s = 1.5 \cdot 10^{-4}$ ft. Thus,

$$\frac{k_s}{D} = \frac{1.5 \cdot 10^{-4} \text{ ft}}{1/12 \text{ ft}} = 0.0018$$

From the Moody diagram, the friction factor, f, is

$$f = 0.025$$

Therefore, the head loss is

$$h_L = f\frac{\overline{u}^2}{2g}\frac{L}{D} = 0.025\left(\frac{100 \text{ ft}^2/\text{sec}^2}{2 \cdot 32.174 \text{ ft/sec}^2}\right)\left(\frac{100 \text{ ft}}{1/12 \text{ ft}}\right) = 46.6 \text{ ft}$$

As an alternative to using the Moody diagram to determine the friction factor, substituting the values above for f, k_s/D and Re_D into the Colebrook formula [Equation (6.60)] yields

$$\frac{1}{\sqrt{f}} = 6.3245553 \quad \text{and} \quad -2\log_{10}\left(\frac{k_s/D}{3.7} + \frac{2.51}{Re_D\sqrt{f}}\right) = 6.3195079$$

Thus, the left-hand and right-hand sides of Equation (6.60) differ by 0.08%. Trial and error shows that we can reduce this difference to 0.003% by using

$$f = 0.02504$$

This difference in the value of f is much smaller than measurement errors (cf. Figure 6.6).

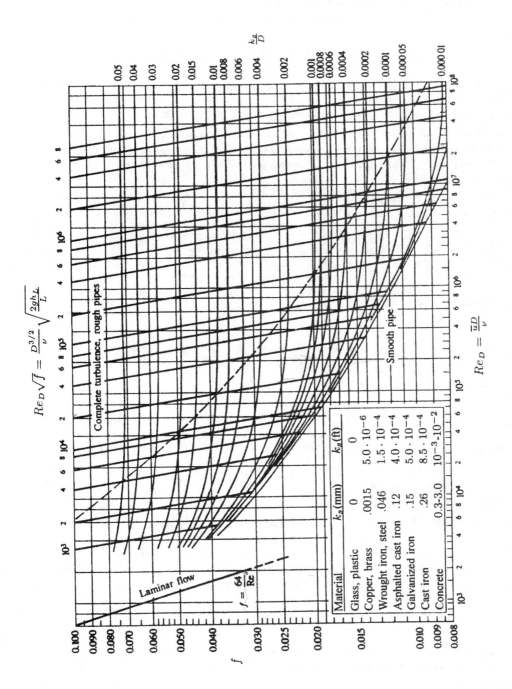

Figure 6.7: *The Moody diagram [From Moody (1944) used with permission of the ASME].*

The second family of curves in the Moody diagram represents the loci of constant head loss, which can be determined from the product of Re_D and $\sqrt{f}$. That is, from Equation (6.56),

$$f\bar{u}^2 = D\frac{2gh_L}{L} \quad \Longrightarrow \quad Re_D\sqrt{f} = \frac{D^{3/2}}{\nu}\sqrt{\frac{2gh_L}{L}} \qquad (6.61)$$

This family of curves consists of straight lines. The constant value of $Re_D\sqrt{f}$ on each line is the value at the intersection with the upper horizontal axis. We can use this family to solve for the flow rate when h_L and D are known.

Example 6.9 *The head loss per kilometer of a 20-cm diameter, asphalted cast-iron pipe is 12.2 meters. The fluid is water, and the temperature is $20°$ C. Determine the flow velocity.*

Solution. First, we compute the dimensionless head-loss parameter defined in Equation (6.61). Reference to Table A.7 shows that the kinematic viscosity is $\nu = 1.0 \cdot 10^{-6}$ m²/sec, so that

$$\frac{D^{3/2}}{\nu}\sqrt{\frac{2gh_L}{L}} = \frac{(0.2 \text{ m})^{3/2}}{1.0 \cdot 10^{-6} \text{ m}^2/\text{sec}}\sqrt{\frac{2 \cdot (9.807 \text{ m/sec}^2) \cdot (12.2 \text{ m})}{1000 \text{ m}}} = 4.38 \cdot 10^4$$

The inset in Figure 6.7 tells us the roughness height is 0.12 mm = 0.00012 m. Thus, the dimensionless roughness height for this pipe is

$$\frac{k_s}{D} = \frac{0.00012 \text{ m}}{0.2 \text{ m}} = 0.0006$$

Reference to the Moody diagram shows that the head-loss line with head-loss parameter equal to $4.38 \cdot 10^4$ crosses the friction curve with $k_s/D = 0.0006$ when

$$f = 0.019$$

Finally, we can solve for the velocity using Equation (6.56), i.e.,

$$\bar{u} = \sqrt{\frac{2gh_L D}{fL}} = \sqrt{\frac{2 \cdot (9.807 \text{ m/sec}^2) \cdot (12.2 \text{ m}) \cdot (0.2 \text{ m})}{0.019 \cdot (1000 \text{ m})}} = 1.59 \frac{\text{m}}{\text{sec}}$$

6.4.2   Minor Losses and Non-Circular Cross Sections

Up to this point, we have considered head loss in a straight pipe of constant cross-sectional area. In real piping systems, we have several complicating factors such as inlets, bends, abrupt changes in area, valves, elbows, tees and other fittings, that create additional head loss. Losses in straight sections discussed in the preceding subsection are referred to as **major losses**. By contrast, losses attending inlets, bends, etc. are termed **minor losses**, and are usually of secondary importance. There are applications where minor losses dominate, however, typically when the straight sections are short.

In general, we compute the total head loss in a complex pipe system according to the following equation.

$$h_L = \sum_i h_{L_i}^{major} + \sum_i h_{L_i}^{minor} \qquad (6.62)$$

On each straight section, $h_{L_i}^{major}$ is given by the Darcy-Weisbach formula, Equation (6.56). Empirical correlations have been developed to describe minor head losses. They are usually expressed in terms of a dimensionless **loss coefficient**, K, viz.,

$$h_L^{minor} = K\frac{\bar{u}^2}{2g} \tag{6.63}$$

Head loss is also expressed in terms of an **equivalent length** of straight pipe, L_e, so that the head loss becomes

$$h_L^{minor} = f\frac{\bar{u}^2}{2g}\frac{L_e}{D} \tag{6.64}$$

Thus, for example, in a constant-diameter pipe system of total length L and diameter D, the total head loss is

$$h_L = \frac{\bar{u}^2}{2g}\left[f\frac{L}{D} + \sum_i K_i\right] \tag{6.65}$$

where we express the minor losses in terms of loss coefficients.

In normal practice, the loss coefficient is used much more often than the equivalent length, although there is no fundamental reason to choose either. Both are empirical in nature and are related to head loss in a dimensionally consistent manner. At a minimum, K and L_e should be expected to depend upon geometry, Reynolds number and roughness. Unfortunately, more often than not, data reported in pipe-flow literature have been correlated with neither Reynolds number nor roughness, and are usually reported only for a given pipe diameter. Measurements are most commonly expressed in terms of equivalent length for bends (and fittings) primarily because $L_e/\mathcal{R}$ tends toward a constant value, where $\mathcal{R}$ is the radius of curvature of the bend.

Measurements have been made to establish loss coefficients and equivalent lengths for pipes with circular cross sections. They can be used as approximations for noncircular geometries as well. To do this, we simply introduce the **hydraulic diameter**, D_h. If A is cross-sectional area and P is perimeter, D_h is defined by

$$D_h = \frac{4A}{P} \tag{6.66}$$

Note that the perimeter, P, includes all surfaces exposed to the fluid, i.e., the so-called **wetted area**. If we have flow between two concentric cylinders, for example, P is the sum of the perimeters of both cylinders. The hydraulic diameter is defined so that it is equal to the physical diameter for a circular cross section. This approximation yields satisfactory results provided cross sections do not depart too much from circular shape. In the case of rectangular geometries, for example, the height to width ratio should not exceed 4.

Example 6.10 *Consider a rectangular cross section of width w and height h, typical of air-conditioning, heating and ventilating ducts. Compute the hydraulic diameter.*

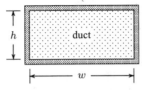

Solution. The area is $A = hw$, while the perimeter is $P = 2(w+h)$. Thus, the hydraulic diameter is

$$D_h = \frac{4A}{P} = \frac{4hw}{2(w+h)} = \frac{2h}{1+h/w}$$

In the limiting case of a square, for which $w = h$, D_h is equal to h.

It might seem questionable that any reasonable degree of accuracy can be attained by what appears to be a completely ad hoc approximation, i.e., using the hydraulic diameter to adapt results for a circular cross section to general shapes. However, the approximation is far less crude than it appears for two key reasons.

First, as long as we deal with fully-developed flow so that properties do not change in the streamwise direction, flow in any pipe or duct involves a balance between the pressure force and the viscous force on the surface [cf. Equation (1.56)]. The control-volume approach tells us that the pressure force is proportional to the cross-sectional area, A, while the viscous force is proportional to the perimeter, P, of the wetted area. Hence, the viscous force adjusts according to A and P independent of the detailed geometry.

Second, for high-Reynolds-number flow characteristic of many applications, the most significant viscous effects contributing to head loss are confined mainly to a very thin region near the surface. Thus, regardless of the actual shape, the walls of the duct or pipe look planar on the scale of the thin viscous region, and are more-or-less independent of the precise geometry. This is a quintessential example of how powerful the control-volume method can be in developing useful engineering methods.

All of the following correlations are limited to incompressible flows, and are generally reliable only for steady flow. Because the correlations rest on an empirical foundation, there is no guarantee that they can be used far beyond their established data base. This was a symptomatic weakness of the hydraulics era in fluid mechanics characterized by the work of Chézy, Weber, Hagen, Poiseuille, Darcy and Weisbach during the nineteenth century. That is not to say that empiricism is entirely without merit, as it is often the only approach practicable with the analytical techniques available. Much of the twentieth century work on predicting turbulence, for example, was based on little more than experimentation, dimensional analysis and empirical relationships [cf. Wilcox (1998)].

Inlet. Figure 6.8 depicts an inlet, sometimes called an entrance, with a bend radius, $\mathcal{R}$. The inset includes values of K for various ratios of $\mathcal{R}$ to pipe diameter, D. Note that the limiting case $\mathcal{R}/D \to 0$ corresponds to a square inlet. There is considerable scatter in measured values of K for inlets (and other minor-loss configurations). The listed values of K are for very smooth inlets and high Reynolds number, and probably represent lower bounds for the specified values of $\mathcal{R}/D$.

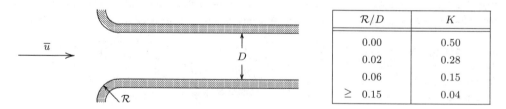

$\mathcal{R}/D$	K
0.00	0.50
0.02	0.28
0.06	0.15
$\geq$ 0.15	0.04

Figure 6.8: *Head loss coefficient at an inlet.*

Figure 6.9 illustrates how much variation there is in measured values. Data are included from several sources [e.g., ASHRAE (1981)], and no attempt has been made to correlate effects of Reynolds number or surface roughness. As shown, values of K differ by as much as 25% for small $\mathcal{R}/D$, i.e., for slight rounding. Differences amongst measurements become smaller as the radius of curvature of the inlet increases. The figure includes an approximate curve fit to the experimental data.

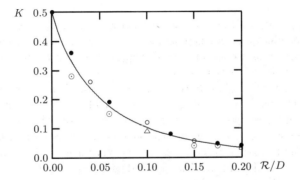

Figure 6.9: *Loss coefficient at an inlet:* ∘, •, ⊙, △, □ *measurements;* —— *approximate curve fit to measurements.*

Figure 6.10 schematically depicts flow at a square inlet. The primary reason for head loss is the inertia of the fluid, which prevents it from following the exact shape of the inlet. Rather, the streamlines are curved as shown and a small region of reverse, or separated, flow exists near the inlet. The fluid swirls around in the reverse-flow region, producing nontrivial viscous losses.

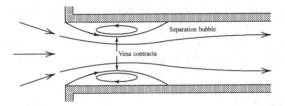

Figure 6.10: *Schematic of flow near a square inlet.*

Because of the obstruction presented by the **separation bubble**, the primary flow in the core of the pipe has a reduced forward-flow area, the minimum area being called the **vena contracta**. The size of the separated region and the vena contracta effect can be reduced significantly by rounding the inlet. As shown in Figures 6.8 and 6.9, the loss coefficient can be reduced to an insignificant value of 0.04 for a radius of curvature, $\mathcal{R}$, as small as $0.15D$.

Contraction. In a contraction, the diameter of the pipe changes from an initial value of D_1 to a smaller value of D_2. Figure 6.11 illustrates such a change in diameter. The inset includes the loss coefficient for a gradual (60°) contraction and a sudden (180°) contraction. The head loss is based on the velocity in the smaller section of the pipe.

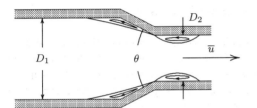

D_2/D_1	$K_{\theta=60°}$	$K_{\theta=180°}$
0.20	0.08	0.49
0.40	0.07	0.42
0.60	0.06	0.32
0.80	0.05	0.18

Figure 6.11: *Head loss coefficient at a contraction.*

Table 6.1 includes additional loss coefficient data [ASHRAE (1981)] valid for both round and rectangular ducts as a function of A_2/A_1. The areas A_1 and A_2 denote cross-sectional area upstream and downstream of the contraction, respectively.

Table 6.1: *Loss Coefficients for Gradual Contractions*

A_2/A_1	Included angle, θ						
	10°	15°- 40°	50°- 60°	90°	120°	150°	180°
0.10	0.05	0.05	0.08	0.19	0.29	0.37	0.43
0.25	0.05	0.04	0.07	0.17	0.27	0.35	0.41
0.50	0.05	0.05	0.06	0.12	0.18	0.24	0.26

As with inlet flow, the primary mechanism for head loss in a contraction is flow separation (at both corners) and the vena contracta. The head loss can be reduced by rounding the corners to permit more gradual turning of the flow. For a sudden contraction, correlation of measurements shows that

$$K \approx \frac{1}{2}\left[1 - \left(\frac{D_2}{D_1}\right)^2\right] \qquad \text{(Sudden contraction)} \qquad (6.67)$$

Expansion. In an expansion, the diameter of the pipe changes from an initial value of D_1 to a larger value of D_2. Figure 6.12 illustrates such a change in diameter. The inset includes the loss coefficient for a gradual (10°) expansion and a sudden (180°) expansion. The head loss is based on the velocity in the smaller section of the pipe.

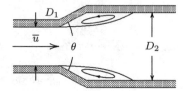

D_1/D_2	$K_{\theta=10°}$	$K_{\theta=180°}$
0.20	0.13	0.92
0.40	0.11	0.72
0.60	0.06	0.42
0.80	0.03	0.16

Figure 6.12: *Head loss coefficient at an expansion.*

As with inlet and contraction flow, the fluid cannot negotiate the sudden turn in an expansion, unless the angle is less than about 10°. Its inertia carries it from the first corner along the trajectory sketched in Figure 6.12, known as the **dividing streamline**, which joins with the pipe surface downstream of the second corner as shown. The flow recirculates below the dividing streamline with significant head loss due to viscous effects.

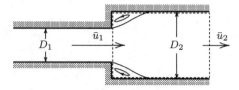

Figure 6.13: *Flow into a sudden expansion.*

We can derive a remarkably accurate expression for the head loss in a sudden expansion (Figure 6.13) using the control-volume method. The only approximations we make are that the pressure on the vertical face within the separation bubble is equal to the inlet pressure and that the kinetic-energy correction factor, α, is 1. For steady flow, conservation of mass, momentum and energy for the control volume indicated by the dashed contour in Figure 6.13 yield:

$$\bar{u}_1 D_1^2 = \bar{u}_2 D_2^2 \tag{6.68}$$

$$\rho \left(\bar{u}_2^2 D_2^2 - \bar{u}_1^2 D_1^2 \right) \approx (p_1 - p_2) D_2^2 \tag{6.69}$$

$$\frac{p_1}{\rho g} + \frac{\bar{u}_1^2}{2g} = \frac{p_2}{\rho g} + \frac{\bar{u}_2^2}{2g} + h_L \tag{6.70}$$

Noting the definition of the loss coefficient, K, given in Equation (6.63), we find

$$K \approx \left[1 - \left(\frac{D_1}{D_2} \right)^2 \right]^2 \qquad \text{(Sudden expansion)} \tag{6.71}$$

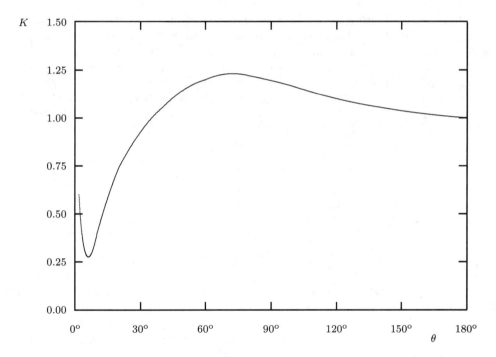

Figure 6.14: *Loss coefficient for a typical conical diffuser with fixed* $D_1/D_2 \ll 1$.

The gradual expansion geometry is usually called a **diffuser**. The loss coefficient varies substantially with the included angle, θ. Figure 6.14, for example, shows the loss coefficient for a typical conical diffuser [cf. White (1994)]. For small angles, the flow is attached and most of the head loss is due to the surface shear stress, similar to the constant-diameter case. For larger angles, the flow separates, and the head loss increases to values that can actually exceed those of a sudden expansion (i.e., $\theta = 180°$). For the case shown, the optimum diffuser angle, θ, is about 5°, at which K achieves a minimum value. For $\theta > 5°$, the boundary layer thickens, increasing the losses. For $\theta > 10°$, the boundary layer separates, as already remarked in the discussion of Figure 6.12.

90° Smooth Bend. In a smooth bend, the loss coefficient varies with the radius of the bend, $\mathcal{R}$. Figure 6.15 illustrates a 90° bend in a pipe of constant diameter, D. The inset includes values of K for various ratios of $\mathcal{R}$ to pipe diameter, D. As $\mathcal{R}/D$ increases, K first falls because "secondary flow" (see cross-section inset) losses decrease but then rises because the "bend" contains a greater length of pipe. For very large $\mathcal{R}/D$, the equivalent length, L_e, [see Equation (6.64)] is just $\pi\mathcal{R}/2$.

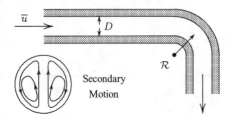

$\mathcal{R}/D$	K
1	0.35
2	0.19
4	0.16
6	0.21
8	0.28
10	0.32

Figure 6.15: *Head loss coefficient at a 90° smooth bend.*

Flow separation also plays a role in the head loss for a bend. Interestingly, the type of secondary motion depicted in Figure 6.15 occurs only for turbulent flow. Secondary motion also occurs for turbulent flow with noncircular cross sections, even when the duct is straight.

Figure 6.16 [cf. White (1994)] can be used to determine the head loss for 90° bends, including effects of surface roughness. The figure shows the increment in loss coefficient, ΔK, that must be added to the contribution due to bend length, so that

$$K = \frac{\pi}{2}f\frac{\mathcal{R}}{D} + \Delta K \qquad (6.72)$$

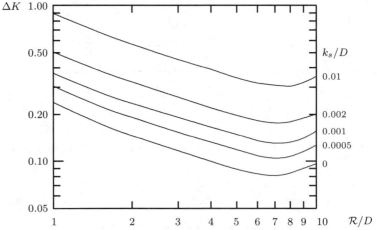

Figure 6.16: *Effect of roughness on head-loss coefficient for a 90° bend.*

Extensive measurements have been made to determine minor loss data for various elements present in typical pipe systems. Figure 6.17 provides an assortment of loss-coefficient and equivalent-length data. The figure includes a sampling of the types of geometries for which loss coefficients have been empirically determined, collected from various sources including Streeter (1961), Cleveland Hydraulic Institute (1979) and Crane Company (1982). If accurate head losses are needed, manufacturer's data should be used.

Reentrant Inlet	ℓ/D	$K_{t/D=0}$	$K_{t/D=0.02}$	
	0.0	0.50	0.50	
	0.1	0.83	0.60	
	0.2	0.92	0.66	
	0.3	0.97	0.69	
	0.4	1.00	0.71	

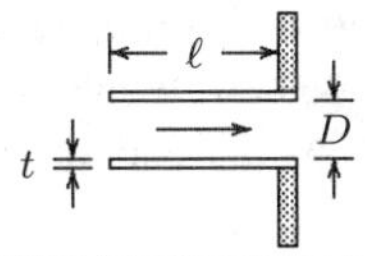

Beveled Inlet	L/D	$K_{\theta=10°}$	$K_{\theta=30°}$	$K_{\theta=50°}$
	0.00	0.50	0.50	0.50
	0.05	0.39	0.30	0.35
	0.10	0.32	0.19	0.28
	0.15	0.27	0.14	0.24
	0.20	0.24	0.12	0.23

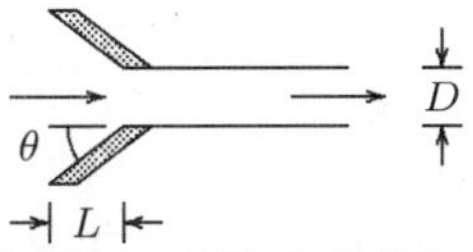

90° Miter Bend

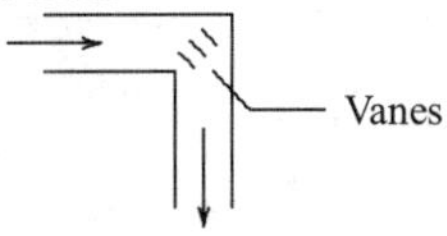

Vanes

	K
Without vanes	1.1
With vanes	0.2

General Miter Bend

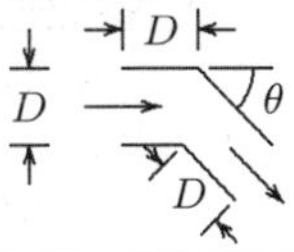

θ	L_e/D
0°	2
30°	8
60°	24
90°	58

Elbow

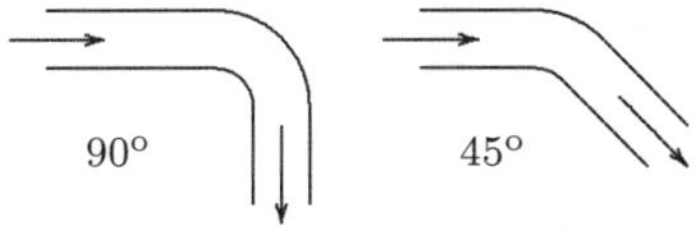

Regular:	K	Long radius:	K
90° – flanged	0.3	90° – flanged	0.2
90° – threaded	1.5	90° – threaded	0.7
45° – threaded	0.4	45° – flanged	0.2

90°   45°

Return Bend

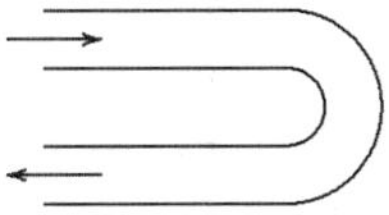

	K
Flanged	0.2
Threaded	1.5

Tee

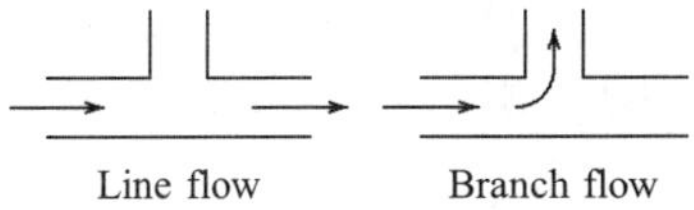

Line flow:	K	Branch flow:	K
Flanged	0.2	Flanged	1.0
Threaded	0.9	Threaded	2.0

Line flow   Branch flow

Threaded Union

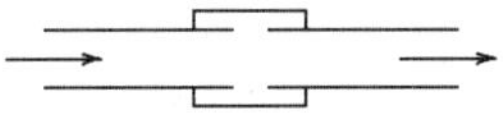

$K = 0.08$

Assorted Valves

	K		K		K
Globe – fully open	10	Gate – fully open	$0.15 - 0.20$	Ball – fully open	0.05
Angle – fully open	$2 - 5$	Gate – $\frac{1}{4}$ closed	0.26	Ball – $\frac{1}{3}$ closed	5.5
Swing check – forward flow	2	Gate – $\frac{1}{2}$ closed	$2.1 - 5.6$	Ball – $\frac{2}{3}$ closed	210
Swing check – backward flow	∞	Gate – $\frac{3}{4}$ closed	17		

Figure 6.17: *Minor loss data for various pipe-system elements.*

Example 6.11 *For the sake of aesthetics, a plumber installs a 90° bend between two points in the pipe system for a swimming pool [Figure (a)]. The same functionality can be accomplished with a straight pipe as shown in Figure (b). What penalty is paid, as measured by head loss and the attendant added burden on the pool's pump, to satisfy the plumber's desire for aesthetics. The pipe is smooth and the flow Reynolds number is $Re_D = 1.75 \cdot 10^6$. Assume that the flow velocity is the same with or without the bend, and that $L = 8D$.*

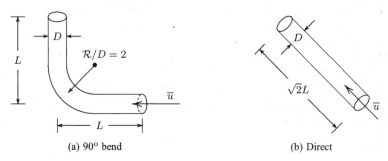

(a) 90° bend   (b) Direct

Solution. Using either Equation (6.59) or the Moody diagram (Figure 6.7), the friction factor is

$$f = 0.0106$$

Without the bend, Figure (b) shows that the length of the pipe connecting the two points is $\sqrt{2}L$. Hence the head loss is

$$h_L = f \frac{\overline{u}^2}{2g} \frac{\sqrt{2}L}{D} = \left(\sqrt{2}f\frac{L}{D}\right)\frac{\overline{u}^2}{2g}$$

Then, in terms of the given properties, we have

$$h_L = \left(\sqrt{2} \cdot 0.0106 \cdot 8\right)\frac{\overline{u}^2}{2g} = 0.12\frac{\overline{u}^2}{2g}$$

By contrast, with the bend [Figure (a)], reference to the inset in Figure 6.15 shows that the loss factor for $R/D = 2$ is

$$K = 0.19$$

Thus, the total head loss is

$$h_L = f \frac{\overline{u}^2}{2g} \frac{2L}{D} + K \frac{\overline{u}^2}{2g} = \left(2f\frac{L}{D} + K\right)\frac{\overline{u}^2}{2g}$$

Again, substituting the given values we find

$$h_L = (2 \cdot 0.0106 \cdot 8 + 0.19)\frac{\overline{u}^2}{2g} = 0.36\frac{\overline{u}^2}{2g}$$

Therefore, for the sake of a pretty pipe arrangement, our plumber has tripled the head loss through this particular section of the pool's pipe system!

6.4.3   Multiple-Pipe Systems

All of the discussion above focuses on a single pipe. However, there are many important applications that involve two or more pipes. Examples are the pipes in the water-distribution system for a housing development and the complex series of bronchial and other tubes in a human lung. The way we analyze complex pipe networks is similar to the method used by electrical engineers to analyze electrical circuits.

To understand the similarity, recall that we can express head loss, h_L, in terms of either the loss coefficient, K, or the equivalent length, L_e, defined in Equations (6.63) and (6.64). Hence, we can say

$$h_L = \bar{u}^2 \hat{R}, \qquad \hat{R} = \frac{K}{2g} \text{ or } \hat{R} = \frac{f}{2g} \frac{L_e}{D} \qquad (6.73)$$

where $\hat{R}$ is an effective resistance. Now, Ohm's law for electrical circuits tells us $V = iR$, where V is voltage, i is current and R is resistance. Thus, head loss is analogous to voltage drop, $\hat{R}$ is analogous to resistance and $\bar{u}^2$ is analogous to current. Note that we don't have a formal mathematical analogy since velocity does not appear linearly in the energy equation. Thus, some, but not all, of the standard circuit-theory methods can be applied to pipe systems.

The foundation of our approach is conservation of mass and energy, which is analogous to conservation of charge and voltage. To illustrate how pipe systems can be analyzed, we consider pipes attached in series and in parallel. Figure 6.18 shows typical arrangements of both types.

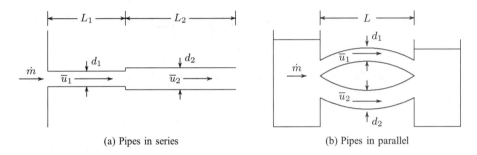

(a) Pipes in series   (b) Pipes in parallel

Figure 6.18: *Pipes in series and in parallel.*

Pipes in Series. Figure 6.18(a) shows two pipes of different diameters, i.e., $d_1 < d_2$, connected in series. Conservation of mass and energy tell us that the mass flow is constant, while the overall head loss is the sum of the head losses in each pipe. Thus, we have

$$\dot{m} = \dot{m}_1 = \dot{m}_2 \qquad (6.74)$$

$$h_L = h_{L1} + h_{L2} \qquad (6.75)$$

Now, for the first pipe, we have a minor loss at the entrance and a major loss proportional to the friction factor, f_1, pipe length, L_1, etc. Pipe 2 has a similar major loss and a minor loss due to sudden expansion. Thus, using Equation (6.71) to approximate the loss at a sudden expansion, we have

$$h_{L1} = \frac{1}{2} \frac{\bar{u}_1^2}{2g} + f_1 \frac{L_1}{d_1} \frac{\bar{u}_1^2}{2g}, \qquad h_{L2} = \left[1 - \left(\frac{d_1}{d_2}\right)^2\right]^2 \frac{\bar{u}_2^2}{2g} + f_2 \frac{L_2}{d_2} \frac{\bar{u}_2^2}{2g} \qquad (6.76)$$

Also, using the fact that $\dot{m} = \frac{\pi}{4}\rho\bar{u}d^2$, we find from Equation (6.74)

$$\bar{u}_2 = \bar{u}_1 \left(\frac{d_1}{d_2}\right)^2 \qquad (6.77)$$

To be specific, suppose the ratio of length to diameter ratio, L_i/d_i, is 10 for both pipes, $d_1 = d$, $d_2 = \sqrt{2}d$ and $\bar{u}_1 = U$. Then, Equation (6.77) shows that

$$\bar{u}_1 = U, \qquad \bar{u}_2 = \frac{1}{2}U \tag{6.78}$$

Substituting these values into (and combining) Equations (6.75) and (6.76) yields

$$h_L = \left[\frac{9}{16} + 10\left(f_1 + \frac{1}{4}f_2\right)\right]\frac{U^2}{2g} \tag{6.79}$$

To complete the solution, all that remains is to compute the friction factor in each pipe, which, in turn, depends upon the Reynolds number. Since we have not specified enough information to compute Reynolds number, this is as far as we can take the solution here.

Note that because mass flow is linear in $\bar{u}$ while head loss is quadratic in $\bar{u}$, we do not have the analogy to electrical circuit theory that would say the total resistance is equal to the sum of the resistances in the two pipes. This becomes very clear if we substitute Equations (6.77) and (6.78) into Equation (6.73), wherefore

$$h_L = \left[\hat{R}_1 + \hat{R}_2\left(\frac{d_1}{d_2}\right)^4\right]\frac{\bar{u}_1^2}{2g} = \left[\hat{R}_1 + \frac{1}{4}\hat{R}_2\right]\frac{U^2}{2g} \tag{6.80}$$

showing that head loss in the first pipe is the most significant. This is true because the velocity is twice as large in the first pipe, and the loss varies as $\bar{u}^2$.

Pipes in Parallel. Figure 6.18(b) shows two pipes of different diameters connected in parallel. As above, we assume $d_1 < d_2$. Conservation of mass and energy tell us that the mass flow is equal to the sum of the mass flows in each pipe, while the overall head loss is the same for each pipe. Therefore, we say

$$\dot{m} = \dot{m}_1 + \dot{m}_2 \tag{6.81}$$

$$h_L = h_{L1} = h_{L2} \tag{6.82}$$

To simplify the analysis, assume that both pipes are such that the product of the equivalent length, L_e, and friction factor, f, is invariant. Then,

$$h_{L1} = \left(\frac{fL_e}{2g}\right)\frac{\bar{u}_1^2}{d_1}, \quad h_{L2} = \left(\frac{fL_e}{2g}\right)\frac{\bar{u}_2^2}{d_2} \quad \Longrightarrow \quad \bar{u}_2 = \bar{u}_1\sqrt{\frac{d_2}{d_1}} \tag{6.83}$$

Also, using the fact that $\dot{m} = \frac{\pi}{4}\rho\bar{u}d^2$, we find

$$\bar{u}_1 d_1^2 + \bar{u}_2 d_2^2 = \frac{4\dot{m}}{\pi\rho} \tag{6.84}$$

Substituting Equation (6.83) into Equation (6.84),

$$\bar{u}_1 d_1^2 + \bar{u}_1\sqrt{\frac{d_2}{d_1}}d_2^2 = \frac{4\dot{m}}{\pi\rho} \quad \Longrightarrow \quad \bar{u}_1\left[1 + \left(\frac{d_2}{d_1}\right)^{5/2}\right] = \frac{4\dot{m}}{\pi\rho d_1^2} \tag{6.85}$$

Therefore, the velocities in the pipes are

$$\left.\begin{array}{rcl}
\bar{u}_1 & = & \dfrac{4\dot{m}}{\pi\rho d_1^2}\left[1 + \left(\dfrac{d_2}{d_1}\right)^{5/2}\right]^{-1} \\[4ex]
\bar{u}_2 & = & \dfrac{4\dot{m}}{\pi\rho d_1^2}\left[1 + \left(\dfrac{d_2}{d_1}\right)^{5/2}\right]^{-1}\sqrt{\dfrac{d_2}{d_1}}
\end{array}\right\} \tag{6.86}$$

As a final comment, these ideas can be extended to a network of pipes in order to analyze the water-distribution system for a housing development or even an entire city. A segment of a pipe network would appear as in Figure 6.19. There are two basic rules that must be applied to solve such a network, and they are identical to the famous **Kirchhoff's rules** for electric circuits. Specifically:

1. The sum of the mass-flow rates entering any junction point is equal to the sum of the mass-flow rates leaving this point.

2. The algebraic sum of the head loss around any closed loop of the network equals zero.

Rule 1 simply means that whatever flows in to a given point must also flow out. Rule 2 means the total enthalpy can assume only one value at a given point. Unlike classical electric circuits, pipe-network equations are nonlinear and require iterative solutions. Such solutions are readily obtained with a computer. Jepson (1976), for example, describes numerical techniques used in solving pipe-network problems.

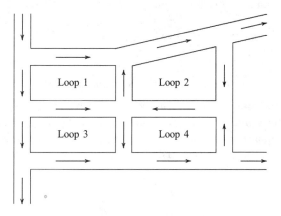

Figure 6.19: *Pipe network.*

6.5 Open-Channel Flow

Our final topic of this chapter is the flow of a liquid (usually water) that occurs in channels with a surface that interfaces with a gas (usually air), i.e., when the flow has a **free surface**. Open-channel flows occur naturally, with typical examples being streams and rivers. Man-made open channels are very common, some of the most familiar being irrigation canals, sewers, aqueducts, drainage ditches and spillways. More subtle examples are the rain gutters on your home and the water running down a driveway when you wash your car.

Open-channel flow, while similar in some ways to pipe flow, includes the extra complication of the free surface. The boundary condition appropriate for a free surface is quite different from that imposed at a solid boundary, and gives rise to interesting phenomena that do not appear in pipe flow. We will see that the gravitational force, rather than the pressure gradient, drives open-channel flows and that the Froude number plays a significant role. Because there is a free surface, we must address the question of the role that surface tension might play. The latter point is easily addressed. For virtually all applications of interest, the Weber number is so large that surface tension can be ignored.

The primary equations used to describe flow in an open channel are conservation of mass and energy. The momentum-conservation principle plays a secondary role. Focusing on steady, incompressible flow, we treat the flow in a **quasi-one-dimensional** sense. That is, we use the control-volume method to relate conditions on cross-sectional planes. Consider Figure 6.20, which is an idealized section of a typical open channel.

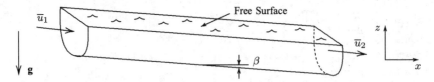

Figure 6.20: *A section of a typical open channel.*

The reason the momentum equation is of secondary importance is because we address only flows for which changes in the streamwise direction occur gradually. This will be true if cross-sectional areas and depth do not change rapidly with streamwise distance and if the inclination of the channel to the horizontal direction is small, i.e., $\beta \ll 1$. When these conditions are satisfied, the vertical velocity is very small compared to the horizontal velocity. Now, the pressure at a free surface is equal to the atmospheric pressure, p_a, even when the fluid at the surface is moving, i.e.,

$$p = p_a \qquad \text{(at a free surface)} \qquad (6.87)$$

Thus, a streamwise pressure gradient cannot occur at the free surface. In fact, straight open-channel flows have no significant streamwise pressure gradient on any part of the cross sections. The primary balance of forces is between the gravitational force and the viscous force on the solid boundaries of the channel. Thus, for steady flow, the vertical (z) component of the momentum equation is as follows.

$$\underbrace{\oiint_S \rho w \, (\mathbf{u} \cdot \mathbf{n}) \, dS}_{\text{Negligible}} = -\mathbf{k} \cdot \oiint_S p \, \mathbf{n} \, dS - \iiint_V \rho g \, dV + \underbrace{\mathbf{k} \cdot \oiint_S \mathbf{f}_v \, dS}_{\text{Negligible}} \qquad (6.88)$$

Note that the vector $\mathbf{f}_v$ is the viscous-force vector introduced in Equation (6.29). The most important conclusion we can draw from the momentum equation comes from applying Equation (6.88) to an infinitesimal element of volume ΔV (cf. Section 3.1). Using Equation (3.8) and the fact that $\int_V \rho g \, dV \approx \rho g \, \Delta V$ as $\Delta V \to 0$, we have

$$0 \approx -\frac{\partial p}{\partial z} \Delta V - \rho g \, \Delta V \qquad \Longrightarrow \qquad \frac{\partial p}{\partial z} \approx -\rho g \qquad (6.89)$$

Hence, we see that the pressure satisfies the hydrostatic relation. To use the streamwise component of the momentum equation, we require some information about the viscous force. Because the flow is incompressible, we can use either the mass and momentum equations or the energy equation. As with pipe flows, it is more convenient to use the energy equation for open-channel flows.

Since we assume the flow is steady, the mass-flow rate, $\dot{m}$, across any cross section is constant. Thus, we can say

$$Q = \frac{\dot{m}}{\rho} = \bar{u} A = \text{constant} \qquad (6.90)$$

where Q is the **volume-flow rate**, $\bar{u}$ is the average streamwise flow velocity and A is the cross-sectional area.

Turning to the energy-conservation equation, we can use the approximate equation derived earlier for pipe-flow applications, viz., Equation (6.52). We make the following three simplifications to the basic conservation equation.

1. Since none are present for the intended applications, we drop turbine and pump terms, h_t and h_p.

2. Because of the large uncertainties associated with the irregular geometries of rivers, streams, etc., we assume the kinetic-energy correction factor, α, is unity throughout. Any errors attending this approximation will be minor relative to typical assumptions made to idealize a real application.

3. Since open channels have negligible streamwise pressure gradient, necessarily $p_1 \approx p_2$.

Thus, Equation (6.52) simplifies to a balance amongst kinetic energy, potential energy, and viscous effects as represented by a head loss, i.e.,

$$\frac{\bar{u}_1^2}{2g} + z_1 = \frac{\bar{u}_2^2}{2g} + z_2 + h_L \tag{6.91}$$

If the depth of the channel is denoted by y, then for a channel segment of length ℓ, reference to Figure 6.21 shows that depth and altitude are related as follows.

$$z_1 = y_1 + \ell \tan \beta \quad \text{and} \quad z_2 = y_2 \tag{6.92}$$

So, denoting the slope of the bottom by

$$S_o \equiv \tan \beta \tag{6.93}$$

the energy-conservation equation becomes

$$y_1 + \frac{\bar{u}_1^2}{2g} + S_o \ell = y_2 + \frac{\bar{u}_2^2}{2g} + h_L \tag{6.94}$$

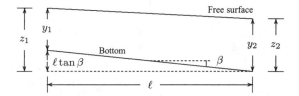

Figure 6.21: *Open-channel geometry—side view.*

To make further progress with these equations, we must determine the head loss. Experience has shown that we can use the results for pipe flow by introducing the **hydraulic radius**, R_h. By definition, the hydraulic radius is

$$R_h = \frac{A}{P} \tag{6.95}$$

where A is cross-sectional area and P is the perimeter of the wetted, solid surface. R_h differs from the hydraulic diameter used for pipe-flow analysis to the extent that it excludes the free surface, while D_h is based on the perimeter bounding the cross-sectional area. This reflects

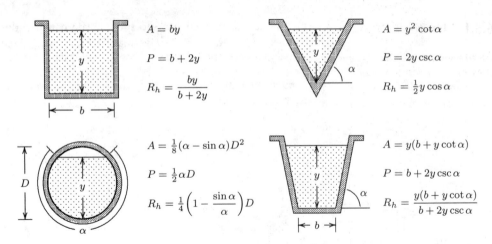

Figure 6.22: *Areas, perimeters and hydraulic radii for common geometries.*

the fact that the free surface cannot support a shear stress. Figure 6.22 includes A, P and R_h for several simple geometries.

An infinitely wide rectangular channel, which might approximate a thin sheet of water flowing down an incline or in a brook (Figure 6.23), is an interesting limiting case. For such a "channel," which is simply a two-dimensional flow, the hydraulic radius is equal to the depth, viz.,

$$R_h = \lim_{b \to \infty} \left(\frac{by}{b + 2y} \right) = y \tag{6.96}$$

For a circular cross section, the information in Figure 6.22 shows that if the channel is full ($\alpha = 2\pi$), then $R_h = \frac{1}{4}D$. This reflects the factor of 4 difference between the definition of hydraulic diameter for pipes [Equation (6.66)] and hydraulic radius for open channels [Equation (6.95)]. We can use the Moody diagram and the correlation formulas for friction factor, f, and head loss, h_L, provided we compute the hydraulic diameter according to

$$D_h = 4R_h \tag{6.97}$$

Figure 6.23: *Preacher's Brook in West Franklin, PA—its hydraulic radius is equal to its depth.*

6.5.1   Uniform Flow

As an example of how we can use the formulas developed above to solve a typical open-channel flow problem, consider a straight channel with a very small inclination to the horizontal. Many rivers and canals have slopes of order $S_o \approx 0.0001$, and this is the value we will use. For this small a slope, the flow will maintain constant depth provided cross-sectional shape remains unchanged. This is referred to as **uniform flow** or **normal flow**.

We will analyze the triangular cross section shown in Figure 6.24. The volume-flow rate is $Q = 5$ m^3/sec and the channel is constructed of brass for which the roughness height is $k_s = 0.6$ mm. Using a value for kinematic viscosity of $\nu = 10^{-6}$ m^2/sec, we wish to determine the normal depth of the water in the channel, y.

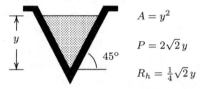

$$A = y^2$$
$$P = 2\sqrt{2}\,y$$
$$R_h = \tfrac{1}{4}\sqrt{2}\,y$$

Figure 6.24: *Flow in an open channel with triangular cross section.*

The governing equations are mass and energy conservation. From mass conservation applied to the specified triangular cross section, we know that

$$Q = \bar{u}A, \quad A = y^2 \quad \Longrightarrow \quad \bar{u} = \frac{Q}{y^2} \tag{6.98}$$

Since the volume flux and the cross-sectional area are constant, so are the average flow velocity and the depth. That is, at any two streamwise sections, we have $\bar{u}_1 = \bar{u}_2$ and $y_1 = y_2$. Hence, Equation (6.94) simplifies to

$$S_o \ell = h_L = f \frac{\ell}{4R_h} \frac{\bar{u}^2}{2g} \quad \Longrightarrow \quad \bar{u}^2 = \frac{8gS_oR_h}{f} \tag{6.99}$$

Observe that we have used the Darcy-Weisbach Equation (6.56) with $D = 4R_h$ to evaluate the head loss.

Now, reference to Figure 6.24 shows that the hydraulic radius for our triangular cross section is

$$R_h = \frac{\sqrt{2}}{4}y \tag{6.100}$$

Substituting Equations (6.98) and (6.100) into Equation (6.99) yields

$$\frac{Q^2}{y^4} = \frac{8gS_o}{f}\frac{\sqrt{2}}{4}y \quad \Longrightarrow \quad y^5 = \frac{Q^2 f}{2\sqrt{2}\,gS_o} \tag{6.101}$$

Therefore, the depth in the channel is

$$y = \left[\frac{Q^2}{2\sqrt{2}\,gS_o}\right]^{1/5} f^{1/5} \tag{6.102}$$

We cannot solve directly for y because the friction coefficient, f, is not yet known. It depends upon Reynolds number and the ratio of k_s to D_h which, in turn, depend upon flow velocity and depth. Thus, we will have to solve with an iterative procedure. As we will see, however, our solution will converge very rapidly, mainly because most open-channel flows occur at high Reynolds number and involve very rough wetted surfaces, so the variation of f with Reynolds number is very slow.

For the given values, the equation relating y and f becomes

$$y = \left[\frac{\left(5 \text{ m}^3/\text{sec}\right)^2}{2\sqrt{2}\left(9.807 \text{ m/sec}^2\right)(0.0001)} \right]^{1/5} f^{1/5} = 6.18 f^{1/5} \qquad (6.103)$$

Omitting the detailed arithmetic calculations for the sake of brevity, the Reynolds number, Re_{D_h}, and dimensionless roughness-height, k_s/D_h, are

$$Re_{D_h} = \frac{\bar{u}D_h}{\nu} = \frac{7.07 \cdot 10^6}{y} \qquad \text{and} \qquad \frac{k_s}{D_h} = \frac{4.25 \cdot 10^{-4}}{y} \qquad (6.104)$$

To solve, we first make a guess for the friction factor and then use Equations (6.103) and (6.104) to determine Reynolds number and the roughness parameter. Reference to the Moody diagram (Figure 6.7) provides a new value for the friction factor. When the guessed and computed values of f agree to within some desired margin of error, the solution has converged.

Table 6.2: *Computational Details for a Triangular Open Channel*

f_{guess}	y (m)	Re_{D_h}	k_s/D_h	$f_{computed}$
0.0100	2.46	$2.87 \cdot 10^6$	$1.73 \cdot 10^{-4}$	0.0135
0.0135	2.61	$2.71 \cdot 10^6$	$1.63 \cdot 10^{-4}$	0.0135

Table 6.2 shows a two-step iteration that yields an acceptable solution for f. The initial guess for the friction factor is $f = 0.0100$. As shown, using the corresponding values of Re_{D_h} and k_s/D_h implies $f \approx 0.0135$. Using $f = 0.0135$ as the guess in the second iteration again yields a computed value of $f = 0.0135$. Substitution of the values for the second iteration into the Colebrook formula [Equation (6.60)] shows that it is satisfied to within 0.5%. Thus, we conclude that the depth of the fluid in the triangular channel is

$$y = 2.61 \text{ m} \qquad (6.105)$$

As mentioned above, the iterative procedure converges very rapidly because typical open-channel flows occur at high Reynolds number, and the wetted surface is generally quite rough. Inspection of the Moody diagram shows that, under such conditions, the friction factor is independent of Reynolds number. Correspondingly, the Colebrook formula simplifies to

$$\frac{1}{\sqrt{f}} \approx -2\log_{10}\left(\frac{k_s/D_h}{3.7}\right) \qquad \text{(Completely-rough flow)} \qquad (6.106)$$

This equation defines a very slow variation of f with k_s/D_h. For example, increasing k_s/D_h by a factor of 1000 from 10^{-4} to 10^{-1} corresponds to f increasing by less than a factor of 10 (from 0.012 to 0.102). Additionally, Equation (6.103) shows that the depth is a slowly varying function of f. Hence, any sensible first guess for f will lead to a close approximation to the actual value of f after a single iteration.

6.5.2 The Equations of Chézy and Manning

Perhaps the most famous formulation for uniform open-channel flow was developed by French engineer Antoine Chézy (1718-1798) in experimental studies of the River Seine and the Courpalet Canal. We can solve for average channel velocity, $\bar{u}$, from Equation (6.99), wherefore

$$\bar{u} = \sqrt{\frac{8g}{f}} \, \sqrt{R_h S_o} \tag{6.107}$$

When the channel shape and roughness are constant, so is the factor $\sqrt{8g/f}$. Calling this constant C, we have the famous **Chézy equation**, viz.,

$$\bar{u} = C\sqrt{R_h S_o} \tag{6.108}$$

The coefficient C is known as the **Chézy coefficient**. It is a dimensional quantity, having dimensions $L^{1/2}/T$, and will thus be different for each new application. In his experiments, Chézy determined that it varies from 60 ft$^{1/2}$/sec (33 m$^{1/2}$/sec) for small rough channels to about 160 ft$^{1/2}$/sec (88 m$^{1/2}$/sec) for large smooth channels. The use of such dimensional, empirically determined, coefficients limits the usefulness of these formulas to the single fluid for which they have been determined, namely, water.

Through the late nineteenth and early twentieth centuries, a great deal of research focused on developing correlations of the Chézy coefficient with roughness, slope and channel shape. Chow (1959), for example, gives a lucid discussion of many correlations devised during this period. By far, the most popular correlation is attributed to Robert Manning (1816-1897). The **Manning correlation** is

$$C = \frac{\chi}{n} R_h^{1/6}, \qquad \chi = 1.49 \text{ ft}^{1/3}/\text{sec} = 1.00 \text{ m}^{1/3}/\text{sec} \tag{6.109}$$

where the dimensionless coefficient n is the **Manning roughness coefficient**. We can arrive at this formula by noting that for completely-rough pipes and channels, the limiting form of the Colebrook formula, Equation (6.106), is closely approximated by

$$f \approx 0.113 \left(\frac{k_s}{R_h}\right)^{1/3}, \qquad 0.001 < \frac{k_s}{R_h} < 0.05 \tag{6.110}$$

Substituting this formula for f into $C = \sqrt{8g/f}$ yields

$$C \approx \left(8.4 g^{1/2}/k_s^{1/6}\right) R_h^{1/6} \tag{6.111}$$

which reproduces the proportionality between C and $R_h^{1/6}$ deduced empirically by Manning. Comparison with Equation (6.109) implies that, for a range of roughness heights, the Manning roughness coefficient is given by

$$n \approx \chi \left(\frac{k_s^{1/6}}{8.4 g^{1/2}}\right), \qquad 0.001 < \frac{k_s}{R_h} < 0.05 \tag{6.112}$$

Finally, combining Equations (6.108) and (6.109), we arrive at the famous **Chézy-Manning equation**, i.e.,

$$\bar{u} = \frac{\chi}{n} S_o^{1/2} R_h^{2/3}, \qquad \chi = 1.49 \text{ ft}^{1/3}/\text{sec} = 1.00 \text{ m}^{1/3}/\text{sec} \tag{6.113}$$

Table 6.3 lists values of n determined by Manning for a variety of open channels with roughness height varying from 0.3 mm to 5 m. Observe that all of the values for n include an error band. This reflects the fact that this empirical parameter depends upon Reynolds number, channel geometry, and other factors that have been glossed over in this rather simplistic formulation.

Table 6.3: *Experimental Values of the Manning Roughness Coefficient*

Type of Channel	n	k_s (ft)	k_s (mm)
Artificially lined channels:			
Glass	0.010 ± 0.002	0.0011	0.3
Brass	0.011 ± 0.002	0.0019	0.6
Smooth steel	0.012 ± 0.002	0.0032	1.0
Finished cement	0.012 ± 0.002	0.0032	1.0
Planed wood	0.012 ± 0.002	0.0032	1.0
Cast iron	0.013 ± 0.003	0.0051	1.6
Painted steel	0.014 ± 0.003	0.0080	2.4
Unfinished cement	0.014 ± 0.002	0.0080	2.4
Clay tile	0.014 ± 0.003	0.0080	2.4
Riveted steel	0.015 ± 0.002	0.0120	3.7
Brickwork	0.015 ± 0.002	0.0120	3.7
Asphalt	0.016 ± 0.003	0.0180	5.4
Corrugated metal	0.022 ± 0.005	0.1200	37.0
Rubble masonry	0.025 ± 0.005	0.2600	80.0
Excavated earth channels:			
Clean	0.022 ± 0.004	0.12	37
Gravelly	0.025 ± 0.005	0.26	80
Weedy	0.030 ± 0.005	0.80	240
Stony, large cobbles	0.035 ± 0.010	1.50	500
Natural channels:			
Clean and straight	0.030 ± 0.005	0.80	240
Major rivers	0.035 ± 0.010	1.50	500
Sluggish, deep pools	0.040 ± 0.010	3.00	900
Floodplains:			
Pasture, farmland	0.035 ± 0.010	1.5	500
Light brush	0.050 ± 0.020	6.0	2000
Heavy brush	0.075 ± 0.025	15.0	5000
Trees	0.150 ± 0.050	—	—

Example 6.12 *Using Equation (6.112), compute the roughness coefficient, n, for sluggish, deep pools.*

Solution. Substituting for χ and g, the roughness coefficient is

$$n = \left(1.00 \ \frac{\mathrm{m}^{1/3}}{\mathrm{sec}} \right) \frac{k_s^{1/6}}{8.4 \left(9.807 \ \mathrm{m/sec^2} \right)^{1/2}} = \frac{k_s^{1/6}}{26.3 \ \mathrm{m}^{1/6}}$$

Reference to Table 6.3 shows that the effective roughness height for a sluggish, deep pool is $k_s = 900$ mm $= 0.90$ m. Thus, the roughness coefficient is

$$n = \frac{(0.90 \ \mathrm{m})^{1/6}}{26.3 \ \mathrm{m}^{1/6}} = 0.037$$

According to the table, the measured roughness coefficient lies between 0.030 and 0.050. Hence, Equation (6.112) gives a realistic value.

In general, the Manning approximation is accurate for a limited range of roughness heights. Equation (6.112) provides an indication as to why its range of applicability might be limited in this manner. In practice, it yields unrealistically low friction, and thus high flow rate, for both deep smooth channels (small k_s/R_h) and shallow rough channels (large k_s/R_h). The Chézy-Manning equation is nevertheless a useful preliminary design tool because of the ease with which it can be applied, which the following example illustrates.

Example 6.13 *Consider the triangular channel discussed in the preceding subsection. Use the Chézy-Manning equation to determine the channel depth.*

Solution. Recall from Equations (6.98) and (6.100) that the velocity and hydraulic radius are

$$\bar{u} = \frac{Q}{y^2}, \qquad R_h = \frac{\sqrt{2}}{4}y = 2^{-3/2}y$$

where $Q = 5$ m^3/sec is the volume-flow rate and y is depth. So, since the slope is $S_o = 0.0001$, the Chézy-Manning Equation (6.113) becomes

$$\frac{Q}{y^2} = \frac{1}{n}(0.0001)^{1/2}\left(2^{-3/2}y\right)^{2/3} = \frac{y^{2/3}}{200n}$$

Thus, solving for y in terms of Q and n, we find

$$y = (200Qn)^{3/8}$$

Since the channel walls are made of brass, reference to Table 6.3 shows that the Manning friction coefficient is $n = 0.011 \pm 0.002$, i.e.,

$$0.009 < n < 0.013$$

Therefore, according to the Chézy-Manning equation, the depth of the water in the channel lies in the range

$$2.28 \text{ m} < y < 2.62 \text{ m}$$

Recall that, using the more-accurate Moody diagram/Colebrook formula values for the friction factor, we found $y = 2.61$ m. Had we used the nominal value of $n = 0.011$, our predicted depth would have been 2.46 m, which is about 6% lower than 2.61 m. Thus, the Chézy-Manning formula is reasonably accurate for this example.

6.5.3 Surface Wave Speed and Flow Classification

While uniform flow in open channels is of interest for practical design of irrigation systems, flood-control studies, etc., there are other flow regimes that are of practical interest. There are two primary ways for classifying open-channel flow, viz., in terms of depth variation and in terms of wave-propagation properties.

Focusing first on depth variation, we have already examined the simplest case of uniform flow in which depth is constant. If the flow encounters an obstacle or if the cross-sectional area changes, then the depth will vary. When this occurs, we say the flow is **varied**. If the flow changes gradually so that it can still be treated analytically in a quasi-one-dimensional manner, we say the flow is **gradually varied**. By contrast, if depth (and other) variations occur so rapidly that a more-complicated multidimensional analysis is required, we say the flow is **rapidly varied**. Summarizing, there are three classes of open-channel flows based on depth variation as follows.

1. *Uniform Flow:* Constant depth and slope, which can be analyzed using simple algebraic equations.

2. *Gradually-Varied Flow:* Depth and/or slope varying sufficiently slowly that the flow can be described with a straightforward one-dimensional, ordinary differential equation [see Equation (6.137) below].

3. *Rapidly-Varied Flow:* Depth and/or slope varying so fast that a two- or three-dimensional analysis is required.

Turning to the second way of classifying open-channel flows, we note that one of the most interesting consequences of having a free surface is the presence of waves that propagate both along and below the surface. Waves on the surface are most easily observed, as anyone who has visited a beach or taken a ride on a boat can attest. An interesting question to pose is, what is the speed of a small wave traveling along a free surface? We might also want to know if the speed depends upon the depth, and if so, what the dependence is. We refer to such waves as **gravity waves**. The speed of propagation, which we denote as c, is often called the **wave celerity**.

We can compute c by analyzing a weak wave traveling along a free surface of an infinitely-wide open channel of undisturbed depth y. As shown in Figure 6.25(a), after the wave passes, it imparts a velocity, Δu, to the fluid. For a weak wave, we assume the changes in flow properties are very small, i.e.,

$$|\Delta y| \ll y, \qquad |\Delta u| \ll c \tag{6.114}$$

We can replace this unsteady problem with an equivalent steady problem by making a Galilean transformation so that the wave is stationary and the flow is from left to right, as shown in Figure 6.25(b). In this coordinate system, the velocity to the left of the wave is c, while the velocity to the right of the wave is $c - \Delta u$.

To analyze the wave, we select the control volume indicated by the dashed contour in Figure 6.25(b). Because we are using a one-dimensional geometry, our analysis can be viewed as applying to a volume of unit width out of the page. Since the flow is steady, conservation of mass tells us the net flux of fluid out of the control volume vanishes. Thus,

$$\iint_S \rho \mathbf{u} \cdot \mathbf{n} \, dS = 0 \tag{6.115}$$

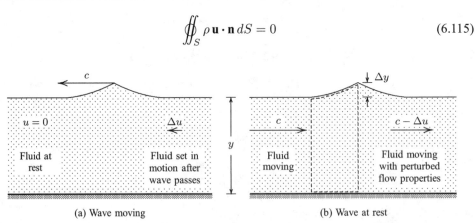

(a) Wave moving   (b) Wave at rest

Figure 6.25: *Analysis of a gravity wave in an open channel; a Galilean transformation produces a steady flow with fluid passing through the stationary wave.*

Hence, since $\mathbf{n}$ is the *outer* unit normal, evaluating the closed-surface integral gives

$$-\rho cy + \rho(c - \Delta u)(y + \Delta y) = 0 \qquad (6.116)$$

Expanding the terms in parentheses, there follows

$$-\rho cy + \rho cy + \rho c\Delta y - \rho y\Delta u - \rho \Delta y\Delta u = 0 \qquad (6.117)$$

wherefore

$$y\Delta u = c\Delta y - \Delta u\Delta y \qquad (6.118)$$

Since we are assuming the surface wave causes an infinitesimally small change in all flow properties (in particular, since $|\Delta u| \ll c$), we can drop the last term in Equation (6.118) as it is quadratic in changes while the other two terms are linear. Hence, we conclude from mass conservation that the changes in velocity and depth are related as follows.

$$\Delta u \approx \frac{c}{y}\Delta y \qquad (6.119)$$

If we ignore viscous effects, the x component of the momentum-conservation equation for steady flow is

$$\oiint_S \rho u(\mathbf{u} \cdot \mathbf{n}) \, dS = -\mathbf{i} \cdot \oiint_S (p - p_a)\mathbf{n} \, dS + \mathbf{i} \cdot \iiint_V \rho g\, \mathbf{k} \, dV \qquad (6.120)$$

where we take the dot product of $\mathbf{i}$ with the net pressure and gravity integrals to extract their x components. Now, the gravity force is in the vertical direction and has no effect on x momentum. Also, as argued above, the pressure is hydrostatic [see Equation (6.89)] so that

$$p - p_a = -\rho gz \qquad (6.121)$$

Expanding the various integrals, with $z = 0$ at the free surface, we have

$$\rho c(-cy) + \rho(c - \Delta u)[(c - \Delta u)(y + \Delta y)] = -\int_{-y}^0 \rho gz \, dz + \int_{-y-\Delta y}^0 \rho gz \, dz \qquad (6.122)$$

We can simplify the momentum equation by combining the two pressure integrals and changing integration variables according to $\zeta = -z$, so that

$$\rho c^2 y = [\rho(c - \Delta u)(y + \Delta y)](c - \Delta u) + \int_y^{y+\Delta y} \rho g\zeta \, d\zeta \qquad (6.123)$$

Then, using Equation (6.116), and evaluating the pressure integral, we can simplify further:

$$
\begin{aligned}
\rho c^2 y &= \rho cy(c - \Delta u) + \frac{1}{2}\rho g\left[(y + \Delta y)^2 - y^2\right] \\
&= \rho c^2 y - \rho cy\Delta u + \frac{1}{2}\rho g(2y + \Delta y)\Delta y \qquad (6.124)
\end{aligned}
$$

Therefore, dropping the term quadratic in Δy, conservation of momentum yields the following relation between the changes in velocity and depth.

$$\rho cy\Delta u \approx \rho gy\Delta y \qquad \Longrightarrow \qquad \Delta u \approx \frac{g}{c}\Delta y \qquad (6.125)$$

Finally, we can combine Equations (6.119) and (6.125) to arrive at

$$c^2 \approx gy \qquad \Longrightarrow \qquad c = \sqrt{gy} \tag{6.126}$$

We can now reveal the second way of classifying open-channel flows. If we compute the ratio of the average flow velocity, $\bar{u}$, to the celerity, c, we have the Froude number, Fr, i.e.,

$$Fr = \frac{\bar{u}}{\sqrt{gy}} \tag{6.127}$$

In terms of the Froude number, there are three types of flow:

1. $Fr < 1$...Subcritical Flow;

2. $Fr = 1$...Critical Flow;

3. $Fr > 1$...Supercritical Flow.

The difference between subcritical and supercritical flow is in the way the flow is affected by an obstruction, a cross-sectional change, a disturbance, etc. In subcritical flow, waves are created by a change in flow conditions. The waves move faster than the oncoming flow, and can propagate upstream at a speed $\tilde{u} \approx \sqrt{gy} - \bar{u}$. Thus, the oncoming flow has advance warning that flow conditions are changing, and it can adjust accordingly in a gradual manner. For example, consider the waves created by a small pebble dropped into a stream of depth y moving at velocity $\bar{u}$. Using a Galilean transformation so that the oncoming stream is at rest, the motion of the wave fronts is as shown in Figure 6.26.

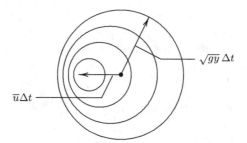

Figure 6.26: *Subcritical wave pattern for a pebble dropped into a stream—a Galilean transformation has been done to put the incident stream at rest.*

In this frame of reference, the point where the pebble strikes advances to the left at a speed $\bar{u}$. Each circle corresponds to the wave front created by the pebble as it moves. The largest circle is the wave front formed at an initial time—at a subsequent time Δt, its radius is $\sqrt{gy}\,\Delta t$. The other three circles denote the wave fronts corresponding to times $\frac{1}{4}\Delta t$, $\frac{1}{2}\Delta t$ and $\frac{3}{4}\Delta t$ with respective radii of $\frac{3}{4}\sqrt{gy}\,\Delta t$, $\frac{1}{2}\sqrt{gy}\,\Delta t$ and $\frac{1}{4}\sqrt{gy}\,\Delta t$.

In supercritical flow, the oncoming stream moves faster than surface waves, which are swept downstream. Thus, the flow has no advance warning that conditions are changing, and the transition caused by the pebble is much more abrupt. Figure 6.27 illustrates the wave pattern—the wave fronts are confined to a triangular region as shown. The sides of the triangle are tangent to the family of circular wave fronts. The half angle of the triangle, μ, is equal to the arcsine of the ratio of distance traveled by the wave front, $\sqrt{gy}\,\Delta t$, to the distance traveled by the pebble, $\bar{u}\Delta t$. Thus,

$$\mu = \sin^{-1}\left(\frac{\sqrt{gy}}{\bar{u}}\right) = \sin^{-1}\left(\frac{1}{Fr}\right) \tag{6.128}$$

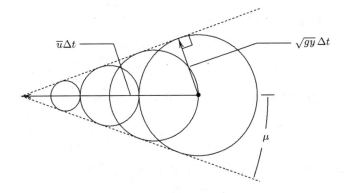

Figure 6.27: *Supercritical wave pattern for a pebble dropped into a stream—a Galilean transformation has been done to put the incident stream at rest.*

Example 6.14 *Compute the Froude number for the triangular channel considered in the two preceding subsections.*

Solution. Based on results obtained, the depth is $y = 2.61$ m and the average velocity is

$$\overline{u} = \frac{Q}{y^2} = \frac{5 \text{ m}^3/\text{sec}}{(2.61 \text{ m/sec})^2} = 0.734 \frac{\text{m}}{\text{sec}}$$

Also, the wave speed is

$$c = \sqrt{gy} = \sqrt{\left(9.807 \frac{\text{m}}{\text{sec}^2}\right)(2.61 \text{ m})} = 5.06 \frac{\text{m}}{\text{sec}}$$

Therefore, for the triangular open channel, the Froude number is

$$Fr = \frac{\overline{u}}{\sqrt{gy}} = 0.145$$

so that the flow is subcritical.

The Froude number appears explicitly in the equation of motion for *gradually-varied* flow, i.e., for flow in which depth and/or cross sections change gradually. To see the variation, recall that for a segment of an open channel of length ℓ, the energy-conservation principle [see Equation (6.94) and Figure 6.21] is

$$y_1 + \frac{\overline{u}_1^2}{2g} + S_o \ell = y_2 + \frac{\overline{u}_2^2}{2g} + h_L \tag{6.129}$$

We can derive a differential equation by considering a channel segment of differential length $\ell = \Delta x$. Then, the various quantities appearing in Equation (6.129) can be written in terms of differential changes, viz.,

$$\left.\begin{array}{ll} \overline{u}_1 = \overline{u}, & \overline{u}_2 = \overline{u} + \Delta\overline{u} \\[2mm] y_1 = y, & y_2 = y + \Delta y \\[2mm] h_L = f\dfrac{\Delta x}{D_h}\dfrac{\overline{u}^2}{2g} \end{array}\right\} \tag{6.130}$$

Substituting Equation (6.130) into Equation (6.129), we obtain

$$y + \frac{\bar{u}^2}{2g} + S_o\,\Delta x = y + \Delta y + \frac{\bar{u}^2}{2g} + \frac{\bar{u}}{g}\Delta\bar{u} + \frac{(\Delta\bar{u})^2}{2g} + f\frac{\Delta x}{D_h}\frac{\bar{u}^2}{2g} \qquad (6.131)$$

Canceling like terms, dropping the term quadratic in $\Delta\bar{u}$ and rearranging a little, there follows

$$S_o \approx \frac{\Delta y}{\Delta x} + \frac{\bar{u}}{g}\frac{\Delta\bar{u}}{\Delta x} + \frac{f}{D_h}\frac{\bar{u}^2}{2g} \qquad (6.132)$$

Taking the limit $\Delta x \to 0$, the differential equation governing gradually-varied flow is

$$S_o = \frac{dy}{dx} + \frac{\bar{u}}{g}\frac{d\bar{u}}{dx} + \frac{f}{D_h}\frac{\bar{u}^2}{2g} \qquad (6.133)$$

We can simplify this equation further by appealing to mass conservation, which permits us to eliminate $d\bar{u}/dx$. For the sake of simplicity, assume the channel has rectangular cross section with width b. Then, mass conservation is $\bar{u}by = $ constant, so that

$$by\frac{d\bar{u}}{dx} + b\bar{u}\frac{dy}{dx} = 0 \qquad\Longrightarrow\qquad \frac{d\bar{u}}{dx} = -\frac{\bar{u}}{y}\frac{dy}{dx} \qquad (6.134)$$

Finally, it is customary to denote the head-loss term by S, i.e.,

$$S \equiv \frac{f}{D_h}\frac{\bar{u}^2}{2g} \qquad (6.135)$$

Note that S is the slope corresponding to uniform flow with the same volume-flow rate, Q [cf. Equation (6.99)]. Substituting Equations (6.134) and (6.135) into Equation (6.133) gives

$$\left[1 - \frac{\bar{u}^2}{gy}\right]\frac{dy}{dx} = S_o - S \qquad (6.136)$$

Recognizing the fact that the quantity in brackets is simply $[1 - Fr^2]$, the final form of the differential equation for gradually-varied flow in a rectangular open channel is

$$\frac{dy}{dx} = \frac{S_o - S}{1 - Fr^2} \qquad (6.137)$$

As discussed by White (1994), for example, there are actually 12 different sub-cases of gradually-varied open-channel flow that can be discerned from Equation (6.137), depending upon the Froude number and upon $(S_o - S)$. As an example, consider a bottom slope, S_o, which exceeds the slope appropriate for uniform flow, S, so that $S_o - S > 0$. If the flow is subcritical, i.e., if $Fr < 1$, then the depth increases with distance, and the velocity decreases. By contrast, the depth decreases and the velocity increases when the flow is supercritical.

6.5.4   Specific Energy

We can gain some qualitative feel for the dynamics of open-channel flows by introducing the **specific energy**, E, defined by

$$E = y + \frac{\bar{u}^2}{2g} \qquad (6.138)$$

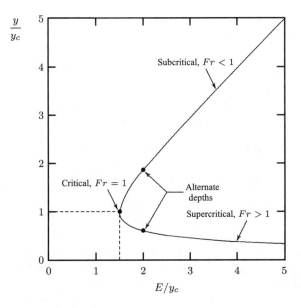

Figure 6.28: *Specific energy for flow in an open channel; y_c is critical depth.*

Focusing on a rectangular channel of width b for simplicity, we know from mass conservation that the volume-flow rate is $Q = \bar{u}by$. Thus, we can rewrite the specific energy as

$$E = y + \frac{Q^2}{2gb^2y^2} \tag{6.139}$$

Figure 6.28 graphically displays specific energy for a fixed value of $Q^2/(gb^2)$. Inspection of the graph shows that E has a minimum value. We can locate this minimum by setting dE/dy equal to zero, i.e.,

$$\frac{dE}{dy} = 1 - \frac{Q^2}{gb^2y^3} = 0 \quad \Longrightarrow \quad y = \left(\frac{Q^2}{gb^2}\right)^{1/3} \equiv y_c \tag{6.140}$$

For reasons that will become obvious below, the quantity y_c is known as the **critical depth**. Dividing Equation (6.139) through by y_c, there follows:

$$\frac{E}{y_c} = \frac{y}{y_c} + \frac{1}{2}\left(\frac{y_c}{y}\right)^2 \tag{6.141}$$

Hence, the minimum value of E is

$$E_{min} = E(y_c) = \frac{3}{2}y_c \tag{6.142}$$

Using this value in the original definition of E, Equation (6.138), we can solve for the velocity at the point where E achieves its minimum value, viz.,

$$y_c + \frac{\bar{u}_c^2}{2g} = \frac{3}{2}y_c \quad \Longrightarrow \quad \bar{u}_c = \sqrt{gy_c} \tag{6.143}$$

The specific-energy curve is the locus of all states possible for a given flow rate and channel width [so that $Q^2/(2gb^2)$ is constant]. Now, since the volume-flow rate is constant on the curve, we know that

$$Q = \overline{u}_c b y_c = \overline{u} b y \quad \Longrightarrow \quad \overline{u} = \overline{u}_c \frac{y_c}{y} \tag{6.144}$$

Hence, the Froude number at any point on the curve is

$$Fr = \frac{\overline{u}}{\sqrt{gy}} = \frac{\overline{u}_c y_c}{\sqrt{gy}\, y} = \frac{\sqrt{gy_c}\, y_c}{\sqrt{gy}\, y} = \left(\frac{y_c}{y}\right)^{3/2} \tag{6.145}$$

Therefore, the Froude number at the critical depth is 1, corresponding to critical flow, which is the reason we call y_c the *critical depth*. This equation also shows that the Froude number is less than 1 on the upper branch of the curve, wherefore the flow is subcritical. By contrast, it is greater than 1 on the lower branch, corresponding to supercritical flow.

As shown in Figure 6.28, for a given volume-flow rate and specific energy, there are two possible depths. These are referred to as **alternate depths**. Since the total energy is the same, the subcritical branch corresponds to a state with high potential energy and low kinetic energy, while the distribution of energy is the opposite on the supercritical branch.

Flow under a **sluice gate** provides an example of flow in which the alternate depths occur for a fixed value of E. A sluice gate, depicted in Figure 6.29, is a control structure used to regulate flow rate. As shown, the subcritical fluid upstream of the gate has high potential energy and low kinetic energy. After passing through the gate, the volume-flow rate is unchanged and, in the absence of head losses, so is the specific energy. The flow becomes supercritical with low potential energy and high kinetic energy. For fixed values of E and Q, the depth upstream of the gate is alternate to the depth downstream.

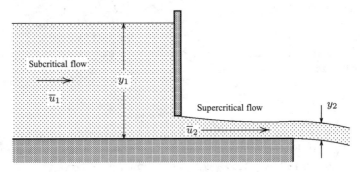

Figure 6.29: *Flow through a sluice gate.*

Example 6.15 *Consider flow in a rectangular open channel that is regulated by a sluice gate. The channel has width b = 10 ft, depth y_1 = 5 ft and volume-flow rate Q = 100 ft^3/sec. Compute the depth downstream of the sluice gate.*

Solution. Conservation of mass tells us the average flow velocity is $\overline{u} = Q/(by_1) = 2$ ft/sec. Also, using Equation (6.140), a short calculation shows that the critical depth is $y_c = 1.46$ ft, wherefore $y_1/y_c = 3.42$. From Figure 6.28, the alternate depth is $y_2/y_c \approx 0.40$. Therefore, the depth downstream of the sluice gate is

$$y_2 = 0.40 y_c = 0.40(1.46 \text{ ft}) = 0.584 \text{ ft} = 7 \text{ inches}$$

6.5.5 Hydraulic Jump

A sluice gate is an example of a flow that changes from subcritical to supercritical, with a negligibly small head loss. When the opposite change occurs with the upstream flow being supercritical and the downstream flow being subcritical, we have what is known as a **hydraulic jump**. This situation is sometimes imposed as a design feature, by forcing an abrupt change in depth, or may occur naturally because of the prevailing depth downstream. The hydraulic jump is analogous to a shock wave in a compressible fluid, which we will discuss in Chapter 8.

As shown in Figure 6.30, the flow upstream of the jump is fast and shallow. After the jump, the flow becomes slow and deep, and there is a substantial head loss. Because they are very effective in dissipating energy and in mixing, hydraulic jumps are commonly used in spillways and in sewage-treatment plants.

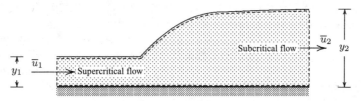

Figure 6.30: *Hydraulic jump.*

We can analyze a hydraulic jump using the control volume indicated by the dashed contour in Figure 6.30. For simplicity, we consider a horizontal bottom and an infinitely-wide channel. We also neglect the viscous force in the momentum equation. Our approach is the same as that used to compute the velocity of a surface wave in Subsection 6.5.3, with the exception that we do not assume changes in velocity and depth are small. Referring to Equations (6.116) and (6.124), we replace c by $\overline{u}_1$, $c - \Delta u$ by $\overline{u}_2$, y by y_1 and $y + \Delta y$ by y_2, wherefore

$$\rho \overline{u}_1 y_1 = \rho \overline{u}_2 y_2 \tag{6.146}$$

$$\rho \overline{u}_1^2 y_1 = \rho \overline{u}_2^2 y_2 + \frac{1}{2} \rho g \left(y_2^2 - y_1^2 \right) \tag{6.147}$$

$$y_1 + \frac{\overline{u}_1^2}{2g} = y_2 + \frac{\overline{u}_2^2}{2g} + h_L \tag{6.148}$$

We include the energy Equation (6.148) to permit solving for the head loss across the jump.

Omitting details of the algebra for the sake of brevity, we can solve for the depth ratio, y_2/y_1, velocity ratio, $\overline{u}_2/\overline{u}_1$, Froude number after the jump, Fr_2, and head loss, h_L, as a function of the Froude number ahead of the jump, Fr_1. The results are as follows.

$$Fr_2 = Fr_1 \left[\frac{2}{\sqrt{1 + 8Fr_1^2} - 1} \right]^{3/2} \tag{6.149}$$

$$\frac{\overline{u}_2}{\overline{u}_1} = \frac{2}{\sqrt{1 + 8Fr_1^2} - 1} \tag{6.150}$$

$$\frac{y_2}{y_1} = \frac{1}{2} \left[\sqrt{1 + 8Fr_1^2} - 1 \right] \tag{6.151}$$

$$h_L = \frac{(y_2 - y_1)^3}{4 y_1 y_2} \tag{6.152}$$

Since, by definition, head loss is always positive, necessarily a hydraulic jump causes depth to increase. Inspection of Equation (6.151) shows that we have $y_2 \geq y_1$ when $Fr_1 \geq 1$, while $y_2 < y_1$ when $Fr_1 < 1$. *Thus, a hydraulic jump can only cause a transition from supercritical to subcritical flow.*

Example 6.16 *Water flows in a wide channel with an incident Froude number of $Fr_1 = 2$ and a volume-flow rate per unit width of $Q = 11.4 \, ft^2/sec$, in which a hydraulic jump occurs. Determine flow properties in the channel.*

Solution. First, we determine the incident depth and velocity by noting that

$$\bar{u}_1 = \frac{Q}{y_1} \quad \text{and} \quad \bar{u}_1 = Fr_1\sqrt{gy_1} \quad \Longrightarrow \quad y_1 = \left(\frac{Q^2}{4g}\right)^{1/3}$$

For the given numerical values, we find

$$y_1 = 1.00 \text{ ft} \quad \text{and} \quad \bar{u}_1 = 11.4 \, \frac{\text{ft}}{\text{sec}}$$

Using Equations (6.149) through (6.152), a straightforward arithmetic exercise shows that:

$$Fr_2 = 0.55, \quad \frac{\bar{u}_2}{\bar{u}_1} = 0.42, \quad \frac{y_2}{y_1} = 2.37, \quad \frac{h_L}{y_1} = 0.27$$

Therefore, downstream of the hydraulic jump, we have

$$y_2 = 2.37 \text{ ft} \quad \text{and} \quad \bar{u}_2 = 4.8 \, \frac{\text{ft}}{\text{sec}}$$

Finally, note that the specific energy in the flow upstream of the hydraulic jump is given by $E_1 = y_1 + \bar{u}_1^2/(2g) = 3.0$ ft, so that the relative loss of energy is

$$\frac{h_L}{E_1} = 0.09$$

For this low a Froude number, a hydraulic jump is usually thought of as being relatively weak. Even so, 9% of the energy in the flow is dissipated into heat. For Froude numbers in excess of 9, the head loss can be as high as 85% of the incident specific energy.

Figure 6.31: *Hydraulic jump—flow is from right to left.*

Problems

6.1 A container filled with 2 slugs of carbon dioxide at $p_1 = 20$ psi and $T_1 = 88°$ F is compressed isothermally to a pressure $p_2 = 60$ psi. How much work, W_c, is done in compressing the gas, where $W_c \equiv \int_{V_1}^{V_2} p\, dV$?

6.2 A container filled with 5 kg of hydrogen at $p_1 = 0.2$ MPa and $T_1 = 0°$ C is compressed isentropically to a pressure $p_2 = 0.4$ MPa. Determine the final temperature, T_2, and the work done, W_c, in compressing the gas.

6.3 In a *polytropic* process, the pressure and density of a perfect gas are related by $p = p_1(\rho/\rho_1)^n$, where n is a constant and subscript '1' denotes initial condition.

  (a) In a reversible, polytropic process 3 slugs of air initially at $p_1 = 14.7$ psi and $T_1 = 60°$ F change to $p_2 = 44.7$ psi and volume $V_2 = 500$ ft^3. What is the polytropic exponent, n?

  (b) Determine the work done by the surroundings on the gas, W, where $W \equiv \int_{V_1}^{V_2} p\, dV$. Express your answer in foot pounds.

  (c) Determine the heat transferred from the surroundings to the gas, Q, in Btu's.

6.4 In a *polytropic* process, the pressure and density of a perfect gas are related by $p = p_1(\rho/\rho_1)^n$, where n is a constant and subscript '1' denotes initial condition.

  (a) In a reversible, polytropic process 50 kg of air initially at $p_1 = 100$ kPa and $T_1 = 20°$ C change to $p_2 = 200$ kPa and $T_2 = 59°$ C. What is the polytropic exponent, n?

  (b) Determine the work done by the surroundings on the gas, W, where $W \equiv \int_{V_1}^{V_2} p\, dV$. Express your answer in Joules.

  (c) Determine the heat transferred from the surroundings to the gas, Q, in Joules.

6.5 Beginning with Gibbs' equation, viz., $T ds = de + p d(1/\rho)$, determine entropy as a function of pressure in an *isothermal* flow of a perfect, calorically-perfect gas.

6.6 Beginning with Gibbs' equation, viz., $T ds = de + p d(1/\rho)$, determine entropy as a function of temperature in an *isobaric* flow of a perfect, calorically-perfect gas.

6.7 Beginning with Gibbs' equation, viz., $T ds = de + p d(1/\rho)$, first introduce the enthalpy, h, and show that $T ds = dh - dp/\rho$. Then, prove that for a perfect gas that is also calorically perfect, the equation of state can be written as

$$\frac{p}{p_r} = \left(\frac{T}{T_r}\right)^{\gamma/(\gamma-1)} \exp\left[\frac{(s_r - s)}{R}\right]$$

where subscript 'r' denotes reference-state conditions.

6.8 Find the compressibility, τ_s, for isentropic flow of a perfect gas defined by

$$\tau_s \equiv \frac{1}{\rho}\left(\frac{\partial \rho}{\partial p}\right)_s$$

where subscript s means entropy is held constant. Compare your result with the isothermal compressibility discussed in Section 1.6.

6.9 For a *thermally-perfect* gas, by definition $e = e(T)$ and $h = h(T)$. Noting that

$$c_p = \left(\frac{\partial h}{\partial T}\right)_p \quad \text{and} \quad c_v = \left(\frac{\partial e}{\partial T}\right)_v$$

verify that $c_p - c_v = R$ for a thermally-perfect gas.

6.10 The internal energy of a perfect, *thermally-perfect* gas is

$$e(T) = e_a + 12R\sqrt{T_a T}$$

where e_a and T_a are constants. Noting that $de = c_v dT$ and $c_p = c_v + R$ for a thermally-perfect gas, determine c_v and c_p. What is the specific-heat ratio, γ, when $T = 9T_a$?

6.11 The pump shown supplies energy to the flow such that the upstream pressure is p and the downstream pressure is $2p$. The steady mass-flow rate is $\dot{m}$ and the pipe diameters are as shown. What is the power, P, delivered by the pump to the flow? The temperature increases by ΔT. Also, you may neglect effects of gravity. Express your answer for the power in terms of $\dot{m}$, p, d, ΔT, water density, ρ, and specific-heat coefficient, c_v.

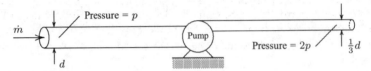

Problem 6.11

6.12 Your yard-drainage system is backing up. To relieve the problem, you install a pump to help drain the water from the primary pipe whose diameter is d. As shown below, the outlet pipe has a diameter of $2d$. The pump supplies energy to the flow such that the upstream pressure is p and the downstream pressure is $4p$, the steady mass-flow rate is $\dot{m}$, and the temperature increases by ΔT. Also, you may neglect effects of gravity. What is the power, P, delivered by the pump to the flow? Express your answer for the power in terms of $\dot{m}$, p, d, ΔT, water density, ρ, and specific-heat coefficient, c_v.

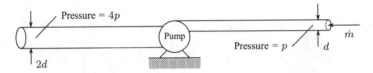

Problem 6.12

6.13 A power plant uses a river to discharge waste heat. Heat is transferred from the plant to the river through a heat exchanger at a rate $\dot{Q}$. The river's volume-flow rate is $\dot{V} = 33$ ft^3/sec and the difference between the temperature downstream and upstream of the power plant is $\Delta T = T_f - T_i = 1.9°$ F. The flow is steady and the river has constant cross-sectional area. Also, you can neglect frictional work from the river, ground and atmosphere. What is the heat-transfer rate from the power plant to the river? Note that $\rho c_p = 62.6$ Btu/(ft$^3 \cdot °$R) for the river water.

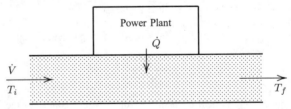

Problems 6.13, 6.14

6.14 A power plant uses a river to discharge waste heat. Heat is transferred from the plant to the river through a heat exchanger at a rate $\dot{Q} = 56$ MW. The river's volume-flow rate is $\dot{V} = 8.9$ m^3/sec and the river temperature upstream of the power plant is $T_i = 18°$ C. The flow is steady and the river has constant cross-sectional area. Also, you can neglect frictional work from the river, ground and atmosphere. What is the temperature downstream of the power plant, T_f? Note that $c_p = 4200$ J/(kg·K) and $\rho = 998$ kg/m^3 for the river water.

6.15 Air flows steadily in a pipe of diameter $d = 8$ cm at a rate of $\dot{m} = 0.5$ kg/sec. The pressure and temperature at the upstream end of the pipe are $p_1 = 150$ kPa and $T_1 = 20°$ C. Ignoring heat transfer, viscous effects and body forces, find the velocity, u_2, and temperature, T_2, at a point in the pipe where the pressure is $p_2 = 100$ kPa. Develop a quadratic equation for T_2/T_1 before determining numerical values for T_2 and u_2.

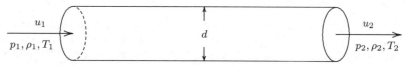

Problems 6.15, 6.16

6.16 Helium flows steadily in a pipe of diameter $d = 8$ cm at a rate of $\dot{m} = 0.5$ kg/sec. The pressure and temperature at the upstream end of the pipe are $p_1 = 150$ kPa and $T_1 = 20°$ C. Ignoring heat transfer, viscous effects and body forces, find the velocity, u_2, and temperature, T_2, at a point in the pipe where the pressure is $p_2 = 100$ kPa. Develop a quadratic equation for T_2/T_1 before determining numerical values for T_2 and u_2.

6.17 A laser is used to energize the flow of air through a channel of height H and width $5H$ (out of the page). The design objective is to increase the velocity in the channel by 5%. Ignoring viscous losses and effects of gravity, what heat-transfer rate, $\dot{Q}$, must be supplied by the laser if $u_1 = 100$ ft/sec and $T_1 = -150°$ F? Assume the flow is steady, pressure is constant and equal to 1 atm throughout the channel and $H = 1$ ft. If an affordable laser delivers 10-20 Btu/sec, can the design objective be realized?

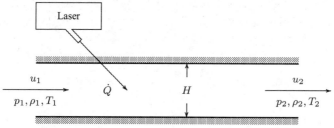

Problem 6.17

6.18 A centrifugal water pump is sufficiently insulated to prevent any heat transfer from the surroundings. The power required to drive the pump is $\dot{W}_s$, and the mass-flow rate through the pump is $\dot{m}$. The flow is steady with water entering normal to the xy plane at Point 1 and exiting to the atmosphere at Point 2. Gravitational effects are negligible and $\Delta h = h_1 - h_2 = 33.5\mathbf{u}_1 \cdot \mathbf{u}_1$.

(a) Compute the force in the xy plane required to hold the pump in place as a function of $\dot{m}$ and $\dot{W}_s$. No other variables should appear in your answer.

(b) What is the force if $\dot{m} = 19.67$ slug/sec and $\dot{W}_s = 52$ hp?

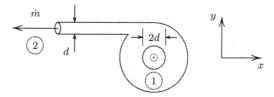

Problems 6.18, 6.19

6.19 The mass-flow rate is $\dot{m}$ for a centrifugal water pump. The flow is steady with water entering normal to the xy plane at Point 1 and exiting at Point 2 as shown. Gravitational effects are negligible, the water exhausts to the atmosphere and $\Delta h = h_1 - h_2 = 9\mathbf{u}_1 \cdot \mathbf{u}_1$. If the power required to drive the pump is $\dot{W}_s = 2\dot{m}^3/(\pi^2 \rho^2 d^4)$, what is the heat transfer rate, $\dot{Q}$?

6.20 The mass-flow rate of water through a turbine is $\dot{m}$ and the pressure drop through the turbine is $p_2 - p_1 = -\Delta p$. Heat-transfer effects can be neglected and the flow is steady.

(a) Determine the power, P, delivered to the turbine from the water. Neglect internal-energy changes.

(b) Compute P for $\dot{m} = 100$ kg/sec, $\Delta p = 300$ kPa, $\rho = 1000$ kg/m^3, $d = 0.2$ m and $\Delta z = 3$ m. Express your answer in kilowatts and in horsepower.

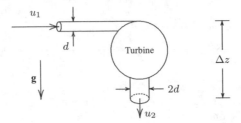

Problem 6.20

6.21 Consider flow through a constant diameter pipe that carries water from an elevation z_1 to another elevation z_2. The average velocity at elevation z_1 is $\bar{u}_1 = U$ and the overall head loss through the pipe is $h_L = 0.03U^2/(2g)$. It turns out that, in terms of the averaged velocities $\bar{u}_1$ and $\bar{u}_2$, Bernoulli's equation is satisfied between elevations z_1 and z_2. Determine the difference between the kinetic-energy correction factors, $\alpha_1 - \alpha_2$.

6.22 The pressure, p, and average velocity, U, remain constant throughout a pipe with a bend lying in a horizontal plane. The fluid is water and the kinetic-energy correction factor upstream of the bend, α_1, exceeds the downstream value by 0.03, i.e., $\alpha_1 = \alpha_2 + 0.03$. If the head loss through the bend is $h_L = 0.11U^2/(2g)$, determine the head, h_p, that the pump must deliver.

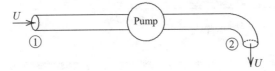

Problem 6.22

6.23 The radius of the pipe section shown below and the inlet velocity are constant and equal to R and U, respectively. At the outlet, the velocity is given by $u = u_o(r)$, where r is distance from the pipe centerline. The difference between inlet pressure, p_i, and outlet pressure, p_o, is $p_i - p_o = 1.20\rho U^2$, where ρ is the density of the (incompressible) fluid in the pipe. The pipe lies in a horizontal plane so that potential energy is constant.

(a) Determine U_{max} as a function of U.

(b) Compute α_1 and α_2.

(c) Compute the head loss, h_L, as a function of U and the acceleration of gravity, g.

Problem 6.23

6.24 For turbulent channel flow, a useful approximation for the velocity profile between the channel wall and centerline is $u/u_m = (y/H)^n$, where $0 \leq y \leq H$, u_m is centerline velocity, y is distance from the channel wall, H is channel half height and n is a constant. Determine the kinetic-energy correction factor, α, as a function of n. Evaluate α for $n = 1/5$, $n = 1/7$ and $n = 1/9$.

6.25 A useful approximation for the velocity profile in turbulent pipe flow is $u/u_m = (y/r_o)^n$, where $y = r_o - r$, u_m is centerline velocity, y is distance from the surface of the pipe, r_o is pipe radius, r is distance from the centerline and n is a constant. Determine the kinetic-energy correction factor, α, as a function of n. Evaluate α for $n = 1/5$, $n = 1/7$ and $n = 1/9$.

6.26 Consider steady, laminar flow of an incompressible fluid with density ρ through the circular pipe of radius R shown below. At the inlet the velocity is uniform and equal to U. At the outlet, the velocity is $U_{out} = U_{max}(1 - r^2/R^2)$, where r is radial distance from the pipe's centerline. The change in elevation, $\Delta z = z_{out} - z_{in} = U^2/(12g)$, and the change in pressure between inlet and outlet, $\Delta p = p_{out} - p_{in} = -\frac{2}{3}\rho U^2$. Determine the maximum velocity, U_{max}, the kinetic-energy correction factors, α_{in} and α_{out}, and the head loss, h_L.

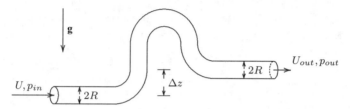

Problem 6.26

6.27 An incompressible fluid of kinematic viscosity $\nu = 10^{-4}$ m^2/sec flows through a pipe of diameter $D = 10$ cm.

(a) If the average flow velocity, $\bar{u}$, is 2 m/sec, is the flow laminar or turbulent? What is the approximate value of the friction coefficient, f?

(b) If the pipe is perfectly smooth and $\bar{u}$ increases to 20 m/sec, will the flow most likely be laminar or turbulent? What is the friction coefficient, f?

(c) Suppose, however, that the pipe surface is very rough. Suppose further that increasing $\bar{u}$ to 200 m/sec yields no change in the value of f determined in Part (a) above. What must the relative roughness, k_s/D, be?

6.28 Consider a pipe of diameter $D = 1$ in and length $L = 280$ ft. A liquid of kinematic viscosity $\nu = 1.25 \cdot 10^{-6}$ ft^2/sec flows through the pipe with average velocity $\bar{u} = 7.5$ ft/sec. Determine the friction factor, f, and head loss, h_L, for pipes made of plastic, steel and galvanized iron.

6.29 Consider a pipe of diameter $D = 25$ cm and length $L = 120$ m. A liquid of kinematic viscosity $\nu = 5 \cdot 10^{-7}$ m^2/sec flows through the pipe with average velocity $\bar{u} = 10$ m/sec. Determine the friction factor, f, and head loss, h_L, for pipes made of glass, wrought iron and cast iron.

6.30 The head loss for a steel pipe of length $L = 2$ km is $h_L = 96.68$ m. Water with kinematic viscosity $\nu = 10^{-6}$ m^2/sec flows through the pipe, whose diameter, D, is 75 mm. What is the average flow velocity, $\bar{u}$?

6.31 The head loss for a perfectly-smooth pipe of length $L = 1$ mile is $h_L = 31.13$ ft. Water with kinematic viscosity $\nu = 10^{-5}$ ft^2/sec flows through the pipe, whose diameter, D, is 0.75 ft. What is the average flow velocity, $\bar{u}$?

6.32 In designing a piping system, you have a choice of pipes made of copper, steel and cast iron. You also have of choice of using a pipe diameter of either $D = 6$ cm or 10 cm. The system calls for a straight section of length $L = 50$ m, through which a liquid of kinematic viscosity $\nu = 2.25 \cdot 10^{-7}$ m^2/sec will flow. Also, the average flow velocity will be $\bar{u} = 7.5$ m/sec, regardless of the pipe diameter. Using the Moody diagram, determine the friction factors for these six different pipes. Also, compute the head loss for the pipes with the minimum and maximum friction factors.

6.33 For plastic piping, what diameter, D, is needed to convey $\dot{V} = 0.025$ m^3/sec of fuel oil if the system can support a head loss of $h_L = 4$ m over a length $L = 280$ m? The kinematic viscosity of the fuel oil is $\nu = 2 \cdot 10^{-4}$ m^2/sec. Assume and verify that the flow is laminar.

6.34 With great care, laminar pipe flow can be realized for Reynolds number, Re_D, as high as 50,000. For water ($\nu = 10^{-5}$ ft^2/sec) flowing in a 3-in diameter, 100-ft long pipe, compute the head loss, in inches, for laminar and turbulent flow when $Re_D = 50,000$.

6.35 A liquid of density ρ is pumped from a tank to the atmosphere. The diameter of the pipe connecting the tank and pump is $4D$. After leaving the pump the pipe diameter is D. The head supplied by the pump is $h_p = 3.1\Delta z$, where Δz is as indicated in the figure. You may assume that throughout this flow the kinetic-energy correction factor, α, is 1.05.

(a) What is the dimensionless head loss, $2gh_L/U^2$? Express your answer in terms of Δz, U and g.

(b) What is $2gh_L/U^2$ if $\Delta z = 5$ ft and $U = 15$ ft/sec?

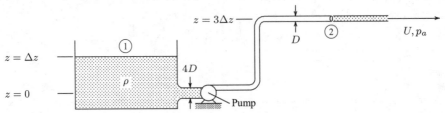

Problems 6.35, 6.36, 6.37

6.36 A liquid of density ρ is pumped from a tank to the atmosphere. The diameter of the pipe connecting the tank and pump is $4D$. After leaving the pump the pipe diameter is D. The head loss is $h_L = \frac{5}{2}U^2/(2g)$, where U is the velocity in the smaller pipe. You may assume that throughout this flow the kinetic-energy correction factor, α, is 1.0.

(a) What head, h_p, must the pump supply? Express your answer in terms of Δz, U and g.

(b) What is h_p if $\Delta z = 2$ m and $U = 8$ m/sec?

6.37 A liquid of density ρ is pumped from a tank to the atmosphere. The diameter of the pipe connecting the tank and pump is $4D$. After leaving the pump the pipe diameter is D. The head supplied by the pump is $h_p = 2.5\Delta z$, where Δz is as indicated in the figure. You may assume that throughout this flow the kinetic-energy correction factor, α, is 1.

(a) What is the velocity of the water as it enters the pump?

(b) What is the dimensionless head loss, $2gh_L/U^2$? Express your answer in terms of Δz, U and g.

6.38 A pump connects two pipes as shown. The head loss between Points 1 and 2 is given by $h_L = 0.65U^2/g$, where U is the flow speed in the inlet pipe and g is gravitational acceleration. The inlet pressure is $p_1 = p_o + 0.1\rho U^2$, where p_o is atmospheric pressure, and the difference in elevation is $\Delta z = U^2/(2g)$. The kinetic-energy correction factor in the inlet pipe is given by $\alpha_1 = 1.1$, while the outlet pipe has $\alpha_2 = 1.125$. The fluid exhausts to the atmosphere. Determine the outlet velocity, U_2, and the head supplied by the pump, h_p.

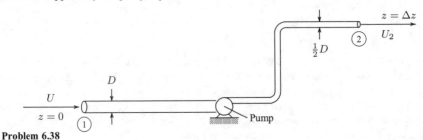

Problem 6.38

6.39 Incompressible, steady flow in a pipe undergoes a sudden expansion as shown. The pipe lies in a horizontal plane so that potential-energy changes can be ignored. The kinetic-energy correction factor, α, is 1.03 throughout the pipe, and the pressure change is found from measurements to be given by $C_p \equiv (p_2 - p_1)/(\frac{1}{2}\rho\bar{u}_1^2) = 0.45$. Measurements indicate that the head loss for a sudden expansion is of the form $h_L = K\bar{u}_1^2/(2g)$. Determine the velocity $\bar{u}_2$ as a function of $\bar{u}_1$ and the coefficient K.

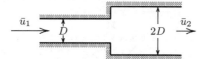

Problem 6.39

6.40 Consider the hydroelectric plant shown. The head loss between Points 1 and 2 is $h_L = \frac{1}{8}H$. If the mass-flow rate through the turbine is $\dot{m}$, how much power is transferred from the water to the turbine? In developing your answer, assume the kinetic-energy correction factor, α, is 1.05 throughout the flow, and that the water emitted by the turbine exhausts to the atmosphere.

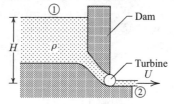

Problems 6.40, 6.41, 6.42

6.41 Consider the hydroelectric plant shown. The head loss between Points 1 and 2 is $h_L = \frac{1}{10}H$. If the mass-flow rate through the turbine is $\dot{m}$, how much power is transferred from the water to the turbine? In developing your answer, assume the kinetic-energy correction factor, α, is 1.06 throughout the flow, and that the water emitted by the turbine exhausts to the atmosphere. Also, the Froude number (based on H) of the water emitted by the turbine is 0.40.

6.42 Consider the hydroelectric plant shown. The head loss between Points 1 and 2 is $h_L = \frac{1}{25}H$. If the head supplied to the turbine is $h_t = \frac{9}{10}H$, what is the Froude number (based on H) of the water emitted by the turbine? In developing your answer, assume the kinetic-energy correction factor, α, is 1.08 throughout the flow, and that the water emitted by the turbine exhausts to the atmosphere.

6.43 For a hydroelectric plant the head loss between Points 1 and 2 is $h_L = \frac{1}{10}H$. The mass-flow rate through the turbine is $\dot{m}$. In order to improve the turbine's performance to counter a power shortage caused by failure of another power plant, Superman heats the water upstream of the turbine with his x-ray vision. In developing your answers below, assume the kinetic-energy correction factor, α, is 1.06 throughout the flow, and that the water emitted by the turbine exhausts to the atmosphere.

(a) Determine the power delivered by the turbine, $\dot{W}_t$, before Superman heats the water.

(b) If the heat transfer rate from Superman is $\dot{Q} = \frac{1}{10}\dot{m}gH$, what is the power delivered by the turbine after Superman heats the water? **HINT:** Check the definition of head loss.

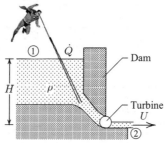

Problem 6.43

6.44 The volume-flow rate through a hydroturbine operating between two large bodies of water of density ρ is $\dot{V}$, and the head loss between Points 1 and 2 is $h_L = \frac{1}{6}H$. If the reference points lie at the free surfaces far from the turbine, what power, $\dot{W}_t$, is generated? Express your answer in terms of ρ, $\dot{m}$, H and gravitational acceleration, g. Compute $\dot{W}_t$ in kW for $\dot{V} = 108$ m³/sec, $\rho = 1000$ kg/m³ and $H = 9.5$ m.

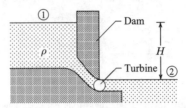

Problem 6.44

6.45 What head, h_p, must be supplied by the pump in order to pump water at a rate $\dot{m}$ from the lower to the upper reservoir? Assume $\alpha_1 = \alpha_2 = 1$ and that the head loss in a pipe of length L and diameter D is given by $h_L = 0.0184 \frac{U^2}{2g} \frac{L}{D}$. For the two pipes, we have $D_1 = D$, $D_2 = D/2$, $L_1/D_1 = 100$ and $L_2/D_2 = 9900/16$. Express your solution for h_p in terms of ρ, Δz, $\dot{m}$, g and D.

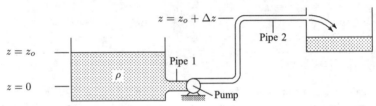

Problems 6.45, 6.46

6.46 What head, h_p, must be supplied by the pump in order to pump water at a rate $\dot{m}$ from the lower to the upper reservoir? Assume $\alpha_1 = \alpha_2 = 1$ and that the head loss in a pipe of length L and diameter D is given by $h_L = 0.028 \frac{U^2}{2g} \frac{L}{D}$. For the two pipes, we have $D_1 = D$, $D_2 = D/2$, $L_1/D_1 = 20$ and $L_2/D_2 = 300$. Express your solution for h_p in terms of ρ, Δz, $\dot{m}$, g and D.

6.47 A pump moves water from Reservoir 1 to Reservoir 2 as shown. Additionally, Reservoir 2 supplies water to a small turbine, and the water exhausts to the atmosphere through a pipe that extends beneath the turbine. The head loss through the entire system of pipes is given by $h_L = 0.20U^2/g$, where U is the flow speed at the outlet and g is gravitational acceleration. Reservoir 1 is extremely large and is open to the atmosphere, while reservoir 2 is closed. The head supplied by the pump is h_p and the head needed to drive the turbine is $h_t = \frac{1}{3}h_p$.

(a) If there were no losses in the system and the pump and turbine were not present, what would the exhaust velocity, $U = U_{ideal}$, be? Assume $\alpha = 1$ in arriving at your answer.

(b) Determine the head supplied by the pump needed to have $U = U_{ideal}$ for the system with the losses, pump, turbine and with the kinetic-energy correction factor throughout the pipes given by $\alpha = 1.1$.

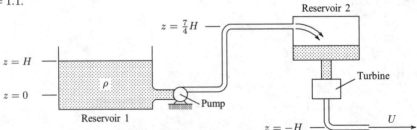

Problem 6.47

6.48 Consider the concentric annulus shown. Determine the hydraulic diameter, noting that the perimeter is that of the *wetted area*, i.e., it is the sum of the perimeters of both circles. Does your answer make sense in the limiting cases $\alpha = 0$ and $\alpha = 1$?

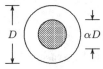

Problem 6.48

6.49 A square duct of side h has a cylindrical obstruction of diameter d centered as shown. If the hydraulic diameter is $D_h = \frac{1}{2}d$, what is h? Note that the perimeter is that of the *wetted area*, i.e., it is the sum of the duct and obstruction perimeters.

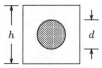

Problem 6.49

6.50 A straight duct is required to carry a given mass flux, $\dot{m}$, of a fluid of density ρ over a distance L. The flow is laminar and the duct cross-sectional area is A. We wish to determine the head loss for several alternative cross sections. Show that the head loss, h_L, is given by $h_L = P^2 h_{Lo}/(4\pi A)$, where P is perimeter and h_{Lo} is the head loss for a circular cross section. Compute h_L/h_{Lo} for the following:

(a) A square of side h;

(b) A rectangle of sides $2h \times h$;

(c) An equilateral triangle of side h.

6.51 Water flows from a large reservoir through a pipe of constant diameter D as shown. The inlet to the pipe is square (i.e., $\mathcal{R}/D \to 0$) and the radius of curvature of each $90°$ bend is $\mathcal{R} = 4D$. The kinetic-energy correction factor is $\alpha = 1.06$ throughout.

(a) If the pipe exit velocity is U and $h/D = 100$, develop an equation for the friction factor, f, as a function of D, U and gravitational acceleration, g.

(b) If $U = 7.13$ m/sec, $D = 11.2$ cm and kinematic viscosity, $\nu = 10^{-6}$ m²/sec, determine the roughness height, k_s, of the pipe.

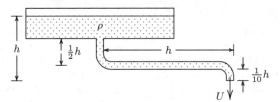

Problems 6.51, 6.52

6.52 Water flows from a large reservoir through a pipe of constant diameter D as shown. The inlet to the pipe is square (i.e., $\mathcal{R}/D \to 0$) and the radius of curvature of each $90°$ bend is $\mathcal{R} = 2D$. The kinetic-energy correction factor is $\alpha = 1.05$ throughout.

(a) If $h/D = 100$, develop an equation for the pipe exit velocity, U, as a function of friction factor, f, h and gravitational acceleration, g.

(b) If the pipe is perfectly smooth, $h = 10$ ft and kinematic viscosity, $\nu = 10^{-5}$ ft²/sec, determine U.
    HINT: You can solve this problem iteratively as follows: (1) make a guess for U; (2) compute Re_D; (3) determine f from the Moody diagram; (4) compute U from the formula derived in Part (a); (5) return to step 1 and repeat until U is determined to within 0.1 ft/sec.

6.53 Water is extracted from a reservoir through a pipe of diameter $D_1 = D$ and length $L_1 = L$ connected to a second pipe of diameter $D_2 = \frac{3}{5}D$ and length $L_2 = \frac{2}{5}L$. The pipe is a distance Δz below the reservoir surface. The friction factor in the large pipe is $f_1 = 0.0120$, while that of the small pipe is $f_2 = 0.0135$. The pipe exit velocity is U.

(a) Ignoring minor losses and assuming $\alpha = 1$, what is D? Express your answer as a function of L, Δz, U and gravitational acceleration, g.

(b) Determine D for $L = 410$ m, $U = 4$ m/sec and $\Delta z = 40$ m.

(c) Now, include the inlet and sudden-contraction losses and repeat Parts (a) and (b).

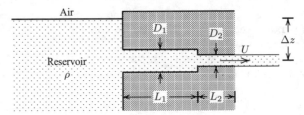

Problems 6.53, 6.54, 6.55

6.54 Water is extracted from a reservoir through a pipe of diameter $D_1 = D$ and length $L_1 = L$ connected to a second pipe of diameter $D_2 = \frac{1}{2}D$ and length $L_2 = \frac{1}{2}L$. The pipe is a distance Δz below the reservoir surface. The friction factor in the large pipe is $f_1 = 0.042$, while that of the small pipe is $f_2 = 0.050$. The desired pipe exit velocity is U.

(a) Ignoring minor losses and assuming $\alpha = 1$, what is Δz? Express your answer as a function of D, L, U and gravitational acceleration, g.

(b) Determine Δz for $D = 6$ in, $L = 120$ ft and $U = 10$ ft/sec.

(c) Now, include the inlet and sudden-contraction losses and repeat Parts (a) and (b).

6.55 Water is extracted from a reservoir through a pipe of diameter $D_1 = D$ and length $L_1 = L$ connected to a second pipe of diameter $D_2 = \frac{1}{2}D$ and length $L_2 = \frac{1}{2}L$. The pipe is a distance Δz below the reservoir surface. The friction factor in the large pipe is $f_1 = 0.021$, while that of the small pipe is $f_2 = 0.025$.

(a) Ignoring minor losses and assuming $\alpha = 1$, what is the exit velocity, U? Express your answer as a function of D, L, Δz and gravitational acceleration, g.

(b) Determine U for $D = 8$ cm, $L = 200$ m and $\Delta z = 20$ m.

(c) Now, include the inlet and sudden-contraction losses and repeat Parts (a) and (b).

6.56 Determine the head loss in the following section of pipe. The pipe has a 50° beveled inlet with a length-to-diameter ratio of $L/D = 0.10$. Also, it has a threaded union and a sudden expansion where the pipe diameter becomes $\frac{3}{2}D$. The length $\ell = 25D$, the surface roughness is $k_s/D = 0.001$ and the Reynolds number is $Re_D = \bar{u}D/\nu = 10^6$. What percentages of the total head loss are due to major and minor losses? Use the Moody diagram to determine the friction factor.

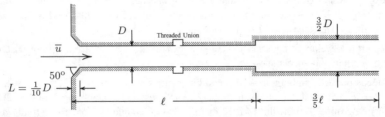

Problem 6.56

6.57 A circular pipe with a 60° gradual contraction followed by a 10° gradual expansion is shown below. Compute the dimensionless head loss, gh_L/U^2 for $d/D = 0.2, 0.4, 0.6$ and 0.8, assuming $L = 100D$, $k_s = 0.006D$ and $Re_D > 10^7$.

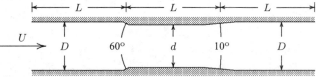

Problem 6.57

6.58 Water flows into a 90° bend with a radius of curvature $\mathcal{R}$. The pipe is circular with diameter D, has surface roughness of $k_s/D = 0.001$, and the Reynolds number is $Re_D = \bar{u}D/\nu = 10^6$. If the radius of curvature is $\mathcal{R} = 7D$ and the lengths of pipe before and after the bend are $\ell = 10D$, what is the head loss? Use the Colebrook formula to determine the friction factor.

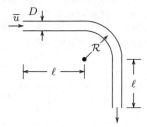

Problem 6.58

6.59 A pipe and nozzle are attached to a reservoir as shown. The inlet to the pipe is not rounded, nor is the entrance to the final pipe section leading into the nozzle. The radius of curvature of both 90° bends is $\mathcal{R} = 0.06D$. The only loss in the nozzle is due to the 60° contraction. The friction factor in the pipe is $f = 0.0255$ when the diameter is D, and $f = 0.0200$ when the diameter is $\frac{5}{2}D$. Determine the nozzle exit velocity, U, as a function of L and gravitational acceleration, g. Assume $\alpha = 1$ throughout.

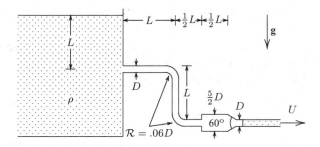

Problem 6.59

6.60 For turbulent flow through three pipes in series as shown, the equivalent length is $L_e = 3L$. If $L/D = 165$, what is the Reynolds number, Re_D? Use the *Blasius formula* for the friction factor, $f \approx 0.3164 Re_D^{-1/4}$.

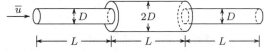

Problems 6.60, 6.61

6.61 For laminar flow through three pipes in series as shown, the equivalent length is $L_e = 4.5L$. If $L/D = 12$, what is the Reynolds number, Re_D?

6.62 SAE30 oil flows through a parallel pipe system with $\rho = 1.71$ slug/ft^3 and $\nu = 1.18 \cdot 10^{-3}$ ft^2/sec. The pressure difference is $p_1 - p_2 = 3.9$ psi, while the pipe diameters and lengths are $D_a = 1$ in, $D_b = 2$ in, $L_a = 10$ ft and $L_b = 20$ ft. The diameter, altitude, z, and α are the same at Points 1 and 2. Ignoring minor losses, compute the volume-flow rate, Q. **HINT:** Assume the flow is laminar. Confirm this assumption after solving.

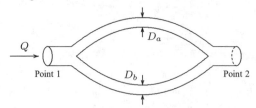

Problems 6.62, 6.63

6.63 Glycerin flows through a parallel pipe system with $\rho = 1260$ kg/m^3 and $\nu = 1.19 \cdot 10^{-3}$ m^2/sec. The volume-flow rate is $Q = 0.25$ m^3/sec, while the pipe diameters and lengths are $D_a = D_b = 10$ cm, $L_a = 20$ m and $L_b = 10$ m. The diameter, altitude, z, and α are the same at Points 1 and 2. Ignoring minor losses, compute the pressure drop, $p_1 - p_2$. **HINT:** Assume the flow is laminar. Confirm this assumption after solving.

6.64 Two reservoirs are connected by a parallel piping system. Water flows between them at 20° C. All pipes have circular cross section and are made of cast iron.

(a) Ignoring minor losses, compute the flow velocity in each pipe as a function of g, d, h, L and the three friction factors, f_1, f_2 and f_3.

(b) Assuming Reynolds numbers are high enough to achieve completely-rough flow so that the friction factors approach their high-Reynolds-number asymptotic values, determine the flow velocities for $D = 1.3$ cm, $h = 200$ m and $L = 400$ m. Use the Moody diagram to estimate the friction factors.

(c) Using your results from Part (b), compute the Reynolds number in each pipe. Compare the left- and right-hand sides of the Colebrook equation for the friction factors used in Part (b).

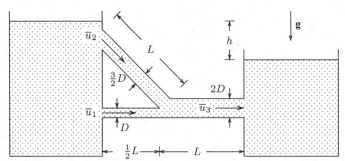

Problems 6.64, 6.65

6.65 Water flows at 68° F between two reservoirs with a parallel piping system as shown. All pipes have circular cross section and are made of galvanized iron.

(a) Ignoring minor losses, compute the flow velocity in each pipe as a function of g, d, h, L and the three friction factors, f_1, f_2 and f_3.

(b) Assuming Reynolds numbers are high enough to achieve completely-rough flow so that the friction factors approach their high-Reynolds-number asymptotic values, determine the flow velocities for $D = 2$ in, $h = 100$ ft and $L = 1000$ ft. Use the Moody diagram to estimate the friction factors.

(c) Using your results from Part (b), compute the Reynolds number in each pipe. Compare the left- and right-hand sides of the Colebrook equation for the friction factors used in Part (b).

6.66 Water at 68° F flows from one large reservoir to another through a pipe system as shown. Both reservoirs are open to the atmosphere and the pipes are made of steel. If $d = 4$ inches and $\bar{u}_2 = 16$ ft/sec, what must the value of $\bar{u}_1$ be? Ignore minor losses.

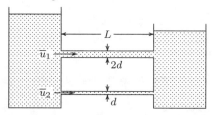

Problem 6.66

6.67 Consider steady flow of an incompressible fluid of density ρ through a multiple nozzle from which fluid discharges to the atmosphere as shown. You may assume flow properties are uniform across the nozzle and that the nozzle lies in a horizontal plane so that potential energy effects are unimportant. If the head loss between the inlet and the largest outlet pipe (the one with diameter $D/2$) is $h_L = 0.2U^2/g$ and the kinetic-energy correction factor is $\alpha = 1$ throughout, determine the force required to hold the pipe in place. Express your answer in terms of ρ, U and D.

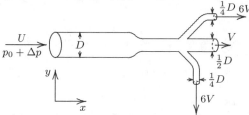

Problem 6.67

6.68 Consider steady flow of water through the *Black Box* shown below. The inlet-pipe pressure at a distance L from the box is $(p_a + \Delta p)$, where p_a denotes atmospheric pressure. The two identical outlet pipes, also of length L, exhaust to the atmosphere. The entire geometry lies in a horizontal plane, so that gravitational effects can be neglected. Assume there is no net reaction force from the *Black Box*.

(a) Use conservation of mass and momentum to determine U_o as a function of U_i, and to derive a relation between U_i, $\Delta p/\rho$, L, D, and the friction coefficients f_i and f_o for the inlet and outlet pipes, respectively.

(b) Assuming the total head loss as the fluid flows from the inlet pipe, through the Black Box and through each of the outlet pipes is h_L, express $\Delta p/\rho$ as a function of U_i, g and h_L. Assume $\alpha = 1$ throughout.

(c) Combining results from Parts (a) and (b), show that if $h_L = 0.1U_i^2/(2g)$ and $L/D = 11$, the friction factors are related by $f_i + 4f_o = 0.1$.

(d) Assuming flow throughout the system occurs at Reynolds number based on pipe diameter in excess of 10^7 and also that the roughness height in the inlet pipe is given by $k_s/D = 0.001$, determine the approximate value of k_s in the outlet pipes. Express your answer in terms of D, the diameter of the inlet pipe.

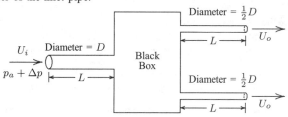

Problem 6.68

6.69 A wide, 1.2 cm sheet of water at $20°$ C runs down a $6°$ glass surface with $k_s = 0.4$ mm. Compute the average flow velocity, $\overline{u}$, first by using the Chézy-Manning equation and then by using the Moody diagram and the Colebrook formula.

6.70 A wide, 1.5 inch sheet of water at $68°$ F runs down an $8°$ asphalt surface with $k_s = 0.02$ ft. Compute the average flow velocity, $\overline{u}$, first by using the Chézy-Manning equation and then by using the Moody diagram and the Colebrook formula.

6.71 A natural channel can be approximated as having trapezoidal cross section with $b = 1000$ ft, $y = 300$ ft and $\alpha = 14°$. The bottom slope is $S_o = 0.0001$ and the Froude number is $Fr = 0.15$. What is the Manning roughness coefficient, n?

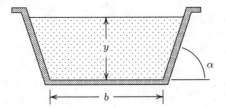

Problems 6.71, 6.72, 6.73

6.72 Water flows in a canal with trapezoidal cross section as shown. The bottom slope is $S_o = 0.0016$, while the canal dimensions are $b = 4$ m, $y = 2$ m and $\alpha = 30°$. Using the Chézy-Manning approach, determine the volume-flow rate, Q, and the Froude number, Fr, if the canal is lined with brickwork and also if it is stony.

6.73 Water flows in a canal with trapezoidal cross section as shown. The bottom slope is $S_o = 0.0008$, while the canal dimensions are $b = 12$ ft, $y = 7$ ft and $\alpha = 36°$. Using the Chézy-Manning approach, determine the volume-flow rate, Q, the Froude number, Fr, and the Reynolds number Re_{D_h}, if the canal is lined with smooth steel and also if it has a weedy surface. Assume $\nu = 10^{-5}$ ft^2/sec.

6.74 A *flume* must be designed to deliver water from a mountain lake to a small hydroelectric power plant. The required volume-flow rate is $Q = 2.5$ m^3/sec, the bottom slope is $S_o = 0.0016$ and the Manning friction coefficient is $n = 0.016$. Determine flume size for a rectangular cross section with $b = 3y$, a semi-circular cross section with $\alpha = \pi$ and a triangular cross section with $\alpha = \frac{1}{3}\pi$ (see Figure 6.22 for geometric details). Make a table summarizing area, A, perimeter, P, hydraulic radius, R_h, and Froude number for all three cases.

6.75 We wish to maximize the volume-flow rate in an open channel of triangular cross section under uniform-flow conditions. The optimum shape is called the *best hydraulic cross section*. Our goal is to find the angle, α, that corresponds to the best hydraulic cross section.

(a) Verify that the channel area, A, satisfies $A^5 y^2 = 4\lambda \left(A^2 + y^4\right)$ where λ is a constant depending on the bottom slope, S_o, the volume-flow rate, Q, and the Manning friction coefficient, n.

(b) Find the minimum value of $A(y)$ and the corresponding angle α. **HINT:** After setting $dA/dy = 0$, eliminate λ by using the equation derived in Part (a).

6.76 We wish to maximize the volume-flow rate in an open channel of rectangular cross section under uniform-flow conditions. The optimum shape is called the *best hydraulic cross section*. Our goal is to find the *aspect ratio*, b/y, (b = width, y = depth) that corresponds to the best hydraulic cross section.

(a) Verify that the channel area, A, satisfies $A^{5/2} y = \lambda \left(A + 2y^2\right)$ where λ is a constant depending on the bottom slope, S_o, the volume-flow rate, Q, and the Manning friction coefficient, n.

(b) Find the minimum value of $A(y)$ and the corresponding aspect ratio. **HINT:** After setting $dA/dy = 0$, eliminate λ by using the equation derived in Part (a).

6.77 A drainage pipe with diameter $D = 3$ inches is half full. The pipe's bottom slope is $S_o = 0.0002$ and the Manning roughness coefficient is $n = 0.012$. What is the Froude number, Fr?

6.78 A triangular open channel has depth $y = 1.25$ m. The bottom slope is $S_o = 0.005$ and the channel is made of cast iron. Using the Chézy-Manning formula, determine the angle α that will give a volume-flow rate, Q, of 3 m^3/sec. Solve for α by trial and error (or Newton's iterations if you are familiar with the method) to the nearest tenth of a degree.

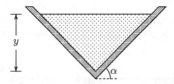

Problem 6.78

6.79 From an overhead bridge, you drop a stone into a wide canal that is 10 m deep. You notice that the resulting ripples on the surface of the water propagate only in the downstream direction. What must be the minimum speed of flow in the canal and the corresponding speed of the ripples traveling in the downstream direction? **HINT:** Waves travel upstream and downstream at velocity $u_{up} = \overline{u} - \sqrt{gy}$ and $u_{down} = \overline{u} + \sqrt{gy}$, respectively.

6.80 A small stone is dropped into a brook. After one second, the ripple created by the stone has traveled 3 ft upstream and 7 ft downstream. Assuming the brook has constant depth, what are the flow speed and the depth? **HINT:** Waves travel upstream and downstream at velocity $u_{up} = \overline{u} - \sqrt{gy}$ and $u_{down} = \overline{u} + \sqrt{gy}$, respectively.

6.81 A small stone is dropped into a brook. After one second, the ripple created by the stone has traveled 3 m downstream. Assuming the brook has a constant depth $y = 40$ cm, what is the flow speed, $\overline{u}$? Also, how far upstream will the ripple travel in one second? **HINT:** Waves travel upstream and downstream at velocity $u_{up} = \overline{u} - \sqrt{gy}$ and $u_{down} = \overline{u} + \sqrt{gy}$, respectively.

6.82 A small surface disturbance is created in a uniform open-channel flow in a very wide channel. The flow velocity is U and the channel depth is y. There are two observers, one a distance ℓ upstream of the point of disturbance and the other a distance ℓ downstream. How much time, Δt, passes after the wave reaches the downstream observer before the upstream observer senses the wave? Express your answer as a function of ℓ, U and Froude number, Fr. Make a graph of the dimensionless time, $U\Delta t/\ell$, versus Fr for $0 \leq Fr < 1$ and discuss what happens when $Fr > 1$. **HINT:** Waves travel upstream and downstream at velocity $u_{up} = \overline{u} - \sqrt{gy}$ and $u_{down} = \overline{u} + \sqrt{gy}$, respectively.

6.83 At two different times, a pebble is dropped into an open channel that is very wide. Each pebble creates a circular ripple that is convected downstream as shown. The channel has constant depth y and the channel flow speed is U.

(a) Noting that the wave speed for weak radial waves is the same as for planar waves, verify that U is given by $U = \ell\sqrt{gy}/(r_1 - r_2)$. **HINT:** Waves travel upstream and downstream at velocity $u_{up} = \overline{u} - \sqrt{gy}$ and $u_{down} = \overline{u} + \sqrt{gy}$, respectively.

(b) If the upstream wavefronts are coincident, what is the Froude number, Fr?

(c) Compute U and Fr for $y = 50$ cm, $r_1 = 8$ m, $r_2 = 5$ m and $\ell = 8$ m.

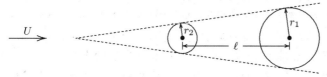

Problem 6.83

6.84 The rectangular open channel shown has width $b = 3$ ft and depth $y = 5$ ft. The bottom slope is $S_o = 0.0005$ and the channel is made of brass. Using the Chézy-Manning formula, determine the specific energy and whether the flow is subcritical or supercritical.

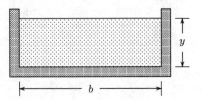

Problem 6.84

6.85 A flood occurs over a pasture and the volume-flow rate per unit width, Q/b, is 8 m²/sec. If the depth is $y = 1$ m, is the flow subcritical or supercritical? What is the ratio of the specific energy to its minimum value?

6.86 Consider a gravelly rectangular channel of width $b = 10$ ft that has a constant volume-flow rate of $Q = 600$ ft³/sec. Determine the critical depth, y_c, and the *critical slope*, S_c, i.e., the slope given by the Chézy-Manning equation with critical depth and critical velocity. What angle, $\theta_c = \tan^{-1} S_c$, corresponds to this bottom slope?

6.87 The Froude number downstream of a sluice gate shown in the figure is $Fr = 8$. What is the Froude number upstream of the gate? **HINT:** Show that y_1/y_c satisfies a cubic equation of the form

$$(y_1/y_c)^3 - N(y_1/y_c)^2 + \frac{1}{2} = 0$$

for which the desired root is $y_1/y_c \approx N$.

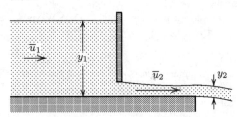

Problems 6.87, 6.88

6.88 Consider a sluice gate in a rectangular channel of width b. The volume-flow rate is Q, and head loss can be neglected.

(a) Determine Q as a function of b, y_1, g and y_2/y_1.

(b) If $y_2 = \frac{1}{2}y_c$, where y_c is the critical depth, what is y_2/y_1?

(c) For the conditions of Part (b), what are the Froude numbers upstream and downstream of the sluice gate?

6.89 A hydraulic jump occurs in a wide rectangular flume. The volume-flow rate per unit width is $Q/b = 25$ ft²/sec and the depth before the jump is $y_1 = 18$ in. Determine the depth after the jump, y_2, and the head loss, h_L.

6.90 Show for a weak hydraulic jump ($Fr_1 \approx 1$) that the Froude number downstream of the jump is approximated by

$$Fr_2 \approx 2 - Fr_1$$

Compare values for Fr_2 obtained using this equation with values for Fr_2 from the exact solution for $Fr_1 = 1, 1.05, 1.1$ and 1.2. **HINT:** Let $Fr_1 = 1 + \epsilon$ where $\epsilon \ll 1$ and expand the exact relationship between Fr_2 and Fr_1.

6.91 Verify that the ratio of the head loss across a hydraulic jump to the initial specific energy is

$$\frac{h_L}{E_1} = \frac{\left(\sqrt{1+8Fr_1^2} - 3\right)^3}{8\left(\sqrt{1+8Fr_1^2} - 1\right)(2+Fr_1^2)}$$

Compute this ratio, as a percentage, for $Fr_1 = 2, 5, 10$ and 20.

6.92 Water flows in a wide rectangular open channel, without head loss, over a bump. Because of the prevailing depth downstream of the bump, a hydraulic jump occurs.

(a) Determine $\bar{u}_0, \bar{u}_1$ and Fr_1 as functions of y_0, y_1 and gravitational acceleration, g.

(b) Compute Fr_1 for $y_0 = 3$ m and $y_1 = 1.1$ m.

(c) Compute y_2 for the Froude number of Part (b).

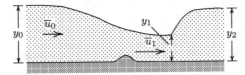

Problem 6.92

6.93 Water flows down a steep hill in a wide rectangular open channel as shown. A weak hydraulic jump occurs, increasing the channel depth from y_1 to y_2. The volume-flow rate is $Q = 60$ m³/sec, the channel width is $b = 25$ m and the depth after the jump is $y_2 = 89$ cm. Compute the depth before the jump, and the Froude numbers and velocities upstream and downstream of the jump. **HINT:** Use the fact that, for a weak hydraulic jump, $Fr_2 \approx 2 - Fr_1$.

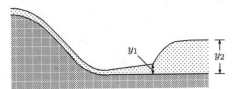

Problems 6.93, 6.94

6.94 Water flows down a steep hill in a wide rectangular open channel as shown. A hydraulic jump occurs, abruptly increasing the channel depth from y_1 to y_2. The volume-flow rate is $Q = 650$ ft³/sec, the channel width is $b = 14$ ft and the depth before the jump is $y_1 = 2.2$ ft. Compute the depth after the jump, and the Froude numbers and velocities upstream and downstream of the jump.

Chapter 7

Turbomachinery

In this chapter, we will investigate some of the features of flow through turbomachines that are commonly found in practical applications. A turbomachine is a device that can either add energy to a flowing fluid or extract energy from it. The origin of the prefix *turbo* is found in Latin, with the translation of "that which spins." Turbomachinery is utilized in our modern world in myriad contexts such as: (a) propulsion systems for airplanes, automobiles, ships and liquid rockets; (b) power-generating plants; (c) gas, oil and water-pumping stations; and (d) numerous other settings ranging from heart-assist pumps to refrigeration facilities.

Flow through a turbomachine is generally three-dimensional, viscous and unsteady (actually periodic). Consequently, accurate analytical predictions can only be made with large-scale computations [see, for example, Lakshminarayana (1996)]. Nevertheless, useful insight can be obtained using the control-volume method and reducing the analysis to essentially one-dimensional computations. That is the purpose of this chapter.

We begin by classifying turbomachines in terms of their geometry and how they affect the flow's energy and direction. Then, to aid in analysis of turbomachinery, we develop an equation for conservation of angular momentum. From the general integral-conservation laws, we derive the **Euler turbomachine equations**. We focus on incompressible flow, for simplicity, with emphasis on water pumps. The analysis applies equally to low-speed gas flows (e.g., ventilation blowers, wind-tunnel fans and windmills). The balance of the chapter applies these equations to turbomachines. We discuss efficiency factors and performance curves to quantify how practical pumps and turbines behave. Above all, this chapter shows how the combination of dimensional analysis (Chapter 2) and the control-volume method (Chapter 5) lays the foundation for developing an understanding of complex fluid flows.

7.1 Classification

There are several ways in which we can classify turbomachines based on their function and geometry. For our purposes, four ways will suffice, viz.,

1. Energy Considerations—how they affect the energy of a given flow

2. Geometrical Considerations—whether or not the turbomachine is enclosed

3. Exit-Flow Considerations—the direction of the outlet flow

4. Number of stages—single-stage and multistage machines

Energy Considerations: When a turbomachine adds energy to the flow, it is commonly referred to as a **pump** or a **compressor**. A turbomachine that extracts energy from the flow is called a **turbine**. We made this distinction in Chapter 6 when we developed an approximate form of the energy-conservation equation [see Equation (6.52)] containing changes in total head due to pumps and turbines.

Geometrical Considerations: Many of the most important turbomachines of engineering interest are enclosed by a permanent housing called the **shroud**. Examples of such turbomachines are jet-engine compressors, gas or steam turbines and centrifugal pumps. For the obvious reason, these machines are described as **enclosed turbomachines**. Not all turbomachines are enclosed, with the most important examples being airplane propellers, ship screws and wind turbines (see Figure 7.1). These devices are called **extended turbomachines**.

Figure 7.1: *ERDA/NASA wind turbine (left) and blade rows in an axial compressor (right).*

Exit-Flow Considerations: Figure 7.2 illustrates the three types of flow exiting from common turbomachines. When the primary flow direction remains parallel to the axis as in Figure 7.2(a), we describe the machine as an **axial** turbomachine. By contrast, if the flow turns and exits the machine in the radial direction as depicted in Figure 7.2(b), we have a **radial** or **centrifugal** turbomachine. If the exit flow is partially axial and partially radial as in Figure 7.2(c), we call the device a **mixed-flow** turbomachine.

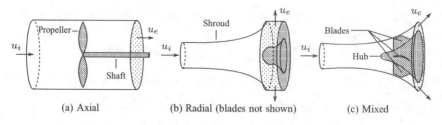

Figure 7.2: *Exit-flow directions for turbomachines.*

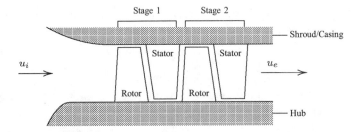

Figure 7.3: *Schematic of two stages in an axial-flow compressor.*

Number of Stages: The total-head (or total-pressure) change for a simple, **single-stage**, turbomachine is limited. To achieve larger changes, conceptually, we can put several single-stage machines in series. In practice, we call such machines **multistage** compressors (or pumps) and multistage turbines. Figure 7.3 depicts two stages in a section of a compressor. Each stage consists of a row of **rotor** blades attached to the hub, followed by a row of **stator** blades attached to the shroud or **casing**. There will be several rows of rotor and stator blades in a multistage arrangement, e.g. Figure 7.1. The function of the rotating rotor blades is to add energy to the flow. The stationary stator blades then remove the angular momentum induced by the torque on the rotor, with a corresponding increase in static pressure.

7.2   Conservation of Angular Momentum

The mass- and momentum-conservation principles are sufficient to solve incompressible flow problems. However, it is sometimes convenient to work with **angular momentum** rather than conventional **linear momentum**. This is especially true in the context of turbomachinery where, for example, we wish to compute the net torque as part of our analysis. In this section, we derive an integral-conservation law for angular momentum and apply the principle to a simple rotating, turbine-like device, viz., a lawn sprinkler.

7.2.1   Derivation

We can derive an integral conservation law for angular momentum in the same manner we have used to derive the principles discussed in Chapters 5 and 6. To motivate the form of the conservation principle, consider Newton's second law for a single particle, i.e.,

$$\mathbf{F} = \frac{d}{dt}(m\mathbf{u}) \tag{7.1}$$

where m is mass. Now, the torque, $\boldsymbol{\tau}$, and angular momentum, $\mathbf{H}$, relative to a fixed point are defined by

$$\boldsymbol{\tau} = \mathbf{r} \times \mathbf{F} \quad \text{and} \quad \mathbf{H} = m\mathbf{r} \times \mathbf{u} \tag{7.2}$$

where $\mathbf{r}$ is the position vector from the fixed point to the point at which $\mathbf{F}$ is applied. Differentiating the angular-momentum vector with respect to time shows that

$$\frac{d\mathbf{H}}{dt} = \frac{d\mathbf{r}}{dt} \times (m\mathbf{u}) + \mathbf{r} \times \frac{d}{dt}(m\mathbf{u}) \tag{7.3}$$

Noting that $\mathbf{u} = d\mathbf{r}/dt$ and substituting from Equation (7.1), we can rewrite the rate of change of angular momentum as

$$\frac{d\mathbf{H}}{dt} = m(\mathbf{u} \times \mathbf{u}) + \mathbf{r} \times \mathbf{F} \tag{7.4}$$

Finally, the first term on the right-hand side of Equation (7.4) vanishes since the cross product of a vector with itself is identically zero. Noting the definition of torque, we conclude that the classical equation for angular-momentum conservation is

$$\frac{d\mathbf{H}}{dt} = \boldsymbol{\tau} \tag{7.5}$$

We recast this equation in a form appropriate for a control volume by using the Reynolds Transport Theorem to express $d\mathbf{H}/dt$ in terms of the absolute[1] angular momentum per unit volume, $\rho\, \mathbf{r} \times \mathbf{u}$. Letting $\mathbf{w}$ denote velocity relative to the rotating control volume to be consistent with classical turbomachine notation, we have the following.

$$\frac{d}{dt} \iiint_V \rho\, \mathbf{r} \times \mathbf{u}\, dV + \oiint_S \rho\, \mathbf{r} \times \mathbf{u}\, (\mathbf{w} \cdot \mathbf{n})\, dS = \boldsymbol{\tau} \tag{7.6}$$

where

$$\mathbf{w} = \mathbf{u}^{rel} \quad \text{and} \quad \mathbf{u} = \boldsymbol{\Omega} \times \mathbf{r} + \mathbf{w} \tag{7.7}$$

The quantity $\boldsymbol{\Omega}$ is the control volume's angular velocity. Before proceeding to application of this principle, it is helpful to pause and discuss two of its interesting subtleties.

1. All of the forces acting (pressure, body forces, viscous stresses) contribute to the torque and should be included where appropriate.

2. This principle has been derived from the linear momentum law, and is therefore not an independent source of equations.

7.2.2   Lawn Sprinkler

As an example of how we implement this principle, consider the lawn sprinkler depicted in Figure 7.4. The sprinkler rotates in the counterclockwise direction with angular velocity (measured in radians/second = sec^{-1})

$$\boldsymbol{\Omega} = \Omega\, \mathbf{k} \tag{7.8}$$

Water enters with velocity U from a vertical pipe through the center and exits with velocity w_e relative to the nozzle. The nozzle area is A_e and is located a distance ℓ from the center. We would like to compute the torque produced by the lawn sprinkler, assuming the flow is steady and incompressible within the sprinkler. We also assume body forces can be ignored.

By symmetry, the torque vector must act about the z axis. Thus, denoting the magnitude of the torque by τ_r, our integral conservation law simplifies to

$$-\tau_r = \mathbf{k} \cdot \oiint_S \rho\, \mathbf{r} \times \mathbf{u}\, (\mathbf{w} \cdot \mathbf{n})\, dS \tag{7.9}$$

where our control volume is coincident with the sprinkler surface and rotates with the same angular velocity. We include a minus sign with τ_r because we have chosen a control volume

---

[1]See Subsection 5.7.1 for a discussion of absolute and relative momentum.

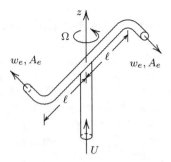

Figure 7.4: *A lawn sprinkler.*

that completely surrounds the sprinkler. Consequently, the torque appearing in our conservation law is the torque exerted by the surroundings on the sprinkler (i.e., it is a "reaction torque"). We take moments about the z axis so that

$$\mathbf{r} = r\mathbf{e}_r \tag{7.10}$$

where $\mathbf{e}_r$ is a unit vector in the radial direction and r is distance from the z axis. Clearly, the angular momentum at the inlet is zero since $r = 0$. The only other parts of the bounding surface that have nonzero angular-momentum flux are the two nozzles. Now, at the nozzles, we have

$$\mathbf{r} = \ell\mathbf{e}_r, \quad \mathbf{w} = -w_e\mathbf{e}_\theta, \quad \mathbf{u} = (\Omega\ell - w_e)\,\mathbf{e}_\theta \tag{7.11}$$

where $\mathbf{e}_\theta$ is a unit vector in the circumferential direction. Thus, the absolute angular momentum per unit mass at the nozzles is

$$\mathbf{r} \times \mathbf{u} = \ell\,(\Omega\ell - w_e)\,(\mathbf{e}_r \times \mathbf{e}_\theta) = \ell\,(\Omega\ell - w_e)\,\mathbf{k} \tag{7.12}$$

Thus, substituting Equation (7.12) into Equation (7.9), we have

$$-\tau_r = \mathbf{k} \cdot \oiint_S \underbrace{\rho\,\ell\,(\Omega\ell - w_e)\,\mathbf{k}}_{\rho\,\mathbf{r}\,\times\,\mathbf{u}}\,\underbrace{w_e}_{(\mathbf{w}\cdot\mathbf{n})}\,dS = \rho\,\ell\,(\Omega\ell - w_e)\,w_e\,(2A_e) \tag{7.13}$$

Therefore, the reaction torque on the sprinkler is

$$\tau_r = 2\rho w_e A_e \ell\,(w_e - \Omega\ell) \tag{7.14}$$

We can rearrange our result in a more interesting form by first using conservation of mass. Clearly, the mass flux, $\dot{m}$, passing through the sprinkler is equal to the combined mass flux through the two exits, wherefore

$$\dot{m} = 2\rho w_e A_e \tag{7.15}$$

Then, the torque produced by the sprinkler, $\tau_{\text{sprinkler}} = -\tau_r$, is given by

$$\tau_{\text{sprinkler}} = \dot{m}\ell u_\theta \tag{7.16}$$

Thus, the torque produced by the sprinkler is the product of the mass-flow rate and the absolute angular momentum per unit mass, $\ell u_\theta = \ell\,(\Omega\ell - w_e)$, at the nozzles. Notably, details of the axial inlet flow do not appear explicitly in Equation (7.16). We will see in the next section that this result is true for a centrifugal pump and, in fact, is valid for turbomachines in general.

7.3   The Euler Turbomachine Equations

The geometry of turbomachines varies greatly, even for the same type of application. Regardless of the type of machine, there is always a rotating component whose purpose is to do work on the fluid. This is accomplished by a rotor (Figure 7.3) which, in the context of a pump, is called an **impeller** (Figure 7.5). In addition, there is a stationary member such as the stator of Figure 7.3 (or a **volute** for a centrifugal pump) whose purpose is to guide the flow after it passes through the rotor or impeller. There is sometimes an additional row of stationary blades known as **guide vanes** that help direct the flow prior to entering the rotor or impeller.

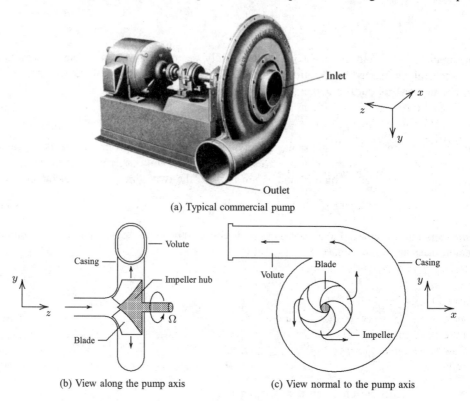

(a) Typical commercial pump

(b) View along the pump axis   (c) View normal to the pump axis

Figure 7.5: *A centrifugal pump; the clearance between the impeller and the casing is exaggerated—it is actually very small in practice.*

To make our discussion more specific, consider a centrifugal pump as sketched in Figure 7.5. The figure presents a side view (yz plane) and a view normal to the pump axis (xy plane). As shown, fluid enters parallel to the z axis, and flows toward the impeller. The impeller rotates at angular velocity Ω, imparting angular momentum to the flow and increasing its pressure.[2] The impeller depicted has what are called **backward-curved** blades. This is not a universal feature as **radial** (not-curved) and **forward-curved** blades are also used.

The impeller causes the flow to rotate and directs it radially outward. The fluid eventually leaves the impeller and passes into the volute. Since the volute area increases as the flow advances, mass conservation dictates a decrease in velocity, accompanied by an additional increase in pressure.

---

[2]The angular rotation rate is often denoted by n or N in turbomachinery literature with units of rpm or Hz.

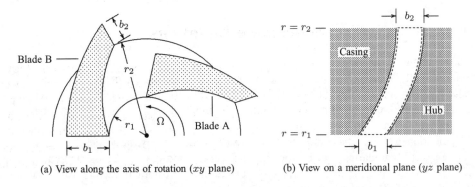

(a) View along the axis of rotation (xy plane)

(b) View on a meridional plane (yz plane)

Figure 7.6: *Closeup view of impeller geometry.*

In general, a detailed analysis requires a computer solution tailored to the specific turbomachine geometry. There are, nevertheless, important global properties of even the most complicated turbomachines that can be gleaned from our trustworthy control-volume method. Since rotation is an integral part of turbomachine flows, we will use the angular-momentum conservation principle developed in the preceding section along with the mass- and energy-conservation principles developed in Chapters 5 and 6, respectively.

The most interesting part of the flow through a centrifugal pump is between the impeller blades. Thus, we select a control volume that lies between two successive blades. Figure 7.6(a) illustrates the impeller geometry from a view along the axis of rotation, which includes the blade shapes. Figure 7.6(b) is a side view in the yz plane, usually referred to as the **meridional plane**. The blade widths at the inlet and outlet are b_1 and b_2, respectively. Also, the radial distance from the axis to the hub at the inlet is r_1, while its value at the outlet is r_2. Our control volume is the region bounded by the hub, the casing [omitted in Figure 7.6(a) to avoid clutter] and two adjacent blades. Control-volume boundaries in the yz and xy planes are indicated by dashed lines in Figures 7.6(b) and 7.7, respectively.

Now, we shift our attention to Figure 7.7, which is a true axial view similar to Figure 7.6(a) with the blade shapes removed. In developing the **Euler turbomachine equations**, we assume the flow is steady and incompressible. We also assume that the blade width is small compared to the hub radius, i.e., that

$$b_1 \ll r_1 \quad \text{and} \quad b_2 \ll r_2 \qquad (7.17)$$

When this is true, we can neglect variations in radial distance between the hub and the casing.

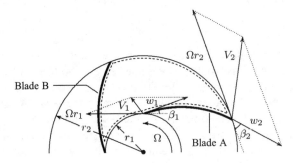

Figure 7.7: *Control volume between two impeller blades with flow directions (xy plane).*

Fluid enters the impeller with velocity relative to the rotating blade of magnitude w_1. We assume this velocity is parallel to the blade at its leading edge, i.e., at $r = r_1$. The blade lies at an angle β_1 to the local tangent. Since the impeller velocity is Ωr_1 at the hub, the absolute velocity is the vector sum of these two velocities, and we denote its magnitude by V_1. Similarly, we assume the flow is parallel to the blade at its trailing edge ($r = r_2$), which makes an angle β_2 with the local tangent to the hub. The impeller and absolute-velocity magnitude at $r = r_2$ are Ωr_2 and V_2, respectively.

As mentioned above, the control volume chosen is enclosed by the blades, hub and casing, which are solid boundaries, and by the surfaces $r = r_1$ and $r = r_2$, across which fluid can flow. The radius vectors, absolute velocity vectors and unit normal vectors at the inlet and outlet portions of the control volume are as follows.

$$\textbf{Inlet:} \quad \mathbf{r} = r_1\mathbf{e}_r, \quad \mathbf{u} = \mathbf{V}_1 = \mathbf{w}_1 + \Omega r_1 \mathbf{e}_\theta, \quad \mathbf{n} = -\mathbf{e}_r \tag{7.18}$$

$$\textbf{Outlet:} \quad \mathbf{r} = r_2\mathbf{e}_r, \quad \mathbf{u} = \mathbf{V}_2 = \mathbf{w}_2 + \Omega r_2 \mathbf{e}_\theta, \quad \mathbf{n} = \mathbf{e}_r \tag{7.19}$$

The vectors $\mathbf{e}_r$ and $\mathbf{e}_\theta$ are the standard radial and circumferential unit vectors for a cylindrical coordinate system (see Appendix D, Section D.2). Note that the impeller rotates in the counterclockwise direction, which corresponds to positive rotation in a mathematical sense. That is, the angular velocity vector, $\mathbf{\Omega}$, points in the positive z direction so that

$$\mathbf{\Omega} = \Omega\mathbf{k} \tag{7.20}$$

This is consistent with the convention that the angular coordinate θ increases in the counterclockwise direction.

Mass Conservation: Because the flow is steady, the integral form of mass conservation simplifies to

$$\oiint_S \rho\mathbf{w} \cdot \mathbf{n} \, dS = 0 \tag{7.21}$$

Expanding the closed-surface integral, we find

$$
\begin{aligned}
0 &= \left\{ \iint_{r=r_1} + \iint_{r=r_2} + \iint_{\text{Blade A}} + \iint_{\text{Blade B}} + \iint_{\text{Hub}} + \iint_{\text{Casing}} \right\} \rho\mathbf{w} \cdot \mathbf{n} \, dS \\
&= \iint_{r=r_1} \rho\mathbf{w}_1 \cdot (-\mathbf{e}_r) \, dS + \iint_{r=r_2} \rho\mathbf{w}_2 \cdot (\mathbf{e}_r) \, dS = -\rho w_{r1} A_1 + \rho w_{r2} A_2
\end{aligned} \tag{7.22}
$$

We have made use of the fact that there is no flow through the blades, hub and casing in the second line of Equation (7.22), and assumed the inlet and outlet areas between the blades are A_1 and A_2, respectively. Hence, denoting the volume-flow rate by Q, conservation of mass tells us

$$\rho w_{r1} A_1 = \rho w_{r2} A_2 = \rho Q \tag{7.23}$$

where w_{r1} and w_{r2} denote the radial components of $\mathbf{w}_1$ and $\mathbf{w}_2$, respectively. Since the flow between all blades is the same (unless, for some reason, the blade spacing is nonuniform), we can regard Q as either the volume-flow rate between two successive blades or the total volume-flow rate through the pump. In the latter case, the areas A_1 and A_2 are the total areas at the inlet and outlet of the impeller. If the blade widths (out of the page) at the inlet and outlet are b_1 and b_2, respectively [Figure 7.6(b)], then $A_1 \approx 2\pi r_1 b_1$ and $A_2 \approx 2\pi r_2 b_2$. Hence, the radial components of the inlet and outlet velocity vectors are

$$w_{r1} = \frac{Q}{2\pi r_1 b_1} \quad \text{and} \quad w_{r2} = \frac{Q}{2\pi r_2 b_2} \tag{7.24}$$

Angular-Momentum Conservation: Clearly, the torque that develops on the impeller is aligned with the z axis. So, calling the torque exerted by the fluid on the impeller τ, the z component of the angular-momentum conservation principle [Equation (7.6)] for steady flow simplifies to

$$\tau = \mathbf{k} \cdot \oiint_S \rho \mathbf{r} \times \mathbf{u} \, (\mathbf{w} \cdot \mathbf{n}) \, dS \qquad (7.25)$$

Appealing to Equations (7.18) and (7.19) and again using the fact that there is no flow through the blades, hub and casing, the closed-surface integral expands as follows.

$$
\begin{aligned}
\tau &= \mathbf{k} \cdot \iint_{r=r_1} \rho \, r_1 \mathbf{e}_r \times \mathbf{V}_1 \, [\mathbf{w}_1 \cdot (-\mathbf{e}_r)] \, dS + \mathbf{k} \cdot \iint_{r=r_2} \rho \, r_2 \mathbf{e}_r \times \mathbf{V}_2 \, [\mathbf{w}_2 \cdot (\mathbf{e}_r)] \, dS \\
&= \mathbf{k} \cdot [\rho r_1 \, (V_{\theta 1} \mathbf{k}) \, (-Q) + \rho r_2 \, (V_{\theta 2} \mathbf{k}) \, (Q)] = \mathbf{k} \cdot [-\rho r_1 V_{\theta 1} Q \mathbf{k} + \rho r_2 V_{\theta 2} Q \mathbf{k}] \quad (7.26)
\end{aligned}
$$

where $V_{\theta 1}$ and $V_{\theta 2}$ are the circumferential components of the absolute velocity vectors, $\mathbf{V}_1$ and $\mathbf{V}_2$, respectively. Therefore, taking the dot product and regrouping terms, the torque is

$$\tau = \rho Q \, [r_2 V_{\theta 2} - r_1 V_{\theta 1}] \qquad (7.27)$$

Finally, the rate at which work is done by this torque is $\dot{W}_p = \boldsymbol{\Omega} \cdot \boldsymbol{\tau}$. Since both $\boldsymbol{\Omega}$ and $\boldsymbol{\tau}$ point in the positive z direction, the power is

$$\dot{W}_p = \rho \Omega Q \, [r_2 V_{\theta 2} - r_1 V_{\theta 1}] \qquad (7.28)$$

Equations (7.27) and (7.28) are known as the **Euler turbomachine equations.** They are based on averaged properties on cross sections, and thus represent a global view. One of the most interesting properties of these equations is the absence of axial inlet-flow properties. As with the simple lawn-sprinkler example of Subsection 7.2.2, we see that the torque is the product of the mass-flow rate, $\dot{m} = \rho Q$, and the change in absolute angular momentum per unit mass of the fluid as it moves through the impeller.

While our derivation has focused on a pump, there is nothing that precludes application of the Euler turbomachine equations to devices that extract energy from the flow, such as turbines. The only difference between pump and turbine applications occurs in the sign of $\dot{W}_p$. That is, when $\boldsymbol{\Omega}$ and $\boldsymbol{\tau}$ are in the same direction, $\dot{W}_p$ is positive so that work is being done on the fluid. This corresponds to a pump or a compressor, which adds energy to the fluid. By contrast, if $\boldsymbol{\Omega}$ and $\boldsymbol{\tau}$ are in opposite directions, $\dot{W}_p$ is negative so that work is being done by the fluid on the rotor. This corresponds to a turbine.

Energy Conservation: Although we can obtain complete details of the solution from the Euler turbomachine equations, it is instructive to see how our analysis relates to the energy-conservation principle developed in Chapter 6. First, note that we can determine the head supplied by the pump, $h_p = \Delta p / (\rho g)$, by combining the definition of head in Equation (6.51) with the second of the Euler relations, Equation (7.28), wherefore

$$h_p \equiv \frac{\dot{W}_p}{\dot{m} g} = \frac{\Omega}{g} \, [r_2 V_{\theta 2} - r_1 V_{\theta 1}] \qquad (7.29)$$

Inspection of Equations (7.18) and (7.19) shows that, in general, the absolute and relative velocities are given by

$$\mathbf{V} = \mathbf{w} + \Omega r \mathbf{e}_\theta \qquad \Longrightarrow \qquad V_r = w_r, \quad V_\theta = w_\theta + \Omega r \qquad (7.30)$$

Thus, we can rewrite the head supplied by the pump as

$$h_p = \frac{\Omega}{g} \left[r_2 \left(w_{\theta 2} + \Omega r_2 \right) - r_1 \left(w_{\theta 1} + \Omega r_1 \right) \right] \qquad (7.31)$$

Now, recall that for steady, incompressible flow, the approximate form of the energy-conservation principle is [cf. Equation (6.52)]:

$$\frac{p_1}{\rho g} + \alpha_1 \frac{\mathbf{V}_1 \cdot \mathbf{V}_1}{2g} + z_1 + h_p = \frac{p_2}{\rho g} + \alpha_2 \frac{\mathbf{V}_2 \cdot \mathbf{V}_2}{2g} + z_2 + h_t + h_L \qquad (7.32)$$

where all notation is as in Section 6.3. Since our flow involves only a pump, obviously the head supplied to a turbine, h_t, is zero. Then, if we assume we have a perfect (inviscid) fluid, we can say the kinetic-energy correction factors, α_1 and α_2, are unity and the head loss, h_L, is zero. Therefore, energy conservation simplifies to

$$\frac{p_1}{\rho g} + \frac{V_{r1}^2 + V_{\theta 1}^2}{2g} + z_1 + h_p = \frac{p_2}{\rho g} + \frac{V_{r2}^2 + V_{\theta 2}^2}{2g} + z_2 \qquad (7.33)$$

Combining Equations (7.30), (7.31) and (7.33), after a bit of straightforward algebra, we find

$$\frac{p_1}{\rho g} + \frac{w_{r1}^2 + w_{\theta 1}^2 - \Omega^2 r_1^2}{2g} + z_1 = \frac{p_2}{\rho g} + \frac{w_{r2}^2 + w_{\theta 2}^2 - \Omega^2 r_2^2}{2g} + z_2 \qquad (7.34)$$

Finally, multiplying through by ρg, we conclude that

$$p + \frac{1}{2} \rho \, \mathbf{w} \cdot \mathbf{w} + \rho g z - \frac{1}{2} \rho \Omega^2 r^2 = \text{constant} \qquad (7.35)$$

This is often referred to as Bernoulli's equation in a rotating coordinate system. We will derive a similar equation in Subsection 9.7.1 as an exact solution to the Euler equation for rigid-body rotation [cf. Equation (9.53)].

7.4   Application to a Centrifugal Pump

While the Euler turbomachine equations are relatively simple, their application is complicated somewhat by the three-dimensional geometry characteristic of virtually all turbomachines. In this section, we apply the equations to the centrifugal pump depicted in Figure 7.5. While their range of applicability extends to most turbomachines, it is convenient at this point to limit our scope to a centrifugal pump, which served as our focus in developing the theory. We will apply the equations to other turbomachines in Section 7.6.

7.4.1   Velocity Triangles

Inspection of the Euler turbomachine Equations (7.27) and (7.28) shows that the individual velocity components of the rotor and the blade at both the impeller inlet and outlet appear explicitly. Thus, to analyze a given pump, sufficient information must be available to define the velocities. To help clarify things, it is useful to discuss a concept known as the **velocity triangle**. This is simply a vector diagram based on the velocity vectors at the inlet and outlet, with sufficient information to determine all components.

Figure 7.8 shows a velocity triangle for the centrifugal pump (see Figure 7.7 for notation). We assume flow relative to the rotor enters and leaves tangent to the blade profile. Recall

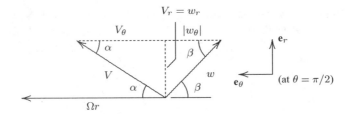

Figure 7.8: *Velocity triangle for a centrifugal pump (xy plane).*

that β is the blade angle relative to the impeller, and Figure 7.8 applies to any part of the blade from leading to trailing edge. Also, α is the angle the absolute velocity makes with the impeller.

In order to use the Euler turbomachine equations to determine the performance of a given machine, we require three types of input data. *First*, the geometry of the device must be specified. For our idealized formulation, the geometry is prescribed in terms of the impeller radii (r_1, r_2), the inlet and outlet blade widths (b_1, b_2) and the inlet and outlet blade angles (β_1, β_2). *Second*, we must know the angular rotation rate, Ω. *Third*, the inlet-flow direction is quantified in terms of the angle α_1. From this information, combined with conservation of mass, we can determine the volume-flow rate, Q, all velocity components of the relative and absolute velocities $(\mathbf{w}_1, \mathbf{V}_1)$ and the angle α_2.

Volume-Flow Rate: From the velocity triangle, clearly

$$|w_\theta| = w_r \cot \beta \quad \text{and} \quad V_\theta = V_r \cot \alpha \tag{7.36}$$

Now, since $\mathbf{V} = \mathbf{w} + \Omega r \mathbf{e}_\theta$ while $\mathbf{V} = V_r \mathbf{e}_r + V_\theta \mathbf{e}_\theta$ and $\mathbf{w} = w_r \mathbf{e}_r - |w_\theta| \mathbf{e}_\theta$, necessarily the radial and circumferential components for this velocity triangle are related by

$$V_\theta + |w_\theta| = \Omega r \quad \text{and} \quad V_r = w_r \tag{7.37}$$

Note that there is no inconsistency with Equation (7.30), which made no explicit assumption about the sign of w_θ—here, we explicitly assume it is negative. Thus, at the inlet where all quantities have subscript 1, we find

$$w_{r1} \cot \alpha_1 + w_{r1} \cot \beta_1 = \Omega r_1 \tag{7.38}$$

which yields the following solution for w_{r1}.

$$w_{r1} = \frac{\Omega r_1}{\cot \alpha_1 + \cot \beta_1} \tag{7.39}$$

Finally, combining Equations (7.39) and (7.24), the volume-flow rate through the pump is

$$Q = \frac{2\pi b_1 \Omega r_1^2}{\cot \alpha_1 + \cot \beta_1} \tag{7.40}$$

As a final comment, the **design flow rate**, Q^*, is the maximum volume-flow rate through the pump. It usually occurs when the inlet flow is purely radial as it enters the impeller, i.e., for $\alpha_1 = \pi/2$.

Velocity Components: At the inlet, we have already shown that the radial and circumferential components of the absolute velocity are $V_{r1} = w_{r1}$ and $V_{\theta 1} = V_{r1} \cot \alpha_1$, respectively. Therefore, making use of Equation (7.39), the circumferential velocity at the inlet is

$$V_{\theta 1} = \frac{\Omega r_1 \cot \alpha_1}{\cot \alpha_1 + \cot \beta_1} \tag{7.41}$$

Conservation of mass, Equation (7.24), shows that the radial velocity components at the inlet and outlet [see Figure 7.6(b)] are related by

$$\frac{w_{r2}}{w_{r1}} = \left(\frac{r_1}{r_2}\right)\left(\frac{b_1}{b_2}\right) \tag{7.42}$$

Therefore, using the value of w_{r1} from Equation (7.39), the radial velocity at the outlet is

$$w_{r2} = \left(\frac{r_1}{r_2}\right)\left(\frac{b_1}{b_2}\right)\frac{\Omega r_1}{\cot \alpha_1 + \cot \beta_1} \tag{7.43}$$

Again referring to the velocity triangle in Figure 7.8, we see that

$$|w_{\theta 2}| = w_{r2} \cot \beta_2 \tag{7.44}$$

Thus, the circumferential component of the velocity relative to the blade at the outlet is

$$|w_{\theta 2}| = \left(\frac{r_1}{r_2}\right)\left(\frac{b_1}{b_2}\right)\frac{\Omega r_1 \cot \beta_2}{\cot \alpha_1 + \cot \beta_1} \tag{7.45}$$

Finally, as discussed above, $V_{\theta 2} + |w_{\theta 2}| = \Omega r_2$, wherefore the circumferential component of the absolute velocity is

$$\begin{aligned}
V_{\theta 2} &= \Omega r_2 - |w_{\theta 2}| \\
&= \Omega r_2 - \left(\frac{r_1}{r_2}\right)\left(\frac{b_1}{b_2}\right)\frac{\Omega r_1 \cot \beta_2}{\cot \alpha_1 + \cot \beta_1} \\
&= \Omega r_2 \left[1 - \left(\frac{r_1}{r_2}\right)^2 \left(\frac{b_1}{b_2}\right)\frac{\cot \beta_2}{\cot \alpha_1 + \cot \beta_1}\right]
\end{aligned} \tag{7.46}$$

Outlet Flow Angle: The final quantity of interest is the angle between the absolute velocity and the impeller at the outlet, α_2. Again referring to the velocity triangle in Figure 7.8, the angle is given by

$$\tan \alpha_2 = \frac{V_{r2}}{V_{\theta 2}} = \frac{w_{r2}}{V_{\theta 2}} \tag{7.47}$$

Combining Equations (7.43), (7.46) and (7.47), after a short calculation we find

$$\alpha_2 = \tan^{-1}\left[\frac{\left(\dfrac{r_1}{r_2}\right)^2 \left(\dfrac{b_1}{b_2}\right)}{\cot \alpha_1 + \cot \beta_1 - \left(\dfrac{r_1}{r_2}\right)^2 \left(\dfrac{b_1}{b_2}\right)\cot \beta_2}\right] \tag{7.48}$$

In summary, the volume-flow rate, relative and absolute velocity vectors at the inlet and outlet, and the outlet flow angle for a centrifugal pump are as follows.

$$
\left.
\begin{aligned}
Q &= \frac{2\pi b_1 \Omega r_1^2}{\cot \alpha_1 + \cot \beta_1} \\[2mm]
\mathbf{w}_1 &= \frac{\Omega r_1}{\cot \alpha_1 + \cot \beta_1} \left[\mathbf{e}_r - \cot \beta_1 \, \mathbf{e}_\theta \right] \\[2mm]
\mathbf{V}_1 &= \frac{\Omega r_1}{\cot \alpha_1 + \cot \beta_1} \left[\mathbf{e}_r + \cot \alpha_1 \, \mathbf{e}_\theta \right] \\[2mm]
\mathbf{w}_2 &= \frac{\Omega r_1}{\cot \alpha_1 + \cot \beta_1} \left(\frac{r_1}{r_2} \right) \left(\frac{b_1}{b_2} \right) \left[\mathbf{e}_r - \cot \beta_2 \, \mathbf{e}_\theta \right] \\[2mm]
\mathbf{V}_2 &= \frac{\Omega r_1}{\cot \alpha_1 + \cot \beta_1} \left(\frac{r_1}{r_2} \right) \left(\frac{b_1}{b_2} \right) \mathbf{e}_r \\[2mm]
&+ \Omega r_2 \left[1 - \left(\frac{r_1}{r_2} \right)^2 \left(\frac{b_1}{b_2} \right) \frac{\cot \beta_2}{\cot \alpha_1 + \cot \beta_1} \right] \mathbf{e}_\theta \\[3mm]
\alpha_2 &= \tan^{-1} \left[\frac{\left(\dfrac{r_1}{r_2} \right)^2 \left(\dfrac{b_1}{b_2} \right)}{\cot \alpha_1 + \cot \beta_1 - \left(\dfrac{r_1}{r_2} \right)^2 \left(\dfrac{b_1}{b_2} \right) \cot \beta_2} \right]
\end{aligned}
\right\} \qquad (7.49)
$$

This completes our formulation of an idealized turbomachine theory. Table 7.1 lists the most useful results of the theory, including the equation numbers for the various properties of interest. In practice, these equations provide estimates of pump performance that can differ from actual values by as much as 25%. There are a variety of reasons why our predictions are so far from the true performance of a real device, all of which stem from the complexity of turbomachine flows. Remember that we are dealing with a three-dimensional flowfield with complicated geometry. Thus, achieving this level of accuracy with a simple one-dimensional formulation is actually remarkable!

Some of the most significant complicating effects are leakage, mismatch, circulation loss and friction loss. **Leakage** is the movement of fluid between the impeller and the casing. Although the gap is small, any pressure difference between upper and lower blade surfaces causes flow in the gaps, with an attendant loss in flow rate and head. This is a three-dimensional effect that has been ignored in our simplified approach. **Mismatch** pertains to

Table 7.1: *Summary of the Euler Turbomachine Equations*

Equation(s)	Turbomachine Property
(7.27)	Torque (Euler turbomachine equation)
(7.28)	Power (Euler turbomachine equation)
(7.29)	Head change
(7.35)	Pressure (Bernoulli's-equation equivalent)
(7.49)	Volume-flow rate and flow velocities

the relation between the inlet flow and the blade passages. Ideally, the flow enters normal to the impeller and tangent to the blades. When these conditions are not met, the pump's performance is overestimated by our idealized model. Similarly, **circulation loss** occurs when the flow does not exit tangent to the blades. This can be caused by separation or stall, a condition where the flow actually reverses direction near the blade trailing edge. This condition reduces the increase in pressure achieved by the pump. There are two types of **friction losses**, viz., losses in the bearings supporting rotating metallic parts and losses caused by the fluid viscosity. Our theory ignores viscous effects, which can alter the flow, especially as the fluid approaches the blade trailing edge. Lakshminarayana (1996) addresses all of these, and other, issues in great detail.

7.4.2   Computing Pump Properties

To illustrate how the Euler turbomachine equations can be used to predict pump performance, consider the following. A typical commercial centrifugal water pump has the following characteristics.

- **Blade Geometry:** $r_1 = 10$ cm, $r_2 = 18$ cm, $b_1 = b_2 = 4.5$ cm, $\beta_1 = 30°$, $\beta_2 = 20°$

- **Angular-Rotation Rate:** $\Omega = 24$ Hz

- **Inlet-Flow Direction:** $\alpha_1 = 90°$

We would like to compute the volume-flow rate, Q, the power required to drive the pump, $\dot{W}_p$, and the increase in head as the water moves through the pump, h_p. Also, to assess how accurate our theory is, we can compare results of our computations with the actual pump properties as determined by the manufacturer, viz.,

$$Q = 0.22 \ \frac{\text{m}^3}{\text{sec}}, \quad \dot{W}_p = 98 \text{ kW}, \quad h_p = 46 \text{ m} \qquad \text{(Measured Pump Properties)} \quad (7.50)$$

The first thing we must do is convert Ω from cycles per second to radians per second. Since there are 2π radians in a complete cycle, we multiply by 2π. Therefore,

$$\Omega = \left(2\pi \ \frac{\text{radians}}{\text{cycle}}\right)\left(24 \ \frac{\text{cycles}}{\text{sec}}\right) = 150.8 \text{ sec}^{-1} \qquad (7.51)$$

Then, from the first of Equations (7.49), the volume-flow rate is

$$Q = \frac{2\pi b_1 \Omega r_1^2}{\cot \alpha_1 + \cot \beta_1} = \frac{2\pi(0.045 \text{ m})\left(150.8 \text{ sec}^{-1}\right)(0.1 \text{ m})^2}{\cot 90° + \cot 30°} = 0.246 \ \frac{\text{m}^3}{\text{sec}} \qquad (7.52)$$

Comparing to the first of Equations (7.50), we see that the predicted value for Q is 12% percent higher than the measured value. While this difference is modest, differences between predicted and measured flow rate are generally larger for off-design conditions, i.e., for $\alpha_1 \neq \pi/2$.

Before we can compute the power and change in head, we must determine the velocities at the inlet and outlet. Using the rest of Equations (7.49), a little arithmetic shows that the velocities (in m/sec) are:

$$
\mathbf{w}_1 = \underbrace{8.71}_{w_{r1}}\mathbf{e}_r - \underbrace{15.08}_{|w_{\theta 1}|}\mathbf{e}_\theta
$$

$$
\mathbf{V}_1 = \underbrace{8.71}_{V_{r1}}\mathbf{e}_r
$$

$$
\mathbf{w}_2 = \underbrace{4.84}_{w_{r2}}\mathbf{e}_r - \underbrace{13.28}_{|w_{\theta 2}|}\mathbf{e}_\theta
$$

$$
\mathbf{V}_2 = \underbrace{4.84}_{V_{r2}}\mathbf{e}_r + \underbrace{13.86}_{V_{\theta 2}}\mathbf{e}_\theta
$$

(7.53)

Also, the angle α_2 follows from the last of Equations (7.49), which yields

$$
\alpha_2 = 19.2^\circ \tag{7.54}
$$

Figure 7.9 shows the velocity vectors at the inlet and outlet for this pump.

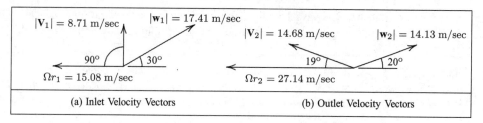

| (a) Inlet Velocity Vectors | (b) Outlet Velocity Vectors |

Figure 7.9: *Velocity vectors for a commercial centrifugal water pump.*

Next, we determine the power required to drive the pump from Equation (7.28). Thus, using the fact that the density of water is $\rho = 998$ kg/m^3 at 20° C,

$$
\begin{aligned}
\dot{W}_p &= \rho\Omega Q\,[r_2 V_{\theta 2} - r_1 V_{\theta 1}] \\
&= \left(998\,\frac{\text{kg}}{\text{m}^3}\right)(150.8\,\text{sec}^{-1})\left(0.246\,\frac{\text{m}^3}{\text{sec}}\right)\left[(0.18\,\text{m})\left(13.86\,\frac{\text{m}}{\text{sec}}\right) - 0\right] \\
&= 92364\,\frac{\text{kg}\cdot\text{m}^2}{\text{sec}^3} = 92.4\,\text{kW}
\end{aligned} \tag{7.55}
$$

Reference to the second of Equations (7.50) shows that our predicted value is just 6% lower than the measured power.

Finally, the increase in head that the fluid receives after passing through the pump is

$$
h_p = \frac{\dot{W}_p}{\dot{m}g} = \frac{\dot{W}_p}{\rho Q g} \tag{7.56}
$$

From the values computed above, we have

$$
h_p = \frac{92364\,\text{kg}\cdot\text{m}^2/\text{sec}^3}{\left(998\,\text{kg/m}^3\right)(0.246\,\text{m}^3/\text{sec})\left(9.807\,\text{m/sec}^2\right)} = 38.4\,\text{m} \tag{7.57}
$$

The last of Equations (7.50) tells us the actual head increase is 46 m, so that our predicted gain is 17% less than measured.

7.4.3   Efficiency

The rate at which the blades do work on the fluid for a pump (and vice versa for a turbine), $\dot{W}_p$, is referred to as the **hydraulic power**. The power required to drive a pump (or power extracted from a turbine), $\Omega\tau$, is known as the **mechanical power**. According to the Euler turbomachine equations, the hydraulic power is equal to the mechanical power. In practice, they are not equal because of the various losses occurring in real flow through the machine.

In the case of a pump, the hydraulic power is less than the mechanical power. Conversely, a turbine absorbs more hydraulic power than it delivers in the form of mechanical power. To describe this mathematically, we introduce the **efficiency**. For a pump, the efficiency, η_p, is defined by

$$\eta_p \equiv \frac{\dot{W}_p}{\Omega\tau} = \frac{\rho Q g h_p}{\Omega\tau} \qquad \text{(Pump)} \qquad (7.58)$$

Similarly, we define the turbine efficiency, η_t, according to

$$\eta_t \equiv \frac{\Omega\tau}{\dot{W}_p} = \frac{\Omega\tau}{\rho Q g h_p} \qquad \text{(Turbine)} \qquad (7.59)$$

In either case, efficiency is always less than one. The turbomachine designer's goal is to achieve as large a value of η_p or η_t over as wide a range of volume-flow rates as possible

---

Example 7.1 *The power, $\dot{W}_p$, for the centrifugal pump discussed in the preceding subsection was quoted in Equation (7.50) as 98 kW. If the manufacturer states that the mechanical power required to drive the pump is $\Omega\tau = 115$ kW, what is its efficiency?*

Solution. The quantity $\dot{W}_p$ is the actual power transferred from the pump to the fluid, while $\Omega\tau$ is required to accomplish the transfer. Hence, the efficiency is

$$\eta_p = \frac{\dot{W}_p}{\Omega\tau} = \frac{98 \text{ kW}}{115 \text{ kW}} = 0.85$$

7.4.4   Performance Curves

The ultimate indicator of a turbomachine's performance comes from extensive testing over a range of operating conditions. A standard format for quantifying turbomachine performance is a family of graphs, known as **performance curves**. Figure 7.10 shows typical centrifugal-pump performance curves. By convention, the curves are drawn for a constant rotation speed, Ω, which is usually given in revolutions per minute (rpm). The three curves shown are the dimensionless efficiency, η_p, the head, h_p, in feet and the mechanical power, $\Omega\tau$, in horsepower. The latter curve is usually identified as the **brake horsepower**. Finally, the volume-flow rate, Q, is generally quoted in gallons per minute for liquids and cubic feet per minute for gases.

There are several interesting points about the performance curves worthy of mention. Focusing first on the head, it is nearly constant at small flow rates and falls to zero at a point where a maximum flow rate, Q_{max}, is reached. This is the maximum flow rate possible for the pump. The dashed part of the curve near $Q = 0$ labeled *unstable* involves a complete breakdown of the flow characterized by oscillations in pressure and mass flow. This breakdown is called **surge**, and can occur whenever h_p increases with Q. While it causes irregular operation for a liquid pump, it can lead to catastrophic failure of a gas compressor or even an entire jet engine.

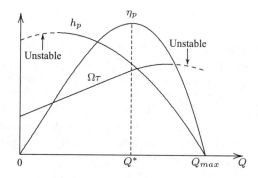

Figure 7.10: *Typical performance curves for a centrifugal pump.*

The mechanical power increases slowly with flow rate and then drops off slightly as the flow rate approaches Q_{max}. The dashed part of the curve near $Q = Q_{max}$ marked *unstable* corresponds to another type of flow breakdown that is especially serious for axial-flow pumps. A key design objective is to minimize the range of flow rates over which unstable conditions can occur.

Finally, as the flow rate increases from zero, the efficiency increases to a peak value at the *design flow rate*, Q^*. Recall that assuming the inlet flow enters the impeller radially in our idealized turbomachine theory yields a reasonably accurate estimate of Q^*. At higher flow rates the efficiency drops off and ultimately falls to zero when the maximum flow rate, Q_{max}, is reached. The point where the efficiency achieves its maximum value is referred to as the **best-efficiency point**.

7.5   Dimensional Considerations

We saw in Subsection 7.4.4 that performance curves for common turbomachines are sometimes plotted in terms of dimensional quantities. While this practice is convenient for designers, it has one important disadvantage. In doing this, the information provided is valid *only for the fluid used in establishing the curves*. As we learned in Chapter 2, dimensional analysis provides a method for correlating experimental data in terms of dimensionless groupings that permits testing with one fluid and extrapolating the test results to a different fluid. It also provides scaling laws between machines of different sizes.

7.5.1   Primary Dimensionless Groupings

As we have seen, there are four key quantities commonly used to characterize centrifugal-pump (and other turbomachine) performance, viz., the change in head, h_p, the mechanical power, $\dot{W}_m = \Omega\tau$, the volume-flow rate, Q, and the efficiency, η_p (or η_t for a turbine). Of these four quantities, only the efficiency is dimensionless. An interesting question to pose is, what are the relevant dimensionless groupings for turbomachines?

To answer this question, we must first decide on the relevant dimensional quantities for flow through a turbomachine. Of course, the rotation speed, Ω, and geometry of the machine are significant. The latter is best characterized for a centrifugal pump by the impeller diameter, D. Also the key fluid properties that should be considered are density, ρ, and viscosity, μ.

Since gravitational acceleration, g, plays a role only when the working fluid is a liquid, it is customary to view the quantity gh_p, which is the hydraulic power per unit mass-flow rate, as a dimensional output quantity rather than h_p. Equivalently, we could work directly with the hydraulic power, $\dot{W}_p = \rho Q g h_p = Q \Delta p$, and arrive at the same results. Note that by dismissing g in this manner, we are making the judgment that buoyancy effects and gravity waves are unimportant in the dynamics of a turbomachine. We will also ignore the effects of surface roughness in our dimensional analysis. Although it will play a bit of a role, it is usually of secondary importance to the overall performance of a properly functioning turbomachine.

Thus, we postulate that gh_p and $\dot{W}_m$ are each functions of five dimensional quantities:

$$gh_p = f_1(Q, D, \Omega, \rho, \mu) \quad \text{and} \quad \dot{W}_m = f_2(Q, D, \Omega, \rho, \mu) \tag{7.60}$$

From dimensional analysis, we wish to correlate gh_p and $\dot{W}_m$ with Q, D, Ω, ρ and μ. In both cases, using the techniques developed in Chapter 2, we find that there are six dimensional quantities and three independent dimensions. The Buckingham Π Theorem tells us there are three dimensionless groupings for each problem. Specifically, we find

$$gh_p = \Omega^2 D^2 \mathcal{F}_1\left(\frac{Q}{\Omega D^3}, \frac{\rho \Omega D^2}{\mu}\right) \quad \text{and} \quad \dot{W}_m = \rho \Omega^3 D^5 \mathcal{F}_2\left(\frac{Q}{\Omega D^3}, \frac{\rho \Omega D^2}{\mu}\right) \tag{7.61}$$

Thus, we conclude that the change in the head and the mechanical power are both functions of the dimensionless flow rate, $Q/(\Omega D^3)$, and the Reynolds (or inverse Ekman) number, $Re = \rho \Omega D^2 / \mu$. In general, Reynolds numbers are very large for turbomachines—typically, $Re > 10^7$. At large Reynolds numbers, fluid flows tend to be relatively insensitive to its magnitude. Thus, it is customary to ignore the Reynolds number, which leaves the following three dimensionless groupings.

$$
\begin{aligned}
C_Q &= \frac{Q}{\Omega D^3} \qquad \text{(Capacity coefficient)} \\[1em]
C_H &= \frac{gh_p}{\Omega^2 D^2} \qquad \text{(Head coefficient)} \\[1em]
C_P &= \frac{\dot{W}_m}{\rho \Omega^3 D^5} \qquad \text{(Power coefficient)}
\end{aligned}
\tag{7.62}
$$

Finally, we can now determine the efficiency as a function of these dimensionless groupings by appealing to its definition in Equation (7.58), viz.,

$$\eta_p = \frac{\rho Q g h_p}{\dot{W}_m} = \frac{\rho \left(\Omega D^3 C_Q\right)\left(\Omega^2 D^2 C_H\right)}{\rho \Omega^3 D^5 C_P} = \frac{C_Q C_H}{C_P} \tag{7.63}$$

Note that this explicitly demonstrates the interdependence of η, h_p, $\dot{W}_m$ and Q, which was obscured by presenting the performance curves in terms of dimensional quantities.

As an example of how the performance curves appear when plotted in terms of dimensionless groupings, we consider another popular centrifugal water pump. The pump comes in two sizes, one with an impeller diameter of 38 inches and a smaller model with an impeller diameter of 32 inches. All other geometrical parameters (casing, blade-to-casing clearance, etc.) are scaled in the same ratio. When conventional performance curves are used, two separate sets of curves must be provided, and each will be valid for just one rotation rate. However, using dimensionless groupings, the performance curves all collapse close to a single set, regardless of pump size and rotation rate!

Table 7.2: *Flow Properties for a Commercial Centrifugal Pump*

38-inch pump, 710 rpm				32-inch pump, 1170 rpm			
Q (gal/min)	h_p (ft)	$\dot{W}_m$ (hp)	η_p	Q (gal/min)	h_p (ft)	$\dot{W}_m$ (hp)	η_p
180	273	-	-	180	498	-	-
4215	267	-	-	4170	496	-	-
5560	266	607	0.617	7310	491	1413	0.643
6725	265	694	0.650	10270	465	1611	0.750
8430	263	805	0.697	14440	431	1862	0.846
12645	255	959	0.851	18565	395	2095	0.886
16860	236	1108	0.909	22735	344	2265	0.874
21075	210	1225	0.914	26545	293	2360	0.834
25290	176	1289	0.874				
26635	169	1311	0.869				

Table 7.2 lists Q, h_p, $\dot{W}_m$ and η_p for the 38-inch pump operating at 710 rpm and for the 32-inch pump operating at 1170 rpm. While the range of flow rates is the same, the smaller pump yields roughly twice the head of the larger pump, and is somewhat less efficient. Using Equations (7.62) and (7.63), we can recast the data of Table 7.2 in dimensionless form. Figure 7.11 shows the dimensionless performance curves for the two pumps, with the open circles corresponding to the larger pump and the closed circles to the smaller pump. The curves are least-squares fits to the data, which fall within 6% of the measurements for all three flow properties, η_p, h_p and $\dot{W}_m$. The minor departures from a single universal curve are attributable to differences in Reynolds number. Even so, the differences are sufficiently small to demonstrate the power of dimensional analysis in correlating measurements.

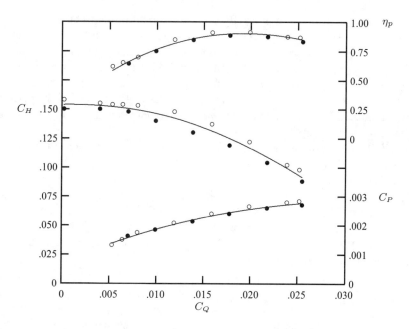

Figure 7.11: *Dimensionless performance curves for a centrifugal pump;* ○ *38-inch pump;* ● *32-inch pump;* —— *least-squares curve fit.*

7.5.2   Specific Speed

There is another dimensionless grouping known as the **specific speed** commonly used in turbomachinery design, that is independent of pump size. It is called N_s and is based on properties at the *best-efficiency point*, Q^* and h_p^*. It can be formed as follows.

$$N_s = \frac{C_{Q^*}^{1/2}}{C_{H^*}^{3/4}} = \left(\frac{Q^*}{\Omega D^3}\right)^{1/2} \left(\frac{\Omega^2 D^2}{gh_p^*}\right)^{3/4} = \frac{\Omega\,(Q^*)^{1/2}}{\left(gh_p^*\right)^{3/4}} \tag{7.64}$$

As will be discussed in the next section, the specific speed is useful for selecting the type of turbomachine for a given application, especially for systems where volume-flow rate, head and rotation speed are dictated by the overall design requirements. In general, the specific speed for centrifugal pumps ranges from about 0.13 to 1.00.

Example 7.2 *For the centrifugal pumps described in the performance curves of Figure 7.11, the best-efficiency point occurs when $C_{Q^*} = 0.0183$ and $C_{H^*} = 0.123$. Determine the specific speed.*

Solution. By definition, the specific speed for these pumps is

$$N_s = \frac{C_{Q^*}^{1/2}}{C_{H^*}^{3/4}} = \frac{(0.0183)^{1/2}}{(0.123)^{3/4}} = 0.65$$

7.6   Common Turbomachines

In this concluding section, we will briefly discuss common turbomachines found in practical applications. The discussion is, of necessity, mostly qualitative. While the Euler turbomachine equations reveal general properties of most pumps, compressors and turbines, they are far too simplistic to describe details of the complex flowfields involved. Rather, we must turn to computational methods if we wish to gain insight into detailed flow through the machine under investigation. Powerful computer programs are available to predict the performance of complex turbomachines, but they require intensive computational resources and are beyond the scope of this text.

Our use of the control-volume method and dimensional analysis lays the foundation for understanding some aspects of turbomachinery. There are aspects we have omitted, such as the shapes of rotor and stator blades and their importance in a machine's performance. The interested reader can learn more about the physical and engineering aspects of turbomachines from the extensive literature on the topic. Some of the most noteworthy books are those by Shepherd (1956), Csanady (1964), Betz (1966), Dixon (1978), Logan (1981), Bathe (1984) and Lakshminarayana (1996). The text by Lakshminarayana cites the most up-to-date methods for analyzing turbomachines.

7.6.1   Pumps and Compressors

While most of our discussion thus far has focused on the centrifugal pump, there are other types of pumps that find practical application. Recall from Figure 7.2 that in addition to centrifugal pumps, which have purely radial outflow, there are also axial-flow pumps and mixed-flow pumps. The latter type of pump has both axial and radial velocity components at the outlet. The performance of pumps is most conveniently summarized in terms of the *specific speed*, N_s. Table 7.3 lists the specific-speed range for the three types of pumps as classified by outflow type.

Table 7.3: *Specific-Speed Ranges for Common Liquid Pumps*

Type of Pump	Specific-Speed Range
Radial-Flow (Centrifugal) Pumps	$0.13 < N_s < 1.00$
Mixed-Flow Pumps	$1.00 < N_s < 3.00$
Axial-Flow Pumps	$N_s > 3.00$

In general, centrifugal pumps are regarded as *high-head, low-flow* machines. They are very effective for pumping in many applications such as the water distribution system in a multi-story building. However, there are situations where an increase in head is of secondary importance relative to the volume-flow rate. A typical example would be pumping water between two reservoirs of slightly different depths. In such an application, we require a *low-head, high-flow* machine. By definition, the specific speed is small for the former type of pump and large for the latter. When we wish to pump large amounts of fluid and the increase in head is less important, mixed-flow and axial-flow pumps are more appropriate.

By convention, when the working fluid is a liquid, we categorize a turbomachine that adds energy to the flow as a "pump." When the working fluid is a gas, we use three different terms depending upon how large a pressure increase is imparted to the gas. By convention, when the pressure increase is 40 psi or more, we call the turbomachine a **compressor**. A machine that imparts a pressure rise of 1 psi or less is called a **fan**. We describe a machine with pressure increase between 1 and 40 psi as a **blower**. Table 7.4 summarizes typical specific-speed ranges for fans, blowers and compressors.

Table 7.4: *Specific-Speed Ranges for Common Gas "Pumps"*

Type of Pump	Specific-Speed Range
Fans	$0.50 < N_s < 2.00$
Blowers	$1.00 < N_s < 3.00$
Compressors	$N_s > 3.00$

While we have not discussed the shapes of the blades in the impeller of a pump, there is one interesting feature that can be discerned from our analysis. In particular, recall that the change in head for the fluid is given by [cf. Equation (7.29)]

$$h_p = \frac{\Omega}{g} \left[r_2 V_{\theta 2} - r_1 V_{\theta 1} \right] \tag{7.65}$$

Also, using Equation (7.40) to eliminate $1/(\cot \alpha_1 + \cot \beta_1)$ in Equation (7.46), the absolute velocity at the outlet, $V_{\theta 2}$, is

$$V_{\theta 2} = \Omega r_2 - \frac{Q \cot \beta_2}{2\pi r_2 b_2} \tag{7.66}$$

Now, ignoring the initial angular momentum, $r_1 V_{\theta 1}$, for simplicity, substituting Equation (7.66) into Equation (7.65) shows that we can write the head as

$$h_p = h_{po} + \frac{dh_p}{dQ} Q \qquad (7.67)$$

where

$$h_{po} \equiv \frac{\Omega^2 r_2^2}{g} \quad \text{and} \quad \frac{dh_p}{dQ} \equiv -\frac{\Omega \cot \beta_2}{2 \pi b_2 g} \qquad (7.68)$$

This is a useful result for the following reason. The angle β_2 is the angle between the impeller and the blade at the trailing edge. When $\beta_2 < 90°$, we have *backward-curved* blades as depicted in Figures 7.5, 7.6 and 7.7. Since $\cot \beta_2 > 0$, necessarily $dh_p/dQ < 0$. By contrast, when $\beta_2 > 90°$, the blades are *forward curved*, and we have $dh_p/dQ > 0$. The special case of $\beta_2 = 90°$ corresponds to *radial* blades and $dh_p/dQ = 0$. Therefore, for small volume-flow rates, the h_p curve has negative slope for backward-curved blades, zero slope for radial blades, and positive slope for forward-curved blades. As discussed by Lakshminarayana (1996), a positive slope can lead to surge, so that backward-curved or radial blades are generally preferred, especially for compressors.

7.6.2 Reaction and Impulse Turbines

The purpose of a turbine is extract power from the fluid that can be used for things such as power generation and propulsion. There are two primary types of turbines known as the **reaction turbine** and the **impulse turbine**.

The reaction turbine is a *low-head, high-flow* turbomachine. It is similar to the pumps we have studied in the sense that fluid flows continuously through the passages between rotor (and stator) blades. As with pumps, one of the ways of classifying reaction turbines is in terms of the exit-flow direction. Thus, there are radial-flow, mixed-flow and axial-flow turbines, whose behavior is adequately described by the Euler turbomachine equations. Radial-flow and mixed-flow turbines are called **Francis turbines** in honor of James B. Francis, an early American pioneer in turbine design. Axial-flow turbines such as the one sketched in Figure 7.2(a) use a propeller and are appropriately called **propeller turbines**.

By contrast, an impulse turbine is a *high-head, low-flow* device. An example is the **Pelton wheel**, sketched in Figure 7.12, which uses a nozzle to create a high-speed jet that strikes a series of "buckets" at the same position as they pass by. Thus, the blade passages, i.e., the regions between the buckets, are not filled with fluid and the pressure is essentially constant

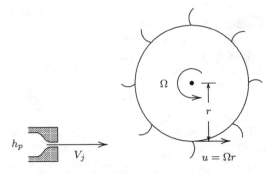

Figure 7.12: *Schematic of a Pelton wheel.*

as it flows past the blades. The pressure is constant since the pressure in a jet is equal to the pressure of the surrounding fluid.

The general principle upon which the Pelton wheel works is straightforward. Fluid from a reservoir with high potential energy flows through a nozzle. Its potential energy is thus transformed to kinetic energy in the jet issuing from the nozzle. The jet strikes the buckets at the wheel rim, and its kinetic energy is transmitted to the wheel to sustain the rotation. Designs exist that use more than one jet to enhance the power of the Pelton wheel. For simplicity, we will analyze a single-jet configuration.

We can formulate a simple theory by using a rotating control volume. We assume the buckets are sufficiently close that the jet always strikes one of the buckets. When this is true, we can approximate the motion as being steady. As shown in Figure 7.13, the fluid from the jet exits the bucket at an angle ϕ to the direction of the incident jet stream. We assume the area of the jet remains constant as it is turned by the bucket. We also assume the bucket width is very small compared to the wheel radius, so that the radial coordinate is approximately equal to the wheel radius, r, everywhere.

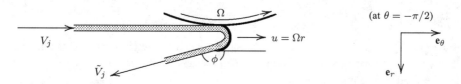

Figure 7.13: *Closeup view of a Pelton-wheel bucket.*

Since the flow is steady, conservation of mass becomes

$$\oiint_S \rho \mathbf{w} \cdot \mathbf{n} \, dS = 0 \tag{7.69}$$

where $\mathbf{w}$ is the fluid velocity relative to the control volume. We select a control volume that is coincident with the jet boundaries. Hence, the only places where fluid crosses the bounding surface of the control volume is through the incident- and deflected-jet cross sections. Now, the incident flow in the jet has absolute velocity, $\mathbf{u} = V_j \mathbf{e}_\theta$. Since the control volume moves with velocity $\mathbf{u}^{cv} = \Omega r \mathbf{e}_\theta$, we have the following at $\theta = -\pi/2$.

Incident Jet: $\mathbf{r} = r\mathbf{e}_r, \quad \mathbf{u} = V_j \mathbf{e}_\theta, \quad \mathbf{w} = (V_j - \Omega r)\,\mathbf{e}_\theta, \quad \mathbf{n} = -\mathbf{e}_\theta$ \hfill (7.70)

Similarly, for the deflected jet, the vectors of interest are

Deflected Jet: $\mathbf{r} = r\mathbf{e}_r, \quad \mathbf{u}- = \tilde{V}_j \mathbf{n}, \quad \mathbf{w} = \tilde{V}_j \mathbf{n} - \Omega r\,\mathbf{e}_\theta, \quad \mathbf{n} = \mathbf{e}_\theta \cos\phi + \mathbf{e}_r \sin\phi$ \hfill (7.71)

Denoting the cross-sectional area of the jet by A, the mass-conservation integral simplifies to

$$\rho\left[-(V_j - \Omega r)\right]A + \rho\left[\tilde{V}_j - \Omega r \cos\phi\right]A = 0 \quad\Longrightarrow\quad \tilde{V}_j = V_j - \Omega r(1 - \cos\phi) \tag{7.72}$$

Since the flow is steady, conservation of angular momentum tells us the torque, τ_b, exerted by the fluid on the bucket is given by

$$\tau_b = \mathbf{k} \cdot \oiint_S \rho \mathbf{r} \times \mathbf{u}\,(\mathbf{w} \cdot \mathbf{n})\, dS \tag{7.73}$$

Therefore, using Equations (7.70) and (7.71) with subscripts *inc* and *def* denoting incident and deflected, respectively, we find

$$\tau_b = \mathbf{k} \cdot \left\{ \underbrace{\rho \left(r V_j \mathbf{k} \right)}_{\rho(\mathbf{r} \times \mathbf{u})_{inc}} \underbrace{\left[-\left(V_j - \Omega r \right) A \right]}_{(\mathbf{w} \cdot \mathbf{n})_{inc} A} + \underbrace{\rho \left(r \tilde{V}_j \cos \phi \mathbf{k} \right)}_{\rho(\mathbf{r} \times \mathbf{u})_{def}} \underbrace{\left[\left(\tilde{V}_j - \Omega r \cos \phi \right) A \right]}_{(\mathbf{w} \cdot \mathbf{n})_{def} A} \right\} \qquad (7.74)$$

So, using the fact that the volume-flow rate is given by $Q = (\mathbf{w} \cdot \mathbf{n})A$, and introducing Equation (7.72) to eliminate $\tilde{V}_j$, the torque on the bucket is

$$\begin{aligned} \tau_b &= -\rho Q r \left(V_j - \tilde{V}_j \cos \phi \right) \\ &= -\rho Q r \left[V_j (1 - \cos \phi) + \Omega r \cos \phi (1 - \cos \phi) \right] \\ &= -\rho Q r (1 - \cos \phi) \left(V_j + \Omega r \cos \phi \right) \end{aligned} \qquad (7.75)$$

The sum of the torque on the Pelton-wheel shaft, τ, and the torque on the bucket, τ_b, is zero so that $\tau = -\tau_b$. Thus,

$$\tau = \rho Q r (1 - \cos \phi) \left(V_j + \Omega r \cos \phi \right) \qquad (7.76)$$

Finally, typical Pelton-wheel buckets are designed so that the angle ϕ is as close as possible to 180°, with a value of 165° being common. Consequently, we can approximate $\cos \phi \approx -1$, so that the final expression for the torque is

$$\tau \approx 2\rho Q r \left(V_j - \Omega r \right) \qquad (7.77)$$

The corresponding power output from the Pelton wheel is

$$\Omega \tau \approx 2\rho Q \Omega r \left(V_j - \Omega r \right) \qquad (7.78)$$

At this point, we can derive an interesting result regarding the maximum power. Letting $U = \Omega r$, we can rewrite the power as

$$\Omega \tau = 2\rho Q U \left(V_j - U \right) \qquad \Longrightarrow \qquad \frac{d\Omega \tau}{dU} = 2\rho r Q \left(V_j - 2U \right) \qquad (7.79)$$

Setting $d\Omega \tau / dU = 0$ and noting that $d^2 \Omega \tau / dU^2 < 0$, we can locate the maximum value of $\Omega \tau$. It occurs when $U = V_j / 2$, or, in terms of Ωr,

$$\Omega r = \frac{1}{2} V_j \qquad \Longrightarrow \qquad \Omega \tau = 2\rho Q \Omega^2 r^2 = \text{maximum} \qquad (7.80)$$

Based on our idealized model of the Pelton wheel, we can develop an expression for the efficiency, η_t. First, however, we must determine how the jet velocity is related to the head in the reservoir, h_p. From energy conservation, we know that

$$\frac{p_{res}}{\rho g} + \frac{\bar{u}_{res}^2}{2g} + h_p = \frac{p_{jet}}{\rho g} + \frac{V_j^2}{2g} \qquad (7.81)$$

Ideally, all of the total head in the reservoir is converted to kinetic energy as the fluid passes through the nozzle and forms the jet. Assuming the reservoir pressure, $p_{res} \approx p_{jet}$ and that the reservoir velocity, $\bar{u}_{res}$, is negligible, the jet velocity is given by

$$V_j \approx \sqrt{2g h_p} \qquad (7.82)$$

We now have sufficient information to compute the efficiency of the Pelton wheel. Substituting Equations (7.78) and (7.82) into Equation (7.59), the efficiency is

$$\eta_t = \frac{\Omega \tau}{\rho Q g h_p} = 4 \frac{\Omega r}{\sqrt{2gh_p}} \left(1 - \frac{\Omega r}{\sqrt{2gh_p}} \right) \tag{7.83}$$

Figure 7.14 compares Equation (7.83) with measurements [White (1994)]. Consistent with Equation (7.80) and the fact that h_p = constant in our idealized theory, the Pelton wheel operates at maximum efficiency when $\Omega r / \sqrt{2gh_p} = \Omega r / V_j = \frac{1}{2}$. This corresponds to fluid exiting with no momentum or energy left and without vertical motion (in absolute terms). Since we have ignored all losses, the theoretical peak efficiency is 1. The measurements indicate a peak efficiency of 80% when $\Omega r / \sqrt{2gh_p} = 0.47$. In general, Pelton wheels are not as efficient as Francis or propeller turbines.

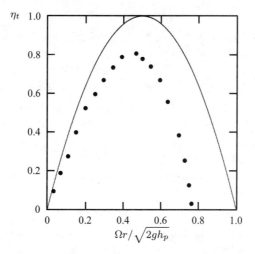

Figure 7.14: *Comparison of theoretical and measured efficiency for a Pelton wheel;* —— *Theoretical;* • *Measured [from White (1994)].*

We conclude our discussion of turbines by summarizing specific-speed ranges for the machines mentioned in this section. Table 7.5 lists the ranges for Pelton wheels, Francis turbines and propeller turbines. Note that a **Kaplan turbine** is a special type of propeller turbine.

Table 7.5: *Specific-Speed Ranges for Common Turbines*

Type of Turbine	Specific-Speed Range
Pelton wheel, single jet	$N_s < 0.13$
Pelton wheel, multijet	$0.10 < N_s < 0.25$
Francis turbines	$0.25 < N_s < 2.10$
Propeller turbines	$1.80 < N_s < 3.50$
Kaplan turbines	$3.00 < N_s < 5.00$

Problems

7.1 The figure shows a lawn sprinkler which ejects water through two nozzles at an angle ϕ to the circumferential direction. The mass-flow rate through the device is $\dot{m} = 7$ kg/sec, and the area of each nozzle is $A_e = 1.2$ cm^2. The water density is $\rho = 1000$ kg/m^3, the sprinkler-arm length is $\ell = 12$ cm, while the sprinkler rotates at $\Omega = 600$ rpm.

   (a) Neglecting all friction losses, compute the power, P, required to drive the sprinkler. Express your answer as a function of $\dot{m}$, ℓ, Ω, w_e and ϕ.

   (b) What is the value of w_e in m/sec?

   (c) Determine the angle ϕ if the power is $P = 210$ W.

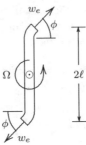

Problems 7.1, 7.2

7.2 Water is ejected through two nozzles at an angle ϕ as shown for a lawn sprinkler.

   (a) Compute the reaction torque on the sprinkler as a function of water density, ρ, ejection velocity, w_e, nozzle area, A_e, rotation rate, Ω, sprinkler-arm length, ℓ, and ϕ.

   (b) Assume the nozzle diameter is $d = 0.24$ in, $\rho = 1.94$ slug/ft^3, $\ell = 3$ in and $\Omega = 300$ rpm. Also, assume the volume-flow rate is $Q = 0.052$ ft^3/sec. Compute the torque for $\phi = 0°$, $30°$ and $60°$.

   (c) At what angle does the torque vanish for the conditions of Part (b)?

7.3 Using the solution developed in this chapter for a centrifugal pump, we would like to determine the conditions under which the outlet flow from the rotor or impeller of a turbomachine is purely radial. Assuming $b_1 = b_2$ while $\alpha_1 = \pi/2$, derive an expression for the exit blade angle, β_2, as a function of r_1, r_2 and β_1 that yields $\alpha_2 = \pi/2$. Make a graph of your results for $\beta_1 = 30°$ and $0 \leq r_1/r_2 \leq 1$.

7.4 The head in a pump can be approximated by $h_p = h_{po} - AQ^2$, where h_{po} is the limiting value of the head as $Q \rightarrow 0$ and A is a constant of dimensions T^2/L^5. For this pump, you can assume the inlet flow is radial so that $\alpha_1 = \pi/2$. Using the solution to the Euler turbomachine equations developed in this chapter, determine h_{po} and A.

7.5 As a first estimate, we can use the solution to the Euler turbomachine equations developed in this chapter to determine the change in pressure across the impeller of a centrifugal pump, $\Delta p = p_2 - p_1$.

   (a) Beginning with Bernoulli's equation in rotating coordinates [Equation (7.35)], determine Δp as a function of ρ, Ω, r_1, r_2, b_1, b_2, β_1 and β_2. Assume $\alpha_1 = \pi/2$.

   (b) Compute the pressure change for the pump discussed in Subsection 7.4.2.

7.6 The maximum volume-flow rate, Q_{max}, for a centrifugal pump occurs when $h_p \rightarrow 0$. Verify that, according to the Euler turbomachine equations, if $\alpha_1 = 90°$ then

$$Q_{max} = 2\pi\Omega r_2^2 b_2 \tan \beta_2$$

7.7 Develop an alternative to Equation (7.68) for dh_p/dQ, taking account of the initial angular momentum, $r_1 V_{\theta 1}$. If the inlet flow angle is $\alpha_1 = 80°$, determine the critical angle, β_{2c}, at which $dh_p/dQ = 0$ for $0 \leq b_1/b_2 \leq 2$. Make a graph of your results and discuss the meaning for the case $b_1/b_2 = 2$.

7.8 The blade geometry of a centrifugal pump is given by $r_1 = 15$ cm, $r_2 = 25$ cm, $b_1 = 4$ cm, $b_2 = 6$ cm, $\beta_1 = 25°$ and $\beta_2 = 90°$. If the angular-rotation rate is $\Omega = 1020$ rpm and $\alpha_1 = 90°$, determine the volume-flow rate and power required to drive the pump if the working fluid is ethyl alcohol at $20°$ C.

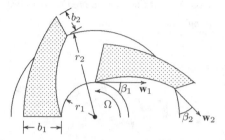

Problems 7.8 through 7.15

7.9 The blade geometry of a centrifugal pump is given by $r_1 = 4$ in, $r_2 = 8$ in, $b_1 = 4$ in, $b_2 = 4$ in, $\beta_1 = 40°$ and $\beta_2 = 90°$. If the angular-rotation rate is $\Omega = 1440$ rpm and $\alpha_1 = 90°$, determine the volume-flow rate and power required to drive the pump if the working fluid is kerosene at $68°$ F.

7.10 A small laboratory pump is used to move mercury between two tanks. It is a centrifugal pump with $r_1 = 3$ cm, $r_2 = 6$ cm, $b_1 = 3$ cm, $b_2 = 5$ cm, $\beta_1 = 35°$ and $\beta_2 = 45°$. If the angular-rotation rate is $\Omega = 600$ rpm and $\alpha_1 = 90°$, determine the volume-flow rate and power required to drive the pump. Express your answer for Q in liters per second (L/sec).

7.11 A commercial water pump is used to empty an overflowing septic tank. Due to suspended solid material, the density of the fluid is 2.15 slug/ft^3. The pump is of the centrifugal type with $r_1 = 4$ in, $r_2 = 12$ in, $b_1 = 2$ in, $b_2 = 1$ in, $\beta_1 = 25°$ and $\beta_2 = 15°$. If the angular-rotation rate is $\Omega = 1600$ rpm and $\alpha_1 = 90°$, determine the volume-flow rate (in gal/min) and power (in hp) required to drive the pump.

7.12 Consider a centrifugal water pump whose impeller geometry is $r_2 = 10$ cm, $b_1 = 3$ cm, $b_2 = 5$ cm, $\beta_1 = 15°$, and $\beta_2 = 10°$. Also, the impeller rotates at an angular velocity of $\Omega = 300$ rpm, and the inlet absolute velocity is purely radial. Neglecting gravitational effects, determine the value of the inlet impeller radius, r_1, for which the pressure rise is a maximum.

7.13 You have just finished painting your living room and you are using a radial fan to dry the walls. Curious about its operation, you find the owner's manual and discover that it includes details of the impeller geometry, which are $r_1 = 7$ in, $r_2 = 15$ in, $b_1 = 6$ in, $b_2 = 2$ in, $\beta_1 = 10°$ and $\beta_2 = 40°$. Also, the impeller rotates at an angular velocity of $\Omega = 2450$ rpm, the inlet absolute velocity is purely radial and the density of air is $\rho = 0.00234$ slug/ft^3. Determine the volume-flow rate and the power required to run the fan.

7.14 Q is designing a miniature radial fan for Agent 007, who will use it to discharge a very potent sleep-inducing gas into the air-conditioning ducts at Spectre headquarters. The density of the gas is $\rho = 1.08$ kg/m^3 and the fan must deliver a volume-flow rate of 40 cm^3/sec with a power consumption of 150 μW (microWatts). The blade geometry must be such that $r_1 = 6$ mm, $b_1 = \frac{1}{2}$ mm, $b_2 = \frac{1}{2}$ mm, $\beta_1 = 30°$ and $\beta_2 = 20°$, while $\alpha_1 = 90°$.

  (a) Determine the angular-rotation speed, Ω, and the outlet radius, r_2.

  (b) Determine the velocity triangles at the inlet and outlet. Be sure to compute the angle α_2.

7.15 A low-speed (incompressible-flow) wind tunnel is powered by a centrifugal fan. The impeller geometry is described by $r_1 = 24$ in, $r_2 = 36$ in, $b_1 = 2$ in, $b_2 = 2$ in, $\beta_1 = 30°$ and $\beta_2 = 16°$. The angular-rotation rate is 600 rpm and $\alpha_1 = 90°$. The working fluid is air with $\rho = 0.00234$ slug/ft^3.

  (a) Compute the volume-flow rate and the power required to run the fan.

  (b) Determine the velocity triangles at the inlet and outlet. Be sure to compute the angle α_2.

7.16 A low-speed (incompressible-flow) wind tunnel is powered by a centrifugal fan. The impeller geometry is described by $r_1 = 60$ cm, $r_2 = 90$ cm, $b_1 = 5$ cm, $b_2 = 4$ cm, $\beta_1 = 27°$ and $\beta_2 = 18°$. The angular-rotation rate is 600 rpm and $\alpha_1 = 80°$. The working fluid is air with $\rho = 1.20$ kg/m^3.

(a) Compute the volume-flow rate and the power required to run the fan.

(b) Determine the velocity triangles at the inlet and outlet. Be sure to compute the angle α_2.

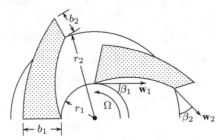

Problems 7.16, 7.17, 7.18

7.17 In a mine shaft a radial fan circulates fresh air with $\rho = 0.00234$ slug/ft^3. The fan operates at angular velocity $\Omega = 6000$ rpm, has a volume-flow rate of 600 ft^3/min, and requires 1 hp to run. The blade geometry is given by $r_1 = 6$ in, $r_2 = 9$ in, $b_1 = 1$ in and $b_2 = 1$ in. Also, the inlet flow has $\alpha_1 = 90°$. Determine the inlet and outlet blade angles, β_1 and β_2.

7.18 In a warehouse, a radial fan circulates fresh air with $\rho = 1.20$ kg/m^3. The fan operates at angular velocity $\Omega = 4000$ rpm, has a volume-flow rate of 0.8 m^3/sec, and requires 2 kW to run. The blade geometry is given by $r_1 = 14$ cm, $r_2 = 20$ cm, $b_1 = 4$ cm and $b_2 = 5$ cm. Also, the inlet flow has $\alpha_1 = 90°$. Determine the inlet and outlet blade angles, β_1 and β_2.

7.19 In the pump industry, it is common practice to compute a dimensional specific speed according to $\tilde{N}_s = \Omega(Q^*)^{1/2}/(h_p^*)^{3/4}$, with Ω in rpm, Q^* in gal/min and h_p^* in ft. Determine the dimensions of $\tilde{N}_s$ and compute the ratio of $\tilde{N}_s$ to N_s.

7.20 The head, h_p, and power, $\Omega\tau$, for an axial-flow pump vary with volume-flow rate according to $h_p = h_o[1 - Q^2/30]$ and $\Omega\tau = P_o[1 + Q^2/15]$, where h_o and P_o are constants and Q is given in ft^3/sec. Determine the volume-flow rate at the *best-efficiency point, Q^**.

7.21 For fans, blowers and compressors, performance curves are customarily presented in terms of pressure rise, Δp, expressed in "inches of H$_2$O." In terms of these units, the hydraulic power is given by $\dot{W}_p = \rho_w Qg\Delta p$ where ρ_w is the density of water (1.94 slug/ft^3), Q is volume-flow rate and g is gravitational acceleration. If the pressure rise is $\Delta p = 10$ inches of H$_2$O, what is the change in head, h_p, if the working fluid is water, air ($\rho = 0.00234$ slug/ft^3), or helium ($\rho = 0.000323$ slug/ft^3)? Express your answers in feet.

7.22 The following data describe the measured performance of a pump operating at 1750 rpm. Complete the pump performance-curve tables by computing the power in hp. Above what volume-flow rate should we be concerned about possible unstable behavior? The working fluid is water with $\rho = 1.94$ slug/ft^3.

Q (gal/min)	h_p (ft)	η_p (%)
0	124	0
400	120	49
800	112	64
1000	104	68
1200	96	71
1400	84	70
1500	76	69
1600	67	67
1800	41	50

Problem 7.22

7.23 The figure shows the performance curves for a centrifugal pump. The head, h_p, and power, $\Omega\tau$, can be approximated by

$$h_p \approx 140 \left[1 - \left(\frac{Q - 400}{2800} \right)^2 \right] \quad \text{and} \quad \Omega\tau \approx 60 - 40 \left(\frac{Q - 2400}{2400} \right)^2$$

with h_p in ft, $\Omega\tau$ in hp and Q in gal/min.

(a) Compute the volume-flow rate of water at the *best-efficiency point*, Q^*, and the corresponding efficiency, η_p. **HINT:** Solve by trial and error, noting that 1600 gal/min $< Q^* <$ 1800 gal/min.

(b) For what range of Q is this pump likely to operate without any type of instability?

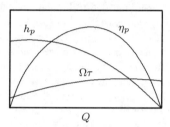

Problems 7.23, 7.24, 7.25, 7.26

7.24 The figure shows performance curves for a centrifugal water pump. For small volume-flow rates, the head varies as $h_p \approx 133 + Q/64$ where h_p is given in ft and Q in gal/min.

(a) Show that, according to the Euler turbomachine equations, if $\alpha_1 = \pi/2$ then

$$h_p = \frac{\Omega^2 r_2^2}{g} - \frac{\Omega Q \cot \beta_2}{2\pi g b_2}$$

(b) Now, assume $b_2 = \frac{1}{5} r_2$ and $\Omega = 1500$ rpm. What are the values of β_2, r_2 and b_2 for this pump?

7.25 The figure shows the performance curves for a centrifugal pump. For the practical operating range of the pump, the head, h_p, and power, $\Omega\tau$, can be approximated by

$$h_p \approx 167 - \frac{3}{80} Q \quad \text{and} \quad \Omega\tau \approx 45 + \frac{1}{160} Q$$

where $800 < Q < 2400$, with h_p in ft, $\Omega\tau$ in hp and Q in gal/min. Determine the volume-flow rate of water at the *best-efficiency point*, Q^*, and the corresponding efficiency, η_p.

7.26 The figure shows the performance curves for a turbomachine. For the practical operating range of the machine, the head, h_p, and power, $\Omega\tau$, can be approximated by

$$h_p \approx \frac{1}{26} (5125 - Q) \quad \text{and} \quad \Omega\tau \approx \frac{1}{137} (Q + 6974)$$

with h_p in m, $\Omega\tau$ in MW (Megawatts) and Q in m^3/sec. Determine the type of turbomachine (i.e., is it a fan, blower or compressor) if the torque is $\tau = 2.75 \cdot 10^5$ N·m when the working fluid is air.

7.27 We know from the manufacturer's specifications that the minimum specific speed at which a centrifugal pump can operate is $(N_s)_{min} = 0.18$. The pump must deliver a volume-flow rate of $Q^* = 3300$ gal/min against a head of $h_p^* = 1025$ ft. What is the minimum rotation speed, Ω_{min}, in rpm?

7.28 A centrifugal pump of diameter 40 cm operating at 2400 rpm has the following performance figures: $Q^* = 0.2$ m^3/sec; $h_p^* = 100$ m; $\dot{W}_m^* = 210$ kW. Compute C_{Q^*}, C_{H^*}, C_{P^*}, η_p and N_s. The working fluid is water with $\rho = 998$ kg/m^3.

7.29 A centrifugal pump of diameter 14.62 in operating at 2134 rpm has the following performance figures: $Q^* = 6$ ft^3/sec; $h_p^* = 350$ ft; $\dot{W}_m^* = 270$ hp. Compute C_{Q^*}, C_{H^*}, C_{P^*}, η_p and N_s. The working fluid is water with $\rho = 1.94$ slug/ft^3.

7.30 A compressor of diameter $D = 1$ m operating at $\Omega = 6000$ rpm has the following performance figures: $Q^* = 140$ m^3/sec; $h_p^* = 1750$ m; $\dot{W}_m^* = 4$ MW. Compute C_{Q^*}, C_{H^*}, C_{P^*}, η_p and N_s for this compressor if the working fluid is air with $\rho = 1.20$ kg/m^3.

7.31 A blower of diameter $D = 24$ in operating at $\Omega = 2000$ rpm has the following performance figures: $Q^* = 335$ ft^3/sec; $h_p^* = 430$ ft; $\dot{W}_m^* = 26$ hp. Compute C_{Q^*}, C_{H^*}, C_{P^*}, η_p and N_s for this blower if the working fluid is air with $\rho = 0.00234$ slug/ft^3.

7.32 The capacity and head coefficients for a 36-inch pump operating at 720 rpm are $C_{Q^*} = 0.02$ and $C_{H^*} = 0.125$ at its *best-efficiency point*, where $\eta_p = 0.83$. We want to design a pump with the same dimensionless performance characteristics that will have a volume-flow rate of $Q^* = 50000$ gal/min at $\Omega = 1200$ rpm. What must the pump diameter be (to the nearest inch) and how much power will be needed to drive it? Assume the efficiency is the same for both pumps. The working fluid is water with $\rho = 1.94$ slug/ft^3.

7.33 Measurements for a model mixed-flow pump with a discharge opening of diameter $D = 1.8$ m operating at $\Omega = 225$ rpm are as listed in the table. Determine the size and rotation speed of a pump with the same dimensionless performance parameters in order to have $Q^* = 6$ m^3/sec and $h_p^* = 18$ m. Make a table of h_p as a function of Q for this pump.

Q (m^3/sec)	h_p (m)	η_p (%)
5.66	18.3	69
7.25	16.8	80
8.58	15.2	86
9.77	13.7	88
11.21	11.4	84
13.00	7.6	64

Problem 7.33

7.34 Larger pumps and turbines are generally more efficient than smaller models of the same design. Moody developed the following empirical formula for estimating the effect:

$$\frac{1 - \eta_2}{1 - \eta_1} = \left(\frac{D_1}{D_2}\right)^{1/4}$$

where η is efficiency, D is diameter, and subscripts identify the two turbomachines. An 88% efficient Francis turbine has an output of 440 kW. A smaller, geometrically-similar turbine has half the volume-flow rate with twice the drop in pressure (head). How much smaller is the turbine and what is the power output, taking account of the reduced efficiency?

7.35 Larger pumps and turbines are generally more efficient than smaller models of the same design. Moody developed the following empirical formula for estimating the effect:

$$\frac{1 - \eta_2}{1 - \eta_1} = \left(\frac{D_1}{D_2}\right)^{1/4}$$

where η is efficiency, D is diameter, and subscripts identify the two turbomachines. Consider a 70% efficient pump with a volume-flow rate of 1000 gal/min against a head of 81 ft. What is the efficiency of a larger, geometrically-similar pump delivering 4000 gal/min against a head of 36 ft?

7.36 A secret research laboratory uses a 95% efficient turbine that develops 25000 hp at 480 rpm under a head of 125 ft. Determine the type of turbine that it is. Assume the working fluid is water with $\rho = 2$ slug/ft^3.

7.37 A power plant uses a 90% efficient turbine that develops 40 MW at 82 rpm under a head of 30 m. What type of turbine is it? Assume the working fluid is water with $\rho = 1000$ kg/m³.

7.38 At its best-efficiency point, the volume-flow rate of a 6-ft radius water turbine is $Q^* = 400$ ft³/sec. Also, it has a power output of $\Omega\tau = 6500$ hp, a best-efficiency-point head of $h_p^* = 615$ ft, and a torque of $\tau = 5 \cdot 10^5$ ft·lbs. The angular-rotation rate is $\Omega = 250$ rpm and the density of water is $\rho = 1.94$ slug/ft³.

(a) What type of turbine is it?

(b) Determine the efficiency of the turbine.

(c) Calculate the capacity and head coefficients.

7.39 A family named Robinson uses a 75% efficient turbine that develops 13 kW at 25 rpm under a head of 5 m. Assuming the working fluid is water with $\rho = 1000$ kg/m³, determine the turbine type.

7.40 For turbines, a useful dimensionless measure of the output power relative to the available head is called the *power-specific speed*, N_{sp}. It increases linearly with rotation speed, Ω, and is independent of rotor diameter, D. Starting with the power coefficient, C_{P*}, and the head coefficient, C_{H*}, deduce N_{sp} as a function of Ω, $\dot{W}_m^*$, ρ, g and h_p^*.

7.41 A Pelton wheel has a diameter $D = 5.05$ m and delivers 15 MW of power. It has a specific speed of $N_s = 0.10$ and operates at 82% efficiency. The diameter of the jet driving the wheel is $d = 35$ cm. You may assume that $\Omega r/V_j = \frac{1}{2}$, where Ω is the rotation rate, V_j is the jet velocity and $r = D/2$. Determine the rotation rate in rpm, assuming the fluid density is $\rho = 1000$ kg/m³.

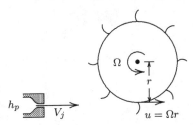

Problems 7.41, 7.42, 7.43, 7.44

7.42 A 10-ft diameter Pelton wheel rotates at 200 rpm. The volume-flow rate of water is 120 ft³/sec, the head is 775 ft and the power delivered is 8400 hp.

(a) Determine the specific speed, N_s and the efficiency, η_t.

(b) If the torque on the wheel is given by Equation (7.77), what is the jet velocity, V_j?

7.43 We have shown from angular-momentum conservation that the torque on a Pelton wheel is given by Equation (7.76). Some researchers use a less-rigorous derivation to conclude that the torque is

$$\tau \approx \rho Q r (1 - \cos\phi)(V_j - \Omega r)$$

For the case $\Omega r = \frac{1}{2}V_j$, we wish to determine the angle ϕ for which τ is a maximum.

(a) Determine the angle and τ_{max} using Equation (7.76)

(b) Determine the angle and τ_{max} using the equation that results from the less-rigorous derivation.

7.44 Imagine that you have been employed as a consultant to design a single-jet Pelton wheel that is 90% efficient with a jet velocity, $V_j = 90$ m/sec. You must use a motor that runs with a torque of $\Omega\tau = 10^7$ N·m. The jet is driven by a water (density $\rho = 998$ kg/m³) reservoir with a head of $h_p = 325$ m. If you arrive at a design with a specific speed of $N_s = 0.11$, determine the angular-rotation rate, Ω, the volume-flow rate, Q, and the radius of the wheel, r. Assume this is the best efficiency point.

7.45 Nozzle losses in a Pelton wheel are accounted for through the empirical formula $V_j = C_v \sqrt{2gh_p}$ where $0.92 \leq C_v \leq 0.98$. The quantity C_v is called the *velocity coefficient*.

(a) Assuming there are no other losses, determine the wheel velocity, Ωr, and maximum efficiency, $(\eta_t)_{max}$, as a function of C_v.

(b) What value of C_v corresponds to $\Omega r / \sqrt{2gh_p} = 0.47$ at the point of maximum efficiency?

(c) How small would C_v have to be to match the value of $(\eta_t)_{max} = 0.8$ indicated in Figure 7.14?

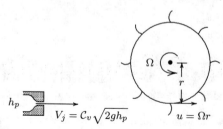

Problem 7.45

Chapter 8

One-Dimensional Compressible Flow

In this chapter, we take up the study of flows in which density variations cannot be ignored. Thus far, most of the fluid-flow applications we have dealt with have been for incompressible fluids. We have tentatively chosen to characterize such flows as having constant density, albeit with a warning in Section 1.6 that an incompressible flow is formally the limiting case of vanishing compressibility, τ. We will show that, for a gas, this occurs in the limit of very small Mach number. The first section provides a basis for classifying the primary compressible-flow regimes in terms of Mach number.

Compressible flow is a fascinating branch of fluid mechanics that introduces us to interesting new physical phenomena. In contrast to incompressible flow, where the energy equation is uncoupled from the equations for mass and momentum conservation, we must now include all three conservation principles to arrive at a closed set of equations. We will also need the equation of state to relate the various thermodynamic properties.

Despite the increased number of equations we have to deal with, we will find the mathematics to be far simpler than in most other branches of fluid mechanics. The latter point is true because a great deal of insight can be gained by analyzing steady, inviscid, one-dimensional flows, most notably **isentropic flow** and **normal shock waves**. Such applications are the primary focus of this chapter. The chapter also includes details of two classical one-dimensional applications, **Fanno flow** and **Rayleigh flow**, which illustrate how viscosity and heat-transfer affect compressible flows.

Examples of compressible flows abound in everyday life. Common compressible-flow applications are an aircraft cruising faster than three-tenths the speed of sound (i.e., approximately 200 mph at sea level) and flow within the standard combustion reciprocating engine. The flow of traffic on a large highway behaves much like flow in a compressible medium. Equally important for the modern comforts we enjoy is flow through rocket engines (to launch communication satellites), gas- or steam-turbine engines (to generate electricity), pipelines at high pressure (to deliver natural gas for heating and cooking) and supersonic wind tunnels (to help design and improve performance of these devices).

There is even an interesting analogy between compressible flows and open-channel flows (Section 6.5). Flow through a sluice gate and a hydraulic jump are analogs of isentropic flow and a shock wave, respectively. Although details are beyond the scope of this chapter, the differential equations governing wave propagation in open channels are identical to those for a compressible fluid with a specific-heat ratio, γ, of 2.

Figure 8.1: *Schlieren photograph of supersonic flow past an airplane model with a piggy-backed vehicle. The photo depicts the shock waves present in the flow. [Courtesy of National Aerospace Laboratories (NLR), Amsterdam, Holland—Used with permission.]*

8.1 Classification

With regard to compressibility, we classify flows according to Mach number, M, where

$$M = \frac{U}{a} = \frac{\text{Flow Speed}}{\text{Sound Speed}} \tag{8.1}$$

There are five different Mach-number regimes, each of which is summarized in Table 8.1. As we will see later in this chapter, most analytical expressions for compressible flows involve the square of the Mach number. Thus, for example, incompressible flow extends all the way to a Mach number of about 0.3. This is true because, to a first approximation, the density in the flow of a gas, ρ, differs from its incompressible value, ρ_t, by a fraction equal to $\frac{1}{2}M^2$ [see Equation (8.60) below]:

$$\rho \approx \rho_t \left(1 - \frac{1}{2}M^2\right) \tag{8.2}$$

When the Mach number is 0.3, this corresponds to a difference of less than 5%. Because the speed of sound is generally very large relative to typical flow velocities, virtually all liquid flows are incompressible. This is not true for gases, however. We will further amplify and quantify this point in Section 9.10 when we develop Bernoulli's equation from the energy-conservation law. In the sections to follow, we will deal mostly with air.

Table 8.1: *Compressible-Flow Regimes*

Mach Number Range	Type of Flow	M^2
0.0 - 0.3	Incompressible	0.00 - 0.09
0.3 - 0.8	Subsonic	0.09 - 0.64
0.8 - 1.2	Transonic	0.64 - 1.44
1.2 - 5.0	Supersonic	1.44 - 25
>5	Hypersonic	>25

The primary difference amongst the various Mach number regimes is in the way waves propagate. Briefly, for incompressible and subsonic flow, waves propagate at a speed much greater than characteristic flow velocities. Consequently, any disturbance in the flow causes waves to move in both the upstream and downstream directions. By contrast, waves move at speeds less than flow velocities in supersonic and hypersonic flow, and flow disturbances can only propagate downstream. The transonic-flow regime is mathematically complicated because it includes regions of both subsonic and supersonic flow. We will illuminate some of the differences amongst the Mach-number regimes in Section 8.5.

Note that the Mach number ranges in Table 8.1 are not intended to mark precise boundaries, but are simply ranges consistent with common nomenclature. Some engineers, for example, consider a Mach 4 flow to be in the low hypersonic range, and many of the approximations that have been developed from hypersonic flow theory are quite accurate for Mach numbers as small as 3. At the lower end of the Mach number spectrum, many Computational Fluid Dynamicists often consider Mach 0.1 to be the upper bound for an accurate incompressible-flow computation.

Example 8.1 *For density variations to become important, the Mach number must be at least 0.3. What flow speed, in mph, must be attained to reach Mach 0.3 in air and in water at 60° F?*

Solution. From Table A.4, the speeds of sound in air and water at 60° F are

$$a_{air} = 1119 \text{ ft/sec} = 763 \text{ mph}$$

$$a_{H_2O} = 4859 \text{ ft/sec} = 3312 \text{ mph}$$

Now the velocity is
$$U = Ma = 0.3a$$

Therefore, the flow speeds required to reach a Mach number of 0.3 in air and in water are

$$U = \begin{cases} 229 \text{ mph}, & \text{air} \\ 994 \text{ mph}, & \text{water} \end{cases}$$

8.2 Thermodynamic Properties of Air

We have discussed thermodynamics in some detail in Section 6.1. We also summarized some common thermodynamic properties of gases in Section 1.5. Because we will be dealing primarily with air in this chapter, it is worthwhile to list its thermodynamic properties here for convenience.

- We assume the thermodynamic properties of air are given by the perfect-gas law, which relates pressure, p, density, ρ, and temperature, T, as follows:

$$p = \rho RT \qquad (8.3)$$

where R = 1716 ft·lb/(slug·°R) [287 J/(kg·K)] is the perfect-gas constant.

- For temperatures up to about 3600° R (2000 K), air is well approximated as being calorically perfect. Thus, its specific-heat coefficients, c_p and c_v, are constant, so that the internal energy, e, and enthalpy, h, are given by

$$e = c_v T \quad \text{and} \quad h = c_p T \qquad (8.4)$$

- As shown in Subsection 6.1.4, the specific-heat coefficients can be expressed in terms of γ and R as follows.

$$c_p = \frac{\gamma}{\gamma - 1}R \quad \text{and} \quad c_v = \frac{1}{\gamma - 1}R \tag{8.5}$$

- Because c_p and c_v are constant, so is the specific-heat ratio, γ. Its value for air is

$$\gamma = \frac{c_p}{c_v} = 1.4 \tag{8.6}$$

- Using the combined first and second laws of thermodynamics (Subsection 6.1.4), we have shown that the entropy, s, for a perfect, calorically-perfect gas is

$$s = c_v \, \ell n \left(\frac{p}{\rho^\gamma} \right) + \text{constant} \tag{8.7}$$

which, for isentropic flow, tells us that

$$p = A\rho^\gamma, \quad A = \text{constant} \qquad \text{(Isentropic flow)} \tag{8.8}$$

Another useful isentropic relation follows from Equation (6.19), which tells us $T/\rho^{\gamma-1}$ is constant for isentropic flow. Thus, $\rho/T^{1/(\gamma-1)} = \text{constant}$, whence

$$p = BT^{\gamma/(\gamma-1)}, \quad B = \text{constant} \qquad \text{(Isentropic flow)} \tag{8.9}$$

8.3   Speed of Sound

We can compute the speed of sound for a perfect gas by analyzing an **acoustic wave**, i.e., a weak sound wave traveling through the gas. We denote the sound speed by a, while the pressure and density of the undisturbed medium are p and ρ. As shown in Figure 8.2(a), after the wave passes, the pressure and density are $p + \Delta p$ and $\rho + \Delta \rho$. Also, the wave imparts a velocity, Δu, to the fluid. For an acoustic wave, we assume the changes in flow properties are very small, i.e.,

$$|\Delta p| \ll p, \quad |\Delta \rho| \ll \rho, \quad |\Delta u| \ll a \tag{8.10}$$

We can simplify our analysis a bit by making a Galilean transformation so that the wave is stationary and the flow is from left to right, as shown in Figure 8.2(b). In this coordinate frame, the velocity upstream (to the left of the wave) is a. Similarly, the velocity downstream (to the right of the wave) is $a - \Delta u$.

To make further progress, we can implement our tried and proven control-volume method. We select a stationary control volume whose boundaries are just upstream and downstream of the wave. Because we are using a one-dimensional geometry, we can select unit area. Since the flow is steady, conservation of mass tells us the net flux of fluid out of the control volume vanishes. Thus,

$$\oiint_S \rho \mathbf{u} \cdot \mathbf{n} \, dS = 0 \tag{8.11}$$

Hence, remembering that $\mathbf{n}$ is always an *outer* unit normal, doing the integrals yields

$$-\rho a + (\rho + \Delta \rho)(a - \Delta u) = 0 \tag{8.12}$$

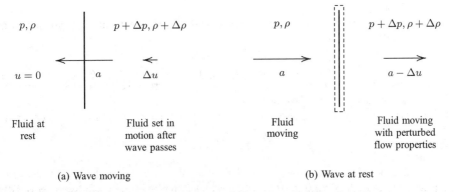

$$p, \rho \qquad p + \Delta p, \rho + \Delta \rho \qquad p, \rho \qquad p + \Delta p, \rho + \Delta \rho$$

$$u = 0 \qquad a \qquad \Delta u \qquad a \qquad a - \Delta u$$

| Fluid at rest | Fluid set in motion after wave passes | Fluid moving | Fluid moving with perturbed flow properties |

(a) Wave moving (b) Wave at rest

Figure 8.2: *Analysis of an acoustic wave; a Galilean transformation produces a steady flow with fluid passing through the stationary wave.*

Now, expanding the terms in parentheses, we have

$$-\rho a + \rho a - \rho \Delta u + a \Delta \rho - \Delta \rho \Delta u = 0 \tag{8.13}$$

so that

$$\rho \Delta u = a \Delta \rho - \Delta \rho \Delta u \tag{8.14}$$

Since we have assumed the wave causes an infinitesimally small change in all flow properties (in particular, since $|\Delta u| \ll a$), we can ignore the last term in Equation (8.14) as it is quadratic in changes while the other two terms are linear. Thus, we conclude from conservation of mass that the changes in velocity and density are related as follows.

$$\rho \Delta u \approx a \Delta \rho \tag{8.15}$$

If we ignore body forces and viscous effects,[1] momentum conservation simplifies to

$$\oiint_S \rho u(\mathbf{u} \cdot \mathbf{n}) \, dS = -\mathbf{i} \cdot \oiint_S p \mathbf{n} \, dS \tag{8.16}$$

where we take the dot product of $\mathbf{i}$ with the net pressure integral to extract its x component. Therefore, expanding the various integrals, we have

$$-\rho a^2 + (\rho + \Delta \rho)(a - \Delta u)^2 = p - (p + \Delta p) \tag{8.17}$$

We can simplify the momentum equation immediately by canceling p, wherefore

$$\rho a^2 = \Delta p + (\rho + \Delta \rho)(a - \Delta u)^2 \tag{8.18}$$

Then, using Equation (8.12), we can simplify this result further, viz.,

$$\rho a^2 = \Delta p + \rho a(a - \Delta u) = \Delta p + \rho a^2 - \rho a \Delta u \tag{8.19}$$

Therefore, conservation of momentum yields the following (exact) relation between the changes in velocity and pressure.

$$\rho a \Delta u = \Delta p \tag{8.20}$$

[1]The latter are in fact zero even in a viscous fluid provided the flow is uniform ahead of and behind the wave.

Finally, we can combine Equations (8.15) and (8.20) to arrive at

$$a^2 \Delta\rho \approx \Delta p \tag{8.21}$$

If we take the limit $\Delta\rho \to 0$, neglecting the higher order term in Equation (8.14) is now seen to be of no consequence. Thus, we can say

$$a^2 = \lim_{\Delta\rho\to 0} \frac{\Delta p}{\Delta\rho} \tag{8.22}$$

The indicated operation must be a partial derivative since all thermodynamic state variables such as sound speed depend upon two other state variables. So, we have to specify what thermodynamic variable is held constant when we differentiate.

Detailed analysis of the viscous-flow equations of motion [cf. Wilcox (2000)] shows that both heat transfer and viscous effects (dissipation) cause entropy changes. Nevertheless, even when we include heat transfer and viscous effects, they are negligible because acoustic signals travel slow enough that large gradients in T and u cannot arise [cf. Liepmann and Roshko (1963)]. This means that entropy changes caused by heat transfer and viscous dissipation, which depend on temperature and velocity gradients, are negligible for an acoustic wave. This is not true for stronger waves, but we are assuming that we have an infinitesimal disturbance. Thus, as we will verify in Chapter 9, the rate of change of entropy following a fluid particle in an adiabatic, inviscid flow is

$$\frac{ds}{dt} = 0 \tag{8.23}$$

This means that an acoustic wave is isentropic, and we have

$$a^2 = \left(\frac{\partial p}{\partial \rho}\right)_s \tag{8.24}$$

where subscript s means the partial derivative is computed with entropy held constant. Equation (8.24) provides the general definition of the speed of sound in a gas.

We can use the isentropic relations for a gas to compute the speed of sound in terms of thermodynamic state variables. Recalling Equation (8.8), differentiation yields

$$a^2 = \gamma A \rho^{\gamma-1} = \gamma \frac{p}{\rho} = \gamma RT \tag{8.25}$$

Therefore, the speed of sound for a perfect gas is

$$a = \sqrt{\gamma RT} \tag{8.26}$$

where the temperature appearing in Equation (8.26) is absolute temperature.

Example 8.2 *Compute the speed of sound on a hot summer day when the temperature is 95° F, and in the middle of winter in Massachusetts when the temperature dips to 5° F.*

Solution. Using Equation (8.26) with $\gamma = 1.4$ and $R = 1716$ ft·lb/(slug·°R), the speed of sound in air under these conditions is

$$a = \begin{cases} 1154 \text{ ft/sec}, & \text{Summer}: \quad (T = 554.67°\text{R}) \\ 1057 \text{ ft/sec}, & \text{Winter}: \quad (T = 464.67°\text{R}) \end{cases}$$

As a final comment, recall from Section 1.6 that the compressibility of a fluid, τ, is defined as follows

$$\tau = \frac{1}{\rho}\frac{\partial \rho}{\partial p} \qquad (8.27)$$

Comparison with Equation (8.24) shows that, for isentropic flow, the compressibility is

$$\tau = \frac{1}{\rho a^2} \qquad (8.28)$$

Thus, since incompressible flow corresponds to the limit $\tau \to 0$, we have shown that incompressible flow also corresponds to the limiting case of very large sound speed. That is, for finite flow velocity, U, the Mach number approaches 0 for $a \gg U$.

8.4 Subsonic Versus Supersonic Flow

There is a fundamental difference between the flow attending motion below and above the speed of sound. Consider a small body moving with velocity U. As the body moves, it collides with molecules that are then set into motion. The molecules transmit information to their neighboring molecules acoustically so that information that the object is coming is transmitted in all directions at the speed of sound. Thus, if the motion is **subsonic** so that $U < a$, the motion appears as shown in Figure 8.3.

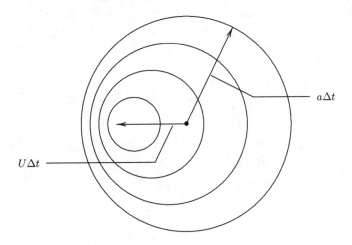

Figure 8.3: *Subsonic motion of a small object.*

Each circle (which is the planar projection of a spherical wave) corresponds to the acoustic wave front of a signal emitted by the object as it moves. The outermost circle is the wave front formed at the beginning of our observations, so that if we have followed the motion for a time Δt, its radius is $a\Delta t$. The inner three circles denote the wave fronts formed at times $\frac{1}{4}\Delta t$, $\frac{1}{2}\Delta t$ and $\frac{3}{4}\Delta t$ with corresponding radii of $\frac{3}{4}a\Delta t$, $\frac{1}{2}a\Delta t$ and $\frac{1}{4}a\Delta t$, respectively. Clearly, because the acoustic wave fronts move faster than the object, the fluid upstream has advance warning that the object is coming. As a result, the fluid particles move aside to allow smooth passage of the object.

Now consider what happens when the velocity of the body is **supersonic** so that $U > a$. As shown in Figure 8.4, when the body collides with a molecule, there can be no advance warning. This is true because the body is moving faster than the velocity at which molecules transmit information. Only molecules within a conical region behind the body have any knowledge that the body is passing through the fluid.

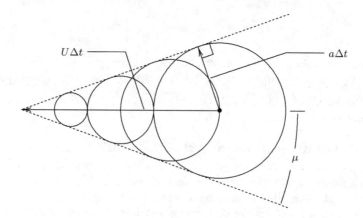

Figure 8.4: *Supersonic motion of a small object.*

The sides of the cone, whose planar projection is a wedge, are tangent to the family of spherical wave fronts. They correspond to a propagating wave front known as a **Mach wave**. The region downstream of the conical wave front is referred to as the **Mach cone**. The half angle of the Mach cone, μ, called the **Mach angle**, is readily computed from its geometry. Inspection of Figure 8.4 shows that the sine of μ is the ratio of distance traveled by the wave front, $a\Delta t$, to the distance traveled by the object, $U\Delta t$. Therefore,

$$\mu = \sin^{-1}\left(\frac{a}{U}\right) = \sin^{-1}\left(\frac{1}{M}\right) \tag{8.29}$$

Example 8.3 *If the Mach angle is $\mu = 30°$, what is the Mach number?*

Solution. From Equation (8.29), the Mach number is

$$M = \frac{1}{\sin \mu} = 2$$

8.5   Analysis of a Streamtube

We can discover some of the interesting differences amongst the various Mach-number ranges by analyzing adiabatic, inviscid flow of a compressible fluid in a **streamtube** of varying area. By definition, a streamtube is a tube whose sides are streamlines (Figure 8.5). Note that because we have assumed the flow is inviscid, the sides of the tube could even be solid boundaries. The object of the following control-volume analysis is to express differential changes in velocity, du, and density, $d\rho$, in terms of the Mach number, M, and the change in area, dA. When we have developed the desired relations, which are generally referred to as the **streamtube equations**, we will be in a position to draw some useful conclusions, and to illuminate some of the interesting properties and idiosyncrasies of compressible flows.

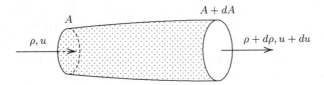

Figure 8.5: *Schematic of a streamtube.*

For steady flow, focusing on a stationary control volume coincident with the streamtube, we know from the integral form of the mass-conservation law that the net mass flux out of the streamtube vanishes, i.e.,

$$\oiint_S \rho\, \mathbf{u} \cdot \mathbf{n}\, dS = 0 \tag{8.30}$$

Expanding the closed-surface integral into its three contributions at the inlet, the outlet and on the perimeter, we have

$$\iint_A \rho\, \mathbf{u} \cdot \mathbf{n}\, dS + \iint_{A+dA} \rho\, \mathbf{u} \cdot \mathbf{n}\, dS + \iint_{Perimeter} \rho\, \mathbf{u} \cdot \mathbf{n}\, dS = 0 \tag{8.31}$$

By definition of a streamline, no fluid flows across the perimeter. Also, since $\mathbf{n}$ is an outer unit normal, $\mathbf{u} \cdot \mathbf{n} = -u$ at the inlet while $\mathbf{u} \cdot \mathbf{n} = u$ at the outlet. Thus, we conclude that

$$\iint_A \rho\, u\, dS = \iint_{A+dA} \rho\, u\, dS \tag{8.32}$$

Finally, we treat flow through our streamtube as a **quasi-one-dimensional** flow, i.e., we regard all flow properties as being constant on cross sections. In practice, this requires A to vary only slowly with x. Hence, Equation (8.32) tells us

$$\rho u A = (\rho + d\rho)(u + du)(A + dA) \tag{8.33}$$

Expanding the right-hand side yields

$$\rho u A = \rho u A + u A d\rho + \rho A du + \rho u dA + \text{(higher order terms)} \tag{8.34}$$

where the "higher order terms" are quadratic and cubic in differential changes. Dropping higher order terms, canceling $\rho u A$ and dividing through by $\rho u A$ gives

$$\frac{d\rho}{\rho} + \frac{du}{u} + \frac{dA}{A} = 0 \tag{8.35}$$

Now, consider momentum conservation. We have a balance between the net flux of x momentum and the x component of the net pressure integral so that

$$\oiint_S \rho u(\mathbf{u} \cdot \mathbf{n})\, dS = -\mathbf{i} \cdot \oiint_S p\, \mathbf{n}\, dS \tag{8.36}$$

Hence, expanding the closed-surface integrals and noting the proper orientation of outer unit normals, we have

$$-\iint_A \rho\, u^2\, dS + \iint_{A+dA} \rho\, u^2\, dS + \iint_{Perimeter} \rho\, u(\mathbf{u} \cdot \mathbf{n})\, dS$$
$$= \iint_A p\, dS - \iint_{A+dA} p\, dS - \iint_{Perimeter} p\, (\mathbf{i} \cdot \mathbf{n})\, dS \tag{8.37}$$

Hence, again regarding the flow as quasi-one-dimensional, we have

$$-\rho u^2 A + (\rho + d\rho)(u + du)^2(A + dA) = pA - (p + dp)(A + dA)$$
$$- \iint_{Perimeter} p\,(\mathbf{i} \cdot \mathbf{n})\,dS \qquad (8.38)$$

In order to compute the integral of $p\,(\mathbf{i} \cdot \mathbf{n})$ on the perimeter, it is instructive to examine a side view of the streamtube as shown in Figure 8.6. Inspection of the geometry shows that the quantity $-(\mathbf{i} \cdot \mathbf{n})dS$ is just the projection of the streamline on the vertical plane. However, this projection is also the differential area change, so that

$$dA = -(\mathbf{i} \cdot \mathbf{n})\,dS \qquad (8.39)$$

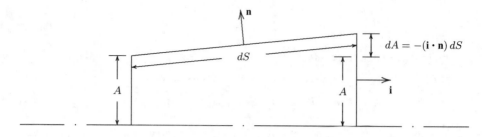

Figure 8.6: *Side view of the streamtube.*

Thus, noting Equation (8.39), we can evaluate the integral on the right-hand side of Equation (8.38). There follows

$$- \iint_{Perimeter} p\,(\mathbf{i} \cdot \mathbf{n})\,dS = pdA \qquad (8.40)$$

Also, note that by making use of Equation (8.33), we have

$$(\rho + d\rho)(u + du)^2(A + dA) = \rho u A(u + du) \qquad (8.41)$$

Substituting Equations (8.40) and (8.41) into Equation (8.38), momentum conservation tells us

$$-\rho u^2 A + \rho u A(u + du) = pA - (p + dp)(A + dA) + pdA \qquad (8.42)$$

Expanding all terms in parentheses and canceling like terms yields

$$\rho u A du = -A dp - dp dA \qquad (8.43)$$

Finally, dropping the $dpdA$ term, which is of higher order, and dividing through by A, we arrive at the desired differential form of the momentum equation, viz.,

$$\rho u du = -dp \qquad (8.44)$$

We now have sufficient information to derive equations for the differential changes in density and velocity. First, we must eliminate the pressure. From momentum conservation [Equation (8.44)], we have

$$udu = -\frac{dp}{\rho} = -\frac{dp}{d\rho}\frac{d\rho}{\rho} = -a^2\frac{d\rho}{\rho} \qquad (8.45)$$

Note that the final equality in Equation (8.45) holds because the flow is adiabatic and inviscid, and must therefore be isentropic. Hence, $dp/d\rho = (\partial p/\partial \rho)_s = a^2$. We can thus express $d\rho/\rho$ as a function of du/u. Using $M^2 = u^2/a^2$, we have

$$\frac{d\rho}{\rho} = -\frac{u}{a^2}du = -M^2\frac{du}{u} \tag{8.46}$$

Substituting this result into Equation (8.35) then yields

$$-M^2\frac{du}{u} + \frac{du}{u} + \frac{dA}{A} = 0 \tag{8.47}$$

Hence, using Equations (8.46) and (8.47), we arrive at the desired **streamtube equations**, viz.,

$$\frac{du}{u} = -\frac{1}{1-M^2}\frac{dA}{A} \quad \text{and} \quad \frac{d\rho}{\rho} = \frac{M^2}{1-M^2}\frac{dA}{A} \tag{8.48}$$

We can use these differential relations to analyze the various compressible-flow regimes.

Because there is a fundamental difference between subsonic and supersonic flow, we will examine behavior predicted by the streamtube equations in these two cases. Additionally, we will analyze the limiting behavior for incompressible flow ($M \to 0$) and for sonic conditions ($M \to 1$).

Incompressible Flow, $M \to 0$: When the speed of sound is so large that the Mach number becomes very small, we have incompressible flow. Then, Equation (8.48) tells us the relation between changes in velocity and area is

$$\frac{du}{u} = -\frac{dA}{A} \tag{8.49}$$

Because the differential changes in u and A differ in sign, a decrease in area yields an increase in velocity, and vice versa. Clearly, this must occur in order to have constant mass flow through each cross section.

Subsonic Flow, $M < 1$: Inspection of Equation (8.48) shows that du still follows $-dA$, although the change in u is greater than in the incompressible case because $1/(1-M^2) > 1$. The velocity must change more than it does for $M \to 0$ because the density also changes. For example, when the area decreases, the density decreases. Thus, to conserve mass, the additional increase in velocity (relative to the incompressible case) attends the reduction in fluid density. A similar argument holds for increasing area.

Supersonic Flow, $M > 1$: When the Mach number exceeds 1, it is illuminating to rewrite Equation (8.48) as follows.

$$\frac{du}{u} = \frac{1}{M^2-1}\frac{dA}{A} \quad \text{and} \quad \frac{d\rho}{\rho} = -\frac{M^2}{M^2-1}\frac{dA}{A} \tag{8.50}$$

Now we find that du has the same sign as dA so that a decrease in area yields a decrease in velocity, while an increase in area corresponds to an increase in velocity. This is quite different from what occurs for $M < 1$. On the one hand, for decreasing area, ρ rises so rapidly that u must decrease in order to conserve mass. This is very similar to our everyday observations. Consider what happens when an accident causes the loss of a lane on a highway, resulting in "bumper-to-bumper" traffic. The density of automobiles increases so much that the same flux of automobiles can occur only at reduced speed. On the other hand, for increasing area, ρ drops so rapidly that u must increase. This is similar to what happens once the automobiles pass beyond the accident and the lost lane is regained.

The Sonic Point, M = 1: We are left with an interesting question of what happens when the Mach number is exactly 1. Both of Equations (8.48) have $1 - M^2$ in the denominator. Hence, since singular behavior does not occur in real fluid flows, the numerators must also vanish. Thus, we can only have $M = 1$ at a point where $dA = 0$, i.e., a point of minimum or maximum area. When it's a minimum, we call such a point a **throat**. However, the inverse is not true. We do not necessarily have $M = 1$ at a throat. If $M \neq 1$ at a throat, then $du = 0$, so that u attains either a minimum or a maximum value at the throat.

8.6 Total Conditions

Again, we consider steady flow of a frictionless gas with no heat transfer. In the absence of body forces, energy conservation for the streamtube depicted in Figure 8.5 simplifies to

$$\oiint_S \rho \left(h + \frac{1}{2}\mathbf{u} \cdot \mathbf{u} \right) (\mathbf{u} \cdot \mathbf{n}) \, dS = 0 \quad \Longrightarrow \quad \rho(h + \frac{1}{2}u^2) u A = \text{constant} \tag{8.51}$$

That is, the total enthalpy flowing in where the area is A equals the total enthalpy flowing out where the area is $A + dA$. Since mass conservation tells us $\rho u A = \text{constant}$, we conclude immediately that

$$h + \frac{1}{2}u^2 = \text{constant} \tag{8.52}$$

Assuming the gas is calorically perfect,

$$h = c_p T \quad \text{and} \quad c_p = \text{constant} \quad \Longrightarrow \quad T + \frac{u^2}{2c_p} = \text{constant} \tag{8.53}$$

We can simplify further by recalling Equations (8.5) and (8.26), so that

$$c_p = \frac{\gamma R}{\gamma - 1} = \frac{\gamma R T}{(\gamma - 1)T} = \frac{a^2}{(\gamma - 1)T} \tag{8.54}$$

So, our energy-conservation equation becomes

$$T + \frac{\gamma - 1}{2} \frac{u^2}{a^2} T = \text{constant} \tag{8.55}$$

Thus, introducing Mach number, we conclude finally that

$$T \left[1 + \frac{\gamma - 1}{2} M^2 \right] = \text{constant} \tag{8.56}$$

The constant appearing in Equation (8.56) is so important in compressible-flow theory that we give it the special name **total temperature**, and denote it by T_t. Therefore, we define the total temperature by

$$T_t = T \left[1 + \frac{\gamma - 1}{2} M^2 \right] \tag{8.57}$$

Total temperature is also referred to as **stagnation temperature** and **reservoir temperature**. It can be thought of as the stagnation temperature in the sense that the fluid temperature would be T_t if it were brought to rest isentropically. This is not true at a stagnation point in a real fluid however, as viscous effects cause losses that reduce the temperature somewhat. Hence, calling T_t stagnation temperature is a bit of a misnomer. It can be thought of as the reservoir temperature in a typical supersonic wind tunnel (Figure 8.7). Gas is pressurized and maintained at a temperature T_t in a large reservoir. The flow then moves through a small nozzle and accelerates from rest to the wind-tunnel design Mach number.

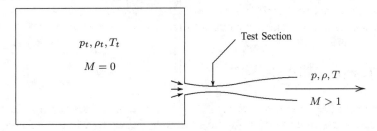

Figure 8.7: *Schematic of a supersonic wind tunnel.*

We can also define **total pressure**, p_t, and **total density**, ρ_t, from the isentropic relations [see Equations (8.8) and (8.9)]. That is,

$$\frac{p_t}{p} = \left(\frac{\rho_t}{\rho}\right)^\gamma = \left(\frac{T_t}{T}\right)^{\gamma/(\gamma-1)} \tag{8.58}$$

Hence, substituting Equation (8.57) into Equation (8.58), there follows

$$p_t = p\left[1 + \frac{\gamma-1}{2}M^2\right]^{\gamma/(\gamma-1)} \tag{8.59}$$

$$\rho_t = \rho\left[1 + \frac{\gamma-1}{2}M^2\right]^{1/(\gamma-1)} \tag{8.60}$$

Like the total temperature, the total pressure and total density are constant in isentropic flow. Table B.1 in Appendix B includes p/p_t, ρ/ρ_t and T/T_t as a function of Mach number for $\gamma = 1.4$.

Example 8.4 *Air is flowing at a Mach number of 0.7, and the freestream temperature, T, is $20°\ C = 293.16\ K$. Compute the temperature the fluid would have if we brought it to rest isentropically.*

Solution. If we bring the fluid to rest isentropically, its temperature will be equal to the total temperature, T_t. From Isentropic Flow Table B.1, for $M = 0.7$ we find

$$\frac{T}{T_t} = 0.9107$$

Thus, the total temperature is

$$T_t = \frac{T}{0.9107} = \frac{293.16\ K}{0.9107} = 321.9\ K = 48.7°\ C$$

8.7   Normal Shock Waves

In Section 8.3 we analyzed an acoustic wave, which is very weak and produces infinitesimally small changes in fluid properties. Not all waves in a compressible medium are weak. The **sonic boom** produced by a supersonic aircraft, for example, is caused by a strong pressure wave known as a **shock wave**. Shock waves produce finite changes in flow properties and, as we will discuss in this section, the changes occur over an extremely thin region.

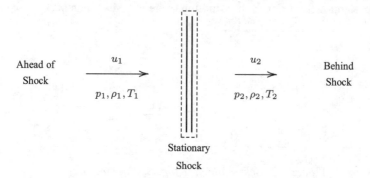

Figure 8.8: *Shock wave and associated terminology.*

Consider a one-dimensional wave propagating into an undisturbed fluid as illustrated in Figure 8.8 (see also Figure 8.9). The pressure, density and temperature of the undisturbed fluid are p_1, ρ_1 and T_1, respectively. If the wave propagates to the left with a wave speed of u_1, we can use a Galilean transformation to put the wave at rest, and the fluid **ahead of the shock** (i.e., the undisturbed fluid) moves to the right with speed u_1. After the fluid passes through the wave, the flow properties are p_2, ρ_2, T_2 and u_2. We refer to the latter region as lying **behind the shock**.

We will assume that the fluid is inviscid and no heat transfer or body forces are present. Because the flow is steady, conservation of mass, momentum and energy for the indicated stationary control volume simplify to the following.

$$\rho_1 u_1 = \rho_2 u_2 \tag{8.61}$$

$$\rho_1 u_1^2 + p_1 = \rho_2 u_2^2 + p_2 \tag{8.62}$$

$$h_1 + \frac{1}{2}u_1^2 = h_2 + \frac{1}{2}u_2^2 \tag{8.63}$$

We can solve these equations with a little algebra and we find two solutions. The first solution has $p_2 = p_1$, $\rho_2 = \rho_1$, $T_2 = T_1$ and $u_2 = u_1$, which corresponds to no wave, whereby the solution is continuous in a mathematical sense. By contrast, the second solution has all flow properties changing from Region 1 to Region 2. This is a rather profound conclusion because we have specified nothing regarding the width of our control volume or the wave. That is, the second solution allows for solution discontinuities in all flow properties. This corresponds to a shock wave, and is consistent with measurements.

The width of a shock wave is a few **mean free paths**, where the mean free path is the average distance a molecule travels before suffering a collision with another molecule. Since the continuum approximation regards molecular dimensions as negligibly small, a shock wave has zero width from a continuum point of view. The second solution is most conveniently

Figure 8.9: *Shadowgraph of a normal shock wave moving from right to left in a shock tube. Shock thickness is a few mean free paths. [Courtesy of H. Liepmann and A. Roshko (1963).]*

written in terms of the Mach number ahead of the shock, M_1. The solution gives what are known as the **normal-shock relations**, viz.,

$$M_2^2 = \frac{1 + \frac{\gamma - 1}{2}M_1^2}{\gamma M_1^2 - \frac{\gamma - 1}{2}} \tag{8.64}$$

$$\frac{\rho_2}{\rho_1} = \frac{u_1}{u_2} = \frac{(\gamma + 1)M_1^2}{2 + (\gamma - 1)M_1^2} \tag{8.65}$$

$$\frac{p_2}{p_1} = 1 + \frac{2\gamma}{\gamma + 1}\left(M_1^2 - 1\right) \tag{8.66}$$

$$\frac{T_2}{T_1} = 1 + \frac{2(\gamma - 1)}{(\gamma + 1)^2}\left(\frac{1 + \gamma M_1^2}{M_1^2}\right)\left(M_1^2 - 1\right) \tag{8.67}$$

Shock waves occur in nature whenever we have supersonic flow past objects. They also occur for supersonic flow within practical fluid-mechanical devices such as jet engines and rocket nozzles. As a final comment, note that the dimensionless grouping

$$\frac{\Delta p}{p_1} \equiv \frac{p_2 - p_1}{p_1} = \frac{2\gamma}{\gamma + 1}\left(M_1^2 - 1\right) \tag{8.68}$$

is known as the **shock strength**. Clearly, the strength of a shock wave increases as the Mach number increases.

Example 8.5 *Suppose we have a normal shock wave in a Mach 3 flow of carbon dioxide with an ambient pressure of 1 atm and an ambient temperature of 50° F = 509.67° R. Calculate the Mach number, M_2, static pressure, p_2, and static temperature, T_2, behind the shock.*

Solution. Using the normal-shock relations given in Equations (8.64) through (8.67) with carbon dioxide's specific-heat ratio of $\gamma = 1.3$, we find

$$M_2 = 0.4511, \quad \frac{p_2}{p_1} = 10.043 \quad \text{and} \quad \frac{T_2}{T_1} = 2.280$$

Therefore, behind the shock wave, we have (with a little rounding):

$$M_2 = 0.451, \quad p_2 = 10.0 \text{ atm}, \quad T_2 = 1162°\text{R} = 702°\text{F}$$

For gases with a γ of 1.4, we can use the Shock Tables of Section B.1 in Appendix B to determine conditions behind a shock wave. The tables include the normal-shock relations as a function of Mach number.

Example 8.6 *Suppose we have a normal shock wave in a Mach 3 airflow with an ambient pressure of 1 atm and an ambient temperature of 50° F = 509.67° R. Calculate the Mach number, M_2, static pressure, p_2, and static temperature, T_2, behind the shock.*

Solution. Reference to the shock tables shows that for $M_1 = 3.00$, conditions behind the shock are

$$M_2 = 0.4752, \quad \frac{p_2}{p_1} = 10.333 \quad \text{and} \quad \frac{T_2}{T_1} = 2.679$$

Therefore, behind the shock wave, we have (with a little rounding):

$$M_2 = 0.475, \quad p_2 = 10.3 \text{ atm}, \quad T_2 = 1365°\text{R} = 906°\text{F}$$

8.8   Directionality

Equation (8.64) has an interesting property. Specifically, a simple algebraic exercise shows that it is still valid if we interchange indices 1 and 2. This suggests that if we have a solution that causes the Mach number, M_1, to change to M_2, then there is also a solution with M_2 ahead of the shock and M_1 behind the shock. Figure 8.10 illustrates the variation of M_2 with M_1. When the flow ahead of the shock is supersonic ($M_1 > 1$), the flow behind the shock is subsonic ($M_2 < 1$), and vice versa.

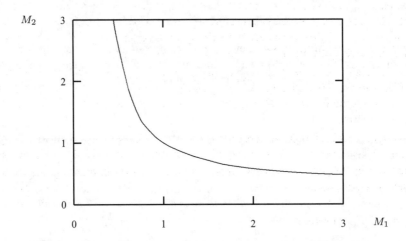

Figure 8.10: *Variation of M_2 with M_1 for $\gamma = 1.4$.*

Thus, there are two possibilities regarding the nature of the shock wave. The first possibility is a transition from supersonic flow ahead of the shock to subsonic flow behind the shock. Since the shock relations show that the density increases for $M_1 > 1$, we call this a **compression shock**. The second possibility is a transition from subsonic flow ahead of the shock to supersonic flow behind the shock so that $M_1 < 1$. In this case, the density decreases and we call it an **expansion shock**.

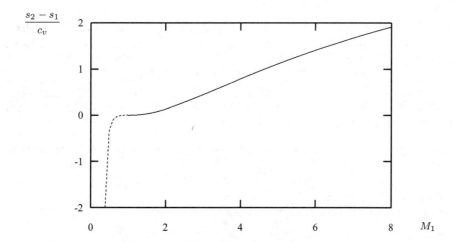

Figure 8.11: *Entropy change across a shock wave.*

The second case, the expansion shock, does not occur in nature because it would result in a decrease in entropy. This is, of course, in conflict with the second law of thermodynamics, which tells us that, in a closed system, entropy is always increasing. Substituting Equations (8.65) and (8.66) into Equation (8.7) yields the entropy change across a shock, viz.,

$$\frac{s_2 - s_1}{c_v} = \ell n \left[1 + \frac{2\gamma}{\gamma + 1} \left(M_1^2 - 1 \right) \right] - \gamma \ell n \left[\frac{(\gamma + 1)M_1^2}{2 + (\gamma - 1)M_1^2} \right] \qquad (8.69)$$

Examination of Equation (8.69) shows that $s_2 = s_1$ when $M_1 = 1$ and $s_2 > s_1$ for large values of M_1. However, because of its algebraic complexity, it is not obvious how the entropy changes when $M_1 < 1$. We can make further progress by examining a graph of entropy change across a shock as a function of Mach number.

Figure 8.11 shows the dimensionless change in entropy as a function of Mach number ahead of the shock, M_1. Clearly, $M_1 < 1$ yields $s_2 - s_1 < 0$, which is forbidden by the second law of thermodynamics. By contrast, $s_2 - s_1 > 0$ for $M_1 > 1$, which is consistent with the second law.

These arguments establish the **directionality** of a shock wave. Obviously it always causes a transition from supersonic flow ahead of the shock to subsonic flow behind the shock. Knowing that shocks always have $M_1 > 1$, we can now determine how all flow properties change across a shock:

- p, ρ and T increase

- T_t is constant

- u, p_t and ρ_t decrease

Because total enthalpy is conserved across a shock, necessarily the total temperature is constant. However, the shock converts some of the flow's kinetic energy into thermal energy in such a way that the total pressure and total density decrease.

8.9   Laval Nozzle

We now turn our focus to the **Laval nozzle**, which is the basis of the supersonic wind tunnel. Figure 8.12 illustrates a nozzle with subsonic flow upstream of the throat. Our goal is to arrive at a nozzle that achieves sonic flow ($M = 1$) at the throat and permits acceleration to supersonic flow. Since expansion shocks do not exist, we must find a way to accelerate the flow isentropically.

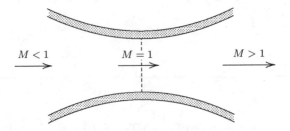

Figure 8.12: *Flow near the throat of a Laval nozzle.*

We can analyze this flow using control-volume methods, just as we did in analyzing streamtube flow in Section 8.5. However, we have no need to deal with differential forms. Rather, we deal directly with mass conservation in its integrated form, which tells us that

$$\rho u A = \rho^* u^* A^* \tag{8.70}$$

where superscript * denotes conditions at the throat. So, the ratio of the nozzle area to the throat area is

$$\frac{A}{A^*} = \frac{\rho^* u^*}{\rho u} = \frac{\rho^*}{\rho} \frac{a^* M^*}{a M} \tag{8.71}$$

Then, noting that $a = \sqrt{\gamma R T}$ for a perfect gas [Equation (8.26)], we have

$$\frac{A}{A^*} = \frac{\rho^*}{\rho} \sqrt{\frac{T^*}{T}} \frac{M^*}{M} \tag{8.72}$$

If our nozzle achieves sonic conditions at the throat, necessarily

$$M^* = 1 \qquad \text{(Sonic conditions at throat)} \tag{8.73}$$

Therefore, the area ratio is given by

$$\frac{A}{A^*} = \frac{\rho^*}{\rho} \sqrt{\frac{T^*}{T}} \frac{1}{M} \tag{8.74}$$

At this point, our formulation is specific to the case of a nozzle that indeed achieves sonic conditions at the throat. However, Equation (8.74) still applies even if the nozzle is operating at off-design conditions (i.e., for $M^* < 1$). In this case we cannot say that A^* is the physical area of the throat. Rather, it is a **fictitious area** whose value represents the area that would exist in the nozzle if the flow were accelerated isentropically to Mach 1.

Now, since the flow is isentropic, the total temperature and total density are both constant, wherefore

$$\left.\begin{aligned}
T^* \left[1 + \frac{\gamma - 1}{2} M^{*2}\right] &= T \left[1 + \frac{\gamma - 1}{2} M^2\right] \\[2ex]
\rho^* \left[1 + \frac{\gamma - 1}{2} M^{*2}\right]^{\frac{1}{\gamma - 1}} &= \rho \left[1 + \frac{\gamma - 1}{2} M^2\right]^{\frac{1}{\gamma - 1}}
\end{aligned}\right\} \tag{8.75}$$

Then, using the fact that, by hypothesis, $M^* = 1$, we have

$$\left. \begin{array}{l} \dfrac{T^*}{T} = \dfrac{2}{\gamma + 1}\left[1 + \dfrac{\gamma - 1}{2}M^2\right] \\[3mm] \dfrac{\rho^*}{\rho} = \left\{\dfrac{2}{\gamma + 1}\left[1 + \dfrac{\gamma - 1}{2}M^2\right]\right\}^{\frac{1}{\gamma - 1}} \end{array} \right\} \qquad (8.76)$$

Substituting Equations (8.76) into Equation (8.74) yields the following equation for the area ratio.

$$\frac{A}{A^*} = \frac{1}{M}\left\{\frac{2}{\gamma + 1}\left[1 + \frac{\gamma - 1}{2}M^2\right]\right\}^{\frac{\gamma + 1}{2(\gamma - 1)}} \qquad (8.77)$$

This equation yields the area ratio as a function of Mach number for *isentropic* flow. As with the isentropic relations for p/p_t, ρ/ρ_t and T/T_t, Table B.1 in Appendix B includes A/A^* as a function of Mach number. Inspection of Equation (8.77) reveals some interesting limiting values of the reciprocal of A/A^*.

$$A^*/A = 0 \text{ when } M = 0, \quad A^*/A = 1 \text{ when } M = 1, \quad A^*/A \to 0 \text{ as } M \to \infty \quad (8.78)$$

Figure 8.13 presents the variation of A^*/A with M for a specific-heat ratio, $\gamma = 1.4$. As shown, the curve is double valued, featuring a subsonic and a supersonic branch.

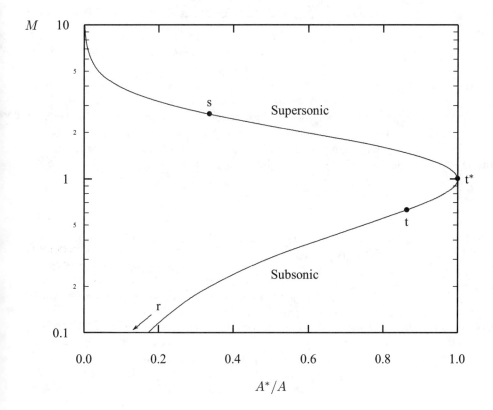

Figure 8.13: *Area-ratio variation with Mach number for $\gamma = 1.4$.*

Subsonic Case. Let Point r denote reservoir conditions and Point t the throat. If the pressure at the nozzle exit plane is too high relative to the reservoir (total) pressure, sonic conditions will not be attained at the throat. Flow will remain subsonic throughout the nozzle. With reference to Figure 8.13, we move along the subsonic branch from r up to t and return along the same branch. Cases a and b in Figure 8.14 correspond to the subsonic case. Note that the Mach number at the throat increases as the exit pressure decreases.

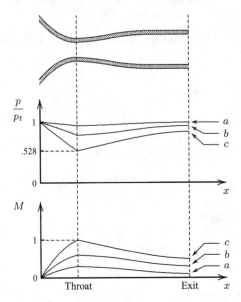

Figure 8.14: *Pressure and Mach number for subsonic flow;* $\gamma = 1.4$.

Clearly, the flow throughout the nozzle is controlled by the exit pressure. Also, the mass flow through the nozzle, $\dot{m} = \rho u A$, increases in a monotone fashion with decreasing exit pressure. As we continue to reduce the pressure at the nozzle exit, we eventually reach a value, p_c, for which the flow is sonic at the throat. In this case, we reach Point t* in Figure 8.13. We can compute the pressure at the throat when sonic conditions are realized by substituting $M = 1$ into Equation (8.59). We thus have

$$\frac{p^*}{p_t} = \left(\frac{2}{\gamma+1}\right)^{\frac{\gamma}{\gamma-1}} = 0.528 \qquad (\gamma = 1.4) \tag{8.79}$$

When the flow at the throat of a Laval nozzle is sonic, we have a condition referred to as **choked flow**. Once the flow is choked, any further decrease in the exit pressure yields no change in the mass-flow rate, $\dot{m}$, through the nozzle! This is true because the mass-flow rate in the nozzle is given by

$$\dot{m} = \rho u A = \rho M \sqrt{\gamma R T}\, A \tag{8.80}$$

For fixed total conditions, changing the exit pressure, p_{exit}, will not change the Mach number at the throat, where the flow must remain at $M^* = 1$. Now, the density and temperature are functions only of Mach number [Equation (8.76)]. Hence, when the flow is choked, ρ and T are fixed at sonic-flow values, regardless of the exit pressure. Since the throat area is a fixed feature of the geometry, we conclude that mass flow is unaffected by p_{exit} once the flow is choked, i.e.,

$$d\dot{m}/dp_{exit} = 0 \qquad \text{for choked flow} \tag{8.81}$$

Sonic Case. For the case where the throat just goes sonic, a second solution exists, albeit with a much lower exit pressure, p_e. Figure 8.15(a) includes the two solutions. As shown, for the supersonic solution, the flow continues expanding after the throat, and we follow the supersonic branch to Point s in Figure 8.13. This is known as the **design case**, and the flow is completely shock free.

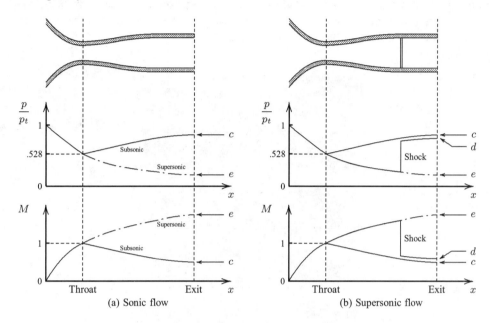

Figure 8.15: *Pressure and Mach number for sonic and supersonic flow;* $\gamma = 1.4$.

Supersonic Case. For any other exit pressures below p_c and above p_e, no isentropic solutions exist. Rather, as illustrated in Figure 8.15(b), either a normal shock wave appears at some point in the nozzle (Case d) or an oblique shock (i.e., a shock inclined at a slanting angle to the flow direction) appears outside the nozzle. This case is referred to as **overexpanded**. Finally, if the pressure is less than p_e, a series of nonisentropic expansion waves occur at the nozzle exit, and the flow is referred to as **underexpanded**. For further details, see any of the excellent texts on compressible flow such as Shapiro (1953), Liepmann and Roshko (1963) or Anderson (1990).

Example 8.7 *A supersonic wind tunnel is designed to have Mach 2 airflow in the test section with conditions corresponding to the U. S. Standard Atmosphere at sea level, i.e., $p_{exit} = 1$ atm and $T_{exit} = 288$ K. Determine the exit-area ratio and the reservoir conditions needed to achieve this design.*

Solution. From the Isentropic Flow and Shock Tables in Section B.1 of Appendix B, for Mach 2 we find

$$\frac{A_{exit}}{A^*} = 1.6875, \quad \frac{p_{exit}}{p_t} = 0.1278, \quad \frac{T_{exit}}{T_t} = 0.5556$$

Thus, rounding to three significant figures:

$$A_{exit} = 1.69A^*, \quad p_t = 7.82 \text{ atm}, \quad T_t = 518 \text{ K}$$

Figure 8.16 includes "schlieren" photographs of flow from a supersonic nozzle at a series of exit pressures. The schlieren system is an optical technique that responds to density changes in a flow, and provides a sharp image on a photographic plate. Because density changes abruptly across a shock wave, a schlieren photograph provides an excellent flow-visualization tool for illustrating shocks.

Figure 8.17 illustrates the diamond-shaped shock-wave patterns emitted from the nozzles of missiles. In a painting, the artist has accurately captured this characteristic feature of the flow in the wake of the nozzle.

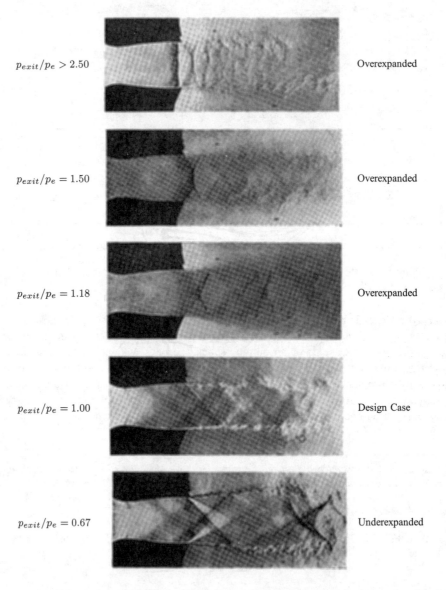

$p_{exit}/p_e > 2.50$ — Overexpanded

$p_{exit}/p_e = 1.50$ — Overexpanded

$p_{exit}/p_e = 1.18$ — Overexpanded

$p_{exit}/p_e = 1.00$ — Design Case

$p_{exit}/p_e = 0.67$ — Underexpanded

Figure 8.16: *Schlieren photographs of flow from a supersonic nozzle for a series of exit pressures. p_e is the design-case exit pressure. [Courtesy of H. Liepmann and A. Roshko]*

Figure 8.17: *Joint Strike Fighter firing missiles with diamond-shaped shock-wave patterns in their exhaust streams. [Painting courtesy of Aviation Week and Space Technology]*

8.10 Fanno Flow

As a simple example of a compressible flow in which viscous effects play a nontrivial role, we consider steady, adiabatic flow in a pipe or duct with constant area, A. This is known as **Fanno flow**, and is depicted schematically in Figure 8.18. To formulate a simplified one-dimensional set of differential equations for averaged properties on cross sections, it is most convenient to begin with the integral form of the conservation laws. The boundary of the control volume we will use is shown as the dashed rectangle in the figure. The control volume is fixed in space and is coincident with the pipe/duct walls.

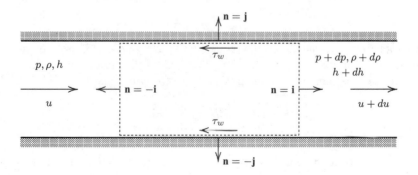

Figure 8.18: *Stationary control volume for Fanno flow; cross-sectional area, A, is constant.*

8.10.1   Fanno-Flow Equations

Mass Conservation. Since we are considering steady flow through a stationary control volume, the mass-conservation principle tells us the rate of change of mass contained in the control volume is zero. Hence, the mass-conservation equation simplifies to

$$\oiint_S \rho\mathbf{u}\cdot\mathbf{n}\,dS = 0 \tag{8.82}$$

Because no fluid flows through the walls of the pipe or duct, the mass flux into the control volume balances the mass flux out. So,

$$-\rho u A + (\rho + d\rho)(u + du)A = 0 \quad\Longrightarrow\quad \rho u = (\rho + d\rho)(u + du) \tag{8.83}$$

Expanding the right-hand side and dropping the higher-order term, $d\rho du$, we find

$$\rho du + u d\rho = d(\rho u) = 0 \tag{8.84}$$

Thus, the differential form of the mass-conservation equation assumes the following form.

$$\frac{d}{dx}(\rho u) = 0 \tag{8.85}$$

Momentum Conservation. Since we have one-dimensional motion, only the x component of the momentum equation is of interest. Appealing to the integral-conservation principle with body forces omitted and including a viscous force that we denote by $\mathbf{f}_v$ [see Equation (6.29)], we have

$$\oiint_S \rho u\,(\mathbf{u}\cdot\mathbf{n})\,dS = -\mathbf{i}\cdot\oiint_S p\mathbf{n}\,dS + \mathbf{i}\cdot\oiint_S \mathbf{f}_v\,dS \tag{8.86}$$

It is most convenient to proceed term by term in evaluating the various surface integrals in this equation. The net flux of momentum is

$$\begin{aligned}
\oiint_S \rho u\,(\mathbf{u}\cdot\mathbf{n})\,dS &= \rho u(-u)A + \underbrace{(\rho + d\rho)(u + du)}_{=\rho u}(u + du)A \\
&= -\rho u^2 A + \rho u(u + du)A = \rho u\,du\,A
\end{aligned} \tag{8.87}$$

For the net-pressure integral, only the contributions from the inlet and outlet affect momentum conservation in the x direction. Hence,

$$\begin{aligned}
-\mathbf{i}\cdot\oiint_S p\mathbf{n}\,dS &= -\mathbf{i}\cdot[-p\,\mathbf{i}A] - \mathbf{i}\cdot[(p + dp)\mathbf{i}A] \\
&= pA - (p + dp)A = -A\,dp
\end{aligned} \tag{8.88}$$

Finally, turning to the viscous term, the viscous force acting on the control volume is due to the viscous shear stress at the pipe/duct wall, τ_w. If P is the distance measured around the perimeter, the viscous force on a differential section of length dx is

$$\mathbf{i}\cdot\oiint_S \mathbf{f}_v\,dS \approx -\tau_w P\,dx \tag{8.89}$$

The net viscous force acting on the control volume must be negative because friction opposes the motion. So, substituting Equations (8.87), (8.88) and (8.89) into Equation (8.86) yields

$$\rho u\,du\,A = -A\,dp - \tau_w P\,dx \tag{8.90}$$

Dividing through by $A dx$ yields the momentum equation for Fanno flow, viz.,

$$\rho u \frac{du}{dx} + \frac{dp}{dx} = -(P/A)\tau_w \tag{8.91}$$

Energy Conservation. In the absence of body forces, energy-conservation Equation (6.34) simplifies to

$$\oiint_S \rho \left(h + \frac{1}{2}u^2 \right) (\mathbf{u} \cdot \mathbf{n}) \, dS = \dot{Q} + \oiint_S \mathbf{f}_v \cdot \mathbf{u} \, dS \tag{8.92}$$

Although a formal proof is beyond the scope of this text, the viscous and heat-transfer contributions on cross sections are negligible relative to the cross-sectional flux of total enthalpy, so that only contributions from the pipe or duct walls are important. Because we assume the flow is adiabatic, the heat flux is exactly zero on the perimeter, wherefore

$$\dot{Q} = 0 \tag{8.93}$$

As discussed in Section 1.9.1, viscous flows satisfy the no-slip boundary condition at solid boundaries. So, on the pipe/duct perimeter, $\mathbf{u} = \mathbf{0}$ so that the friction force does no work at the pipe or duct walls. Therefore,

$$\oiint_S \mathbf{f}_v \cdot \mathbf{u} \, dS = 0 \tag{8.94}$$

This means the net flux of total enthalpy out of the control volume is zero, i.e.,

$$\rho u \left(h + \frac{1}{2}u^2 \right) = \text{constant} \quad \Longrightarrow \quad h + \frac{1}{2}u^2 = \text{constant} \tag{8.95}$$

where we make use of Equation (8.85) to conclude that ρu is also constant. In differential form, energy conservation becomes

$$\frac{dh}{dx} + u\frac{du}{dx} = 0 \tag{8.96}$$

We can arrive at a more illuminating form of the energy equation by appealing to Gibbs' equation [Equation (6.12)]. This will permit us to derive an equation for the entropy as follows.

$$
\left.
\begin{aligned}
T\frac{ds}{dx} &= \frac{de}{dx} + p\frac{d(1/\rho)}{dx} && \text{[from Gibbs' equation]} \\[2mm]
&= \frac{dh}{dx} - \frac{1}{\rho}\frac{dp}{dx} && \text{[since } h = e + p/\rho \text{]} \\[2mm]
&= -\frac{1}{\rho}\left[\rho u\frac{du}{dx} + \frac{dp}{dx}\right] && \text{[from Equation (8.96)]} \\[2mm]
&= -\frac{1}{\rho}\left[-(P/A)\tau_w\right] && \text{[from Equation (8.91)]}
\end{aligned}
\right\} \tag{8.97}
$$

Therefore, the entropy equation for Fanno flow is

$$\rho T\frac{ds}{dx} = (P/A)\tau_w \tag{8.98}$$

Equations (8.96) and (8.98) provide an immediate glimpse of the physical process occurring in Fanno flow. Since the right-hand side of the entropy equation is positive, the entropy increases in a pipe or duct flow with friction. However, friction changes the entropy without changing the total enthalpy.[2] The primary reason total enthalpy is conserved rests in the fact that the viscous stresses do no work at the pipe/duct surface where the stresses are largest.

To close our system of equations, we must determine the shear stress, τ_w. Following our analysis of pipe flow in Chapter 6, we introduce the conventional **friction factor**, f, and Darcy's **hydraulic diameter**, D_h, viz.,

$$f = \frac{\tau_w}{\frac{1}{8}\rho u^2} \quad \text{and} \quad D_h = 4A/P \tag{8.99}$$

Combining these relations, the right-hand side of Equation (8.98) can be written as

$$(P/A)\tau_w = \frac{1}{2}\rho u^2 f/D_h \tag{8.100}$$

Summarizing, the equations for conservation of mass, momentum and energy appropriate for Fanno flow are as follows.

$$\left.\begin{array}{c} \dfrac{d}{dx}(\rho u) = 0 \\[2ex] \dfrac{dp}{dx} + \rho u \dfrac{du}{dx} = -\dfrac{1}{2}\rho u^2 f/D_h \\[2ex] \rho T \dfrac{ds}{dx} = \dfrac{1}{2}\rho u^2 f/D_h \end{array}\right\} \tag{8.101}$$

8.10.2   Solution of the Fanno-Flow Equations

Because we have derived these equations by integrating over cross sections, flow properties such as ρ, u, h, etc. represent average conditions on each cross section. Since we have not explicitly introduced the constitutive relation for τ_w, our solution applies to either laminar or turbulent flow. Of course, the type of flow we have determines the value we use for the friction factor.

Since we are considering compressible flow, we must raise the question of the effect Mach number has on f. There is indeed an effect, and it is stronger for turbulent flow than for laminar flow. At Mach 10, for example, Schlichting-Gersten (2000) indicates that the skin friction on a flat plate is about 2/3 of the incompressible value for laminar flow and 1/4 for turbulent flow. In both cases, there is only a minor reduction for subsonic flow. Similar trends should be expected for flow in pipes and ducts.

Based on these observations, for subsonic Mach numbers, the value of f can be determined from either the Moody diagram (Figure 6.7) or Equations (6.59) and (6.60). When the flow is supersonic, a compressibility correction can be applied to the incompressible friction factor. Based on measurements presented by Schlichting-Gersten (2000), for turbulent flow we can compute f according to

$$f \approx \frac{f_{inc}}{\sqrt{1 + \frac{\gamma-1}{2}rM^2}} \tag{8.102}$$

---

[2]Although the physical processes are different, this also occurs for flow through a normal shock wave.

where f_{inc} is the incompressible friction factor, and r is the recovery factor. Note that for turbulent flow, r is only weakly dependent on the fluid and has been found from measurements to be

$$r = 0.89 \qquad \text{(turbulent flow)} \tag{8.103}$$

With a bit more algebra, we can obtain a closed-form solution to Equations (8.101). Assuming a perfect, calorically-perfect gas, we proceed in a manner similar to what we did for the Laval nozzle in Section 8.9. That is, we recast the problem in terms of Mach number and reference all conditions to the sonic point. We can solve for the thermodynamic properties of the flow using basic definitions along with conservation of energy and mass, as given in Equations (8.96) and the first of (8.101). The primary relations we need to accomplish this are $a^2 = \gamma p/\rho$, $M^2 = u^2/a^2$, and $p = \rho RT$. Also, the mass flux and total enthalpy are constant so that we have $\rho u = \text{constant}$ and $h + u^2/2 = \text{constant}$.

While the algebra is straightforward, it is a bit tedious and is addressed in the Problems section. Table B.2 in Appendix B includes all pertinent flow properties as a function of Mach number. Letting superscript * denote sonic-flow $(M = 1)$ reference conditions, the primary thermodynamic properties are given by

$$\left.\begin{aligned}
\frac{p}{p^*} &= \frac{1}{M}\left[\frac{\gamma+1}{2+(\gamma-1)M^2}\right]^{1/2} \\[2mm]
\frac{\rho}{\rho^*} &= \frac{1}{M}\left[\frac{2+(\gamma-1)M^2}{\gamma+1}\right]^{1/2} \\[2mm]
\frac{T}{T^*} &= \frac{\gamma+1}{2+(\gamma-1)M^2} \\[2mm]
\frac{p_t}{p_t^*} &= \frac{1}{M}\left[\frac{2+(\gamma-1)M^2}{\gamma+1}\right]^{(\gamma+1)/[2(\gamma-1)]}
\end{aligned}\right\} \tag{8.104}$$

Turning to the momentum equation, rearrangement in terms of Mach number yields the following differential equation.

$$\frac{dM^2}{dx} = \gamma M^4 \left[\frac{1+\frac{\gamma-1}{2}M^2}{1-M^2}\right]\frac{f}{D_h} \tag{8.105}$$

Hence, integrating from $x = 0$ to $x = L^*$, where L^* corresponds to the sonic point, we have

$$\int_0^{L^*}\frac{f}{D_h}dx = \int_{M^2}^1 \frac{(1-M^2)\,dM^2}{\gamma M^4\left[1+\frac{\gamma-1}{2}M^2\right]} = \frac{1-M^2}{\gamma M^2} + \frac{\gamma+1}{2\gamma}\ell n\left[\frac{(\gamma+1)M^2}{2+(\gamma-1)M^2}\right] \tag{8.106}$$

Finally, we define an average friction factor, $\overline{f}$, by

$$\overline{f} \equiv \frac{1}{L^*}\int_0^{L^*} f\,dx \tag{8.107}$$

Then, substituting Equation (8.107) into Equation (8.106), we arrive at the following equation, which relates the local Mach number and the distance from the point of interest to a reference point at which sonic conditions occur.

$$\frac{\overline{f}L^*}{D_h} = \frac{1-M^2}{\gamma M^2} + \frac{\gamma+1}{2\gamma}\ell n\left[\frac{(\gamma+1)M^2}{2+(\gamma-1)M^2}\right] \tag{8.108}$$

To understand how we use this relationship, it is instructive to consider two cases distinguished by the nature of the inlet flow. The two possibilities are a subsonic inlet and a supersonic inlet. For simplicity, the primary difference between the two cases we will consider is in the inlet Mach number. The temperature is the same for both cases, so that the difference in Mach number results from changing the pressure.

8.10.3   Subsonic Inlet

We consider the flow of air in a smooth cylindrical pipe of diameter $D = 0.2$ m and length $L = 20$ m. At the inlet, the Mach number is $M_1 = 0.3$, the pressure is $p_1 = 1$ atm and the temperature is $T_1 = 293$ K. Using our Fanno-flow solution, we would like to determine the Mach number, M_2, pressure, p_2, and temperature, T_2, at the exit. Because we have referenced all flow properties to sonic-flow conditions, we imagine that our pipe extends beyond the exit as shown in Figure 8.19.

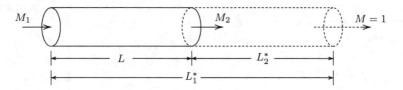

Figure 8.19: Adiabatic flow in a pipe with friction.

As our first step, we must compute the Reynolds number and determine the friction factor. For the inlet conditions, the speed of sound, a_1, velocity, u_1, and density, ρ_1 are:

$$\left.\begin{aligned}
a_1 &= \sqrt{\gamma R T_1} &&= 343 \text{ m/sec} \\[2mm]
u_1 &= M_1 a_1 &&= 103 \text{ m/sec} \\[2mm]
\rho_1 &= \frac{p_1}{R T_1} &&= 1.201 \text{ kg/m}^3
\end{aligned}\right\} \tag{8.109}$$

The viscosity, μ_1, follows from Sutherland's law, Equation (1.47), which gives

$$\mu_1 = \frac{1.46 \cdot 10^{-6} T^{3/2}}{T + 110.3} = 1.816 \cdot 10^{-5} \text{ kg/(m} \cdot \text{sec)} \tag{8.110}$$

Thus, the Reynolds number at the inlet, Re_D, is

$$Re_D = \frac{\rho_1 u_1 D}{\mu_1} = 1.36 \cdot 10^6 \tag{8.111}$$

Reference to the Moody diagram in Figure 6.7 shows that the flow is turbulent and the friction factor is

$$f \approx 0.011 \tag{8.112}$$

Now, the friction factor will vary in our pipe, and conceptually we must use the average value in Equation (8.108). For the sake of simplicity, it is customary to use the inlet friction factor as an approximation to $\overline{f}$. Also, note that the perimeter of the pipe is $P = \pi D$ and the cross-sectional area is $A = \pi D^2/4$. Hence, the hydraulic diameter is

$$D_h = 4A/P = D \tag{8.113}$$

At this point, we refer to Table B.2 in Appendix B to determine the inlet properties in relation to the sonic-flow reference conditions. For Mach 0.3, we find

$$\frac{\overline{f}L_1^*}{D_h} = 5.2993, \quad \frac{p_1}{p^*} = 3.6191, \quad \frac{T_1}{T^*} = 1.1788 \tag{8.114}$$

Now, as shown in Figure 8.19, the length from the exit to the sonic point is $L_2^* = L_1^* - L$, wherefore

$$\frac{\overline{f}L_2^*}{D_h} = \frac{\overline{f}L_1^*}{D_h} - \frac{\overline{f}L}{D_h} = 5.2993 - \frac{(.011)(20 \text{ m})}{(.2 \text{ m})} = 4.1993 \tag{8.115}$$

Again referring to Table B.2, this value of $\overline{f}L_2^*/D_h$ lies 60% of the way between the values listed for Mach numbers of 0.30 and 0.35. Hence, interpolation tells us the exit Mach number, pressure and temperature are

$$M_2 \approx 0.33, \quad \frac{p_2}{p^*} \approx 3.3030, \quad \frac{T_2}{T^*} \approx 1.1743 \tag{8.116}$$

Finally, we can compute the actual values of p_2 and T_2 according to

$$p_2 = p_1 \cdot \frac{p_2/p^*}{p_1/p^*} = 0.91 \text{ atm} \quad \text{and} \quad T_2 = T_1 \cdot \frac{T_2/T^*}{T_1/T^*} = 292 \text{ K} \tag{8.117}$$

8.10.4 Supersonic Inlet

We consider the same pipe as in the subsonic case with one change. We now increase the inlet Mach number to 3. The inlet speed of sound, density and viscosity are still given by Equations (8.109) and (8.110). However, the velocity and Reynolds number both increase by a factor of 10, so that

$$u_1 = 1030 \text{ m/sec} \quad \text{and} \quad Re_D = 1.36 \cdot 10^7 \tag{8.118}$$

We cannot use the Moody diagram to determine f_{inc} as the appropriate value is off scale for a smooth pipe. Rather, we use the turbulent-flow value for f given in Equation (6.59) to show that

$$f_{inc} \approx 0.00775 \tag{8.119}$$

Then, using Equation (8.102) to correct for effects of compressibility, we obtain

$$f = \frac{f_{inc}}{\sqrt{1 + \frac{\gamma-1}{2}rM^2}} \approx 0.0048 \tag{8.120}$$

Turning to the Fanno-flow tables, at Mach 3 the relevant properties for our problem are

$$\frac{\overline{f}L_1^*}{D_h} = 0.5222, \quad \frac{p_1}{p^*} = 0.2182, \quad \frac{T_1}{T^*} = 0.4286 \tag{8.121}$$

Hence, the effective distance of the exit from the sonic point is

$$\frac{\overline{f}L_2^*}{D_h} = \frac{\overline{f}L_1^*}{D_h} - \frac{\overline{f}L}{D_h} = 0.5222 - \frac{(.0048)(20 \text{ m})}{(.2 \text{ m})} = 0.0422 \tag{8.122}$$

Reference to Table B.2, shows that this value of $\overline{f}L_2^*/D_h$ lies 57% of the way between the values listed for Mach numbers of 1.20 and 1.25. Hence, interpolation tells us the exit Mach number, pressure and temperature for our supersonic inlet case are

$$M_2 \approx 1.23, \quad \frac{p_2}{p^*} \approx 0.7819, \quad \frac{T_2}{T^*} \approx 0.9218 \tag{8.123}$$

and the dimensional values of p_2 and T_2 become

$$p_2 = p_1 \cdot \frac{p_2/p^*}{p_1/p^*} = 3.58 \text{ atm} \quad \text{and} \quad T_2 = T_1 \cdot \frac{T_2/T^*}{T_1/T^*} = 630 \text{ K} \tag{8.124}$$

8.10.5   Mollier Diagram

There is an additional equation for Fanno flow that is particularly illuminating, and reveals the primary effect of friction on the flow. As a direct consequence of the fact that Fanno flow has constant total enthalpy, a straightforward derivation shows that the rate of change of the enthalpy is given by

$$\frac{dh}{dx} = \frac{\gamma M^2}{M^2 - 1} T \frac{ds}{dx} \tag{8.125}$$

Since the second law of thermodynamics tells us the entropy must always increase, enthalpy must decrease with x when the flow is subsonic and vice versa. Since total enthalpy, $h + \frac{1}{2}u^2$, is constant, necessarily u varies in the opposite direction to h. Thus, the cooling associated with a decrease in h for subsonic flow results in an increase in the flow's kinetic energy per unit mass, $\frac{1}{2}u^2$. Conversely, under supersonic conditions, friction heats the flow at the expense of kinetic energy. Inspection of Equation (8.125) shows that when the Mach number is unity, necessarily $ds/dx = 0$ (to preclude the physically impossible condition that $dh/dx \to \infty$). This means s achieves its maximum value at the sonic point.

These facts are illustrated in Figure 8.20, which shows a typical variation of h with s. This type of graph is called a **Mollier diagram**. The curve is known as the **Fanno line** or, more appropriately, the **Fanno curve**. Entropy is a maximum at point a where the flow is sonic. All states with higher enthalpy lie on the subsonic, or upper, branch of the curve, while states with lower enthalpy lie on the supersonic, or lower, branch.

As indicated in the figure, the flow is always driven toward the sonic point. When the inlet flow is supersonic, for example, we begin at point 1. The flow downstream of the inlet approaches point a, with an attendant decrease in Mach number towards one. Each point,

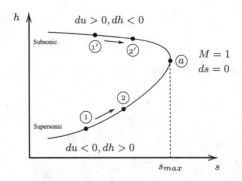

Figure 8.20: *Mollier diagram for Fanno flow.*

say 2, on the supersonic branch corresponds to a specific pipe/duct length, L. Increasing the length L causes point 2 to move closer to point a. When the length is selected so that the flow just becomes sonic at the exit, points 2 and a are coincident and we have $L = L^*$. The flow is then said to be **choked**, and no further increase in L can occur without modifying the inlet conditions. That is, we must move point 1 further back on the branch to a position corresponding to the desired length. If this is not done, a shock will form, causing the flow to jump from point 2 to $2'$.

Similarly, when the inlet flow is subsonic, we begin at point $1'$ and the flow is driven toward point a. The pipe/duct length can be increased until the flow becomes choked. Further increases in L cause a reduction in the mass-flow rate and the inlet Mach number.

For the two examples above, it is a straightforward matter to compute the length of pipe required to choke the flow. Specifically, the length is given by

$$L^* = \left(\frac{\overline{f}L^*}{D_h}\right)\left(\frac{D_h}{\overline{f}}\right) = \begin{cases} 96.4 \text{ m}, & \text{Subsonic inlet} \\ 21.8 \text{ m}, & \text{Supersonic inlet} \end{cases} \tag{8.126}$$

As a final comment, it is worthwhile to summarize how flow properties vary in Fanno flow. The direction in which most flow properties change depends upon whether the inlet flow is subsonic or supersonic. Table 8.2 lists the various flow properties and the directions in which they vary.

Table 8.2: *Variation of Fanno-Flow Properties*

Subsonic Inlet	Supersonic Inlet
M and u increase	p, ρ and T increase
T_t is constant	T_t is constant
p, ρ, T, p_t and ρ_t decrease	M, u, p_t and ρ_t decrease

8.11 Rayleigh Flow

Rayleigh flow is a straightforward, one-dimensional example that illustrates effects of heat transfer in a compressible medium. We again consider steady flow of a gas in a duct or pipe with constant cross-sectional area A. In this application, we ignore viscous effects and focus rather on heat transfer as indicated in Figure 8.21. This analysis is valid for flows heated by processes such as combustion, nuclear radiation and laser irradiation.

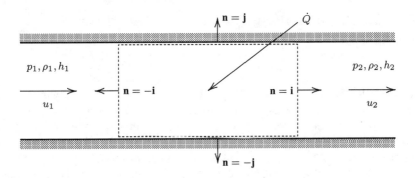

Figure 8.21: *Stationary control volume for Rayleigh flow; cross-sectional area, A, is constant.*

8.11.1   Rayleigh-Flow Equations

Mass and Momentum Conservation. It is most convenient to apply the conservation principles to a control volume relating the inlet and a specific point downstream. Our geometry and control volume are the same as for Fanno flow. Thus, the mass- and momentum-conservation equations are nearly the same for both Fanno and Rayleigh flow. Noting that we are neglecting effects of friction, the integral conservation principles for mass [Equation (8.82)] and momentum [Equation (8.86) with $\mathbf{f}_v = \mathbf{0}$] simplify immediately to

$$\rho_1 u_1 = \rho_2 u_2 \tag{8.127}$$

$$p_1 + \rho_1 u_1^2 = p_2 + \rho_2 u_2^2 \tag{8.128}$$

Energy Conservation. As with Fanno flow we ignore body forces. Consequently, the integral form of the energy-conservation principle is

$$\oiint_S \rho \left(h + \frac{1}{2} u^2 \right) (\mathbf{u} \cdot \mathbf{n}) \, dS = \dot{Q} \tag{8.129}$$

where $\dot{Q}$ is the heat transfer rate to the control volume. We define the **heat added per unit mass**, q, according to

$$\dot{Q} = \dot{m} q = \rho u A q \tag{8.130}$$

So, evaluating the total enthalpy-flux integral yields the following equation for energy conservation.

$$h_1 + \frac{1}{2} u_1^2 + q = h_2 + \frac{1}{2} u_2^2 \tag{8.131}$$

8.11.2   Solution of the Rayleigh-Flow Equations

The first important observation we can make regarding Rayleigh flow concerns the total temperature. Noting that the enthalpy is given by $h = c_p T$ and that $c_p = \gamma R / (\gamma - 1)$, the total enthalpy is given by

$$h + \frac{1}{2} u^2 = c_p T + \frac{1}{2} u^2 = c_p T \left(1 + \frac{\gamma - 1}{2} M^2 \right) = c_p T_t \tag{8.132}$$

where T_t is total temperature. Therefore, we can rewrite Equation (8.131) as

$$T_{t2} = T_{t1} + \frac{q}{c_p} \tag{8.133}$$

This equation tells us that the heat added to the duct or pipe serves to directly increase the total temperature of the flow. This is especially interesting as total temperature has been constant in virtually all of the flows we have discussed thus far, including flows in which a shock wave is present.

Using the perfect-gas law, $p = \rho R T$, along with Equations (8.127), (8.128) and (8.131), we can solve for all flow properties as a function of Mach number. As with Fanno flow, we select our reference state as the sonic point. The solution for the primary thermodynamic

variables and total conditions is tabulated in Table B.3 of Appendix B. Letting superscript *
denote sonic conditions, the algebraic form of the solution is as follows.

$$\left.\begin{array}{rcl} \dfrac{p}{p^*} & = & \dfrac{\gamma + 1}{1 + \gamma M^2} \\[2.5ex] \dfrac{\rho}{\rho^*} & = & \dfrac{1}{M^2}\left[\dfrac{1 + \gamma M^2}{\gamma + 1}\right] \\[2.5ex] \dfrac{T}{T^*} & = & M^2\left[\dfrac{\gamma + 1}{1 + \gamma M^2}\right]^2 \\[2.5ex] \dfrac{p_t}{p_t^*} & = & \dfrac{\gamma + 1}{1 + \gamma M^2}\left[\dfrac{2 + (\gamma - 1)M^2}{\gamma + 1}\right]^{\gamma/(\gamma - 1)} \\[2.5ex] \dfrac{T_t}{T_t^*} & = & \dfrac{(\gamma + 1)M^2}{(1 + \gamma M^2)^2}\left[2 + (\gamma - 1)M^2\right] \end{array}\right\} \qquad (8.134)$$

As with Fanno flow, the solution depends upon whether the flow is subsonic or supersonic
at the inlet. Furthermore, our solution is valid for both heating and cooling, and the direction
of changes in flow properties differs for the four possible conditions, i.e., subsonic inlet with
heating, subsonic inlet with cooling, supersonic inlet with heating and supersonic inlet with
cooling.

8.11.3   Subsonic Inlet With Heating

Consider a duct with air entering at a Mach number $M_1 = 0.3$. The static pressure and
temperature are $p_1 = 1$ atm and $T_1 = 293$ K, respectively. The heat added per unit mass is
$q = 3 \cdot 10^5$ J/kg. We would like to determine M_2, p_2 and p_{t2}. Figure 8.22 schematically
illustrates such a flow, including an imaginary extension to a length at which sonic flow exists.
Note that beginning with the conditions at the duct exit, the heat required to bring the flow to
sonic conditions is q_2^*. Thus, we infer that the heat required to take the flow from the inlet
conditions to sonic conditions, q_1^*, must be

$$q_1^* = q_1 + q_2^* \qquad (8.135)$$

This is different from Fanno flow in which sonic conditions correspond to an imaginary point
that can be reached *adiabatically*. Here, we have *non-adiabatic* flow, so that p^*, T^*, etc. are
the conditions that would exist if sufficient heat were added to reach Mach 1.

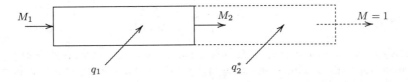

Figure 8.22: *Frictionless flow in a duct with heat addition.*

As our first step in solving this problem, we must determine the total pressure and total
temperature at the inlet. We need these quantities in order to compute the magnitude of p_t

and T_t at the exit. To find the inlet total conditions, we can use the isentropic flow Table B.1 in Appendix B. For an inlet Mach number, $M_1 = 0.3$,

$$\frac{p_1}{p_{t1}} = 0.9395 \quad \text{and} \quad \frac{T_1}{T_{t1}} = 0.9823 \qquad (8.136)$$

Also, we need the specific heat, c_p, to compute T_{t2} from Equation (8.133).

$$c_p = \frac{\gamma R}{\gamma - 1} = \frac{(1.4)[287 \text{ J}/(\text{kg} \cdot \text{K})]}{0.4} = 1004.5 \text{ J}/(\text{kg} \cdot \text{K}) \qquad (8.137)$$

Therefore, the inlet total pressure and total temperature are

$$\left.\begin{aligned} T_{t1} &= \frac{T_1}{T_1/T_{t1}} = \frac{293 \text{ K}}{0.9823} = 298.3 \text{ K} \\[2mm] p_{t1} &= \frac{p_1}{p_1/p_{t1}} = \frac{1 \text{ atm}}{0.9395} = 1.0644 \text{ atm} \end{aligned}\right\} \qquad (8.138)$$

From Equation (8.133), the total temperature at the exit is

$$T_{t2} = T_{t1} + \frac{q}{c_p} = 298.3 \text{ K} + \frac{3 \cdot 10^5 \text{ J}/\text{kg}}{1004.5 \text{ J}/(\text{kg} \cdot \text{K})} = 298.3 \text{ K} + 298.7 \text{ K} = 597.0 \text{ K} \quad (8.139)$$

Now, we turn to the Rayleigh-flow Table B.3 in Appendix B. We require the total temperature ratio, T_{t1}/T_t^*, which we will use to determine the Mach number at the exit. We also require the static and total pressure ratios, p_1/p^* and p_{t1}/p_t^*, which will eventually be needed to determine the magnitudes of p_2 and p_{t2}. Reference to the tables shows that for $M_1 = 0.3$,

$$\frac{T_{t1}}{T_t^*} = 0.3469, \quad \frac{p_1}{p^*} = 2.1314, \quad \frac{p_{t1}}{p_t^*} = 1.1985 \qquad (8.140)$$

In order to find the Mach number at the exit, M_2, we first compute T_{t2}/T_t^* as follows.

$$\frac{T_{t2}}{T_t^*} = \frac{T_{t2}}{T_{t1}}\frac{T_{t1}}{T_t^*} = \left(\frac{597.0 \text{ K}}{298.3 \text{ K}}\right)(0.3469) = 0.6943 \qquad (8.141)$$

Again referring to Table B.3, we find that $T_t/T_t^* = 0.6943$ lies 4% of the way between the values listed for Mach numbers of 0.50 and 0.55. So, linear interpolation tells us that

$$M_2 \approx 0.50, \quad \frac{p_2}{p^*} \approx 1.7739, \quad \frac{p_{t2}}{p_t^*} \approx 1.1132 \qquad (8.142)$$

Finally, the static and total pressure at the exit are

$$p_2 = p_1 \cdot \frac{p_2/p^*}{p_1/p^*} = 0.83 \text{ atm} \quad \text{and} \quad p_{t2} = p_{t1} \cdot \frac{p_{t2}/p_t^*}{p_{t1}/p_t^*} = 0.99 \text{ atm} \qquad (8.143)$$

8.11.4 Supersonic Inlet With Heating

We now increase the inlet Mach number to $M_1 = 3$, leaving all other conditions the same as in the subsonic case. Reference to the isentropic flow Table B.1 shows that

$$\frac{p_1}{p_{t1}} = 0.02722 \quad \text{and} \quad \frac{T_1}{T_{t1}} = 0.3571 \qquad (8.144)$$

So, the inlet total pressure and total temperature are

$$T_{t1} = \frac{T_1}{T_1/T_{t1}} = 820.5 \text{ K} \quad \text{and} \quad p_{t1} = \frac{p_1}{p_1/p_{t1}} = 36.74 \text{ atm} \tag{8.145}$$

The total temperature at the exit follows from Equation (8.133), viz.,

$$T_{t2} = T_{t1} + \frac{q}{c_p} = 820.5 \text{ K} + \frac{3 \cdot 10^5 \text{ J/kg}}{1004.5 \text{ J/(kg} \cdot \text{K)}} = 820.5 \text{ K} + 298.7 \text{ K} = 1119.2 \text{ K} \tag{8.146}$$

From the Rayleigh flow Table B.3 for an inlet Mach number, $M_1 = 3$,

$$\frac{T_{t1}}{T_t^*} = 0.6540, \quad \frac{p_1}{p^*} = 0.1765, \quad \frac{p_{t1}}{p_t^*} = 3.4245 \tag{8.147}$$

Therefore, the total temperature ratio at the duct exit is

$$\frac{T_{t2}}{T_t^*} = \frac{T_{t2}}{T_{t1}} \frac{T_{t1}}{T_t^*} = \left(\frac{1119.2 \text{ K}}{820.5 \text{ K}} \right) (0.6540) = 0.8921 \tag{8.148}$$

Referring again to the Rayleigh flow Table B.3, we see that $T_t/T_t^* = 0.8921$ lies 37% of the way between values for Mach numbers of 1.55 and 1.60. Using interpolation there follows:

$$M_2 \approx 1.57, \quad \frac{p_2}{p^*} \approx 0.5403, \quad \frac{p_{t2}}{p_t^*} \approx 1.1577 \tag{8.149}$$

Thus, the static and total pressure at the exit are

$$p_2 = p_1 \cdot \frac{p_2/p^*}{p_1/p^*} = 3.06 \text{ atm} \quad \text{and} \quad p_{t2} = p_{t1} \cdot \frac{p_{t2}/p_t^*}{p_{t1}/p_t^*} = 12.42 \text{ atm} \tag{8.150}$$

8.11.5 Mollier Diagram

The two examples above show that Rayleigh flow has some similarity to Fanno flow on a couple of counts. It appears that, at least in the case of heat addition, the flow is driven toward sonic conditions. This raises the question of the possibility that sufficient heating can choke the flow. As with Fanno flow, the Mollier diagram is useful for illustrating these features, along with another curious nuance of this flow. The entropy for a perfect gas is given by Equation (8.7), i.e.,

$$s = c_v \, \ell n \left(\frac{p}{\rho^\gamma} \right) + \text{constant} \tag{8.151}$$

Using the Rayleigh-flow solution for p and ρ [Equations (8.134)], we can express the entropy as a function of Mach number.

$$s = c_v \, \ell n \left[M^{2\gamma} \left(\frac{\gamma + 1}{1 + \gamma M^2} \right)^{\gamma + 1} \right] + \text{constant} \tag{8.152}$$

Since the enthalpy is $h = c_p T$, clearly $h/h^* = T/T^*$, which is given by the third of Equations (8.134). So, differentiating s and h with respect to x, a bit of algebra leads to the following equation relating dh/dx and ds/dx.

$$\frac{dh}{dx} = \frac{\gamma M^2 - 1}{M^2 - 1} T \frac{ds}{dx} \tag{8.153}$$

Equation (8.153) defines the **Rayleigh line** or, more appropriately, the **Rayleigh curve**, which Figure 8.23 illustrates.

There are several similarities between the Mollier diagrams for Fanno flow (Figure 8.20) and Rayleigh flow. Both have supersonic and subsonic branches. Similar to Fanno flow, when Rayleigh flow has heat addition the flow is driven toward the sonic point a, where the entropy is a maximum. As discussed earlier, when we add the critical amount of heat, q^*, the flow will be choked. If we add heat in excess of q^*, a shock will form if the inlet is supersonic. If the inlet is subsonic, excess heating will cause the inlet conditions to change. Again, this behavior is analogous to what we found for Fanno flow.

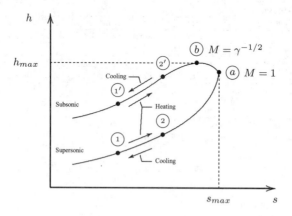

Figure 8.23: *Mollier diagram for Rayleigh flow.*

There is an interesting subtlety of Rayleigh flow that has no analog in Fanno flow. Inspection of Equation (8.153) shows that on the supersonic branch, the enthalpy, and hence the temperature, increases when we heat the flow. This is obvious since $ds/dx > 0$ corresponds to $dh/dx > 0$ when $M > 1$. However, on the subsonic branch, the factor $(\gamma M^2 - 1)/(M^2 - 1)$ changes sign at $M = \gamma^{-1/2}$. As a result, the enthalpy increases for Mach numbers less than $\gamma^{-1/2}$ and decreases for $M > \gamma^{-1/2}$.

In other words, for Mach numbers between $\gamma^{-1/2}$ (0.845 for air) and 1, the temperature decreases as a result of heating the flow! Because total temperature increases, this means the kinetic energy increases so rapidly in this range that the fluid must cool slightly to conserve energy.

When we have heat extraction, all of these trends are reversed. While the entropy of the duct or pipe decreases because of the heat extracted, the energy of the surroundings must increase so that the entropy of the entire system increases. This is dictated by the second law of thermodynamics. When heat is extracted, the flow is driven away from the sonic point on both the subsonic and supersonic branches of the Rayleigh curve. Table 8.3 summarizes the directions in which flow properties change in Rayleigh flow.

For the two Rayleigh-flow examples discussed above, we can compute the amount of heat added per unit mass that will choke the flow. For the subsonic and supersonic cases, the inlet total temperature is $T_{t1} = 298.3$ K and 820.5 K, respectively. Reference to Table B.3 shows that T_{t1}/T_t^* is 0.3469 for subsonic flow and 0.6540 for the supersonic case. Hence, a bit of arithmetic shows that the reference total temperature, T_t^*, is 860 K for the subsonic inlet and 1255 K for the supersonic inlet. Finally, noting that the flow is choked when $T_{t2} = T_t^*$, we use Equation (8.133) to compute the required heat added per unit mass, which we call q^*. A little more arithmetic yields the following.

$$q^* = \begin{cases} 5.64 \cdot 10^5 \text{ J/kg}, & \text{Subsonic inlet} \\ 4.36 \cdot 10^5 \text{ J/kg}, & \text{Supersonic inlet} \end{cases} \tag{8.154}$$

Table 8.3: *Variation of Rayleigh-Flow Properties*

Subsonic Inlet $0 < M < \gamma^{-1/2}$	Subsonic Inlet $\gamma^{-1/2} < M < 1$	Supersonic Inlet $M > 1$
With Heat Addition:		
M, u, T and T_t increase	M, u and T_t increase	p, ρ, T and T_t increase
p, ρ, p_t and ρ_t decrease	p, ρ, T, p_t and ρ_t decrease	M, u, p_t and ρ_t decrease
With Heat Extraction:		
p, ρ, p_t and ρ_t increase	p, ρ, T, p_t and ρ_t increase	M, u, p_t and ρ_t increase
M, u, T and T_t decrease	M, u and T_t decrease	p, ρ, T and T_t decrease

Problems

8.1 For density variations to become important, the Mach number must be at least 0.3. If the temperature is 60° F, what flow speed, in mph, must be attained to reach Mach 0.3 in helium and in mercury? See Table A.4 for speed of sound values.

8.2 Compute the velocity required to reach the transonic flow regime ($M \approx 0.8$) in air and water at 15.6° C. Express your answers in km/hr. See Table A.4 for speed of sound values.

8.3 Determine the Mach numbers and corresponding flow classification for the following. Use Table A.1 to determine appropriate gas properties.

   (a) Carbon dioxide at 50° F flowing with a velocity of 1000 ft/sec.

   (b) Hydrogen at 50° F flowing with a velocity of 1000 ft/sec.

   (c) Air at 0° F flowing with a velocity of 1000 ft/sec.

   (d) Helium at −300° F flowing with a velocity of 9500 ft/sec.

8.4 Determine the Mach numbers and corresponding flow classification for the following. Use Table A.1 to determine appropriate gas properties.

   (a) Methane at 26° C flowing with a velocity of 410 m/sec.

   (b) Helium at 26° C flowing with a velocity of 610 m/sec.

   (c) Air at 0° C flowing with a velocity of 994 m/sec.

   (d) Air at −49.16° C flowing with a velocity of 300 m/sec.

8.5 According to the U. S. Standard Atmosphere, the temperature in the atmosphere is $T = T_a - \alpha z$, where $\alpha = 6.50$ K/km, z is altitude and T_a is surface temperature. On a day when $T_a = 10°$ C, a missile flies at Mach 0.5. Determine its speed in m/sec if it is flying at $z = 25$ m and at $z = 2$ km.

8.6 Imagine that you launch a small rocket from a beach somewhere in the Los Angeles area. At the same time, a friend launches a similar rocket in Denver, which is a mile above sea level. If both rockets have a Mach number of 0.7, whose rocket is traveling faster? Assume conditions in the atmosphere are given by the U. S. Standard Atmosphere with a sea-level temperature of $T_o = 78°$ F.

8.7 According to the U. S. Standard Atmosphere, the temperature in the atmosphere is $T = T_a - \alpha z$, where $\alpha = 18.85°$ R/mi, z is altitude and T_a is surface temperature. At an altitude of 4 miles, a rocket is flying at 1340 mph, which corresponds to Mach 2. What is the surface temperature to the nearest ° F?

8.8 Consider a liquid of density ρ_ℓ that has gas bubbles uniformly distributed. A fraction λ of the total volume, V, is occupied by the air. The fluid mixture is compressed isentropically and $\lambda \ll 1$.

   (a) Show that differential pressure and volume changes are related by

$$\frac{dp}{p} = -\frac{\gamma}{\lambda}\frac{dV}{V}$$

   where γ is the specific-heat ratio of the gas. Neglect any volume change for the liquid.

   (b) Explain why $d\rho/\rho = -dV/V$ and $\rho \approx (1 - \lambda)\rho_\ell$.

   (c) Combine results of Parts (a) and (b) to show that the speed of sound, a, is

$$a = \sqrt{\frac{\gamma}{\lambda(1 - \lambda)}\frac{p}{\rho_\ell}}$$

   (d) Suppose the gas is air and the liquid is water with $\rho_\ell = 1000$ kg/m³. Compute a/a_o for the mixture, where $a_o = 1481$ m/sec is the sound speed in pure water, when $\lambda = 0.01, 0.02$ and 0.03.

8.9 An empirical formula relating pressure and density for seawater is

$$p/p_a \approx (\alpha + 1)(\rho/\rho_a)^7 - \alpha$$

where p_a and ρ_a are surface values and $\alpha = 3000$. If $p_a = 1$ atm and $\rho_a = 1030$ kg/m^3, what is the speed of sound at a depth where $\rho = 1040$ kg/m^3? Compare with the fresh-water value of 1481 m/sec.

8.10 A classical result of the *kinetic theory of gases* tells us that the internal energy for each degree of freedom is $\frac{1}{2}RT$, where R is the perfect-gas constant and T is absolute temperature. A molecule of a monatomic gas is viewed as a sphere and thus has 3 translational degrees of freedom. A diatomic molecule appears as a 'barbell' with 5 degrees of freedom, 3 translational and 2 rotational (the rotational mode about the axis can be ignored). At very high temperatures, two additional vibrational modes appear for diatomic molecules.

(a) According to this theory, what is γ for a perfect, calorically-perfect gas with n degrees of freedom?

(b) Compare with measured values (Table A.1) for helium ($n = 3$), air ($n = 5$), carbon dioxide at room temperature ($n = 5$) and carbon dioxide at very high temperature ($n = 7$).

8.11 If we account for *quantum-mechanical* effects, we find that the specific heat for carbon dioxide is

$$c_v = \frac{5}{2}R + \frac{(\Theta_{vib}/T)^2\, e^{\Theta_{vib}/T}}{(e^{\Theta_{vib}/T} - 1)^2}R, \qquad \Theta_{vib} = 954 \text{ K}$$

(a) Using the fact that $c_p - c_v = R$, compute the specific-heat ratio, γ, for $T = 10°$ C, $100°$ C, $1000°$ C and $5000°$ C.

(b) What temperature does the nominal value, $\gamma = 1.30$, correspond to? **HINT:** Solve by trial and error—it lies between $300°$ C and $400°$ C.

8.12 Gas flows with velocity $U = 1000$ m/sec, pressure $p = 125$ kPa and temperature $T = 250°$ C. We want to determine the pressure and temperature the gas would reach if it were brought to rest isentropically. Do your computations for air and for methane. See Table A.1 for properties of methane.

8.13 Gas flows with velocity $U = 1500$ ft/sec, pressure $p = 20$ psi and temperature $T = 94°$ F. We want to determine the pressure and temperature the gas would reach if it were brought to rest isentropically. Do your computations for air and for hydrogen. See Table A.1 for properties of hydrogen.

8.14 Air undergoes an isentropic expansion through a nozzle from a stagnation temperature of $60°$ C. What is the Mach number at the point where the flow velocity is 150 m/sec?

8.15 Helium ($\gamma = 5/3$) experiences an isentropic expansion from a stagnation temperature of $100°$ F. What is the Mach number at points where the flow velocity is 100 ft/sec, 500 ft/sec and 1000 ft/sec?

8.16 Hypersonic experiments are proposed for a wind tunnel with a reservoir temperature of $300°$ F. If the test-section Mach number is 6, determine the temperature in the test section if the working gas is air ($\gamma = 1.40$) and if it is helium ($\gamma = 5/3$).

8.17 Hypersonic experiments are proposed for a wind tunnel with a reservoir temperature of $200°$ C. If the test-section Mach number is 10, determine the temperature in the test section if the working gas is air ($\gamma = 1.40$) and if it is helium ($\gamma = 5/3$).

8.18 Wind-tunnel tests are being planned for Mach 1.6 airflow past a scale model of an airplane. The reservoir temperature will be $20°$ C and the pressure in the wind-tunnel test section must be 1 atm. Assuming the flow will be isentropic, determine the required reservoir pressure and the flow velocity in the test section.

8.19 Helium ($\gamma = 5/3$) is brought to rest isentropically. If the static pressure increases from 8 kPa to 256 kPa, at what Mach number is the helium flowing?

8.20 Wind-tunnel tests are being planned for Mach 2.5 airflow past a scale model of a cruise missile. The reservoir pressure will be 280 psi and the temperature in the wind-tunnel test section must be 50° F. Assuming the flow will be isentropic, determine the required reservoir temperature and the density in the test section.

8.21 The Mach numbers at two points in an isentropic flow of air are $M_A = 0.3$ and $M_B = 0.6$. Determine p_A/p_B and T_A/T_B. Also, compute T_{tA}/T_{tB} for this flow.

8.22 The Mach numbers at two points in an isentropic flow of air are $M_A = 3$ and $M_B = 6$. Determine p_A/p_B and T_A/T_B. Also, compute T_{tA}/T_{tB} for this flow.

8.23 Consider subsonic flow past an airplane wing. The flow speed far ahead of the wing is U_∞. The flow accelerates over the wing, reaching a maximum speed of $\frac{7}{5}U_\infty$. The freestream density and temperature are $\rho_\infty = 1.20$ kg/m^3 and $T_\infty = 300$ K, respectively. Compute M_∞ and the percentage change in temperature at the point where the maximum velocity is achieved for the following.

   (a) $U_\infty = 100$ m/sec

   (b) $U_\infty = 200$ m/sec

8.24 Consider subsonic flow past an airplane wing. The flow speed far ahead of the wing is U_∞. The flow accelerates over the wing, reaching a maximum speed of $\frac{3}{2}U_\infty$. The freestream density and temperature are $\rho_\infty = 0.00234$ slug/ft^3 and $T_\infty = 519°$ R, respectively. Compute M_∞ and the percentage change in density at the point where the maximum velocity is achieved for the following.

   (a) $U_\infty = 100$ mph

   (b) $U_\infty = 450$ mph

8.25 Consider subsonic flow of helium ($\gamma = 5/3$) approaching a Pitot-static tube. The flow is essentially isentropic. Determine the Mach number when the measured total and static pressures are as follows.

   (a) $p = 101$ kPa, $p_t = 126$ kPa

   (b) $p = 14.7$ psi, $p_t = 26.5$ psi

   (c) $p = 1.06$ atm, $p_t = 1.11$ atm

8.26 In Chapter 3 we learned that, for incompressible flow, the flow speed, U_p, can be obtained from a Pitot-static tube according to $U_p = \sqrt{2(p_t - p)/\rho}$, where ρ is density, p_t is total pressure and p is static pressure.

   (a) Assuming isentropic, subsonic flow, show that U_p/U is the following function of Mach number, M, where U is the true flow speed:

$$\frac{U_p}{U} = \frac{1}{M}\sqrt{\frac{2}{\gamma}\left[\left(1 + \frac{\gamma-1}{2}M^2\right)^{\gamma/(\gamma-1)} - 1\right]}$$

   (b) Verify that $U_p \to U$ as $M \to 0$.

   (c) Compute U_p/U for $M = 0.2, 0.4, 0.6, 0.8$ and 1.0. Below what Mach number is the difference between U_p and U less than 2%?

8.27 Consider subsonic flow of air approaching a Pitot-static tube. The flow is essentially isentropic. Determine the Mach number when the measured total and static pressures are as follows.

   (a) $p = 101$ kPa, $p_t = 126$ kPa

   (b) $p = 14.7$ psi, $p_t = 26.5$ psi

   (c) $p = 1.06$ atm, $p_t = 1.11$ atm

8.28 The ratio of the total density to the static density of a gas is $\rho_t/\rho = 5.02$. The Mach number is $M = 2.4$ and the flow is isentropic.

(a) Assuming the gas is one of those listed in Table A.1, which gas is it?

(b) If the total temperature is $T_t = 600°$ C, what is the static temperature, T?

(c) If the total pressure is $p_t = 1050$ kPa, find the static pressure, p.

8.29 You have no calculator available, but you do have your shock tables (see Section B.1 of Appendix B). You also know that

$$M_2^2 = \frac{1 + \frac{1}{2}(\gamma - 1)M_1^2}{\gamma M_1^2 - \frac{1}{2}(\gamma - 1)}$$

In your shock tube you have air with $M_1 = 2.65$. You wish to change your gas to helium ($\gamma = 5/3$), but your application requires having the same value for M_2 that you have with air. What must M_1 be when you use helium? **NOTE**: This problem is intended to encourage you to make rational engineering approximations. In this spirit, any interpolation you do in the shock tables should be done to two significant figures, and no more.

8.30 Superman is traveling faster than a speeding bullet. In fact, he is traveling at 7 times the speed of sound in the ambient atmosphere (temperature $72°$ F at 14.7 psi). Estimate the maximum temperature on his head (stagnation temperature). Also, determine the maximum pressure on his head, taking into consideration the shock wave standing in front of him. Note that, although the shock is curved, the part of the wave just upstream of Superman's head is normal to the direction of his motion.

Shock
Wave

Problem 8.30

8.31 The static pressure increases from 100 kPa to 600.5 kPa across a shock wave in air. If the static temperature of the fluid ahead of the shock is 150 K, what is the temperature behind the shock?

8.32 Wind-tunnel experiments are being planned to simulate Mach 1.5 flow past a blunt-nosed object. We require the static pressure just behind the normal shock standing in front of the object to be 1 atm. What must the pressure be in the wind-tunnel reservoir for air and for carbon dioxide ($\gamma = 1.30$)?

M_1

$p_2 = 1$ atm

Problems 8.32, 8.33

8.33 Wind-tunnel experiments are being planned to simulate Mach 3.3 flow past a blunt-nosed object. We require the static pressure just behind the normal shock standing in front of the object to be 1 atm. What must the pressure be in the wind-tunnel reservoir for air and for helium ($\gamma = 5/3$)?

8.34 Just ahead of a normal shock, conditions are given by $M_1 = 3$, $T_1 = 550°$ R and $p_1 = 21$ psi. Determine the values of M_2, T_{t2} and p_2 (i.e., conditions behind the shock) for air and helium ($\gamma = 5/3$). Use the shock tables for the parts of this problem to which they apply.

8.35 Just ahead of a normal shock, conditions are given by $M_1 = 4$, $T_1 = 350$ K and $p_1 = 8$ kPa. Determine the values of M_2, T_{t2} and p_2 (i.e., conditions behind the shock) for air and carbon dioxide ($\gamma = 9/7$). Use the shock tables for the parts of this problem to which they apply.

8.36 Just behind a normal shock, the Mach number is $M_2 = 0.7$ and the temperature is $T_2 = 87.4°$ C. Determine the values of M_1 and T_1 (i.e., conditions ahead of the shock) for air and for helium ($\gamma = 5/3$). Use the shock tables for the parts of this problem to which they apply.

8.37 A Pitot-static tube in a Mach 2 air flow indicates a total pressure, $p_t = 99$ kPa. Taking account of the *detached shock* standing upstream of the tube, determine the static pressure measured by the tube.

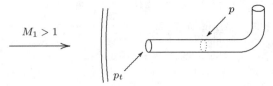

Problems 8.37, 8.38, 8.39

8.38 A Pitot-static tube is inserted in the test section of a supersonic wind tunnel with a reservoir pressure of 10 atm. The test-section Mach number is 3 and the working gas is helium ($\gamma = 5/3$). Accounting for the *detached shock* upstream of the tube, what pressure, p_t, does the stagnation-pressure tap indicate?

8.39 A Pitot-static tube is inserted in the test section of a supersonic wind tunnel with a reservoir pressure of 9 atm. The stagnation-pressure tap indicates $p_t = 1.17$ atm. Accounting for the *detached shock* standing ahead of the tube, what is the test-section Mach number if the working gas is air?

8.40 A supersonic airplane is equipped with a Pitot-static tube in order to measure its speed. If the tube indicates that the static pressure is $p = 13.5$ atm and the total pressure is $p_t = 20.1$ atm, at what Mach number is the airplane moving?

8.41 In the process of calibrating a wind tunnel, an engineer inadvertently places her Pitot tube and other instrumentation so far into the test section that the devices lie downstream of a normal shock. She realizes her oversight when the measured flow properties correspond to subsonic flow. Rather than move the instruments and rerun the test, she opts to simply use the normal-shock tables. The flow downstream of the shock has Mach number, $M_2 = 0.45$, static temperature, $T_2 = 15.8°$ C and static pressure, $p_2 = 40$ kPa. Find the Mach number, M_1, total temperature, T_{t_1}, and total pressure, p_{t_1}, ahead of the shock.

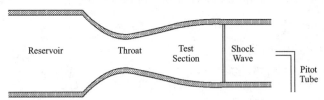

Problem 8.41

8.42 The normal-shock relations give flow properties as a function of Mach number ahead of the shock. Alternatively, we can express properties as functions of the pressure ratio, p_2/p_1. In this form, they are called the *Rankine-Hugoniot relations*. Verify that the Rankine-Hugoniot relations for density and temperature are

$$\frac{\rho_2}{\rho_1} = \frac{1 + \dfrac{\gamma+1}{\gamma-1}\dfrac{p_2}{p_1}}{\dfrac{p_2}{p_1} + \dfrac{\gamma+1}{\gamma-1}} \quad \text{and} \quad \frac{T_2}{T_1} = 1 + \frac{\left(\dfrac{p_2}{p_1}\right)^2 - 1}{\dfrac{\gamma+1}{\gamma-1}\dfrac{p_2}{p_1} + 1}$$

8.43 Beginning with the normal-shock relation for Mach number, viz.,

$$M_2^2 = \frac{1 + \frac{1}{2}(\gamma - 1)M_1^2}{\gamma M_1^2 - \frac{1}{2}(\gamma - 1)}$$

solve for M_1^2 as a function of M_2^2. Compare your result with the equation you are starting with.

8.44 Show for a weak shock wave ($M \approx 1$) that the Mach number downstream of the shock is approximated by $M_2^2 \approx 2 - M_1^2$. Compare values for M_2 obtained using this equation with values for M_2 from the shock tables, with $M_1 = 1, 1.05, 1.1$ and 1.2. **HINT:** Let $M_1^2 = 1 + \epsilon$, where $\epsilon \ll 1$, and expand the exact relationship between M_1 and M_2.

8.45 Show that the lowest Mach number and the largest density ratio possible downstream of a normal shock wave are

$$M_2 = \sqrt{\frac{(\gamma - 1)}{2\gamma}} \quad \text{and} \quad \frac{\rho_2}{\rho_1} = \frac{(\gamma + 1)}{(\gamma - 1)}$$

What are the limiting values of M_2 and ρ_2/ρ_1 for air?

8.46 A shock wave is moving to the left with a velocity W as shown. It is advancing into a uniform freestream with velocity U_∞ and Mach number M_∞. The specific-heat ratio of the gas is γ. If the shock velocity is γU_∞, what is M_∞?

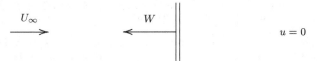

    Problem 8.46

8.47 If the test-section Mach number of a supersonic wind tunnel using air is 2.8 and the test-section area is 70 cm^2, what is the throat area?

8.48 If the test-section Mach number of a supersonic wind tunnel using helium is 3 and the test-section area is 48 cm^2, what is the throat area?

8.49 A wind tunnel is designed to have a Mach number of 2.5, a static pressure of 2 psi, and a static temperature of $-5°$ F in the test section. Determine the area ratio of the nozzle required and the reservoir conditions that must be maintained if air is to be used.

8.50 Consider a supersonic wind tunnel designed by an eager young research assistant.

  (a) If the test-section Mach number is 3 and the test-section area is 50 cm^2, what is the throat area?

  (b) If, as a result of poor design, a shock forms upstream of the test section and the test-section Mach number is only 0.504, what is the throat area?

8.51 What is the pressure, p^*/p_t, at the throat of a Laval nozzle if the working gas is methane ($\gamma = 1.31$)?

8.52 What is the pressure, p^*/p_t, at the throat of a Laval nozzle if the working gas is carbon dioxide ($\gamma = 1.3$)?

8.53 What is the pressure, p^*/p_t, at the throat of a Laval nozzle if the working gas is helium ($\gamma = 5/3$)?

8.54 A perfect gas experiences an isentropic expansion through a Laval nozzle to a point where the area is 1.787 times the throat area. The Mach number at this point is 2. What is the specific-heat ratio, γ, of the gas? **HINT:** Assume $\gamma = 1 + 2/n$ where $n \geq 3$ is an integer.

8.55 Air flows through a Laval nozzle. The exit pressure is sufficiently high to cause a normal shock to form inside the nozzle at a point where the nozzle area is 30 cm^2. Determine the pressure just downstream of the shock if the reservoir pressure is 600 kPa and the throat area is 15 cm^2.

8.56 Air flows from a reservoir with total pressure 1600 kPa through a supersonic nozzle. The ratio of the nozzle-exit area to its throat area is $A_e/A^* = 3.5$. If the exit pressure is $p_e = 101$ kPa, will a shock be present in the nozzle?

8.57 The ratio of exit to throat area for a Laval nozzle is $A_e/A_t = 10$. A normal shock stands inside the nozzle at a point where $A/A_t = 2.5$. The total pressure ahead of the shock is 750 kPa. What is the nozzle-exit Mach number?

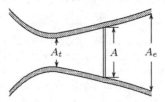

Problems 8.57, 8.58, 8.59, 8.60

8.58 You have just finished assembling your *Laval Nozzle at Home Kit* and you are ready to try it out. Due to external constraints, the nozzle-exit Mach number can be no greater than 1/2. The ratio of the throat area, A_t, to the exit area, A_e, is 1/4. Also, a normal shock wave lies at a point where $A = 2A_t$. Will you be able to satisfy the external constraints?

8.59 The ratio of exit to throat area for a Laval nozzle is $A_e/A_t = 5$. A normal shock stands inside the nozzle at a point where $A/A_t = 3$. The total pressure ahead of the shock is 900 kPa. What is the nozzle-exit Mach number?

8.60 A gas with a specific-heat ratio $\gamma = 1.28$ flows through a nozzle and achieves supersonic conditions. A normal shock stands in the nozzle at a point where the cross-sectional area of the shock is A. The area at the throat is A_t, while the nozzle-exit area is A_e. The static pressure ahead of the shock is $p_1 = 16.25$ psi and the Mach number behind the shock is $M_2 = 0.7$.

(a) Determine the Mach number ahead of the shock, M_1, and the pressure behind the shock, p_2.

(b) Find the total pressure ahead of the shock, p_t, and the static pressure at the throat, p^*.

(c) Compute A_e/A_t and A/A_t for a nozzle-exit Mach number of $M_e = 1/2$.

8.61 Verify that the mass-flow rate, $\dot{m}$, for flow of a perfect, calorically-perfect gas through a choked nozzle is

$$\dot{m} = \frac{p_t A^*}{\sqrt{T_t}}\sqrt{\frac{\gamma}{R}\left(\frac{2}{\gamma+1}\right)^{(\gamma+1)/(\gamma-1)}}$$

8.62 One problem in creating high Mach number flows is condensation of the oxygen component in air when the temperature drops below 50 K. Consider isentropic flow through a nozzle in which the reservoir temperature is 300 K.

(a) At what Mach number will condensation of oxygen occur?

(b) If the radius of the nozzle is given by $r = r_o\sqrt{(1+30x^2)/(x^2+1)}$, where the throat is located at $x = 0$, at what value of x will the oxygen first begin to condense?

8.63 Air flows from a reservoir through a supersonic nozzle. The ratio of the nozzle-exit area to its throat area is $A_e/A^* = 2.5$. If the exit pressure is $p_e = 1$ atm, what must the reservoir pressure be for the *design case*, i.e., for a shock-free flow?

8.64 Transonic wind-tunnel tests are complicated by the fact that *model blockage* can cause significant changes in the flow. To see why, consider a tunnel with a test-section Mach number $M = 1.10$, area $A = 10$ ft^2 and reservoir temperature $T_t = 68°$ F. The working gas is air.

(a) Compute the test-section flow speed, U, and the nozzle-throat area, A^*.

(b) If a model of cross-sectional area $A_m = 0.006A$ is placed in the tunnel, what is the new test-section Mach number?

(c) Compute the test-section flow speed, $\tilde{U}$, with the model of Part (b) present. What is the percentage difference from the flow speed for the unblocked tunnel?

8.65 Compute the area ratio, A/A^*, for isentropic flow at Mach 2 as a function of γ. Make a graph of your results. What are the limiting values of A/A^* for $\gamma \to 1$ and $\gamma \to \infty$?

8.66 In 1985, Cadillac introduced a digital display of the temperature outside the automobile. Assume the temperature sensor actually measures stagnation temperature. If the temperature outside your Cadillac is $70°$ F and, thanks to the protection of your radar jammer, you are traveling at 80 mph, determine the percentage error in the indicated temperature, where Error $= 100 \cdot (T_{\text{actual}} - T_{\text{indicated}})/T_{\text{actual}}$. Suppose the Supersonic Transport (SST) were to use the same device to measure outside temperature. If the SST moves at Mach 2, compute the percentage error, assuming a *detached normal shock wave* is present just upstream of the sensor.

8.67 Verify that the pressure coefficient, C_p, for a calorically-perfect gas can be written as

$$C_p = \frac{2}{\gamma M_\infty^2} \left(\frac{p}{p_\infty} - 1 \right)$$

where subscript ∞ denotes freestream conditions.

8.68 A shock wave propagates into a quiescent fluid in which the static pressure is p_1 and the speed of sound is a_1.

(a) If the pressure behind the shock is p_2, show that the shock moves at a speed U_s, where

$$U_s = a_1 \sqrt{\frac{\gamma - 1}{2\gamma} + \frac{\gamma + 1}{2\gamma} \frac{p_2}{p_1}}$$

(b) A nuclear explosion can generate a pressure of 5000 psi. Using the result of Part (a), compute the speed of the shock created by such an explosion on a day when $p = 14.7$ psi and $T_1 = 50°$ F.

8.69 A blow-down supersonic wind tunnel is supplied with air from a large reservoir as shown. The Mach number in the test section is $M_2 = 2$, and the pressure is below atmospheric so that a shock wave is formed just at the exit. The pressure at Point 3 immediately behind the shock is 14.7 psi.

(a) Find p_0, p_1, p_2, and M_3.

(b) A Pitot tube is placed in the exit jet as shown. What is the pressure p_4? Why is $p_4 < p_0$? How does T_4 compare with T_0?

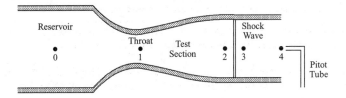

Problem 8.69

8.70 A Pitot-static tube mounted on an airplane indicates the static pressure is 96 kPa and the stagnation pressure is 120 kPa. Determine the two possible Mach numbers at which the plane might be moving.

8.71 A Pitot-static tube mounted on an airplane indicates the static pressure is 95 kPa and the stagnation pressure is 130 kPa. Determine the two possible Mach numbers at which the plane might be moving.

8.72 Consider the flow of air through a duct of varying area. At the inlet (Station 1), the Mach number is $M_1 = 1.7$. A normal shock lies at Station 2. If the cross-sectional areas stand in the ratio $A_1 : A_2 : A_3 = 5 : 4 : 5$, what is the Mach number, M_3, at Station 3?

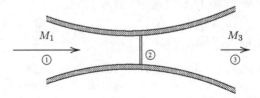

Problems 8.72, 8.73

8.73 Consider the flow of air through a duct of varying area. At the inlet (Station 1), the Mach number is $M_1 = 3$. A normal shock lies at Station 2. If the cross-sectional areas stand in the ratio $A_1 : A_2 : A_3 = 4 : 3 : 6$, what is the Mach number, M_3, at Station 3?

8.74 Air from a large reservoir flows steadily and subsonically through a nozzle. The subsonic jet impinges on a plate. The force required to hold the plate in place is F, the reservoir pressure is p_t and the nozzle-exit area is A_e.

(a) Letting p_a denote atmospheric pressure, verify that

$$F = \frac{2\gamma}{\gamma - 1} \left[(p_t/p_a)^{(\gamma-1)/\gamma} - 1 \right] p_a A_e$$

(b) If $F = 50$ lb, $A_e = 5$ in^2 and $p_a = 14.7$ psi, determine p_t and the nozzle-exit Mach number.

(c) If the working gas is changed to helium at the same total pressure determined in Part (b), what is the force? What is the Mach number at the nozzle exit?

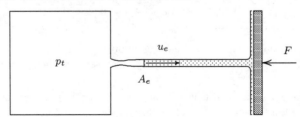

Problems 8.74, 8.75

8.75 Air from a large reservoir flows steadily and subsonically through a nozzle. The subsonic jet impinges on a plate. The force required to hold the plate in place is F, the reservoir pressure is p_t and the nozzle-exit area is A_e.

(a) Letting p_a denote atmospheric pressure, verify that

$$F = \frac{2\gamma}{\gamma - 1} \left[(p_t/p_a)^{(\gamma-1)/\gamma} - 1 \right] p_a A_e$$

(b) If $F = 150$ N, $A_e = 30$ cm^2 and $p_a = 101$ kPa, determine p_t and the nozzle-exit Mach number.

(c) For the conditions of Part (b), what is the force if p_t increases by 25%? Verify that the flow will still be subsonic.

8.76 Oxygen begins to dissociate at a temperature of 2000 K. When this happens, air cannot be approximated as a calorically-perfect gas.

(a) At what Mach number will this temperature be reached for flow past a blunt-nosed body if the freestream temperature is $T_\infty = 20°$ C?

(b) For an insulated body, the temperature at the stagnation point is actually a bit less than the total temperature. It is $T_{aw} = T_\infty \left[1 + \frac{\gamma-1}{2}rM^2\right]$, where $r = 0.85$ is the *recovery factor*. How does this change your estimate of Part (a)?

(c) A common approximation for hypersonic flows is to treat air as a perfect, calorically-perfect gas with $\gamma = 1.2$. Repeat your estimate of Part (b) with $\gamma = 1.2$.

8.77 The Newtonian approximation is often used in hypersonic flow theory. In this approximation we set the specific-heat ratio $\gamma = 1$. What are the normal shock relations, i.e., M_2^2, p_2/p_1, ρ_2/ρ_1 and T_2/T_1 as functions of M_1, according to this approximation?

8.78 Air flows adiabatically through a pipe of diameter $D = 6$ inches. The flow enters with a pressure $p_i = 100$ psi, average flow velocity $\bar{u} = 1650$ ft/sec and temperature $T_i = -4°$ F. If the pipe length is $L = 45$ inches and the exit pressure is $p_e = 145$ psi, what is the average friction factor, $\bar{f}$, and the exit Mach number, M_e?

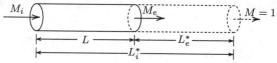

Problems 8.78 through 8.82

8.79 Find the maximum length, in the absence of choking, for adiabatic flow of air in a 20-cm diameter pipe if the friction factor is $\bar{f} = 0.022$. The inlet-flow conditions are $\bar{u} = 250$ m/sec, $T = 44°$ C and $p = 2.5$ atm. Also, determine the temperature at the exit.

8.80 An experiment is designed so that gas flows into a 5-cm diameter pipe with $p = 600$ kPa, $\bar{u} = 300$ m/sec and $T = 20°$ C. The friction factor is $\bar{f} = 0.025$. How far from the entrance, L^*, is the flow choked for hydrogen and for helium?

8.81 Air flows adiabatically through a short pipe of diameter $D = 12$ cm. The flow enters with $p = 500$ kPa, $\bar{u} = 500$ m/sec and $T = -20°$ C. If the pipe length is $L = 1$ m and the exit pressure is $p_e = 720$ kPa, what is the average friction factor, $\bar{f}$, and the exit Mach number?

8.82 Carbon dioxide flows adiabatically through a smooth cylindrical pipe of diameter $D = 10$ cm and length $L = 10$ m. At the inlet, the pressure is $p = 1$ atm, $\bar{u} = 798$ m/sec and $T = 15°$ C. Determine the Mach number, pressure and temperature at the exit.

8.83 Air enters a 5 cm by 5 cm square duct at a velocity $\bar{u} = 900$ m/sec and a temperature $T = 300$ K. The friction factor, $\bar{f}$, is 0.018.

(a) For what duct length will the flow exactly decelerate to Mach 1?

(b) If the duct length is 2 m, will there be a normal shock in the duct?

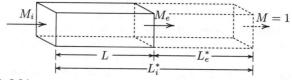

Problems 8.83, 8.84

8.84 Find the maximum length, in the absence of choking, for adiabatic flow of air in a 1-in square duct if the friction factor is $\bar{f} = 0.0145$. The inlet-flow conditions are $\bar{u} = 955$ ft/sec, $T = 9°$ F and $p = 18.1$ psi. Also, determine the pressure at the exit.

8.85 Helium flows adiabatically through a smooth rectangular channel of sides 1 in by 2 in and length $L = 4$ ft. At the inlet, the pressure is $p = 14.7$ psi, $\bar{u} = 8175$ ft/sec and $T = 59°$ F. Determine the Mach number, pressure and temperature at the exit.

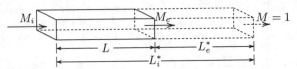

Problems 8.85, 8.86

8.86 Carbon dioxide flows adiabatically through a smooth rectangular channel of sides 9 cm by 20 cm and length $L = 10$ m. At the inlet, the pressure is $p = 100$ kPa, the velocity is $\bar{u} = 665$ m/sec and the temperature is $T = 15°$ C. Determine the Mach number, pressure and temperature at the exit.

8.87 Derive the Fanno-flow Equations (8.104).

8.88 Air flows adiabatically through a short duct whose cross section is an equilateral triangle of side $s = 4$ in. The flow enters with $p = 20$ psi, $\bar{u} = 2000$ ft/sec and $T = 100°$ F. If the duct length is $L = 2$ ft and the exit temperature is $T_e = 210°$ F, what is the average friction factor, $\bar{f}$, and the exit Mach number?

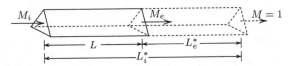

Problem 8.88

8.89 Show that for Fanno flow

$$\frac{dh}{dx} = \frac{\gamma M^2}{M^2 - 1} T \frac{ds}{dx}$$

8.90 Air with velocity $\bar{u} = 330$ m/sec enters a 6 cm by 12 cm rectangular duct at a pressure of $p = 300$ kPa and a temperature of $T = 300$ K. How much heat per unit mass (in J/kg) must be added to achieve sonic conditions at the exit? What is the exit temperature?

8.91 Air with velocity $\bar{u} = 250$ ft/sec enters a circular pipe at a pressure of $p = 30$ psi and a temperature of $T = 30°$ F. How much heat per unit mass (in Btu/slug) must be added to achieve sonic conditions at the exit? What is the exit pressure?

8.92 Helium enters a frictionless duct at a Mach number of 0.8 and a total temperature of $T = -50°$ F. How much heat per unit mass (in Btu/slug) must be extracted to reduce the Mach number to 0.2?

8.93 An experiment is set up with gas entering a frictionless duct with $\bar{u} = 500$ m/sec, $p = 3$ atm and $T = 350$ K. How much heat per unit mass (in J/kg) must be added in order to choke the flow for air and for hydrogen? What is the exit temperature in each case?

8.94 Hydrogen enters a frictionless duct at a Mach number of 4 and a total temperature of $T = 12°$ F. How much heat per unit mass (in Btu/slug) must be extracted to achieve Mach numbers of 3, 2 and 1?

8.95 Air enters a frictionless duct at a Mach number of 0.3 and a total temperature of $T = 293$ K. How much heat per unit mass (in J/kg) must be added to achieve Mach numbers of 0.5, 0.7 and 0.9?

8.96 Gas is flowing through a frictionless pipe, the exit Mach number is 2 and the static temperature is $20°$ C. If the heat extracted from the pipe is 100 kJ/kg, determine the inlet Mach number and static temperature for hydrogen and for air.

8.97 Methane enters a frictionless duct at a Mach number of 1.3 and a total temperature of $T_t = 200°$ F. How much heat per unit mass (in Btu/slug) must be extracted to reduce the total temperature to $0°$ F, and what is the Mach number when this condition is reached?

8.98 Air with velocity $\bar{u} = 75$ m/sec enters a pipe of circular cross section at a pressure $p = 200$ kPa and a temperature $T = -10°$ C. Find the heat per unit mass that must be added to achieve sonic conditions at the exit, and compute the corresponding exit pressure.

8.99 Derive the Rayleigh-flow Equations (8.134).

8.100 You have just purchased a *Blast-Gas Dispenser*, which shoots any loaded gas out of a tube at high velocity. The gas you have chosen to work with is helium. The dispenser includes a heat-control unit that helps regulate the exit temperature. The gas exhausts to the atmosphere for which the pressure is 14.7 psi. The speed of the exit gas is 4500 ft/sec, while the static and total temperature are $70°$ F and $90°$ F, respectively. The static temperature of the helium as it enters the heat-control unit is $110°$ F. You may assume the flow through the heat-control unit can be approximated as Rayleigh flow, and that the inlet flow is supersonic.

(a) Find the flow speed at the inlet to the heat-control unit.

(b) Compute the heat transfer rate as the gas flows through the heat-control unit and determine if the unit cools or heats the gas.

8.101 Show that for Rayleigh flow

$$\frac{dh}{dx} = \frac{\gamma M^2 - 1}{M^2 - 1} T \frac{ds}{dx}$$

Chapter 9

Conservation Laws: Differential Form

In this chapter we derive the differential equations governing conservation of mass, momentum and energy for a fluid. To understand why we might care about the differential equations, consider the following.

On the one hand, we have seen in Chapters 5 through 8 that the integral forms of the conservation laws provide an excellent tool for analyzing fluid-flow problems. When they are applied to a finite control volume, they give a *global view* that permits accurate computation of things such as the overall force on an object in terms of known or measurable flow properties entering and leaving the chosen volume.

On the other hand, using the integral forms for a finite control volume yields no information about flow details within. We might desire such information in order to determine why a given vehicle design fails to meet or, better yet, exceeds desired design objectives. Only with a *detailed view* can we examine fine details of the flow about the vehicle to discover what might happen to be deficient or surprisingly proficient about our design. The differential forms of the conservation laws provide such a detailed view.

By focusing on a differential-sized control volume, we can deduce the differential equations governing motion of a fluid directly from the integral forms. We thus begin the chapter by applying the Reynolds Transport Theorem to an infinitesimal control volume. Confining our focus to inviscid fluids, we then use the results to derive the differential equations for conservation of mass, momentum and energy.

After deriving the differential forms, we examine a few properties of fluid motion, including a rigorous derivation of **Bernoulli's equation**. We show how Bernoulli's equation, which represents conservation of mechanical energy, differs from exact energy conservation where internal energy is included in the overall balance.

We also demonstrate **Galilean invariance** of the momentum equation. This is a nontrivial consideration for two key reasons. First, a Galilean transformation is a linear operation while the Eulerian description makes the momentum equation quasi-linear. Thus, it is not obvious that the fluid mechanics equations of motion are invariant under a Galilean transformation. Second, if our equations are not invariant, we cannot use measurements in a wind tunnel for a stationary model to infer forces on a prototype moving into a fluid that is at rest. This would greatly complicate the job of the experimenter who would have to design models that are in motion in a wind tunnel.

9.1   Reynolds Transport Theorem Revisited

In developing the differential equations governing motion of a fluid, we shift our focus from a finite control volume to a specific point in the flowfield. As we will see, this is equivalent to taking the limit as the size of the control volume goes to zero.

Recall that in Section 4.7 we stated the Reynolds Transport Theorem in two forms. The first [Equation (4.69)] is appropriate for general control-volume applications. The second, equally valid form is Equation (4.73), which we repeat here for convenience.

$$\frac{dB}{dt} = \iiint_V \frac{\partial(\rho\beta)}{\partial t}\, dV + \oiint_S \rho\beta\, \mathbf{u}\cdot\mathbf{n}\, dS \tag{9.1}$$

The quantity B denotes an extensive variable while β denotes the corresponding intensive variable. This is the form that is most suitable for deducing the differential equations of fluid motion.

As our first step in developing the differential conservation laws, we apply the Reynolds Transport Theorem to the differential element shown in Figure 9.1. The element is a rectangular parallelepiped and fluid passes through its six faces. We assume that the fluid element's center lies at the point (x, y, z) and that the lengths of the three sides are Δx, Δy and Δz in the x, y and z directions, respectively.

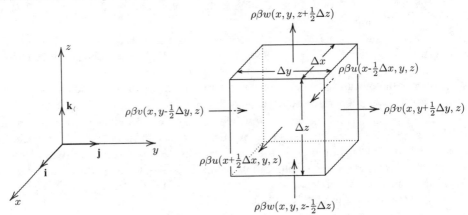

Figure 9.1: *Fluid property β flowing through an infinitesimally small control volume centered at the point (x, y, z). $\mathbf{i}$, $\mathbf{j}$ and $\mathbf{k}$ are unit vectors in the x, y and z directions, respectively.*

Clearly, as the volume $\Delta V = \Delta x \Delta y \Delta z$ becomes infinitesimally small, the volume integral can be approximated as

$$\iiint_V \frac{\partial(\rho\beta)}{\partial t}\, dV \approx \frac{\partial(\rho\beta)}{\partial t}\Delta V \quad \text{as} \quad \Delta V \to 0 \tag{9.2}$$

where $\partial(\rho\beta)/\partial t$ is evaluated at the center of the rectangular parallelepiped.[1]

Turning now to the integral of $\rho\beta\, \mathbf{u}\cdot\mathbf{n}$ over the surface bounding the fluid element, we observe that the net flux of property β across the control surface, $\dot{B}_{cs}$, is

$$\dot{B}_{cs} \equiv \oiint_S \rho\beta\, \mathbf{u}\cdot\mathbf{n}\, dS = \sum_{m=1}^{6} \iint_{A_m} \rho\beta\, \mathbf{u}\cdot\mathbf{n}\, dA_m \tag{9.3}$$

---

[1]This will be exact in the limit as Δx, Δy and Δz approach zero because the ratio of the volume integral of any quantity Φ to ΔV is exactly $\Phi(x, y, z)$.

where the six A_m are the areas of the six faces of the element. For this geometry, the outer unit normal vectors are:

- Face at $x+\frac{1}{2}\Delta x$: $\mathbf{n} = \mathbf{i} \implies \mathbf{u} \cdot \mathbf{n} = u$, Face at $x-\frac{1}{2}\Delta x$: $\mathbf{n} = -\mathbf{i} \implies \mathbf{u} \cdot \mathbf{n} = -u$

- Face at $y+\frac{1}{2}\Delta y$: $\mathbf{n} = \mathbf{j} \implies \mathbf{u} \cdot \mathbf{n} = v$, Face at $y-\frac{1}{2}\Delta y$: $\mathbf{n} = -\mathbf{j} \implies \mathbf{u} \cdot \mathbf{n} = -v$

- Face at $z+\frac{1}{2}\Delta z$: $\mathbf{n} = \mathbf{k} \implies \mathbf{u} \cdot \mathbf{n} = w$, Face at $z-\frac{1}{2}\Delta z$: $\mathbf{n} = -\mathbf{k} \implies \mathbf{u} \cdot \mathbf{n} = -w$

So, for our infinitesimal control volume, we have

$$\dot{B}_{cs} = \underbrace{\iint_A (\rho\beta u)\, dydz}_{\text{Face at } x+\frac{1}{2}\Delta x} - \underbrace{\iint_A (\rho\beta u)\, dydz}_{\text{Face at } x-\frac{1}{2}\Delta x} + \underbrace{\iint_A (\rho\beta v)\, dxdz}_{\text{Face at } y+\frac{1}{2}\Delta y}$$

$$- \underbrace{\iint_A (\rho\beta v)\, dxdz}_{\text{Face at } y-\frac{1}{2}\Delta y} + \underbrace{\iint_A (\rho\beta w)\, dxdy}_{\text{Face at } z+\frac{1}{2}\Delta z} - \underbrace{\iint_A (\rho\beta w)\, dxdy}_{\text{Face at } z-\frac{1}{2}\Delta z} \qquad (9.4)$$

Because we are dealing with infinitesimally small quantities, we can treat $\rho\beta\,\mathbf{u}\cdot\mathbf{n}$ as being constant on each face. Our result will be exact in the limit as Δx, Δy and Δz approach zero. Thus,

$$\begin{aligned}\dot{B}_{cs} &\approx \left[\rho\beta u(x+\tfrac{1}{2}\Delta x, y, z) - \rho\beta u(x-\tfrac{1}{2}\Delta x, y, z)\right]\Delta y\Delta z \\ &+ \left[\rho\beta v(x, y+\tfrac{1}{2}\Delta y, z) - \rho\beta v(x, y-\tfrac{1}{2}\Delta y, z)\right]\Delta x\Delta z \\ &+ \left[\rho\beta w(x, y, z+\tfrac{1}{2}\Delta z) - \rho\beta w(x, y, z-\tfrac{1}{2}\Delta z)\right]\Delta x\Delta y \end{aligned} \qquad (9.5)$$

Next, we rearrange terms to arrive at

$$\begin{aligned}\dot{B}_{cs} &\approx \frac{\rho\beta u(x+\tfrac{1}{2}\Delta x, y, z) - \rho\beta u(x-\tfrac{1}{2}\Delta x, y, z)}{\Delta x}\Delta V \\ &+ \frac{\rho\beta v(x, y+\tfrac{1}{2}\Delta y, z) - \rho\beta v(x, y-\tfrac{1}{2}\Delta y, z)}{\Delta y}\Delta V \\ &+ \frac{\rho\beta w(x, y, z+\tfrac{1}{2}\Delta z) - \rho\beta w(x, y, z-\tfrac{1}{2}\Delta z)}{\Delta z}\Delta V \end{aligned} \qquad (9.6)$$

where, as noted above, $\Delta V = \Delta x\Delta y\Delta z$. Finally, as Δx, Δy and Δz approach zero, each line in Equation (9.6) contains a partial derivative of a component of the vector $\rho\beta\,\mathbf{u}$. Therefore, the net flux of property β across the control surface is

$$\dot{B}_{cs} \approx \left[\frac{\partial(\rho\beta u)}{\partial x} + \frac{\partial(\rho\beta v)}{\partial y} + \frac{\partial(\rho\beta w)}{\partial z}\right]\Delta V \qquad (9.7)$$

The quantity in square brackets is the familiar divergence of $\rho\beta\,\mathbf{u}$, which we write symbolically as

$$\nabla \cdot (\rho\beta\,\mathbf{u}) = \frac{\partial(\rho\beta u)}{\partial x} + \frac{\partial(\rho\beta v)}{\partial y} + \frac{\partial(\rho\beta w)}{\partial z} \qquad (9.8)$$

Therefore, we have proven that

$$\oiint_S \rho\beta\,\mathbf{u}\cdot\mathbf{n}\, dS \approx \nabla \cdot (\rho\beta\,\mathbf{u})\,\Delta V \quad \text{as} \quad \Delta V \to 0 \qquad (9.9)$$

Substituting Equations (9.2) and (9.9) into Equation (9.1), dividing through by ΔV and letting $\Delta V \to 0$, we arrive at the limiting form of the Reynolds Transport applied to an infinitesimal control volume.

$$\lim_{\Delta V \to 0} \frac{1}{\Delta V} \frac{dB}{dt} = \frac{\partial(\rho\beta)}{\partial t} + \nabla \cdot (\rho\beta\,\mathbf{u}) \tag{9.10}$$

With just a bit more manipulation we can cast this result in an especially useful form that will minimize the number of mathematical operations needed to derive the differential conservation laws. Specifically, we can differentiate the terms in parentheses to obtain

$$\frac{\partial(\rho\beta)}{\partial t} = \rho\frac{\partial\beta}{\partial t} + \beta\frac{\partial\rho}{\partial t} \tag{9.11}$$

and (see Appendix C for vector differential identities)

$$\nabla \cdot (\rho\beta\,\mathbf{u}) = \rho\,\mathbf{u} \cdot \nabla\beta + \beta\,\nabla \cdot (\rho\,\mathbf{u}) \tag{9.12}$$

Consequently, we find

$$\lim_{\Delta V \to 0} \frac{1}{\Delta V} \frac{dB}{dt} = \rho\left[\frac{\partial\beta}{\partial t} + \mathbf{u} \cdot \nabla\beta\right] + \beta\left[\frac{\partial\rho}{\partial t} + \nabla \cdot (\rho\,\mathbf{u})\right] \tag{9.13}$$

The first term in square brackets is just the Eulerian derivative of β. Thus, the limiting form of the Reynolds Transport Theorem that we will use in the next three sections to derive the differential equations governing the motion of a fluid is:

$$\lim_{\Delta V \to 0} \frac{1}{\Delta V} \frac{dB}{dt} = \rho\frac{d\beta}{dt} + \beta\left[\frac{\partial\rho}{\partial t} + \nabla \cdot (\rho\,\mathbf{u})\right] \tag{9.14}$$

As we will discover in the next section, this equation simplifies even further when we account for mass conservation.

9.2   Conservation of Mass

We deduce a differential equation for mass conservation by applying the limiting form of the Reynolds Transport Theorem, Equation (9.14), to an infinitesimal control volume. Just as in our derivation for a finite-sized control volume in Section 5.2, our extensive variable B is the mass of the system that is coincident with our chosen control volume at time t. The corresponding intensive variable is $\beta = 1$.

Since the mass of a system and the intensive variable β are both constant, necessarily $dB/dt = 0$ and $d\beta/dt = 0$. Hence, inspection of Equation (9.14) shows that the differential equation governing mass conservation for a fluid *at every point within the control volume* is

$$\frac{\partial\rho}{\partial t} + \nabla \cdot (\rho\,\mathbf{u}) = 0 \tag{9.15}$$

This equation is often referred to as the **continuity equation**. Equation (9.15) is in what is known as **conservation form**. By definition, this means the differential equation consists of the sum of the time derivative of one quantity and the divergence of another. We can expand $\nabla \cdot (\rho\,\mathbf{u})$ and rewrite Equation (9.15) as

$$\frac{\partial\rho}{\partial t} + \mathbf{u} \cdot \nabla\rho + \rho\nabla \cdot \mathbf{u} = 0 \tag{9.16}$$

The sum of the first two terms on the left-hand side of Equation (9.16) is the Eulerian derivative of ρ. Thus, we arrive at the continuity equation in **primitive-variable form**.

$$\frac{d\rho}{dt} + \rho\nabla \cdot \mathbf{u} = 0 \qquad (9.17)$$

Before moving on to momentum and energy conservation, there is a very important simplification we can make to the limiting form of the Reynolds Transport Theorem for an infinitesimal control volume. Specifically, substituting the continuity equation in conservation form [Equation (9.15)] into the Reynolds Transport Theorem yields a remarkable result, i.e.,

$$\lim_{\Delta V \to 0} \frac{1}{\Delta V}\frac{dB}{dt} = \rho\frac{d\beta}{dt} \qquad (9.18)$$

In addition to simplifying our derivations in the next two subsections, this illustrates the connection between the Reynolds Transport Theorem and the Eulerian derivative, both of which focus on the rate of change following fluid particles. It tells us, in words, that in the limit as a control volume approaches a single point in the flow, the rate of change of a specified extensive variable per unit volume for a system equals the fluid density times the Eulerian derivative of the corresponding intensive variable within the control volume.

9.3 Conservation of Momentum

As with mass conservation, we deduce a differential equation for momentum conservation by applying the limiting form of the Reynolds Transport Theorem, Equation (9.18), to an infinitesimal control volume. Our extensive variable is the momentum, $\mathbf{P}$, and the corresponding intensive variable is the velocity vector, $\mathbf{u}$. Confining our attention to inviscid fluids, the only surface force acting is pressure. Then, if we have a body force present, Newton's second law of motion tells us that for a moving system (see Section 5.3):

$$\frac{d\mathbf{P}}{dt} = -\oiint_S p\,\mathbf{n}\,dS + \iiint_V \rho\mathbf{f}\,dV \qquad (9.19)$$

where $\mathbf{f}$ is the specific body force vector. Proceeding term by term from left to right, the Reynolds Transport Theorem [Equation (9.18)] tells us that for the momentum equation,

$$\lim_{\Delta V \to 0} \frac{1}{\Delta V}\frac{d\mathbf{P}}{dt} = \rho\frac{d\mathbf{u}}{dt} \qquad (9.20)$$

Next, recall that in Section 3.1 we evaluated the net pressure force acting on an infinitesimal control volume and demonstrated that

$$-\oiint_S p\,\mathbf{n}\,dS \approx -\nabla p\,\Delta V \quad \text{for} \quad \Delta V \to 0 \quad \Longrightarrow \quad -\lim_{\Delta V \to 0}\frac{1}{\Delta V}\oiint_S p\,\mathbf{n}\,dS = -\nabla p \quad (9.21)$$

Also, for the obvious reason,

$$\lim_{\Delta V \to 0}\frac{1}{\Delta V}\iiint_V \rho\mathbf{f}\,dV = \rho\mathbf{f} \qquad (9.22)$$

Collecting all of this, the resulting differential equation governing momentum conservation at each point in a flowfield is

$$\rho\frac{d\mathbf{u}}{dt} = -\nabla p + \rho\mathbf{f} \qquad (9.23)$$

This equation, valid for a perfect fluid, is known as **Euler's equation**. It is in primitive-variable form, and can be used to compute the details of general fluid motion at every point in a flow. In words, this equation says that *mass per unit volume times acceleration equals the sum of forces per unit volume*, i.e., it is Newton's second law of motion per unit volume.

9.4   Conservation of Energy

Energy conservation for a fluid is based upon the first law of thermodynamics, which says

$$\frac{dE}{dt} = \dot{Q} - \dot{W} \tag{9.24}$$

where E is total energy of the system, $\dot{Q}$ is the rate at which heat is transferred from the surroundings to the system and $\dot{W}$ is work done by the system on the surroundings. Because details of the rate of heat transfer, $\dot{Q}$, depend upon the process and have not been specified, it is impossible at this point to write them as either a surface or a volume integral. We thus confine our attention to a differential form of the energy-conservation equation in the special case of adiabatic flow (for which $\dot{Q} = 0$). We restrict our formulation to a frictionless fluid, so that the only force doing work is the pressure, and we assume the body force acting on our system and control volume is conservative (and independent of time) so that $\mathbf{f} = -\nabla \mathcal{V}$. To further simplify our analysis, we will also assume our infinitesimal control volume is stationary.

Thus, the extensive variable for our infinitesimal system is total energy and the corresponding intensive variable is specific total energy, viz., $\beta = \mathcal{E}$, where

$$\mathcal{E} \equiv e + \frac{1}{2}\mathbf{u} \cdot \mathbf{u} + \mathcal{V} \tag{9.25}$$

In words, the specific total energy is the sum of specific internal energy, specific kinetic energy and the force potential.

As with the momentum equation, we proceed term by term beginning with the left-hand side of Equation (9.24). Using the Reynolds Transport Theorem for an infinitesimal control volume [Equation (9.18)], we have

$$\lim_{\Delta V \to 0} \frac{1}{\Delta V} \frac{dE}{dt} = \rho \frac{d\mathcal{E}}{dt} \tag{9.26}$$

Since we are excluding heat transfer, only the work term remains to be evaluated. Recalling Equation (6.30), the work term for an inviscid fluid in a stationary control volume is

$$\dot{W} = \oiint_{S} p\,(\mathbf{u} \cdot \mathbf{n})\,dS \tag{9.27}$$

We can evaluate this integral for an infinitesimal control volume exactly as in Section 9.1. Clearly, if we make the formal substitution of p for $\rho\beta$ in Equation (9.9), we have[2]

$$\lim_{\Delta V \to 0} \frac{\dot{W}}{\Delta V} = \nabla \cdot (p\,\mathbf{u}) \tag{9.28}$$

wherefore the differential equation for energy in terms of total energy, $\mathcal{E}$ is

$$\rho \frac{d\mathcal{E}}{dt} = -\nabla \cdot (p\,\mathbf{u}) \tag{9.29}$$

---

[2]We are not redefining the intensive variable β. We are simply making use of the fact that all of the steps leading to Equation (9.1) would be unchanged had we started with p rather than $\rho\beta$.

It is customary to recast this equation in terms of total enthalpy, $\mathcal{H}$, defined by

$$\mathcal{H} \equiv h + \frac{1}{2}\mathbf{u}\cdot\mathbf{u} + \mathcal{V} \tag{9.30}$$

where $h = e + p/\rho$ is specific enthalpy. Thus, specific total energy and specific total enthalpy are related by

$$\mathcal{H} = \mathcal{E} + \frac{p}{\rho} \tag{9.31}$$

So, we have

$$\rho\frac{d\mathcal{H}}{dt} = \rho\frac{d\mathcal{E}}{dt} + \rho\frac{d}{dt}\left(\frac{p}{\rho}\right) = \rho\frac{d\mathcal{E}}{dt} + \frac{dp}{dt} - \frac{p}{\rho}\frac{d\rho}{dt} \tag{9.32}$$

Appealing to the continuity equation [Equation (9.17)] to eliminate $d\rho/dt$ and expanding dp/dt, Equation (9.33) simplifies to the following.

$$\rho\frac{d\mathcal{H}}{dt} = \rho\frac{d\mathcal{E}}{dt} + \frac{\partial p}{\partial t} + \mathbf{u}\cdot\nabla p + p\,\nabla\cdot\mathbf{u} = \rho\frac{d\mathcal{E}}{dt} + \frac{\partial p}{\partial t} + \nabla\cdot(p\,\mathbf{u}) \tag{9.33}$$

Finally, combining Equations (9.29) and (9.33), the energy equation becomes

$$\rho\frac{d\mathcal{H}}{dt} = \frac{\partial p}{\partial t} \tag{9.34}$$

where the total enthalpy, $\mathcal{H}$, is defined in Equation (9.30). We can make an immediate observation that follows from Equation (9.34). On the one hand, if a flow is steady, the total enthalpy will remain constant along streamlines. On the other hand, if the flow is unsteady, the $\partial p/\partial t$ term causes the total enthalpy of a fluid particle to vary as it moves through the flowfield. That is, because kinetic energy is not Galilean invariant, neither is $\mathcal{H}$.

9.5 Summary of the Conservation Equations

Before proceeding to an investigation of the conservation laws in differential form, it is worthwhile to summarize the equations developed in Sections 9.2, 9.3 and 9.4. We have shown that the differential equations of motion for an inviscid, adiabatic fluid are the continuity equation, Euler's equation and the energy equation. These equations apply to both liquids and gases. All we need to complete our set of equations is the equation of state. In summary, remembering that total enthalpy is defined by $\mathcal{H} \equiv h + \frac{1}{2}\mathbf{u}\cdot\mathbf{u} + \mathcal{V}$, the equations are:

Equation of State:

$$\rho = \begin{cases} \dfrac{p}{RT}, & \text{gases} \\[2mm] \text{constant}, & \text{liquids} \end{cases} \tag{9.35}$$

Continuity Equation:

$$\frac{d\rho}{dt} + \rho\nabla\cdot\mathbf{u} = 0 \tag{9.36}$$

Euler's Equation:

$$\rho\frac{d\mathbf{u}}{dt} = -\nabla p + \rho\mathbf{f} \tag{9.37}$$

Energy Equation:

$$\rho\frac{d\mathcal{H}}{dt} = \frac{\partial p}{\partial t} \tag{9.38}$$

9.6   Mass Conservation for Incompressible Flows

In Cartesian coordinates, the continuity equation [Equation (9.36)] is

$$\frac{\partial \rho}{\partial t} + u\frac{\partial \rho}{\partial x} + v\frac{\partial \rho}{\partial y} + w\frac{\partial \rho}{\partial z} + \rho\frac{\partial u}{\partial x} + \rho\frac{\partial v}{\partial y} + \rho\frac{\partial w}{\partial z} = 0 \qquad (9.39)$$

Continuity assumes an especially simple form for incompressible flows. Specifically, if the density, ρ, is constant, we have

$$\nabla \cdot \mathbf{u} = \frac{\partial u}{\partial x} + \frac{\partial v}{\partial y} + \frac{\partial w}{\partial z} = 0 \qquad \text{(Incompressible)} \qquad (9.40)$$

Equation (9.40) has the remarkable property that it holds for both steady and unsteady incompressible flows. Given the velocity vector for a flowfield, we can use this equation to determine whether or not the flow is incompressible. We can also use it to determine necessary conditions for a flow to be incompressible. The following two examples illustrate how the incompressible continuity equation can be used.

Example 9.1 *Consider two-dimensional flow approaching a stagnation point. In the immediate vicinity of the stagnation point, the velocity vector is given by* $\mathbf{u} = A(x\,\mathbf{i} - y\,\mathbf{j})$*, where x and y are tangent to and normal to the surface, respectively. The quantity A is a constant. Is this flow incompressible?*

Solution. Taking the divergence of this vector, we find

$$\nabla \cdot \mathbf{u} = \left(\mathbf{i}\frac{\partial}{\partial x} + \mathbf{j}\frac{\partial}{\partial y} \right) \cdot (Ax\,\mathbf{i} - Ay\,\mathbf{j}) = A - A = 0$$

Thus, we conclude that this flow is incompressible.

Example 9.2 *An unsteady flow has the following velocity field:*

$$\mathbf{u} = \Omega(\Omega x t + y)\,\mathbf{i} + \Omega(Ax + B\Omega y t)\,\mathbf{j}$$

where Ω is a constant of dimension 1/time. The quantities A and B are dimensionless constants. The flow is incompressible and irrotational. Find the values of A and B necessary to guarantee these conditions.

Solution. Since the flow is incompressible and two-dimensional, necessarily

$$\frac{\partial u}{\partial x} + \frac{\partial v}{\partial y} = \Omega\frac{\partial}{\partial x}(\Omega x t + y) + \Omega\frac{\partial}{\partial y}(Ax + B\Omega y t) = \Omega^2(t + Bt) = 0 \qquad \Longrightarrow \qquad B = -1$$

Also, since the flow is irrotational, the vorticity is $(\partial v/\partial x - \partial u/\partial y)\mathbf{k} = \mathbf{0}$, so that

$$\frac{\partial v}{\partial x} - \frac{\partial u}{\partial y} = \Omega\frac{\partial}{\partial x}(Ax + B\Omega y t) - \Omega\frac{\partial}{\partial y}(\Omega x t + y) = \Omega(A - 1) = 0 \qquad \Longrightarrow \qquad A = 1$$

We can gain some physical insight into the incompressible form of the continuity equation by substituting specific volume, $v = 1/\rho$, into Equation (9.36). With a little algebra, we obtain the following equation.

$$\frac{1}{v}\frac{dv}{dt} = \nabla \cdot \mathbf{u} \qquad (9.41)$$

Now, consider a fluid particle whose initial mass and volume are M and V, respectively. When we compute the Eulerian derivative of a flow property, we are computing the rate of change following the fluid particle, i.e., we are computing the rate of change of a *system* coincident with our fluid particle at a specified instant in time. But, the mass of a system is constant, wherefore the volume is $V = Mv$ and its rate of change is $dV/dt = M\,dv/dt$. Therefore,

$$\frac{1}{V}\frac{dV}{dt} = \nabla \cdot \mathbf{u} \tag{9.42}$$

This result tells us that when we have incompressible flow, for which $\nabla \cdot \mathbf{u}$ vanishes, the volume of a fluid particle remains constant.

Figure 9.2 illustrates two examples of volume distortions that are possible in an incompressible flow. The initial shape of the fluid particle [Figure 9.2(a)] is assumed to be a cube. If the particle undergoes compression in the xy plane, it must expand in the z direction. Figure 9.2(b), drawn exactly to scale, shows what happens when the x side is reduced by 10% and the y side by 40%. To have the same volume as the original cube, the z side must increase by 85% ($0.90 \times 0.60 \times 1.85 = 1.00$). Similarly, for the expansion case shown in Figure 9.2(c), doubling the length of the x side with the y side unchanged requires halving the z side to maintain constant volume.

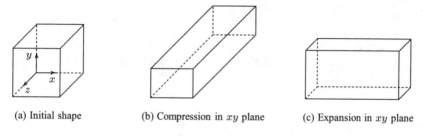

(a) Initial shape   (b) Compression in xy plane   (c) Expansion in xy plane

Figure 9.2: *Incompressible volume-deformation examples.*

9.7 Euler's Equation

Although Euler's equation [Equation (9.37)] looks relatively simple in vector form, our shorthand notation for the Eulerian derivative conceals the complexity of this vector partial differential equation. Expanding the differential operator d/dt into its unsteady and convective parts, the three components of Euler's equation in Cartesian coordinates are

$$\left. \begin{array}{l} \rho\dfrac{\partial u}{\partial t} + \rho u\dfrac{\partial u}{\partial x} + \rho v\dfrac{\partial u}{\partial y} + \rho w\dfrac{\partial u}{\partial z} = -\dfrac{\partial p}{\partial x} + \rho f_x \\[3mm] \rho\dfrac{\partial v}{\partial t} + \rho u\dfrac{\partial v}{\partial x} + \rho v\dfrac{\partial v}{\partial y} + \rho w\dfrac{\partial v}{\partial z} = -\dfrac{\partial p}{\partial y} + \rho f_y \\[3mm] \rho\dfrac{\partial w}{\partial t} + \rho u\dfrac{\partial w}{\partial x} + \rho v\dfrac{\partial w}{\partial y} + \rho w\dfrac{\partial w}{\partial z} = -\dfrac{\partial p}{\partial z} + \rho f_z \end{array} \right\} \tag{9.43}$$

Inspection of Equations (9.43) tells us that Euler's equation is not linear. That is, the convective acceleration terms, i.e., terms such as $u\partial u/\partial x$, involve products of the velocity and its derivative. These terms make the equation **quasi-linear**.[3] Unlike linear equations, we cannot use superposition or even prove that our solution exists and is unique.

On the one hand, these coupled, quasi-linear, partial differential equations are not easy to solve, even for simple, idealized geometries. One noteworthy exception is for incompressible, irrotational flow, known as **potential flow**. As we will see in Chapter 10, under these conditions the continuity equation provides a linear partial differential equation and all of the nonlinearity is confined to the relation between pressure and velocity, viz., through Bernoulli's equation. Even in this special case, solution of the equations of motion requires complex computer programs for general fluid-flow problems.

On the other hand, after several decades of research and development, such programs are readily available, not only for potential flows, but for rotational flows in a compressible medium. This is one of the key advances in fluid mechanics attributable to the special branch known as Computational Fluid Dynamics (CFD).

We will not attempt to solve Euler's equation for any but the simplest applications in this text, typically where the velocity or acceleration is known in advance. Aside from a cursory examination of *potential-flow theory* in Chapter 10, we leave the difficult task of analyzing and solving these equations to advanced fluid mechanics books such as those written by Landau and Lifshitz (1966), Panton (1996) and Wilcox (2000).

Example 9.3 *The average velocity of water flowing through a nozzle increases from $u_1 = 5$ m/sec to $u_2 = 20$ m/sec. Assuming the average velocity varies linearly with distance along the nozzle, x, and that the length of the nozzle is $\ell = 1$ m, estimate the pressure gradient, dp/dx, at a point midway through the nozzle. The density of water is $\rho = 1000$ kg/m³. You may assume the flow can be approximated as one dimensional.*

Solution. From the one-dimensional Euler equation,

$$\rho u \frac{du}{dx} = -\frac{dp}{dx}$$

The velocity varies linearly from u_1 to u_2 as x increases from 0 to ℓ. Thus,

$$u(x) = u_1 + (u_2 - u_1)\frac{x}{\ell} \quad \Longrightarrow \quad \frac{du}{dx} = \frac{u_2 - u_1}{\ell}$$

Halfway through the nozzle, we thus have

$$u = \frac{1}{2}(u_1 + u_2) \quad \text{and} \quad \frac{du}{dx} = \frac{u_2 - u_1}{\ell}$$

Therefore, the pressure gradient is

$$\frac{dp}{dx} = -\rho\left(\frac{u_1 + u_2}{2}\right)\left(\frac{u_2 - u_1}{\ell}\right) = -\rho\frac{u_2^2 - u_1^2}{2\ell}$$

For the given values,

$$\frac{dp}{dx} = -\left(1000\,\frac{\text{kg}}{\text{m}^3}\right)\frac{(20\,\text{m/sec})^2 - (5\,\text{m/sec})^2}{2(1\,\text{m})}$$

$$= -1.875 \cdot 10^5\,\frac{\text{N}}{\text{m}^3} = -187.5\,\frac{\text{kPa}}{\text{m}}$$

---

[3]We stop short of calling the equation **nonlinear** because, in strict mathematical terms, an equation is nonlinear when the highest derivative in the equation appears in other than a linear form.

9.7.1 Rotating Tank

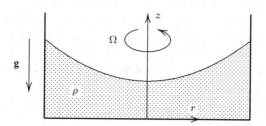

Figure 9.3: *Rotating tank of incompressible fluid—cross-sectional view.*

Incompressible flow in a rotating cylindrical tank (Figure 9.3) is an interesting flow to analyze using Euler's equation. We assume the tank has been rotating for a long time, so that the fluid all moves with constant angular velocity, $\mathbf{\Omega} = \Omega\mathbf{k}$. That is, the fluid is in a state of **rigid-body rotation** (recall our discussion of flow in a rotating cylinder in Section 4.3), so that the velocity at any point in the fluid is

$$\mathbf{u} = \mathbf{\Omega} \times \mathbf{r} = \Omega r \mathbf{e}_\theta \qquad (9.44)$$

where r is radial distance from the center of the tank, and $\mathbf{e}_\theta$ is a unit vector in the circumferential direction. Hence, as shown in Section 4.3, the vorticity is

$$\boldsymbol{\omega} = \nabla \times \mathbf{u} = 2\mathbf{\Omega} \qquad (9.45)$$

Because the vorticity is nonvanishing, we cannot use Bernoulli's equation to help analyze this flow. Rather, we must solve Euler's equation if we wish to determine, for example, the pressure in the fluid.

Since the geometry is symmetric about the z axis, we use the axisymmetric form of Euler's equation. From Appendix D, we find that the three components are as follows.

$$\left.\begin{array}{rcl}
\rho u_r \dfrac{\partial u_r}{\partial r} + \rho \dfrac{u_\theta}{r}\dfrac{\partial u_r}{\partial \theta} + \rho w \dfrac{\partial u_r}{\partial z} - \rho\dfrac{u_\theta^2}{r} & = & -\dfrac{\partial p}{\partial r} \\[2ex]
\rho u_r \dfrac{\partial u_\theta}{\partial r} + \rho \dfrac{u_\theta}{r}\dfrac{\partial u_\theta}{\partial \theta} + \rho w \dfrac{\partial u_\theta}{\partial z} + \rho\dfrac{u_r u_\theta}{r} & = & -\dfrac{1}{r}\dfrac{\partial p}{\partial \theta} \\[2ex]
\rho u_r \dfrac{\partial w}{\partial r} + \rho \dfrac{u_\theta}{r}\dfrac{\partial w}{\partial \theta} + \rho w \dfrac{\partial w}{\partial z} & = & -\dfrac{\partial p}{\partial z} - \rho g
\end{array}\right\} \qquad (9.46)$$

Now, for rigid-body rotation, we know that the radial (u_r) and axial (w) velocity components vanish and the circumferential component is given by $u_\theta = \Omega r$. Hence, Equations (9.46) simplify to

$$\left.\begin{array}{rcl}
-\rho\Omega^2 r & = & -\dfrac{\partial p}{\partial r} \\[2ex]
0 & = & -\dfrac{\partial p}{\partial \theta} \\[2ex]
0 & = & -\dfrac{\partial p}{\partial z} - \rho g
\end{array}\right\} \qquad (9.47)$$

To solve this coupled set of equations, we begin by integrating the first of the three with respect to r. Note that when we perform an integration of a function of r, θ and z with respect to r, we introduce a *function of integration*, $f(\theta, z)$. This is the analog of the *constant of integration* that appears when we integrate a function of a single variable. Therefore, we have

$$p(r, \theta, z) = \frac{1}{2}\rho\Omega^2 r^2 + f(\theta, z) \qquad (9.48)$$

Next, we differentiate Equation (9.48) with respect to θ and substitute into the second of Equations (9.47), viz.,

$$\frac{\partial p}{\partial \theta} = \frac{\partial f}{\partial \theta} = 0 \quad \Longrightarrow \quad f(\theta, z) = F(z) \qquad (9.49)$$

That is, we have shown that, at most, our function of integration is a function only of z. This is consistent with the axial symmetry of the flow. So, the pressure is now given by

$$p(r, \theta, z) = \frac{1}{2}\rho\Omega^2 r^2 + F(z) \qquad (9.50)$$

Finally, to determine the function $F(z)$, we differentiate Equation (9.50) with respect to z and substitute into the last of Equations (9.47). This yields

$$\frac{\partial p}{\partial z} = \frac{dF}{dz} = -\rho g \quad \Longrightarrow \quad F(z) = -\rho gz + \text{constant} \qquad (9.51)$$

wherefore the pressure is given by

$$p(r, \theta, z) = \frac{1}{2}\rho\Omega^2 r^2 - \rho gz + \text{constant} \qquad (9.52)$$

Thus, we conclude that the pressure in the rotating tank satisfies the following equation.

$$p + \rho gz - \frac{1}{2}\rho\Omega^2 r^2 = \text{constant} \qquad (9.53)$$

Although similar in form, this is not Bernoulli's equation. The kinetic-energy term appears with a minus sign, which is a result of the rotational nature of this particular flow. Rather, it is a solution to Euler's equation. *This example illustrates that Euler's equation applies to all inviscid flows while Bernoulli's equation does not.*

Example 9.4 *Determine the shape of the free surface for the rotating tank of incompressible fluid in Figure 9.3.*

Solution. Denoting the depth of the fluid at the center of the tank (where $r = 0$) by $z = z_{min}$, Equation (9.53) tells us that

$$p + \rho gz - \frac{1}{2}\rho\Omega^2 r^2 = p_a + \rho gz_{min}$$

Since the pressure is atmospheric at the liquid-air interface, i.e., at the free surface, we have

$$p_a + \rho gz - \frac{1}{2}\rho\Omega^2 r^2 = p_a + \rho gz_{min}$$

Simplifying, the shape of the free surface is

$$z = z_{min} + \frac{\Omega^2 r^2}{2g}$$

9.7.2 Galilean Invariance of Euler's equation

Suppose we have a body advancing into a quiescent fluid with constant velocity $\mathbf{u} = -U\mathbf{i}$ as illustrated in Figure 9.4(a). Clearly, this flow is unsteady for an observer in the main body of fluid since the geometry looks different at each instant. Thus, we cannot use Bernoulli's equation to relate pressure and velocity, even if the flow is inviscid, incompressible and irrotational, with conservative body forces.

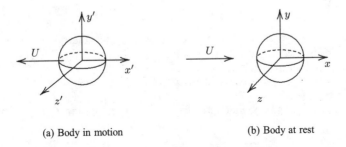

(a) Body in motion   (b) Body at rest

Figure 9.4: *Motion of a body in different coordinate frames.*

Now, suppose we observe the motion in a coordinate frame translating with the same velocity as the body [Figure 9.4(b)]. In this coordinate frame, the flow geometry does not change with time, and the motion is in fact steady. This is a dramatic improvement from an analytical point of view. In addition to reducing the number of independent variables by one (time is no longer of any consequence), we can use the powerful Bernoulli equation provided, of course, that the flow is also inviscid, irrotational, incompressible and subject only to conservative body forces.

This is a smooth move provided the Euler equation is invariant under this transformation. What we have done is made the classical **Galilean transformation**. When we say the Euler equation is invariant under such a transformation, we mean the equation holds when we write the equation in terms of all transformed velocities and pertinent flow properties. The physical meaning of invariance is straightforward. From elementary physics we know that Newton's laws of motion are Galilean invariant for motion of discrete particles, and there is no reason for this to change for fluids. However, there is some cause for concern as the Galilean transformation is a linear operation and the Eulerian description introduces nonlinear terms in the time derivative. Hence, the purpose of this section is to demonstrate Galilean invariance of the Euler equation.

Before proceeding to the proof, it is worthwhile to pause and discuss the reason why Galilean invariance matters. Very simply, if the equations of fluid mechanics were not Galilean invariant, any measurements made in a wind tunnel with fluid moving past a stationary model could not be used to predict forces on a full-scale model moving through a fluid at rest. Rather, the model would have to move through the wind tunnel to simulate the full-scale object's motion, with all the difficulties attending acceleration from rest, attainment of steady flow, and deceleration to rest prior to crashing into the end of the tunnel. Clearly, it is preferable to have a stationary model, and Galilean invariance of the equations of motion guarantees applicability of the measurements to a moving object.

Letting primed quantities denote conditions in the frame where the fluid is at rest and the body moves, the Euler equation is given by

$$\frac{\partial \mathbf{u}'}{\partial t'} + \mathbf{u}' \cdot \nabla' \mathbf{u}' = -\frac{\nabla' p}{\rho} + \mathbf{f} \qquad (9.54)$$

Clearly, since p and ρ are thermodynamic properties of the fluid, they cannot depend upon the coordinate frame from which we make our observations. Likewise, the body force, $\mathbf{f}$, will be independent of the coordinate frame provided it doesn't depend explicitly upon velocity or position. Body forces that are coordinate-system dependent do exist, e.g., Coriolis force (see Appendix D, Section D.4), but their presence signifies a noninertial frame for which Galilean invariance does not hold.

By definition, in a Galilean transformation, the coordinates and time transform according to (see Figure 9.4)

$$x = x' + Ut, \quad y = y', \quad z = z', \quad t = t' \tag{9.55}$$

where unprimed quantities correspond to the frame in which the body is at rest. Also, the velocity vectors in the two coordinate frames are related by

$$\mathbf{u} = \mathbf{u}' + U\mathbf{i} \tag{9.56}$$

Clearly, spatial differentiation is unaffected by a Galilean transformation, so that

$$\frac{\partial}{\partial x} = \frac{\partial}{\partial x'}, \quad \frac{\partial}{\partial y} = \frac{\partial}{\partial y'}, \quad \frac{\partial}{\partial z} = \frac{\partial}{\partial z'} \quad \Longrightarrow \quad \nabla = \nabla' \tag{9.57}$$

By contrast, temporal differentiation is affected. From the chain rule, the time derivative transforms as follows.

$$\left(\frac{\partial}{\partial t'}\right)_{x'} = \left(\frac{\partial t}{\partial t'}\right)_{x'} \left(\frac{\partial}{\partial t}\right)_x + \left(\frac{\partial x}{\partial t'}\right)_{x'} \left(\frac{\partial}{\partial x}\right)_t = \frac{\partial}{\partial t} + U\frac{\partial}{\partial x} \tag{9.58}$$

Note that we have omitted y and z from Equation (9.58) for the sake of brevity as all attending derivatives vanish (only x depends upon t). So, transforming the unsteady term first,

$$\frac{\partial \mathbf{u}'}{\partial t'} = \left(\frac{\partial}{\partial t} + U\frac{\partial}{\partial x}\right) (\mathbf{u} - U\mathbf{i}) = \frac{\partial \mathbf{u}}{\partial t} + U\frac{\partial \mathbf{u}}{\partial x} \tag{9.59}$$

Thus, the unsteady term is most certainly not invariant under a Galilean transformation.[4] Now consider the convective acceleration. We have

$$\mathbf{u}' \cdot \nabla' \mathbf{u}' = (\mathbf{u} - U\mathbf{i}) \cdot \nabla (\mathbf{u} - U\mathbf{i}) = \mathbf{u} \cdot \nabla \mathbf{u} - U\frac{\partial \mathbf{u}}{\partial x} \tag{9.60}$$

which shows that the convective acceleration is not Galilean invariant either. However, when we sum the unsteady and convective acceleration terms, we find

$$\frac{\partial \mathbf{u}'}{\partial t'} + \mathbf{u}' \cdot \nabla' \mathbf{u}' = \frac{\partial \mathbf{u}}{\partial t} + U\frac{\partial \mathbf{u}}{\partial x} + \mathbf{u} \cdot \nabla \mathbf{u} - U\frac{\partial \mathbf{u}}{\partial x} = \frac{\partial \mathbf{u}}{\partial t} + \mathbf{u} \cdot \nabla \mathbf{u} \tag{9.61}$$

In other words, the sum of the unsteady and convective acceleration terms, which is the Eulerian derivative, is invariant under a Galilean transformation. Therefore, the Euler equation in the transformed coordinate frame is

$$\frac{\partial \mathbf{u}}{\partial t} + \mathbf{u} \cdot \nabla \mathbf{u} = -\frac{\nabla p}{\rho} + \mathbf{f} \tag{9.62}$$

which is identical to Equation (9.54) with all primes omitted. Thus, the Euler equation is invariant under a Galilean transformation.

---

[4]As discussed at the beginning of this section, a term is invariant under the transformation if and only if transforming the term is equivalent to dropping primes.

Example 9.5 *A body moves at constant velocity $U = 30$ ft/sec through water ($\rho = 1.94$ slug/ft^3). The difference between the pressure at the front stagnation point on the body and at Point A is $p_{stag} - p_A = 4.91$ psi. What is the velocity at Point A?*

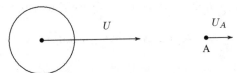

Solution. We must make a Galilean transformation in order to arrive at a steady flow, wherefore we can use Bernoulli's equation. The velocities transform as indicated in the figure below.

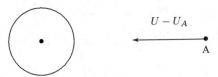

By subtracting U from the velocity of the cylinder, it is now at rest. Subtracting U from the velocity at Point A shows that the flow moves to the left at a speed $U - U_A$. Because the flow is steady in this coordinate frame, we can use Bernoulli's equation, viz.,

$$p_{stag} = p_A + \frac{1}{2}\rho\,(U - U_A)^2 \qquad \Longrightarrow \qquad U - U_A = \sqrt{\frac{2}{\rho}\,(p_{stag} - p_A)}$$

Hence, the velocity at Point A is

$$U_A = U - \sqrt{\frac{2}{\rho}\,(p_{stag} - p_A)}$$

For the given values,

$$U_A = 30\,\frac{\text{ft}}{\text{sec}} - \sqrt{\frac{2}{1.94\,\text{slug/ft}^3}\left(4.91\,\frac{\text{lb}}{\text{in}^2}\right)\left(144\,\frac{\text{in}^2}{\text{ft}^2}\right)} = 3\,\frac{\text{ft}}{\text{sec}}$$

9.8 Entropy Generation

It is instructive to develop the equation for entropy in an adiabatic, inviscid medium, subject to conservative body forces implied by the energy-conservation law developed in Section 9.4. In addition to illustrating an important implication of the energy-conservation law, our analysis will yield a result we will need to analyze the connection between Bernoulli's equation and the exact energy-conservation law in Section 9.10. As noted in Subsection 6.1.4, we can appeal to the combined first and second laws of thermodynamics, i.e., to Gibbs' equation, to determine the entropy for a given system or control volume. First, note that

$$
\begin{aligned}
T ds &= de + p\,d(1/\rho) \\
&= d(e + p/\rho) - \frac{dp}{\rho} \\
&= dh - \frac{dp}{\rho}
\end{aligned}
\tag{9.63}
$$

Hence, the entropy must satisfy the following differential equation.

$$\rho T \frac{ds}{dt} = \rho \frac{dh}{dt} - \frac{dp}{dt} \tag{9.64}$$

Now, the momentum equation for an inviscid fluid is simply Euler's equation, which is

$$\rho \frac{d\mathbf{u}}{dt} = -\nabla p + \rho \mathbf{f} = -\nabla p - \rho \nabla \mathcal{V} \tag{9.65}$$

where we note that the body force is assumed to be conservative. Taking the dot product of Euler's equation with the velocity and noting from the chain rule that $\mathbf{u} \cdot d\mathbf{u}/dt = d(\frac{1}{2}\mathbf{u} \cdot \mathbf{u})/dt$, we have

$$\rho \frac{d}{dt}\left(\frac{1}{2}\mathbf{u} \cdot \mathbf{u}\right) = -\mathbf{u} \cdot \nabla p - \rho \mathbf{u} \cdot \nabla \mathcal{V} \tag{9.66}$$

Hence, since we have also assumed that the body force (and thus its potential function, $\mathcal{V}$) is independent of time, then

$$\frac{\partial \mathcal{V}}{\partial t} = 0 \quad \Longrightarrow \quad \mathbf{u} \cdot \nabla \mathcal{V} = \frac{d\mathcal{V}}{dt} \tag{9.67}$$

wherefore Equation (9.66) can be rearranged to read

$$0 = \rho \frac{d}{dt}\left(\frac{1}{2}\mathbf{u} \cdot \mathbf{u} + \mathcal{V}\right) + \mathbf{u} \cdot \nabla p \tag{9.68}$$

This equation represents conservation of mechanical energy. Now, adding the respective sides of Equations (9.64) and (9.68) yields

$$\rho T \frac{ds}{dt} = \underbrace{\rho \frac{d}{dt}\left(h + \frac{1}{2}\mathbf{u} \cdot \mathbf{u} + \mathcal{V}\right)}_{\rho \, d\mathcal{H}/dt} - \underbrace{\left(\frac{dp}{dt} - \mathbf{u} \cdot \nabla p\right)}_{\partial p/\partial t} \tag{9.69}$$

Finally, reference to the differential form of the energy-conservation law [Equation (9.34)] tells us the right-hand side of Equation (9.69) vanishes. Thus, we arrive at the important result that for inviscid, adiabatic flow with conservative body forces, the rate of change of entropy following a fluid particle is zero:

$$\rho T \frac{ds}{dt} = 0 \tag{9.70}$$

Consistent with our naming conventions discussed in Subsection 6.1.1, we refer to a constant-entropy flow (or process) as **isentropic**.

9.9   Derivation of Bernoulli's Equation

We introduced Bernoulli's equation in Section 3.5 without proof, although we promised to derive the equation later on. We now have sufficient background to prove the validity of Bernoulli's equation. To derive it, we will integrate Euler's equation subject to a few limiting conditions, which we will identify below.

Our first step in our derivation is to note that the convective acceleration, $\mathbf{u} \cdot \nabla \mathbf{u}$, which appears in Euler's equation can be rewritten as follows (see Appendix C).

$$\mathbf{u} \cdot \nabla \mathbf{u} = \nabla\left(\frac{1}{2}\mathbf{u} \cdot \mathbf{u}\right) - \mathbf{u} \times (\nabla \times \mathbf{u}) \tag{9.71}$$

This is a general vector identity that will help us arrive at the desired result with a minimum of algebraic operations. Hence, since total acceleration is the sum of instantaneous and convective contributions, i.e., $d\mathbf{u}/dt = \partial\mathbf{u}/\partial t + \mathbf{u} \cdot \nabla\mathbf{u}$, in terms of vorticity, $\omega = \nabla \times \mathbf{u}$, Euler's equation becomes

$$\rho\frac{\partial\mathbf{u}}{\partial t} + \rho\nabla\left(\frac{1}{2}\mathbf{u}\cdot\mathbf{u}\right) - \rho\mathbf{u}\times\omega = -\nabla p + \rho\mathbf{f} \tag{9.72}$$

As with all of our analysis in this chapter, we presume that the fluid with which we are working is inviscid. As noted when we first introduced Bernoulli's equation, having an inviscid fluid is one of the constraints we must impose. We must require four more conditions in order to arrive at this most famous of all equations in the field of fluid mechanics. So, our list of five constraints on the flow under consideration is as follows.

1. Inviscid fluid (Euler's equation is valid);

2. Steady flow ($\partial\mathbf{u}/\partial t = \mathbf{0}$);

3. Incompressible flow (ρ = constant);

4. Conservative body force ($\mathbf{f} = -\nabla\mathcal{V}$);

5. Irrotational flow ($\omega = \mathbf{0}$).

Under these conditions, Equation (9.72) simplifies to

$$\nabla\left(p + \frac{1}{2}\rho\mathbf{u}\cdot\mathbf{u} + \rho\mathcal{V}\right) = \mathbf{0} \tag{9.73}$$

Since this holds for all points in the flow, necessarily the quantity in parentheses is constant. Therefore, we conclude that

$$p + \frac{1}{2}\rho\mathbf{u}\cdot\mathbf{u} + \rho\mathcal{V} = \text{constant} \tag{9.74}$$

Equation (9.74) is, of course, **Bernoulli's equation**. Although we arrived at this result by integrating the momentum equation, it is actually an equation for *conservation of mechanical energy*. This is especially clear when the body force is gravity for which the body-force potential is $\mathcal{V} = gz$. Note that in this spirit, we can regard pressure as the pressure-force potential. Equation (9.74) says the sum of the pressure, p, kinetic energy per unit volume (also known as the **dynamic pressure** or **dynamic head**), $\frac{1}{2}\rho\mathbf{u}\cdot\mathbf{u}$, and potential energy per unit volume, $\rho\mathcal{V}$, is constant. There is no contradiction here. A constant-density fluid flow is specified completely by the momentum and continuity equations, so any other property can be deduced from them.

As a final comment on the conditions required for Bernoulli's equation to hold, note that we can actually relax the irrotationality condition. When we do this, Equation (9.74) holds along a streamline, although the constant is different on each streamline. To see this, note first that for steady, incompressible flow with a conservative body force, $\mathbf{f} = -\nabla\mathcal{V}$, Euler's equation simplifies to

$$\mathbf{u}\cdot\nabla\mathbf{u} + \nabla\left(\frac{p}{\rho} + \mathcal{V}\right) = \mathbf{0} \tag{9.75}$$

Using natural coordinates (see Appendix D), we can write the component of Equation (9.75) along a streamline as

$$u\frac{\partial u}{\partial s} + \frac{\partial}{\partial s}\left(\frac{p}{\rho} + \mathcal{V}\right) = 0 \tag{9.76}$$

Then, noting that $u \partial u / \partial s = \partial(\frac{1}{2} u^2)/\partial s$, we have

$$\frac{\partial}{\partial s}\left(\frac{p}{\rho} + \frac{1}{2}u^2 + \mathcal{V}\right) = 0 \qquad (9.77)$$

This shows that the quantity in parentheses is constant only along a streamline as opposed to being constant everywhere. By contrast, we found above that when the flow is irrotational, the gradient of this quantity [cf. Equation (9.73)] vanishes, which is a much stronger condition. Nevertheless, we conclude that even when the flow has nonzero vorticity,

$$p + \frac{1}{2}\rho \mathbf{u} \cdot \mathbf{u} + \rho \mathcal{V} = \text{constant on streamlines for } \nabla \times \mathbf{u} \neq \mathbf{0} \qquad (9.78)$$

where we note that $u^2 = \mathbf{u} \cdot \mathbf{u}$. This result is interesting from a conceptual point of view, but is not very helpful in general practice. That is, we don't know where the streamlines are until we have solved the equations of motion. Hence, this form of Bernoulli's equation is more limited than the form for irrotational flow.

It is always a major simplification if we can use Bernoulli's equation to relate pressure and velocity, rather than having to solve Euler's equation. The reason is obvious, i.e., Bernoulli's equation is a solution to Euler's equation so that no additional computation is needed. With the exception of certain idealized flows, solutions to Euler's equation are difficult to obtain and usually require a numerical solution. For this reason, Bernoulli's equation, whenever valid, provides a major simplification for determining flow properties.

Example 9.6 *Determine the pressure in the rotating tank of incompressible fluid in Figure 9.3.*

Solution. As we saw in Subsection 9.7.1, this flow is rotational for a stationary observer and that precludes using Bernoulli's equation. However, the fluid is at rest for an observer rotating with the tank. Thus, the flow in a tank-fixed rotating coordinate frame is steady, irrotational and we are given that the fluid is incompressible. If we can show that the body forces acting are conservative, we can apply Bernoulli's equation in this coordinate frame.

There are two body forces acting, namely, the gravitational force and the centrifugal force. We already know that gravity is a conservative body force and its force potential function is $\mathcal{V}_g = gz$. By definition, the centrifugal force in a coordinate frame rotating with circumferential velocity $u_\theta = \Omega r$ is

$$\mathbf{f}_c = \frac{u_\theta^2}{r} \mathbf{e}_r = \Omega^2 r\, \mathbf{e}_r$$

where r is radial distance in a horizontal plane measured from the center of the tank and $\mathbf{e}_r$ is a unit vector pointing radially outward. Therefore,

$$\mathbf{f}_c = -\nabla \mathcal{V}_c = -\frac{\partial \mathcal{V}_c}{\partial r} \mathbf{e}_r \quad \text{where} \quad \mathcal{V}_c = -\frac{1}{2}\Omega^2 r^2$$

So, since both of the body forces acting are conservative and the other necessary conditions are satisfied, an observer rotating with the tank can use Bernoulli's equation. Consequently, the pressure is given by

$$p + \rho \mathcal{V}_g + \rho \mathcal{V}_c = \text{constant}$$

Substituting for the force potentials, we conclude that

$$p + \rho gz - \frac{1}{2}\rho \Omega^2 r^2 = \text{constant}$$

9.10 Bernoulli's Equation and Energy Conservation

We can use the results of the two preceding sections to illustrate the connection between exact energy conservation and Bernoulli's equation. We derived the equation in the preceding section by manipulating the momentum equation. Hence, it represents conservation of mechanical energy, and excludes thermal effects. We escaped the need to consider any but mechanical forms of energy with the postulate that the flow is incompressible. In this section, we will see exactly what the limits of Bernoulli's equation are regarding effects of compressibility.

For *steady flow* of a perfect fluid with no heat transfer, Equation (9.34) simplifies to

$$\frac{d\mathcal{H}}{dt} = 0 \tag{9.79}$$

where we note that $d/dt = \mathbf{u} \cdot \nabla$ for steady flow. Therefore, under these conditions, we conclude that

$$h + \frac{1}{2}\mathbf{u} \cdot \mathbf{u} + \mathcal{V} = \text{constant on a streamline} \tag{9.80}$$

We call this condition **particle isenthalpic**.

9.10.1 Flow of a Liquid

The density, ρ, is constant for a liquid. Entropy is also constant in the absence of heat transfer and viscous effects, and we infer from Equation (9.63) that

$$dh = dp/\rho \quad \Longrightarrow \quad h = p/\rho + \text{constant} \quad \text{(for isentropic flow of a liquid)} \tag{9.81}$$

Therefore, the exact energy-conservation principle for a steady, adiabatic, inviscid flow of a liquid simplifies immediately to Bernoulli's equation, viz.,

$$\frac{p}{\rho} + \frac{1}{2}\mathbf{u} \cdot \mathbf{u} + \mathcal{V} = \text{constant} \tag{9.82}$$

9.10.2 Flow of a Gas

We cannot neglect density changes in a gas. Thus, demonstrating that the energy-conservation principle is consistent with Bernoulli's equation under incompressible, or nearly incompressible, conditions is a bit more subtle. We begin by noting that, if we select a reference point at infinity (i.e., very far from the point of interest), then the energy-conservation principle tells us

$$h + \frac{1}{2}\mathbf{u} \cdot \mathbf{u} + \mathcal{V} = h_\infty + \frac{1}{2}U_\infty^2 + \mathcal{V}_\infty \tag{9.83}$$

Now, noting that adiabatic, inviscid flows with conservative body forces are isentropic, one thermodynamic state variable (entropy) is fixed. Thus, all other thermodynamic state variables can be expressed as a function of a single state variable. So, if we regard enthalpy as a function of pressure, we can expand in Taylor series according to

$$h = h_\infty + \left(\frac{dh}{dp}\right)_\infty (p - p_\infty) + \frac{1}{2}\left(\frac{d^2h}{dp^2}\right)_\infty (p - p_\infty)^2 + \cdots \tag{9.84}$$

For a perfect, calorically-perfect gas in an isentropic flow, we know from Equation (6.22) that

$$p = A\rho^\gamma, \quad A = \text{constant} \quad \Longrightarrow \quad \rho = (p/A)^{1/\gamma} \tag{9.85}$$

Also, the enthalpy can be written as

$$h = c_p T = \frac{c_p}{R}\frac{p}{\rho} = \frac{\gamma}{\gamma - 1}\frac{p}{\rho} \tag{9.86}$$

where we make use of Equation (6.15) to replace the ratio of c_p to R by the factor $\gamma/(\gamma - 1)$. Then, combining Equations (9.85) and (9.86), the enthalpy for a perfect, calorically-perfect gas in an isentropic flow is given by

$$h = \frac{\gamma}{\gamma - 1}A^{1/\gamma}p^{(\gamma-1)/\gamma} \tag{9.87}$$

We can differentiate this equation to show that

$$\frac{dh}{dp} = \frac{\gamma}{\gamma - 1}A^{1/\gamma}\frac{\gamma - 1}{\gamma}p^{-1+(\gamma-1)/\gamma} = A^{1/\gamma}p^{-1/\gamma} \tag{9.88}$$

$$\frac{d^2h}{dp^2} = -\frac{1}{\gamma}A^{1/\gamma}p^{-1-1/\gamma} = -\frac{1}{\gamma p}\frac{dh}{dp} \tag{9.89}$$

Now, from Equation (9.85) we know that $\rho = (p/A)^{1/\gamma}$. Also, as shown in Section 8.3, the speed of sound for a perfect gas is given by $a = \sqrt{\gamma p/\rho}$. Consequently, Equations (9.88) and (9.89) simplify to

$$\frac{dh}{dp} = \frac{1}{\rho}, \qquad \frac{d^2h}{dp^2} = -\frac{1}{\rho a}\frac{dh}{dp} \tag{9.90}$$

Therefore, evaluating the first and second derivatives in the farfield, we have

$$\left(\frac{dh}{dp}\right)_\infty = \frac{1}{\rho_\infty}, \qquad \left(\frac{d^2h}{dp^2}\right)_\infty = -\frac{1}{\rho_\infty^2 a_\infty^2} \tag{9.91}$$

Substituting these results into Equation (9.84), we conclude that the enthalpy is given by the following Taylor series.

$$\begin{aligned}
h &= h_\infty + \frac{(p - p_\infty)}{\rho_\infty} - \frac{1}{2}\frac{(p - p_\infty)^2}{\rho_\infty^2 a_\infty^2} + \cdots \\
&= h_\infty + \frac{(p - p_\infty)}{\rho_\infty} - \frac{1}{2}U_\infty^2 M_\infty^2\left(\frac{p - p_\infty}{\rho_\infty U_\infty^2}\right)^2 + \cdots
\end{aligned} \tag{9.92}$$

where $M_\infty = U_\infty/a_\infty$ is the Mach number. Hence, energy-conservation Equation (9.83) can be approximated by

$$h_\infty + \frac{(p - p_\infty)}{\rho_\infty} - \frac{1}{2}U_\infty^2 M_\infty^2\left(\frac{p - p_\infty}{\rho_\infty U_\infty^2}\right)^2 + \frac{1}{2}\mathbf{u}\cdot\mathbf{u} + \mathcal{V} \approx h_\infty + \frac{1}{2}U_\infty^2 + \mathcal{V}_\infty \tag{9.93}$$

Canceling h_∞ and regrouping terms, we have

$$\frac{p}{\rho_\infty} + \frac{1}{2}\mathbf{u}\cdot\mathbf{u} + \mathcal{V} \approx \frac{p_\infty}{\rho_\infty} + \frac{1}{2}U_\infty^2\left[1 + M_\infty^2\left(\frac{p - p_\infty}{\rho_\infty U_\infty^2}\right)^2\right] + \mathcal{V}_\infty \tag{9.94}$$

Thus, in the limit of small Mach number, we can say (with ρ_∞ replaced by ρ)

$$\frac{p}{\rho} + \frac{1}{2}\mathbf{u}\cdot\mathbf{u} + \mathcal{V} \approx \frac{p_\infty}{\rho} + \frac{1}{2}U_\infty^2 + \mathcal{V}_\infty \tag{9.95}$$

which is again Bernoulli's equation. Therefore, as with a liquid, we see that the limiting form of the exact energy-conservation law (as Mach number approaches zero) is Bernoulli's equation. This also suggests that the proper definition of incompressible flow is the limiting case $M_\infty \to 0$.

Example 9.7 *For low-speed flow past an object, measurements show that the pressure coefficient is $C_p = (p - p_\infty)/(\frac{1}{2}\rho_\infty U_\infty^2) = -0.1$ and the Mach number is $M_\infty = 0.25$. Ignoring body forces, use Equations (9.94) and (9.95) to determine $|\mathbf{u}|/U_\infty$.*

Solution. From Equation (9.94), a little rearrangement yields

$$C_p + \frac{\mathbf{u} \cdot \mathbf{u}}{U_\infty^2} = 1 + \frac{1}{4}M_\infty^2 C_p^2$$

where we have dropped the body-force potential. Solving for the velocity, we find

$$\frac{|\mathbf{u}|}{U_\infty} = \sqrt{1 - C_p + \frac{1}{4}M_\infty^2 C_p^2}$$

So, for the given conditions,

$$\frac{|\mathbf{u}|}{U_\infty} = \sqrt{1 + 0.1 + \frac{1}{4}(0.25)^2(0.1)^2} = 1.04888$$

By contrast, Equation (9.95) yields

$$\frac{|\mathbf{u}|}{U_\infty} = \sqrt{1 - C_p} = \sqrt{1 + 0.1} = 1.04881$$

The difference is 0.007%.

Problems

9.1 Consider steady, compressible flow of a gas through a nozzle. The velocity can be approximated as $\mathbf{u} = U_o(1+x/x_o)\,\mathbf{i}$, where U_o and x_o are constant reference velocity and length, respectively. Determine the density, ρ, if its value at $x = 0$ is ρ_a. At what point does the density fall to 60% of ρ_a?

9.2 For steady flow of a compressible fluid, the velocity vector is

$$\mathbf{u} = u_o \left(\frac{x}{x_o}\right)^2 \mathbf{i}$$

where u_o and x_o are reference velocity and position, respectively. The fluid density is ρ_o at $x = x_o$. Determine the density, ρ, as a function of ρ_o, x and x_o.

9.3 In a one-dimensional, compressible flow, the density decreases exponentially from ρ_a to ρ_b, i.e., $\rho = \rho_a - (\rho_a - \rho_b)\,e^{-t/\tau}$, where τ is a constant of dimensions time. If the velocity at $x = 0$ is $u(0, t) = u_o$, where u_o is constant, determine $u(x, t)$ as a function of u_o, ρ_a, ρ, x, t and τ.

9.4 Appendix D includes the divergence of the velocity in cylindrical and spherical coordinates. Determine the most general form of the velocity for incompressible flow if the following conditions hold.

(a) Axially-symmetric flow with $u_\theta = w = 0$.

(b) Spherically-symmetric flow with $u_\theta = u_\phi = 0$.

9.5 The velocity vector for a flow is

$$\mathbf{u} = \frac{U}{h^3}\left[6x^2 y\,\mathbf{i} + 2x^3\mathbf{j} + 10h^3\mathbf{k}\right]$$

where U and h are constants. Is the flow incompressible? Is the flow irrotational?

9.6 The Cartesian velocity components for a two-dimensional flow are

$$u = \frac{UD^2 y}{(x^2 + y^2)^{3/2}}, \qquad v = -\frac{UD^2 x}{(x^2 + y^2)^{3/2}}$$

where U and D are constants. Is the flow incompressible? Is the flow irrotational?

9.7 The velocity vector for a flow is

$$\mathbf{u} = \frac{U_o}{L^3}\left[x^2 y\,\mathbf{i} + xy^2\,\mathbf{j} - 4xyz\,\mathbf{k}\right]$$

where U_o and L are constants. Is the flow incompressible? Is the flow irrotational?

9.8 A flowfield has the following velocity vector

$$\mathbf{u} = \frac{x^3 z}{y^2}\mathbf{i} - \frac{3x^2 z}{y}\mathbf{j} - \frac{3x^2 z^2}{y^2}\mathbf{k}$$

where all quantities are dimensionless. Is the flow incompressible? Is the flow irrotational?

9.9 The Cartesian velocity components for a two-dimensional flow are

$$u = \frac{UR^2(y^2 - x^2)}{(x^2 + y^2)^2}, \qquad v = -\frac{2UR^2 xy}{(x^2 + y^2)^2}$$

where U and R are constants. Is the flow incompressible? Is the flow irrotational?

9.10 The velocity for a two-dimensional flow in cylindrical coordinates is

$$\mathbf{u} = U\left(\frac{r}{R}\right)\cos\theta\,\mathbf{e}_r - 2U\left(\frac{r}{R}\right)\sin\theta\,\mathbf{e}_\theta$$

where U and R are constants. Is the flow incompressible? Is the flow irrotational?

9.11 The velocity for a two-dimensional flow in cylindrical coordinates is

$$u_r = U\left(1 - \frac{R^2}{r^2}\right)\cos\theta, \qquad u_\theta = U\frac{R}{r} - U\left(1 + \frac{R^2}{r^2}\right)\sin\theta$$

where U and R are constants. Is the flow incompressible? Is the flow irrotational?

9.12 The y component of the velocity vector for a two-dimensional, incompressible, irrotational flow is $v(x,y) = Bx/(x^2 + y^2)$, where B is a constant. Determine the x component, $u(x,y)$.

9.13 The y component of the velocity vector for a two-dimensional, incompressible, irrotational flow is $v(x,y) = -U(y/h - 1)$ where U and h are constants. Determine the x component, $u(x,y)$.

9.14 The x component of the velocity vector for a two-dimensional, incompressible, irrotational flow is

$$u(x,y) = U\left[1 - e^{-\lambda x}\cos\lambda y\right]$$

where U and λ are constant velocity and length scales, respectively. Determine the y component of the velocity vector, $v(x,y)$, assuming there is a stagnation point at $x = y = 0$.

9.15 The x component of the velocity vector for a two-dimensional, incompressible, irrotational flow is $u(x,y) = Cxy$ where C is a constant. Determine the y component, $v(x,y)$.

9.16 The circumferential component of the velocity vector for a two-dimensional, incompressible, irrotational flow is $u_\theta(r,\theta) = -3Ar^2\sin 3\theta$ where A is a constant. Determine the radial component, $u_r(r,\theta)$.

9.17 The radial component of the velocity vector for a two-dimensional, incompressible, irrotational flow is $u_r(r,\theta) = (S/r^2)\cos\theta$ where S is a constant. Determine the circumferential component, $u_\theta(r,\theta)$.

9.18 A flowfield has the velocity vector $\mathbf{u} = Ar\cos\theta\,\mathbf{e}_r - r\sin\theta\,\mathbf{e}_\theta$, where A is a constant and all quantities are dimensionless.

    (a) Is there any value of A for which this flow is irrotational?

    (b) Is there any value of A for which this flow is incompressible?

9.19 An unsteady flow has the velocity vector $\mathbf{u} = xe^{-t/\tau}\mathbf{i} + Cye^{-t/\tau}\mathbf{j}$, where C and τ are constants and all quantities are dimensionless. The flow is incompressible and irrotational. Find the values of C and τ necessary to guarantee these conditions.

9.20 Consider unsteady flow of an incompressible fluid with negligible body forces. The velocity vector is $\mathbf{u} = U\mathbf{i} + Ue^{-t/\tau}\mathbf{j}$, where U and τ are constants. Determine the pressure, $p(x,y,z,t)$ for this flow.

9.21 Consider an unsteady flow in an incompressible fluid in which gravitational effects are important. The velocity and gravitational vectors are $\mathbf{u} = U\mathbf{i} + U\cosh[\kappa(x - Ut)]\mathbf{k}$ and $\mathbf{g} = -g\,\mathbf{k}$, where U and κ are constants and g is gravitational acceleration. Determine the pressure, $p(x,y,z,t)$ for this flow.

9.22 Consider a pipe whose cross-sectional area is $A(x) = A_oF(x)$, where A_o is the area at $x = 0$, $F(0) = 1$ and $F(x)$ is an obscure function you've never heard of. Assume the flow is inviscid, incompressible, can be approximated as one dimensional and that body forces are negligible. Using the fact that mass flux, $\dot{m} = \rho uA$, is constant and $p(0) = p_a$, compute the pressure throughout the pipe. How does $p(x)$ vary with increasing area? How does it vary with decreasing area? Explain your results.

9.23 The average velocity of water in a nozzle increases from $u_1 = 4$ m/sec to $u_2 = 10$ m/sec. Assuming the average velocity varies linearly with distance along the nozzle, x, and that the length of the nozzle is $\ell = 1$ m, estimate the pressure gradient, dp/dx, at a point midway through the nozzle. The density of water is $\rho = 1000$ kg/m^3. You may assume the flow can be approximated as one dimensional.

9.24 The velocity vector for a steady, incompressible flow is $\mathbf{u} = (U/L^2)(yz\,\mathbf{i} + xz\,\mathbf{j} + xy\,\mathbf{k})$, where U and L are constant velocity and length scales, respectively. The fluid density is ρ and the pressure at $x = y = z = 0$ is p_o. Verify that this flow is irrotational and incompressible. Also, using Euler's equation, determine the pressure as a function of ρ, U, L, p_o, x, y and z. Assume there are no body forces acting. Compare your result with the pressure according to Bernoulli's equation.

9.25 We wish to analyze an incompressible, two-dimensional flow with velocity vector $\mathbf{u} = Ax\,\mathbf{i} - Ay\,\mathbf{j}$, where A is a constant and body forces are negligible.

(a) Does this velocity field satisfy the continuity equation?

(b) Is the flow irrotational?

(c) If the flow is inviscid, what is the pressure, $p(x, y)$, if $p(0, 0) = p_t$?

9.26 The constant-diameter U-tube shown rotates about axis O-O at angular velocity Ω.

(a) Determine the new positions of the water surfaces, ℓ_1 and ℓ_2. Neglect the diameter of the U-tube in your computations and assume the tubes are open to the atmosphere.

(b) What does your answer for Part (a) predict for rotation rates in excess of the critical value defined by $\Omega_{crit} = 2\sqrt{gL_2}/L_3$? Explain how to reformulate the problem with a diagram of the fluid in the U-tube when $\Omega > \Omega_{crit}$.

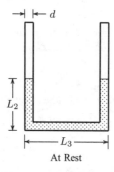
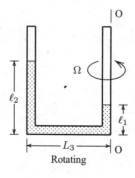

At Rest   Rotating

Problem 9.26

9.27 Determine the new positions of the water surfaces, ℓ_1 and ℓ_2 when the U-tube shown rotates about axis O-O at angular velocity Ω. Assume the diameter of the thickest part of the U-tube, $2d$, is very small compared to L_3, and that the tubes are open to the atmosphere.

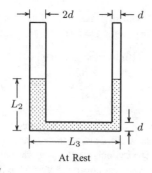
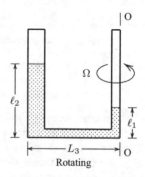

At Rest   Rotating

Problem 9.27

9.28 Fluid in a cylindrical tank of radius R rotates about the z axis with angular velocity Ω. The fluid has been rotating for a time sufficient to establish rigid-body rotation. The initial fluid level (indicated by the dashed line) is h, the fluid density is ρ and atmospheric pressure is p_a. As shown in the text, the equation of the free surface is $z = \zeta(r) = h_{min} + \frac{1}{2}\Omega^2 r^2/g$. Appealing to mass conservation, compute h_{max} and h_{min} as functions of h, Ω, R and g.

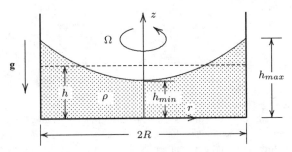

Problems 9.28, 9.29, 9.30

9.29 Fluid in a cylindrical tank of radius R rotates about the z axis with angular velocity Ω. The fluid has been rotating for a time sufficient to establish rigid-body rotation. The initial fluid level (indicated by the dashed line) is h, the fluid density is ρ and atmospheric pressure is p_a. As shown in the text, the equation of the free surface is $z = \zeta(r) = h_{min} + \frac{1}{2}\Omega^2 r^2/g$. Appealing to mass conservation, find the rotation rate for which the center of the container just becomes exposed, i.e., $h_{min} = 0$.

9.30 A cylindrical tank partially filled with a liquid of density ρ rotates about the z axis with angular velocity Ω. The tank is open to the atmosphere and has been rotating for a time sufficient to establish rigid-body rotation. Determine the shape of the free surface on the plane passing through the axis of rotation if the minimum and maximum depths are $h_{min} = 0.08R$ and $h_{max} = 0.4R$. Also, determine the angular-rotation rate, Ω, as a function of g and R.

9.31 Imagine you are rushing to the university to avoid being late for your fluid-mechanics class. Your coffee cup is resting next to you in your car. In your haste to get to class, you accelerate at λ g's, i.e., your acceleration is $a = \lambda g$ where λ is a constant. What is the maximum height, h_o, to which the cup can be filled to avoid spilling any coffee? Assume the cup is a cylinder of height $h = 10$ cm and diameter $d = 8$ cm. **HINT:** To conserve mass in this geometry, necessarily $h_o = \frac{1}{2}(h + h_{min})$. As a percentage, determine how full the cup can be if you are driving your Volkswagen Bug ($\lambda = 1/6$) or your Corvette ($\lambda = 2/5$).

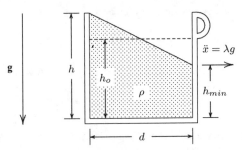

Problems 9.31, 9.32

9.32 Imagine you are rushing to the university to avoid being late for your fluid-mechanics class. Your coffee cup is resting next to you in your car. In your haste to get to class, you accelerate at λ g's, i.e., your acceleration is $a = \lambda g$ where λ is a constant. What is the maximum value of λ possible to avoid spilling if the cup is initially 75% full? Assume the cup is a cylinder of height $h = 4$ inches and diameter $d = 3.5$ inches. **HINT:** To conserve mass in this geometry, necessarily $h_o = \frac{1}{2}(h + h_{min})$.

9.33 A truck carries a tank of water that is open at the top with length ℓ, width w and depth h. Assuming the driver will not accelerate the truck at a rate greater than a, what is the maximum depth, h_o, to which the tank may be filled to prevent spilling any water? Assume constant acceleration, and note that the pressure is constant and equal to its atmospheric value at the free surface.

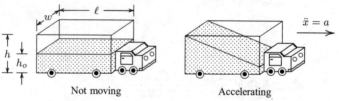

Not moving   Accelerating

Problems 9.33, 9.34

9.34 A truck carries a tank of water that is open at the top with length ℓ, width w and depth h. If the truck is half full and $\ell = 5h$, what is the maximum acceleration, a, that can be sustained without spilling any water? Assume constant acceleration, and note that the pressure is constant and equal to its atmospheric value at the free surface.

9.35 A small car containing an incompressible fluid of density ρ is rolling down an inclined plane. Determine the location and value of the maximum pressure in the car. Ignore any friction in the wheels. **HINT:** Use a coordinate system with x and z parallel to and normal to the inclined plane, respectively.

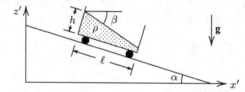

Problems 9.35, 9.36

9.36 A small car containing an incompressible fluid of density ρ is rolling down an inclined plane. Show that the free surface is planar and determine the angle it makes with the horizontal, β. Ignore any friction in the wheels. **HINT:** Do your work in a coordinate system for which x and z are parallel to and normal to the inclined plane, respectively.

9.37 In terms of natural coordinates (see Appendix D, Section D.5), the s and n components of Euler's equation are

$$\rho u \frac{\partial u}{\partial s} = -\frac{\partial p}{\partial s} - \rho \frac{\partial \mathcal{V}}{\partial s} \quad \text{and} \quad -\rho \frac{u^2}{\mathcal{R}} = -\frac{\partial p}{\partial n} - \rho \frac{\partial \mathcal{V}}{\partial n}$$

As discussed in Section 3.5, integrating the streamwise, or s, component yields

$$p + \frac{1}{2}\rho u^2 + \rho \mathcal{V} = F(n)$$

where $F(n)$ is constant along a streamline (because $n =$ constant on a streamline). Verify that, with $\omega = u/\mathcal{R} + \partial u/\partial n$ denoting the vorticity,

$$F'(n) = \rho \omega u$$

9.38 For hurricanes, the Coriolis acceleration is much larger than the convective acceleration. Reference to Appendix D, Section D.4 shows that the Euler equation simplifies to

$$2\rho \, \Omega \times \mathbf{u} = -\nabla \hat{p}$$

where $\hat{p}$ is a reduced pressure that includes the centrifugal acceleration and Ω is Earth's angular velocity. Based on this equation, explain why hurricanes rotate counterclockwise in the northern hemisphere and clockwise in the southern hemisphere. **HINT:** To simplify your explanation, consider hurricanes centered at the north pole and the south pole.

9.39 A Pitot-static tube is placed in a flow of air with $\rho = 1.20$ kg/m^3. The stagnation- and static-pressure taps read 103.16 kPa and 101 kPa, respectively. What is the velocity of the air? If the velocity changes to 80 m/sec and the static pressure is unchanged, what is the corresponding stagnation pressure?

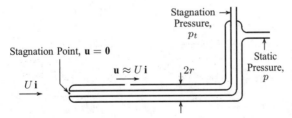

Problems 9.39, 9.40, 9.41

9.40 Consider a poorly designed Pitot-static tube with a single static-pressure hole at the top of the tube as shown. If the tube radius is r, develop a formula for the true velocity, U_{true}, as a function of gravitational acceleration, g, radius, r, and the velocity inferred from $U = \sqrt{2(p_t - p)/\rho}$. If $r = 5$ mm, determine the percentage error in velocity for an indicated value of $U = 1$, 10 and 100 m/sec.

9.41 A Pitot-static tube is placed in a flow of helium with $\rho = 3.2 \cdot 10^{-4}$ slug/ft^3. The static-pressure tap reads 1725 psf and the flow velocity is 300 ft/sec. What is the stagnation pressure? If the stagnation pressure changes to 1760 psf and the static pressure is unchanged, what is the corresponding velocity?

9.42 A *Venturi meter* is a device used to measure fluid velocities and flow rates for incompressible, steady flow. As shown, the pressure is measured at two sections of a pipe with different cross-sectional areas. A straightforward derivation shows that the volume-flow rate $Q = A_2\sqrt{2(p_1 - p_2)/[\rho(1 - A_2^2/A_1^2)]}$, where p is pressure, ρ is density and A is cross-sectional area. Consider a Venturi meter that has $A_1 = 6$ ft^2 and $A_2 = 5$ ft^2. If the volume-flow rate is $Q = 50$ ft^3/sec, what is the pressure difference, $p_1 - p_2$, if the fluid flowing is air, water or mercury?

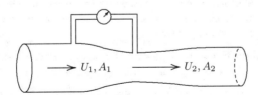

Problems 9.42, 9.43, 9.44

9.43 A *Venturi meter* is a device used to measure fluid velocities and flow rates for incompressible, steady flow. As shown, the pressure is measured at two sections of a pipe with different cross-sectional areas. A straightforward derivation shows that the volume-flow rate $Q = A_2\sqrt{2(p_1 - p_2)/[\rho(1 - A_2^2/A_1^2)]}$, where p is pressure, ρ is density and A is cross-sectional area. Consider a Venturi meter that has $A_1 = 0.1$ m^2 and $A_2 = 0.075$ m^2. The attached pressure gage can accurately measure pressures no smaller than 0.1 Pa. If air of density $\rho = 1.20$ kg/m^3 is flowing, what is the minimum flow rate that can be accurately measured?

9.44 A *Venturi meter* is a device used to measure fluid velocities and flow rates for incompressible, steady flow. As shown, the pressure is measured at two sections of a pipe with different cross-sectional areas. You may assume the flow is irrotational and that effects of body forces can be ignored. Noting that for steady flow the mass-flow rate, $\dot{m}$, is constant and equal to $\rho U A$, where ρ is density, U is average velocity and A is cross-sectional area, verify that $\dot{m}$ is

$$\dot{m} = \rho A_2 \sqrt{\frac{2(p_1 - p_2)}{\rho(1 - A_2^2/A_1^2)}}$$

9.45 A *barotropic fluid* is one for which the pressure is a function only of density, i.e., $p = p(\rho)$. For a steady, irrotational flow of a barotropic fluid with a conservative body force $\mathbf{f} = -\nabla \mathcal{V}$, derive the following replacement for Bernoulli's equation.

$$\int \frac{dp}{\rho} + \frac{1}{2}\mathbf{u} \cdot \mathbf{u} + \mathcal{V} = \text{constant}$$

HINT: Introduce a function $F(\rho)$ defined by

$$\frac{\nabla p}{\rho} = \frac{1}{\rho}\frac{dp}{d\rho}\nabla \rho = F'(\rho)\nabla \rho$$

9.46 For incompressible, irrotational flows, even when the flow is unsteady, the velocity vector, $\mathbf{u}$, can be written as $\mathbf{u} = \nabla \phi$, where $\phi(x, y, z, t)$ is the *velocity potential*. Beginning with the vector identity

$$\mathbf{u} \cdot \nabla \mathbf{u} = \nabla \left(\frac{1}{2}\mathbf{u} \cdot \mathbf{u}\right) - \mathbf{u} \times (\nabla \times \mathbf{u})$$

derive an unsteady-flow replacement for Bernoulli's equation. Assume a conservative body force is present so that $\mathbf{f} = -\nabla \mathcal{V}$.

9.47 A cylinder moves at constant velocity U through water of density $\rho = 1000$ kg/m³. The difference between the pressure at the front stagnation point on the cylinder and at Point P in the cylinder's wake is $p_t - p_P = 4.5$ kPa. If the velocity at Point P is $U_P = 5$ m/sec, what is U?

Problems 9.47, 9.48

9.48 A cylinder moves at constant velocity $U = 9$ m/sec through water of density $\rho = 1000$ kg/m³. The difference between the pressure at the front stagnation point on the cylinder and at Point P in the cylinder's wake is $p_t - p_P = 18$ kPa. What is the velocity at Point P?

9.49 Body A travels through water at a constant speed of $U_A = 20$ ft/sec. Velocities at points B and C are induced by the moving body and have magnitudes of $U_B = 7$ ft/sec and $U_C = 3$ ft/sec. If the density of water is 1.94 slug/ft³ and effects of gravity can be ignored, what is $p_B - p_C$ in psi?

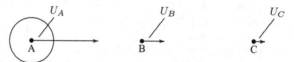

Problems 9.49, 9.50

9.50 Body A travels through water at a constant speed of $U_A = 15$ m/sec. Velocities at points B and C are induced by the moving body and have magnitudes of $U_B = 4$ m/sec and $U_C = 2$ m/sec. If the density of water is 1000 kg/m³ and effects of gravity can be ignored, what is $p_B - p_C$ in kPa?

9.51 An object is traveling through water at a speed $U_1 = 20$ m/sec. The flow speeds at Points 2 and 3 are $U_2 = 12$ m/sec and $U_3 = 7$ m/sec. If the pressure difference between Points 3 and 4 is half the difference between Points 2 and 3, what is the flow speed at Point 4, U_4?

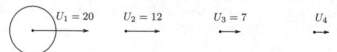

Problem 9.51

9.52 Reference to Appendix D, Section D.4 shows that for inviscid flow in a coordinate system rotating about the z axis with angular velocity $\mathbf{\Omega} = \Omega\mathbf{k}$, the momentum equation assumes the following form.

$$\frac{\partial \mathbf{u}'}{\partial t'} + \mathbf{u}' \cdot \nabla' \mathbf{u}' = -\frac{\nabla' p}{\rho} - \mathbf{\Omega} \times \mathbf{\Omega} \times \mathbf{r}' - 2\mathbf{\Omega} \times \mathbf{u}'$$

The last two terms on the right-hand side of this equation are the centrifugal and Coriolis forces, respectively. Taking advantage of the results developed in Section 9.7.2, determine the form this equation assumes under a Galilean transformation. Is this equation Galilean invariant?

Chapter 10

Potential Flow

Now that we've derived the differential form of the conservation laws for fluid motion, we can shift from the control-volume oriented *global view* of a given flow to a *detailed view*. Achieving a detailed view requires solving, subject to appropriate boundary conditions, the differential equations that we inferred from the integral conservation laws. This permits determining fluid velocity, pressure and other flow properties throughout a given control volume. Note that, since we will actually be dropping the control-volume concept, in describing the flow domain we will replace the term *control volume* with the commonly-used term of *flowfield*.

Recall from our discussion in Section 9.7 that, because we use the Eulerian description, the differential equations describing fluid motion are *quasi-linear*. Since the basic differential equations are not linear, their solution can be very difficult, and usually must be carried out numerically. There is one specialized branch of fluid mechanics where the primary equations of motion, with no approximation, are linear. Specifically, if we confine our focus to inviscid, incompressible flows with no vorticity, there is a dramatic simplification of the fluid-flow equations. This branch of fluid mechanics is known as **potential-flow theory**.

Lest you think this might constitute a pathological case consider that, for the high Reynolds numbers typical of most practical flows, viscous effects are confined to extremely thin regions known as boundary layers. Thus, since viscous effects are confined to extremely thin layers, it should be reasonable to ignore viscosity for the rest of the flowfield, which often accounts for most of the flow domain. And, indeed, we find that the flow outside the boundary layers behaves just as the inviscid equations predict, at least for streamlined bodies.

We know that incompressible flow is very common in nature, so only the assumption of irrotational flow might be a cause for concern. In practice, vorticity is often confined to small regions such as the boundary layers close to a solid surface. Although effects of vorticity can extend over great distances, the flow away from the concentrated vorticity is usually irrotational. For example, we will see that flow past a rotating cylinder can be adequately described as irrotational. The circulation required to achieve the side force attending the rotation results from a fictitious vortex that we embed within the cylinder. This fictitious vortex is separate from the physical flowfield. The chapter includes a home experiment that demonstrates the connection between vorticity and lift.

Some of the earliest work in Computational Fluid Dynamics focused on potential-flow theory. Under the guidance of A. M. O. Smith [see Hess and Smith (1966)] of the Douglas Aircraft Company, for example, potential-flow computations were routinely done for complete airplane fuselages in the early 1960's. A great deal of effort in CFD focused on potential-flow theory well into the 1980's, and some activity continues today.

10.1 Differential Equations of Motion

We begin with the differential form of the mass- and momentum-conservation principles. As summarized in Chapter 9, the appropriate equations are the continuity and Euler equations. For incompressible flow, conservation of mass simplifies to [see Equation (9.40)]

$$\nabla \cdot \mathbf{u} = 0 \tag{10.1}$$

For a flow with a conservative body force given by $\mathbf{f} = -\nabla \mathcal{V}$, Euler's equation [Equation (9.23)] is

$$\rho \frac{d\mathbf{u}}{dt} = -\nabla p - \rho \nabla \mathcal{V} \tag{10.2}$$

Taking advantage of the vector identity for $\mathbf{u} \cdot \nabla \mathbf{u}$ given in Appendix C, we can rewrite the acceleration as

$$\frac{d\mathbf{u}}{dt} = \frac{\partial \mathbf{u}}{\partial t} + \nabla \left(\frac{1}{2} \mathbf{u} \cdot \mathbf{u} \right) - \mathbf{u} \times (\nabla \times \mathbf{u}) \tag{10.3}$$

Then, for irrotational flow, because $\nabla \times \mathbf{u} = \mathbf{0}$, Euler's equation for incompressible, irrotational flow with a conservative body force is

$$\frac{\partial \mathbf{u}}{\partial t} + \nabla \left(\frac{1}{2} \mathbf{u} \cdot \mathbf{u} \right) = -\nabla \left(\frac{p}{\rho} \right) - \nabla \mathcal{V} \tag{10.4}$$

Finally, we can move all terms to the left-hand side of this equation. The result is the desired form of Euler's equation that we will work with in this chapter, viz.,

$$\frac{\partial \mathbf{u}}{\partial t} + \nabla \left(\frac{p}{\rho} + \frac{1}{2} \mathbf{u} \cdot \mathbf{u} + \mathcal{V} \right) = \mathbf{0} \tag{10.5}$$

10.2 Mathematical Foundation

To understand the origin of the name, *potential flow*, consider the following. For an incompressible, irrotational flow of a perfect (inviscid) fluid, we can make use of the fact that the velocity field is curl and divergence free. That is, by definition of irrotational flow, we have

$$\nabla \times \mathbf{u} = \mathbf{0} \tag{10.6}$$

As discussed in the preceding section, conservation of mass for an incompressible fluid tells us

$$\nabla \cdot \mathbf{u} = 0 \tag{10.7}$$

First consider the irrotationality condition, Equation (10.6). As noted in Appendix C, for any scalar function ϕ,

$$curl \; grad \; \phi = \nabla \times \nabla \phi \equiv \mathbf{0} \tag{10.8}$$

Hence, if a function ϕ exists such that

$$\mathbf{u} = \nabla \phi \tag{10.9}$$

then Equation (10.6) is automatically satisfied. By definition, the quantity ϕ is the **velocity potential**, and this is the basis of the name potential-flow theory. Second, consider the incompressibility condition, Equation (10.7). For any vector $\mathbf{\Psi}$, reference to Appendix C shows that

$$div \ curl \ \mathbf{\Psi} = \nabla \cdot (\nabla \times \mathbf{\Psi}) \equiv 0 \qquad (10.10)$$

So, if we can find a vector $\mathbf{\Psi}$ such that

$$\mathbf{u} = \nabla \times \mathbf{\Psi} \qquad (10.11)$$

then the incompressibility condition is automatically satisfied.

As we will see in the following subsections, by virtue of Equations (10.6) and (10.7), the functions ϕ and $\mathbf{\Psi}$ exist and, to within an additive constant, are unique. Although potential-flow theory is not confined to two-dimensional flow, we will limit our study of the topic to two dimensions to simplify the analysis. One advantage of specializing to two-dimensional flow is that the vector $\mathbf{\Psi}$ has only one component, which is normal to the plane of flow. Thus, $\mathbf{\Psi}$ can be written as

$$\mathbf{\Psi} = \psi \mathbf{k} \qquad (10.12)$$

where ψ is a scalar and $\mathbf{k}$ is a unit vector normal to the xy plane. The quantity ψ is known as the **streamfunction**.

Use of the streamfunction is not confined to potential-flow theory, as the condition for its existence is simply that the flow be incompressible. It is frequently used in viscous-flow problems as well. By contrast, the velocity potential exists only if the flow is irrotational. Since viscous flows are inherently rotational, ϕ is confined to inviscid-flow applications.

10.2.1 Cylindrical Coordinates

Often the symmetry of a problem makes it convenient to work in cylindrical coordinates, r and θ, for potential-flow problems. In fact, we will make continuous use of both Cartesian and cylindrical coordinates in this chapter, often mixing both in the same problem. Thus, it is worthwhile to summarize key relations between the two systems, with regard to both coordinate axes and velocity components. Additional details can be found in Appendix D.

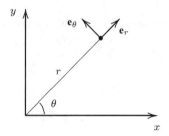

Figure 10.1: *Cylindrical coordinates.*

As illustrated in Figure 10.1, in two dimensions the cylindrical polar coordinate system is related to the rectangular Cartesian coordinate system by

$$x = r \cos \theta, \qquad y = r \sin \theta \qquad (10.13)$$

where

$$r = \sqrt{x^2 + y^2}, \qquad \theta = \tan^{-1} \left(\frac{y}{x} \right) \qquad (10.14)$$

Also, in cylindrical coordinates, we have unit vectors, $\mathbf{e}_r$ and $\mathbf{e}_\theta$, in the radial and circumferential directions, respectively. In terms of the Cartesian system unit vectors $\mathbf{i}$ and $\mathbf{j}$, we have

$$\mathbf{e}_r = \mathbf{i}\cos\theta + \mathbf{j}\sin\theta, \qquad \mathbf{e}_\theta = -\mathbf{i}\sin\theta + \mathbf{j}\cos\theta \tag{10.15}$$

The velocity vector has a component u_r in the radial direction and a component u_θ in the circumferential direction, so that

$$\mathbf{u} = u_r\,\mathbf{e}_r + u_\theta\,\mathbf{e}_\theta \tag{10.16}$$

The Cartesian (u, v) and cylindrical (u_r, u_θ) velocity components are related to each other as follows (see Appendix D).

$$\left.\begin{array}{ll} u = u_r\cos\theta - u_\theta\sin\theta & \text{and} \quad v = u_r\sin\theta + u_\theta\cos\theta \\[2mm] u_r = u\cos\theta + v\sin\theta & \text{and} \quad u_\theta = -u\sin\theta + v\cos\theta \end{array}\right\} \tag{10.17}$$

10.2.2 Velocity-Potential Representation

For simplicity, we begin our discussion in terms of rectangular Cartesian coordinates. In component form, the velocity potential, ϕ, is related to the velocity components by

$$u = \frac{\partial\phi}{\partial x} \quad \text{and} \quad v = \frac{\partial\phi}{\partial y} \tag{10.18}$$

Then, from the continuity Equation (10.7), we have

$$\frac{\partial u}{\partial x} + \frac{\partial v}{\partial y} = \frac{\partial^2\phi}{\partial x^2} + \frac{\partial^2\phi}{\partial y^2} = \nabla^2\phi \tag{10.19}$$

Hence, the velocity potential satisfies **Laplace's equation**, viz.,

$$\nabla^2\phi = 0 \tag{10.20}$$

As noted above, Laplace's equation is a linear partial differential equation about which a great deal is known [e.g., Kellogg (1953)]. We will generate numerous solutions to this equation in this chapter.

For problems in which cylindrical symmetry is present, we use cylindrical coordinates. Reference to Appendix D shows that the velocity potential is related to the velocity by

$$\nabla\phi = \mathbf{e}_r\frac{\partial\phi}{\partial r} + \mathbf{e}_\theta\frac{1}{r}\frac{\partial\phi}{\partial\theta} \tag{10.21}$$

Therefore, the relation between the velocity components and the velocity potential becomes

$$u_r = \frac{\partial\phi}{\partial r} \quad \text{and} \quad u_\theta = \frac{1}{r}\frac{\partial\phi}{\partial\theta} \tag{10.22}$$

Finally, Laplace's equation for ϕ transforms to the following.

$$r\frac{\partial}{\partial r}\left(r\frac{\partial\phi}{\partial r}\right) + \frac{\partial^2\phi}{\partial\theta^2} = 0 \tag{10.23}$$

10.2.3 Streamfunction Representation

For two-dimensional flow, the streamfunction, ψ, is related to the rectangular Cartesian velocity components by

$$u = \frac{\partial \psi}{\partial y} \quad \text{and} \quad v = -\frac{\partial \psi}{\partial x} \tag{10.24}$$

Then, from Equations (10.6), (10.11) and (10.12), there follows immediately

$$\nabla \times \mathbf{u} = \mathbf{k}\left(\frac{\partial v}{\partial x} - \frac{\partial u}{\partial y}\right) = -\mathbf{k}\left(\frac{\partial^2 \psi}{\partial x^2} + \frac{\partial^2 \psi}{\partial y^2}\right) = -\mathbf{k}\nabla^2\psi = \mathbf{0} \tag{10.25}$$

Hence, the streamfunction also satisfies Laplace's equation, viz.,

$$\nabla^2\psi = 0 \tag{10.26}$$

For cylindrical coordinates, we know from Appendix D that

$$\nabla \times (\psi\mathbf{k}) = \mathbf{e}_r\frac{1}{r}\frac{\partial \psi}{\partial \theta} - \mathbf{e}_\theta\frac{\partial \psi}{\partial r} \tag{10.27}$$

so that the relation between the velocity components and the streamfunction is

$$u_r = \frac{1}{r}\frac{\partial \psi}{\partial \theta} \quad \text{and} \quad u_\theta = -\frac{\partial \psi}{\partial r} \tag{10.28}$$

Finally, Laplace's equation for the streamfunction in cylindrical coordinates is

$$r\frac{\partial}{\partial r}\left(r\frac{\partial \psi}{\partial r}\right) + \frac{\partial^2 \psi}{\partial \theta^2} = 0 \tag{10.29}$$

Table 10.1: *Velocity-Potential and Streamfunction Representations*

	Velocity Potential	Streamfunction
General	$\mathbf{u} = \nabla\phi$	$\mathbf{u} = \nabla \times \psi\mathbf{k}$
	$\nabla^2\phi = 0$ (from $\nabla \cdot \mathbf{u} = 0$)	$\nabla^2\psi = 0$ (from $\nabla \times \mathbf{u} = \mathbf{0}$)
2 Dimensional	$u = \dfrac{\partial \phi}{\partial x}, \quad v = \dfrac{\partial \phi}{\partial y}$	$u = \dfrac{\partial \psi}{\partial y}, \quad v = -\dfrac{\partial \psi}{\partial x}$
(Cartesian)	$\dfrac{\partial^2 \phi}{\partial x^2} + \dfrac{\partial^2 \phi}{\partial y^2} = 0$	$\dfrac{\partial^2 \psi}{\partial x^2} + \dfrac{\partial^2 \psi}{\partial y^2} = 0$
Axisymmetric	$u_r = \dfrac{\partial \phi}{\partial r}, \quad u_\theta = \dfrac{1}{r}\dfrac{\partial \phi}{\partial \theta}$	$u_r = \dfrac{1}{r}\dfrac{\partial \psi}{\partial \theta}, \quad u_\theta = -\dfrac{\partial \psi}{\partial r}$
(Cylindrical)	$r\dfrac{\partial}{\partial r}\left(r\dfrac{\partial \phi}{\partial r}\right) + \dfrac{\partial^2 \phi}{\partial \theta^2} = 0$	$r\dfrac{\partial}{\partial r}\left(r\dfrac{\partial \psi}{\partial r}\right) + \dfrac{\partial^2 \psi}{\partial \theta^2} = 0$

As a convenience for future reference, Table 10.1 summarizes the velocity potential and streamfunction representations in both Cartesian and cylindrical coordinates. The velocity-potential and streamfunction representations form the basis of potential-flow theory. As noted above, one of the key advantages to this approach is the fact that Laplace's equation is linear. Thus, we can use features such as **superposition** and the various uniqueness and existence theorems for linear partial differential equations.

10.3   Streamlines and Equipotential Lines

The most useful feature of the streamfunction representation is its relation to streamlines. Consider the curve defined by

$$\psi(x, y) = \text{constant} \tag{10.30}$$

On this curve, an incremental change in ψ is given by:

$$d\psi = \frac{\partial \psi}{\partial x} dx + \frac{\partial \psi}{\partial y} dy = 0 \tag{10.31}$$

Then using Equation (10.24), we can rewrite this equation as

$$-v dx + u dy = 0 \tag{10.32}$$

or,

$$\frac{dx}{u} = \frac{dy}{v} \tag{10.33}$$

But, as discussed in Section 4.4, this is the equation of a streamline, i.e., a curve that is tangent to the local flow velocity. Therefore, $\psi(x, y) = \text{constant}$ defines a streamline.

Furthermore, the mass flux between two streamlines is proportional to the difference between the values of ψ on the two streamlines. To see this, consider flow across an arc ab of arc length Δs (see Figure 10.2) with the control volume shown. Clearly, the mass flowing across the arc, $\Delta \dot{m}_{ab}$, is

$$\Delta \dot{m}_{ab} = \rho \mathbf{u} \cdot \mathbf{n}_{ab} \Delta s \tag{10.34}$$

Now, for steady flow, we know that the net mass flux out of the control volume vanishes, wherefore

$$\oiint_S \rho \mathbf{u} \cdot \mathbf{n} \, dS = 0 \tag{10.35}$$

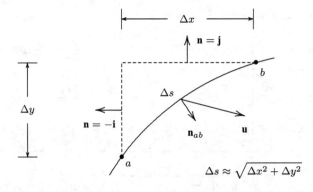

Figure 10.2: *Mass flux across an arc of length Δs.*

Hence, for the chosen control volume,

$$\rho \mathbf{u} \cdot \mathbf{n}_{ab} \, \Delta s + \rho \underbrace{[\mathbf{u} \cdot \mathbf{j}]}_{=v} \Delta x + \rho \underbrace{[\mathbf{u} \cdot (-\mathbf{i})]}_{=-u} \Delta y = 0 \qquad (10.36)$$

or,

$$\Delta \dot{m}_{ab} + \rho v \Delta x - \rho u \Delta y = 0 \qquad \Longrightarrow \qquad \Delta \dot{m}_{ab} = \rho \left[u \Delta y - v \Delta x \right] \qquad (10.37)$$

In differential form, we thus have

$$d\dot{m}_{ab} = \rho \left[u\,dy - v\,dx \right] \qquad (10.38)$$

Now, using Equation (10.24), we can rewrite this equation in terms of the streamfunction, so that the differential mass flux across arc ab becomes

$$d\dot{m}_{ab} = \rho \left[\frac{\partial \psi}{\partial y} dy + \frac{\partial \psi}{\partial x} dx \right] = \rho d\psi \qquad (10.39)$$

Therefore, $d\psi$ is the differential volume flux, viz.,

$$d\psi = \frac{1}{\rho} d\dot{m}_{ab} \qquad (10.40)$$

In a similar way, we can compute a differential change in the velocity potential.

$$d\phi = \frac{\partial \phi}{\partial x} dx + \frac{\partial \phi}{\partial y} dy = u\,dx + v\,dy \qquad (10.41)$$

So, on lines for which $\phi = $ constant, which are called the **equipotential lines**, there follows

$$\left(\frac{dy}{dx} \right)_{\phi=constant} = -\frac{u}{v} \qquad (10.42)$$

As shown above, on lines for which $\psi = $ constant (the streamlines), we have

$$\left(\frac{dy}{dx} \right)_{\psi=constant} = \frac{v}{u} \qquad (10.43)$$

Thus, the slopes of the equipotential lines and the streamlines are negative reciprocals, which means these lines are mutually perpendicular. In general, we can construct families of curves as illustrated in Figure 10.3.

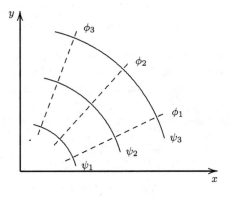

Figure 10.3: *Equipotential lines and streamlines.*

10.4   Bernoulli's Equation

Thus far, the equations we have discussed are the kinematical constraint of irrotationality, Equation (10.6), and mass conservation, Equation (10.7). However, aside from specializing Euler's equation for incompressible, irrotational flow [see Equation (10.5)], we have not discussed momentum conservation. The astute reader will observe that since the original mass-conservation equation is linear, it is no surprise that the governing equation for the velocity potential, $\nabla^2\phi = 0$, is also linear. Similarly, the irrotationality condition, from which the streamfunction equation, $\nabla^2\psi = 0$, is derived, is a linear equation. But, what about Euler's equation, which is where the nonlinearity associated with the Eulerian derivative lies? The answer to this question reflects the central role played by vorticity in determining the structure of an incompressible flow.

To explain this point, consider the following. If we introduce the velocity potential into Equation (10.5), we find

$$\nabla\left(\frac{\partial\phi}{\partial t} + \frac{p}{\rho} + \frac{1}{2}\mathbf{u}\cdot\mathbf{u} + \mathcal{V}\right) = \mathbf{0} \tag{10.44}$$

where we note that $\partial/\partial t$ and ∇ commute. Therefore, since its gradient, and thus all spatial derivatives are zero, the quantity within parentheses can be, at most, a function only of time, t. We thus have an unsteady-flow equivalent of Bernoulli's equation, viz.,

$$\frac{\partial\phi}{\partial t} + \frac{p}{\rho} + \frac{1}{2}\mathbf{u}\cdot\mathbf{u} + \mathcal{V} = F(t) \tag{10.45}$$

where $F(t)$ must be determined from initial and/or boundary conditions. This equation serves as a relation between the pressure and the velocity field for a given flow. Most importantly, if the velocity field is known, we can use Equation (10.45) to compute the pressure.

The question thus becomes, can we compute the velocity independent of the unsteady equivalent of Bernoulli's equation? Because the equation for ϕ is independent of pressure, we need only determine whether the initial and boundary conditions are independent of pressure. Regarding initial conditions, the strongest dependence on pressure they can have is upon the initial pressure. Hence, initial conditions are obviously independent of the instantaneous value of p. Regarding boundary conditions, the most common boundary condition imposed is at a solid boundary, through which there can be no flow. That is,

$$\mathbf{u}\cdot\mathbf{n} = 0 \qquad \text{(at solid boundaries)} \tag{10.46}$$

In terms of the velocity potential, we therefore have

$$\frac{\partial\phi}{\partial n} = 0 \qquad \text{(at solid boundaries)} \tag{10.47}$$

where n is the direction normal to the solid boundary. This is referred to as a *Neumann-type boundary condition*. We also impose farfield conditions, which are typically either undisturbed flow or uniform flow. Hence, the most common boundary conditions are completely independent of pressure.

Thus, in the context of potential-flow theory, we can pose a problem with an equation independent of the pressure, viz.,

$$\frac{\partial^2\phi}{\partial x^2} + \frac{\partial^2\phi}{\partial y^2} = 0 \tag{10.48}$$

Further, it is usually solved subject to boundary conditions of Neumann type that are also independent of pressure. The extensive theory developed by mathematicians for Laplace's equation tells us that a unique solution to this problem exists which is, of necessity, independent of pressure. In other words, we can solve a potential-flow problem with no prior knowledge of the pressure. Once we have this solution in hand, Equation (10.45) provides the pressure.

Most importantly, this shows that all of the nonlinearity appears in the equation for pressure, which is not required to determine the velocity. That is, the mass-conservation principle uncouples from the momentum-conservation equation, thus permitting a solution for $\mathbf{u}$ independent of p.

This shows that, at least in a mathematical sense, the pressure is a passive quantity. In terms of the physics of fluid motion, this is consistent with the notion that the vorticity is the driving force in an incompressible flow. The condition of irrotationality places a strong kinematic constraint on the way fluid particles move. In general, we can envision the logical sequence of cause and effect as follows.

1. The vorticity field dictates the local acceleration in the flow.

2. The pressure adjusts to balance the acceleration.

The mathematical structure of the potential-flow problem is completely consistent with this chronology. When the vorticity is non-vanishing, the continuity and momentum equations are coupled and this interpretation is less obvious.

As a final comment, while time appears in Equation (10.45), it does not appear in Laplace's equation. Hence, we can generate the solution for an unsteady flow problem by simply reflecting the time dependence in the farfield and/or surface boundary conditions. This mirrors the physical fact that incompressible flow is the limiting case of infinite sound speed, so that changes in the flow are communicated throughout the flowfield instantaneously. Thus, even when flow properties are changing with time, we are effectively treating the flow in a quasi-steady manner.

10.5 Fundamental Solutions

Because Laplace's equation is linear, we can take advantage of the powerful concept of superposition. Specifically, if $\phi_1(x, y)$ and $\phi_2(x, y)$ are both solutions to Laplace's equation, then

$$\phi(x, y) = c_1 \phi_1(x, y) + c_2 \phi_2(x, y) \tag{10.49}$$

is also a solution, where c_1 and c_2 are arbitrary constants. Furthermore, we can even develop a solution of the form

$$\phi(x, y) = \iint \left[c_1(\xi, \eta) \phi_1(x - \xi, y - \eta) + c_2(\xi, \eta) \phi_2(x - \xi, y - \eta) \right] d\xi d\eta \tag{10.50}$$

where $c_1(\xi, \eta)$ and $c_2(\xi, \eta)$ are arbitrary functions.

By using the superposition principle, we can construct solutions to complex problems based on some very simple solutions known as **fundamental solutions**. The fundamental solutions would correspond to ϕ_1 and ϕ_2. Generating a solution to a complex problem would then require solving for c_1 and c_2, for example. In this section, we will examine commonly used fundamental solutions to Laplace's equation, including both the velocity potential and the streamfunction. We will use fundamental solutions in the following sections to solve some interesting problems.

10.5.1   Uniform Flow

Description. The first fundamental solution we will consider is that of a uniform flow, i.e., a flow with constant velocity. Assuming the flow is parallel to the x axis, the constant velocity components are

$$u = U \quad \text{and} \quad v = 0 \tag{10.51}$$

Velocity Potential. To determine the velocity potential, $\phi(x, y)$, we note that

$$\frac{\partial \phi}{\partial x} = U \quad \text{and} \quad \frac{\partial \phi}{\partial y} = 0 \tag{10.52}$$

Integrating the first of Equations (10.52) over x, we find

$$\phi(x, y) = Ux + f(y) \tag{10.53}$$

where $f(y)$ is a function of integration. Now, we differentiate with respect to y and use the second of Equations (10.52), wherefore

$$\frac{\partial \phi}{\partial y} = \frac{df}{dy} = 0 \quad \Longrightarrow \quad f(y) = \text{constant} \tag{10.54}$$

Therefore, the velocity potential for uniform flow is

$$\phi(x, y) = Ux + \text{constant} \tag{10.55}$$

Clearly this potential function satisfies Laplace's equation since, by inspection, $\partial^2 \phi / \partial x^2$ and $\partial^2 \phi / \partial y^2$ both vanish.

Streamfunction. We follow a similar procedure to compute the streamfunction. First, by definition of the streamfunction, ψ, we know that

$$\frac{\partial \psi}{\partial y} = U \quad \text{and} \quad \frac{\partial \psi}{\partial x} = 0 \tag{10.56}$$

Integrating the first of Equations (10.56) over y, we find

$$\psi(x, y) = Uy + g(x) \tag{10.57}$$

where $g(x)$ is a function of integration. Differentiating with respect to x and using the second of Equations (10.56), we arrive at

$$\frac{\partial \psi}{\partial x} = \frac{dg}{dx} = 0 \quad \Longrightarrow \quad g(x) = \text{constant} \tag{10.58}$$

Thus, the streamfunction for uniform flow is

$$\psi(x, y) = Uy + \text{constant} \tag{10.59}$$

As with the potential function, the streamfunction obviously satisfies Laplace's equation since $\partial^2 \psi / \partial x^2$ and $\partial^2 \psi / \partial y^2$ both vanish.

Streamlines and Equipotential Lines. From Equation (10.59), the streamlines are simply a family of straight lines parallel to the x axis, i.e., lines along which y is constant. Similarly, from Equation (10.55), the equipotentials are a family of straight lines parallel to the y axis. Figure 10.4 shows the streamlines and equipotentials for uniform flow.

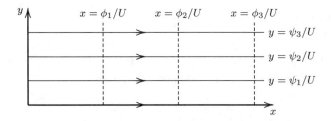

Figure 10.4: *Streamlines (——) and equipotential lines (- - -) for uniform flow.*

10.5.2   Potential Vortex

Description. We discussed the potential vortex in Section 4.3 when we introduced the concepts of vorticity and circulation. By definition, the velocity field for this flow is

$$u_r = 0 \quad \text{and} \quad u_\theta = \frac{\Gamma}{2\pi r} \tag{10.60}$$

where Γ is the circulation, and r is distance from the origin. Note that the circumferential velocity is singular at the origin, i.e., $u_\theta \to \infty$ as $r \to 0$. This is characteristic of all the fundamental solutions we will discuss, with the exception of uniform flow. To constitute part of the solution to a physical flowfield, such singular points must lie outside the solution domain—we will see how this is accomplished in subsequent sections. As a final comment, recall that the circulation for the potential vortex is the line integral of velocity on a contour enclosing the vortex. Selecting a circle of radius r, we have

$$\oint_C \mathbf{u} \cdot d\mathbf{s} = \int_0^{2\pi} (u_\theta \mathbf{e}_\theta) \cdot (r d\theta \mathbf{e}_\theta) = \int_0^{2\pi} \frac{\Gamma}{2\pi r} r d\theta = \Gamma \tag{10.61}$$

Velocity Potential. To determine the velocity potential, $\phi(r, \theta)$, we begin with

$$\frac{\partial \phi}{\partial r} = 0 \quad \text{and} \quad \frac{1}{r} \frac{\partial \phi}{\partial \theta} = \frac{\Gamma}{2\pi r} \tag{10.62}$$

Integrating the second of Equations (10.62) over θ, we find

$$\phi(r, \theta) = \frac{\Gamma}{2\pi} \theta + f(r) \tag{10.63}$$

where $f(r)$ is a function of integration. Now, we differentiate with respect to r and use the first of Equations (10.62), so that

$$\frac{\partial \phi}{\partial r} = \frac{df}{dr} = 0 \quad \Longrightarrow \quad f(r) = \text{constant} \tag{10.64}$$

Therefore, the velocity potential for the potential vortex is

$$\phi(r, \theta) = \frac{\Gamma}{2\pi} \theta + \text{constant} \tag{10.65}$$

To verify that ϕ satisfies Laplace's equation, observe that it is a function only of θ. Hence, because it is linear in θ, both terms in Laplace's equation vanish.

Streamfunction. In terms of cylindrical coordinates, the streamfunction, ψ, satisfies the following equations.

$$\frac{1}{r}\frac{\partial\psi}{\partial\theta} = 0 \quad \text{and} \quad \frac{\partial\psi}{\partial r} = -\frac{\Gamma}{2\pi r} \tag{10.66}$$

Integrating the second of Equations (10.66) over r, we find

$$\psi(r,\theta) = -\frac{\Gamma}{2\pi}\ell nr + g(\theta) \tag{10.67}$$

where $g(\theta)$ is a function of integration. Differentiating with respect to θ yields

$$\frac{1}{r}\frac{\partial\psi}{\partial\theta} = \frac{1}{r}\frac{dg}{d\theta} = 0 \quad \Longrightarrow \quad g(\theta) = \text{constant} \tag{10.68}$$

Thus, the streamfunction for the potential vortex is

$$\psi(r,\theta) = -\frac{\Gamma}{2\pi}\ell nr + \text{constant} \tag{10.69}$$

To verify that the streamfunction satisfies Laplace's equation, note first that it is a function of r only. Also, $r\partial\psi/\partial r = -\Gamma/(2\pi) = \text{constant}$, so that Laplace's equation (see Table 10.1) is trivially satisfied.

Streamlines and Equipotential Lines. From Equation (10.69), the streamlines are simply a family of concentric circles centered at the origin. The equipotentials are a family of radial lines radiating from the origin. Figure 10.5 shows the streamlines and equipotentials for the potential vortex.

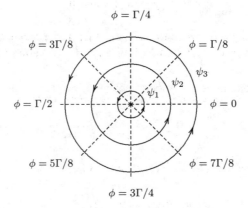

Figure 10.5: *Streamlines (——) and equipotential lines (- - -) for a potential vortex; the radii of the circular streamlines are $r_i = \exp(-2\pi\psi_i/\Gamma)$.*

10.5.3   Source and Sink

Description. The source is another singular solution. It represents a continuous source of fluid emanating from the origin. The source provides a constant volume flux of fluid, Q. When the value of Q is negative, we refer to this fundamental solution as a *sink*. The velocity components for a source are

$$u_r = \frac{Q}{2\pi r} \quad \text{and} \quad u_\theta = 0 \tag{10.70}$$

where Q is the source strength (volume flux), and r is distance from the origin. The mass flowing across any boundary enclosing a source is constant. For a circle of radius r centered at the origin, we have

$$\dot{m} = \oiint_S \rho \mathbf{u} \cdot \mathbf{n} \, dS = \int_0^{2\pi} \rho \left(u_r \mathbf{e}_r \right) \cdot \mathbf{e}_r \, r d\theta = \int_0^{2\pi} \rho \frac{Q}{2\pi r} r d\theta = \rho Q \tag{10.71}$$

Velocity Potential. The velocity potential, $\phi(r, \theta)$, satisfies

$$\frac{\partial \phi}{\partial r} = \frac{Q}{2\pi r} \quad \text{and} \quad \frac{1}{r}\frac{\partial \phi}{\partial \theta} = 0 \tag{10.72}$$

Integrating the first of Equations (10.72) over r yields

$$\phi(r, \theta) = \frac{Q}{2\pi} \ell n r + f(\theta) \tag{10.73}$$

where $f(\theta)$ is a function of integration. Hence, differentiating with respect to θ and using the second of Equations (10.72), there follows

$$\frac{\partial \phi}{\partial \theta} = \frac{df}{d\theta} = 0 \quad \Longrightarrow \quad f(\theta) = \text{constant} \tag{10.74}$$

Thus, the velocity potential is

$$\phi(r, \theta) = \frac{Q}{2\pi} \ell n r + \text{constant} \tag{10.75}$$

To verify that ϕ satisfies Laplace's equation, observe that it is a function only of r, and that $r\partial\phi/\partial r = Q/(2\pi) = \text{constant}$. Substitution into Laplace's equation (see Table 10.1) thus shows that it is satisfied.

Streamfunction. Turning to the streamfunction, ψ, we know that

$$\frac{1}{r}\frac{\partial \psi}{\partial \theta} = \frac{Q}{2\pi r} \quad \text{and} \quad \frac{\partial \psi}{\partial r} = 0 \tag{10.76}$$

Integrating the first of Equations (10.76) over θ shows that

$$\psi(r, \theta) = \frac{Q}{2\pi} \theta + g(r) \tag{10.77}$$

where $g(r)$ is a function of integration. Differentiation with respect to r and using the second of Equations (10.76) gives

$$\frac{\partial \psi}{\partial r} = \frac{dg}{dr} = 0 \quad \Longrightarrow \quad g(r) = \text{constant} \tag{10.78}$$

Hence, the streamfunction for the source is

$$\psi(r, \theta) = \frac{Q}{2\pi} \theta + \text{constant} \tag{10.79}$$

Because $\psi(r, \theta)$ is a linear function of θ only, it obviously satisfies Laplace's equation.

Streamlines and Equipotential Lines. Equation (10.79) shows that the streamlines are a family of radial lines with origin at $r = 0$. Similarly, Equation (10.75) shows that the equipotential lines are a family of concentric circles centered at the origin. Figure 10.6 depicts the streamlines and equipotentials for a source and a sink.

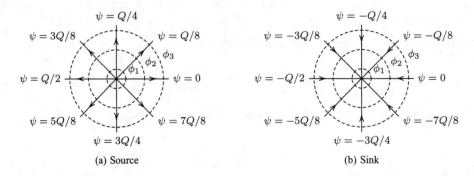

(a) Source   (b) Sink

Figure 10.6: *Streamlines (——) and equipotential lines (- - -) for a source and a sink; the radii of the circular equipotential lines are $r_i = \exp(-2\pi\phi_i/Q)$.*

10.5.4   Doublet

Description. Consider a source of strength Q located on the x axis at $x = -a$, and a sink of strength $-Q$ located on the same axis at $x = a$. Figure 10.7(a) shows the geometry for such a source/sink superposition. A doublet is formed when the distance between the source and sink goes to zero in such a way that the product of the spacing and source strength is constant,

$$\lim_{a \to 0} 2aQ = \mathcal{D} \tag{10.80}$$

where $\mathcal{D}$ is a constant referred to as the doublet strength. The doublet is a useful fundamental solution that finds extensive use in potential-flow theory. There is no net flow out of a contour enclosing a doublet. Rather, fluid flows from the source to the sink with the streamlines approaching circular shapes tangent to the x axis as shown in Figure 10.7(b). The flow directions indicated on the streamlines correspond to positive doublet strength.

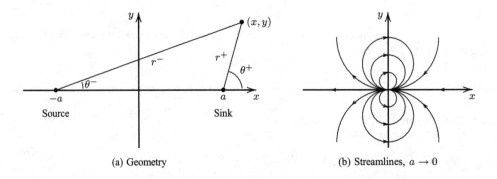

(a) Geometry   (b) Streamlines, $a \to 0$

Figure 10.7: *Source-sink geometry and streamlines.*

Velocity Potential. Using the superposition principle, we can construct the velocity potential as the sum of the potentials for a source and a sink. Hence, in terms of Cartesian coordinates,

$$\phi(x,y) = \lim_{a \to 0} \frac{Q}{2\pi} \left(\ell nr^- - \ell nr^+ \right) \tag{10.81}$$

The quantities r^+ and θ^+ are the cylindrical coordinates corresponding to coordinate axes coincident with the sink, while r^- and θ^- are cylindrical coordinates centered at the location of the source. They are defined as follows.

$$\left. \begin{array}{ll} \text{Sink}: & r^+ = \sqrt{(x-a)^2 + y^2}, \quad \theta^+ = \tan^{-1}\left(\dfrac{y}{x-a}\right) \\[3mm] \text{Source}: & r^- = \sqrt{(x+a)^2 + y^2}, \quad \theta^- = \tan^{-1}\left(\dfrac{y}{x+a}\right) \end{array} \right\} \tag{10.82}$$

Hence, we can rewrite the velocity potential as

$$\phi(x,y) = \lim_{a \to 0} \left(\frac{2aQ}{2\pi} \right) \frac{\ell n\sqrt{(x+a)^2 + y^2} - \ell n\sqrt{(x-a)^2 + y^2}}{2a} \tag{10.83}$$

Then, noting the definition of the doublet strength quoted in Equation (10.80), we can rearrange this equation as follows.

$$\phi(x,y) = \frac{D}{2\pi} \lim_{a \to 0} \frac{\ell n\sqrt{(x+a)^2 + y^2} - \ell n\sqrt{(x-a)^2 + y^2}}{2a} \tag{10.84}$$

But, by definition, the indicated limit is simply the partial derivative with respect to x of the function $\ell n\sqrt{x^2 + y^2} = \ell nr$. Therefore, we have

$$\phi(x,y) = \frac{D}{2\pi} \frac{\partial}{\partial x} \ell nr \tag{10.85}$$

We can simplify this further by taking the derivative, i.e.,

$$\frac{\partial}{\partial x} \ell nr = \frac{1}{r} \frac{\partial r}{\partial x} = \frac{1}{r} \frac{\partial}{\partial x} \sqrt{x^2 + y^2} = \frac{1}{r} \frac{x}{r} = \frac{x}{r^2} \tag{10.86}$$

Finally, since the Cartesian coordinate x is related to cylindrical coordinates by $x = r\cos\theta$, we conclude that

$$\frac{\partial}{\partial x} \ell nr = \frac{\cos\theta}{r} \tag{10.87}$$

Thus, substituting Equation (10.87) into Equation (10.85), we arrive at the velocity potential for a doublet, viz.,

$$\phi(r,\theta) = \frac{D}{2\pi r} \cos\theta \tag{10.88}$$

Streamfunction. We can derive the streamfunction using the same procedure. Again using superposition, the streamfunction for the source-sink pair is

$$\psi(x,y) = \lim_{a \to 0} \frac{Q}{2\pi} \left(\theta^- - \theta^+ \right) = \lim_{a \to 0} \left(\frac{2aQ}{2\pi} \right) \frac{\tan^{-1}\left(\dfrac{y}{x+a}\right) - \tan^{-1}\left(\dfrac{y}{x-a}\right)}{2a} \tag{10.89}$$

Taking the limit, the streamfunction thus becomes

$$\psi(x, y) = \frac{\mathcal{D}}{2\pi} \frac{\partial}{\partial x} \tan^{-1} \left(\frac{y}{x} \right) = -\frac{\mathcal{D}}{2\pi} \frac{y}{x^2 + y^2} \tag{10.90}$$

Then noting that $y = r \sin\theta$ and $r^2 = x^2 + y^2$, the streamfunction for a doublet is

$$\psi(r, \theta) = -\frac{\mathcal{D}}{2\pi r} \sin\theta \tag{10.91}$$

Streamlines and Equipotential Lines. We can most conveniently determine the equations of the streamlines and equipotentials by using Cartesian coordinates. Considering the streamlines first, Equation (10.91) can be rewritten as

$$r^2 = -\frac{\mathcal{D}}{2\pi\psi} r \sin\theta \quad \Longrightarrow \quad x^2 + y^2 = -2 \left(\frac{\mathcal{D}}{4\pi\psi} \right) y \tag{10.92}$$

or,

$$x^2 + \left(y + \frac{\mathcal{D}}{4\pi\psi} \right)^2 = \left(\frac{\mathcal{D}}{4\pi\psi} \right)^2 \tag{10.93}$$

Similarly, the equipotentials follow from Equation (10.88), viz.,

$$r^2 = \frac{\mathcal{D}}{2\pi\phi} r \cos\theta \quad \Longrightarrow \quad x^2 + y^2 = 2 \left(\frac{\mathcal{D}}{4\pi\phi} \right) x \tag{10.94}$$

so that

$$\left(x - \frac{\mathcal{D}}{4\pi\phi} \right)^2 + y^2 = \left(\frac{\mathcal{D}}{4\pi\phi} \right)^2 \tag{10.95}$$

Therefore, the streamlines are a family of circles of radii $\mathcal{D}/(4\pi|\psi|)$ tangent to the x axis, and the equipotentials are a family of circles of radii $\mathcal{D}/(4\pi|\phi|)$ tangent to the y axis. Figure 10.7(b)—presented earlier in this subsection—shows the streamlines.

10.5.5   Comments on Use of Fundamental Solutions

In the following sections we will use the fundamental solutions described above to construct solutions to some interesting fluid-flow problems. Table 10.2 summarizes the velocity potentials and streamfunctions for the four fundamental solutions covered in this section. Note that we recast the uniform-flow results in cylindrical coordinates for the sake of consistency.[1] Before proceeding, it is worthwhile to pause and summarize a few salient observations regarding potential-flow theory, and to describe what our solution strategy is.

Simulating Solid Boundaries. Because a streamline is a contour to which the velocity is everywhere tangent, there can be no flow normal to a streamline. Thus, when we build a solution based on fundamental solutions that contains a closed streamline, we can always consider it to be the boundary of a solid object. We call such a boundary a **dividing streamline**. Closed bodies will result, for example, when the net source strength is zero, such as with the *doublet*. In general, we introduce *sources* and *sinks* to determine the shape of the body of interest. Any streamline, closed or not, can always be regarded as a solid boundary.

---

[1] All of the singular potentials and streamfunctions in the Table assume the origin is located at $(x, y) = (0, 0)$. If we place the source, vortex or doublet at some other location, say $(x, y) = (\xi, \eta)$, the quantities r and θ are given by $r = \sqrt{(x - \xi)^2 + (y - \eta)^2}$ and $\theta = \tan^{-1}[(y - \eta)/(x - \xi)]$.

Table 10.2: *Fundamental-Solution Velocity Potentials and Streamfunctions*

Flow	Velocity Potential, $\phi(r,\theta)$	Streamfunction, $\psi(r,\theta)$
Uniform Flow	$Ur\cos\theta$	$Ur\sin\theta$
Potential Vortex	$\dfrac{\Gamma}{2\pi}\theta$	$-\dfrac{\Gamma}{2\pi}\ell nr$
Source	$\dfrac{Q}{2\pi}\ell nr$	$\dfrac{Q}{2\pi}\theta$
Doublet	$\dfrac{\mathcal{D}}{2\pi r}\cos\theta$	$-\dfrac{\mathcal{D}}{2\pi r}\sin\theta$

Suppressing Singular Behavior. As noted earlier, with the one exception of *uniform flow*, all of the fundamental solutions discussed in this section contain a singular point, i.e., a point where the velocity is infinite. Because such points do not occur in nature, the question naturally arises about how we can use fundamental solutions to construct a physically realistic flowfield solution. The answer is simple. The solutions we construct confine the singularities to a region external to the flowfield. In the case of a closed body, for example, the singularity lies within the interior of the dividing streamline. Note that the region within the dividing streamline is a separate flowfield according to potential-flow theory, although the singularities preclude using it as a description of a physically realizable flow.

Simulating Forces on Objects. We can simulate a lifting force in our solution by adding potential vortices. Although the potential vortex is irrotational, it has constant circulation that corresponds to having all of the vorticity concentrated at the origin, where the solution is singular. Again, provided we locate the vortex within a dividing streamline, the solution will correspond to a physically realizable flow with lift.

Direct and Indirect Solutions. Laplace's equation is one of the basic equations of mathematical physics, and its properties and methods of solution are well established. Entire books have been written about Laplace's equation [cf. Kellogg (1953)] and its application in fluid mechanics [e.g., Robertson (1965) and Milne-Thompson (1960)]. In the following sections, we will construct what are known as **indirect solutions**, in which we distribute fundamental solutions and infer the shape generated by our selection of source, vortex, or doublet strengths. While this provides algebraically manageable solutions, we have virtually no control over the geometrical shapes that we obtain the solution for. To generate a solution for a given geometry, we require a **direct solution**. Such solutions are far more pertinent in general engineering practice. However, they usually require solution of an integral equation such as Equation (10.50), and fall beyond the scope of this text. Nevertheless, the reader should be aware that such methods exist [cf. Hess (1975)] and are generally implemented numerically.

Complex Variables. As a final comment, all of the fundamental solutions can be recast in terms of complex-variable theory [cf. Churchill and Brown (1990), Carrier, Krook and Pearson (1966), or Whittaker and Watson (1963)]. In this context, which is inherently limited to two-dimensional flows, there are many powerful advanced concepts from complex variables such as contour integration [Hildebrand (1976), Wilcox (1995)] and conformal mapping [Carrier, Krook and Pearson (1966)] that greatly enhance the range of problems that can be solved. Many texts describe the use of complex variables for potential-flow problems. Panton (1996) provides a particularly lucid discussion. The Problems section includes a few examples.

10.6   Flow Past a Cylinder

Constructing the solution for flow past a cylinder is especially illuminating as it illustrates many of the most important features of potential-flow theory. The solution involves superposition of three of the fundamental solutions of the preceding section, and shows many of the subtleties of the method. We begin by superposing a uniform flow and a doublet to generate the basic solution. We will find that there is no force on the cylinder. Then, we add a vortex at the center of the cylinder so that its singularity is isolated from the main flow. We will find that the vortex gives rise to a lifting force.

10.6.1   The Basic Solution

Statement of the Problem. We consider a cylinder of radius R immersed in a flow that has velocity $\mathbf{u} = U\,\mathbf{i}$ far upstream (Figure 10.8). Hence, assuming the flow is steady, inviscid, incompressible and irrotational, the problem at hand can therefore be stated as

$$\mathbf{u} = \nabla\phi, \qquad \nabla^2\phi = 0 \tag{10.96}$$

which must be solved subject to

$$\mathbf{u} \to U\,\mathbf{i} \quad \text{as} \quad r \to \infty, \qquad \mathbf{u}\cdot\mathbf{n} = 0 \quad \text{on} \quad r = R \tag{10.97}$$

The first of Equations (10.97) is the farfield condition that the flow be uniform far from the cylinder. The second is the condition that there be no flow across the cylinder surface. In terms of cylindrical coordinates, Equations (10.97) become

$$\left.\begin{array}{ll} u_r \to U\cos\theta, & r \to \infty \\[4pt] u_\theta \to -U\sin\theta, & r \to \infty \\[4pt] u_r = 0, & r = R \end{array}\right\} \tag{10.98}$$

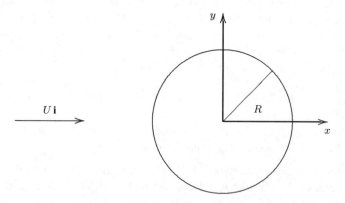

Figure 10.8: *Flow past a cylinder.*

There are ways to solve this linear problem directly. We could use a separation of variables, for example, i.e., assume $\phi(r, \theta) = R(r)\Theta(\theta)$, substitute into Laplace's equation and the boundary conditions, and solve. We will not pursue a direct solution however, as our purpose here is to examine the role of fundamental solutions in potential-flow theory. Thus, we will simply make a guess about what the solution is, with an adjustable parameter (the doublet strength), and verify that we have solved the correct problem with proper selection of the parameter. Alternatively, we can view what we are doing as follows. First, we construct a solution by superposition of fundamental solutions. Then, we determine the problem we have solved. From either point of view, we now proceed to the solution.

Velocity Potential and Streamfunction. If we superpose a uniform flow with freestream velocity U and a doublet of strength $\mathcal{D}$, reference to Table 10.2 shows that the velocity potential and streamfunction are as follows.

$$\left.\begin{aligned}
\phi(r, \theta) &= U r \cos\theta + \frac{\mathcal{D}}{2\pi r} \cos\theta \\[2mm]
\psi(r, \theta) &= U r \sin\theta - \frac{\mathcal{D}}{2\pi r} \sin\theta
\end{aligned}\right\} \tag{10.99}$$

Note that, with no loss of generality, we have set the constants that appear with ϕ and ψ equal to zero. Thus, the streamlines for this flow are given by

$$\left(U r - \frac{\mathcal{D}}{2\pi r}\right) \sin\theta = \text{constant} \tag{10.100}$$

In the special case where the constant is zero, there are two solutions, viz.,

$$\left.\begin{aligned}
\sin\theta = 0 \quad &\Longrightarrow \quad y = r\sin\theta = 0 \\[2mm]
U r - \frac{\mathcal{D}}{2\pi r} = 0 \quad &\Longrightarrow \quad r^2 = \frac{\mathcal{D}}{2\pi U}
\end{aligned}\right\} \tag{10.101}$$

Therefore, if we define

$$R = \sqrt{\frac{\mathcal{D}}{2\pi U}} \tag{10.102}$$

then the last of Equations (10.101) shows that this streamline corresponds to flow about a cylinder of radius R. Also, the solution with $y = 0$ corresponds to the parts of the dividing streamline upstream and downstream of the cylinder. Eliminating $\mathcal{D}$, the velocity potential and streamfunction are

$$\left.\begin{aligned}
\phi(r, \theta) &= U \left(r + \frac{R^2}{r}\right) \cos\theta \\[2mm]
\psi(r, \theta) &= U \left(r - \frac{R^2}{r}\right) \sin\theta
\end{aligned}\right\} \tag{10.103}$$

Because both the uniform flow and the doublet satisfy Laplace's equation, so must their sum. Hence, we know that our solution satisfies the basic equations of motion. To be certain we have satisfied the boundary conditions, we should compute the velocity and verify that Equations (10.98) are satisfied.

Velocity Components. The corresponding velocity components are

$$\left.\begin{aligned}
u_r &= \frac{\partial \phi}{\partial r} = U\left(1 - \frac{R^2}{r^2}\right)\cos\theta \\[2mm]
u_\theta &= \frac{1}{r}\frac{\partial \phi}{\partial \theta} = -U\left(1 + \frac{R^2}{r^2}\right)\sin\theta
\end{aligned}\right\} \tag{10.104}$$

Clearly, the velocity approaches the freestream velocity as $r \to \infty$, where the contribution from the doublet approaches zero as $1/r^2$. Inspection of the solution for u_r shows that it vanishes on the surface of the cylinder. Therefore, the boundary conditions are satisfied. Since the solution satisfies Laplace's equation and the boundary conditions, we know that this is the (unique) solution.

Note that, because the doublet lies within the cylinder, there is no singular behavior for $r > R$. Figure 10.9 shows the streamlines for the flow both inside and outside the cylinder. The streamlines within the cylinder (shown as dashed lines) resemble those of pure doublet flow. For our purposes, we can ignore the flow for $r < R$ as the surface, $r = R$, is a closed streamline, so that there is no mixing of fluid from the two regions.

Stagnation Points. To locate any stagnation points that might be present in the flow, we seek points where *both* u_r and u_θ vanish. For the case at hand, appeal to Equations (10.104) yields

$$\left.\begin{aligned}
\left(1 - \frac{R^2}{r^2}\right)\cos\theta = 0 \\[2mm]
\left(1 + \frac{R^2}{r^2}\right)\sin\theta = 0
\end{aligned}\right\} \tag{10.105}$$

Clearly, since $(1 + R^2/r^2)$ is nonzero at all points in the flowfield ($r > R$), if any stagnation points are present, they must occur at points where $\sin\theta = 0$. Now, since $\cos\theta \neq 0$ at points where $\sin\theta = 0$, necessarily the factor $(1 - R^2/r^2) = 0$, so that $r = R$. Finally, $\sin\theta = 0$ when $\theta = 0$ and $\theta = \pi$. Therefore, the flow has two stagnation points located at

$$(r, \theta) = (R, \pi) \quad \text{and} \quad (r, \theta) = (R, 0) \tag{10.106}$$

corresponding to the points where the dividing streamline ($y = 0$ upstream of the cylinder) splits and rejoins (to $y = 0$ downstream of the cylinder), respectively. Figure 10.9 shows the stagnation points.

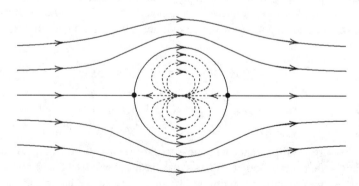

Figure 10.9: *Streamlines for flow past a cylinder;* • *denotes a stagnation point.*

Another interesting observation about the velocity field is its variation on the surface of the cylinder. When $r = R$, we have

$$u_r = 0, \qquad u_\theta = -2U \sin \theta \tag{10.107}$$

where the minus sign reflects the fact that u_θ is positive in the counterclockwise direction. Thus, the magnitude of the velocity on the cylinder increases from zero at the leading stagnation point ($\theta = \pi$) to twice the freestream velocity at the midpoint ($\theta = \pi/2$), and drops back to zero at the trailing stagnation point. Because of the high velocity at the top of the cylinder, the spacing between the streamlines must decrease from the upstream spacing in order to conserve mass (see Figure 10.9).

10.6.2   Force on the Cylinder

We have sufficient information to compute the force on the cylinder. We can compute it by integrating the pressure over the cylinder surface. Letting $\mathbf{n}$ denote the outer unit normal to the cylinder, the net force exerted by the fluid on the cylinder is

$$\mathbf{F} = - \oiint_S p\, \mathbf{n}\, dS \tag{10.108}$$

Now, taking advantage of the fact that we can replace p by $(p - p_\infty)$ [see Subsection 5.4.2], and noting that

$$\mathbf{n} = \mathbf{e}_r = \mathbf{i} \cos \theta + \mathbf{j} \sin \theta, \qquad dS = R d\theta \tag{10.109}$$

the force (per unit length) is thus given by

$$\mathbf{F} = - \int_0^{2\pi} (p - p_\infty)\, (\mathbf{i} \cos \theta + \mathbf{j} \sin \theta)\, R d\theta \tag{10.110}$$

For steady flow with no body force, Bernoulli's equation tells us the pressure satisfies

$$p + \frac{1}{2} \rho\, \mathbf{u} \cdot \mathbf{u} = p_\infty + \frac{1}{2} \rho\, U^2 \tag{10.111}$$

On the surface of the cylinder, we have

$$\mathbf{u} \cdot \mathbf{u} = u_r^2(R, \theta) + u_\theta^2(R, \theta) = 4U^2 \sin^2 \theta \tag{10.112}$$

Therefore, the pressure difference for flow on the cylinder surface is

$$p - p_\infty = \frac{1}{2} \rho\, U^2 \left(1 - 4 \sin^2 \theta\right) \tag{10.113}$$

Then,

$$\mathbf{F} = - \frac{1}{2} \rho\, U^2 R \int_0^{2\pi} \left(1 - 4 \sin^2 \theta\right) (\mathbf{i} \cos \theta + \mathbf{j} \sin \theta)\, d\theta \tag{10.114}$$

Finally, we define the lift, L, and drag, D, (per unit cylinder height) as follows.

$$\mathbf{F} = D\mathbf{i} + L\mathbf{j} \tag{10.115}$$

Hence, for our cylinder, the lift and drag according to the potential-flow solution are

$$\left. \begin{array}{l} L = -\dfrac{1}{2}\rho U^2 R \displaystyle\int_0^{2\pi} \left(1 - 4\sin^2\theta\right)\sin\theta d\theta = 0 \\[3mm] D = -\dfrac{1}{2}\rho U^2 R \displaystyle\int_0^{2\pi} \left(1 - 4\sin^2\theta\right)\cos\theta d\theta = 0 \end{array} \right\} \tag{10.116}$$

Therefore, the net force on the cylinder is zero. This result is expected of course, as the flow, by construction of our solution, has zero vorticity. Examination of the pressure variation over the cylinder surface clearly illustrates why the net force is zero. Figure 10.10 shows the **pressure coefficient,** C_p, defined by

$$C_p \equiv \frac{p - p_\infty}{\frac{1}{2}\rho U^2} = 1 - 4\sin^2\theta \tag{10.117}$$

on the upper surface as a function of θ, which is measured from the positive x axis.

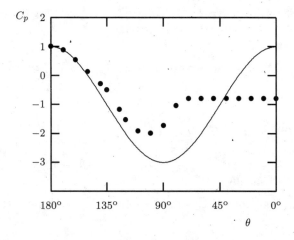

Figure 10.10: *Pressure coefficient for flow past a cylinder;* —— *Potential-flow theory;* ● *Measured [Patel (1968)].*

By symmetry, the pressure on the lower surface must be identical to that on the upper surface. Because the unit normal has opposite directions on the upper and lower surfaces, clearly there can be no lift. This conclusion would be true even for a viscous flow. It is less obvious for the drag, however. According to our potential-flow solution, the pressure is also symmetric about the midsection of the cylinder, i.e., about the y axis. Thus, for any pressure force exerted on the front face of the cylinder, there is an equal and opposite pressure force exerted on the back face. Figure 10.11 depicts the balance of pressure forces schematically for a given angle θ_o.

Figure 10.10 also includes measured C_p values [Patel (1968)]. As shown, except within about 75° of the leading stagnation point, the pressure on a cylinder in a real fluid is quite different from the potential-flow prediction. The large difference is caused by a phenomenon known as **separation**. In a real flow, the no-slip boundary condition (Section 11.3) tells us the velocity vanishes on the cylinder surface. This means the fluid very close to the surface, i.e., within the boundary layer, has low momentum. As we move beyond the top of the cylinder,

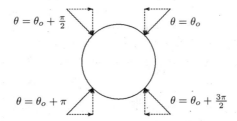

Figure 10.11: *Pressure-force balance for flow past a cylinder.*

the pressure begins to increase and decelerates the flow. The fluid in the boundary layer lacks sufficient momentum to overcome the increasing pressure, and leaves the surface before reaching the rear stagnation point predicted by our potential-flow solution. Rather, it enters a viscous region known as the **wake**. The flow in the wake near the cylinder has much lower pressure, and the asymmetric pressure distribution (Figure 10.10) yields a drag force.

This is illustrated schematically in Figure 10.12. Points where the flow leaves the surface are known as **separation points**. For very high Reynolds numbers, measurements show that separation occurs at about $\theta = 60°$ (120° from the forward stagnation point). For lower Reynolds numbers, it can occur as far forward as $\theta = 98°$.

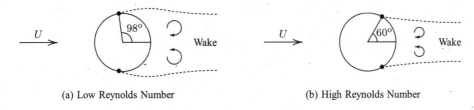

(a) Low Reynolds Number   (b) High Reynolds Number

Figure 10.12: *Flow separation for real flow past a cylinder;* • *denotes a separation point.*

As a final comment, a cylinder is not a thin, streamlined body. Thus, the fact that the predicted and measured pressure distributions differ so greatly is unsurprising, and should not be regarded as a failure of potential-flow theory. Thick, non-streamlined bodies are simply beyond the range of applicability of the theory. By contrast, for streamlined objects such as airplane wings, which are designed to pass with minimum resistance through the air, the theory provides an excellent prediction of most features of the flow.

However, if a streamlined object is inclined to the flow at too large an angle to the oncoming flow, separation occurs and, again, the potential-flow prediction will differ greatly from observations. In general, potential-flow methods fail to provide a physically realistic solution when increasing pressure is encountered near a solid boundary.

10.6.3 Adding a Vortex

To illustrate how we can add a lifting force to a potential-flow solution, we pose the following questions. If we add a potential vortex at the origin, what happens to our flow? Does the geometry of the dividing streamline change? What physical flow does the new solution correspond to? What is the magnitude and direction of the force predicted by the theoretical solution? To answer these questions, we will add a vortex and examine the resulting solution.

Velocity Potential and Streamfunction. To add a vortex, appeal to Table 10.2 and Equations (10.103) tells us the new velocity potential and streamfunction assume the following form:

$$\left.\begin{array}{l} \phi(r,\theta) = U\left(r + \dfrac{R^2}{r}\right)\cos\theta + \dfrac{\Gamma}{2\pi}\theta \quad + \text{(constant)}_\phi \\[3mm] \psi(r,\theta) = U\left(r - \dfrac{R^2}{r}\right)\sin\theta - \dfrac{\Gamma}{2\pi}\ell nr + \text{(constant)}_\psi \end{array}\right\} \qquad (10.118)$$

With no loss of generality, we can select $\text{(constant)}_\phi = 0$. However, it is advantageous to choose

$$\text{(constant)}_\psi = \frac{\Gamma}{2\pi}\ell nR \qquad (10.119)$$

for then, we still have $\psi(R,\theta) = 0$, i.e., the original cylinder remains on the streamline for which ψ vanishes. Thus, we propose the following forms for ϕ and ψ.

$$\left.\begin{array}{l} \phi(r,\theta) = U\left(r + \dfrac{R^2}{r}\right)\cos\theta + \dfrac{\Gamma}{2\pi}\theta \\[3mm] \psi(r,\theta) = U\left(r - \dfrac{R^2}{r}\right)\sin\theta - \dfrac{\Gamma}{2\pi}\ell n\left(\dfrac{r}{R}\right) \end{array}\right\} \qquad (10.120)$$

Velocity Components. Working with the streamfunction, the velocity components for this flow are

$$\left.\begin{array}{lclcl} u_r & = & \dfrac{1}{r}\dfrac{\partial\psi}{\partial\theta} & = & U\left(1 - \dfrac{R^2}{r^2}\right)\cos\theta \\[3mm] u_\theta & = & -\dfrac{\partial\psi}{\partial r} & = & -U\left(1 + \dfrac{R^2}{r^2}\right)\sin\theta + \dfrac{\Gamma}{2\pi r} \end{array}\right\} \qquad (10.121)$$

Inspection of the velocity components shows that our solution still satisfies the original boundary conditions of Equation (10.98). Since the $\psi = 0$ streamline still defines the original circular shape and the farfield and surface boundary conditions are satisfied, our solution still corresponds to flow past a cylinder. Regarding the velocity field, the circumferential velocity on the cylinder surface is

$$u_\theta(R,\theta) = -2U\sin\theta + \frac{\Gamma}{2\pi R} \qquad (10.122)$$

The second term on the right-hand side is the contribution from the potential vortex. Because it is a constant, it corresponds to the cylinder rotating at an angular velocity of $\Gamma/(2\pi R^2)$. Thus, the problem we have solved with our superposition of a uniform flow, a doublet and a potential vortex is flow past a rotating cylinder.

Stagnation Points. Interestingly, as illustrated in Figure 10.13, there are two possibilities for the location of the stagnation points in this flow, depending on the strength of the circulation, Γ. The first possibility has the stagnation points lying on the cylinder surface. Since $r = R$ guarantees that $u_r = 0$, we examine u_θ to determine their locations. Thus, we have

$$u_\theta(R,\theta) = 0 \quad \Longrightarrow \quad -2U\sin\theta + \frac{\Gamma}{2\pi R} = 0 \qquad (10.123)$$

Solving for θ, we find

$$r = R, \quad \theta = \sin^{-1}\left(\frac{\Gamma}{4\pi RU}\right) \qquad (\Gamma \le 4\pi RU) \qquad (10.124)$$

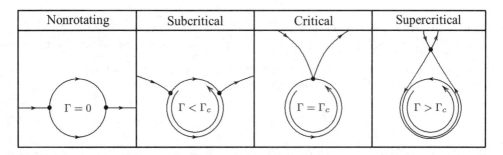

| Nonrotating | Subcritical | Critical | Supercritical |

Figure 10.13: *Dividing streamlines for rotating-cylinder flow;* $\Gamma_c = 4\pi RU$ *is the critical circulation;* • *denotes a stagnation point.*

As long as $\Gamma < 4\pi RU$, there will be two stagnation points on the cylinder surface with the values of θ given by Equation (10.124). When $\Gamma = 4\pi RU$, there is a single stagnation point at $\theta = 90°$. For larger values of Γ, the stagnation point moves above the cylinder. To solve for the location of the stagnation point, note first that u_r always vanishes along the y axis where $\theta = \pi/2$, so that $\cos\theta = 0$. Hence, since $\sin\theta$ must equal 1, u_θ is zero when

$$0 = -U\left(1 + \frac{R^2}{r^2}\right) + \frac{\Gamma}{2\pi r} \tag{10.125}$$

which can be rearranged to read

$$\left(\frac{r}{R}\right)^2 - 2\left(\frac{\Gamma}{4\pi RU}\right)\left(\frac{r}{R}\right) + 1 = 0 \tag{10.126}$$

Solving, the location of the stagnation point is given by

$$r = \frac{\Gamma}{4\pi U}\left[1 + \sqrt{1 - \left(\frac{4\pi RU}{\Gamma}\right)^2}\right], \quad \theta = \frac{\pi}{2} \qquad (\Gamma > 4\pi RU) \tag{10.127}$$

Figure 10.13 shows the dividing streamline for several values of circulation relative to the **critical circulation**, $\Gamma_c = 4\pi RU$. For the supercritical case, i.e., for $\Gamma > \Gamma_c$, the cylinder lies within the dividing streamline, so that fluid is trapped within the two closed streamlines. Alternatively, we could view our supercritical solution as flow past the tear-drop shaped geometry defined by the dividing streamline.

Observe that the dividing streamlines presented in Figure 10.13 are shown only in the immediate vicinity of the cylinder. Although not obvious from the figure, which is based on exact streamline coordinates, even in the supercritical case, the streamlines ultimately are parallel to the x axis far from the body. However, the effects of the vortex are much stronger than those of the doublet. This is evident by inspection of the velocity field. While the doublet contribution to the velocity falls off as $1/r^2$ far from the cylinder, the vortex contribution falls off as $1/r$. Hence, nontrivial effects of the vortex persist much farther than those of the doublet. This explains the strongly distorted streamline shapes near our rotating cylinder.

10.6.4 Force Computation Redone

Examination of the velocity components shows that, because the vortex contribution adds a counterclockwise circumferential component at the cylinder surface (provided $\Gamma > 0$), it increases the velocity at the bottom of the cylinder. Bernoulli's equation then tells us the pressure is reduced. Similarly, the velocity decreases at the top of the cylinder, corresponding to an increase in pressure. Hence, we should expect to find a downward, or negative lift, force on the cylinder. To compute the magnitude of the force, we again note that the force per unit length of the cylinder is given by Equation (10.110), viz.,

$$\mathbf{F} = -\int_0^{2\pi} (p - p_\infty) (\mathbf{i}\cos\theta + \mathbf{j}\sin\theta) \, Rd\theta \qquad (10.128)$$

The pressure follows from Bernoulli's equation, which simplifies to

$$p - p_\infty = \frac{1}{2}\rho \left(U^2 - \mathbf{u}\cdot\mathbf{u}\right) = \frac{1}{2}\rho U^2 \left[1 - \left(2\sin\theta - \frac{\Gamma}{2\pi RU}\right)^2\right] \qquad (10.129)$$

where we use the fact that the radial velocity component, u_r, vanishes on the cylinder surface, so that $\mathbf{u}\cdot\mathbf{u} = u_\theta^2(R,\theta)$. Hence, expanding the quadratic term, we have

$$p - p_\infty = \frac{1}{2}\rho U^2 \left\{\left[1 - \left(\frac{\Gamma}{2\pi RU}\right)^2\right] + \frac{2\Gamma}{\pi RU}\sin\theta - 4\sin^2\theta\right\} \qquad (10.130)$$

Now, the first and third terms on the right-hand side of this expression for $p - p_\infty$ are even functions of θ. Hence, when multiplied by the unit normal, which is an odd function of θ, they must integrate to zero (integrating from 0 to 2π is the same as integrating from $-\pi$ to π). This means the force on the cylinder is simply

$$\begin{aligned}\mathbf{F} &= -\frac{1}{2}\rho U^2 \int_0^{2\pi} \frac{2\Gamma}{\pi RU}\sin\theta\,(\mathbf{i}\cos\theta + \mathbf{j}\sin\theta)\, Rd\theta \\ &= -\frac{\rho U\Gamma}{\pi}\int_0^{2\pi} \left(\mathbf{i}\sin\theta\cos\theta + \mathbf{j}\sin^2\theta\right)\, d\theta\end{aligned} \qquad (10.131)$$

Finally, we make use of the fact that, from elementary calculus,

$$\left.\begin{aligned}\int_0^{2\pi} \sin\theta\cos\theta\, d\theta &= 0 \\ \int_0^{2\pi} \sin^2\theta\, d\theta &= \pi\end{aligned}\right\} \qquad (10.132)$$

Therefore the force on a counterclockwise rotating cylinder, according to potential-flow theory, is

$$\mathbf{F} = -\rho U\Gamma\,\mathbf{j} \qquad (10.133)$$

As expected, our solution predicts a negative lift force and zero drag. This force is generally referred to as the **Magnus force**, and has the interesting feature of being perpendicular to the freestream flow direction. Thus, the lift force per unit length of the cylinder, L, is

$$L = -\rho U\Gamma \qquad (10.134)$$

Equation (10.134) is the famous result known as the **Kutta-Joukowski law** or the **Kutta-Joukowski lift theorem**. This result was established by Kutta and Joukowski, two researchers who developed Equation (10.134) independent of each other's efforts in 1902 and 1906, respectively.

In words, Kutta and Joukowski stated their theorem as follows.

"According to inviscid theory, the lift per unit depth of any cylinder of any shape immersed in a uniform stream equals $\rho U\Gamma$, where Γ is the total net circulation contained within the body shape. The direction of the lift is $90°$ from the stream direction, rotating opposite to the circulation."

Figure 10.14 compares computed and measured [Rouse (1946) and Hoerner and Borst (1975)] lift and drag coefficients, C_L and C_D, for flow past a rotating cylinder. By definition, these dimensionless coefficients are

$$C_L \equiv \frac{\tilde{L}}{\frac{1}{2}\rho U^2 (2Rb)}, \qquad C_D \equiv \frac{\tilde{D}}{\frac{1}{2}\rho U^2 (2Rb)} \qquad (10.135)$$

where b is cylinder height. The quantities $\tilde{L} = Lb$ and $\tilde{D} = Db$ are total lift and drag, respectively. In the experiments, the cylinder rotates in the clockwise direction with angular velocity ω, so that the circulation is

$$\Gamma = -2\pi\omega R^2 \qquad (10.136)$$

Results are plotted as a function of the Strouhal number, St, defined by

$$St = \frac{\omega R}{U} \qquad (10.137)$$

In terms of our potential-flow solution, the lift and drag coefficients are

$$C_L = 2\pi St, \qquad C_D = 0 \qquad \text{(Potential-flow theory)} \qquad (10.138)$$

As shown, measurements indicate that less than half the potential-flow value of C_L is realized, and there is a significant drag force. The discrepancies are caused by flow separation. Interestingly, the lift force is much larger than that on a typical airfoil of similar size. Hence, generating lift with a rotating cylinder can be a useful aerodynamic device in spite of flow separation. This is remarkable because separation is generally regarded as catastrophic failure for most aerodynamic devices, especially for an airfoil.

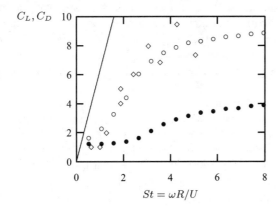

Figure 10.14: *Lift and drag coefficients for flow past a cylinder rotating with angular velocity ω; —— potential-flow theory; $\circ$, $\bullet$ Measured C_L, C_D [Rouse (1946)]; $\diamond$ Measured C_L [Hoerner and Borst (1975)].*

To make it easier to determine the direction of the Magnus force, we can rewrite the Kutta-Joukowski lift theorem in vector form as

$$\mathbf{F} = \rho\, \mathbf{U} \times \mathbf{\Gamma} \qquad (10.139)$$

where the circulation vector, $\mathbf{\Gamma}$, is parallel to the direction of the bound vorticity. Using this generalized form of the Kutta-Joukowski law, we can apply what we have learned to help understand the motion of a baseball. Figure 10.15 illustrates three of the most common ingredients of a successful pitcher's arsenal, viz., a fastball, a sinker and a curveball. While the ball is a sphere rather than a cylinder, the Magnus force is present, and acts in the direction indicated by Equation (10.139).

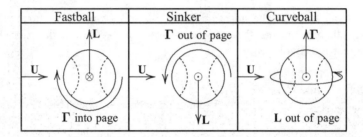

Figure 10.15: *Various motions of a baseball.*

As illustrated, a pitcher's fastball is thrown with "backspin" corresponding to clockwise circulation. This is done to counter the effect of gravity and to permit the ball to travel in a more nearly straight line. This permits a talented pitcher to more precisely control the point where the ball passes through the strike zone. Countering gravity is just as important for long throws, where the object is to get the ball to its desired destination as quickly as possible. An example is an outfielder attempting to throw out a runner trying to advance a base. Clearly, the more level the throw, the shorter the distance it must travel. Imparting backspin produces lift, which permits a flatter (and thus shorter) trajectory.

By contrast, some pitchers have mastered the so-called sinker. With this pitch, a counterclockwise spin is imparted to the ball as it is thrown, thus producing a downward force. As the ball approaches the batter, the downward force causes the ball to drop. If the batter succeeds in making contact with the ball, it is usually hit downward, i.e., on the ground. This has given rise to the colorful description of the plight of batters facing a pitcher with an effective sinker, i.e., that they spend the afternoon "killing worms" (presumably living in the ground near home plate).

With the curveball, the pitcher provides spin in the vertical direction. This causes a sideways force that makes the ball move continually farther away from the batter as it approaches home plate, if he bats with the same hand as the pitcher throws. If the batter bats with the opposite hand, the ball will "curve" toward him as it approaches. In either case, the pitch is usually more difficult to hit than a pitch thrown in a straight line.

As it turns out, the most effective curveball is a combination sinker and pure curveball. That is, the orientation of the spin should be at an angle to the vertical so that it not only moves away from (or toward) the batter, but also drops. This is true because changing the elevation of the bat is physically more difficult than simply extending or pulling in the arms to compensate for a curveball. For the reader interested in exploring the fluid dynamics of a baseball further, the book by Adair (1990) provides an excellent starting point.

Perhaps the most unusual application of the Magnus force is the famous **Flettner-Rotor ship** built in 1927. The ship had two rotating cylinders in place of sails. This ship was quite successful, and its builder sailed it across the Atlantic Ocean. Figure 10.16 shows a side view of the Flettner-Rotor ship.

Figure 10.16: *The Flettner-Rotor ship. Its builder sailed the ship across the Atlantic Ocean.*

AN EXPERIMENT YOU CAN DO AT HOME
Excerpted from Shames (1992)

You can observe the Magnus force on a rotating cylinder by assembling the following materials:

- An $8\frac{1}{2}$ inch by 11 inch sheet of paper

- A pair of scissors or an exacto knife

- Some tape and a three-foot long piece of string

To minimize weight, cut the sheet of paper down to an $8\frac{1}{2}$ inch by 4 inch segment. Now, roll this segment into a cylinder (of height $8\frac{1}{2}$ inches) and tape it in a way that conceals the seam, and permits it to roll freely. Allowing a little overlap to make handling easy while taping, you will have a cylinder that is roughly 1 inch in diameter. Tie the string around the center of the cylinder and wrap it around until about 4 inches are free. You are now ready to do the experiment.

Place the cylinder three feet from the edge of a table top with the free end of the string looping beneath the cylinder and lying flat on the surface. Jerk the string horizontally causing it to unwind and let go. When done properly, the translating cylinder will lift above the plane of the table, thus demonstrating the Magnus force.

10.7   Other Interesting Solutions

This section includes several interesting potential-flow solutions to problems that illustrate a few more subtle features of the theory. First, we use potential-flow theory to solve for accelerating flow past a cylinder. Then, we examine flow past a classic body known as a Rankine oval. Next, we analyze flow past a wedge. Another application is flow near a stagnation point. We conclude by showing how general solid boundaries can be simulated by suitable placement of fundamental solutions, viz., with the method of images.

10.7.1   Accelerating Cylinder

We noted at the end of Section 10.4 that Laplace's equation does not explicitly include time. This means we can solve an unsteady-flow problem in a quasi-steady sense, simply substituting time-dependent boundary conditions into what would otherwise be a steady-flow solution. The unsteady equivalent of Bernoulli's equation, Equation (10.45), shows that the pressure is directly affected by unsteadiness. This opens the possibility that forces might develop in unsteady potential flow. As we will see in the present example, this is indeed the case.

Basic Solution. We focus on potential flow past a cylinder of radius R where the freestream velocity, $U(t)$, varies with time, t. The preceding section tells us the velocity potential is

$$\phi(r, \theta, t) = U(t) \left(r + \frac{R^2}{r} \right) \cos \theta \qquad (10.140)$$

The unsteady Bernoulli's equation tells us that, in the absence of body forces,

$$\frac{\partial \phi}{\partial t} + \frac{p}{\rho} + \frac{1}{2} \mathbf{u} \cdot \mathbf{u} = F(t) \qquad (10.141)$$

where $F(t)$ is a function that must be determined from the boundary conditions.

Surface Pressure. Noting that

$$\frac{\partial \phi}{\partial t} = \frac{dU}{dt} \left(r + \frac{R^2}{r} \right) \cos \theta \qquad (10.142)$$

we see that because $\cos \theta = 0$ on the y axis, necessarily $\partial \phi / \partial t = 0$. Hence, we know every term on the left-hand side of Equation (10.141) on the y axis as $|y| \to \infty$. Therefore, we have

$$F(t) = \frac{p_\infty}{\rho} + \frac{1}{2} U^2 \qquad (10.143)$$

Substituting Equation (10.143) into Equation (10.141) yields the pressure:

$$p = p_\infty + \frac{1}{2} \rho \left(U^2 - u_r^2 - u_\theta^2 \right) - \rho \frac{\partial \phi}{\partial t} \qquad (10.144)$$

Force on the Cylinder. Our immediate goal in this exercise is to compute any force that might develop on the cylinder. So, noting that on the cylinder, $u_r = 0$ and $u_\theta = -2U \sin \theta$, the pressure simplifies to

$$p = p_\infty + \frac{1}{2} \rho U^2 \left(1 - 4 \sin^2 \theta \right) - \rho \frac{\partial \phi}{\partial t} \qquad (10.145)$$

Also, reference to Equation (10.142) shows that on the surface of the cylinder,

$$\frac{\partial \phi}{\partial t} = 2R \frac{dU}{dt} \cos \theta \qquad (10.146)$$

Therefore, the pressure on the surface of the cylinder is

$$p = p_\infty + \frac{1}{2} \rho U^2 \left(1 - 4 \sin^2 \theta\right) - 2\rho R \frac{dU}{dt} \cos \theta \qquad (10.147)$$

As with steady-flow, the force on the cylinder is given by Equation (10.110). The only difference between the steady-flow pressure [Equation (10.113)] and the pressure for the present case is the term proportional to dU/dt. Since we know the other terms yield no force, we can omit all but this term and write

$$
\begin{aligned}
\mathbf{F} &= -\int_0^{2\pi} \left(-2\rho R \frac{dU}{dt} \cos \theta\right) (\mathbf{i} \cos \theta + \mathbf{j} \sin \theta) \, R d\theta \\
&= 2\rho R^2 \frac{dU}{dt} \int_0^{2\pi} \left(\mathbf{i} \cos^2 \theta + \mathbf{j} \sin \theta \cos \theta\right) d\theta = 2\pi \rho R^2 \frac{dU}{dt} \mathbf{i}
\end{aligned}
\qquad (10.148)
$$

Thus, the cylinder develops a drag force proportional to the acceleration of the freestream flow. Consistent with the symmetry of the flow, the lift is zero.

Virtual Mass. When an object moves through an incompressible fluid, it pushes a finite mass of fluid out of its way. When the object accelerates, as in the present example, the surrounding fluid that is set in motion must also accelerate. This is the reason the cylinder experiences a drag force. A classical result of potential-flow theory is the following [cf. Milne-Thompson (1960) or Saffman (1993)]. For an accelerating body, the force that develops on the body is

$$\mathbf{F} = (m + m_h) \frac{d\mathbf{U}}{dt} \qquad (10.149)$$

where m is the actual mass of the body and m_h is the **virtual mass**[2] defined in terms of the total relative kinetic energy of the fluid by

$$\frac{1}{2} m_h \mathbf{U} \cdot \mathbf{U} \equiv \iiint_V \frac{1}{2} \rho \mathbf{U}_{rel} \cdot \mathbf{U}_{rel} \, dV \qquad (10.150)$$

where $\mathbf{U}_{rel}$ is the fluid velocity relative to the object, and the integration is performed for the entire volume surrounding the object.

For the case at hand, since the fluid within the dividing streamline has density ρ, the mass of the body (per unit length) is the product of ρ and the area of the cylinder, so that

$$m = \pi \rho R^2 \qquad (10.151)$$

The relative velocity is simply the doublet contribution, so that

$$\mathbf{U}_{rel} = -U \frac{R^2}{r^2} \cos \theta \, \mathbf{e}_r - U \frac{R^2}{r^2} \sin \theta \, \mathbf{e}_\theta \implies \mathbf{U}_{rel} \cdot \mathbf{U}_{rel} = U^2 \frac{R^4}{r^4} \qquad (10.152)$$

The virtual mass per unit length thus becomes

$$m_h = \int_0^{2\pi} \int_R^\infty \rho \left(\frac{R}{r}\right)^4 r \, dr \, d\theta = 2\pi \rho R^4 \int_R^\infty \frac{dr}{r^3} = \pi \rho R^2 \qquad (10.153)$$

[2]Virtual mass is also referred to as **added mass** and **hydrodynamic mass**.

which, coincidentally, is equal to the mass of fluid displaced by the cylinder. So, according to the virtual mass concept, the force on the cylinder should be

$$\mathbf{F} = \left(\pi\rho R^2 + \pi\rho R^2\right)\frac{dU}{dt}\mathbf{i} = 2\pi\rho R^2 \frac{dU}{dt}\mathbf{i} \tag{10.154}$$

Comparison with Equation (10.148) shows that the force is identical to the result obtained by integrating the pressure directly.

10.7.2 Rankine Oval

Thus far, none of our applications has made direct use of the source or the sink. In this application, we superpose a uniform stream with a source and a sink of equal strengths to obtain a closed body known as the **Rankine oval** (Figure 10.17). In general, when we construct a potential-flow solution from fundamental solutions, the resulting body will be closed provided the sum of the strengths of all sources matches the sum of the strengths of all sinks.

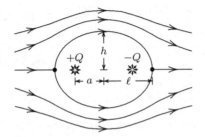

Figure 10.17: *Streamlines for flow past a Rankine oval; • denotes a stagnation point.*

We place a source of strength Q on the x axis at $x = -a$ and a sink of the same strength on the x axis at $x = a$. That is, we select the same geometry considered when we introduced the doublet (see Figure 10.7). However, the spacing between the source and the sink remains finite for the Rankine oval.

Streamfunction. As shown when we formulated the doublet relationships, the streamfunction for this flow is

$$\psi = Uy + \frac{Q}{2\pi}\left(\theta^- - \theta^+\right); \qquad \theta^- = \tan^{-1}\left(\frac{y}{x+a}\right), \quad \theta^+ = \tan^{-1}\left(\frac{y}{x-a}\right) \tag{10.155}$$

We can simplify the algebra by making use of the trigonometric identity for the tangent of the difference between two angles, i.e.,

$$\tan\left(\theta^- - \theta^+\right) = \frac{\tan\theta^- - \tan\theta^+}{1 + \tan\theta^- \tan\theta^+} \tag{10.156}$$

Thus, for the Rankine oval angles, θ^- and θ^+,

$$\tan\left(\theta^- - \theta^+\right) = \frac{\dfrac{y}{x+a} - \dfrac{y}{x-a}}{1 + \dfrac{y^2}{x^2 - a^2}} = -\frac{2ay}{x^2 + y^2 - a^2} \tag{10.157}$$

Substituting into Equation (10.155), we arrive at a somewhat simplified form of the stream-function for flow past a Rankine oval.

$$\psi = Uy - \frac{Q}{2\pi}\tan^{-1}\left(\frac{2ay}{x^2 + y^2 - a^2}\right) \tag{10.158}$$

Determining Body Shape. Before proceeding with our analysis, it is convenient to recast the streamfunction and coordinates in nondimensional form. Using the half-spacing distance, a, as the appropriate length scale, we find

$$\overline{\psi} = \overline{y} - \overline{Q}\tan^{-1}\left(\frac{2\overline{y}}{\overline{x}^2 + \overline{y}^2 - 1}\right) \tag{10.159}$$

where

$$\overline{\psi} \equiv \frac{\psi}{Ua}, \quad \overline{x} \equiv \frac{x}{a}, \quad \overline{y} \equiv \frac{y}{a}, \quad \overline{Q} \equiv \frac{Q}{2\pi Ua} \tag{10.160}$$

The body lies on the $\psi = 0$ streamline. Hence, the body shape is defined, in terms of dimensionless quantities, by the following equation.

$$0 = \overline{y} - \overline{Q}\tan^{-1}\left(\frac{2\overline{y}}{\overline{x}^2 + \overline{y}^2 - \overline{a}^2}\right) \implies \tan\left(\frac{\overline{y}}{\overline{Q}}\right) = \frac{2\overline{y}}{\overline{x}^2 + \overline{y}^2 - 1} \tag{10.161}$$

Equation (10.161) is a transcendental equation that cannot be solved in closed form if we regard $\overline{y}$ as a function of $\overline{x}$. However, we can solve this equation for $\overline{x}$ as an explicit function of $\overline{y}$. The solution is

$$\overline{x} = \pm\sqrt{1 + 2\overline{y}\cot\left(\overline{y}/\overline{Q}\right) - \overline{y}^2} \tag{10.162}$$

The overall shape of the Rankine oval is best characterized by its height, $2h$, and length, 2ℓ. To determine the length, we must find the value of $\overline{x}$ that corresponds to $\overline{y} = 0$. Similarly, the height is the value of $\overline{y}$ corresponding to $\overline{x} = 0$. Considering the length first,

$$\overline{x} = \lim_{\overline{y}\to 0}\sqrt{1 + 2\overline{y}\cot\left(\overline{y}/\overline{Q}\right) - \overline{y}^2} = \sqrt{1 + 2\overline{Q}} \tag{10.163}$$

Returning to dimensional parameters, we have

$$\frac{\ell}{a} = \sqrt{1 + \frac{Q}{\pi Ua}} \tag{10.164}$$

Now, we turn to computation of the height. Setting $\overline{x} = 0$ in Equation (10.161), the value of $\overline{y}$ at the top of the oval satisfies

$$\tan\left(\frac{\overline{y}}{\overline{Q}}\right) = \frac{2\overline{y}}{\overline{y}^2 - 1} \tag{10.165}$$

We can simplify this equation by taking advantage of the trigonometric half-angle formula for the tangent, i.e.,

$$\tan\alpha = \frac{2\tan\left(\frac{\alpha}{2}\right)}{1 - \tan^2\left(\frac{\alpha}{2}\right)} = \frac{2\cot\left(\frac{\alpha}{2}\right)}{\cot^2\left(\frac{\alpha}{2}\right) - 1} \tag{10.166}$$

Hence, making the identification $\alpha = \overline{y}/\overline{Q}$, we have

$$\frac{2\cot\left(\dfrac{\overline{y}}{2\overline{Q}}\right)}{\cot^2\left(\dfrac{\overline{y}}{2\overline{Q}}\right) - 1} = \frac{2\overline{y}}{\overline{y}^2 - 1} \tag{10.167}$$

By inspection, it must be the case that

$$\overline{y} = \cot\left(\frac{\overline{y}}{2\overline{Q}}\right) \tag{10.168}$$

Although this is still a transcendental equation that must be solved numerically, it is nevertheless simpler than the original equation. Rewriting the relationship in terms of dimensional quantities, we have

$$\frac{h}{a} = \cot\left(\frac{\pi U a}{Q}\frac{h}{a}\right) \tag{10.169}$$

Table 10.3 includes values for the height and width of the Rankine oval as a function of the dimensionless source strength. Note that as the source strength becomes very large, the spacing between the source and sink becomes less and less significant. Both $h/a \to \infty$ and $\ell/a \to \infty$ so that, in effect, the limit $Q/(\pi U a) \to \infty$ is the same as the limit $a \to 0$. In other words, as the source strength increases, our source/sink pair approaches a doublet. Hence, we should expect the solution to asymptotically approach the flow past a circular cylinder (for which $h = \ell$). This is borne out in Table 10.3.

Table 10.3: *Dimensions of a Rankine Oval*

$Q/(\pi U a)$	h/a	ℓ/a	h/ℓ
0	0.000	1.000	0.000
0.01	0.016	1.005	0.015
0.1	0.143	1.049	0.136
1	0.860	1.414	0.608
10	3.111	3.317	0.938
100	9.983	10.050	0.993
∞	∞	∞	1.000

10.7.3   Wedge

Any function that satisfies Laplace's equation can be regarded as a possible solution to some potential-flow problem. Hence, given such a function, it is interesting to see what the streamlines look like, determine any useful dividing streamlines, locate stagnation points, etc. For example, consider the function

$$\psi(r, \theta) = \psi_o r^n \sin n\theta \tag{10.170}$$

where ψ_o and n are constants. The exponent n is dimensionless and the coefficient ψ_o has dimensions $(\text{length})^{2-n}/(\text{time})$. This is indeed a solution to Laplace's equation, which can be verified by direct substitution.

Velocity Components. Before determining the shape of the streamlines implied by this stream-function, it is worthwhile to first compute the velocity components. This will permit us to determine flow direction on the streamlines and to locate any stagnation points that might exist. Hence, differentiating the streamfunction, we find

$$\left.\begin{array}{ccccc} u_r & = & \dfrac{1}{r}\dfrac{\partial\psi}{\partial\theta} & = & n\psi_o r^{n-1}\cos n\theta \\[3mm] u_\theta & = & -\dfrac{\partial\psi}{\partial r} & = & -n\psi_o r^{n-1}\sin n\theta \end{array}\right\} \qquad (10.171)$$

Determining Body Shape. Although there is no guarantee that the $\psi = 0$ streamline defines a meaningful dividing streamline [recall the rotating cylinder and Equation (10.119)], it usually serves as a good starting point. For the streamfunction at hand,

$$\psi = 0 \quad\Longrightarrow\quad \theta = \frac{m\pi}{n}, \quad m = \text{integer} \qquad (10.172)$$

If we confine our attention to the cases $m = 0$ and $m = 1$, this defines the wedge-shaped region between $\theta = 0$ and $\theta = \pi/n$. Along the sides of the wedge, the circumferential velocity, u_θ, is zero, while the radial velocity varies with radial distance according to

$$u_r(r,0) = n\psi_o r^{n-1} \quad\text{and}\quad u_r(r,\pi/n) = -n\psi_o r^{n-1} \qquad (10.173)$$

Inspection of Equations (10.171) shows that, as long as $n > 1$, there is a single stagnation point at $r = 0$. Figure 10.18 shows several wedge solutions for angles ranging from 45° to 225° ($\pi/4$ to $5\pi/4$).

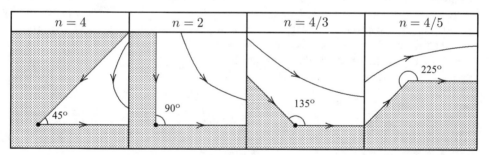

Figure 10.18: *Dividing streamlines for wedge flow;* • *denotes a stagnation point.*

The potential-flow wedge solution is valid only in the immediate vicinity of the origin when $n > 1$. This is true because the velocity becomes infinite as $r \to \infty$. Similarly, because the solution is infinite at the origin for $n < 1$, it is valid provided we exclude the immediate vicinity of the origin. This is true of all potential-flow solutions. That is, as can be shown from complex-variable theory, the only function that is completely bounded for the entire region $0 \le r < \infty$ is a constant. We see this with the fundamental solutions, for example, where the vortex, source and doublet velocities are singular as $r \to 0$, and otherwise well behaved.

If we continue to $m = 2, 3, \ldots$ in Equation (10.172), we find additional wedges, each with an included angle of π/n. They may overlap, of course, depending upon the value of n.

We can also use more than a single wedge in a flowfield. For example, consider the case $n = 4/3$. If we use the three streamlines defined by setting $m = -1, 0, +1$, assuming $\psi_o > 0$, we would have flow past the trailing edge of a body with a sharp trailing edge as illustrated in

Figure 10.19(a). The primary flow direction is from left to right. The body is defined by the streamlines with $m = \pm 1$. The $m = 0$ streamline lies along the plane of symmetry between the upper and lower streams.

Similarly, if we have $\psi_o < 0$, the flow direction is reversed, and the solution corresponds to flow near the leading edge of a sharp-nosed body. This is illustrated in Figure 10.19(b). For this case, flow is from right to left.

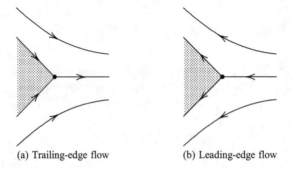

(a) Trailing-edge flow   (b) Leading-edge flow

Figure 10.19: *Dividing streamlines for flow near sharp leading and trailing edges; • denotes a stagnation point.*

10.7.4   Stagnation Point

One especially interesting case of wedge flow worthy of additional discussion occurs when we select $n = 2$. If we focus on a single wedge, Figure 10.18 shows that this corresponds to flow into a 90° corner. However, if we use the streamlines with $m = 0, 1, 2$ in Equation (10.172), then we can regard the x axis as the body surface with the y axis being an axis of symmetry for the flow. Figure 10.20 shows the resulting flow.

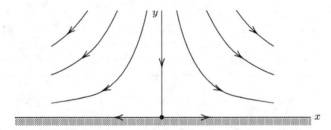

Figure 10.20: *Stagnation-point flow; • denotes the stagnation point.*

Because it assumes a simple form, it is worthwhile to recast the streamfunction in terms of Cartesian coordinates. That is,

$$\psi = \psi_o r^2 \sin 2\theta = 2\psi_o r^2 \sin \theta \cos \theta = 2\psi_o xy \tag{10.174}$$

Clearly, the streamlines are a family of hyperbolas given by $xy = $ constant. The velocity components are

$$u = \frac{\partial \psi}{\partial y} = 2\psi_o x \quad \text{and} \quad v = -\frac{\partial \psi}{\partial x} = -2\psi_o y \tag{10.175}$$

We noted in the preceding section that, because the velocity becomes infinite far from the stagnation point at $x = y = 0$, the solution can be valid only in the "immediate vicinity" of the origin. We can make a quantitative statement about what this phrase means in the special case of flow past a cylinder. For consistency in notation, consider flow past a cylinder with the freestream flow oriented along the y axis as shown in Figure 10.21.

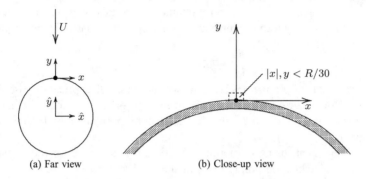

(a) Far view   (b) Close-up view

Figure 10.21: *Flow past a cylinder revisited.*

Since we have rotated our coordinate system by 90° relative to that used in Section 10.6, we simply replace the circumferential angle θ by $\theta + \pi/2$. Thus, the radial velocity component follows from the first of Equations (10.104), and becomes

$$u_r(r, \theta) = U\left(1 - \frac{R^2}{r^2}\right)\cos\left(\theta + \frac{\pi}{2}\right) = -U\left(1 - \frac{R^2}{r^2}\right)\sin\theta \qquad (10.176)$$

Hence, along the positive y axis, which corresponds to $\theta = \pi/2$, we see that

$$u_r(r, \pi/2) = -U\left(1 - \frac{R^2}{\hat{y}^2}\right) \qquad (10.177)$$

where $\hat{y}$ is measured from the center of the cylinder. Then, changing to the coordinate system whose origin lies at the forward stagnation point [Figure 10.21(a)], and noting that the Cartesian velocity component, v, is equal to u_r, we have

$$v(x, y) = -U\left[1 - \frac{R^2}{(R + y)^2}\right] = -U\left[1 - \frac{1}{(1 + y/R)^2}\right] \qquad (10.178)$$

Finally, expanding in Taylor series, we know that

$$\frac{1}{(1 + y/R)^2} = 1 - 2\left(\frac{y}{R}\right) + 3\left(\frac{y}{R}\right)^2 + \cdots \qquad (10.179)$$

Thus, substituting Equation (10.179) into Equation (10.178), the vertical velocity approaching the stagnation point is

$$v = -2U\left(\frac{y}{R}\right)\left[1 - \frac{3}{2}\left(\frac{y}{R}\right) + \cdots\right] \qquad (10.180)$$

Hence, for $y \ll R$, the velocity is closely approximated by

$$v \approx -2\frac{U}{R}y \qquad (10.181)$$

Comparison with the second of Equations (10.175) shows that our stagnation-point solution is valid with ψ_o chosen to be

$$\psi_o = \frac{U}{R} \tag{10.182}$$

To quantify the terminology "immediate vicinity," we simply examine the Taylor series expansion for v. Clearly, our solution is accurate as long as the first term in the series that we neglect is small compared to the last term retained. Specifically, if we insist that our solution be valid to within 5%, then the immediate vicinity of the stagnation point means

$$\frac{3}{2}\frac{y}{R} < \frac{1}{20} \quad \Longrightarrow \quad y < \frac{R}{30} \tag{10.183}$$

The dashed rectangle in Figure 10.21(b) shows the range of applicability of the stagnation-point solution. As an example, if we consider flow past a $2\frac{1}{2}$ foot circular cylinder, the stagnation-point solution will be valid in the region approximately 1 inch away from the stagnation point.

As a final comment, using Bernoulli's equation, we can solve for the pressure. Denoting the pressure at the stagnation point by p_a, then,

$$p + \frac{1}{2}\rho\left(2\psi_o\right)^2\left(x^2 + y^2\right) = p_a \quad \Longrightarrow \quad p = p_a - 2\rho\psi_o^2\left(x^2 + y^2\right) \tag{10.184}$$

10.7.5   The Method of Images

All of the flows discussed thus far have occurred in a fluid that extends infinitely far from the object of interest. There are important applications that involve the presence of a solid boundary at a finite distance from an object. For example, we might want to compute flow past an object close to a plane such as an airfoil or an automobile-like shape to simulate motion on a runway or highway, respectively. We might also want to simulate the constraining walls of a wind tunnel to see how much difference there is between flow past an object in a finite and an infinite fluid, thus helping establish error bounds for measured flow properties.

A useful technique for simulating a planar boundary is known as the **method of images**. This method is best explained by example. Consider a source of strength Q. If we place the source at the point $(x, y) = (0, h)$ and another source of strength Q at $(x, y) = (0, -h)$, the streamfunction is

$$\psi(x, y) = \underbrace{\frac{Q}{2\pi}\tan^{-1}\left(\frac{y - h}{x}\right)}_{\text{Source at }(0,h)} + \underbrace{\frac{Q}{2\pi}\tan^{-1}\left(\frac{y + h}{x}\right)}_{\text{Source at }(0,-h)} \tag{10.185}$$

Differentiation shows that the velocity components for this flow are

$$\left.\begin{aligned} u &= \frac{\partial\psi}{\partial y} = \frac{Q}{2\pi}\left[\frac{x}{x^2 + (y - h)^2} + \frac{x}{x^2 + (y + h)^2}\right] \\[2mm] v &= -\frac{\partial\psi}{\partial x} = \frac{Q}{2\pi}\left[\frac{y - h}{x^2 + (y - h)^2} + \frac{y + h}{x^2 + (y + h)^2}\right] \end{aligned}\right\} \tag{10.186}$$

Hence, the velocity along the x axis is given by

$$u(x, 0) = \frac{Qx}{\pi\left(x^2 + h^2\right)}, \quad v(x, 0) = 0 \quad \Longrightarrow \quad \mathbf{u}\cdot\mathbf{n} = 0 \text{ on } y = 0 \tag{10.187}$$

where $\mathbf{n} = \mathbf{j}$ is the unit normal to the x axis. Therefore, we have a symmetry plane aligned with the x axis. The source below the axis is known as the **mirror image** of the source above the axis. Figure 10.22(a) shows a few streamlines for a source and its mirror image.

In a similar way, the mirror image of a vortex of strength Γ a distance h above the x axis is a vortex of *opposite strength*, $-\Gamma$, a distance h below the axis. The streamfunction for a vortex and its image is

$$\psi(x,y) = \underbrace{-\frac{\Gamma}{2\pi}\ell n\sqrt{x^2 + (y-h)^2}}_{\text{Counterclockwise Vortex at } (0,h)} + \underbrace{\frac{\Gamma}{2\pi}\ell n\sqrt{x^2 + (y+h)^2}}_{\text{Clockwise Vortex at } (0,-h)} \qquad (10.188)$$

while the velocity components are

$$\left.\begin{aligned} u &= \frac{\partial\psi}{\partial y} = \frac{\Gamma}{2\pi}\left[\frac{y+h}{x^2+(y+h)^2} - \frac{y-h}{x^2+(y-h)^2}\right] \\ v &= -\frac{\partial\psi}{\partial x} = \frac{\Gamma}{2\pi}\left[\frac{x}{x^2+(y-h)^2} - \frac{x}{x^2+(y+h)^2}\right] \end{aligned}\right\} \qquad (10.189)$$

Again, we see that the velocity along the x axis has only an x component, i.e.,

$$u(x,0) = \frac{\Gamma h}{\pi\,(x^2+h^2)}, \qquad v(x,0) = 0 \qquad (10.190)$$

Figure 10.22(b) shows a few streamlines for a vortex and its mirror image.

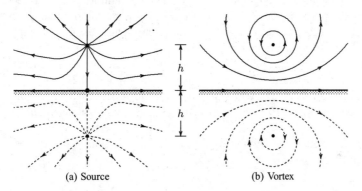

(a) Source   (b) Vortex

Figure 10.22: *Method of images for a source and a vortex.*

Note that for both the source and the vortex, because the velocity on the x axis is nonuniform, Bernoulli's equation tells us the pressure is also nonuniform. Hence, it is reasonable to expect that there is a nonzero force on the axis caused by the presence of the source and vortex pairs. Equivalently, we can view this as the force exerted by the image source or vortex on the other. To determine the force, we observe first that along the x axis, the pressure is

$$p = p_\infty - \frac{1}{2}\rho u^2(x,0) \qquad (10.191)$$

where we note that the fluid velocity goes to zero far from the source or vortex. The force per unit depth exerted on the wall from the source or vortex above the x axis is given by

$$\mathbf{F}_w = \int_{-\infty}^{+\infty} (p - p_\infty)\,\mathbf{j}\,dx \qquad (10.192)$$

Thus, appealing to Equations (10.187) and (10.190) for $u(x,0)$, the force on the x axis is

$$
\mathbf{F}_w = \begin{cases} -\mathbf{j}\dfrac{\rho Q^2}{2\pi^2}\displaystyle\int_{-\infty}^{+\infty}\dfrac{x^2\,dx}{(x^2+h^2)^2}, & \text{Source} \\[4mm] -\mathbf{j}\dfrac{\rho\Gamma^2}{2\pi^2}\displaystyle\int_{-\infty}^{+\infty}\dfrac{h^2\,dx}{(x^2+h^2)^2}, & \text{Vortex} \end{cases}
\tag{10.193}
$$

Finally, reference to a standard table of integrals (or through a substitution such as $x = h\tan\phi$) shows that

$$
\int_{-\infty}^{+\infty}\frac{x^2\,dx}{(x^2+h^2)^2} = \frac{\pi}{2h} \quad\text{and}\quad \int_{-\infty}^{+\infty}\frac{h^2\,dx}{(x^2+h^2)^2} = \frac{\pi}{2h}
\tag{10.194}
$$

Therefore, the force exerted on the wall is

$$
\mathbf{F}_w = \begin{cases} -\dfrac{\rho Q^2}{4\pi h}\mathbf{j}, & \text{Source} \\[4mm] -\dfrac{\rho\Gamma^2}{4\pi h}\mathbf{j}, & \text{Vortex} \end{cases}
\tag{10.195}
$$

In both cases, the force is directed downward, increases with the strength of the source or vortex, and increases as the distance from the x axis decreases.

Figure 10.22 shows that, in addition to producing a symmetry plane, the introduction of images induces changes throughout the flow. The streamlines are modified even in the immediate vicinity of the source and the vortex. Furthermore, the distortion increases as we move the source or vortex closer to the symmetry plane. Thus, for more complex flows, we can expect similar distortion. For example, consider flow past a cylinder of radius R. Suppose we wish to construct a solution for flow past a cylinder whose center is a distance nR above a plane surface, where n is a number greater than 1 (see Figure 10.23).

As a first approximation, we will superpose a uniform flow with a doublet of strength $\mathcal{D} = 2\pi R^2 U$ [see Equation (10.102)] at $(x,y) = (0, nR)$ and an image doublet of the same strength at $(x,y) = (0, -nR)$. We know that both the doublet and its image must have the same sign since the doublet is simply the limiting case of a source superposed with a sink. So, the streamfunction is

$$
\psi \;=\; \underbrace{Uy}_{\text{Uniform flow}} \;-\; \underbrace{U\frac{R^2}{r^+}\sin\theta^+}_{\text{Doublet at }(0,nR)} \;-\; \underbrace{U\frac{R^2}{r^-}\sin\theta^-}_{\text{Doublet at }(0,-nR)}
\tag{10.196}
$$

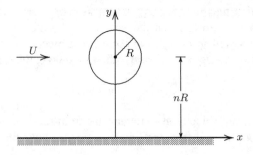

Figure 10.23: *Flow past a cylinder above a solid boundary.*

where r^+ and r^- are the radial distances measured from the doublets at $(0, nR)$ and $(0, -nR)$, respectively, while θ^+ and θ^- are the angles measured from $y = nR$ and $y = -nR$, i.e.,

$$\left. \begin{array}{ll} r^+ = \sqrt{x^2 + (y - nR)^2}, & \theta^+ = \tan^{-1}\left(\dfrac{y - nR}{x}\right) \\[2ex] r^- = \sqrt{x^2 + (y + nR)^2}, & \theta^- = \tan^{-1}\left(\dfrac{y + nR}{x}\right) \end{array} \right\} \qquad (10.197)$$

Because of the influence of the mirror-image doublet, the shape of the dividing streamline must differ from that given by a superposition of the original doublet and the uniform stream. Although noticeable, the effect is relatively small, even for $n = 2$. To illustrate the magnitude of the effect, Table 10.4 lists the velocities at the original stagnation-point locations for several values of n. As shown, for $n > 6$, deviations from the solution for flow past a cylinder are less than 0.2%.

Table 10.4: *Induced Velocities at Original Stagnation Points*

Stagnation Point	$n = 2$		$n = 3$		$n = 6$	
Location	u/U	v/U	u/U	v/U	u/U	v/U
Forward	0.052	-0.028	0.026	-0.009	0.002	-0.001
Rearward	0.052	0.028	0.026	0.009	0.002	0.001

The velocities in the table suggest that the upstream portion of the dividing streamline shifts upward slightly, and vice versa for the downstream portion. This turns out to be true for this flow. Also, the effective body shape is very nearly circular, although the center is shifted a bit to the right. The modified geometry is sketched in Figure 10.24. The distortion can be removed by adding so-called "corrective images" [cf. Robertson (1965)]. For general problems, this consists of adding an (often infinite) series of imaged sources, doublets, etc. to restore the original dividing streamline. Numerical solution, implementing an iterative procedure, is usually required for all but the simplest problems.

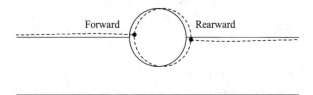

Figure 10.24: *Schematic showing the effect of the mirror doublet on the dividing streamline. The solid contour (——) is the original dividing streamline and the dashed line (- - -) is the shifted streamline; • denotes a stagnation point.*

Problems

10.1 Verify by direct substitution that the velocity potential and streamfunction for a doublet, Equations (10.88) and (10.91), satisfy Laplace's equation.

10.2 Determine the dimensions of the following quantities.

(a) The velocity potential, ϕ, and the streamfunction, ψ.

(b) The strength of a source, Q.

(c) The strength of a potential vortex, Γ.

(d) The strength of a doublet, $\mathcal{D}$.

10.3 Compute the mass flux from a potential vortex.

10.4 Compute the circulation of a source.

10.5 Compute the circulation of a doublet.

10.6 Compute the mass flux from a doublet.

10.7 Show that for an incompressible, irrotational flow, the equation of motion

$$\nabla \left[\frac{\partial \phi}{\partial t} + \frac{p}{\rho} + \frac{1}{2} \left(u^2 + v^2 \right) \right] = 0$$

is identical to the Euler equation in two dimensions. **HINT:** This problem is most conveniently done in component form.

10.8 Consider the streamfunction given by $\psi(x, y) = Ax^2 + Bxy - Ay^2$, where A and B are constants and $B^2 \neq 4A^2$.

(a) Compute the velocity components and locate all stagnation points.

(b) In what part of the xy plane are the x and y velocity components equal?

10.9 For steady, compressible, two-dimensional flows, we can define a "streamfunction" according to

$$u = \frac{1}{\rho} \frac{\partial \tilde{\psi}}{\partial y}, \qquad v = -\frac{1}{\rho} \frac{\partial \tilde{\psi}}{\partial x}$$

where ρ is the (variable) density.

(a) What are the dimensions of $\tilde{\psi}$?

(b) Verify that conservation of mass is satisfied.

(c) If the flow is irrotational, verify that

$$\nabla^2 \tilde{\psi} = \mathbf{u} \times \nabla \rho \qquad \text{where} \qquad \tilde{\boldsymbol{\psi}} \equiv \tilde{\psi} \mathbf{k}$$

10.10 Consider the streamfunction given by

$$\psi(r, \theta) = Qr^3 \sin 3\theta$$

where Q is a constant.

(a) Compute the velocity components and locate all stagnation points.

(b) If $\psi = 4$ when $x = 2$ and $y = 2$, what is the value of Q?

(c) Develop an equation of the form $x = f(\psi, y)$ for the streamlines in this flow.

10.11 We can construct another fundamental solution to Laplace's equation, known as the dipole, using two potential vortices as indicated below. Assume the circulation, Γ, is such that the dipole strength, $\mathcal{D}_p$, is given by

$$\mathcal{D}_p = \lim_{a \to 0} 2a\Gamma$$

(a) Determine the velocity potential.

(b) Determine the streamfunction.

(c) Compare your results with those of a doublet.

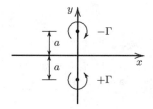

Problem 10.11

10.12 *Complex variables* are especially useful in potential-flow theory. We define the *complex potential*, $F(z)$, as a function of the complex variable, z, according to $F(z) = \phi + i\psi$, where $z = x + iy = re^{i\theta}$ and $i = \sqrt{-1}$ is the imaginary number. The quantities ϕ and ψ are the standard velocity potential and streamfunction, respectively. Determine the complex potential as a function of z for uniform flow, a point source, a potential vortex and a doublet.

10.13 *Complex variables* are especially useful in potential-flow theory. We define the *complex potential*, $F(z)$, as a function of the complex variable, z, according to $F(z) = \phi + i\psi$, where $z = x + iy = re^{i\theta}$ and $i = \sqrt{-1}$ is the imaginary number. The quantities ϕ and ψ are the standard velocity potential and streamfunction, respectively. We want to determine the complex potential for flow past a wedge. First, beginning with the streamfunction, $\psi(r, \theta) = Ar^n \sin n\theta$, where A is a constant, find the conventional velocity potential, $\phi(r, \theta)$. Now, determine the complex potential as a function of z.

10.14 *Complex variables* are especially useful in potential-flow theory. We define the *complex potential*, $F(z)$, as a function of the complex variable, z, according to $F(z)$, as a function of the complex variable, z, according to $F(z) = \phi + i\psi$, where $z = x + iy = re^{i\theta}$ and $i = \sqrt{-1}$ is the imaginary number. The quantities ϕ and ψ are the standard velocity potential and streamfunction, respectively.

(a) Verify that the complex velocity is $w(z) \equiv dF/dz = u - iv$. **HINT:** Assume $F(z)$ is an *analytic function* so that, according to complex-variable theory, the derivative at any point in the complex plane is independent of the direction in which we differentiate.

(b) Letting w^* denote the *complex conjugate* of w, rewrite Bernoulli's equation in terms of the complex velocity. Assume the flow is steady and that $|\mathbf{u}| \to U$ as $r \to \infty$.

10.15 *Complex variables* are especially useful in potential-flow theory. We define the *complex potential*, $F(z)$, as a function of the complex variable, z, according to $F(z)$, as a function of the complex variable, z, according to $F(z) = \phi + i\psi$, where $z = x + iy = re^{i\theta}$ and $i = \sqrt{-1}$ is the imaginary number. The quantities ϕ and ψ are the standard velocity potential and streamfunction, respectively. If the complex potential is $F(z) = Ue^{-i\alpha}z$, where U and α are constants, determine $\phi(x, y)$ and $\psi(x, y)$. What flow does this correspond to?

10.16 *Complex variables* are especially useful in potential-flow theory. We define the *complex potential*, $F(z)$, as a function of the complex variable, z, according to $F(z)$, as a function of the complex variable, z, according to $F(z) = \phi + i\psi$, where $z = x + iy = re^{i\theta}$ and $i = \sqrt{-1}$ is the imaginary number. The quantities ϕ and ψ are the standard velocity potential and streamfunction, respectively.

(a) If the complex potential is $F(z) = U[z + R^2/z]$, where R is a constant, determine $\phi(r, \theta)$ and $\psi(r, \theta)$. What flow does this correspond to? **HINT:** Do your work in cylindrical coordinates.

(b) If we add a point vortex of strength Γ, how does $F(z)$ change?

10.17 Determine the streamlines and geometry for wedge flow with $n = \frac{1}{2}$.

10.18 The streamfunction for an incompressible flow is $\psi(x,y) = x^2y - 2y^2$. Plot the streamlines in the first quadrant for $\psi = -2, 0, 2$, and find any stagnation points. Be sure to indicate flow direction on the streamlines. Is this flow irrotational?

10.19 The streamfunction for an incompressible flow is $\psi(x,y) = 3x^2y - y^3$. Plot the streamlines in the full xy plane, and find any stagnation points. Be sure to indicate flow direction on the streamlines. Is this flow irrotational?

10.20 Investigate the streamfunction $\psi(x,y) = C(x^2 - y^2)$, where C is constant. Plot the streamlines in the full xy plane, find any stagnation points, and interpret what the flow could represent.

10.21 The streamfunction for an incompressible flow is $\psi(x,y) = x^2 + 4y^2$. Plot the streamlines in the full xy plane, and find any stagnation points. Be sure to indicate flow direction on the streamlines. Is this flow irrotational?

10.22 Determine the velocity potential for a flow whose velocity vector is $\mathbf{u} = U\left[\mathbf{i} + (y/a)\mathbf{j}\right]$, where U and a are constant velocity and length scales, respectively. Is this flow incompressible?

10.23 Determine the velocity potential and streamfunction for a uniform flow inclined at an angle α to the x axis, i.e., for $\mathbf{u} = U(\mathbf{i}\cos\alpha + \mathbf{j}\sin\alpha)$. Set all additive constants equal to zero for convenience.

10.24 Determine the velocity potential for an unsteady, three-dimensional flow whose velocity vector is

$$\mathbf{u} = Ue^{-\kappa x}\cos\omega t\left(\kappa z\cos\kappa y\,\mathbf{i} + \kappa z\sin\kappa y\,\mathbf{j} - \cos\kappa y\,\mathbf{k}\right)$$

where U, ω and κ are constant velocity, frequency and wavenumber (dimensions $1/L$), respectively.

10.25 Determine the velocity potential for a flow whose (three-dimensional) velocity vector is

$$\mathbf{u} = \frac{U}{h}\left(x\,\mathbf{i} + y\,\mathbf{j} - 2z\,\mathbf{k}\right)$$

where U and h are constant velocity and length scales, respectively. Set all additive constants equal to zero for convenience. Is this flow incompressible?

10.26 The Cartesian velocity components for stagnation-point flow are $u(x,y) = Ax$ and $v(x,y) = -Ay$, where A is a constant.

    (a) Determine the streamfunction, $\psi(x,y)$.

    (b) Determine the velocity potential, $\phi(x,y)$.

    (c) Now add a source of strength Q. What must Q be in order to move the stagnation point to $x = 0$, $y = y_o$?

    (d) In general, ψ and ϕ are unique to within an additive constant. What must the constant for ψ be if the dividing streamline passes through the displaced stagnation point and is defined by $\psi = 0$?

10.27 Compute the velocity potential, $\phi(x,y)$, for flow in the vicinity of a stagnation point beginning with the streamfunction, $\psi(x,y) = 2\psi_o xy$. Plot three streamlines and equipotential lines in each of the upper quadrants.

10.28 The velocity potential for a two-dimensional flow is

$$\phi(x,y) = U\left[x + \lambda e^{-x/\lambda}\sin(y/\lambda)\right], \qquad x \geq 0$$

where U and λ are constant velocity and length scales, respectively. Determine the corresponding streamfunction, $\psi(x,y)$.

10.29 The streamfunction for a potential flow is $\psi(x, y) = f_1(x) + f_2(y)$. What is the most general form of the functions $f_1(x)$ and $f_2(y)$?

10.30 The streamfunction for a two-dimensional flow is $\psi(r, \theta) = Qr^n (\sin n\theta + \cos n\theta)$, where Q and n are constants. Determine the corresponding velocity potential, $\phi(r, \theta)$.

10.31 Compute the velocity potential for flow past a wedge beginning with $\psi(r, \theta) = r^n \sin n\theta$.

10.32 A two-dimensional potential flow has streamfunction $\psi(r, \theta) = Ar^4 \sin 4\theta$ with A = constant.

(a) Determine the velocity components $u_r(r, \theta)$ and $u_\theta(r, \theta)$.

(b) Determine the corresponding velocity potential, $\phi(r, \theta)$.

(c) Locate any stagnation points. Convert to Cartesian coordinates for the sake of clarity.

(d) The body shape is given by $\psi = 0$. If we confine our interest to the first quadrant excluding the y axis, i.e., $0 \leq \theta < \pi/2$, what is the body shape?

(e) Sketch a few streamlines. Include any stagnation points, body contours and flow direction along the streamlines. **HINT:** Find the flow direction on the body surface.

10.33 A two-dimensional potential flow has streamfunction $\psi(r, \theta) = -Ar^2 \cos 2\theta$ with A = constant.

(a) Determine the velocity components $u_r(r, \theta)$ and $u_\theta(r, \theta)$.

(b) Determine the corresponding velocity potential, $\phi(r, \theta)$.

(c) Locate any stagnation points. Convert your answer to Cartesian coordinates for the sake of clarity.

(d) The body shape is given by $\psi = 0$. If we confine our interest to the first three quadrants, i.e., $0 \leq \theta < 3\pi/2$, what is the body shape? Note that the body contours of interest lie in the first ($0 \leq \theta \leq \pi/2$) and third ($\pi \leq \theta \leq 3\pi/2$) quadrants, and that there is a dividing streamline in the second ($\pi/2 \leq \theta \leq \pi$).

(e) Sketch a few streamlines. Include any stagnation points, body contours and flow direction along the streamlines. **HINT:** Find the flow direction on the body surface and on the dividing streamline.

10.34 Determine the pressure induced by a potential vortex in an otherwise motionless vat of mercury at $20°$ C, assuming $p \to p_\infty$ as $r \to \infty$. If the farfield pressure is $p_\infty = 100$ kPa and the vortex strength is $\Gamma = 10$ m^2/sec, what is the minimum value of r for which the solution is physically meaningful?

10.35 Determine the pressure induced by a point sink in an otherwise motionless vat of ethyl alcohol at $20°$ C, assuming $p \to p_\infty$ as $r \to \infty$. If $p_\infty = 100$ kPa and the sink strength is $Q = -1$ m^3/sec, what is the minimum value of r for which the solution is physically meaningful?

10.36 Determine the pressure induced by a doublet in an otherwise motionless vat of kerosene at $68°$ F, assuming $p \to p_\infty$ as $r \to \infty$. If $p_\infty = 14.7$ psi and the doublet strength is $\mathcal{D} = 325$ ft^3/sec, what is the minimum value of r for which the solution is physically meaningful?

10.37 Very far from a two-dimensional lifting body, we can approximate the velocity potential as

$$\phi \approx Ur \cos \theta + \frac{\Gamma}{2\pi}\theta + \frac{\mathcal{D}}{2\pi r} \cos \theta \qquad (r \to \infty)$$

where U is freestream velocity, Γ is circulation and $\mathcal{D}$ is an effective doublet strength. Verify that as $r \to \infty$, the pressure coefficient is independent of doublet strength and is given by

$$C_p \approx \frac{C_L}{\pi} \left(\frac{R}{r}\right) \sin \theta$$

where $C_L \equiv \Gamma/(UR)$ is the *lift coefficient* and R is a length characteristic of the body size.

10.38 A simplified model of a hurricane approximates the flow as a rigid-body rotation in the inner core and a *potential vortex* outside of the core (cf. Section 4.3). That is, the circumferential velocity, u_θ, is

$$u_\theta \approx \begin{cases} \dfrac{\Gamma r}{2\pi R^2}, & r \leq R \\[2mm] \dfrac{\Gamma}{2\pi r}, & r \geq R \end{cases}$$

where r is radial distance, R is core radius, and Γ is the circulation.

(a) Determine the velocity potential.

(b) Determine the pressure for $p \to p_\infty$ as $r \to \infty$.

(c) What is the minimum pressure if the air density is $\rho = 0.00234$ slug/ft^3, $p_\infty = 14.7$ psi and the peak velocity is 200 mph?

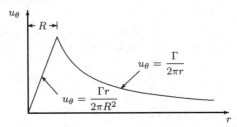

   Problem 10.38

10.39 As an airplane taxis down the runway in preparation for takeoff, the velocity on the fuselage is

$$|\mathbf{u}| = U \cdot \frac{s}{L} \cdot \frac{t}{\tau}$$

where s is arc length and t is time. The quantities U, L and τ are reference velocity, length and time scales related by $L = U\tau$. In this problem, we examine the error in using a Pitot-static tube to infer velocity if effects of the acceleration (unsteadiness) are ignored (this problem had to be corrected to get the DC-10 certified by the FAA).

(a) Neglecting body forces and assuming the flow is incompressible, compute the pressure coefficient, C_p, defined by $C_p \equiv (p - p_s)/(\frac{1}{2}\rho U^2)$ where p_s is the pressure at the stagnation point ($s = 0$). Note that the velocity potential near the stagnation point is $\phi = \frac{1}{2}U(s^2/L)(t/\tau)$.

(b) Starting with the value for C_p computed in Part (a), use the steady-flow version of Bernoulli's Equation to infer the velocity at $s = L$ as a function of t/τ.

(c) Compare your result of Part (b) with the actual velocity given above by making a graph of the percentage error for $0.5 < t/\tau < 4$.

10.40 Find the stagnation-point location for a *spiral vortex* in a uniform freestream of velocity U. A spiral vortex is the superposition of a point source of strength Q and a potential vortex of strength Γ.

10.41 Determine the streamlines for a *spiral vortex*, which is the superposition of a point source of strength Q and a potential vortex of strength Γ. Plot a streamline for $\Gamma = 2Q$.

10.42 Determine the equipotentials for a *spiral vortex*, which is the superposition of a point source of strength Q and a potential vortex of strength Γ. Plot an equipotential for $\Gamma = -\frac{1}{2}Q$.

10.43 We can approximate a tornado by combining a potential vortex of strength Γ with a point sink of strength $-Q$, both centered at the origin. If the pressure is given by $p = p_\infty - \rho Q^2/(2\pi r^2)$, where p_∞ is the farfield pressure, what must Γ be? Also, what is the angle, α, between the velocity vector and the radial direction?

10.44 A uniform stream of velocity $\mathbf{u} = U\mathbf{i}$ flows past sources of strength Q at $x = \pm a$, $y = 0$ and a sink of strength $-2Q$ at $x = y = 0$. What is the thickness of the resulting closed body at $x = 0$?

10.45 Consider superposition of a source of strength $Q = Uh$ and a uniform flow of velocity U from left to right.

  (a) What is the complete streamfunction for this flow? Include an appropriate constant in the streamfunction to insure that the negative x axis is a dividing streamline along which $\psi = 0$.

  (b) Locate any stagnation points that might be present.

  (c) Noting that $y = r\sin\theta$, show that the body shape ($\psi = 0$) is such that $y \rightarrow \pm h/2$ as $x \rightarrow \infty$.

  (d) Make a qualitative sketch of what the body and streamlines look like. Be sure to include some streamlines both inside and outside the body.

10.46 Consider superposition of a sink of strength $Q = -Uh$ and a uniform flow of velocity U from left to right.

  (a) What is the complete streamfunction for this flow? You may choose any extra constant in the streamfunction to be zero.

  (b) Locate any stagnation points that might be present.

  (c) Noting that $y = r\sin\theta$, show that the body shape ($\psi = 0$) is such that $y \rightarrow \pm h/2$ as $\theta \rightarrow \pm\pi$.

  (d) Make a qualitative sketch of what the body and streamlines look like. Be sure to include some streamlines both inside and outside the body.

10.47 A uniform stream flows downward so that $u = 0$ and $v = -V$ where V is constant.

  (a) What are the streamfunction and velocity potential for this flow in Cartesian coordinates?

  (b) What are the streamfunction and velocity potential for this flow in cylindrical coordinates?

  (c) Now add a source of strength Q located at the origin. What are the streamfunction and velocity potential in cylindrical coordinates?

  (d) For the streamfunction of Part (c), locate any stagnation points in the flow. Express your answer in Cartesian coordinates. (Note that most of the work is most conveniently done in cylindrical coordinates—just convert when you're done.)

  (e) Now add a sink of strength Q located at $x = a$, $y = -a$ to the flow of Part (c). What are the streamfunction and velocity potential in Cartesian coordinates?

  (f) Sketch what you think the flow of Part (e) looks like, i.e., draw streamlines and indicate flow direction on the streamlines. Don't worry about precise location of stagnation point(s), etc., just indicate what the body shape might look like and give a rough idea of where the streamlines and stagnation point(s) might be.

10.48 Consider flow past a wall with a source of strength Q at the origin. At infinity, the flow is parallel to the x axis with uniform velocity U. The freestream pressure is p_∞ and the density is ρ.

  (a) Determine the streamfunction and velocity potential for this flow. Be sure to verify that $y = 0$ is a streamline to show that you have solved the right problem. **NOTE:** $y = 0$ is a different streamline (corresponds to a different value of ψ) for $x > 0$ and $x < 0$.

  (b) Determine the pressure, p, along the wall.

  (c) Find the value of x at which the pressure along the wall is a maximum.

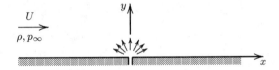

Problem 10.48

10.49 Consider flow past a wall with a sink of strength $-Q$ at the origin. At infinity, the flow is parallel to the x axis with uniform velocity U. The freestream pressure is p_∞ and the density is ρ.

(a) Determine the streamfunction and velocity potential for this flow. Be sure to verify that $y = 0$ is a streamline to show that you have solved the right problem. **NOTE:** $y = 0$ is a different streamline (corresponds to a different value of ψ) for $x > 0$ and $x < 0$.

(b) Determine the pressure, p, along the wall.

(c) Find the value of x at which the pressure along the wall is a maximum.

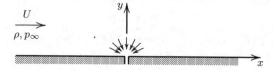

Problem 10.49

10.50 If we superimpose a uniform flow of velocity U with a doublet of strength $\mathcal{D}$, a point vortex of strength Γ and a point source of strength Q, the farfield solution corresponds to a two-dimensional lifting body that emits a stream of fluid. The emitted stream could represent the exhaust of a jet engine. State the streamfunction and velocity potential for this flow, noting that the lift should be upward.

(a) If there is a stagnation point at $r = R$ and $\theta = 210°$, how are the vortex, source and doublet strengths related?

(b) Suppose now that the body is a circular cylinder with a doublet strength $\mathcal{D} = 2\pi U R^2$. Including an additive constant in the streamfunction, explain why $\psi(R, \theta) = 0$ no longer corresponds to the cylinder surface. Also, verify that $\psi(R, \pi) - \psi(R, 0) = \frac{1}{2}Q$.

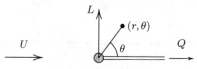

Problem 10.50

10.51 Far from a non-lifting body that is emitting fluid (e.g., from a jet engine), we can approximate the velocity potential as

$$\phi \approx Ur\cos\theta + \frac{Q}{2\pi}\ell nr + \frac{\mathcal{D}}{2\pi r}\cos\theta \qquad (r \to \infty)$$

where U is freestream velocity, Q is the mass-injection rate and $\mathcal{D}$ is an effective doublet strength.

(a) Verify that

$$p - p_\infty \approx -\frac{\rho QU}{2\pi r}\cos\theta \qquad \text{as} \qquad r \to \infty$$

(b) Using a circular control volume of radius r, show that

$$\oint_S \rho\mathbf{u}(\mathbf{u}\cdot\mathbf{n})dS = \frac{3}{2}\rho QU\,\mathbf{i} \qquad \text{and} \qquad \oint_S (p - p_\infty)\,\mathbf{n}\,dS = -\frac{1}{2}\rho QU\,\mathbf{i}$$

(c) Use the results of Part (b) to determine the force on the body.

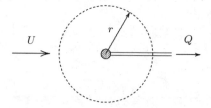

Problem 10.51

10.52 The wind is blowing with velocity U and freestream pressure p_∞ past a quonset hut, which is a half cylinder of radius R and length L (out of the page). The internal pressure is p_i. Using potential-flow theory, derive an expression for the vertical force on the hut due to the pressure difference on the hut.

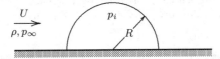

Problems 10.52, 10.53

10.53 In a hurricane, the wind is blowing with velocity $U = 90$ mph, freestream pressure $p_\infty = 14.7$ psi and density $\rho = 0.00234$ slug/ft^3 past a quonset hut, which is a half cylinder of radius $R = 12$ ft and length $L = 50$ ft (out of the page). To help the hut survive the hurricane, the internal pressure, p_i, is regulated so that there is zero net vertical force on the hut. Using potential-flow theory, determine p_i.

10.54 An arctic hut in the shape of a half-circular cylinder has radius R. A wind of velocity U is blowing and creates a substantial aerodynamic force on the hut. This force is due to the difference between the external pressure and the pressure inside the hut, p_i. Making a thin slit at an angle θ_o from the ground level causes the net force on the hut to vanish. Determine $\sin \theta_o$. **NOTE:** Assume the opening is very small compared to R so that it has a negligible effect on the flow about the hut. Assume incompressible potential flow and note that the pressure must be continuous at the slit.

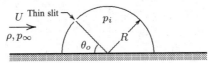

Problem 10.54

10.55 An interesting way to generate an exact solution for flow past a two-dimensional bump is to use a streamline for flow past a cylinder as shown. The desired height of the bump is $R/2$, where R is the cylinder radius. Determine the distance h and the maximum velocity on the bump, u_{max}.

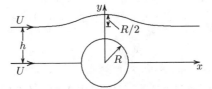

Problem 10.55

10.56 Consider flow past a circular body of radius R with a blunted base as shown below. As a first approximation, assume the base pressure, p_b, is constant across the base and that the flow on the circular part of the body follows potential-flow theory.

(a) Sketch the pressure coefficient, C_p, as a function of θ for the assumed flow, with $0 \le \theta \le 2\pi$. The angle θ is zero on the positive x axis. Explain the basis of your sketch, e.g., maxima and minima, etc.

(b) Compute the drag on the body. (It is not zero because this is not an exact solution.)

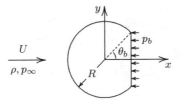

Problem 10.56

10.57 Consider flow past a semi-circular body of radius R with a blunted base as shown below. As a first approximation, assume the base pressure, p_b, is equal to the pressure at the shoulder and that the flow on the circular part of the body follows the potential-flow solution for flow past a circle.

(a) Sketch the pressure coefficient, C_p, as a function of θ for the assumed flow, with $0 \le \theta \le 2\pi$. The angle θ is zero on the positive x axis. Explain the basis of your sketch, e.g., maxima and minima, etc.

(b) Compute the drag on the body. (It is not zero because this is not an exact solution.)

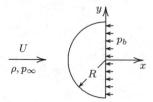

Problem 10.57

10.58 An incompressible fluid of density ρ and velocity U flows past a rotating circular cylinder of diameter D. The cylinder develops lift, L, in the direction shown. In what direction is the cylinder rotating? If the lift coefficient is $C_L \equiv L/(\frac{1}{2}\rho U^2 D) = 2\pi$, determine the circulation, Γ.

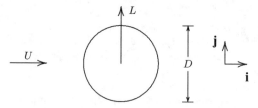

Problem 10.58

10.59 The *Flettner-Rotor* ship was designed with two rotating cylinders of height H acting as sails. The cylinders have diameter D and rotate with angular velocity Ω. The ship moves with velocity $\mathbf{U}_s = U\,\mathbf{i}$, the wind velocity is $\mathbf{U}_w = -V[\mathbf{i}\cos\alpha + \mathbf{j}\sin\alpha]$ and the air density is ρ.

(a) Estimate the force on the ship as a function of Ω, D, H, U, V, α and ρ.

(b) If the side force vanishes, what is the thrust?

(c) Determine the thrust, in tons, for $\Omega = 12$ Hz, $D = 10$ ft, $H = 40$ ft, $U = 4$ knots, $V = 15$ knots and $\rho = 0.00234$ slug/ft^3.

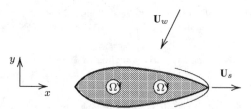

Problems 10.59, 10.60

10.60 The *Flettner-Rotor* ship was designed with two rotating cylinders of height H acting as sails. The cylinders have diameter D and rotate with angular velocity Ω. The ship moves with velocity $\mathbf{U}_s = U\,\mathbf{i}$, the wind velocity is $\mathbf{U}_w = -V\,\mathbf{j}$ and the air density is ρ.

(a) Estimate the force on the ship as a function of Ω, D, H, U, V and ρ.

(b) Determine the thrust for $\Omega = 800$ rpm, $D = 2.85$ m, $H = 16$ m, $U = 3$ knots, $V = 12$ knots and $\rho = 1.2$ kg/m^3.

10.61 Recall the home experiment discussed in this chapter. We would like to estimate the minimum speed, U_{min}, and rotation rate, Ω_{min}, needed to cause the cylinder to rise above the table. Assume that the angular velocity is given by $\Omega = U/R$, where U is the translation velocity and R is the cylinder radius. Let ρ, H and W denote air density, cylinder height and cylinder weight, respectively.

(a) Determine U_{min} and Ω_{min} as functions of W, ρ, R and H.

(b) Compute U_{min} (in ft/sec) and Ω_{min} (in Hz) for $W = 0.005$ lb, $\rho = 0.00234$ slug/ft^3, $R = \frac{1}{2}$ in and $H = 11$ in.

10.62 A circular cylinder of radius $R = 10$ inches is placed in a wind tunnel with a test-section velocity of $U = 10$ ft/sec. The cylinder rotates in the counterclockwise direction about its longitudinal axis with an angular velocity, Ω, which can be varied. Using the potential-flow solution, determine the angular velocities (in rpm) corresponding to critical flow, subcritical flow with $\Gamma = \frac{1}{2}\Gamma_c$ and supercritical flow with $\Gamma = 2\Gamma_c$. Also, determine the location of the stagnation points for each case.

10.63 Consider *steady potential flow* past a cylindrical tube with three radially drilled holes as shown below. Whenever the pressure on the two side holes is equal, the center hole (located halfway between the side holes) will point in the direction of flow. The pressure at the center hole is then the stagnation pressure. This device is called a *direction-finding Pitot tube*.

(a) Since the side holes must measure the freestream static pressure, where must they be located, i.e., what is the angle α?

(b) Now assume the freestream flow is uniform but changes with time, i.e., $U = U(t)$. Taking account of this unsteadiness, determine the difference between the pressure at the centrally located hole, p_1, and the pressure at either one of the side holes, p_2. Express your answer in terms of ρ, R, α and $U(t)$.

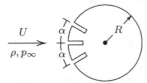

Problem 10.63

10.64 Consider *unsteady potential flow* past a cylinder of radius R where the freestream velocity, $U(t)$, varies with time, t.

(a) Compute the pressure coefficient, C_p, on the surface of the cylinder. Express your answer in terms of R, θ, U and dU/dt. **HINT:** Note that $\partial\phi/\partial t$ is always zero on the y axis.

(b) Compute the lift, L, and drag, D, on the cylinder. You may need to use the following trigonometric identity: $\cos^2\theta = \frac{1}{2}(1 + \cos 2\theta)$.

10.65 To simulate the effect of incoming and reflected waves on a pier, imagine a cylinder of radius R immersed in a fluid with a sinusoidally-varying freestream velocity, viz., $U = U_o \sin \omega t$, where U_o and ω are characteristic velocity and frequency scales, respectively.

(a) Using the potential-flow solution for flow past a cylinder, find the pressure coefficient on the surface of the cylinder, $C_p \equiv (p - p_\infty)/(\frac{1}{2}\rho U_o^2)$. Your answer should involve *Strouhal number*.

(b) Verify that the average value of C_p defined by

$$\overline{C}_p \equiv \frac{\omega}{2\pi} \int_0^{2\pi/\omega} C_p \, dt$$

is half the value of that for a steady flow with $U = U_o$, and is independent of Strouhal number.

10.66 Consider a point source whose strength, $Q(t)$, varies with time. The source is immersed in an infinite fluid of density ρ. The pressure far from the point source (i.e., as $r \to \infty$) is p_∞. Determine the pressure as a function of p_∞, ρ, $Q(t)$ and the coordinates r and θ. Ignore body forces.

10.67 To approximate the motion of a slender object launched from a submarine as it rises to the surface, we approximate the object as an ellipsoid of revolution of volume V. If the object's density is $\rho_m = \frac{1}{2}\rho$, where ρ is the density of water, compute the acceleration, dW/dt, according to potential-flow theory, including effects of virtual mass. Express your answer as a function of $m_h/(\rho V)$ and the acceleration of gravity, g. Make a graph of the error we would make as a function of t/c if we were to ignore the virtual mass. The virtual mass for longitudinal motion of an ellipsoid of revolution is listed below.

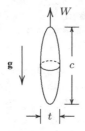

t/c	$m_h/(\rho V)$
0.0	0
0.1	0.0207
0.2	0.0591
0.3	0.1054
0.4	0.1563

Problem 10.67

10.68 The concept of *virtual mass* is valid for both two- and three-dimensional flows. The virtual mass of a sphere of radius R is $m_h = \frac{2}{3}\rho\pi R^3$, where ρ is the fluid density. Ignoring viscous drag, compute the acceleration of a sphere with density $\rho_s = \frac{1}{2}\rho$ as it rises through a fluid in a gravitational field.

10.69 Using the method of images, indicate how you would distribute vortices to simulate walls along $y = 0$ and along $x = 0$, i.e., to bound the first quadrant $(x > 0, y > 0)$.

10.70 Using the method of images, indicate how you would distribute sources to simulate walls along $y = 0$ and along $x = 0$, i.e., to bound the first quadrant $(x > 0, y > 0)$.

10.71 Using the method of images, compute the force exerted by a doublet on a planar wall. The doublet strength is $\mathcal{D}$ and it is located a distance h above the wall. **HINT:** The following integrals may be of use in deriving your answer.

$$\int_{-\pi/2}^{\pi/2} \cos^2 \phi \, d\phi = \frac{1}{2}\pi, \qquad \int_{-\pi/2}^{\pi/2} \cos^4 \phi \, d\phi = \frac{3}{8}\pi, \qquad \int_{-\pi/2}^{\pi/2} \cos^6 \phi \, d\phi = \frac{5}{16}\pi$$

10.72 As an airplane approaches a runway, its lift is altered by the so-called *ground effect*. To determine the effect, approximate the wing and the ground plane by a potential vortex of strength Γ and its image. If the approach velocity is U (parallel to the ground), determine the lift per unit wingspan when the wing is a distance h above the ground. Write your final result in terms of *aerodynamic circulation*, $\Gamma_a = -\Gamma$. Does your answer make sense for $h \to \infty$? Does the lift increase or decrease due to the *ground effect*?

10.73 Use the method of images to compute the flow past a cylinder a distance $4R$ from a wall as shown in Figure 10.23. Compute and compare flow velocities to the solution in an infinite domain at the four points $(x, y) = (0, 0), (0, 4R), (0, 6R), (0, 10R)$.

Chapter 11

Vorticity, Viscosity, Lift and Drag

This chapter can be viewed both as a post script to the material covered in Chapters 1 through 10 and as a preview of what should follow in an intermediate course. In the spirit of a post script to what we've already covered, we have gotten about as much out of inviscid control-volume theory as we can. Dealing with finite-sized control volumes provides important gross flow properties but no details of motion within the control volume. In the spirit of looking ahead, the object of further study should be to examine fluid motion at a more fundamental level. To accomplish this, continued study should do two things.

1. **Turn the focus exclusively to the differential equations of motion:** We took preliminary steps in Chapter 9 by decreasing the size of our control volume to infinitesimal size and applying them to irrotational, incompressible flows in Chapter 10. Future study should include effects of compressibility and vorticity.

2. **Include viscous effects in the equations of motion:** We must revise the differential conservation laws with friction and heat-transfer effects included.

This chapter touches on both of these topics, albeit in a cursory manner. One of the most important aspects of real-fluid behavior is the presence of vorticity, and the role viscosity plays in its creation. Vorticity is perhaps the most important of all fluid-flow properties. For example, without it we would have no lift on an airplane's wing or a sailboat's sail. The dynamics of the gulf stream and of hurricanes are characterized by the strength of the vorticity involved. On the one hand, the classical theory of fluid mechanics and aerodynamics deals with vorticity as an essentially inviscid-flow phenomenon. On the other hand, as we will see in this chapter, inviscid fluids in an unbounded medium have no physical mechanism for developing vorticity. We will also see the dilemma of eighteenth- and nineteenth-century mathematicians and physicists who concluded that viscous effects were too small to have a significant effect on fluid motion.

This dilemma is best characterized by two famous inviscid-flow facts known as the **Helmholtz Theorem** and **d'Alembert's Paradox**. These two mathematical results, which we will discuss in this chapter, leave us with an apparent contradiction between theory and experiment. The theory says it is impossible to develop aerodynamic forces on objects such as an airplane wing, while experimental evidence shows that great forces do in fact develop.

It took Prandtl's pioneering work on viscous effects, first published in 1904, to resolve this dilemma, and to put theoretical fluid mechanics on a solid footing as a rational science. Prandtl argued that, in fluids such as water and air, close to a solid boundary viscous effects are confined to very thin layers known as **boundary layers**. As a result, very large velocity gradients develop—so large that significant viscous forces are present. The large velocity gradients also create vorticity at solid boundaries, thus providing the missing physical mechanism for generating vorticity. This chapter is offered as a quick look at this historical road of discovery, which ultimately revolutionized fluid mechanics as a science. It is also intended to stress the close tie between viscosity and vorticity that is often overlooked in a first course on fluid mechanics. The chapter includes two fascinating home experiments that reveal subtle phenomena illustrating the connection between vorticity and viscosity.

11.1   The Vortex Force

One way to understand the connection between vorticity and the force that develops when fluid flows past an object is to rearrange Euler's equation to reveal the presence of an apparent force known as the **vortex force**. As noted by Saffman (1993), the notion of the vortex force finds its origins in work by Prandtl in the early twentieth century, and was used extensively by von Kármán and associates in developing the theory of aerodynamics.

To see how the vortex force appears in the fluid-mechanical equations of motion, remember that Euler's equation can be written as [see Equation (9.72)]:

$$\rho\frac{\partial \mathbf{u}}{\partial t} + \rho\nabla\left(\frac{1}{2}\mathbf{u}\cdot\mathbf{u}\right) - \rho\mathbf{u}\times\boldsymbol{\omega} = -\nabla p + \rho\mathbf{f} \tag{11.1}$$

where $\boldsymbol{\omega}$ is the vorticity vector given by $\boldsymbol{\omega} = \nabla\times\mathbf{u}$. The term proportional to the cross product of $\mathbf{u}$ and $\boldsymbol{\omega}$ is similar to the Coriolis acceleration that appears in a rotating coordinate frame (see Section D.4 of Appendix D). Just as we often treat the Coriolis term as an apparent force, so we can rewrite Euler's equation as follows.

$$\rho\frac{\partial \mathbf{u}}{\partial t} + \rho\nabla\left(\frac{1}{2}\mathbf{u}\cdot\mathbf{u}\right) = \underbrace{\rho\mathbf{u}\times\boldsymbol{\omega}}_{Vortex\ Force} - \nabla p + \rho\mathbf{f} \tag{11.2}$$

The vortex force acts in a direction normal to the flow direction. For example, in two dimensions, the vorticity vector is

$$\boldsymbol{\omega} = \left(\frac{\partial v}{\partial x} - \frac{\partial u}{\partial y}\right)\mathbf{k} \tag{11.3}$$

Figure 11.1 illustrates the direction of the vortex force for a two-dimensional flowfield. As shown, if the flow has vorticity, a force will be present at 90° to the direction of flow. This is the situation that arises when we move an object such as an airplane wing through the air.

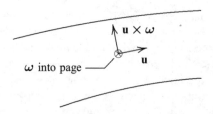

Figure 11.1: *The vortex force in a two-dimensional flow.*

Example 11.1 *Compute the rate at which work is done by the vortex force.*

Solution. In general, the differential work done by a force, **F**, acting over a differential distance d**r** is

$$dW = \mathbf{F} \cdot d\mathbf{r} = \mathbf{F} \cdot \mathbf{u}\, dt$$

where $\mathbf{u} = d\mathbf{r}/dt$ is the velocity. So, for the vortex force, $\mathbf{F} = \rho\, \mathbf{u} \times \omega$, we have

$$dW = \rho\, \mathbf{u} \times \omega \cdot \mathbf{u}\, dt = 0$$

which is true because $\mathbf{u} \times \omega$ is perpendicular to **u**. Therefore,

$$\frac{dW}{dt} = 0$$

11.2 Helmholtz's Theorem and d'Alembert's Paradox

Given the importance of vorticity in developing a force on an object in a fluid, it would be helpful to determine the conditions necessary to create vorticity. There is a profound result known as the **Helmholtz Theorem** that created a major stumbling block for mathematicians and scientists in developing the basic theoretical foundations of fluid mechanics. Specifically, the Helmholtz Theorem states the following.

> *If a fluid particle moving in an unbounded frictionless, incompressible fluid under conservative body forces has zero vorticity initially, it always has zero vorticity.*

Thus, for example, when a flow starts from rest this theorem tells us there is no physical mechanism in an unbounded frictionless, incompressible fluid to develop vorticity.

To prove this theorem, we begin with Euler's equation for an incompressible flow with a conservative body force given by $\mathbf{f} = -\nabla\mathcal{V}$, viz.,

$$\frac{\partial \mathbf{u}}{\partial t} + \nabla\left(\frac{1}{2}\mathbf{u}\cdot\mathbf{u}\right) - \mathbf{u}\times\omega = -\nabla\left(\frac{p}{\rho} + \mathcal{V}\right) \tag{11.4}$$

We can derive an equation for the vorticity by taking the curl of this differential equation. First, we note that the curl of the gradient of a scalar vanishes (see Appendix C). Hence,

$$\nabla \times \nabla\left(\frac{1}{2}\mathbf{u}\cdot\mathbf{u}\right) = \mathbf{0} \quad \text{and} \quad \nabla \times \nabla\left(\frac{p}{\rho} + \mathcal{V}\right) = \mathbf{0} \tag{11.5}$$

Since the curl commutes with time differentiation, the curl of the first term in Equation (11.4) becomes

$$\nabla \times \frac{\partial \mathbf{u}}{\partial t} = \frac{\partial}{\partial t}\left(\nabla \times \mathbf{u}\right) = \frac{\partial \omega}{\partial t} \tag{11.6}$$

Reference to Appendix C shows that the curl of the cross product of **u** and ω is

$$\nabla \times (\mathbf{u} \times \omega) = (\omega \cdot \nabla)\mathbf{u} + \mathbf{u}(\nabla \cdot \omega) - (\mathbf{u} \cdot \nabla)\omega - \omega(\nabla \cdot \mathbf{u}) \tag{11.7}$$

Since the divergence of the curl of any vector is zero, we can say

$$\nabla \cdot \omega = \nabla \cdot (\nabla \times \mathbf{u}) = 0 \tag{11.8}$$

Finally, since the flow is incompressible,

$$\nabla \cdot \mathbf{u} = 0 \tag{11.9}$$

Collecting these vector identities and substituting into Equation (11.4) yields the following differential equation for the vorticity vector.

$$\frac{\partial \omega}{\partial t} + \mathbf{u} \cdot \nabla \omega = \omega \cdot \nabla \mathbf{u} \tag{11.10}$$

Thus, since the sum of the two terms on the left-hand side of this equation is simply the Eulerian derivative of ω, the final form of the differential equation governing the vorticity is:

$$\frac{d\omega}{dt} = \omega \cdot \nabla \mathbf{u} \tag{11.11}$$

The term on the right-hand side of Equation (11.11) is called the **vortex-stretching** term.

We are now in a position to prove the Helmholtz Theorem. Because the proof is very simple for two-dimensional flow, we will confine our attention to two dimensions. Although we omit the proof here, the theorem applies to three-dimensional flows also. By definition, for two-dimensional flow in the xy plane, we know that

$$w = 0 \quad \text{and} \quad \frac{\partial}{\partial z} \to 0 \tag{11.12}$$

Noting that the vorticity has only a z component for two-dimensional flow, i.e.,

$$\omega = \left(\frac{\partial v}{\partial x} - \frac{\partial u}{\partial y} \right) \mathbf{k} = \omega \mathbf{k} \tag{11.13}$$

necessarily the vortex-stretching term vanishes, viz.,

$$\omega \cdot \nabla \mathbf{u} = \omega \frac{\partial \mathbf{u}}{\partial z} = \mathbf{0} \tag{11.14}$$

Therefore, in a two-dimensional flow, we have shown that

$$\frac{d\omega}{dt} = \mathbf{0} \tag{11.15}$$

In words, this equation says *the rate of change of vorticity following a fluid particle is zero.* Hence, if ω is initially zero, then it will remain zero for all time. Thus, we have proven the Helmholtz Theorem for two-dimensional flows. As mentioned above, the theorem holds for three-dimensional flows as well.

Consequently, if we have an object that is initially at rest and begins moving at $t = 0$, its initial vorticity is zero. At all subsequent times its vorticity remains zero, wherefore no force develops on the object. However, this is completely inconsistent with everyday observations. An airplane wing develops a tremendous lifting force, a sail causes a sailboat to be able to sail nearly into the wind, a hand extended out of an automobile's window feels a strong force. Practical experience tells us such objects do develop forces, indicating that the vorticity does not remain zero.

Eighteenth- and nineteenth-century scientists were able to solve Euler's equation for slender objects. They were also able to make crude wind-tunnel measurements late in the nineteenth century that confirmed their solutions for most of the flowfield, with one important exception. The wind-tunnel models developed forces that the theory failed to predict.

These apparently contradictory results are referred to as **d'Alembert's Paradox**. That is, measurements and theory were completely consistent in many details, providing evidence that the theory was, for the most part, correct. However, the theory failed to predict the most important detail—the force! This was indeed a paradox. Complete details regarding the Helmholtz Theorem and d'Alembert's Paradox can be found at an introductory level in the book by Anderson (1989), or in the advanced text by Saffman (1993).

As a final comment, the Helmholtz Theorem and d'Alembert's Paradox left practicing engineers with a sense that theoretical fluid mechanics was of little value for design purposes. Rather, the topic was viewed by many as an intellectual exercise for mathematicians, devoid of practical use. This point of view had a lot to do with the empiricism and lack of rigor in much of the work of hydraulics engineers of the eighteenth and nineteenth centuries. The methods used by hydraulicists often achieved their design objectives, but with little or no understanding of why any new innovation they stumbled upon might function as well as it did.

11.3 Boundary Conditions at a Solid Boundary

To help resolve d'Alembert's Paradox, we must understand what happens in the immediate vicinity of a solid surface past which a fluid is flowing. If the surface has unit normal $\mathbf{n}$, we can always write the velocity vector, $\mathbf{u}$, as

$$\mathbf{u} = u_n\mathbf{n} + u_t\mathbf{t} \tag{11.16}$$

where u_n is the magnitude of the component of $\mathbf{u}$ normal to the surface. The vector $u_t\mathbf{t} = \mathbf{u} - u_n\mathbf{n}$ is the contribution to the velocity parallel to the surface, where $\mathbf{t}$ is a unit vector tangent to the surface (see Figure 11.2). Since there can be no flow through the boundary, if the surface is stationary, we can say that

$$\mathbf{u} \cdot \mathbf{n} = u_n = 0 \qquad \text{(at the surface)} \tag{11.17}$$

This result is valid for any fluid, whether we regard the fluid as real or ideal. For an ideal fluid, i.e., for an inviscid or frictionless fluid, this is the only condition we can specify at a solid boundary. Thus, we can have a non-vanishing tangential velocity component in an ideal fluid.

By contrast, for a real fluid, we find that fluid "sticks" to solid boundaries. That is, we find from observation of flow past stationary solid boundaries that $u_t = 0$. This is known as

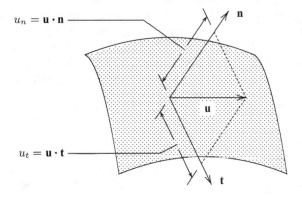

Figure 11.2: *Solid boundary and associated vectors.*

the **no-slip boundary condition**. As we will discover in Section 11.5, including viscosity in the momentum equation increases the order of the momentum equation, thus permitting (and requiring) additional boundary conditions.

If the surface moves with velocity $\mathbf{U}_s$, all of the comments above apply to the velocity relative to the surface. Thus, the boundary condition on the velocity appropriate at a solid boundary is as follows.

$$\left.\begin{array}{ll} (\mathbf{u} - \mathbf{U}_s) \cdot \mathbf{n} = 0, & \text{Ideal Fluid} \\[2mm] (\mathbf{u} - \mathbf{U}_s) = \mathbf{0}, & \text{Real Fluid} \end{array}\right\} \tag{11.18}$$

We will use this information in the next section to help explain how an object moving in a real fluid develops vorticity and, hence, a force.

11.4   Viscous Effects and Vorticity Generation

Consider flow past an *airfoil*, i.e., a two-dimensional cross section of a wing. Figure 11.3 contrasts the streamlines for a perfect fluid and a real fluid. As shown, for a perfect fluid, the streamlines have an unusual kink near the trailing edge of the airfoil, and there is a rear stagnation point on the upper surface upstream of the trailing edge. The streamlines shown are consistent with the Helmholtz Theorem, corresponding to irrotational flow. Assuming the motion started from rest, the irrotational flowfield that develops produces zero vorticity and zero force on the airfoil.

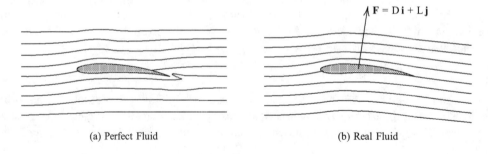

(a) Perfect Fluid   (b) Real Fluid

Figure 11.3: *Flow past an airfoil.*

By contrast, for a real fluid, the stagnation point lies precisely at the trailing edge. The streamlines leave the trailing edge smoothly without any kinks, and without the attending rapid local acceleration that would be required to negotiate the abrupt turning associated with ideal flow. This is a manifestation of the effect of friction. The primary purposes that friction serves in fluid flows are to drive the flow toward equilibrium and to smooth out discontinuities. Most importantly, the real flow develops vorticity even when the motion starts from rest, with an attendant force, $\mathbf{F}$, that is usually written as

$$\mathbf{F} = D\mathbf{i} + L\mathbf{j} \tag{11.19}$$

where D is called **drag** and L is called **lift**. To further underscore the connection between vorticity and the force on the object, a reasonable approximation for the lift per unit width on the object is given by the **Kutta-Joukowski law** that we derived in Chapter 10.

$$L = \rho U_\infty \Gamma_a \tag{11.20}$$

The quantity U_∞ is freestream velocity and Γ_a is the aerodynamic circulation [see Equation (4.27)].

The question remains about how the vorticity develops in the flow. For simplicity, consider a surface element parallel to the freestream flow direction, x. Recall from the preceding section that the no-slip boundary condition means both the tangential- and normal-velocity components, u and v, vanish at the surface. Since we assume for the purpose of our illustrative argument that the surface is parallel to the x axis, necessarily $\partial v / \partial x = 0$ as v vanishes at any two adjacent points on the surface. Hence, in a two-dimensional flow, the vorticity at the surface becomes

$$\boldsymbol{\omega} = -\frac{\partial u}{\partial y}\mathbf{k} \qquad \text{at} \qquad y = 0 \qquad\qquad (11.21)$$

Now, as shown in Figure 11.4, as the distance above the surface, y, increases, u increases from zero to its freestream value. Thus, except at special points known as separation points,[1] the velocity gradient, $\partial u / \partial y$, is nonvanishing at the surface. Hence, in a real fluid, the kinematical constraint imposed by the no-slip boundary condition causes the surface value of the vorticity to be nonzero. Although we have demonstrated this phenomenon for a simplified planar geometry, it is true for arbitrary shapes.

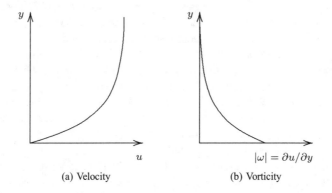

(a) Velocity (b) Vorticity

Figure 11.4: *Velocity and vorticity close to a solid boundary.*

The vorticity that is continually created at the surface diffuses into the flow under the action of the fluid's viscosity.[2] Then, it is swept into the main flow by the inertia of fluid particles passing near the surface. This is the process through which forces develop on an object immersed in a flowing fluid. It should not be surprising that friction plays a central role in controlling the force on an object.

Consider the role friction plays, for example, in creating traction for an automobile's tires. Imagine what would happen when you press the accelerator pedal if your automobile were parked on a frictionless street. The idealization of a frictionless fluid is to the airfoil what that frictionless street is to your motionless automobile with rapidly spinning wheels!

The streamline pattern shown in Figure 11.3(a) is observed at the beginning of the motion of an airfoil. Consistent with the discussion above, because of the action of viscous forces, the stagnation point moves toward the trailing edge. During this transient stage, vorticity is developing as a result of viscous diffusion. When the stagnation point reaches the trailing

---

[1]A separation point is a point where $\partial u / \partial y = 0$.

[2]As we will see in Section 11.5, the physical process of vorticity diffusion is the same as heat diffusing through a substance heated at a boundary.

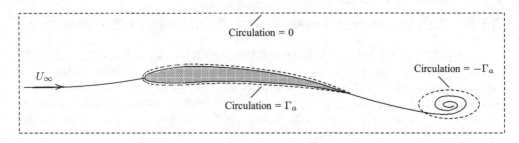

Figure 11.5: *An airfoil and a shed vortex.*

edge, a vortex is shed by the airfoil (Figure 11.5), and it propagates downstream. If we denote the strength of the vortex in terms of its circulation as $-\Gamma_a$, we discover two interesting facts. First, if we compute the circulation for a contour containing the airfoil and the shed vortex, the circulation about the contour is zero. Second, if we compute the circulation for a contour containing only the airfoil, we find that it is Γ_a.

The former fact provides a useful method for including lift in an inviscid computation. There is, in fact, a well-developed procedure for computing flow about lifting airfoils using potential-flow theory [cf. Glauert (1948), Milne-Thompson (1960), Robertson (1965) or Valentine (1967)]. Noting that the Helmholtz Theorem tells us that $d\omega/dt = \mathbf{0}$, if we could somehow inject vorticity into a flow, it would never go away. Thus, it is completely consistent to imagine that we have a vortex of strength Γ_a within the airfoil that is permanently bound to that location. We refer to such a vortex as a **bound vortex**. Then, if we solve the Euler equation, subject to the condition that the flow leaves the trailing edge smoothly as in Figure 11.3(b), we have a solution that includes lift. It turns out that there is a unique value of Γ_a that causes the flow to leave the trailing edge in this manner. We refer to the boundary condition that makes the flow leave the trailing edge smoothly as the **Kutta condition**.

The latter fact is consistent with the Helmholtz Theorem. Although we haven't proven it here, a similar result known as **Kelvin's Circulation Theorem** is written in terms of circulation. It states that the circulation about any contour always containing the same fluid particles is constant for an inviscid, incompressible fluid in an unbounded domain, i.e., $d\Gamma_a/dt = 0$. Thus, the net circulation of the airfoil and the shed vortex must be zero. When we use the Kutta condition, we imagine that the shed vortex lies at infinity, and as noted above, the force on the airfoil is given by $\rho U_\infty \Gamma_a$. Conversely, the force on the shed vortex is $-\rho U_\infty \Gamma_a$, so that *the net force on the airfoil and shed vortex is zero*, as demanded by the Helmholtz Theorem.

Example 11.2 *Compute the aerodynamic circulation for a small, single-engine airplane for which $L = 100$ lb/ft, $U_\infty = 200$ mph and $\rho = 0.00216$ slug/ft^3.*

Solution. First, note that from the Kutta-Joukowski law, the lift, L, and aerodynamic circulation, Γ_a, are related according to

$$L = \rho U_\infty \Gamma_a \quad \Longrightarrow \quad \Gamma_a = \frac{L}{\rho U_\infty}$$

For the given conditions, since $U_\infty = 200$ mph $= 293.4$ ft/sec, we have

$$\Gamma_a = \frac{100 \text{ lb/ft}}{\left(0.00216 \text{ slug/ft}^3\right)(293.4 \text{ ft/sec})} = 157.8 \ \frac{\text{ft}^2}{\text{sec}}$$

11.5 Navier-Stokes Equation and Diffusion of Vorticity

When we include friction in the momentum equation, it appears as an additional surface force. Although details of how the friction force is represented are beyond the scope of this text, it is nevertheless illuminating to discuss how friction alters the equations of motion. We find that the Euler equation is replaced by the most fundamental equation governing the motion of a fluid, viz., the **Navier-Stokes** equation. For incompressible flow, the Navier-Stokes equation is

$$\rho \frac{d\mathbf{u}}{dt} = -\nabla p + \mu \nabla^2 \mathbf{u} + \rho \mathbf{f} \tag{11.22}$$

where μ is viscosity (assumed constant here) and the differential operator ∇^2 is the **Laplacian operator** (see Appendix D) defined by

$$\nabla^2 \equiv \frac{\partial^2}{\partial x^2} + \frac{\partial^2}{\partial y^2} + \frac{\partial^2}{\partial z^2} \tag{11.23}$$

Note first that the Navier-Stokes equation, because of the presence of the viscous term, $\mu\nabla^2\mathbf{u}$, is a second-order differential equation. By contrast, the Euler equation is of first order. Thus, we must specify additional boundary conditions in general, such as the no-slip condition. Note also that taking the curl of this equation involves the same sequence of operations used for Euler's equation (see Section 11.2). The only difference is the viscous term, for which

$$\nabla \times \left(\mu \nabla^2 \mathbf{u} \right) = \mu \nabla^2 \boldsymbol{\omega} \tag{11.24}$$

Therefore, if we have an incompressible, viscous fluid subjected to conservative body forces, the equation for the vorticity becomes (with $\nu = \mu/\rho$):

$$\frac{d\boldsymbol{\omega}}{dt} = \boldsymbol{\omega} \cdot \nabla \mathbf{u} + \nu \nabla^2 \boldsymbol{\omega} \tag{11.25}$$

The last term on the right-hand side of this equation represents the physical process of diffusion. As discussed in the preceding section, this is the physical mechanism through which vorticity is created in flow past solid objects. If the "vortex-stretching" term, $\boldsymbol{\omega} \cdot \nabla \mathbf{u}$, is omitted and d/dt regarded as a pure time derivative, Equation (11.25) is identical with the unsteady heat-conduction equation in any number of spatial directions.

11.6 Prandtl's Boundary Layer

As noted in the introductory comments at the beginning of this chapter, mathematicians and scientists prior to the twentieth century were uncertain about the validity of the Newtonian (or Stokes) model for viscous stresses, i.e., $\tau = \mu \partial u / \partial y$. This uncertainty was based on the observation that for the large Reynolds numbers encountered in practical flows,

$$\left| \mu \nabla^2 \mathbf{u} \right| \ll \left| \rho \mathbf{u} \cdot \nabla \mathbf{u} \right| \tag{11.26}$$

They were left with the dilemma that their best model for effects of friction seemed to provide an unsatisfactory explanation for the generation of vorticity in a real fluid.

In 1904, Prandtl announced his discovery that in slightly viscous fluids (in the sense that $\mu \ll \rho U L$) such as air and water, viscous effects are confined to thin layers near a solid boundary. He originally called this the **transition layer**. Blasius renamed it the **boundary layer** in his 1908 doctoral thesis.

Equation (11.26) implicitly assumes that derivatives of the velocity (velocity gradients) are of the order of U/L where U and L are characteristic flow velocity and length, e.g., L could be the length of the body under consideration, and the same characteristic length applies to derivatives in any direction. However, the thickness of the boundary layer, δ, and not the length of the body, is the characteristic length for y derivatives. Succinctly put, Prandtl argued that the streamwise length scale is $L_x = L$, while the normal length scale is $L_y = \delta$.

Furthermore, as we will see in this section, velocity gradients are so large in boundary layers that $\mu \partial^2 \mathbf{u}/\partial y^2$ and $\rho \mathbf{u} \cdot \nabla \mathbf{u}$ are of comparable order of magnitude (y denotes distance normal to the surface). Prandtl further showed that outside of the boundary layer, the flow behaves as though it is inviscid, in complete agreement with the observations of his predecessors. His pioneering work proved that the Newtonian/Stokes model for representing the shear stress is correct. He also proved that the mechanism of vorticity diffusion is of sufficient magnitude to create the vorticity that leads to lift and drag. Prandtl's discovery of the boundary layer thus resolved d'Alembert's Paradox.

11.6.1   High-Reynolds-Number Flow Near a Surface

As pointed out by Schlichting and Gersten (2000), to illustrate how viscous effects can have such a profound effect at very large Reynolds number, Prandtl discussed the solution to a differential equation of the following form.

$$\epsilon \frac{d^2u}{dy^2} + (1+\epsilon)\frac{du}{dy} + y = 0, \qquad 0 \le y \le 1 \qquad (\epsilon \ll 1) \tag{11.27}$$

This equation is to be solved subject to the boundary conditions

$$u(0) = 0 \quad \text{and} \quad u(1) = 1 \tag{11.28}$$

Prandtl offered this two-point boundary value problem as a simplified analog of the Navier-Stokes equation.[3] The second-derivative term has a small coefficient just as the second-derivative term in the Navier-Stokes equation, in nondimensional form, has the reciprocal of the Reynolds number as its coefficient. An immediate consequence is that only one boundary condition can be satisfied if we set $\epsilon = 0$. This is similar to setting viscosity equal to zero in the Navier-Stokes equation, which yields Euler's equation, and the attendant consequence that only the normal-velocity surface boundary condition can be satisfied. That is, we cannot enforce the no-slip boundary condition for Euler-equation solutions. The exact solution to this equation is

$$u(y; \epsilon) = \frac{e^{1-y} - e^{1-y/\epsilon}}{1 - e^{1-1/\epsilon}} \tag{11.29}$$

If we set $\epsilon = 0$ in Equation (11.27), we have the following first-order equation:

$$\frac{du}{dy} + y = 0 \tag{11.30}$$

and the solution, $u(y; 0)$, is

$$u(y; 0) = e^{1-y} \tag{11.31}$$

where we use the boundary condition at $y = 1$. Provided $y \gg \epsilon$, this solution is the limiting form of Equation (11.29). To see this, simply note that $e^{-y/\epsilon} \to 0$ as $\epsilon \to 0$. However, the

---

[3]Strictly speaking, Prandtl's model equation did not include the $(1 + \epsilon)$ factor – it has been included here to simplify the algebra.

solution obviously fails to satisfy the boundary condition at $y = 0$ because the solution gives $u(0; 0) = e = 2.71828 \cdots$. Figure 11.6 displays the solution to our simplified equation for several values of ϵ. As shown, the smaller the value of ϵ, the more closely $u(y; 0)$ represents the solution throughout the region $0 < y \leq 1$. Only in the immediate vicinity of $y = 0$ is the solution inaccurate. We call the thin layer where $u(y; 0)$ departs from the exact solution a **boundary layer**.

This simple analogy displays a key feature of solutions to the Navier-Stokes equation for **slightly-viscous** fluids, flowing over a solid boundary. We call a fluid slightly viscous whenever its kinematic viscosity, ν, is very small compared to UL, so that the Reynolds number is very large. In such a fluid, we indeed find in practice that viscous effects are confined to thin layers near the boundary.

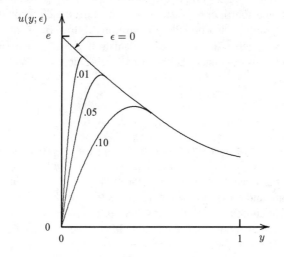

Figure 11.6: *Solutions to the model equation for several values of ϵ.*

11.6.2 Boundary-Layer Equations

As discussed above, Prandtl found that flow properties in a boundary layer vary much more rapidly in the direction normal to the surface, y, than in the direction parallel to the surface, x (Figure 11.7). In fact, variations are so rapid that the convective acceleration, $u\partial u/\partial x$, and the viscous term, $\nu\partial^2 u/\partial y^2$, are of the same order of magnitude. To understand Prandtl's reasoning, we will consider flow past the flat plate depicted in the figure. The dashed line delineates the "edge" of the viscous region, which is assumed to have thickness $\delta(x)$. We do not have to define $\delta(x)$ exactly at this point.

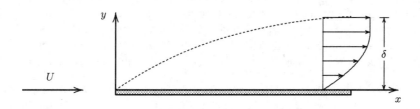

Figure 11.7: *Viscous flow over a solid surface.*

We begin with the continuity and Navier-Stokes equations. Confining our attention to steady, incompressible, two-dimensional flow, we have

$$\frac{\partial u}{\partial x} + \frac{\partial v}{\partial y} = 0 \tag{11.32}$$

$$u\frac{\partial u}{\partial x} + v\frac{\partial u}{\partial y} = -\frac{1}{\rho}\frac{\partial p}{\partial x} + \nu\left(\frac{\partial^2 u}{\partial x^2} + \frac{\partial^2 u}{\partial y^2}\right) \tag{11.33}$$

$$u\frac{\partial v}{\partial x} + v\frac{\partial v}{\partial y} = -\frac{1}{\rho}\frac{\partial p}{\partial y} + \nu\left(\frac{\partial^2 v}{\partial x^2} + \frac{\partial^2 v}{\partial y^2}\right) \tag{11.34}$$

These equations constitute a fourth order system, so that four boundary conditions are needed to have a well-posed problem. Two boundary conditions follow from the no-slip boundary condition at the surface, which tells us u and v vanish at $y = 0$ for a stationary surface. Additionally, the solution must approach the freestream velocity, $u = U$ and $v = 0$ as $y \to \infty$. Thus,

$$\left.\begin{array}{l} u = 0, \quad v = 0, \quad y = 0 \\[2mm] u \to U, v \to 0, \quad y \to \infty \end{array}\right\} \tag{11.35}$$

Prandtl used order-of-magnitude estimates to develop an approximate set of equations valid in the boundary layer. His approach was heuristic, and involved excellent engineering judgment guided by experimental observations. We will follow his arguments here, noting that they are approximate and lacking in rigor. Nevertheless, the equations are well founded, and represent an excellent approximation for **attached**, or unseparated, flows. We will discuss what happens when the flow undergoes separation in Subsection 11.6.6. It is possible to deduce these equations with advanced mathematical techniques. Modern developments in applied mathematics, most notably in **singular perturbation theory** [cf. Van Dyke (1975), Kevorkian and Cole (1981), Neyfeh (1981) or Wilcox (1995)], provide a solid foundation for development of the **boundary-layer equations**.

Prandtl's arguments begin with the assumption that the viscous layer is very thin, which means we can say $\delta(x) \ll x$. Anticipating that properties change over a much shorter distance in the vertical direction, δ, than in the horizontal direction, x, we expect to have $\partial f/\partial y \gg \partial f/\partial x$ for any property f. Additionally, we estimate the order of magnitude of such derivatives to be given by

$$\frac{\partial f}{\partial x} \sim \frac{F}{x} \quad \text{and} \quad \frac{\partial f}{\partial y} \sim \frac{F}{\delta} \tag{11.36}$$

where F is a measure of the change in f. The notation "$\sim$" means order of magnitude, and does not imply equality.[4] The idea behind these estimates is that if a quantity changes by an amount F over a given distance (x or δ here), then its derivative is proportional to the ratio of F to that distance. The closer the function is to being linear, the closer the proportionality coefficient is to one. Although the factor will differ from unity for general functions, these estimates still provide a reasonable estimate of the order of magnitude of a derivative.

Mass Conservation. Since the horizontal velocity, u, changes from 0 to U as we move from the plate surface to the freestream, we can plausibly estimate the order of magnitude of u to

---

[4]In the spirit of singular perturbation theory, the $\sim$ symbol means the ratio of the terms on each side of the $\sim$ approaches a finite, nonzero value in the limit $\delta \to 0$, i.e., their ratio is bounded.

be U, so that the order of magnitude of $\partial u / \partial x$ is

$$\frac{\partial u}{\partial x} \sim \frac{U}{x} \tag{11.37}$$

Similarly, if we denote the change in vertical velocity by V, an estimate of its derivative across the boundary layer is given by

$$\frac{\partial v}{\partial y} \sim \frac{V}{\delta} \tag{11.38}$$

So, we can estimate the order of magnitude of the terms in the continuity Equation (11.32) as follows.

$$\underbrace{\frac{\partial u}{\partial x}}_{U/x} + \underbrace{\frac{\partial v}{\partial y}}_{V/\delta} = 0 \tag{11.39}$$

Therefore, the order of magnitude of the vertical velocity is

$$\frac{V}{\delta} \sim \frac{U}{x} \quad \Longrightarrow \quad V \sim \frac{\delta}{x} U \ll U \tag{11.40}$$

Hence, the vertical velocity is very small relative to the horizontal velocity in a boundary layer. For typical boundary layers, we find that v is positive at the boundary-layer edge, corresponding to fluid being displaced from the surface as fluid moves over the plate in the streamwise direction.[5] Thus, having $v \ll u$ is consistent with having $\delta \ll x$. By contrast, if v were comparable to u, the edge of the boundary layer would spread at a finite angle, which would correspond to $\delta \sim x$. This would contradict our basic assumption that the boundary layer is very thin, which tells us it is plausible to have $v \ll u$.

Horizontal-Momentum Conservation. Using Equation (11.40) for the vertical velocity, and denoting the order of magnitude of the change in pressure by P, the various terms in the x-momentum Equation (11.33) are

$$\underbrace{u\frac{\partial u}{\partial x}}_{U^2/x} + \underbrace{v\frac{\partial u}{\partial y}}_{U^2/x} = \underbrace{-\frac{1}{\rho}\frac{\partial p}{\partial x}}_{P/\rho x} + \underbrace{\nu\frac{\partial^2 u}{\partial x^2}}_{\nu U/x^2} + \underbrace{\nu\frac{\partial^2 u}{\partial y^2}}_{\nu U/\delta^2} \tag{11.41}$$

We can make three immediate observations regarding the relative importance of the terms in this equation.

1. $\nu U / x^2 \ll \nu U / \delta^2$ because $\delta \ll x$. Thus, we can ignore the term $\nu \partial^2 u / \partial x^2$.

2. Since we know from experiments that pressure gradient has a significant effect on a boundary layer, we expect the pressure-gradient term to be of the same order of magnitude as the acceleration terms. Hence, $U^2 / x \sim P / \rho x$, so that $P \sim \rho U^2$. This is consistent with Bernoulli's equation, which is valid above the boundary layer.

3. In order for viscous effects to be significant, the term $\nu \partial^2 u / \partial y^2$ must be of the same order of magnitude as the convective terms. Hence, we must have $U^2 / x \sim \nu U / \delta^2$, wherefore the thickness of the boundary layer is given by

---

[5]In fact, we will find that at the edge of the boundary layer, necessarily $v < u d\delta / dx$ when the layer is growing in thickness. Even when it decreases in thickness, we have $v \sim u d\delta / dx$ at the boundary-layer edge.

$$\delta(x) \sim \sqrt{\frac{\nu x}{U}} \tag{11.42}$$

This result tells us that the smaller the viscosity, ν, the thinner the boundary layer will be. To cast this in quantitative terms, we first rewrite the boundary-layer thickness in terms of Reynolds number based on x, i.e., $Re_x = Ux/\nu$. There follows:

$$\delta \sim \frac{x}{\sqrt{Re_x}} \tag{11.43}$$

Since Reynolds number is typically in excess of 10^6 for most practical applications, this result indicates that the thickness of a boundary layer will be less than a thousandth of the distance along a surface. Consequently, derivatives normal to a solid surface will be several orders of magnitude larger than those parallel to the same surface!

Vertical-Momentum Conservation. Using the estimates for vertical velocity, $V \sim U\delta/x$, and pressure, $P \sim \rho U^2$, the orders of magnitude of the terms in the y-momentum Equation (11.34) are

$$\underbrace{u\frac{\partial v}{\partial x}}_{U^2\delta/x^2} + \underbrace{v\frac{\partial v}{\partial y}}_{U^2\delta/x^2} = \underbrace{-\frac{1}{\rho}\frac{\partial p}{\partial y}}_{U^2/\delta} + \underbrace{\nu\frac{\partial^2 v}{\partial x^2}}_{\nu U\delta/x^3} + \underbrace{\nu\frac{\partial^2 v}{\partial y^2}}_{\nu U/x\delta} \tag{11.44}$$

As with horizontal momentum, we can make three immediate observations regarding the relative importance of the terms in this equation.

1. $\nu U\delta/x^3 \ll \nu U/x\delta$ because $\delta \ll x$. Thus, we can ignore the viscous term $\nu\partial^2 v/\partial x^2$.

2. Since the boundary-layer thickness is given by Equation (11.43), necessarily

$$u\frac{\partial u}{\partial x} \sim v\frac{\partial v}{\partial y} \sim \nu\frac{\partial^2 v}{\partial y^2} \sim \frac{U^2\delta}{x^2}$$

3. Because $\delta \ll x$, necessarily $U^2\delta/x^2 \ll U^2/\delta$. This means the acceleration and viscous terms are of a much smaller order of magnitude than the pressure gradient across the boundary layer. All but the normal pressure-gradient term, $\partial p/\partial y$, are unimportant in Equation (11.34), so that the vertical momentum equation simplifies to

$$0 \approx -\frac{1}{\rho}\frac{\partial p}{\partial y}$$

On reflection, the third point may seem a little odd. After all, we are, in effect, saying that $\partial p/\partial y$ is enormous compared to all other terms in the y-momentum equation. Using this information, we then conclude that it must be zero! There is no contradiction, however, and the physical meaning of this result is the following. On the one hand, the magnitude of the pressure, $p \sim \rho U^2$, is established from the balance of forces in the horizontal-momentum equation. On the other hand, any change in pressure across the boundary layer must be comparable in magnitude to the convective and viscous terms in the vertical momentum equation, which are substantially smaller than U^2/δ. Therefore, as a first approximation, the pressure must be constant across a boundary layer, so that $\partial p/\partial y \approx 0$.

Since mass conservation [Equation (11.32)] remains unaltered, the following approximate set of equations, known as **Prandtl's boundary-layer equations**, can be used to describe the flow in a boundary layer.

$$\frac{\partial u}{\partial x} + \frac{\partial v}{\partial y} = 0 \tag{11.45}$$

$$u\frac{\partial u}{\partial x} + v\frac{\partial u}{\partial y} = -\frac{1}{\rho}\frac{\partial p}{\partial x} + \nu\frac{\partial^2 u}{\partial y^2} \tag{11.46}$$

$$0 = -\frac{1}{\rho}\frac{\partial p}{\partial y} \tag{11.47}$$

Although we have developed the boundary-layer equations in a planar geometry, they hold as written in general body-oriented coordinates. That is, these equations are valid along the surface of a general geometry where x becomes arc length and y becomes distance normal to the surface (see Figure 11.8). This will be true in general because the boundary layer is very thin for most applications relative to the radius of curvature, $\mathcal{R}$, of the surface. There are exceptions, of course, and when the boundary-layer thickness, δ, is comparable in magnitude to $\mathcal{R}$, the boundary layer is said to be "thick." The boundary-layer equations as quoted above must be modified for thick boundary-layer applications [cf. Van Dyke (1975) or Emanuel (1994)]. Equations (11.45) through (11.47) are restricted to the classical case of a **thin shear layer**, which covers the vast majority of everyday applications.

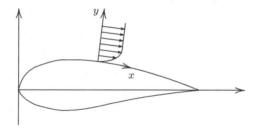

Figure 11.8: *Boundary-layer coordinates; x is arc length and y is normal distance.*

Because Equation (11.47) tells us the pressure is independent of y, necessarily the pressure is constant across a boundary layer. Consequently, the pressure can be determined from an inviscid solution. That is, we can solve Euler's equation or use potential-flow methods to determine the inviscid solution, including the pressure at all points in the flow. Then, the pressure in the boundary layer is given by the surface value of p according to the inviscid solution. The methodology for computing flow past a slender body whose surface can be described in conventional Cartesian coordinates by $y = y_b(x)$ is as follows.

1. Solve Euler's equation for flow past the body. Part of the solution is $p_{Euler}(x, y)$.

2. Compute the boundary-layer pressure from $p(x) = p_{Euler}(x, y_b)$.

3. Solve the boundary-layer equations using $p(x)$.

This is a great simplification from a computational point of view. Inviscid solutions are far less computationally intensive than Navier-Stokes solutions, mainly because of the need to resolve thin boundary layers for the latter. Similarly, because the pressure is known in advance (from the inviscid solution), the boundary-layer equations are also much easier to solve numerically than the full Navier-Stokes equations. The boundary-layer equations are accurate provided the pressure gradient, dp/dx, is not strong enough to cause separation of the boundary layer. As noted earlier, the boundary-layer equations are strictly valid only for **attached flows**, i.e., flows with no separation. When separation occurs, only a complete solution to the Navier-Stokes equations is sufficient.

Since the pressure, $p(x)$, is a known function in the boundary-layer equations, it turns out that the boundary-layer equations are of third order. This stands in contrast to the full, two-dimensional Navier-Stokes system of equations in component form, which we noted earlier is of fourth order [see the discussion just below Equation (11.34)]. Consequently, we must give up one of the boundary conditions of Equation (11.35). We retain the no-slip condition so that u and v still vanish at the surface. Also, we still insist that $u \to U$ as $y \to \infty$. Thus, we impose no condition on the vertical velocity in the freestream, and the appropriate boundary conditions for the boundary-layer equations are

$$\left. \begin{array}{lll} u = 0, & v = 0, & y = 0 \\[2mm] u \to U, & & y \to \infty \end{array} \right\} \tag{11.48}$$

Note that this means we have no control over the vertical velocity as $y \to \infty$ in a boundary-layer computation—it is part of the solution.

Example 11.3 *The velocity,* **u**, *in a two-dimensional, incompressible boundary layer can be approximated by*

$$\mathbf{u} \approx U \frac{y}{\delta} \left(1 - \frac{y}{\delta} \right) \mathbf{i}, \qquad \delta \approx 5 \sqrt{\frac{\nu x}{U}}$$

where U is freestream velocity, y is distance from the surface, δ is boundary-layer thickness, ν is kinematic viscosity and x is streamwise distance. Compute the dimensionless vorticity at the surface, $\nu \omega_w / U^2$, as a function of Reynolds number, $Re_x = Ux/\nu$.

Solution. For the given velocity profile, $v = 0$, wherefore the vorticity is

$$\boldsymbol{\omega} = \omega \mathbf{k} = \left(\frac{\partial v}{\partial x} - \frac{\partial u}{\partial y} \right) \mathbf{k} = -\frac{\partial u}{\partial y} \mathbf{k}$$

Now, differentiating the given u profile, we have

$$\frac{\partial u}{\partial y} = \frac{\partial}{\partial y} \left[\frac{U}{\delta} y - \frac{U}{\delta^2} y^2 \right] = \frac{U}{\delta} - \frac{2U}{\delta^2} y$$

Therefore, the dimensionless value of ω at the surface, $\nu \omega_w / U^2$, is

$$\frac{\nu \omega_w}{U^2} = -\frac{\nu}{U^2} \left(\frac{\partial u}{\partial y} \right)_{y=0} = -\frac{\nu}{U \delta}$$

Finally, since $\delta \approx 5 \sqrt{\nu x / U}$, we find

$$\frac{\nu \omega_w}{U^2} = -\frac{1}{5} \frac{\nu}{U} \sqrt{\frac{U}{\nu x}} = -\frac{1}{5} \sqrt{\frac{\nu}{Ux}} = -\frac{1}{5 \sqrt{Re_x}}$$

Prior to the advent of very fast computers, great advances in aerodynamics and aviation occurred using this procedure, i.e., combined inviscid solutions and boundary-layer theory. Because most aerodynamic and hydrodynamic devices are designed to move smoothly through the fluid with minimal resistance, flow separation is usually avoided at the designated operating condition. Thus, for flow about a wing, the classical inviscid-flow/boundary-layer approach is generally sufficient for design purposes, and is still used at aircraft companies.

11.6.3 Momentum-Integral Equation

In this subsection, we will use the control-volume method to analyze flow in a boundary layer. While returning to the control-volume methodology may appear to run counter to the aims of the differential approach that is now our primary focus, there is method in our madness. The result we obtain, known as the **momentum-integral equation**, is particularly useful in both experimental and theoretical work. Additionally, the equation has played a central role in the development of boundary-layer theory, and helps illuminate some aspects of the nature of boundary layers.

Figure 11.9 shows a segment of a growing boundary layer and suitably chosen control-volume boundaries. We consider two streamwise locations, $x_1 = x$ and $x_2 = x + \Delta x$, where Δx is a small distance. The boundary-layer thickness is $\delta_1 \equiv \delta(x)$ at the upstream boundary of the control volume and $\delta_2 \equiv \delta(x + \Delta x)$ at the downstream boundary. The quantities u_e and v_e are the velocity components at the edge of the boundary layer, and are defined by

$$u_e \equiv u(x, \delta) \quad \text{and} \quad v_e \equiv v(x, \delta) \tag{11.49}$$

The surface shear stress is denoted by τ_w, and will, in general, vary with x. For simplicity, we consider the constant-pressure case. We also assume the flow is steady, incompressible and two dimensional.

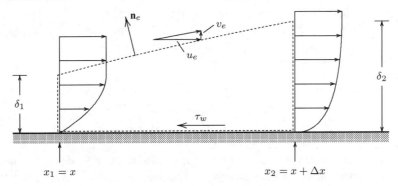

Figure 11.9: *A two-dimensional boundary layer.*

Mass Conservation. Beginning with mass conservation, we know that

$$\oiint_S \rho \mathbf{u} \cdot \mathbf{n} \, dS = 0 \tag{11.50}$$

Noting that there is no flow through the solid boundary, the closed-surface integral consists of contributions from the upstream boundary ($x = x_1$), the downstream boundary ($x = x_2$), and the boundary-layer edge. At the upstream boundary, we have $\mathbf{u} \cdot \mathbf{n} = -u$, and at the downstream boundary, $\mathbf{u} \cdot \mathbf{n} = u$. At the boundary-layer edge, the unit normal is $\mathbf{n}_e$, and the velocity is equal to the freestream value. Consistent with the boundary-layer approximation, $d\delta/dx \ll 1$, the unit normal to the upper boundary is

$$\mathbf{n}_e \approx \mathbf{j} - \mathbf{i} \frac{d\delta}{dx} \tag{11.51}$$

Therefore, the value of $\mathbf{u} \cdot \mathbf{n}$ on the upper boundary of the control volume is

$$(\mathbf{u} \cdot \mathbf{n})_e \approx v_e - u_e \frac{d\delta}{dx} \tag{11.52}$$

This is the **entrainment velocity**, and it is negative when the boundary layer is growing. While v_e is generally positive, the quantity $u_e d\delta/dx$ is larger than v_e so that fluid is drawn into the boundary layer. The opposite is true, of course, if the boundary layer is decreasing in thickness. So, expanding the closed-surface integral into its three contributions, we find

$$- \int_0^{\delta_1} \rho u \, dy + \int_0^{\delta_2} \rho u \, dy + \int_{x_1}^{x_2} \rho \left(v_e - u_e \frac{d\delta}{dx} \right) dx = 0 \tag{11.53}$$

which can be rewritten as

$$\frac{\int_0^{\delta(x+\Delta x)} \rho u \, dy - \int_0^{\delta(x)} \rho u \, dy}{\Delta x} + \frac{1}{\Delta x} \int_x^{x+\Delta x} \rho \left(v_e - u_e \frac{d\delta}{dx} \right) dx = 0 \tag{11.54}$$

Taking the limit as $\Delta x \to 0$, the final form of the mass-conservation principle for the boundary layer is:

$$\frac{d}{dx} \int_0^{\delta} \rho u \, dy + \rho \left(v_e - u_e \frac{d\delta}{dx} \right) = 0 \tag{11.55}$$

The physical interpretation of this equation is obvious. It says the rate of change of mass flux through the boundary layer is balanced by the rate at which fluid is entrained from the freestream. Clearly, when the entrainment velocity, $(\mathbf{u} \cdot \mathbf{n})_e$, is negative, Equation (11.55) tells us the mass flux increases, so that the boundary layer grows in thickness. We will use this equation to simplify the momentum equation below.

Momentum Conservation. The x component of the momentum equation is

$$\oint\!\!\!\oint_S \rho u (\mathbf{u} \cdot \mathbf{n}) dS = -\mathbf{i} \cdot \oint\!\!\!\oint_S (p - p_\infty) \mathbf{n} \, dS - \oint\!\!\!\oint_S \tau_w \, dS \tag{11.56}$$

where p_∞ is the freestream pressure and τ_w is the surface shear stress due to friction. Because pressure is constant across a boundary layer, necessarily $p = p_\infty$ throughout the control volume and on the entire bounding surface. Therefore, the net pressure integral vanishes. Since the entrainment velocity is generally nonvanishing so that $\rho u (\mathbf{u} \cdot \mathbf{n}) = \rho u_e (\mathbf{u} \cdot \mathbf{n})_e$ at the boundary-layer edge, the momentum-flux and viscous-stress integrals expand to yield the following equation.

$$- \int_0^{\delta_1} \rho u^2 dy + \int_0^{\delta_2} \rho u^2 dy + \int_{x_1}^{x_2} \rho u_e \left(v_e - u_e \frac{d\delta}{dx} \right) dx = - \int_{x_1}^{x_2} \tau_w dx \tag{11.57}$$

Hence, regrouping terms, we have

$$\frac{\int_0^{\delta(x+\Delta x)} \rho u^2 dy - \int_0^{\delta(x)} \rho u^2 dy}{\Delta x} + \frac{1}{\Delta x} \int_x^{x+\Delta x} \rho u_e \left(v_e - u_e \frac{d\delta}{dx} \right) dx$$

$$= -\frac{1}{\Delta x} \int_x^{x+\Delta x} \tau_w \, dx \tag{11.58}$$

Thus, taking the limit $\Delta x \to 0$, the momentum equation becomes:

$$\frac{d}{dx} \int_0^\delta \rho u^2 \, dy + \rho u_e \left(v_e - u_e \frac{d\delta}{dx} \right) = -\tau_w \tag{11.59}$$

Again, the physical interpretation of this equation is straightforward. The rate of change of the momentum flux through the boundary layer is balanced by the rate at which momentum is entrained from the freestream, $-\rho u_e(v_e - u_e d\delta/dx)$, and the surface shear stress.

We can rearrange Equation (11.59) to arrive at the classical result known as the momentum-integral equation. To accomplish this end, we begin by substituting for the entrainment velocity from Equation (11.55).

$$\frac{d}{dx} \int_0^\delta \rho u^2 \, dy - u_e \frac{d}{dx} \int_0^\delta \rho u \, dy = -\tau_w \tag{11.60}$$

Now, recall that Euler's equation holds at the edge of the boundary layer, wherefore

$$\rho u_e \frac{du_e}{dx} = -\frac{dp}{dx} \tag{11.61}$$

Hence, since the pressure is constant, necessarily the freestream velocity is also constant. Consequently, we can rewrite Equation (11.60) as

$$\frac{d}{dx} \int_0^\delta \rho \left(u^2 - u_e u \right) dy = -\tau_w \tag{11.62}$$

Dividing Equation (11.62) through by ρu_e^2, there follows:

$$\frac{d}{dx} \int_0^\delta \frac{u}{u_e} \left(1 - \frac{u}{u_e} \right) dy = \frac{\tau_w}{\rho u_e^2} = \frac{c_f}{2} \tag{11.63}$$

Thus, we arrive at the momentum-integral equation for a constant-pressure boundary layer, viz.,

$$\frac{d\theta}{dx} = \frac{c_f}{2} \tag{11.64}$$

where the quantity θ is the **momentum thickness** defined by

$$\theta \equiv \int_0^\delta \frac{u}{u_e} \left(1 - \frac{u}{u_e} \right) dy \tag{11.65}$$

Had we allowed for variable pressure, a more general version of the equation results. The complete incompressible, two-dimensional form of the equation is [see Schlichting and Gersten (2000)]

$$\frac{d\theta}{dx} + (2 + H) \frac{\theta}{u_e} \frac{du_e}{dx} = \frac{c_f}{2} \tag{11.66}$$

where H is the **shape factor** defined by

$$H = \frac{\delta^*}{\theta} \tag{11.67}$$

and δ^* is the **displacement thickness**, i.e.,

$$\delta^* \equiv \int_0^\delta \left(1 - \frac{u}{u_e} \right) dy \tag{11.68}$$

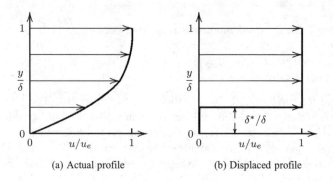

(a) Actual profile   (b) Displaced profile

Figure 11.10: *Physical meaning of displacement thickness, δ^*.*

The quantities c_f, θ, δ^* and H are collectively referred to as **integral parameters**. The displacement thickness is of especial interest in boundary-layer theory. It has a straightforward physical interpretation, which Figure 11.10 illustrates. Figure 11.10(a) shows a typical velocity profile. For incompressible flow, the mass flux per unit width (out of the page) for this profile, $\dot{m}_a$, is

$$\dot{m}_a = \rho \int_0^\delta u \, dy \tag{11.69}$$

Now, consider the velocity profile of Figure 11.10(b), for which the velocity is constant and equal to the freestream value above $y = \delta^*$, and zero below. The mass flux per unit width for this profile, $\dot{m}_b$, is

$$\dot{m}_b = \rho u_e \left(\delta - \delta^*\right) = \rho \int_0^\delta u_e \, dy - \rho u_e \delta^* \tag{11.70}$$

In order to have the same mass flux for both profiles, we conclude that

$$\dot{m}_a = \dot{m}_b \quad \Longrightarrow \quad \rho \int_0^\delta u_e \, dy - \rho u_e \delta^* = \rho \int_0^\delta u \, dy \tag{11.71}$$

With a little rearrangement of terms, the value of δ^* is

$$\delta^* \equiv \int_0^\delta \left(1 - \frac{u}{u_e}\right) dy \tag{11.72}$$

which is identical to the definition of displacement thickness given in Equation (11.68). Therefore, the displacement thickness is the distance viscous effects appear to displace the inviscid flow from the surface. This phenomenon is generally referred to as the **displacement effect** (Figure 11.11). When the displacement thickness becomes large, e.g., near a separation point, the effective shape of the body seen by the inviscid flow differs from the actual shape.

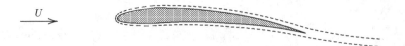

Figure 11.11: *Illustration of the displacement effect; the dashed contour shows the effect of adding δ^* to the actual airfoil shape.*

Even for attached flow, the displacement effect changes the effective shape of an airfoil as illustrated in Figure 11.11. As a result, the actual pressure distribution differs from inviscid-flow theory in a way that is strongly dependent on airfoil shape. As a general rule of thumb, the displacement effect usually **reduces** the lift, relative to the inviscid value, by about 10%.

The momentum thickness defined in Equation (11.65) does not lend itself to quite so straightforward a physical interpretation. However, it should be clear that it is a quantitative measure of the momentum lost to friction.

The momentum-integral equation, Equation (11.66), served as the foundation of boundary-layer computational work until the early 1970's. It has since given way to numerical solution of the boundary-layer equations, which provides more detailed information. Nevertheless, the equation still provides some interesting insight into the dynamics of the boundary layer. Additionally, it is useful in both experimental and computational work. On the one hand, the experimenter can measure velocity profiles at several values of x, compute θ for each profile, and use Equation (11.66) to infer the skin friction. Alternatively, if the skin friction has been measured also, Equation (11.66) can be used as a consistency check. Similarly, a computational fluid dynamicist can use the equation to check for numerical accuracy.

Example 11.4 *The freestream horizontal velocity for stagnation-point flow is $u_e = Ax$. A reasonable approximation to the velocity profile is*

$$\frac{u}{u_e} = \left(2 - \frac{y}{\delta}\right)\frac{y}{\delta}$$

where $\delta(x)$ is the boundary-layer thickness. Use the momentum-integral equation to set up a differential equation for $\delta(x)$, and solve the equation.

Solution. From the velocity profile, we can compute the skin friction directly. A short calculation shows that

$$c_f = \frac{\mu(\partial u/\partial y)_{y=0}}{\frac{1}{2}\rho u_e^2} = \frac{4\nu}{Ax\delta}$$

Using the definitions of δ^*, θ and H, we also find

$$\delta^* = \frac{1}{3}\delta, \qquad \theta = \frac{2}{15}\delta, \qquad H = \frac{5}{2}$$

Substitution of these integral parameters into the general form of the momentum-integral equation [Equation (11.66)], we obtain

$$\frac{2}{15}\frac{d\delta}{dx} + \frac{9}{2}\frac{2}{15}\frac{\delta}{x} = \frac{2\nu}{Ax\delta}$$

Rearranging terms a bit, this equation can be rewritten as

$$x\frac{d\delta^2}{dx} + 9\delta^2 = \frac{30\nu}{A}$$

As can be verified by direct substitution, the general solution to this equation is

$$\delta^2(x) = \frac{10}{3}\frac{\nu}{A} + Cx^{-9}$$

where C is an integration constant. Far downstream, the term proportional to x^{-9} will be negligible so that we can ignore it. Thus, the thickness of this boundary layer is

$$\delta = \sqrt{\frac{10}{3}\frac{\nu}{A}} \approx 1.83\sqrt{\frac{\nu}{A}}$$

AN EXPERIMENT YOU CAN DO AT HOME
Excerpted from Granger (1988)

To observe an interesting interaction of vorticity and viscosity, you must assemble a cup, some water, tea leaves and a spoon. The ideal cup is as nearly cylindrical as possible and should be about 2/3 full of water. Add a few tea leaves and vigorously stir with the spoon in a circular motion for a few seconds. Now, remove the spoon and marvel in wonder as the tea leaves settle in a nice little heap at the center of the cup!

As discussed by Granger (1988), this is one of Prof. Nicholas Rott's favorite experiments. It shows some of the interesting features of what are known as the *spin-up* and *spin-down* processes. The stirring causes the fluid to rotate in a nearly rigid-body state. The centrifugal force causes the tea leaves to assemble along the cylindrical boundary of the cup. Because of the no-slip condition at the cup boundaries, thin boundary layers form along the bottom and sides of the cup. The fluid in these boundary layers flows radially outward along the bottom, up the sides and back into the main body of fluid. The circulation is completed by a cylindrical front that advances from the cup boundary toward the non-rotating core during spin up.

When the stirring ceases, a cylindrical front moves back toward the cup boundaries, leaving the fluid near the center of the cup motionless. During the spin-down process, the overall circulation reverses and the tea leaves are deposited in the motionless core at the center of the cup.

While the motion is primarily inviscid, the flow is strongly influenced by the boundary layer at the bottom of the cup, which is known as an *Ekman layer*. As fluid moves radially outward, the layer draws fluid downward in order to conserve mass. This so-called *Ekman suction* provides the mechanism for the unusual spin-up and spin-down dynamics. Maxworthy (1968) describes this flow in a paper entitled "A Storm in a Teacup."

11.6.4   Blasius Solution

The boundary-layer Equations (11.45) - (11.47) have an exact solution for constant pressure, known as the **Blasius solution**. Constant pressure corresponds to a flat plate that is aligned with the freestream velocity vector (see Figure 11.12). We solve the continuity and boundary-layer equations with zero pressure gradient subject to the no-slip boundary condition and the relevant freestream condition, so that the problem we must solve is

$$\frac{\partial u}{\partial x} + \frac{\partial v}{\partial y} = 0 \tag{11.73}$$

$$u\frac{\partial u}{\partial x} + v\frac{\partial u}{\partial y} = \nu\frac{\partial^2 u}{\partial y^2} \tag{11.74}$$

$$u = v = 0 \quad \text{at} \quad y = 0 \quad \text{and} \quad u \to U \quad \text{as} \quad y \to \infty \tag{11.75}$$

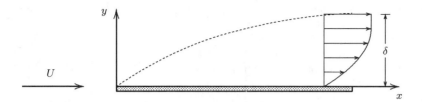

Figure 11.12: *Boundary layer on a flat plate.*

This problem was the first example demonstrating Prandtl's boundary-layer theory, and was solved by H. Blasius in his Ph.D. thesis. Blasius found that the constant-pressure boundary layer is **self similar**. This means that velocity profiles retain their shape at all streamwise positions when displayed in terms of appropriately scaled parameters. Specifically, the streamwise velocity, for example, behaves according to $u(x,y) = u_o(x)f(\eta)$, where $\eta = y/\delta(x)$. The partial differential equations can be transformed to ordinary differential equations and boundary conditions that permit solving for $u_o(x)$, $\delta(x)$ and $f(\eta)$. The Blasius solution is most conveniently obtained in terms of the streamfunction, which assumes the following form.

$$\psi(x,y) = \sqrt{2U\nu x}\, G(\eta), \qquad \eta \equiv y\sqrt{\frac{U}{2\nu x}} \qquad (11.76)$$

Recalling that $u = \partial\psi/\partial y$ and $v = -\partial\psi/\partial x$, substitution into the equations of motion and the boundary conditions yields the following two-point boundary value problem for the dimensionless streamfunction, $G(\eta)$.

$$\frac{d^3 G}{d\eta^3} + G\frac{d^2 G}{d\eta^2} = 0 \qquad (11.77)$$

$$\left.\begin{array}{l} G = 0, \quad G' = 0, \quad \eta = 0 \\[2mm] G' \to 1, \qquad\qquad \eta \to \infty \end{array}\right\} \qquad (11.78)$$

Although the solution for $G(\eta)$ cannot be written in closed form, it is readily obtained by numerical methods. Defining the boundary-layer edge as the point where $u = 0.99u_e$, we find $\eta \approx 3.5$ at this point (which corresponds to $y = \delta$), so that

$$\delta \approx 5\sqrt{\frac{\nu x}{u_e}} = \frac{5x}{\sqrt{Re_x}} \qquad (11.79)$$

This confirms our earlier estimate [Equation (11.43)] of the boundary-layer thickness. Note in passing that this definition is a bit arbitrary, and therefore not worth quoting to high accuracy. Furthermore, it would be quite difficult to determine δ to such precision experimentally. Consequently, δ is less reliable than other boundary-layer length scales such as momentum thickness and displacement thickness.

Forming the appropriate integrals, we can compute the momentum thickness, displacement thickness and shape factor. We find

$$\delta^* = \frac{1.721x}{\sqrt{Re_x}}, \qquad \theta = \frac{0.664x}{\sqrt{Re_x}}, \qquad H = 2.59 \qquad (11.80)$$

Hence, for the constant-pressure boundary layer, the displacement thickness is about a third of the boundary-layer thickness, while momentum thickness is approximately an eighth of δ.

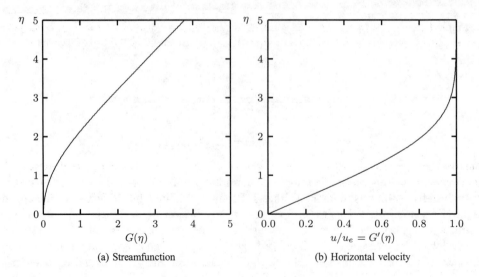

(a) Streamfunction   (b) Horizontal velocity

Figure 11.13: *Blasius solution.*

Because both δ^* and θ grow at the same rate, the shape factor remains constant. Another quantity of interest is the surface shear stress, τ_w. From the Blasius solution, there follows

$$\tau_w = \mu \left(\frac{\partial u}{\partial y} \right)_{y=0} = \frac{\mu U G''(0)}{\sqrt{2\nu x/U}} = 0.332 \frac{\rho U^2}{\sqrt{Re_x}} \tag{11.81}$$

where we use the fact that the numerical solution (see Figure 11.13) gives $G''(0) = 0.332\sqrt{2}$. We can recast this in dimensionless form by introducing the **skin-friction coefficient**, c_f, defined as

$$c_f \equiv \frac{\tau_w}{\frac{1}{2}\rho U^2} \tag{11.82}$$

Thus, for the Blasius boundary layer, the skin friction is

$$c_f = \frac{0.664}{\sqrt{Re_x}} \tag{11.83}$$

To determine the drag force on one side of a plate of length L, we can integrate the surface shear stress [Equation (11.81)] from the leading to trailing edges of the plate. Denoting the drag force by F_D, we have

$$
\begin{aligned}
F_D &= \int_0^L \tau_w \, dx = 0.332 \rho U^2 \int_0^L \frac{dx}{\sqrt{Re_x}} \\
&= 0.332 \rho U^2 \frac{\nu}{U} \int_0^{Re_L} \frac{dRe_x}{\sqrt{Re_x}} = 0.664 \frac{\rho U^2 L}{Re_L} \sqrt{Re_L} \\
&= 0.664 \frac{\rho U^2 L}{\sqrt{Re_L}}
\end{aligned}
\tag{11.84}
$$

Thus, the drag coefficient for a plate of length L is

$$C_D = \frac{F_D}{\frac{1}{2}\rho U^2 L} = \frac{1.328}{\sqrt{Re_L}} \tag{11.85}$$

Example 11.5 *Assuming that the Blasius solution applies along each spanwise location, z, compute the drag coefficient, $C_D = D/(\frac{1}{2}\rho U^2 A)$, where $A = \frac{1}{2}L^2$, for the triangular flat plate shown.*

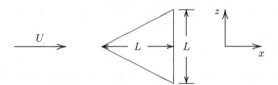

Solution. In order to use the Blasius solution, which is strictly valid for a two-dimensional geometry (i.e., x and y, not x, y and z), we compute the differential drag, dD, on a strip of width dz as shown below. Then, we can integrate the result over z to obtain the drag, D.

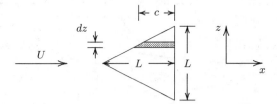

At a point where the chord length is c, the drag from both surfaces for a strip of width dz is

$$
\begin{aligned}
dD &= 2dz \int_0^c \tau_w\, dx = 2dz \int_0^c \frac{1}{2}\rho U^2 c_f dx = \rho U^2 dz \int_0^c \frac{0.664}{\sqrt{Re_x}} dx \\
&= 0.664 \rho U^2 dz \frac{\nu}{U} \int_0^{Re_c} \frac{dRe_x}{\sqrt{Re_x}} = 0.664 \mu U dz \left. \left(2\sqrt{Re_x} \right) \right|_{Re_x=0}^{Re_x=Re_c} \\
&= 1.328 \mu U \sqrt{Re_c}\, dz
\end{aligned}
$$

Now, for this triangular flat plate, the arc length is

$$
c = L(1 - 2z/L)
$$

Hence, the differential drag for a strip of width dz is

$$
dD = 1.328 \mu U \sqrt{Re_L} \sqrt{1 - 2z/L}\, dz
$$

So, taking account of the symmetry about $z = 0$, we have

$$
\begin{aligned}
D &= 2 \int_0^{L/2} dD = 2(1.328)\mu U \sqrt{Re_L} \int_0^{L/2} \sqrt{1 - 2z/L}\, dz \\
&= 1.328 \mu U L \sqrt{Re_L} \int_0^1 \sqrt{1 - \xi}\, d\xi \qquad (\xi \equiv 2z/L) \\
&= 1.328 \mu U L \sqrt{Re_L} \left. \left(-\frac{2}{3} \right) (1 - \xi)^{3/2} \right|_{\xi=0}^{\xi=1} \\
&= 0.885 \mu U L \sqrt{Re_L}
\end{aligned}
$$

Finally, the drag coefficient, C_D, is

$$
C_D = \frac{D}{\frac{1}{4}\rho U^2 L^2} = \frac{3.54 \mu U L \sqrt{Re_L}}{\rho U^2 L^2} = \frac{3.54}{\sqrt{Re_L}}
$$

11.6.5   Effects of Pressure Gradient

In addressing effects of pressure gradient on a boundary layer, Falkner and Skan (1931) found that similarity solutions exist for the boundary-layer equations when the freestream velocity is given by

$$u_e(x) = U_o \left(\frac{x}{L}\right)^m \tag{11.86}$$

where U_o is a characteristic velocity, L is a characteristic length and m is a dimensionless constant. For inviscid flow past a wedge, we found in Subsection 10.7.3 that the velocity on the surface of the wedge varies with distance along the surface raised to a power [Equation (10.173)]. When m lies between 0 and 1, the freestream velocity given in Equation (11.86) corresponds to the surface velocity for flow past a wedge of angle $2m\pi/(m+1)$ as depicted in Figure 11.14.

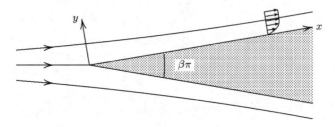

Figure 11.14: *Viscous flow past a wedge.*

The similarity solution for this flow is again most conveniently formulated in terms of the streamfunction. The Falkner-Skan solution is of the form

$$\psi(x,y) = \sqrt{\frac{2\nu U_o L}{m+1}} \left(\frac{x}{L}\right)^{(m+1)/2} G(\eta), \qquad \eta \equiv y\sqrt{\frac{(m+1)U_o}{2\nu L}} \left(\frac{x}{L}\right)^{(m-1)/2} \tag{11.87}$$

Substitution into the boundary-layer equations yields

$$\frac{d^3G}{d\eta^3} + G\frac{d^2G}{d\eta^2} + \beta\left[1 - \left(\frac{dG}{d\eta}\right)^2\right] = 0 \tag{11.88}$$

$$\left.\begin{array}{ll} G = 0, \quad G' = 0, \quad \eta = 0 \\[2mm] G' \to 1, \qquad\qquad \eta \to \infty \end{array}\right\} \tag{11.89}$$

The constant coefficient β is related to the constant m by the simple algebraic relation

$$\beta = \frac{2m}{m+1} \tag{11.90}$$

Equation (11.88) is known as the **Falkner-Skan equation**. Hartree (1937) investigated solutions to the Falkner-Skan equation in detail. His analysis showed that physically realistic solutions exist for $-0.0905 \leq m \leq 2$. There are solutions beyond this range as well that are summarized by Stewartson (1954). Figure 11.15 shows four velocity profiles corresponding to some of the most interesting values of the parameter m.

The solutions for $0 < m < 1$ correspond to the wedge shown in Figure 11.14. The solution is physically relevant for flow near the leading edge of a body with a pointed leading edge.

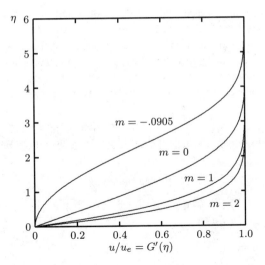

Figure 11.15: *Falkner-Skan solutions.*

When $1 < m < 2$, the wedge angle exceeds $180°$, so that we have what can be described as flow into a corner. As noted by Panton (1996), this type of flow can be very difficult to simulate experimentally. There are also solutions for $m > 2$, but none corresponds to a simple ideal flow.

All of the solutions for $m > 0$ have one thing in common. Specifically, they correspond to decreasing pressure, which we refer to as a **favorable pressure gradient**. The description of decreasing pressure as being favorable has its origin in the effect varying pressure has on a boundary layer. Because decreasing pressure accelerates the flow, especially the low-momentum portion of the boundary layer closest to the surface, favorable pressure gradient energizes the flow, and poses no threat to the boundary layer.

By contrast, for $m < 0$, the pressure increases in the streamwise direction. We describe this type of flow as having an **adverse pressure gradient**. An adverse pressure gradient opposes the motion, and most strongly affects the low-momentum fluid near the surface. If the adverse gradient is too strong, it can cause a reversal in flow direction. When this occurs, we say that the boundary layer **separates** from the surface.

The Falkner-Skan equation has solutions for adverse pressure gradients with m in the range $-0.0905 < m < 0$. This corresponds to flow past a convex corner (Figure 11.16) with an angle as large as $18°$. Hartree originally found **attached-flow** solutions for m in this range, i.e., solutions with positive velocity throughout the boundary layer. When $m = -0.0905$, corresponding to $\beta = -0.199$ (so that $\beta\pi/2 = 18°$), there is a solution with $\partial u/\partial y = 0$ at the surface. This means the skin friction is exactly zero, and we call this incipient separation.

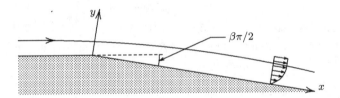

Figure 11.16: *Falkner-Skan flow past a convex corner.*

Stewartson (1954) later discovered that for $-0.0905 < m < 0$, **reverse-flow** solutions also exist. These solutions have a small region close to the surface with negative velocities. It is doubtful that the Stewartson solutions bear any relation to the physical phenomenon of separation as they still entail nearly **parallel flow**, i.e., a flow that has $v \ll u$. By contrast, when boundary-layer separation occurs, we find the vertical velocity to be comparable in magnitude to the horizontal velocity, and the boundary-layer approximation is not valid.

Example 11.6 *Show that $c_f Re_\theta = $ constant for the Falkner-Skan solution, where $c_f \equiv \tau_w/(\frac{1}{2}\rho u_e^2)$ and $Re_\theta \equiv u_e \theta/\nu$.*

Solution. The streamfunction for the Falkner-Skan solution is

$$\psi(x,y) = \sqrt{\frac{2\nu U_o L}{m+1}} \left(\frac{x}{L}\right)^{(m+1)/2} G(\eta), \qquad \eta \equiv y\sqrt{\frac{(m+1)U_o}{2\nu L}} \left(\frac{x}{L}\right)^{(m-1)/2}$$

Hence, the streamwise velocity, u, is

$$u = \frac{\partial \psi}{\partial y} = \frac{\partial \eta}{\partial y}\frac{\partial \psi}{\partial \eta} = \sqrt{\frac{(m+1)U_o}{2\nu L}} \left(\frac{x}{L}\right)^{(m-1)/2} \sqrt{\frac{2\nu U_o L}{m+1}} \left(\frac{x}{L}\right)^{(m+1)/2} G'(\eta) = u_e G'(\eta)$$

where $u_e = U_o(x/L)^m$ is the freestream velocity for the Falkner-Skan solution. Therefore,

$$\begin{aligned}
\theta &= \int_0^\infty \frac{u}{u_e}\left(1 - \frac{u}{u_e}\right) dy = \sqrt{\frac{2\nu L}{(m+1)U_o}} \left(\frac{x}{L}\right)^{(1-m)/2} \int_0^\infty G'(\eta)\left[1 - G'(\eta)\right] d\eta \\
&= \sqrt{\frac{2\nu L}{(m+1)}} \sqrt{\frac{x}{L}} \sqrt{\frac{1}{U_o(x/L)^m}} \int_0^\infty G'(\eta)\left[1 - G'(\eta)\right] d\eta = I\sqrt{\frac{2\nu x}{(m+1)u_e}}
\end{aligned}$$

where I is the numerical value of the integral. Hence, the Reynolds number based on θ is

$$Re_\theta = I\sqrt{\frac{2}{m+1}Re_x}$$

where $Re_x = u_e x/\nu$. Turning to the skin friction, note first that the velocity gradient at $y = 0$ is

$$\left(\frac{\partial u}{\partial y}\right)_{y=0} = u_e \frac{\partial \eta}{\partial y} G''(0) = u_e \sqrt{\frac{(m+1)U_o}{2\nu L}} \left(\frac{x}{L}\right)^{(m-1)/2} G''(0)$$

Therefore, the skin friction is

$$\begin{aligned}
c_f &= \frac{2\mu(\partial u/\partial y)_{y=0}}{\rho u_e^2} = \frac{2\mu}{\rho u_e^2} u_e \sqrt{\frac{(m+1)U_o}{2\nu L}} \left(\frac{x}{L}\right)^{(m-1)/2} G''(0) \\
&= 2\frac{\nu}{u_e} \sqrt{\frac{(m+1)U_o}{2\nu L}} \left(\frac{x}{L}\right)^{(m-1)/2} G''(0) = 2\frac{\nu}{u_e} \sqrt{\frac{(m+1)}{2\nu L}} \sqrt{U_o\left(\frac{x}{L}\right)^m} \sqrt{\frac{L}{x}} G''(0) \\
&= 2\frac{\nu}{u_e} \sqrt{\frac{(m+1)}{2\nu L}} \sqrt{u_e} \sqrt{\frac{L}{x}} G''(0) = \sqrt{\frac{2(m+1)\nu}{u_e x}} G''(0) = \sqrt{\frac{2(m+1)}{Re_x}} G''(0)
\end{aligned}$$

Finally, combining the results obtained for c_f and Re_θ, and noting that I and $G''(0)$ are constants, we conclude that

$$c_f Re_\theta = \sqrt{\frac{2(m+1)}{Re_x}} G''(0) \cdot I\sqrt{\frac{2}{m+1}Re_x} = 2I G''(0) = \text{constant}$$

11.6.6 Boundary-Layer Separation

On the one hand, because of the retarding effect of the viscous stress, the flow within a boundary layer has a smaller velocity than in the inviscid freestream. The stress is largest near the surface where the velocity gradient is largest, so that fluid particles close to the surface have far less momentum than those in the freestream. On the other hand, since the pressure is constant through a boundary layer, the pressure force acts with the same magnitude on all fluid particles within the layer. When the pressure acts to decelerate the flow, i.e., when an adverse pressure gradient exists, the low-momentum fluid particles nearest the surface are most easily stopped. If the adverse pressure gradient is strong enough, the flow will reverse direction. As discussed in the previous subsection, when this occurs we say the boundary layer has **separated**. By definition, the **separation point** in a steady, two-dimensional flow is the point where $\partial u/\partial y = 0$ at the surface. Separation in three dimensions and in unsteady flows is more complex and beyond the scope of this text.

We can use a simple geometrical argument to demonstrate why having an adverse pressure gradient is a necessary (but not a sufficient) condition for separation. We begin by noting that at a solid boundary, the convective terms in the x-momentum equation vanish (because $u = v = 0$), wherefore

$$\mu \left(\frac{\partial^2 u}{\partial y^2} \right)_{y=0} = \frac{dp}{dx} \tag{11.91}$$

For a favorable pressure gradient we have $dp/dx < 0$, so that the second derivative of u is negative at the surface. Since we expect the maximum velocity to occur at the boundary-layer edge, necessarily $\partial^2 u/\partial y^2 < 0$ at the edge as well. Thus, in a favorable pressure gradient, the curvature of the velocity profile, $\partial^2 u/\partial y^2$, is everywhere negative. The Falkner-Skan profiles for $m = 1$ and $m = 2$ shown in Figure 11.15 are examples of uninflected velocity profiles with negative curvature. Similarly, for zero pressure gradient (the $m = 0$ case in Figure 11.15), there is an inflection point at the surface, i.e., a point where $\partial^2 u/\partial y^2 = 0$, while the curvature is negative at all other points in the boundary layer.

By contrast, in an adverse pressure gradient we have $dp/dx > 0$, wherefore the second derivative of u is positive at the surface. Consequently, since $\partial^2 u/\partial y^2 < 0$ at the boundary-layer edge, there must be an inflection point somewhere in the boundary layer. When an inflection point is present, the boundary-layer velocity profile shape changes as illustrated in Figure 11.17. Profile A corresponds to constant pressure. The boundary layer depicted then encounters an adverse pressure gradient and develops an inflection point above the surface as shown in Profile B. A bit farther downstream, the boundary layer separates and assumes the shape of Profile C. Beyond this point, there is reverse flow near the surface corresponding to

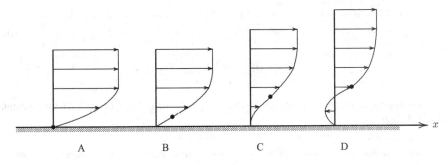

Figure 11.17: *Boundary-layer velocity profiles near separation; • denotes an inflection point.*

Profile D. Clearly the separated-flow profile must have an inflection point, thus demonstrating that adverse pressure gradient is a necessary condition for separation. It is not a sufficient condition, i.e., it does not guarantee separation, as a boundary layer can remain attached for mild adverse gradients (e.g., the Falkner-Skan cases with $-0.0905 < m < 0$).

Separation is generally undesirable as it leads to increased drag for devices such as wings. Separation occurs on a wing for too large an angle of attack, a condition referred to as **stall**. It causes not only an increase in drag, but a loss of lift as well. This condition is a catastrophic failure, and poses a limit on the angle of attack and load carrying capacity of an airplane. Figure 11.18 schematically contrasts unseparated, or **attached**, flow past an airfoil with massively separated flow corresponding to stall. For the attached case [Figure 11.18(a)], the flow leaves the trailing edge smoothly and enters a thin viscous layer known as a wake. When the airfoil stalls [Figure 11.18(b)], the flow separates on the upper surface, and the boundary layer abruptly breaks away. A large region of complicated eddying motions forms in the wake of the airfoil.

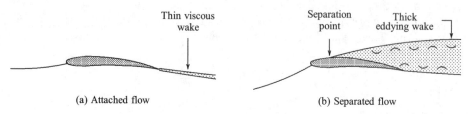

(a) Attached flow   (b) Separated flow

Figure 11.18: *Attached and separated flow past an airfoil.*

The loss of lift is caused by the grossly altered shape presented to the inviscid flow. The effective upper surface is the edge of the wake, which corresponds to a much thicker airfoil. Clearly, the effective angle of attack seen by the upper surface is greatly reduced. This increases the low pressure that prevails on the upper, or **suction surface** of an airfoil that provides most of the lift.[6] Since lift is proportional to the angle of attack, the reduced angle accounts for the dramatic loss of lift. This is an extreme example of the displacement effect discussed at the end of Subsection 11.6.3.

Separation occurs for internal flows also, such as in a rapidly expanding duct or pipe. This is the cause of the head loss for a sudden expansion in ducts and pipes discussed in Subsection 6.4.2. Figure 11.19 illustrates how separation alters the flow. The adverse pressure gradient encountered as the flow expands causes an extended region of separated flow. The presence of the boundary layer, and its susceptibility to separation, thus brings about a severe distortion in the overall flow, including strong viscous losses. This is the source of the head loss.

The contour bounding the separated region in Figure 11.19 is the locus of points, y_o, below which the net mass flux in the streamwise direction is zero, i.e.,

$$\int_{y_w}^{y_o} \rho u \, dy = 0 \qquad (11.92)$$

where y_w is the value of y on the duct wall. It is called the **dividing streamline**, and separates the recirculating flow from the flow in the central portion of the duct or pipe. The point where the dividing streamline intersects the surface downstream of the separation point is called the **reattachment point**. The flow downstream of reattachment has no reverse-flow region, and

---

[6]The high pressure that develops on the lower or **pressure surface** typically provides 20% or less of the lift.

gradually approaches a new equilibrium state. The flow eventually reattaches when the adverse pressure gradient is removed. In our sudden-expansion example, the pressure approaches a nominally constant value downstream of the corner, so that $dp/dx \to 0$.

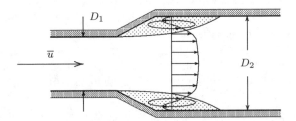

Figure 11.19: *Separation in duct or pipe flow at a sudden expansion; the flow is separated in the shaded region. As illustrated, a streamline in the separation bubble is a closed contour.*

A boundary-layer computation that uses a measured pressure distribution provides an accurate prediction for the location of separation. However, the boundary-layer equations are singular at a separation point, with the vertical velocity and displacement thickness becoming infinite. This is true provided we insist upon specifying the pressure. The singularity is not an inherent property of the boundary-layer equations, but rather is a consequence of specifying dp/dx. Methods have been developed that specify either the displacement thickness or the surface shear stress and determine the pressure as part of the solution [cf. Panton (1996)]. These are called **inverse boundary-layer methods**, and require iteration between the inviscid and boundary-layer solutions to achieve compatible pressure distributions. These methods are limited to small separated regions.

The boundary-layer equations fail completely for massively-separated flows. A stalled airfoil, for example, cannot be treated with a boundary-layer computation. The same is true for flow past a **bluff body** such as an automobile, which has a more-or-less blunt geometry at its "trailing edge."

With the increasing power of computers, approximate procedures have largely given way to full Navier-Stokes solutions. This is appropriate from a theoretical point of view as the boundary-layer equations are unsuitable for describing separation. Recalling the basic premises underlying the boundary-layer approximation, they are inappropriate for two key reasons. First, the velocity components in separated regions are both of the same order of magnitude, while the boundary-layer approximation assumes $v \ll u$. Second, the pressure varies normal to the surface in a separated region wherefore the boundary-layer approximation that $\partial p/\partial y = 0$ is violated.

11.7 Turbulence

For "low enough" velocities, in the sense that the Reynolds number is not too large, the equations of motion for a viscous fluid have well-behaved, steady solutions. We have seen viscous-flow solutions for Couette flow (Subsection 1.9.1), pipe flow (Subsection 1.9.2), constant-pressure boundary-layer flow (Subsection 11.6.4) and boundary-layer flow with pressure gradient (Subsection 11.6.5). For all of these solutions, the motion is termed **laminar** and can be observed both experimentally and in nature.

At larger Reynolds numbers, the viscous stresses are completely overwhelmed by the fluid's inertia, and the laminar motion becomes unstable. Rapid temporal velocity and pressure fluctuations appear and the motion becomes inherently unsteady. When this occurs, we describe the motion as being **turbulent**. In the cases of fully-developed Couette flow and pipe flow, for example, laminar flow is assured only if the Reynolds number based on channel height and pipe radius is less than 1500 and 2300, respectively.

Virtually all flows of practical engineering interest are turbulent. Turbulent flows always occur when the Reynolds number is large. For slightly viscous fluids such as water and air, large Reynolds number corresponds to anything stronger than a small breeze or a puff of wind. Thus, to analyze fluid motion for general applications, we must deal with turbulence. In this section, we will explore a few salient aspects of this phenomenon.

11.7.1   Laminar, Transitional and Turbulent Flow

Prior to 1930, experimentalists lacked instrumentation capable of sensing rapid velocity and pressure fluctuations that are always present in turbulent flows. Rather, they determined mean values (i.e., statistical averages) of flow properties. All flow properties undergo significant changes between laminar and turbulent regimes. In terms of mean measurements, early experimenters found quite different velocity profiles in pipe flow.

Figure 11.20 schematically compares laminar and mean turbulent velocity profiles. As discussed in Section 1.9.2, the laminar profile varies quadratically with pipe radius. By contrast, the turbulent-flow profile is much more filled out, appearing nearly uniform across most of the pipe. The slope of the velocity profile is much steeper at the pipe surface for the turbulent profile, corresponding to a much larger friction factor. Similar differences in velocity profiles are present for boundary layers.

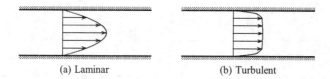

(a) Laminar   (b) Turbulent

Figure 11.20: *Pipe-flow velocity profiles.*

When measurements cover a wide enough range of Reynolds numbers to include both laminar and turbulent flow, dramatic changes in measured properties are observed as the motion changes from laminar to turbulent. The change in the character of the motion is called **transition**, and generally occurs over a finite streamwise distance. In the case of the flat-plate boundary layer, for example, the Blasius laminar-flow solution is observed experimentally for Reynolds number based on plate length, Re_x, of at least $9 \cdot 10^4$. The precise point where transition begins depends on a variety of factors including the level of turbulence in the freestream and how rough the surface is.

Figure 11.21 displays the skin friction for a flat-plate boundary layer with Re_x ranging from 10^3 to 10^8. As shown, for large Reynolds numbers, the skin friction is significantly larger than the Blasius value ($c_f = 0.664 \, Re_x^{-1/2}$) that we determined in Subsection 11.6.4. Also, the skin friction falls off roughly as $Re_x^{-1/5}$ (the exact origin of x depending on the transition point), which is much more gradual than the Blasius variation.

The same kind of abrupt change in properties is observed for internal flows. Examination of the Moody diagram for flow in a pipe, Figure 6.7, shows a similar variation of the friction factor, f, with Reynolds number. At very low Reynolds numbers, the laminar-flow solution matches measured values. As Reynolds number increases, the flow undergoes transition, and

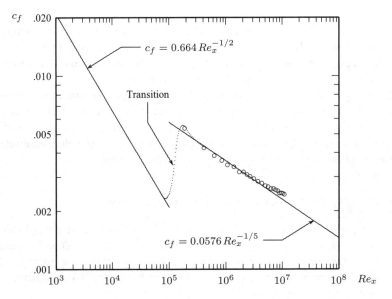

Figure 11.21: *Skin friction for a flat-plate boundary layer; ○ measurements of Wieghardt [see Coles and Hirst (1969)].*

the friction factor asymptotes to values significantly above the laminar prediction. As with the boundary layer, the turbulent-flow rate of decrease of f with Reynolds number is less rapid than the laminar rate.

In 1883 Reynolds performed a simple experiment (see Figure 11.22) that demonstrates the existence of the markedly different behavior of laminar and turbulent flow. Using a transparent pipe, he injected dye along the centerline. He adjusted the flow rate to control the Reynolds number, Re_D, in the pipe. For $Re_D < 2300$, the dye stream is straight and, aside from a minor amount of diffusion, remains unmixed with the surrounding fluid. As Re_D increases beyond 2300, the dye stream suddenly becomes wavy. When the Reynolds number increases still further, the distinct dye stream disappears completely and the dye spreads uniformly through the pipe. Figure 11.22 illustrates this process in sketches similar to those given by Reynolds (1883).

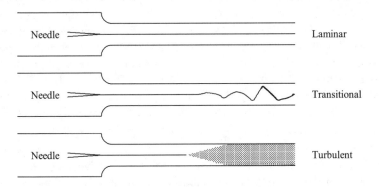

Figure 11.22: *Transition from laminar to turbulent flow in a pipe as revealed by continuous injection of dye.*

Since 1930, instrumentation has been developed that permits detailed measurement of flow properties revealing important aspects of turbulence. Devices such as the *hot-wire anemometer* and, more recently, the *laser-Doppler anemometer* provide instantaneous measurements sensitive enough to record a wide range of frequencies. This is important because turbulence includes oscillatory motions of a very wide range of frequencies, and the range increases with Reynolds number. If we place one of these devices in the pipe flow analyzed by Reynolds, for example, we can measure the velocity as a function of time.

Suppose we make our observations at a point a distance r from the pipe centerline near the downstream end of the pipe shown in Figure 11.22. For the three cases—laminar, transitional and turbulent—the measured velocity as a function of time, t, would be as illustrated in Figure 11.23.

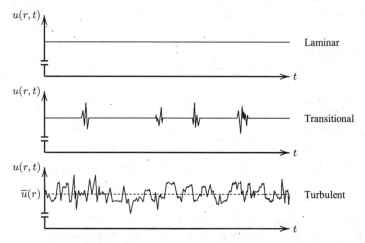

Figure 11.23: *Velocity versus time for laminar, transitional and turbulent pipe flow. Fluctuations are typically less than 10% of the mean value.*

For laminar conditions, the velocity has a constant value at each point across the pipe. The flow is steady and completely consistent with the laminar-flow solution developed in Section 1.9.2. In the transitional case, the velocity shows sudden bursts of unsteadiness separated by finite intervals during which the velocity is constant. When the flow becomes turbulent, the velocity displays rapid fluctuations in time. The fluctuations include frequencies over a wide range, and are quite irregular and aperiodic. The magnitude of the fluctuations is typically within 10% of the average value, $\bar{u}(r)$. As mentioned above, prior to development of modern, high-frequency, measuring devices, experimenters were able to determine only averaged values such as $\bar{u}$ (e.g., Figure 11.20).

Often, as an approximation, turbulent boundary-layer profiles are represented by a **power-law** relationship. That is, we sometimes say

$$\frac{\bar{u}}{u_e} = \left(\frac{y}{\delta}\right)^{1/n} \tag{11.93}$$

where n is typically an integer between 6 and 8. A value of $n = 7$ yields a good approximation at high Reynolds number for the flat-plate boundary layer. Figure 11.24 compares a 1/7 power-law profile with measured values. As shown, the agreement between measured values and the approximate profile is excellent with differences everywhere less than 3%. The figure includes the Blasius profile to show the contrast between laminar- and turbulent-flow velocity profiles.

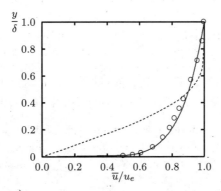

Figure 11.24: *Power-law velocity profile;* —— $\bar{u}/u_e = (y/\delta)^{1/7}$;∘ *Wieghardt data at* $Re_x = 1.09 \cdot 10^7$ *[Coles and Hirst (1969)];* - - - *Blasius (laminar).*

For a constant-pressure boundary layer, the measured boundary-layer thickness, δ, is

$$\delta \approx 0.37 x Re_x^{-1/5} \tag{11.94}$$

Using the 1/7 power-law profile, we can compute the displacement thickness, δ^*, momentum thickness, θ, and shape factor, H. Table 11.1 lists values for constant-pressure laminar and turbulent boundary layers. Based on these results, we can see immediately how much more filled out a turbulent boundary layer is relative to the laminar profile. Using values from the table, the ratio of displacement thickness to boundary-layer thickness is

$$\frac{\delta^*}{\delta} = \begin{cases} 0.344, & \text{Laminar} \\ 0.124, & \text{Turbulent} \end{cases} \tag{11.95}$$

Hence, the effective displacement is reduced from a little more than 1/3 of δ for a laminar boundary layer to 1/8 of δ for the turbulent case.

Table 11.1: *Integral Parameters for Laminar and Turbulent Flat-Plate Boundary Layers*

Property	Laminar Flow	Turbulent Flow
δ	$5.0xRe_x^{-1/2}$	$0.37xRe_x^{-1/5}$
δ^*	$1.721xRe_x^{-1/2}$	$0.046xRe_x^{-1/5}$
θ	$0.664xRe_x^{-1/2}$	$0.036xRe_x^{-1/5}$
H	2.59	1.28
c_f	$0.664Re_x^{-1/2}$	$0.0576Re_x^{-1/5}$

Table 11.1 also includes the skin friction, which follows from the constant-pressure version of the momentum-integral equation [Equation (11.64)]. As shown in Figure 11.21, the power-law inferred skin-friction formula matches measured values for plate-length Reynolds number, Re_x, only up to about 10^6. A more accurate representation of the skin friction is provided by the Kármán-Schoenherr correlation [see Hopkins and Inouye (1971)]:

$$\frac{1}{c_f} = 17.08 \left(\log_{10} Re_\theta\right)^2 + 25.11 \log_{10} Re_\theta + 6.012 \tag{11.96}$$

This correlation more-closely follows the trend of measured skin friction, which exhibits a somewhat reduced rate of decrease with increasing Reynolds number.

Example 11.7 *Compute the skin friction for a turbulent flat-plate boundary layer when $Re_\theta = 10^5$ according to the Kármán-Schoenherr correlation and the formulas from Table 11.1.*

Solution. Using the Kármán-Schoenherr correlation, we have

$$(c_f)_{K-S} = \left(17.08 \cdot 5^2 + 25.11 \cdot 5 + 6.012\right)^{-1} = 0.00179$$

Using the formula for θ in the table, the momentum-thickness Reynolds number is

$$Re_\theta = 0.036 Re_x^{4/5} \quad \Longrightarrow \quad Re_x = \left(\frac{Re_\theta}{0.036}\right)^{5/4} = 1.13 \cdot 10^8$$

Hence, the skin friction is

$$(c_f)_{Table} = 0.0576 Re^{-1/5} = 0.00141 \qquad \text{(21\% lower than Kármán-Schoenherr)}$$

AN EXPERIMENT YOU CAN DO AT HOME
Contributed by Prof. Peter Bradshaw, Stanford University

The curious reader can perform a simple experiment to observe how fast turbulent mixing occurs. Fill a transparent cup or glass with water and allow it to settle. Next, very gently add a few drops of milk or thin cream to the water. Note that non-fat milk won't work—the drops tend to break up. Initially, the milk will sink to the bottom and spread over the bottom of the vessel. After a few seconds, a steady state will be reached in which no motion is evident, and most of the water will be clear. Although not evident to the naked eye, mixing of the milk and water proceeds at the molecular level. If the water is three inches in depth, you would have to wait approximately three weeks for the mixing to be complete. The time required for laminar mixing is proportional to the square of the size of the container, so that adding the milk to a six-inch glass of water would require just over three months to mix.

Now, dip a spoon into the cup or glass and remove it. Make no attempt to stir the mixture. You will observe a rather remarkable phenomenon. Within a few seconds, the milk and water will be completely mixed! The mixing occurs as a result of the turbulence in the wake of the spoon. To observe the turbulent-mixing process in more detail, use a larger container such as a fish bowl, being sure to remove the fish first. The mixing will take longer, and the turbulent eddies should be readily observable.

11.8   Lift and Drag of Common Objects

The forces exerted by a fluid on an object moving through it are due to the pressure, which acts normal to the surface of the object, and the viscous stresses, which act both normal and tangent to the surface. Recall that we can symbolically write the surface force exerted by a fluid on an object, $\mathbf{F}_s$, with outer unit normal $\mathbf{n}$ as

$$\mathbf{F}_s = -p\mathbf{n} + \mathbf{f}_v \tag{11.97}$$

where p is the pressure and $\mathbf{f}_v$ is the viscous force. In general, the viscous contribution to the normal force is very small compared to the pressure, and is exactly zero at the surface. So, we need only consider the tangential, or shear, stress in computing forces. Thus, letting $\mathbf{t}$ denote a unit vector tangent to the surface, we can say $\mathbf{f}_v = \tau_w \mathbf{t}$, where τ_w is the surface shear stress. Then, the net force, $\mathbf{F}$, follows from integrating over the surface of the object, viz.,

$$\mathbf{F} = \oiint_S [-p\mathbf{n} + \mathbf{f}_v]\, dS = \oiint_S [-(p - p_\infty)\mathbf{n} + \tau_w \mathbf{t}]\, dS \qquad (11.98)$$

where we reference the pressure to its freestream value, p_∞. Recall from Subsection 5.4.2 that we can always subtract a constant value from the pressure with no effect on the pressure integral over a closed surface.

As noted in Section 11.4, the component of the net force parallel to the freestream flow direction is the **drag**, D, while the component normal to the freestream flow direction is the **lift**, L. For objects that are symmetric about the xy plane (so that $\mathbf{F}$ has no z component), we can write the net force as $\mathbf{F} = D\,\mathbf{i} + L\,\mathbf{j}$, where we assume the freestream flow moves parallel to the x axis. Therefore, the drag and lift are given by

$$D = \mathbf{i} \cdot \oiint_S [-(p - p_\infty)\mathbf{n} + \tau_w \mathbf{t}]\, dS \qquad (11.99)$$

$$L = \mathbf{j} \cdot \oiint_S [-(p - p_\infty)\mathbf{n} + \tau_w \mathbf{t}]\, dS \qquad (11.100)$$

Figure 11.25 shows the pressure and viscous forces that act on an airfoil. The pressure is less than the freestream value on the upper surface so that the vector $-(p - p_\infty)\mathbf{n}$ points away from the airfoil. Since the low pressure effectively "sucks" the airfoil upward, we refer to this as the **suction surface** or the **suction side**. By contrast, the pressure on the lower surface is greater than the freestream value so that $-(p - p_\infty)\mathbf{n}$ points toward the airfoil, and we call this the **pressure surface** or **pressure side**. The magnitudes shown are typical of a modern airfoil, with both upper and lower surfaces contributing to the lift, the larger contribution coming from the upper surface.

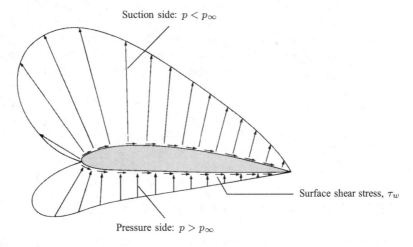

Figure 11.25: *Pressure and viscous forces acting on an airfoil.*

Traditionally, for incompressible flow, we consider the drag on an object to consist of two separate contributions. The first is the **pressure drag**, corresponding to the part of the drag integral [Equation (11.100)] involving only the pressure. The second contribution is the **skin-friction drag** or **friction drag** for short, corresponding to the part of the drag integral involving only the surface shear stress. Their sum is referred to as the **form drag**. Keep in mind that distinguishing between pressure and skin-friction drag in this manner does not imply that the former is present even in an inviscid fluid. That is, the pressure drag is exactly zero in the limit of zero viscosity, i.e., in the limit of an inviscid fluid. The only reason the pressure yields a nonzero drag-force contribution is because viscous effects have altered the flow to the extent that the surface pressures are significantly different from the pressures that would prevail in an inviscid fluid.

Most lift and drag data are presented in terms of the dimensionless lift and drag coefficients, C_L and C_D, respectively, defined as

$$C_L \equiv \frac{L}{\frac{1}{2}\rho U^2 A} \quad \text{and} \quad C_D \equiv \frac{D}{\frac{1}{2}\rho U^2 A} \tag{11.101}$$

where A is the so-called **frontal area**, ρ is the freestream fluid density and U is the freestream velocity. The frontal area is usually the projection of the object on a plane normal to the flow direction. To avoid ambiguity, the frontal area is usually quoted with published data.

Certainly the most important context in which lift arises is the wing and, in two-dimensional flow, the airfoil. Because an airfoil is a two-dimensional object, the conventional manner in which C_L and C_D are defined requires explanation. For an airfoil, which is simply a cross section of a wing, L and D are the lift and drag per unit span. The lift and drag coefficients are formed with respect to the chord length, c, which is the distance from leading to trailing edge. That is, we replace A with c in the definitions of C_L and C_D [Equation (11.101)].

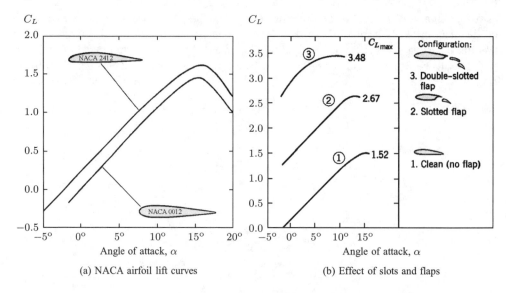

(a) NACA airfoil lift curves   (b) Effect of slots and flaps

Figure 11.26: *Airfoil lift coefficients.*

One of the most comprehensive collections of lift (and drag) data for wings and airfoils is contained in the book by Abbott and von Doenhoff (1959). Figure 11.26, patterned after similar figures in the Abbott-von Doenhoff text, show lift coefficients as a function of airfoil angle of attack, α, i.e., the angle at which the airfoil is inclined to the freestream flow direction. Figure 11.26(a) shows the measured variation of C_L with α for two NACA airfoils.

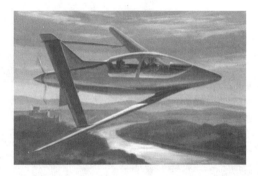

Figure 11.27: *Joined-wing airplane designed by Julian Wolkovich. In addition to being very strong structurally, such aircraft can achieve very high lift to drag ratio.*

In both cases the lift coefficient increases with α up to a maximum value, $C_{L_{max}}$. Any further increase in α causes C_L to decrease, a condition known as **stall**. An airfoil stalls because of flow separation on the upper surface of the airfoil, which also causes an increase in drag. Figure 11.26(b) illustrates the effect of slots and flaps on another NACA airfoil, namely the NACA 23012. Flaps, sketched in the inset, are parts of a wing that can be extended to increase wing area, and hence lift, during takeoff and landing. As shown, the lift coefficient can be more than doubled through the use of a double-slotted flap.

Drag is generally a function of Reynolds number, especially at relatively small values. For high Reynolds number, the drag coefficient on many objects approaches a constant value. The drag coefficient for airfoils is much smaller than the lift coefficient. In fact, the lift-to-drag ratio, C_L/C_D, for modern high-performance airfoils and exotic designs such as the Wolkovich joined-wing airplane shown in Figure 11.27 can actually exceed 100. For comprehensive collections of drag data, see Morrison (1962), Hoerner (1965) or Blevins (1984).

Figure 11.28 shows the **drag coefficient**, C_D, for cylinders and spheres as determined from wind-tunnel experiments. Figure 11.29 includes high-Reynolds-number drag coefficients for an assortment of three-dimensional objects.

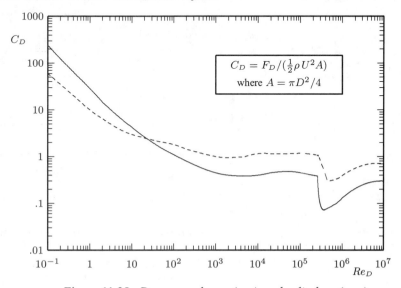

Figure 11.28: *Drag on spheres (——) and cylinders (- - -).*

Cube				
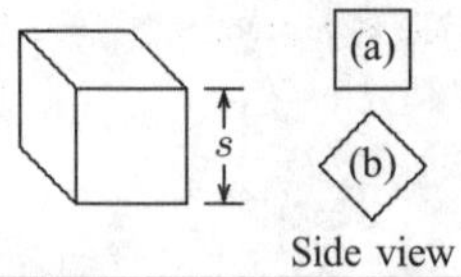 Side view	$A = s^2$		Orientation: (a) Unrotated (b) Rotated 45°	C_D 1.08 0.81

Rectangular plate				
	$A = bh$	b/h 1 5 10 20 ∞	C_D 1.18 1.20 1.30 1.50 2.00	

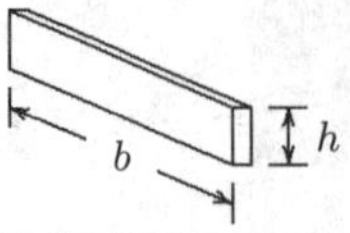

Hemisphere				
(a) (b)	$A = \dfrac{\pi}{4}d^2$		Flat side facing: (a) Upstream (b) Downstream	C_D 1.42 0.38

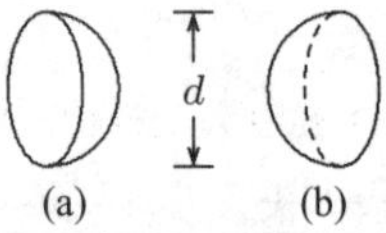

Ellipsoid				
	$A = \dfrac{\pi}{4}d^2$	ℓ/d 0.75 1 2 4 8	C_D 0.20 0.20 0.13 0.10 0.08	

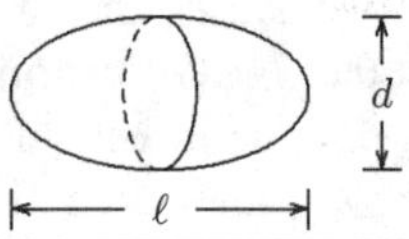

Cone				
	$A = \dfrac{\pi}{4}d^2$	θ 10° 30° 60° 90°	C_D 0.30 0.55 0.80 1.15	

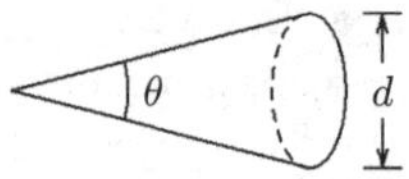

Flat-Faced Cylinder				
	$A = \dfrac{\pi}{4}d^2$	ℓ/d 0.5 1 2 4 8	C_D 1.15 0.90 0.85 0.87 0.99	

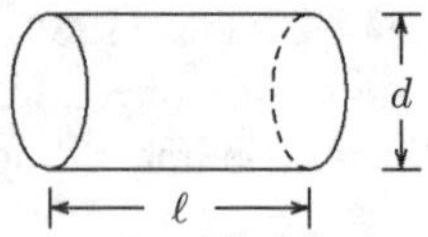

Thin Disk				
	$A = \dfrac{\pi}{4}d^2$		$C_D = 1.10 - 1.17$	

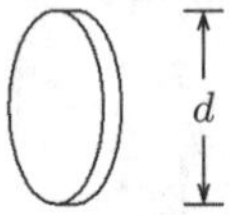

Parachute				
	$A = \dfrac{\pi}{4}d^2$		$C_D = 1.20 - 1.40$	

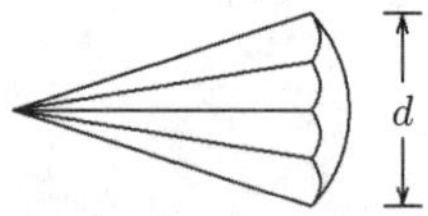

Waving Flag				
	$A = \ell h$	ℓ/h 1 2 3	C_D 0.07 0.12 0.15	

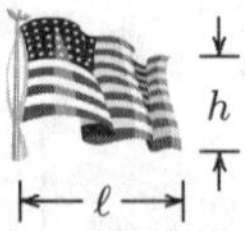

Figure 11.29: *Drag-coefficient data for high-Reynolds-number flow past assorted three-dimensional objects.*

Example 11.8 *A flag of size $\ell = 2$ m by $h = 1$ m is flying on a flagpole that is $H = 10$ m high. Estimate the bending moment at the base of the pole if the wind is blowing at $U = 3$ m/sec.*

Solution. If we assume the force on the flag acts through the center of the flag, then the moment on the flagpole, M, due to the aerodynamic drag, D, is

$$M = \left(H - \frac{h}{2} \right) D$$

where H is the height of the pole and h is the vertical width of the flag. Letting $\rho = 1.20$ kg/m^3 denote the air density, the drag is

$$D = \frac{1}{2} \rho U^2 h \ell C_D, \qquad C_D = 0.12$$

Therefore, the moment is

$$M = \frac{1}{2} \rho U^2 h \ell \left(H - \frac{h}{2} \right) C_D$$

So, for the given values, we find

$$M = \frac{1}{2} \left(1.20 \text{ kg/m}^3 \right) (3 \text{ m/sec})^2 (2 \text{ m})(1 \text{ m})(10 \text{ m} - 0.5 \text{ m})(0.12) = 12.3 \text{ N} \cdot \text{m}$$

Problems

11.1 Derive the compressible-flow vorticity equation by starting with the momentum equation in the following form.

$$\frac{\partial \mathbf{u}}{\partial t} + \nabla \left(\frac{1}{2} \mathbf{u} \cdot \mathbf{u} \right) - \mathbf{u} \times \boldsymbol{\omega} = -\frac{\nabla p}{\rho} - \nabla \mathcal{V}$$

(a) Mimic the steps in the text to arrive at a differential equation for $\boldsymbol{\omega}$ in as simplified a form as possible. Use the continuity equation to replace $\nabla \cdot \mathbf{u}$ by a term proportional to $d\rho/dt$.

(b) Rewrite the equation derived in Part (a) as a differential equation for $\boldsymbol{\omega}/\rho$.

(c) How does your result change if the flow is *barotropic*, i.e., if the pressure depends only upon ρ?

11.2 Are there any conditions under which the vortex force is conservative for inviscid, incompressible flow subjected to conservative body forces? **HINT:** A conservative force has zero curl.

11.3 *Beltrami flow* is an idealized type of flow sometimes used in turbomachinery analysis. In this type of flow, the velocity vector, $\mathbf{u}$, is everywhere parallel to the vorticity vector, $\boldsymbol{\omega}$.

(a) What is the vortex force in Beltrami flow?

(b) If the body force is conservative with $\mathbf{f} = -\nabla \mathcal{V}$, determine the pressure, p, as a function of $\mathcal{V}$, $\mathbf{u}$ and density, ρ, for steady, incompressible Beltrami flow.

11.4 There are physical mechanisms for generating vorticity in an inviscid, compressible flow. To see this, we can develop *Crocco's equation* as follows.

(a) Beginning with Gibbs' equation [Equation (6.12)], verify that the entropy, s, is given by

$$T\nabla s = \nabla h - \frac{\nabla p}{\rho}$$

where p, ρ, T and h are pressure, density, temperature and enthalpy, respectively.

(b) For steady flow with no body forces, use Euler's equation to eliminate $\nabla p/\rho$ and verify that

$$\mathbf{u} \times \boldsymbol{\omega} = \nabla h_t - T\nabla s$$

where $h_t = h + \frac{1}{2}\mathbf{u} \cdot \mathbf{u}$ is the total enthalpy. This is *Crocco's equation*, which shows that variations in either h_t or s can lead to creation of vorticity.

(c) Based on this equation, if h_t is constant, what must be true of irrotational flows? What must be true of isentropic flows? The answers to these two questions constitute *Crocco's theorem*.

11.5 A simplified model of a hurricane approximates the flow as a rigid-body rotation in the inner core and a *potential vortex* outside of the core (cf. Section 4.3). That is, the circumferential velocity, u_θ, is

$$u_\theta \approx \begin{cases} \Omega r, & r \leq R \\ \dfrac{\Gamma}{2\pi r}, & r \geq R \end{cases}$$

where r is radial distance, R is core radius, Ω is the angular-rotation rate in the core, and Γ is the circulation. If a hurricane has peak winds of 100 mph and the core radius is 120 ft, what is the circulation? What is the angular-rotation rate (in rpm) in the core?

11.6 Compute the aerodynamic circulation for the following aircraft.

(a) The *Gossamer Condor*, the first human-powered aircraft, for which $L = 2.2$ lb/ft, $U_\infty = 10$ mph and $\rho = 0.00234$ slug/ft^3.

(b) A *Boeing 747* for which $L = 4000$ lb/ft, $U_\infty = 570$ mph and $\rho = 7.0 \cdot 10^{-4}$ slug/ft^3.

11.7 A *Boeing 747* jumbo jet weighs 750000 lb when fully fueled with 150 passengers aboard. To generate sufficient lift to take off at this weight, the 747 must achieve a runway speed of 140 mph. Assuming the circulation, Γ, increases linearly with speed (as it will be for unchanged flap settings and angle of attack), what runway speed is required to take off with 325 passengers? Assume an average passenger with luggage weighs 200 lb.

11.8 Consider viscous flow close to a solid boundary for which

$$\mathbf{u} = U\left[\frac{y}{h} + K\left(\frac{y}{h}\right)^2\right]\mathbf{i}$$

where y is distance normal to the boundary, U is a constant reference velocity and K is a dimensionless constant. Also, h is the thickness of the viscous layer and is independent of x.

(a) Using the closed contour shown, compute the circulation in terms of K, U and Δx.

(b) Is there a value of K for which the circulation is zero?

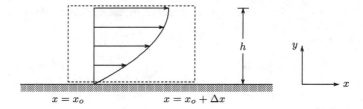

Problem 11.8

11.9 Compute the aerodynamic circulation, Γ_a, for Couette flow, using the dashed rectangular contour shown. Recall from Chapter 1 that the velocity is $u(y) = Uy/h$.

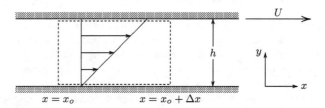

Problem 11.9

11.10 We want to analyze the aerodynamic circulation, Γ_a, for flow in a two-dimensional channel. The velocity profile for channel flow is $u(y) = u_m[1 - 4(y/h)^2]$, where u_m is the maximum velocity.

(a) Using the dashed rectangular contour shown, verify that the circulation is zero.

(b) Now use a rectangular contour that extends from either of the channel walls to the centerline, and compute the nonzero circulation.

(c) Compute the vorticity and, noting that $\Gamma = \iint_A \omega\, dxdy$ where ω is the vorticity component normal to the xy plane, explain the results of Parts (a) and (b).

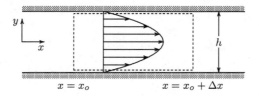

Problem 11.10

11.11 For flow past a flat plate, the boundary layer is as shown. The dotted line depicts the edge of the layer, and the flow is essentially uniform ($\mathbf{u} \approx U\,\mathbf{i}$) and inviscid above the edge. Also, the vertical velocity, v, in the boundary layer is very small compared to the horizontal component, i.e., $v \ll u$. The velocity at $x = L$ is $u(L, y) = Uf(y/\delta)$, where f is a function that has to be determined numerically. The point of this problem is to show that, regardless of the precise details of flow within the boundary layer, the circulation is a function only of the freestream velocity, U, and the plate length, L.

(a) Using the dashed rectangular contour shown, compute the circulation from its definition, $\Gamma = \oint_C \mathbf{u} \cdot d\mathbf{s}$. Neglect any contributions involving v.

(b) Now compute the circulation from $\Gamma = \iint \boldsymbol{\omega} \cdot \mathbf{k}\, dA$, and verify that the answer matches that of Part (a).

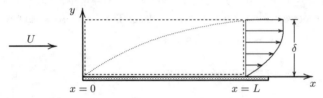

Problem 11.11

11.12 Compute and compare the Reynolds numbers for: (a) a 17-foot long automobile in a traffic jam moving at 7 mph on a day when the temperature is 90° F; and (b) a 25-cm long toy car traveling at 12 m/sec on a day when the temperature is 21° C. Assume the pressure is 1 atm, and use *Sutherland's equation* to determine the viscosity.

11.13 Compute and compare the Reynolds numbers for: (a) a 26-foot long sailboat drifting at 2 knots on a day when the water temperature is 50° F; and (b) a 50-cm long toy motorboat traveling at 10 m/sec on a day when the Jamaican water temperature is 30° C.

11.14 Variations in viscosity with temperature are much more rapid in water than in air. To appreciate just how much of an effect there is and how it might affect viscous forces on an object, compute the Reynolds number on a 100-meter long tanker moving at 15 knots near Hawaii (water temperature 30° C) and near Alaska (water temperature 0° C – but not frozen). Refer to Table A.7 for the kinematic viscosity of water. Repeat your computations with the viscosity of air at the same temperatures. Assume the pressure is 1 atm, and use *Sutherland's equation* to determine the viscosity.

11.15 At what speed must an object of characteristic size 1 foot move to have a Reynolds number of just 10^5 in air and in water? Assume pressure is atmospheric and the temperature is 68° F.

11.16 The phenomenon illustrated in the home experiment described in this chapter is relevant to missiles carrying liquid fuel that spin for directional stability. Detailed analysis shows that during *spin up*, the radial position, r, of the front as it moves from the missile's inner surface to its centerline is given by

$$r = Re^{-t\sqrt{\nu\Omega}/h}$$

where R and h are fuel-compartment radius and height, ν is kinematic viscosity and Ω is rotation rate. Determine the approximate spin-up time, i.e., the time to achieve nearly rigid-body rotation ($r/R \approx .05$), for the following.

(a) A missile with $R = 110$ cm, $h = 200$ cm, $\nu = 1.5 \cdot 10^{-6}$ m²/sec and $\Omega = 10$ sec^{-1}.

(b) The home experiment with $R = 3.5$ in, $h = 3.5$ in, $\nu = 1.08 \cdot 10^{-5}$ ft²/sec and $\Omega = 120$ rpm.

(c) Due to turbulence, the spin-up time for Part (b) is only $t_{\text{true}} = 2.5$ sec. Assuming the equation above is valid with an effective kinematic viscosity, ν_{eff}, replacing ν, what is ν_{eff}/ν?

11.17 The surface shear stress, τ_w, for viscous flow over a solid boundary is $\tau_w = \mu(\partial u/\partial y)_{y=0}$. How is τ_w related to the vorticity at the surface, ω_w, in a two-dimensional flow? For simplicity, you may assume the surface is planar.

11.18 A liquid is flowing with constant pressure and a freestream velocity of $U = 4$ ft/sec over a flat plate. The boundary layer thickness is $\delta = 0.9$ inch at a distance $x = 1.1$ ft from the leading edge of the plate. Assuming the flow is laminar, determine the kinematic viscosity of the fluid. If the liquid is one of those listed in Table A.6 of Appendix A, which is it?

11.19 To see how viscosity affects the thickness of a laminar boundary layer, consider flow past a thin flat plate using fluids of widely varying viscosities. The freestream velocity, U, is the same for both fluids, and we measure the boundary-layer thickness, δ, at the same plate length, x, in both experiments. First, we use a *highly-viscous* gas, helium ($\nu = 1.23 \cdot 10^{-3}$ ft^2/sec), and find $\delta = 4$ in. Next, we use a *slightly-viscous* gas, carbon dioxide ($\nu = 8.44 \cdot 10^{-5}$ ft^2/sec). What is δ when carbon dioxide is used?

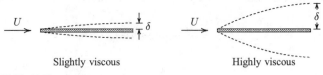

Slightly viscous   Highly viscous

Problems 11.19, 11.20

11.20 To see how viscosity affects the thickness of a laminar boundary layer, consider flow past a thin flat plate with fluids of widely varying viscosities. The freestream velocity, U, is the same for both fluids, and we measure the boundary-layer thickness, δ, at the same plate length, x, in both experiments. First, we use a *slightly-viscous* liquid, water ($\nu = 10^{-6}$ m^2/sec), and find $\delta = 3$ mm. Next, we use a *highly-viscous* liquid, glycerin ($\nu = 1.19 \cdot 10^{-3}$ m^2/sec). What is δ when glycerin is used?

11.21 The boundary-layer thickness is $\delta \approx 5.0 x Re_x^{-1/2}$ for laminar flow over a flat surface, where x is distance along the surface and Re_x is Reynolds number based on x. Using this formula, estimate the thickness of the boundary layer on an 8-meter long submarine that uses surface heating to maintain laminar flow. The submarine is moving at $U_\infty = 30$ knots, and the kinematic viscosity of the water near the submarine surface is $\nu = 8 \cdot 10^{-7}$ m^2/sec^2.

11.22 The boundary-layer thickness is $\delta \approx 5.0 x Re_x^{-1/2}$ for laminar flow over a flat surface, where x is distance along the surface and Re_x is Reynolds number based on x. Using this formula, estimate the thickness of the boundary layer near the trailing edge of a sailboat's sail where $x = 10$ ft. Assume the winds are very light so that sailboat is traveling at a frustrating speed of $U_\infty = 3$ knots. Also, assume the kinematic viscosity of air is $\nu = 1.62 \cdot 10^{-4}$ ft^2/sec^2.

11.23 Consider the flow of kerosene ($\nu = 2.55 \cdot 10^{-5}$ ft^2/sec) past a thin flat plate. The freestream velocity is 2 ft/sec. Assuming sufficient care has been taken to maintain laminar flow, determine the boundary-layer thickness and skin friction at distances $x = 1$ ft and $x = 5$ ft from the plate leading edge.

11.24 Q has designed a flying mirror to permit James Bond to reflect a laser beam around a corner. The mirror has width, $w = 10$ cm, and length, $\ell = 5$ cm. Q's design guarantees laminar flow over the mirror when it flies at $U = 50$ m/sec with its long side normal to the motion. This design feature assures that the mirror can be quickly and silently moved from one place to another. Use Blasius formulas in the computations below, and assume pressure is atmospheric while the temperature is 20° C.

(a) Assuming that transition to turbulence (a source of noise) will not occur on the mirror provided $Re_x < 4 \cdot 10^5$, will the mirror still have laminar flow if, in his haste to avoid capture, 007 makes it fly with its short side normal to the mirror's motion?

(b) What is the total drag on the mirror in the design case and when 007 rotates it as in Part (a)?

11.25 For flow near the entrance to a wide rectangular duct, boundary layers on the lower and upper walls grow and merge on the centerline. Assuming the flow remains laminar, estimate the distance, ℓ_m, at which the boundary layers merge. Compare your answer with the distance needed to achieve fully-developed flow, $\ell_e \approx 0.06 H Re_H$, where $Re_H \equiv UH/\nu$.

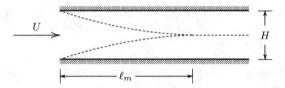

Problem 11.25

11.26 Assuming the Blasius solution applies, compute the drag coefficient, $C_D = D/(\frac{1}{2}\rho U^2 A)$, where $A = \pi d^2/4$, for the circular flat plate shown. **HINT:** Compute the differential drag dD on a strip of width dz and integrate the result over z. The following integral will help in arriving at an answer.

$$\int_0^{\pi/2} \cos^{3/2}\phi \, d\phi \approx 1.57$$

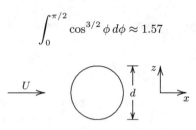

Problem 11.26

11.27 Consider the laminar boundary layer that develops on a flat plate aligned with the freestream flow direction. The flow is incompressible, the freestream flow speed is U and the pressure is constant in the flow direction, i.e., $\partial p/\partial x = 0$. The vertical velocity component is constant and equal to $-v_w$. Determine the horizontal velocity component, $u(x, y)$. Is there any restriction on the value of v_w?

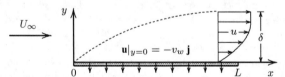

Problem 11.27, 11.28

11.28 For flow over a porous flat plate with *suction*, or surface mass removal, the horizontal velocity, u, is a function only of distance from the surface, y, and is given by

$$u(y) = U_\infty \left[1 - \exp\left(-C_Q U_\infty y/\nu\right)\right], \qquad v_w = C_Q U_\infty$$

where C_Q is the suction coefficient, U_∞ is the freestream velocity and ν is kinematic viscosity.

  (a) Compute the z component of the vorticity, ω, for constant U_∞, C_Q and ν.

  (b) The boundary-layer thickness, δ, is usually defined as the value of y where $u = 0.99U_\infty$. Using this definition, determine δ for this flow.

  (c) Let L denote the length of the plate and let $Re = U_\infty L/\nu$ denote the Reynolds number. Rewrite the velocity and vorticity in dimensionless form.

  (d) Assuming $C_Q = 0.01$, compute δ/L for $Re = 10^3$ and 10^4.

  (e) Make graphs of u/U_∞ and $|\nu\omega/U_\infty^2|$ versus y/L with $0 \le y/L \le 1$ for $C_Q = 0.01$. Make a single graph for each property including results on the same graph for the two Reynolds numbers, $Re = 10^3$ and 10^4.

11.29 To see how much larger variations in velocity normal to a surface are relative to streamwise variations, it is instructive to examine the boundary layer on a $30°$ wedge. The *Falkner-Skan* solution tells us the velocity in the boundary layer is

$$u(x, y) = u_e(x) f\left(\frac{y}{\delta}\right), \quad u_e(x) = u_o \left(\frac{x}{L}\right)^{1/6}, \quad \delta(x) \approx 6 \sqrt{\frac{\nu L}{u_o}} \left(\frac{x}{L}\right)^{5/12}$$

where $u_e(x)$ is the velocity above the boundary layer, f is a function satisfying $f(0) = 0$ and $f(\eta) \to 1$ as $\eta \to \infty$, and $\delta(x)$ is the boundary-layer thickness. Also, u_o and L are characteristic velocity and length scales, while ν is kinematic viscosity.

(a) Develop an estimate of $\partial u / \partial x$ by ignoring the variation of $f(y/\delta)$, i.e., by computing du_e/dx.

(b) Develop an estimate of $\partial u / \partial y$ by approximating it as $\Delta u / \Delta y \approx u_e / \delta$.

(c) Determine the ratio of $\partial u / \partial y$ to $\partial u / \partial x$ at $x/L = 1$ for $Re \equiv u_o L / \nu = 10^4$, 10^5 and 10^6.

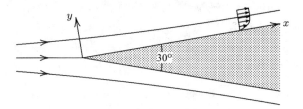

Problem 11.29

11.30 Compute the Reynolds number based on length of a 1-foot long model airplane moving at 8 ft/sec when the temperature is $68°$ F. Assuming the boundary layer experiences a negligible pressure gradient and remains attached along the fuselage, estimate the thickness of the boundary layer near the tail.

11.31 Compute the Reynolds number based on length of an 18-foot long automobile moving at 60 mph when the temperature is $68°$ F. Assuming the boundary layer remains attached along the side of the car and that it experiences a negligible pressure gradient, estimate the thickness of the boundary layer near the tail light.

11.32 Compute the Reynolds number based on length of a 100-meter long ship moving at 14 knots when the water temperature is $10°$ C. Assuming the boundary layer remains attached along the side of the hull and that it experiences a negligible pressure gradient, estimate the thickness of the boundary layer near the stern.

11.33 The point at which a boundary layer experiences transition from laminar to turbulent flow is very sensitive to the freestream turbulence level. Transition has been observed for the flat-plate boundary layer at plate-length Reynolds number, Re_x, ranging from about 10^5 to $3 \cdot 10^6$. Also, measurements show that the length of the transition region is $Re_{\Delta x} \approx 25 Re_{x_t}^{2/3}$, where Re_{x_t} is the Reynolds number at which transition begins. To appreciate how abrupt transition is and how much flow properties change, consider the following two situations. In each case, compute the transition-point location, x_t, the width of the transition region, Δx, the skin friction at the beginning of transition, and the skin friction at the end of transition (i.e., at $x_f = x_t + \Delta x$). Assume that, for a turbulent flat-plate boundary layer, $c_f = 0.0576 Re_x^{-1/5}$.

(a) Air ($\nu = 1.62 \cdot 10^{-4}$ ft^2/sec) flowing over a plate at $U = 100$ ft/sec with transition occurring at $Re_{x_t} = 2 \cdot 10^6$.

(b) Water ($\nu = 1.08 \cdot 10^{-5}$ ft^2/sec) flowing over a plate at $U = 30$ ft/sec with transition occurring at $Re_{x_t} = 5 \cdot 10^5$.

(c) Make a log-log plot of c_f versus Re_x based on your results of Parts (a) and (b) similar to Figure 11.21.

11.34 To appreciate why laminar flow is of minimal importance in many engineering applications, compute the percent of the vehicle over which laminar flow exists for the following situations. In each case, consider the boundary layer on a flat portion of the vehicle and assume transition occurs at a (very high) Reynolds number of $Re_{x_t} = 10^6$.

(a) A 10-meter sailboat moving at 3 knots ($\nu = 1.00 \cdot 10^{-6}$ m²/sec).

(b) A 10-meter sailboat moving at 7.5 knots ($\nu = 1.00 \cdot 10^{-6}$ m²/sec).

(c) A 25-meter yacht moving at 12 knots ($\nu = 0.80 \cdot 10^{-6}$ m²/sec).

(d) A 100-meter tanker moving at 15 knots ($\nu = 1.50 \cdot 10^{-6}$ m²/sec).

11.35 To appreciate why laminar flow is of minimal importance in many engineering applications, compute the percent of the vehicle over which laminar flow exists for the following situations. In each case, consider the boundary layer on a flat portion of the vehicle and assume transition occurs at a (very high) Reynolds number of $Re_{x_t} = 5 \cdot 10^5$.

(a) A 20-foot automobile moving at 25 mph ($\nu = 1.62 \cdot 10^{-4}$ ft²/sec).

(b) A 20-foot automobile moving at 65 mph ($\nu = 1.62 \cdot 10^{-4}$ ft²/sec).

(c) A small aircraft with a wing chord length of 8 feet moving at 150 mph ($\nu = 1.67 \cdot 10^{-4}$ ft²/sec).

(d) A *Boeing 747* with a wing chord length of 30 feet moving at 570 mph ($\nu = 4.27 \cdot 10^{-4}$ ft²/sec).

11.36 A flat plate that is $L = 10$ ft long by $W = 5$ ft wide is immersed in a stream of air moving at $U = 60$ ft/sec with kinematic viscosity $\nu = 1.62 \cdot 10^{-4}$ ft²/sec and density $\rho = 0.00234$ slug/ft³. The boundary layer undergoes transition from laminar to turbulent flow at a plate-length Reynolds number, Re_{x_t}, of 10^6. Ignoring the transition region, i.e., assuming the width of the transition region is negligibly small, compute the total force on the plate.

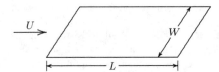

Problem 11.36

11.37 Using Equation (11.94), estimate the thickness of the boundary layer near the trailing edge of the wing on a *Boeing 747* traveling at $U_\infty = 570$ mph with $\nu = 4.27 \cdot 10^{-4}$ ft²/sec². Assume the chord length is 25 ft. To appreciate how thin the boundary layer is, compute the angle $\tan^{-1}(\delta/x)$, which would be the angle measured from the leading edge if the wing were flat.

11.38 Using Equation (11.94), estimate the thickness of the boundary layers above and below the water line near the stern of a 100-meter long ship traveling at $U_\infty = 16$ knots. The kinematic viscosity of air (at 20° C) is $\nu_a = 1.51 \cdot 10^{-5}$ m²/sec², while the value for water (at 10° C) is $\nu_w = 1.31 \cdot 10^{-6}$ m²/sec². To appreciate how thin these boundary layers are, compute the angle $\tan^{-1}(\delta/x)$, for both cases, which would be the angle measured from the leading edge if the side of the hull were flat.

11.39 Assuming the flow is turbulent from the leading edge and that the skin friction can be determined from $c_f \approx 0.0576 Re_x^{-1/5}$, compute the drag coefficient, $C_D = D/(\frac{1}{2}\rho U^2 A)$, where $A = bL$, for the rectangular flat plate shown. For $Re_L = 10^5$, 10^6 and 10^7, compare your answer with the laminar drag coefficient [Equation (11.85)].

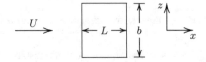

Problem 11.39

11.40 Assuming the flow is turbulent from the leading edge and that the skin friction can be determined from $c_f \approx 0.0576 Re_x^{-1/5}$, compute the drag coefficient, $C_D = D/(\frac{1}{2}\rho U^2 A)$, where $A = \frac{1}{2}L^2$, for the triangular flat plate shown. For $Re_L = 10^5$, 10^6 and 10^7, compare your answer with the laminar drag coefficient, $C_D = 3.54/\sqrt{Re_L}$. **HINT:** Compute the differential drag dD on a strip of width dz and integrate the result over z.

Problem 11.40

11.41 For two-dimensional turbulent flow over a solid boundary, the horizontal velocity component close to the surface is approximately

$$u \approx u_\tau \left[\frac{1}{\kappa} \ell n \frac{u_\tau y}{\nu} + C \right], \qquad y \geq 30 \frac{\nu}{u_\tau}$$

where $u_\tau = \sqrt{\tau_w/\rho}$ is the *friction velocity*, $\tau_w = \mu(\partial u/\partial y)_{y=0}$, y is distance normal to the surface and ν is kinematic viscosity. Also, $\kappa = 0.41$ and $C = 5.0$. Determine the vorticity at $y = 30\nu/u_\tau$ and compute its ratio to the vorticity at $y = 0$.

11.42 Assume that the velocity profile for a laminar, flat-plate boundary layer can be approximated by $u/u_e = (y/\delta)^{1/2}$. Noting that experimental data indicate $\tau_w = 1.66 u_e \mu/\delta$, use the momentum-integral equation and these relations to obtain a differential equation for δ. Solve for δ as a function of x and compare your result with the exact laminar (Blasius) result.

11.43 For a turbulent flat-plate boundary layer, we can estimate the boundary-layer thickness, δ, as follows. Assume the wall shear stress is the same as for pipe flow, i.e.,

$$\tau_w \approx 0.0225 \rho U^2 \left(\frac{\nu}{U\delta} \right)^{1/4}$$

Then, assume that the ratio of momentum thickness, θ, to boundary layer thickness is given by the 1/7-power law velocity profile so that $\theta/\delta = 7/72$. Using the momentum-integral equation, determine the ratio of δ to distance from the plate leading edge, x. Also, compute the skin friction, c_f, and the drag on a plate of length L, taking account of both sides of the plate.

11.44 Correlation of measurements shows that the velocity profile and the boundary-layer thickness for a laminar boundary layer can be approximated as follows:

$$\frac{u}{u_e} = 2 \left(\frac{y}{\delta} \right) - \left(\frac{y}{\delta} \right)^2 \quad \text{and} \quad \frac{\delta}{x} = 2 \left(\frac{\nu}{U_o x} \right)^{3/5}$$

where ν is kinematic viscosity and U_o is a reference velocity. Using the momentum-integral equation, determine the corresponding freestream velocity, $u_e(x)$.

11.45 Consider a simple linear approximation to the velocity profile for an incompressible flat-plate boundary layer.

$$\frac{u}{U} = \frac{y}{\delta}$$

(a) Compute the skin friction, c_f, displacement thickness, δ^*, momentum thickness, θ, and shape factor, H.

(b) Using the momentum-integral equation, determine the boundary-layer thickness, δ.

(c) Quantify the differences between the results of Parts (a) and (b) and the Blasius solution.

11.46 Consider the following approximation to the velocity profile for an incompressible flat-plate boundary layer.

$$\frac{u}{U} = \sin\left(\frac{\pi}{2}\frac{y}{\delta}\right)$$

(a) Compute the skin friction, c_f, displacement thickness, δ^*, momentum thickness, θ, and shape factor, H.

(b) Using the momentum-integral equation, determine the boundary-layer thickness, δ.

(c) Quantify the differences between the results of Parts (a) and (b) and the Blasius solution.

11.47 Consider the following approximation to the velocity profile for an incompressible flat-plate boundary layer.

$$\frac{u}{U} = \frac{3}{2}\left(\frac{y}{\delta}\right) - \frac{1}{2}\left(\frac{y}{\delta}\right)^3$$

(a) Compute the skin friction, c_f, displacement thickness, δ^*, momentum thickness, θ, and shape factor, H.

(b) Using the momentum-integral equation, determine the boundary-layer thickness, δ.

(c) Quantify the differences between the results of Parts (a) and (b) and the Blasius solution.

11.48 Consider the following approximation to the velocity profile for an incompressible flat-plate boundary layer.

$$\frac{u}{U} = \frac{3}{2}\left(\frac{y}{\delta}\right) - \frac{1}{2}\left(\frac{y}{\delta}\right)^{5/2}$$

(a) Compute the skin friction, c_f, displacement thickness, δ^*, momentum thickness, θ, and shape factor, H.

(b) Using the momentum-integral equation, determine the boundary-layer thickness, δ.

(c) Quantify the differences between the results of Parts (a) and (b) and the Blasius solution.

11.49 Consider the following approximation to the velocity profile for an incompressible flat-plate boundary layer.

$$\frac{u}{U} = 2\left(\frac{y}{\delta}\right) - 2\left(\frac{y}{\delta}\right)^3 + \left(\frac{y}{\delta}\right)^4$$

(a) Compute the skin friction, c_f, displacement thickness, δ^*, momentum thickness, θ, and shape factor, H.

(b) Using the momentum-integral equation, determine the boundary-layer thickness, δ.

(c) Quantify the differences between the results of Parts (a) and (b) and the Blasius solution.

11.50 An experimenter has made a series of wind-tunnel measurements to calibrate his tunnel. His measurements correspond to what he believes is a constant-pressure, laminar boundary layer. Based on his data, he has developed the following correlation for the velocity profile.

$$\frac{u}{U} = \frac{3}{2}\left(\frac{y}{\delta}\right) - \frac{2}{5}\left(\frac{y}{\delta}\right)^3 - \frac{1}{10}\left(\frac{y}{\delta}\right)^5$$

(a) Compute the skin friction, c_f, displacement thickness, δ^*, momentum thickness, θ, and shape factor, H.

(b) Using the momentum-integral equation, determine the boundary-layer thickness, δ.

(c) Quantify the differences between the results of Parts (a) and (b) and the Blasius solution. Comment on whether or not the measurements establish the intended purpose of certifying the accuracy of the measurements.

11.51 For a boundary layer with sufficient surface mass removal (suction), transition to turbulence can be prevented. For such a boundary layer, the laminarized velocity profile asymptotes to:

$$u = U \left[1 - e^{-C_Q U y / \nu} \right]$$

where $C_Q > 0$ is the (dimensionless) suction coefficient. Compute skin friction, c_f, displacement thickness, δ^*, momentum thickness, θ, and shape factor, H, for such a boundary layer. **HINT:** Do all integrals from $y = 0$ to $y \to \infty$.

11.52 For an incompressible boundary layer with surface mass removal (suction), the momentum-integral equation is modified slightly and becomes

$$\frac{d\theta}{dx} + (H + 2) \frac{\theta}{u_e} \frac{du_e}{dx} = \frac{c_f}{2} - C_Q$$

where $C_Q = -v_w / u_e$ is the (dimensionless) suction coefficient, and v_w is the suction velocity.

(a) Assuming the horizontal velocity is given by $u(x, y) = u_e(x)\{1 - \exp[-y/a(x)]\}$, compute skin friction, c_f, displacement thickness, δ^*, momentum thickness, θ, and shape factor, H. **HINT:** Do all integrals from $y = 0$ to $y \to \infty$.

(b) If $C_Q Re_\theta = 1/2$, solve for $\theta(x)$ in terms of $u_e(x)$, assuming $\theta(x_o) = \theta_o$ and $u_e(x_o) = u_o$ where x_o is a reference point.

11.53 *Wieghardt* proposed an *energy-integral equation* for incompressible, laminar boundary layers, viz.,

$$\frac{d}{dx} \left(u_e^3 \theta_E \right) = 2\nu \int_0^\delta \left(\frac{\partial u}{\partial y} \right)^2 dy$$

The quantity θ_E is the energy thickness defined by

$$\theta_E \equiv \int_0^\delta \frac{u}{u_e} \left[1 - \left(\frac{u}{u_e} \right)^2 \right] dy$$

Use this equation to determine θ_E and δ for a flat-plate boundary layer assuming $u/u_e = y/\delta$. Compare your result for δ with the Blasius thickness, i.e., $\delta = 5.0\sqrt{\nu x/u_e}$.

11.54 *Wieghardt* proposed an *energy-integral equation* for incompressible, laminar boundary layers, viz.,

$$\frac{d}{dx} \left(u_e^3 \theta_E \right) = 2\nu \int_0^\delta \left(\frac{\partial u}{\partial y} \right)^2 dy$$

The quantity θ_E is the energy thickness defined by

$$\theta_E \equiv \int_0^\delta \frac{u}{u_e} \left[1 - \left(\frac{u}{u_e} \right)^2 \right] dy$$

Use this equation to determine θ_E and δ for a flat-plate boundary layer assuming $u/u_e = \sin(\frac{1}{2}\pi y/\delta)$. Compare your result for δ with the Blasius thickness, i.e., $\delta = 5.0\sqrt{\nu x/u_e}$.

11.55 The Falkner-Skan velocity profile for $m = 2$ can be approximated as $u/u_e = 1 - e^{-5y/\delta}$. Using the momentum-integral equation, solve for δ and c_f. **HINT:** Do all integrals from $y = 0$ to $y = \infty$. Compare your computed c_f with the exact value, viz., $c_f = 4.13/\sqrt{Re_x}$.

11.56 The Falkner-Skan velocity profile for $m = \frac{1}{2}$ can be approximated as $u/u_e = 1 - e^{-\lambda y/\delta}$, where λ is a constant. Using the momentum-integral equation, solve for δ and c_f. **HINT:** Do all integrals from $y = 0$ to $y = \infty$. Compare your computed c_f with the exact value, viz., $c_f = 1.61/\sqrt{Re_x}$. Comment on the optimum choice of λ if the goal is to compute an accurate value for c_f.

11.57 For the *Falkner-Skan* family of velocity profiles [$u_e(x) = U_o(x/L)^m$], show that the profile corresponding to incipient separation, $m \approx -1/11$, has a shape factor of $H = 4$. **HINT:** Since the similarity variable, η, is proportional to $y/(x/L)^{(1-m)/2}$, we expect all boundary-layer thickness parameters for this profile to be proportional to $(x/L)^{(1-m)/2}$.

11.58 *Thwaites' method* is a useful empirical procedure for predicting laminar boundary-layer properties with variable freestream pressure, including separation point. With this method, the momentum thickness, θ, and skin friction, c_f, are given by

$$\theta^2 \approx \frac{0.450\nu}{u_e^6} \int_0^x u_e^5(x)dx, \quad c_f \approx \frac{2(K+0.09)^{0.62}}{Re_\theta} \quad \text{where} \quad K \equiv \frac{\theta^2}{\nu}\frac{du_e}{dx}$$

Also, Re_θ is Reynolds number based on θ. According to this method, boundary-layer separation occurs when $K = -0.09$. Consider a linearly-decreasing velocity, i.e., $u_e(x) = U(1-x/L)$, where U and L are constant velocity and length scales, respectively. At what value of x/L, according to *Thwaites' method*, does the boundary layer separate? Compare your answer with the exact value of $x_{sep}/L = 0.120$.

11.59 *Thwaites' method* is a useful empirical procedure for predicting laminar boundary-layer properties with variable freestream pressure. To develop the method, proceed as follows.

(a) Multiply both sides of the momentum-integral equation by $2\theta/\nu$, and write the equation as

$$\frac{d}{dx}\left(\frac{\theta^2}{\nu}\right) = \frac{F(K)}{u_e} \quad \text{where} \quad K \equiv \frac{\theta^2}{\nu}\frac{du_e}{dx}$$

(b) Now, observe that for laminar boundary layers, correlation of exact results shows that the function $F(K)$ is given by $F(K) \approx 0.450 - 6K$. Substitute this correlation into the equation derived in Part (a), and verify that

$$\frac{u_e\theta^2}{\nu} \approx \frac{0.450}{u_e^5} \int_0^x u_e^5(x)dx$$

(c) Use *Thwaites' method*, i.e., the equation developed in Part (b), to compute θ for constant pressure. Compare your result with the Blasius solution.

11.60 The lift coefficient of an unpowered aircraft is 28.5 times its drag coefficient. Assuming the aircraft moves at constant speed with lift, drag and weight in equilibrium, at what angle to the horizontal is it inclined?

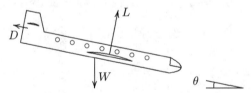

Problems 11.60, 11.61

11.61 The lift coefficient of a commercial airliner is 20 times its drag coefficient. At an altitude of 10 km the airliner loses all engine power. In terms of horizontal distance, how far can it glide before reaching the ground? Assume the airliner glides at constant speed with lift, drag and weight in equilibrium.

11.62 A cube weighing 1 lb falls in a deep tank filled with glycerine at 68° F. The bottom of the cube is always parallel to the bottom of the tank. If its terminal velocity is 6 ft/sec, determine the size of the cube in inches.

11.63 Suppose we choose the wetted area instead of the frontal area in defining the drag coefficient of a cone. The wetted area is the surface area of the cone facing the flow, i.e., excluding the base area. Verify that the revised drag coefficient is half the conventional value for a cone angle $\theta = 60°$. **HINT:** The wetted area of a cone of radius r and height h is $A_{wet} = \pi r\sqrt{r^2 + h^2}$.

11.64 James Bond has just been thrown out of a New York skyscraper. Fortunately, Q has provided 007 with a combination handkerchief/parachute that expands to a diameter, d, of 4 ft. Q has also shown 007 how to fall so that his body achieves $C_D A = 10$ ft^2. Assume Bond weighs 180 pounds, and that the air density is 0.00234 slug/ft^3.

(a) What is the drag coefficient, C_{D_p}, of Q's new handkerchief/parachute if Bond's terminal velocity is 50 ft/sec?

(b) What would Bond's terminal velocity be for a conventional parachute of the same size?

11.65 A baseball is dropped from the Goodyear blimp, which is hovering over Yankee Stadium. Assume, for simplicity, that the ball's drag coefficient, C_D, is constant and equal to 0.4.

(a) Compute the terminal velocity, w_f, as a function of C_D, the mass of the ball, m, the diameter of the ball, D, the density of air, ρ, and gravitational acceleration, g.

(b) Determine the velocity, w, as a function of time, t. Express your answer in terms of g, t and w_f.
HINT: The following integral may be of use in arriving at your answer.

$$\int \frac{dw}{w_f^2 - w^2} = \frac{1}{w_f} \tanh^{-1}\left(\frac{w}{w_f}\right)$$

(c) Make a graph of your solution for $mg = 5$ ounces, $\rho = 0.00234$ slug/ft^3 and $D = 0.25$ ft.

(d) For Reynolds number, Re_D, less than 1000, C_D is greater than 0.4. According to your solution, at what time does the Reynolds number reach 1000 (so that your solution is valid)?

11.66 Due to fatigue from temperatures in excess of 100° F, a construction worker drops a standard 15 inch long 1 by 2 wooden board of nominal cross section $\frac{3}{4}$ inch by $\frac{3}{2}$ inch from the roof of a very tall building. The wood is pine and has a density of $\rho_o = 1$ slug/ft^3. Compute the terminal velocity of the board as it falls with a frontal area as shown in each of the two configurations in the figure. Assume the density of air is $\rho = 0.00220$ slug/ft^3 and that the building is sufficiently high that the board reaches its terminal velocity before striking the ground.

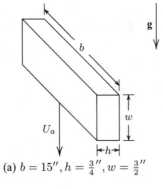

(a) $b = 15''$, $h = \frac{3}{4}''$, $w = \frac{3}{2}''$

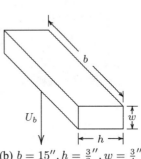

(b) $b = 15''$, $h = \frac{3}{2}''$, $w = \frac{3}{4}''$

Problem 11.66

Appendix A

Fluid Properties

A.1 Perfect-Gas Properties

The pressure, p, density, ρ, and temperature, T, are related by the perfect-gas equation of state,

$$p = \rho R T \tag{A.1}$$

where R is the **perfect-gas constant**. Table A.1 includes values of R for common gases. The table also includes the specific heat coefficient, c_p, and the specific-heat ratio, γ.

Table A.1: *Thermodynamic Properties of Common Gases*

Gas	γ	J/(kg·K)		ft·lb/(slug·°R)	
		R	c_p	R	c_p
Air	1.40	287	1004	1716	6003
Carbon dioxide	1.30	189	841	1130	5028
Helium	1.66	2077	5225	12419	31240
Hydrogen	1.40	4124	14180	24677	84783
Methane	1.31	518	2208	3098	13783
Nitrogen	1.40	297	1039	1776	6212
Oxygen	1.40	260	910	1555	5440

A.2 Pressure

The standard value of the pressure in the atmosphere is

$$p = \begin{cases} 1 & \text{atm} & \text{(atmospheres)} \\ 1 & \text{torr} & \text{(Torr)} \\ 14.7 & \text{psi} & \text{(pounds per square inch)} \\ 101 & \text{kPa} & \text{(kiloPascals)} \\ 760 & \text{mmHg} & \text{(millimeters of mercury)} \\ 2116.8 & \text{psf} & \text{(pounds per square foot)} \end{cases} \tag{A.2}$$

where one Pascal = one Newton per square meter. Pressure is often quoted in terms of **absolute** (psia) and **gage** (psig) values, with the latter relative to the atmospheric pressure. Thus, $p = 14.9$ psi is an absolute pressure of 14.9 psia and a gage pressure of 0.2 psig.

A.3 Density

Values of the density of air and water at atmospheric pressure and a temperature of 20° C (68° F) are as follows.

$$\rho = \begin{cases} 1.20 & \text{kg/m}^3 \quad (0.00234 \text{ slug/ft}^3), \quad \text{Air} \\ 998 & \text{kg/m}^3 \quad (1.94 \text{ slug/ft}^3), \qquad \text{Water} \end{cases} \tag{A.3}$$

Table A.2 includes densities of common liquids, while Table A.3 includes the variation of density with temperature for water.

Table A.2: *Densities of Common Liquids*

Liquid	T (° C)	ρ (kg/m^3)	T (° F)	ρ (slug/ft^3)
Carbon tetrachloride	20	1,590	68	3.09
Ethyl alcohol	20	789	68	1.53
Glycerin	20	1,260	68	2.44
Kerosene	20	814	68	1.58
Mercury	20	13,550	68	26.30
Oil: SAE 10W	38	870	100	1.69
Oil: SAE 10W-30	38	880	100	1.71
Oil: SAE 30	38	880	100	1.71
Seawater	16	1,030	60	1.99

Table A.3: *Density of Water as a Function of Temperature*

T (° C)	ρ (kg/m^3)	T (° F)	ρ (slug/ft^3)
0	1,000	32	1.94
10	1,000	50	1.94
20	998	68	1.94
30	996	86	1.93
40	992	104	1.93
50	988	122	1.92
60	983	140	1.91
70	978	158	1.90
80	972	176	1.89
90	965	194	1.87
100	958	212	1.86

A.4 Compressibility and Speed of Sound

Table A.4 lists compressibility, τ, and the speed of sound, a, for several common fluids. In general, for a perfect gas, the speed of sound is

$$a = \sqrt{\gamma R T} \tag{A.4}$$

Table A.4: *Compressibility and Sound Speed of Common Fluids for a Pressure of 1 atm*

Fluid	T ($^\circ$ C)	τ (m^2/N)	a (m/sec)	T ($^\circ$ F)	τ (ft^2/lb)	a (ft/sec)
Air	15.6	$1.00 \cdot 10^{-5}$	341	60.0	$4.79 \cdot 10^{-4}$	1119
Ether	15.0	$1.63 \cdot 10^{-9}$	1032	59.0	$7.82 \cdot 10^{-8}$	3386
Ethyl alcohol	20.0	$9.43 \cdot 10^{-10}$	1213	68.0	$4.52 \cdot 10^{-8}$	3980
Glycerin	15.6	$2.21 \cdot 10^{-10}$	1860	60.0	$1.06 \cdot 10^{-8}$	6102
Helium	15.6	$1.00 \cdot 10^{-5}$	998	60.0	$4.79 \cdot 10^{-4}$	3274
Mercury	15.6	$0.35 \cdot 10^{-10}$	1450	60.0	$0.17 \cdot 10^{-8}$	4757
Water	15.6	$4.65 \cdot 10^{-10}$	1481	60.0	$2.23 \cdot 10^{-8}$	4859

A.5 Surface Tension

In general, surface tension, σ, depends upon the surface material as well as the nature of the two fluids involved, and it is a function of pressure and temperature. It is also considerably affected by dirt on the surface. Table A.5 lists surface tension for several common liquids.

Table A.5: *Surface Tension of Common Fluids*

Fluid$_1$	Fluid$_2$	T ($^\circ$ C)	σ (N/m)	T ($^\circ$ F)	σ (lb/ft)
Carbon tetrachloride	Air	20	.027	68	.0018
Carbon tetrachloride	Water	20	.045	68	.0030
Ethyl alcohol	Air	20	.023	68	.0016
Gasoline	Air	16	.022	60	.0015
Glycerin	Air	20	.063	68	.0043
Mercury	Air	20	.466	68	.0319
Mercury	Water	20	.375	68	.0257
SAE 10 Oil	Air	16	.036	60	.0025
SAE 30 Oil	Air	16	.035	60	.0024
Seawater	Air	16	.073	60	.0050
Water	Air	0	.076	32	.0052
Water	Air	20	.073	68	.0050
Water	Air	100	.059	212	.0040

A.6   Viscosity

For air and water at atmospheric pressure and $20°$ C ($68°$ F), the kinematic viscosity is

$$\nu = \begin{cases} 1.51 \cdot 10^{-5} \text{ m}^2/\text{sec} & (1.62 \cdot 10^{-4} \text{ ft}^2/\text{sec}), \quad \text{Air} \\ 1.00 \cdot 10^{-6} \text{ m}^2/\text{sec} & (1.08 \cdot 10^{-5} \text{ ft}^2/\text{sec}), \quad \text{Water} \end{cases} \qquad (A.5)$$

Table A.6 includes kinematic viscosities of common liquids and gases.

Table A.6: *Kinematic Viscosities of Common Liquids and Gases*

Fluid	T ($°$ C)	$10^6\nu$ (m^2/sec)	T ($°$ F)	$10^5\nu$ (ft^2/sec)
Carbon tetrachloride	20	0.60	68	0.65
Ethyl alcohol	20	1.51	68	1.62
Gasoline	16	0.46	60	0.49
Glycerin	20	1190.00	68	1280.00
Kerosene	20	2.37	68	2.55
Mercury	20	0.12	68	0.13
Oil: SAE 10W	38	41.00	100	44.00
Oil: SAE 10W-30	38	76.00	100	82.00
Oil: SAE 30	38	110.00	100	118.00
Air	15	14.60	59	15.71
Carbon dioxide	15	7.84	59	8.44
Helium	15	114.00	59	123.00
Hydrogen	15	101.00	59	109.00
Methane	15	15.90	59	17.10
Nitrogen	15	14.50	59	15.60
Oxygen	15	15.00	59	16.10

Table A.7 includes the variation of ν with temperature for water.

Table A.7: *Kinematic Viscosity of Water as a Function of Temperature*

T ($°$ C)	$10^6\nu$ (m^2/sec)	T ($°$ F)	$10^5\nu$ (ft^2/sec)
0	1.79	32	1.93
10	1.31	50	1.41
20	1.00	68	1.08
30	0.80	86	0.86
40	0.66	104	0.71
50	0.55	122	0.59
60	0.47	140	0.51
70	0.41	158	0.44
80	0.36	176	0.39
90	0.33	194	0.36
100	0.29	212	0.31

The viscosity of air and many other common gases is well approximated by **Sutherland's Law**, which is an empirical equation that is quite accurate for a wide range of temperatures. The formula is

$$\mu = \frac{AT^{3/2}}{T + S} \qquad (A.6)$$

where A and S are empirical constants. Note that T is the absolute temperature, and is thus given either in Kelvins (K) or degrees Rankine ($^\circ$R). Table A.8 lists A and S for air and several common gases.

Table A.8: *Sutherland Viscosity Law Coefficients for Common Gases*

Gas	A [kg/(m·sec·K$^{1/2}$)]	S (K)	A [slug/(ft·sec·($^\circ$R)$^{1/2}$)]	S ($^\circ$R)
Air	$1.46 \cdot 10^{-6}$	110.3	$2.27 \cdot 10^{-8}$	198.6
Ammonia	$1.30 \cdot 10^{-6}$	370.0	$2.02 \cdot 10^{-8}$	666.0
Carbon dioxide	$1.57 \cdot 10^{-6}$	240.0	$2.45 \cdot 10^{-8}$	432.0
Carbon monoxide	$1.43 \cdot 10^{-6}$	118.0	$2.23 \cdot 10^{-8}$	212.4
Hydrogen	$0.64 \cdot 10^{-6}$	72.0	$0.99 \cdot 10^{-8}$	129.6
Nitrogen	$1.41 \cdot 10^{-6}$	111.0	$2.19 \cdot 10^{-8}$	199.8
Oxygen	$1.69 \cdot 10^{-6}$	127.0	$2.64 \cdot 10^{-8}$	228.6
Sulfur dioxide	$1.77 \cdot 10^{-6}$	416.0	$2.76 \cdot 10^{-8}$	748.8

A.7 Vapor Pressure

The vapor pressure of a fluid varies with temperature. Table A.9 lists vapor pressure for several common liquids.

Table A.9: *Vapor Pressure of Common Fluids*

Fluid	T ($^\circ$C)	p_v (kPa)	T ($^\circ$F)	p_v (psi)
Carbon tetrachloride	20	13.00	68	1.89
Ethyl alcohol	20	5.90	68	0.86
Gasoline	16	55.00	60	7.98
Glycerin	20	$1.4 \cdot 10^{-5}$	68	$2.0 \cdot 10^{-6}$
Mercury	16	$1.6 \cdot 10^{-4}$	68	$2.3 \cdot 10^{-5}$
Seawater	16	1.82	60	0.26
Water	0	0.61	32	0.09
Water	10	1.23	50	0.18
Water	20	2.33	68	0.34
Water	30	4.23	86	0.61
Water	40	7.35	104	1.07
Water	50	12.29	122	1.78
Water	60	19.84	140	2.88
Water	70	31.06	158	4.51
Water	80	47.27	176	6.86
Water	90	70.12	194	10.17
Water	100	101.61	212	14.74

Appendix B

Compressible Flow Tables

This appendix includes several tables that are valid for gases with a specific-heat ratio, $\gamma = 1.4$.

B.1 Isentropic-Flow and Normal-Shock Relations

Notation used in the following tables is as follows.

Isentropic flow:

M	=	local Mach number
p/p_t	=	ratio of static to total pressure
ρ/ρ_t	=	ratio of static to total density
T/T_t	=	ratio of static to total temperature
A/A^*	=	ratio of local area to reference sonic-flow area

Normal shock waves:

M_1	=	Mach number ahead of the shock wave
M_2	=	Mach number behind the shock wave
p_2/p_1	=	pressure ratio across the shock wave
T_2/T_1	=	temperature ratio across the shock wave
p_{t2}/p_{t1}	=	total pressure ratio across the shock wave

Subsonic Flow of a Perfect Gas with $\gamma = 1.4$

M	p/p_t	ρ/ρ_t	T/T_t	A/A^*	M	p/p_t	ρ/ρ_t	T/T_t	A/A^*
0.00	1.0000	1.0000	1.0000	∞	0.50	0.8430	0.8852	0.9524	1.3398
0.02	0.9997	0.9998	0.9999	28.9421	0.52	0.8317	0.8766	0.9487	1.3034
0.04	0.9989	0.9992	0.9997	14.4815	0.54	0.8201	0.8679	0.9449	1.2703
0.06	0.9975	0.9982	0.9993	9.6659	0.56	0.8082	0.8589	0.9410	1.2403
0.08	0.9955	0.9968	0.9987	7.2616	0.58	0.7962	0.8498	0.9370	1.2130
0.10	0.9930	0.9950	0.9980	5.8218	0.60	0.7840	0.8405	0.9328	1.1882
0.12	0.9900	0.9928	0.9971	4.8643	0.62	0.7716	0.8310	0.9286	1.1656
0.14	0.9864	0.9903	0.9961	4.1824	0.64	0.7591	0.8213	0.9243	1.1451
0.16	0.9823	0.9873	0.9949	3.6727	0.66	0.7465	0.8115	0.9199	1.1265
0.18	0.9776	0.9840	0.9936	3.2779	0.68	0.7338	0.8016	0.9153	1.1097
0.20	0.9725	0.9803	0.9921	2.9635	0.70	0.7209	0.7916	0.9107	1.0944
0.22	0.9668	0.9762	0.9904	2.7076	0.72	0.7080	0.7814	0.9061	1.0806
0.24	0.9607	0.9718	0.9886	2.4956	0.74	0.6951	0.7712	0.9013	1.0681
0.26	0.9541	0.9670	0.9867	2.3173	0.76	0.6821	0.7609	0.8964	1.0570
0.28	0.9470	0.9619	0.9846	2.1656	0.78	0.6691	0.7505	0.8915	1.0471
0.30	0.9395	0.9564	0.9823	2.0351	0.80	0.6560	0.7400	0.8865	1.0382
0.32	0.9315	0.9506	0.9799	1.9219	0.82	0.6430	0.7295	0.8815	1.0305
0.34	0.9231	0.9445	0.9774	1.8229	0.84	0.6300	0.7189	0.8763	1.0237
0.36	0.9143	0.9380	0.9747	1.7358	0.86	0.6170	0.7083	0.8711	1.0179
0.38	0.9052	0.9313	0.9719	1.6587	0.88	0.6041	0.6977	0.8659	1.0129
0.40	0.8956	0.9243	0.9690	1.5901	0.90	0.5913	0.6870	0.8606	1.0089
0.42	0.8857	0.9170	0.9659	1.5289	0.92	0.5785	0.6764	0.8552	1.0056
0.44	0.8755	0.9094	0.9627	1.4740	0.94	0.5658	0.6658	0.8498	1.0031
0.46	0.8650	0.9016	0.9594	1.4246	0.96	0.5532	0.6551	0.8444	1.0014
0.48	0.8541	0.8935	0.9559	1.3801	0.98	0.5407	0.6445	0.8389	1.0003
0.50	0.8430	0.8852	0.9524	1.3398	1.00	0.5283	0.6339	0.8333	1.0000

Supersonic Flow of a Perfect Gas with $\gamma = 1.4$

	Isentropic Flow				Normal Shock Wave				
M_1	p/p_t	ρ/ρ_t	T/T_t	A/A^*	M_2	p_2/p_1	ρ_2/ρ_1	T_2/T_1	p_{t2}/p_{t1}
1.00	0.5283	0.6339	0.8333	1.0000	1.0000	1.0000	1.0000	1.0000	1.0000
1.02	0.5160	0.6234	0.8278	1.0003	0.9805	1.0471	1.0334	1.0132	1.0000
1.04	0.5039	0.6129	0.8222	1.0013	0.9620	1.0952	1.0671	1.0263	0.9999
1.06	0.4919	0.6024	0.8165	1.0029	0.9444	1.1442	1.1009	1.0393	0.9998
1.08	0.4800	0.5920	0.8108	1.0051	0.9277	1.1941	1.1349	1.0522	0.9994
1.10	0.4684	0.5817	0.8052	1.0079	0.9118	1.2450	1.1691	1.0649	0.9989
1.12	0.4568	0.5714	0.7994	1.0113	0.8966	1.2968	1.2034	1.0776	0.9982
1.14	0.4455	0.5612	0.7937	1.0153	0.8820	1.3495	1.2378	1.0903	0.9973
1.16	0.4343	0.5511	0.7879	1.0198	0.8682	1.4032	1.2723	1.1029	0.9961
1.18	0.4232	0.5411	0.7822	1.0248	0.8549	1.4578	1.3069	1.1154	0.9946
1.20	0.4124	0.5311	0.7764	1.0304	0.8422	1.5133	1.3416	1.1280	0.9928
1.22	0.4017	0.5213	0.7706	1.0366	0.8300	1.5698	1.3764	1.1405	0.9907
1.24	0.3912	0.5115	0.7648	1.0432	0.8183	1.6272	1.4112	1.1531	0.9884
1.26	0.3809	0.5019	0.7590	1.0504	0.8071	1.6855	1.4460	1.1657	0.9857
1.28	0.3708	0.4923	0.7532	1.0581	0.7963	1.7448	1.4808	1.1783	0.9827
1.30	0.3609	0.4829	0.7474	1.0663	0.7860	1.8050	1.5157	1.1909	0.9794
1.32	0.3512	0.4736	0.7416	1.0750	0.7760	1.8661	1.5505	1.2035	0.9758
1.34	0.3417	0.4644	0.7358	1.0842	0.7664	1.9282	1.5854	1.2162	0.9718
1.36	0.3323	0.4553	0.7300	1.0940	0.7572	1.9912	1.6202	1.2290	0.9676
1.38	0.3232	0.4463	0.7242	1.1042	0.7483	2.0551	1.6549	1.2418	0.9630
1.40	0.3142	0.4374	0.7184	1.1149	0.7397	2.1200	1.6897	1.2547	0.9582
1.42	0.3055	0.4287	0.7126	1.1262	0.7314	2.1858	1.7243	1.2676	0.9531
1.44	0.2969	0.4201	0.7069	1.1379	0.7235	2.2525	1.7589	1.2807	0.9476
1.46	0.2886	0.4116	0.7011	1.1501	0.7157	2.3202	1.7934	1.2938	0.9420
1.48	0.2804	0.4032	0.6954	1.1629	0.7083	2.3888	1.8278	1.3069	0.9360
1.50	0.2724	0.3950	0.6897	1.1762	0.7011	2.4583	1.8621	1.3202	0.9298
1.52	0.2646	0.3869	0.6840	1.1899	0.6941	2.5288	1.8963	1.3336	0.9233
1.54	0.2570	0.3789	0.6783	1.2042	0.6874	2.6002	1.9303	1.3470	0.9166
1.56	0.2496	0.3710	0.6726	1.2190	0.6809	2.6725	1.9643	1.3606	0.9097
1.58	0.2423	0.3633	0.6670	1.2344	0.6746	2.7458	1.9981	1.3742	0.9026
1.60	0.2353	0.3557	0.6614	1.2502	0.6684	2.8200	2.0317	1.3880	0.8952
1.62	0.2284	0.3483	0.6558	1.2666	0.6625	2.8951	2.0653	1.4018	0.8877
1.64	0.2217	0.3409	0.6502	1.2836	0.6568	2.9712	2.0986	1.4158	0.8799
1.66	0.2151	0.3337	0.6447	1.3010	0.6512	3.0482	2.1318	1.4299	0.8720
1.68	0.2088	0.3266	0.6392	1.3190	0.6458	3.1261	2.1649	1.4440	0.8639
1.70	0.2026	0.3197	0.6337	1.3376	0.6405	3.2050	2.1977	1.4583	0.8557
1.72	0.1966	0.3129	0.6283	1.3567	0.6355	3.2848	2.2304	1.4727	0.8474
1.74	0.1907	0.3062	0.6229	1.3764	0.6305	3.3655	2.2629	1.4873	0.8389
1.76	0.1850	0.2996	0.6175	1.3967	0.6257	3.4472	2.2952	1.5019	0.8302
1.78	0.1794	0.2931	0.6121	1.4175	0.6210	3.5298	2.3273	1.5167	0.8215
1.80	0.1740	0.2868	0.6068	1.4390	0.6165	3.6133	2.3592	1.5316	0.8127

Supersonic Flow of a Perfect Gas with $\gamma = 1.4$ (continued)

	Isentropic Flow				Normal Shock Wave				
M_1	p/p_t	ρ/ρ_t	T/T_t	A/A^*	M_2	p_2/p_1	ρ_2/ρ_1	T_2/T_1	p_{t2}/p_{t1}
1.80	0.1740	0.2868	0.6068	1.4390	0.6165	3.6133	2.3592	1.5316	0.8127
1.82	0.1688	0.2806	0.6015	1.4610	0.6121	3.6978	2.3909	1.5466	0.8038
1.84	0.1637	0.2745	0.5963	1.4836	0.6078	3.7832	2.4224	1.5617	0.7948
1.86	0.1587	0.2686	0.5910	1.5069	0.6036	3.8695	2.4537	1.5770	0.7857
1.88	0.1539	0.2627	0.5859	1.5308	0.5996	3.9568	2.4848	1.5924	0.7765
1.90	0.1492	0.2570	0.5807	1.5553	0.5956	4.0450	2.5157	1.6079	0.7674
1.92	0.1447	0.2514	0.5756	1.5804	0.5918	4.1341	2.5463	1.6236	0.7581
1.94	0.1403	0.2459	0.5705	1.6062	0.5880	4.2242	2.5767	1.6394	0.7488
1.96	0.1360	0.2405	0.5655	1.6326	0.5844	4.3152	2.6069	1.6553	0.7395
1.98	0.1318	0.2352	0.5605	1.6597	0.5808	4.4071	2.6369	1.6713	0.7302
2.00	0.1278	0.2300	0.5556	1.6875	0.5774	4.5000	2.6667	1.6875	0.7209
2.02	0.1239	0.2250	0.5506	1.7160	0.5740	4.5938	2.6962	1.7038	0.7115
2.04	0.1201	0.2200	0.5458	1.7451	0.5707	4.6885	2.7255	1.7203	0.7022
2.06	0.1164	0.2152	0.5409	1.7750	0.5675	4.7842	2.7545	1.7369	0.6928
2.08	0.1128	0.2104	0.5361	1.8056	0.5643	4.8808	2.7833	1.7536	0.6835
2.10	0.1094	0.2058	0.5313	1.8369	0.5613	4.9783	2.8119	1.7705	0.6742
2.12	0.1060	0.2013	0.5266	1.8690	0.5583	5.0768	2.8402	1.7875	0.6649
2.14	0.1027	0.1968	0.5219	1.9018	0.5554	5.1762	2.8683	1.8046	0.6557
2.16	0.9956^{-1}	0.1925	0.5173	1.9354	0.5525	5.2765	2.8962	1.8219	0.6464
2.18	0.9649^{-1}	0.1882	0.5127	1.9698	0.5498	5.3778	2.9238	1.8393	0.6373
2.20	0.9352^{-1}	0.1841	0.5081	2.0050	0.5471	5.4800	2.9512	1.8569	0.6281
2.22	0.9064^{-1}	0.1800	0.5036	2.0409	0.5444	5.5831	2.9784	1.8746	0.6191
2.24	0.8785^{-1}	0.1760	0.4991	2.0777	0.5418	5.6872	3.0053	1.8924	0.6100
2.26	0.8514^{-1}	0.1721	0.4947	2.1153	0.5393	5.7922	3.0319	1.9104	0.6011
2.28	0.8251^{-1}	0.1683	0.4903	2.1538	0.5368	5.8981	3.0584	1.9285	0.5921
2.30	0.7997^{-1}	0.1646	0.4859	2.1931	0.5344	6.0050	3.0845	1.9468	0.5833
2.32	0.7751^{-1}	0.1609	0.4816	2.2333	0.5321	6.1128	3.1105	1.9652	0.5745
2.34	0.7512^{-1}	0.1574	0.4773	2.2744	0.5297	6.2215	3.1362	1.9838	0.5658
2.36	0.7281^{-1}	0.1539	0.4731	2.3164	0.5275	6.3312	3.1617	2.0025	0.5572
2.38	0.7057^{-1}	0.1505	0.4688	2.3593	0.5253	6.4418	3.1869	2.0213	0.5486
2.40	0.6840^{-1}	0.1472	0.4647	2.4031	0.5231	6.5533	3.2119	2.0403	0.5401
2.42	0.6630^{-1}	0.1439	0.4606	2.4479	0.5210	6.6658	3.2367	2.0595	0.5317
2.44	0.6426^{-1}	0.1408	0.4565	2.4936	0.5189	6.7792	3.2612	2.0788	0.5234
2.46	0.6229^{-1}	0.1377	0.4524	2.5403	0.5169	6.8935	3.2855	2.0982	0.5152
2.48	0.6038^{-1}	0.1346	0.4484	2.5880	0.5149	7.0088	3.3095	2.1178	0.5071
2.50	0.5853^{-1}	0.1317	0.4444	2.6367	0.5130	7.1250	3.3333	2.1375	0.4990
2.52	0.5674^{-1}	0.1288	0.4405	2.6865	0.5111	7.2421	3.3569	2.1574	0.4911
2.54	0.5500^{-1}	0.1260	0.4366	2.7372	0.5092	7.3602	3.3803	2.1774	0.4832
2.56	0.5332^{-1}	0.1232	0.4328	2.7891	0.5074	7.4792	3.4034	2.1976	0.4754
2.58	0.5169^{-1}	0.1205	0.4289	2.8420	0.5056	7.5991	3.4263	2.2179	0.4677
2.60	0.5012^{-1}	0.1179	0.4252	2.8960	0.5039	7.7200	3.4490	2.2383	0.4601

n^{-p} is shorthand for $n \cdot 10^{-p}$

Supersonic Flow of a Perfect Gas with $\gamma = 1.4$ (continued)

	Isentropic Flow				Normal Shock Wave				
M_1	p/p_t	ρ/ρ_t	T/T_t	A/A^*	M_2	p_2/p_1	ρ_2/ρ_1	T_2/T_1	p_{t2}/p_{t1}
2.60	0.5012^{-1}	0.1179	0.4252	2.8960	0.5039	7.720	3.449	2.238	0.4601
2.62	0.4859^{-1}	0.1153	0.4214	2.9511	0.5022	7.842	3.471	2.259	0.4526
2.64	0.4711^{-1}	0.1128	0.4177	3.0073	0.5005	7.965	3.494	2.280	0.4452
2.66	0.4568^{-1}	0.1103	0.4141	3.0647	0.4988	8.088	3.516	2.301	0.4379
2.68	0.4429^{-1}	0.1079	0.4104	3.1233	0.4972	8.213	3.537	2.322	0.4307
2.70	0.4295^{-1}	0.1056	0.4068	3.1830	0.4956	8.338	3.559	2.343	0.4236
2.72	0.4165^{-1}	0.1033	0.4033	3.2440	0.4941	8.465	3.580	2.364	0.4166
2.74	0.4039^{-1}	0.1010	0.3998	3.3061	0.4926	8.592	3.601	2.386	0.4097
2.76	0.3917^{-1}	0.9885^{-1}	0.3963	3.3695	0.4911	8.721	3.622	2.407	0.4028
2.78	0.3799^{-1}	0.9671^{-1}	0.3928	3.4342	0.4896	8.850	3.643	2.429	0.3961
2.80	0.3685^{-1}	0.9463^{-1}	0.3894	3.5001	0.4882	8.980	3.664	2.451	0.3895
2.82	0.3574^{-1}	0.9259^{-1}	0.3860	3.5674	0.4868	9.111	3.684	2.473	0.3829
2.84	0.3467^{-1}	0.9059^{-1}	0.3827	3.6359	0.4854	9.243	3.704	2.496	0.3765
2.86	0.3363^{-1}	0.8865^{-1}	0.3794	3.7058	0.4840	9.376	3.724	2.518	0.3701
2.88	0.3263^{-1}	0.8675^{-1}	0.3761	3.7771	0.4827	9.510	3.743	2.540	0.3639
2.90	0.3165^{-1}	0.8489^{-1}	0.3729	3.8498	0.4814	9.645	3.763	2.563	0.3577
2.92	0.3071^{-1}	0.8307^{-1}	0.3696	3.9238	0.4801	9.781	3.782	2.586	0.3517
2.94	0.2980^{-1}	0.8130^{-1}	0.3665	3.9993	0.4788	9.918	3.801	2.609	0.3457
2.96	0.2891^{-1}	0.7957^{-1}	0.3633	4.0763	0.4776	10.055	3.820	2.632	0.3398
2.98	0.2805^{-1}	0.7788^{-1}	0.3602	4.1547	0.4764	10.194	3.839	2.656	0.3340
3.00	0.2722^{-1}	0.7623^{-1}	0.3571	4.2346	0.4752	10.333	3.857	2.679	0.3283
3.02	0.2642^{-1}	0.7461^{-1}	0.3541	4.3160	0.4740	10.474	3.875	2.703	0.3227
3.04	0.2564^{-1}	0.7303^{-1}	0.3511	4.3989	0.4729	10.615	3.893	2.726	0.3172
3.06	0.2489^{-1}	0.7149^{-1}	0.3481	4.4835	0.4717	10.758	3.911	2.750	0.3118
3.08	0.2416^{-1}	0.6999^{-1}	0.3452	4.5696	0.4706	10.901	3.929	2.774	0.3065
3.10	0.2345^{-1}	0.6852^{-1}	0.3422	4.6573	0.4695	11.045	3.947	2.799	0.3012
3.12	0.2276^{-1}	0.6708^{-1}	0.3393	4.7467	0.4685	11.190	3.964	2.823	0.2960
3.14	0.2210^{-1}	0.6568^{-1}	0.3365	4.8377	0.4674	11.336	3.981	2.848	0.2910
3.16	0.2146^{-1}	0.6430^{-1}	0.3337	4.9304	0.4664	11.483	3.998	2.872	0.2860
3.18	0.2083^{-1}	0.6296^{-1}	0.3309	5.0248	0.4654	11.631	4.015	2.897	0.2811
3.20	0.2023^{-1}	0.6165^{-1}	0.3281	5.1210	0.4643	11.780	4.031	2.922	0.2762
3.22	0.1964^{-1}	0.6037^{-1}	0.3253	5.2189	0.4634	11.930	4.048	2.947	0.2715
3.24	0.1908^{-1}	0.5912^{-1}	0.3226	5.3186	0.4624	12.081	4.064	2.972	0.2668
3.26	0.1853^{-1}	0.5790^{-1}	0.3199	5.4201	0.4614	12.232	4.080	2.998	0.2622
3.28	0.1799^{-1}	0.5671^{-1}	0.3173	5.5234	0.4605	12.385	4.096	3.023	0.2577
3.30	0.1748^{-1}	0.5554^{-1}	0.3147	5.6286	0.4596	12.538	4.112	3.049	0.2533
3.32	0.1698^{-1}	0.5440^{-1}	0.3121	5.7358	0.4587	12.693	4.128	3.075	0.2489
3.34	0.1649^{-1}	0.5329^{-1}	0.3095	5.8448	0.4578	12.848	4.143	3.101	0.2446
3.36	0.1602^{-1}	0.5220^{-1}	0.3069	5.9558	0.4569	13.005	4.158	3.127	0.2404
3.38	0.1557^{-1}	0.5113^{-1}	0.3044	6.0687	0.4560	13.162	4.173	3.154	0.2363
3.40	0.1512^{-1}	0.5009^{-1}	0.3019	6.1837	0.4552	13.320	4.188	3.180	0.2322

n^{-p} is shorthand for $n \cdot 10^{-p}$

Supersonic Flow of a Perfect Gas with $\gamma = 1.4$ (continued)

	Isentropic Flow					Normal Shock Wave			
M_1	p/p_t	ρ/ρ_t	T/T_t	A/A^*	M_2	p_2/p_1	ρ_2/ρ_1	T_2/T_1	p_{t2}/p_{t1}
3.4	0.1512^{-1}	0.5009^{-1}	0.3019	6.18	0.4552	13.32	4.188	3.180	0.2322
3.5	0.1311^{-1}	0.4523^{-1}	0.2899	6.79	0.4512	14.13	4.261	3.315	0.2129
3.6	0.1138^{-1}	0.4089^{-1}	0.2784	7.45	0.4474	14.95	4.330	3.454	0.1953
3.7	0.9903^{-2}	0.3702^{-1}	0.2675	8.17	0.4439	15.81	4.395	3.596	0.1792
3.8	0.8629^{-2}	0.3355^{-1}	0.2572	8.95	0.4407	16.68	4.457	3.743	0.1645
3.9	0.7532^{-2}	0.3044^{-1}	0.2474	9.80	0.4377	17.58	4.516	3.893	0.1510
4.0	0.6586^{-2}	0.2766^{-1}	0.2381	10.72	0.4350	18.50	4.571	4.047	0.1388
4.1	0.5769^{-2}	0.2516^{-1}	0.2293	11.71	0.4324	19.44	4.624	4.205	0.1276
4.2	0.5062^{-2}	0.2292^{-1}	0.2208	12.79	0.4299	20.41	4.675	4.367	0.1173
4.3	0.4449^{-2}	0.2090^{-1}	0.2129	13.95	0.4277	21.41	4.723	4.532	0.1080
4.4	0.3918^{-2}	0.1909^{-1}	0.2053	15.21	0.4255	22.42	4.768	4.702	0.9948^{-1}
4.5	0.3455^{-2}	0.1745^{-1}	0.1980	16.56	0.4236	23.46	4.812	4.875	0.9170^{-1}
4.6	0.3053^{-2}	0.1597^{-1}	0.1911	18.02	0.4217	24.52	4.853	5.052	0.8459^{-1}
4.7	0.2701^{-2}	0.1464^{-1}	0.1846	19.58	0.4199	25.60	4.893	5.233	0.7809^{-1}
4.8	0.2394^{-2}	0.1343^{-1}	0.1783	21.26	0.4183	26.71	4.930	5.418	0.7214^{-1}
4.9	0.2126^{-2}	0.1233^{-1}	0.1724	23.07	0.4167	27.84	4.966	5.607	0.6670^{-1}
5.0	0.1890^{-2}	0.1134^{-1}	0.1667	25.00	0.4152	29.00	5.000	5.800	0.6172^{-1}
5.1	0.1683^{-2}	0.1044^{-1}	0.1612	27.07	0.4138	30.18	5.033	5.997	0.5715^{-1}
5.2	0.1501^{-2}	0.9620^{-2}	0.1561	29.28	0.4125	31.38	5.064	6.197	0.5297^{-1}
5.3	0.1341^{-2}	0.8875^{-2}	0.1511	31.65	0.4113	32.60	5.093	6.401	0.4913^{-1}
5.4	0.1200^{-2}	0.8197^{-2}	0.1464	34.17	0.4101	33.85	5.122	6.610	0.4560^{-1}
5.5	0.1075^{-2}	0.7578^{-2}	0.1418	36.87	0.4090	35.13	5.149	6.822	0.4236^{-1}
5.6	0.9643^{-3}	0.7012^{-2}	0.1375	39.74	0.4079	36.42	5.175	7.038	0.3938^{-1}
5.7	0.8663^{-3}	0.6496^{-2}	0.1334	42.80	0.4069	37.74	5.200	7.258	0.3664^{-1}
5.8	0.7794^{-3}	0.6023^{-2}	0.1294	46.05	0.4059	39.08	5.224	7.481	0.3412^{-1}
5.9	0.7021^{-3}	0.5590^{-2}	0.1256	49.51	0.4050	40.44	5.246	7.709	0.3179^{-1}
6.0	0.6334^{-3}	0.5194^{-2}	0.1220	53.18	0.4042	41.83	5.268	7.941	0.2965^{-1}
6.1	0.5721^{-3}	0.4829^{-2}	0.1185	57.08	0.4033	43.24	5.289	8.176	0.2767^{-1}
6.2	0.5173^{-3}	0.4495^{-2}	0.1151	61.21	0.4025	44.68	5.309	8.415	0.2584^{-1}
6.3	0.4684^{-3}	0.4187^{-2}	0.1119	65.59	0.4018	46.14	5.329	8.658	0.2416^{-1}
6.4	0.4247^{-3}	0.3904^{-2}	0.1088	70.23	0.4011	47.62	5.347	8.905	0.2259^{-1}
6.5	0.3855^{-3}	0.3643^{-2}	0.1058	75.13	0.4004	49.13	5.365	9.156	0.2115^{-1}
6.6	0.3503^{-3}	0.3402^{-2}	0.1030	80.32	0.3997	50.65	5.382	9.411	0.1981^{-1}
6.7	0.3187^{-3}	0.3180^{-2}	0.1002	85.80	0.3991	52.20	5.399	9.670	0.1857^{-1}
6.8	0.2902^{-3}	0.2974^{-2}	0.0976	91.59	0.3985	53.78	5.415	9.933	0.1741^{-1}
6.9	0.2646^{-3}	0.2785^{-2}	0.0950	97.70	0.3979	55.38	5.430	10.199	0.1634^{-1}
7.0	0.2416^{-3}	0.2609^{-2}	0.0926	104.14	0.3974	57.00	5.444	10.469	0.1535^{-1}
7.1	0.2207^{-3}	0.2446^{-2}	0.0902	110.93	0.3968	58.64	5.459	10.744	0.1443^{-1}
7.2	0.2019^{-3}	0.2295^{-2}	0.0880	118.08	0.3963	60.31	5.472	11.022	0.1357^{-1}
7.3	0.1848^{-3}	0.2155^{-2}	0.0858	125.60	0.3958	62.00	5.485	11.304	0.1277^{-1}
7.4	0.1694^{-3}	0.2025^{-2}	0.0837	133.52	0.3954	63.72	5.498	11.590	0.1202^{-1}
7.5	0.1554^{-3}	0.1904^{-2}	0.0816	141.84	0.3949	65.46	5.510	11.879	0.1133^{-1}

n^{-p} is shorthand for $n \cdot 10^{-p}$

Supersonic Flow of a Perfect Gas with $\gamma = 1.4$ (continued)

	Isentropic Flow				Normal Shock Wave				
M_1	p/p_t	ρ/ρ_t	T/T_t	A/A^*	M_2	p_2/p_1	ρ_2/ρ_1	T_2/T_1	p_{t2}/p_{t1}
7.5	0.1554^{-3}	0.1904^{-2}	0.8163^{-1}	141.8	0.3949	65.46	5.510	11.88	0.1133^{-1}
8.0	0.1024^{-3}	0.1414^{-2}	0.7246^{-1}	190.1	0.3929	74.50	5.565	13.39	0.8488^{-2}
8.5	0.6898^{-4}	0.1066^{-2}	0.6472^{-1}	251.1	0.3912	84.13	5.612	14.99	0.6449^{-2}
9.0	0.4739^{-4}	0.8150^{-3}	0.5814^{-1}	327.2	0.3898	94.33	5.651	16.69	0.4964^{-2}
9.5	0.3314^{-4}	0.6313^{-3}	0.5249^{-1}	421.1	0.3886	105.13	5.685	18.49	0.3866^{-2}
10.0	0.2356^{-4}	0.4948^{-3}	0.4762^{-1}	535.9	0.3876	116.50	5.714	20.39	0.3045^{-2}

n^{-p} is shorthand for $n \cdot 10^{-p}$

B.2 Fanno Flow

Notation used in the Fanno-flow tables of this section is as follows.

M = local Mach number
fL^*/D = dimensionless length to sonic point
p/p^* = ratio of static to sonic pressure
ρ/ρ^* = ratio of static to sonic density
T/T^* = ratio of static to sonic temperature
p_t/p_t^* = ratio of local total to sonic total pressure

Fanno Flow for $\gamma = 1.4$

M	fL^*/D	p/p^*	ρ/ρ^*	T/T^*	p_t/p_t^*
0.00	∞	∞	∞	1.2000	∞
0.05	280.0203	21.9034	18.2620	1.1994	11.5914
0.10	66.9216	10.9435	9.1378	1.1976	5.8218
0.15	27.9320	7.2866	6.0995	1.1946	3.9103
0.20	14.5333	5.4554	4.5826	1.1905	2.9635
0.25	8.4834	4.3546	3.6742	1.1852	2.4027
0.30	5.2993	3.6191	3.0702	1.1788	2.0351
0.35	3.4525	3.0922	2.6400	1.1713	1.7780
0.40	2.3085	2.6958	2.3184	1.1628	1.5901
0.45	1.5664	2.3865	2.0693	1.1533	1.4487
0.50	1.0691	2.1381	1.8708	1.1429	1.3398
0.55	0.7281	1.9341	1.7092	1.1315	1.2549
0.60	0.4908	1.7634	1.5753	1.1194	1.1882
0.65	0.3246	1.6183	1.4626	1.1065	1.1356
0.70	0.2081	1.4935	1.3665	1.0929	1.0944
0.75	0.1273	1.3848	1.2838	1.0787	1.0624
0.80	0.0723	1.2893	1.2119	1.0638	1.0382
0.85	0.0363	1.2047	1.1489	1.0485	1.0207
0.90	0.0145	1.1291	1.0934	1.0327	1.0089
0.95	0.0033	1.0613	1.0440	1.0165	1.0021
1.00	0.0000	1.0000	1.0000	1.0000	1.0000

Fanno Flow for $\gamma = 1.4$ (continued)

M	fL^*/D	p/p^*	ρ/ρ^*	T/T^*	p_t/p_t^*
1.00	0.0000	1.0000	1.0000	1.0000	1.0000
1.05	0.0027	0.9443	0.9605	0.9832	1.0020
1.10	0.0099	0.8936	0.9249	0.9662	1.0079
1.15	0.0205	0.8471	0.8926	0.9490	1.0175
1.20	0.0336	0.8044	0.8633	0.9317	1.0304
1.25	0.0486	0.7649	0.8367	0.9143	1.0468
1.30	0.0648	0.7285	0.8123	0.8969	1.0663
1.35	0.0820	0.6947	0.7899	0.8794	1.0890
1.40	0.0997	0.6632	0.7693	0.8621	1.1149
1.45	0.1178	0.6339	0.7503	0.8448	1.1440
1.50	0.1361	0.6065	0.7328	0.8276	1.1762
1.55	0.1543	0.5808	0.7166	0.8105	1.2116
1.60	0.1724	0.5568	0.7016	0.7937	1.2502
1.65	0.1902	0.5342	0.6876	0.7770	1.2922
1.70	0.2078	0.5130	0.6745	0.7605	1.3376
1.75	0.2250	0.4929	0.6624	0.7442	1.3865
1.80	0.2419	0.4741	0.6511	0.7282	1.4390
1.85	0.2583	0.4562	0.6404	0.7124	1.4952
1.90	0.2743	0.4394	0.6305	0.6969	1.5553
1.95	0.2899	0.4234	0.6211	0.6816	1.6193
2.00	0.3050	0.4082	0.6124	0.6667	1.6875
2.05	0.3197	0.3939	0.6041	0.6520	1.7600
2.10	0.3339	0.3802	0.5963	0.6376	1.8369
2.15	0.3476	0.3673	0.5890	0.6235	1.9185
2.20	0.3609	0.3549	0.5821	0.6098	2.0050
2.25	0.3738	0.3432	0.5756	0.5963	2.0964
2.30	0.3862	0.3320	0.5694	0.5831	2.1931
2.35	0.3983	0.3213	0.5635	0.5702	2.2953
2.40	0.4099	0.3111	0.5580	0.5576	2.4031
2.45	0.4211	0.3014	0.5527	0.5453	2.5168
2.50	0.4320	0.2921	0.5477	0.5333	2.6367
2.55	0.4425	0.2832	0.5430	0.5216	2.7630
2.60	0.4526	0.2747	0.5385	0.5102	2.8960
2.65	0.4624	0.2666	0.5342	0.4991	3.0359
2.70	0.4718	0.2588	0.5301	0.4882	3.1830
2.75	0.4809	0.2513	0.5262	0.4776	3.3377
2.80	0.4898	0.2441	0.5225	0.4673	3.5001
2.85	0.4983	0.2373	0.5189	0.4572	3.6707
2.90	0.5065	0.2307	0.5155	0.4474	3.8498
2.95	0.5145	0.2243	0.5123	0.4379	4.0376
3.00	0.5222	0.2182	0.5092	0.4286	4.2346

Fanno Flow for $\gamma = 1.4$ (continued)

M	fL^*/D	p/p^*	ρ/ρ^*	T/T^*	p_t/p_t^*
3.00	0.5222	0.2182	0.5092	0.4286	4.2346
3.05	0.5296	0.2124	0.5062	0.4195	4.4410
3.10	0.5368	0.2067	0.5034	0.4107	4.6573
3.15	0.5437	0.2013	0.5007	0.4021	4.8838
3.20	0.5504	0.1961	0.4980	0.3937	5.1210
3.25	0.5569	0.1911	0.4955	0.3855	5.3691
3.30	0.5632	0.1862	0.4931	0.3776	5.6286
3.35	0.5693	0.1815	0.4908	0.3699	5.9000
3.40	0.5752	0.1770	0.4886	0.3623	6.1837
3.45	0.5809	0.1727	0.4865	0.3550	6.4801
3.50	0.5864	0.1685	0.4845	0.3478	6.7896
3.55	0.5918	0.1645	0.4825	0.3409	7.1128
3.60	0.5970	0.1606	0.4806	0.3341	7.4501
3.65	0.6020	0.1568	0.4788	0.3275	7.8020
3.70	0.6068	0.1531	0.4770	0.3210	8.1691
3.75	0.6115	0.1496	0.4753	0.3148	8.5517
3.80	0.6161	0.1462	0.4737	0.3086	8.9506
3.85	0.6206	0.1429	0.4721	0.3027	9.3661
3.90	0.6248	0.1397	0.4706	0.2969	9.7990
3.95	0.6290	0.1366	0.4691	0.2912	10.2496
4.00	0.6331	0.1336	0.4677	0.2857	10.7188
4.05	0.6370	0.1307	0.4663	0.2803	11.2069
4.10	0.6408	0.1279	0.4650	0.2751	11.7147
4.15	0.6445	0.1252	0.4637	0.2700	12.2427
4.20	0.6481	0.1226	0.4625	0.2650	12.7916
4.25	0.6516	0.1200	0.4613	0.2602	13.3622
4.30	0.6550	0.1175	0.4601	0.2554	13.9549
4.35	0.6583	0.1151	0.4590	0.2508	14.5706
4.40	0.6615	0.1128	0.4579	0.2463	15.2099
4.45	0.6646	0.1105	0.4569	0.2419	15.8735
4.50	0.6676	0.1083	0.4559	0.2376	16.5622
4.55	0.6706	0.1062	0.4549	0.2334	17.2767
4.60	0.6734	0.1041	0.4539	0.2294	18.0178
4.65	0.6762	0.1021	0.4530	0.2254	18.7862
4.70	0.6790	0.1001	0.4521	0.2215	19.5828
4.75	0.6816	0.0982	0.4512	0.2177	20.4084
4.80	0.6842	0.0964	0.4504	0.2140	21.2637
4.85	0.6867	0.0946	0.4495	0.2104	22.1497
4.90	0.6891	0.0928	0.4487	0.2068	23.0671
4.95	0.6915	0.0911	0.4480	0.2034	24.0169
5.00	0.6938	0.0894	0.4472	0.2000	25.0000

B.3   Rayleigh Flow

Notation used in the Rayleigh-flow tables of this section is as follows.

M = local Mach number
T_t/T_t^* = ratio of local total to sonic total temperature
p/p^* = ratio of static to sonic pressure
ρ/ρ^* = ratio of static to sonic density
T/T^* = ratio of static to sonic temperature
p_t/p_t^* = ratio of local total to sonic total pressure

Rayleigh Flow for $\gamma = 1.4$

M	T_t/T_t^*	p/p^*	ρ/ρ^*	T/T^*	p_t/p_t^*
0.00	0.0000	2.4000	∞	0.0000	1.2679
0.05	0.0119	2.3916	167.2500	0.0143	1.2657
0.10	0.0468	2.3669	42.2500	0.0560	1.2591
0.15	0.1020	2.3267	19.1019	0.1218	1.2486
0.20	0.1736	2.2727	11.0000	0.2066	1.2346
0.25	0.2568	2.2069	7.2500	0.3044	1.2177
0.30	0.3469	2.1314	5.2130	0.4089	1.1985
0.35	0.4389	2.0487	3.9847	0.5141	1.1779
0.40	0.5290	1.9608	3.1875	0.6151	1.1566
0.45	0.6139	1.8699	2.6409	0.7080	1.1351
0.50	0.6914	1.7778	2.2500	0.7901	1.1141
0.55	0.7599	1.6860	1.9607	0.8599	1.0940
0.60	0.8189	1.5957	1.7407	0.9167	1.0753
0.65	0.8683	1.5080	1.5695	0.9608	1.0582
0.70	0.9085	1.4235	1.4337	0.9929	1.0431
0.75	0.9401	1.3427	1.3241	1.0140	1.0301
0.80	0.9639	1.2658	1.2344	1.0255	1.0193
0.85	0.9810	1.1931	1.1600	1.0285	1.0109
0.90	0.9921	1.1246	1.0977	1.0245	1.0049
0.95	0.9981	1.0603	1.0450	1.0146	1.0012
1.00	1.0000	1.0000	1.0000	1.0000	1.0000

Rayleigh Flow for $\gamma = 1.4$ (continued)

M	T_t/T_t^*	p/p^*	ρ/ρ^*	T/T^*	p_t/p_t^*
1.00	1.0000	1.0000	1.0000	1.0000	1.0000
1.05	0.9984	0.9436	0.9613	0.9816	1.0012
1.10	0.9939	0.8909	0.9277	0.9603	1.0049
1.15	0.9872	0.8417	0.8984	0.9369	1.0109
1.20	0.9787	0.7958	0.8727	0.9118	1.0194
1.25	0.9689	0.7529	0.8500	0.8858	1.0303
1.30	0.9580	0.7130	0.8299	0.8592	1.0437
1.35	0.9464	0.6758	0.8120	0.8323	1.0594
1.40	0.9343	0.6410	0.7959	0.8054	1.0777
1.45	0.9218	0.6086	0.7815	0.7787	1.0983
1.50	0.9093	0.5783	0.7685	0.7525	1.1215
1.55	0.8967	0.5500	0.7568	0.7268	1.1473
1.60	0.8842	0.5236	0.7461	0.7017	1.1756
1.65	0.8718	0.4988	0.7364	0.6774	1.2066
1.70	0.8597	0.4756	0.7275	0.6538	1.2402
1.75	0.8478	0.4539	0.7194	0.6310	1.2767
1.80	0.8363	0.4335	0.7119	0.6089	1.3159
1.85	0.8250	0.4144	0.7051	0.5877	1.3581
1.90	0.8141	0.3964	0.6988	0.5673	1.4033
1.95	0.8036	0.3795	0.6929	0.5477	1.4516
2.00	0.7934	0.3636	0.6875	0.5289	1.5031
2.05	0.7835	0.3487	0.6825	0.5109	1.5579
2.10	0.7741	0.3345	0.6778	0.4936	1.6162
2.15	0.7649	0.3212	0.6735	0.4770	1.6780
2.20	0.7561	0.3086	0.6694	0.4611	1.7434
2.25	0.7477	0.2968	0.6656	0.4458	1.8128
2.30	0.7395	0.2855	0.6621	0.4312	1.8860
2.35	0.7317	0.2749	0.6588	0.4172	1.9634
2.40	0.7242	0.2648	0.6557	0.4038	2.0451
2.45	0.7170	0.2552	0.6527	0.3910	2.1311
2.50	0.7101	0.2462	0.6500	0.3787	2.2218
2.55	0.7034	0.2375	0.6474	0.3669	2.3173
2.60	0.6970	0.2294	0.6450	0.3556	2.4177
2.65	0.6908	0.2216	0.6427	0.3448	2.5233
2.70	0.6849	0.2142	0.6405	0.3344	2.6343
2.75	0.6793	0.2071	0.6384	0.3244	2.7508
2.80	0.6738	0.2004	0.6365	0.3149	2.8731
2.85	0.6685	0.1940	0.6346	0.3057	3.0014
2.90	0.6635	0.1879	0.6329	0.2969	3.1359
2.95	0.6586	0.1820	0.6312	0.2884	3.2768
3.00	0.6540	0.1765	0.6296	0.2803	3.4245

Rayleigh Flow for $\gamma = 1.4$ (continued)

M	T_t/T_t^*	p/p^*	ρ/ρ^*	T/T^*	p_t/p_t^*
3.00	0.6540	0.1765	0.6296	0.2803	3.4245
3.05	0.6495	0.1711	0.6281	0.2725	3.5790
3.10	0.6452	0.1660	0.6267	0.2650	3.7408
3.15	0.6410	0.1612	0.6253	0.2577	3.9101
3.20	0.6370	0.1565	0.6240	0.2508	4.0871
3.25	0.6331	0.1520	0.6228	0.2441	4.2721
3.30	0.6294	0.1477	0.6216	0.2377	4.4655
3.35	0.6258	0.1436	0.6205	0.2315	4.6674
3.40	0.6224	0.1397	0.6194	0.2255	4.8783
3.45	0.6190	0.1359	0.6183	0.2197	5.0984
3.50	0.6158	0.1322	0.6173	0.2142	5.3280
3.55	0.6127	0.1287	0.6164	0.2088	5.5676
3.60	0.6097	0.1254	0.6155	0.2037	5.8173
3.65	0.6068	0.1221	0.6146	0.1987	6.0776
3.70	0.6040	0.1190	0.6138	0.1939	6.3488
3.75	0.6013	0.1160	0.6130	0.1893	6.6314
3.80	0.5987	0.1131	0.6122	0.1848	6.9256
3.85	0.5962	0.1103	0.6114	0.1805	7.2318
3.90	0.5937	0.1077	0.6107	0.1763	7.5505
3.95	0.5914	0.1051	0.6100	0.1722	7.8820
4.00	0.5891	0.1026	0.6094	0.1683	8.2268
4.05	0.5869	0.1002	0.6087	0.1645	8.5853
4.10	0.5847	0.0978	0.6081	0.1609	8.9579
4.15	0.5827	0.0956	0.6075	0.1573	9.3451
4.20	0.5807	0.0934	0.6070	0.1539	9.7473
4.25	0.5787	0.0913	0.6064	0.1506	10.1649
4.30	0.5768	0.0893	0.6059	0.1473	10.5985
4.35	0.5750	0.0873	0.6054	0.1442	11.0486
4.40	0.5732	0.0854	0.6049	0.1412	11.5155
4.45	0.5715	0.0836	0.6044	0.1383	11.9999
4.50	0.5698	0.0818	0.6039	0.1354	12.5023
4.55	0.5682	0.0800	0.6035	0.1326	13.0231
4.60	0.5666	0.0784	0.6030	0.1300	13.5629
4.65	0.5651	0.0767	0.6026	0.1274	14.1223
4.70	0.5636	0.0752	0.6022	0.1248	14.7017
4.75	0.5622	0.0736	0.6018	0.1224	15.3019
4.80	0.5608	0.0722	0.6014	0.1200	15.9234
4.85	0.5594	0.0707	0.6010	0.1177	16.5667
4.90	0.5581	0.0693	0.6007	0.1154	17.2325
4.95	0.5568	0.0680	0.6003	0.1132	17.9213
5.00	0.5556	0.0667	0.6000	0.1111	18.6339

Appendix C

Vector Differential Operator ∇

C.1 Definition of ∇

Much of the analysis in the main text uses rectangular Cartesian coordinates x, y, and z with corresponding unit vectors $\mathbf{i}$, $\mathbf{j}$ and $\mathbf{k}$ as shown in Figure C.1. The vector differential operator ∇ is defined by

$$\nabla = \mathbf{i}\,\frac{\partial}{\partial x} + \mathbf{j}\,\frac{\partial}{\partial y} + \mathbf{k}\,\frac{\partial}{\partial z} \tag{C.1}$$

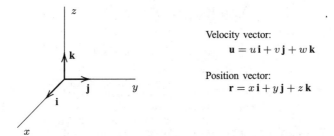

Velocity vector:
$$\mathbf{u} = u\,\mathbf{i} + v\,\mathbf{j} + w\,\mathbf{k}$$

Position vector:
$$\mathbf{r} = x\,\mathbf{i} + y\,\mathbf{j} + z\,\mathbf{k}$$

Figure C.1: *Rectangular Cartesian coordinates.*

C.2 Gradient, Divergence and Curl

The primary ways ∇ can operate on a vector $\mathbf{F} = F_x\mathbf{i} + F_y\mathbf{j} + F_z\mathbf{k}$ or a scalar ϕ are as follows.

$$\nabla \cdot \mathbf{F} = div\,\mathbf{F} = \frac{\partial F_x}{\partial x} + \frac{\partial F_y}{\partial y} + \frac{\partial F_z}{\partial z} \tag{C.2}$$

$$\nabla \phi = grad\,\phi = \mathbf{i}\,\frac{\partial \phi}{\partial x} + \mathbf{j}\,\frac{\partial \phi}{\partial y} + \mathbf{k}\,\frac{\partial \phi}{\partial z} \tag{C.3}$$

$$\nabla \times \mathbf{F} = curl\,\mathbf{F} = \mathbf{i}\left[\frac{\partial F_z}{\partial y} - \frac{\partial F_y}{\partial z}\right] + \mathbf{j}\left[\frac{\partial F_x}{\partial z} - \frac{\partial F_z}{\partial x}\right] + \mathbf{k}\left[\frac{\partial F_y}{\partial x} - \frac{\partial F_x}{\partial y}\right] \tag{C.4}$$

C.3   Directional Derivative

There is a particularly illuminating geometrical interpretation of the differential operator ∇. Specifically, if $\mathbf{n}$ is a unit vector, then

$$\mathbf{n} \cdot \nabla = n_x \frac{\partial}{\partial x} + n_y \frac{\partial}{\partial y} + n_z \frac{\partial}{\partial z} = \frac{\partial}{\partial n} \tag{C.5}$$

That is, $\mathbf{n} \cdot \nabla$ is the **directional derivative** with $\partial/\partial n$ representing the rate of change in the direction of $\mathbf{n}$.

C.4   Operations and Identities

Finally, the following relations hold for any vectors $\mathbf{u}$ and $\mathbf{v}$ and scalars ρ and ϕ. Note that the **Laplacian operator**, ∇^2, is the dot product of ∇ with itself, i.e., $\nabla^2 = \nabla \cdot \nabla$.

$$\nabla \times \nabla \phi = \mathbf{0} \tag{C.6}$$

$$\nabla \cdot (\nabla \times \mathbf{u}) = 0 \tag{C.7}$$

$$\nabla \times (\nabla \times \mathbf{u}) = \nabla(\nabla \cdot \mathbf{u}) - \nabla^2 \mathbf{u} \tag{C.8}$$

$$\nabla \cdot (\rho \mathbf{u}) = \mathbf{u} \cdot \nabla \rho + \rho \nabla \cdot \mathbf{u} \tag{C.9}$$

$$\nabla \times (\rho \mathbf{u}) = \rho \nabla \times \mathbf{u} + \nabla \rho \times \mathbf{u} \tag{C.10}$$

$$\mathbf{u} \cdot \nabla \mathbf{u} = \nabla \left(\frac{1}{2} \mathbf{u} \cdot \mathbf{u} \right) - \mathbf{u} \times (\nabla \times \mathbf{u}) \tag{C.11}$$

$$\nabla \times (\mathbf{u} \times \mathbf{v}) = (\mathbf{v} \cdot \nabla)\mathbf{u} + \mathbf{u}(\nabla \cdot \mathbf{v}) - (\mathbf{u} \cdot \nabla)\mathbf{v} - \mathbf{v}(\nabla \cdot \mathbf{u}) \tag{C.12}$$

Appendix D

Equations of Motion in Various Coordinate Systems

Often, taking advantage of special coordinate systems can simplify the problem we wish to solve. For example, problems with cylindrical geometries such as flow in a pipe are most conveniently solved using cylindrical polar coordinates. The purpose of this Appendix is to list the incompressible continuity and Navier-Stokes[1] equations for several useful coordinate systems.

For a general body force per unit mass, **f**, the Navier-Stokes equation is

$$\frac{d\mathbf{u}}{dt} = -\frac{1}{\rho}\nabla p + \mathbf{f} + \nu\nabla^2\mathbf{u} \tag{D.1}$$

where the acceleration vector, $d\mathbf{u}/dt$, is computed as the Eulerian derivative of velocity, **u**, defined as follows.

$$\frac{d\mathbf{u}}{dt} = \frac{\partial\mathbf{u}}{\partial t} + \mathbf{u}\cdot\nabla\mathbf{u} = \frac{\partial\mathbf{u}}{\partial t} + \left(u\frac{\partial}{\partial x} + v\frac{\partial}{\partial y} + w\frac{\partial}{\partial z}\right)\mathbf{u} \tag{D.2}$$

The convective part of the acceleration, $\mathbf{u}\cdot\nabla\mathbf{u}$, can be expressed in terms of the more familiar *grad* and *curl* operations. Specifically,

$$\mathbf{u}\cdot\nabla\mathbf{u} = \nabla\left(\frac{1}{2}\mathbf{u}\cdot\mathbf{u}\right) - \mathbf{u}\times(\nabla\times\mathbf{u}) \tag{D.3}$$

We can make use of a theorem from tensor calculus that tells us the following. Any vector identity that can be proven in Cartesian coordinates and written strictly in terms of *grad*, *div* and *curl* operations holds in all coordinate systems. Hence, we can rewrite the continuity equation and the Navier-Stokes equation as follows.

$$\nabla\cdot\mathbf{u} = 0 \tag{D.4}$$

$$\frac{\partial\mathbf{u}}{\partial t} + \nabla\left(\frac{1}{2}\mathbf{u}\cdot\mathbf{u}\right) - \mathbf{u}\times(\nabla\times\mathbf{u}) = -\frac{1}{\rho}\nabla p + \mathbf{f} + \nu\nabla^2\mathbf{u} \tag{D.5}$$

Beginning with Equations (D.4) and (D.5), we can write the continuity and Navier-Stokes equations in any coordinate system for which we know how to express *grad*, *div* and *curl*. This appendix lists the equations, in component form, for several useful coordinate systems.

---

[1]Euler's equation follows from setting $\nu = 0$ in the Navier-Stokes equation.

D.1 Rectangular Cartesian Coordinates

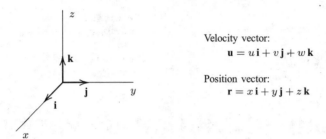

Velocity vector:
$$\mathbf{u} = u\,\mathbf{i} + v\,\mathbf{j} + w\,\mathbf{k}$$

Position vector:
$$\mathbf{r} = x\,\mathbf{i} + y\,\mathbf{j} + z\,\mathbf{k}$$

Figure D.1: *Rectangular Cartesian coordinates.*

We express the velocity vector, $\mathbf{u}$, in terms of its three components u, v and w as shown in Figure D.1. Typical *grad*, *div*, *curl* and Laplacian operations are as follows.

$$\nabla p = \mathbf{i}\,\frac{\partial p}{\partial x} + \mathbf{j}\,\frac{\partial p}{\partial y} + \mathbf{k}\,\frac{\partial p}{\partial z} \tag{D.6}$$

$$\nabla \cdot \mathbf{u} = \frac{\partial u}{\partial x} + \frac{\partial v}{\partial y} + \frac{\partial w}{\partial z} \tag{D.7}$$

$$\nabla \times \mathbf{u} = \mathbf{i}\left[\frac{\partial w}{\partial y} - \frac{\partial v}{\partial z}\right] + \mathbf{j}\left[\frac{\partial u}{\partial z} - \frac{\partial w}{\partial x}\right] + \mathbf{k}\left[\frac{\partial v}{\partial x} - \frac{\partial u}{\partial y}\right] \tag{D.8}$$

$$\nabla^2 \mathbf{u} = \left[\frac{\partial^2}{\partial x^2} + \frac{\partial^2}{\partial y^2} + \frac{\partial^2}{\partial z^2}\right](u\,\mathbf{i} + v\,\mathbf{j} + w\,\mathbf{k}) \tag{D.9}$$

The continuity equation and the three components of the Navier-Stokes equation with a body force $\mathbf{f} = f_x\mathbf{i} + f_y\mathbf{j} + f_z\mathbf{k}$ are:

$$\frac{\partial u}{\partial x} + \frac{\partial v}{\partial y} + \frac{\partial w}{\partial z} = 0 \tag{D.10}$$

$$\frac{\partial u}{\partial t} + u\frac{\partial u}{\partial x} + v\frac{\partial u}{\partial y} + w\frac{\partial u}{\partial z} = -\frac{1}{\rho}\frac{\partial p}{\partial x} + f_x + \nu\left(\frac{\partial^2 u}{\partial x^2} + \frac{\partial^2 u}{\partial y^2} + \frac{\partial^2 u}{\partial z^2}\right) \tag{D.11}$$

$$\frac{\partial v}{\partial t} + u\frac{\partial v}{\partial x} + v\frac{\partial v}{\partial y} + w\frac{\partial v}{\partial z} = -\frac{1}{\rho}\frac{\partial p}{\partial y} + f_y + \nu\left(\frac{\partial^2 v}{\partial x^2} + \frac{\partial^2 v}{\partial y^2} + \frac{\partial^2 v}{\partial z^2}\right) \tag{D.12}$$

$$\frac{\partial w}{\partial t} + u\frac{\partial w}{\partial x} + v\frac{\partial w}{\partial y} + w\frac{\partial w}{\partial z} = -\frac{1}{\rho}\frac{\partial p}{\partial z} + f_z + \nu\left(\frac{\partial^2 w}{\partial x^2} + \frac{\partial^2 w}{\partial y^2} + \frac{\partial^2 w}{\partial z^2}\right) \tag{D.13}$$

D.2 Cylindrical Polar Coordinates

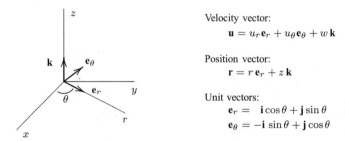

Velocity vector:
$$\mathbf{u} = u_r \mathbf{e}_r + u_\theta \mathbf{e}_\theta + w\,\mathbf{k}$$

Position vector:
$$\mathbf{r} = r\,\mathbf{e}_r + z\,\mathbf{k}$$

Unit vectors:
$$\mathbf{e}_r = \mathbf{i}\cos\theta + \mathbf{j}\sin\theta$$
$$\mathbf{e}_\theta = -\mathbf{i}\sin\theta + \mathbf{j}\cos\theta$$

Figure D.2: *Cylindrical polar coordinates.*

We express the velocity vector, $\mathbf{u}$, in terms of its three components u_r, u_θ and w as shown in Figure D.2. Typical *grad*, *div*, *curl* and Laplacian operations are as follows

$$\nabla p = \mathbf{e}_r \frac{\partial p}{\partial r} + \mathbf{e}_\theta \frac{1}{r}\frac{\partial p}{\partial \theta} + \mathbf{k}\frac{\partial p}{\partial z} \tag{D.14}$$

$$\nabla \cdot \mathbf{u} = \frac{1}{r}\frac{\partial}{\partial r}(r u_r) + \frac{1}{r}\frac{\partial u_\theta}{\partial \theta} + \frac{\partial w}{\partial z} \tag{D.15}$$

$$\nabla \times \mathbf{u} = \mathbf{e}_r\left[\frac{1}{r}\frac{\partial w}{\partial \theta} - \frac{\partial u_\theta}{\partial z}\right] + \mathbf{e}_\theta\left[\frac{\partial u_r}{\partial z} - \frac{\partial w}{\partial r}\right] + \mathbf{k}\left[\frac{1}{r}\frac{\partial}{\partial r}(r u_\theta) - \frac{1}{r}\frac{\partial u_r}{\partial \theta}\right] \tag{D.16}$$

$$\nabla^2 \mathbf{u} = \left[\frac{1}{r}\frac{\partial}{\partial r}\left(r\frac{\partial}{\partial r}\right) + \frac{1}{r^2}\frac{\partial^2}{\partial \theta^2} + \frac{\partial^2}{\partial z^2}\right](u_r \mathbf{e}_r + u_\theta \mathbf{e}_\theta + w\,\mathbf{k}) \tag{D.17}$$

The continuity equation and the three components of the Navier-Stokes equation with a body force $\mathbf{f} = f_r \mathbf{e}_r + f_\theta \mathbf{e}_\theta + f_z \mathbf{k}$ are:

$$\frac{1}{r}\frac{\partial}{\partial r}(r u_r) + \frac{1}{r}\frac{\partial u_\theta}{\partial \theta} + \frac{\partial w}{\partial z} = 0 \tag{D.18}$$

$$\frac{\partial u_r}{\partial t} + u_r \frac{\partial u_r}{\partial r} + \frac{u_\theta}{r}\frac{\partial u_r}{\partial \theta} + w\frac{\partial u_r}{\partial z} - \frac{u_\theta^2}{r} = -\frac{1}{\rho}\frac{\partial p}{\partial r} + f_r$$
$$+ \nu\left(\frac{\partial^2 u_r}{\partial r^2} + \frac{1}{r^2}\frac{\partial^2 u_r}{\partial \theta^2} + \frac{\partial^2 u_r}{\partial z^2} + \frac{1}{r}\frac{\partial u_r}{\partial r} - \frac{2}{r^2}\frac{\partial u_\theta}{\partial \theta} - \frac{u_r}{r^2}\right) \tag{D.19}$$

$$\frac{\partial u_\theta}{\partial t} + u_r \frac{\partial u_\theta}{\partial r} + \frac{u_\theta}{r}\frac{\partial u_\theta}{\partial \theta} + w\frac{\partial u_\theta}{\partial z} + \frac{u_r u_\theta}{r} = -\frac{1}{\rho r}\frac{\partial p}{\partial \theta} + f_\theta$$
$$+ \nu\left(\frac{\partial^2 u_\theta}{\partial r^2} + \frac{1}{r^2}\frac{\partial^2 u_\theta}{\partial \theta^2} + \frac{\partial^2 u_\theta}{\partial z^2} + \frac{1}{r}\frac{\partial u_\theta}{\partial r} + \frac{2}{r^2}\frac{\partial u_r}{\partial \theta} - \frac{u_\theta}{r^2}\right) \tag{D.20}$$

$$\frac{\partial w}{\partial t} + u_r \frac{\partial w}{\partial r} + \frac{u_\theta}{r}\frac{\partial w}{\partial \theta} + w\frac{\partial w}{\partial z} = -\frac{1}{\rho}\frac{\partial p}{\partial z} + f_z$$
$$+ \nu\left(\frac{\partial^2 w}{\partial r^2} + \frac{1}{r^2}\frac{\partial^2 w}{\partial \theta^2} + \frac{\partial^2 w}{\partial z^2} + \frac{1}{r}\frac{\partial w}{\partial r}\right) \tag{D.21}$$

D.3   Spherical Coordinates

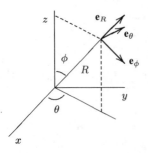

Velocity vector:
$$\mathbf{u} = u_R \mathbf{e}_R + u_\theta \mathbf{e}_\theta + u_\phi \mathbf{e}_\phi$$

Position vector:
$$\mathbf{r} = R \mathbf{e}_R$$

Unit vectors:
$$\mathbf{e}_R = \mathbf{i} \cos\theta \sin\phi + \mathbf{j} \sin\theta \sin\phi + \mathbf{k} \cos\phi$$
$$\mathbf{e}_\theta = -\mathbf{i} \sin\theta + \mathbf{j} \cos\theta$$
$$\mathbf{e}_\phi = \mathbf{i} \cos\theta \cos\phi + \mathbf{j} \sin\theta \cos\phi - \mathbf{k} \sin\phi$$

Figure D.3: *Spherical coordinates.*

We express the velocity vector, **u**, in terms of its three components u_R, u_θ and u_ϕ as shown in Figure D.3. Typical *grad, div, curl* and Laplacian operations are as follows.

$$\nabla p = \mathbf{e}_R \frac{\partial p}{\partial R} + \mathbf{e}_\theta \frac{1}{R\sin\phi}\frac{\partial p}{\partial \theta} + \mathbf{e}_\phi \frac{1}{R}\frac{\partial p}{\partial \phi} \tag{D.22}$$

$$\nabla \cdot \mathbf{u} = \frac{1}{R^2}\frac{\partial}{\partial R}(R^2 u_R) + \frac{1}{R\sin\phi}\frac{\partial u_\theta}{\partial \theta} + \frac{1}{R\sin\phi}\frac{\partial}{\partial \phi}(u_\phi \sin\phi) \tag{D.23}$$

$$\begin{aligned}
\nabla \times \mathbf{u} = {} & \frac{\mathbf{e}_R}{R\sin\phi}\left[\frac{\partial}{\partial \phi}(u_\theta \sin\phi) - \frac{\partial u_\phi}{\partial \theta}\right] \\
& + \frac{\mathbf{e}_\theta}{R}\left[\frac{\partial}{\partial R}(Ru_\phi) - \frac{\partial u_R}{\partial \phi}\right] \\
& + \frac{\mathbf{e}_\phi}{R\sin\phi}\left[\frac{\partial u_R}{\partial \theta} - \frac{\partial}{\partial R}(Ru_\theta \sin\phi)\right]
\end{aligned} \tag{D.24}$$

$$\begin{aligned}
\nabla^2 \mathbf{u} = {} & \left[\frac{1}{R^2}\frac{\partial}{\partial R}\left(R^2\frac{\partial}{\partial R}\right) + \frac{1}{R^2\sin^2\phi}\frac{\partial^2}{\partial \theta^2}\right. \\
& \left. + \frac{1}{R^2\sin\phi}\frac{\partial}{\partial \phi}\left(\sin\phi\frac{\partial}{\partial \phi}\right)\right](u_R\mathbf{e}_R + u_\theta\mathbf{e}_\theta + u_\phi\mathbf{e}_\phi)
\end{aligned} \tag{D.25}$$

The continuity equation and the three components of the Navier-Stokes equation with a body force $\mathbf{f} = f_R\mathbf{e}_R + f_\theta\mathbf{e}_\theta + f_\phi\mathbf{e}_\phi$ are:

$$\frac{1}{R^2}\frac{\partial}{\partial R}(R^2 u_R) + \frac{1}{R\sin\phi}\frac{\partial u_\theta}{\partial \theta} + \frac{1}{R\sin\phi}\frac{\partial}{\partial \phi}(u_\phi \sin\phi) = 0 \tag{D.26}$$

$$\begin{aligned}
& \frac{\partial u_R}{\partial t} + u_R\frac{\partial u_R}{\partial R} + \frac{u_\theta}{R\sin\phi}\frac{\partial u_R}{\partial \theta} + \frac{u_\phi}{R}\frac{\partial u_R}{\partial \phi} - \frac{u_\phi^2 + u_\theta^2}{R} \\
& = -\frac{1}{\rho}\frac{\partial p}{\partial R} + f_R + \nu\left[\frac{1}{R}\frac{\partial^2(Ru_R)}{\partial R^2} + \frac{1}{R^2\sin^2\phi}\frac{\partial^2 u_R}{\partial \theta^2} + \frac{1}{R^2}\frac{\partial^2 u_R}{\partial \phi^2}\right. \\
& \quad \left. + \frac{\cot\phi}{R^2}\frac{\partial u_R}{\partial \phi} - \frac{2}{R^2}\frac{\partial u_\phi}{\partial \phi} - \frac{2}{R^2\sin\phi}\frac{\partial u_\theta}{\partial \theta} - \frac{2u_R}{R^2} - \frac{2u_\phi\cot\phi}{R^2}\right]
\end{aligned} \tag{D.27}$$

$$\frac{\partial u_\theta}{\partial t} + u_R \frac{\partial u_\theta}{\partial R} + \frac{u_\theta}{R \sin \phi} \frac{\partial u_\theta}{\partial \theta} + \frac{u_\phi}{R} \frac{\partial u_\theta}{\partial \phi} + \frac{u_R u_\theta}{R} + \frac{u_\phi u_\theta \cot \phi}{R}$$

$$= -\frac{1}{\rho R \sin \phi} \frac{\partial p}{\partial \theta} + f_\theta + \nu \left[\frac{1}{R} \frac{\partial^2 (R u_\theta)}{\partial R^2} + \frac{1}{R^2 \sin^2 \phi} \frac{\partial^2 u_\theta}{\partial \theta^2} + \frac{1}{R^2} \frac{\partial^2 u_\theta}{\partial \phi^2} \right.$$

$$\left. + \frac{\cot \phi}{R^2} \frac{\partial u_\theta}{\partial \phi} + \frac{2}{R^2 \sin \phi} \frac{\partial u_R}{\partial \theta} + \frac{2 \cos \phi}{R^2 \sin^2 \phi} \frac{\partial u_\phi}{\partial \theta} - \frac{u_\theta}{R^2 \sin^2 \phi} \right] \tag{D.28}$$

$$\frac{\partial u_\phi}{\partial t} + u_R \frac{\partial u_\phi}{\partial R} + \frac{u_\theta}{R \sin \phi} \frac{\partial u_\phi}{\partial \theta} + \frac{u_\phi}{R} \frac{\partial u_\phi}{\partial \phi} + \frac{u_R u_\phi}{R} - \frac{u_\theta^2 \cot \phi}{R}$$

$$= -\frac{1}{\rho R} \frac{\partial p}{\partial \phi} + f_\phi + \nu \left[\frac{1}{R} \frac{\partial^2 (R u_\phi)}{\partial R^2} + \frac{1}{R^2 \sin^2 \phi} \frac{\partial^2 u_\phi}{\partial \theta^2} + \frac{1}{R^2} \frac{\partial^2 u_\phi}{\partial \phi^2} \right.$$

$$\left. + \frac{\cot \phi}{R^2} \frac{\partial u_\phi}{\partial \phi} - \frac{2 \cos \phi}{R^2 \sin^2 \phi} \frac{\partial u_\theta}{\partial \theta} + \frac{2}{R^2} \frac{\partial u_R}{\partial \phi} - \frac{u_\phi}{R^2 \sin^2 \phi} \right] \tag{D.29}$$

D.4 Rotating Coordinate System

For flow in a coordinate system rotating with angular velocity Ω, the absolute velocity (velocity seen by a stationary observer), $\mathbf{u}_{abs}$, is related to the velocity seen by an observer in the rotating frame, $\mathbf{u}$, according to

$$\mathbf{u}_{abs} = \mathbf{u} + \Omega \times \mathbf{r} \tag{D.30}$$

where $\mathbf{r}$ is position vector. Figure D.4 shows a coordinate system rotating about the z axis.

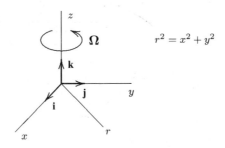

Figure D.4: *Rotating coordinate system.*

A key difference from stationary coordinates is the presence of the Coriolis and centrifugal accelerations, which are usually treated as body forces. For convenience, denote the sum of these two forces by $\mathbf{f}_{Extra}$. Then, we write

$$\mathbf{f}_{Extra} = \mathbf{f}_{Coriolis} + \mathbf{f}_{Centrifugal} \tag{D.31}$$

where

$$\mathbf{f}_{Coriolis} = -2\Omega \times \mathbf{u} \tag{D.32}$$

$$\mathbf{f}_{Centrifugal} = -\Omega \times \Omega \times \mathbf{r} \tag{D.33}$$

The Coriolis force is non-conservative and cannot be simplified. By contrast, the centrifugal force is conservative as can be seen from the following operations.

$$-\mathbf{\Omega} \times \mathbf{\Omega} \times \mathbf{r} = -\Omega \mathbf{k} \times \Omega \mathbf{k} \times (x\,\mathbf{i} + y\,\mathbf{j} + z\,\mathbf{k}) = \Omega^2(x\,\mathbf{i} + y\,\mathbf{j}) \tag{D.34}$$

However, the vector $(x\,\mathbf{i} + y\,\mathbf{j})$ can be written as the gradient of $\frac{1}{2}r^2$, where r is the radial distance from the origin in the xy plane, viz,

$$x\,\mathbf{i} + y\,\mathbf{j} = \frac{1}{2}\nabla\left(x^2 + y^2\right) = \frac{1}{2}\nabla r^2 \tag{D.35}$$

Therefore, we can rewrite the centrifugal force as follows.

$$\mathbf{f}_{Centrifugal} = \nabla\left(\frac{1}{2}\Omega^2 r^2\right) \tag{D.36}$$

Thus, the Navier-Stokes equation in a rotating coordinate system assumes the following form.

$$\frac{d\mathbf{u}}{dt} + 2\mathbf{\Omega} \times \mathbf{u} = -\frac{1}{\rho}\nabla p + \nabla\left(\frac{1}{2}\Omega^2 r^2\right) + \mathbf{f} + \nu\nabla^2\mathbf{u} \tag{D.37}$$

The continuity equation is unaffected. As a final comment, absolute vorticity is

$$\nabla \times \mathbf{u}_{abs} = \nabla \times \mathbf{u} + \nabla \times (\mathbf{\Omega} \times \mathbf{r}) = \nabla \times \mathbf{u} + 2\mathbf{\Omega} \tag{D.38}$$

D.5 Natural (Streamline) Coordinates

Natural, or streamline, coordinates are sometimes useful as a conceptual tool for illustrating nuances of fluid motion. For two-dimensional, steady flow we envision one of the coordinates, s, being everywhere parallel to the fluid velocity, i.e., parallel to a streamline. The other coordinate, n, is chosen to be normal to streamlines (see Figure D.5).

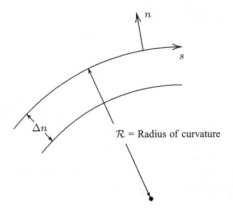

Figure D.5: *Natural coordinates.*

Since there is no flow normal to a streamline, the velocity has one component, u. An additional consequence is that the amount of fluid flowing between two streamlines is constant. Thus, denoting the distance between two streamlines by Δn, conservation of mass becomes

$$\rho u \Delta n = \text{constant} \tag{D.39}$$

The streamwise and normal components of the Euler equation are

$$u\frac{\partial u}{\partial s} = -\frac{1}{\rho}\frac{\partial p}{\partial s} + f_s \qquad \text{and} \qquad -\frac{u^2}{\mathcal{R}} = -\frac{1}{\rho}\frac{\partial p}{\partial n} + f_n \tag{D.40}$$

where f_s and f_n are the body-force components tangent to and normal to the streamline, respectively. If the body force is conservative so that $\mathbf{f} = -\nabla V$, then

$$u\frac{\partial u}{\partial s} = -\frac{1}{\rho}\frac{\partial p}{\partial s} - \frac{\partial V}{\partial s} \qquad \text{and} \qquad -\frac{u^2}{\mathcal{R}} = -\frac{1}{\rho}\frac{\partial p}{\partial n} - \frac{\partial V}{\partial n} \tag{D.41}$$

The vorticity in natural coordinates is normal to the plane of motion, and its value is

$$\omega = \frac{u}{\mathcal{R}} + \frac{\partial u}{\partial n} \tag{D.42}$$

Answers to Selected Problems

SPECIAL NOTE TO STUDENTS

From a student's point of view, of all the places where typographical errors might be lurking in a book, the answers section is undoubtedly the most unwelcome place for them to hide! We have made extensive efforts to assure the accuracy of these answers. However, if you find errors in this section or any other part of the text, report them to DCW Industries' Home Page on the Worldwide Web at **http://www.dcwindustries.com**. We provide an updated list of known typographical errors on the site for all of DCW Industries' publications.

1.1 0.3165 MJ, 205 kW, 7.50 m/sec

1.5 168 lb

1.9 19 molecules

1.13 $d/\ell_{mfp} = 575$, yes

1.15 1.86 hours

1.19 CO_2: 88.2 N, O_2: 64.1 N

1.21 241 kPa

1.23 37.2 atm

1.27 463 J/(kg·K)

1.31 Air: 0.0753 lb/ft^3
Water: 62.4 lb/ft^3

1.35 0.69 atm

1.39 (a) $\lambda = 0.977$
(b) $x/\mathcal{R} = 1.80$

1.43 Air: 0.049 atm, Ether: 302 atm
Glycerin: 2228 atm

1.47 −0.02 liter

1.51 $\sigma = 0.075$ N/m and $T = 7°$ C

1.53 $D_a/D_w = 1.24$

1.57 $\Delta h = 3.46$ mm

1.61 $\sigma = 0.0015$ lb/ft

1.65 $\mu_r = 2.61 \cdot 10^{-7}$, $\omega = 0.62$

1.69 (a) $\tau_w = \mu U_e/\delta$
(b) $\tau_w = \pi\mu U_e/(2\delta)$
(c) $\tau_w = 3\mu U_e/(2\delta)$

1.75 $\overline{u} = U/2$, $\frac{1}{2}\overline{u^2} = U^2/6$

1.79 $U \rightarrow W(R-r)/(2\pi r\ell\mu) = 0.212$ m/sec

1.3 3.42 slugs, 49.9 kg, 1.63 m

1.7 millimeter

1.11 (a) $3.0 \cdot 10^{13}$ molecules
(b) 30 molecules

1.17 Air: 0.779 kg/m^3, CO_2: 1.183 kg/m^3,
H_2: 0.054 kg/m^3, CH_4: 0.431 kg/m^3,
N_2: 0.753 kg/m^3, O_2: 0.860 kg/m^3

1.25 519 J/(kg·K)

1.29 Ethyl alcohol: 0.79, CCl_4: 1.59
Hg: 13.56

1.33 $c_v = R/(\gamma - 1)$, $c_p = \gamma R/(\gamma - 1)$

1.37 (a) 13.0 kPa
(b) 51° C

1.41 $\tau \approx 2.31 \cdot 10^{-8}$ ft^2/lb

1.45 Air: $\Delta\rho/\rho = 0.2151$
Water: $\Delta\rho/\rho = 1.00 \cdot 10^{-5}$

1.49 (a) $\tau = [n(p + 3040p_a)]^{-1}$
(b) $n = 7.00$

1.55 $\Delta p = 4\sigma/R = 58.4$ N/m^2

1.59 $\Delta h = 4\sigma \cos\phi/(\rho g s)$

1.63 $\Delta h_2/\Delta h_1 = 0.856$

1.67 (a) $\mu_r = 2.123 \cdot 10^{-6}$, $T_r = 1819$

1.71 (a) $\mu_e = 0.12/(du/dy)^{1/3}$
(b) $\mu_e/\mu_w \approx 120/(du/dy)^{1/3}$

1.73 (b) $\tau_\ell = \mu U/h$, $\tau_u = -\mu U/h$

1.77 (a) $\mu = 1.2 \cdot 10^{-3}$ kg/(m·sec)
(b) $\rho_{max} = 750$ kg/m^3

1.81 $\mu = MAd/(\pi LUD)$
     $= 2.14 \cdot 10^{-3}$ slug/(ft·sec)
1.87 (a) $u_m = 1.58$ ft/sec
     (b) $Re_R = 1121$
     (c) $F = -4.53 \cdot 10^{-4}$ lb
1.91 $\overline{\frac{1}{2}u^2} = \frac{1}{6}u_m^2$

1.83 $Re_h = 0.03$, Couette flow valid
1.85 (a) $Q = \pi(p_1 - p_2)R^4/(8\mu L)$
     (b)Time $= 0.94$ sec
1.89 (a) $\overline{u} = \frac{1}{2}u_m$
     (b) $f = 64/Re_D$

2.5 $[\dot{h}] = M/(T^3\Theta)$
2.13 $[\omega] = T^{-1}$, α is dimensionless
2.17 $[\chi] = L^{1/3}/T$
2.25 $\nu_t = $ constant $\cdot k^2/\epsilon$
2.29 $R = $ constant $\cdot (E/\rho_a)^{1/5}t^{2/5}$
2.33 $E(\kappa) = $ constant $\cdot \epsilon^{2/3}\kappa^{-5/3}$
2.37 2: Vh/ν, $MV^2/(\dot{m}\nu)$
2.41 2: UR/ν, $\mathcal{R}/R$
2.45 $\omega = (U/D) f(UD/\nu)$
2.49 $T = \rho\Omega^2 d^5 f(Q/[\Omega d^3])$
2.53 3: $\omega r^2/\Gamma_o$, $\Gamma_o t/r^2$, ν/Γ_o
2.57 3: $\Delta T/T_\infty$, $c_p\mu/k$, $U/\sqrt{c_p T_\infty}$
2.61 3: $\Omega^6 F/(\rho g^4)$, $\Omega U/g$, $\Omega^2 \ell/g$
2.65 4: $\omega^2 P/(\rho U^5)$, U/a, $\omega D/U$, $\rho UD/\mu$
2.69 $U_m/U_p = 5$
2.73 (a) $U_m = 30$ ft/sec
     (b) $F_p = 132$ lb
     (c) $P = 6$ hp
2.77 $U_m = 630$ mph, $p_m = 3.95$ atm
2.81 (a) $U_m = 8000$ mph
     (b) $T_{max} = 5380°$ R, model vaporized
2.85 (a) $R/(\rho g L^3)$, $U/\sqrt{gL}$
     (b) $U_p = 15$ knots, $R_p = 93750$ lb
2.89 (a) 3: $\rho LF/\mu^2$, t/L, $\rho UL/\mu$
     (b) $U_m = 1$ m/sec
     (c) $F_p = 5000$ N/m
2.93 (a) Q^2N/ν^3, $\nu D/Q$
     (b) $Q_m = 19.0$ ft³/sec, $N_m = 228$ rpm

2.11 $U_{hull} = 2.43\sqrt{L_w}$
2.15 C is dimensionless
2.23 $u_m = $ constant $\cdot R^2(dp/dx)/\mu$
2.27 $u_m(x) = $ constant $\cdot \sqrt{J/x}$
2.31 $v = $ constant $\cdot (\nu\epsilon)^{1/4}$
2.35 $\mathcal{R} = (\mathcal{M}/\rho)^{1/3} f(\mathcal{M}^{2/3}\rho^{1/3}g/\sigma)$
2.39 $u/U = f(y/h, \rho y^2/[\mu t])$
2.43 2: $Q/(\Omega D^3)$, $gh_p/(\Omega^2 D^2)$
2.47 2: $D^2(dp/dx)/(\mu U)$, $\rho UD/\mu$
2.51 3: $\omega^2 F/(\rho U^4)$, Ω/ω, $\omega L/U$
2.55 3: $D/(\mu\Omega L)$, $\rho UL/\mu$, $\Omega L/U$
2.59 3: $U/\sqrt{\nu\omega}$, $\omega h^2/\nu$, $\omega k_s^2/\nu$
2.63 4: $T/(\rho\Omega^2\ell^4)$, $U/(\Omega\ell)$, $a/(\Omega\ell)$, $V_t/(\Omega\ell)$
2.67 4: $T/(\rho n^2 D^4)$, nD/U, UD/ν, $U/\sqrt{gD}$
2.71 (a) $\omega_m = 18$ sec^{-1}, $\mu_m = \frac{3}{4}\mu_p$
     (b) $T_m = -107°$ F
2.75 (a) $U_p = 8$ m/sec, $N_p = 135$ rpm
     (b) $Re_{D_m} = 3.6 \cdot 10^5$, $Re_{D_p} = 2.3 \cdot 10^7$
2.79 (a) $h_m = 7.65$ mm, $U_m = 0.643$ m/sec
     (b) $\tau = 51$ min
2.83 (a) $U_m = 87.4$ m/sec
     (b) $M_m = 0.239$
2.87 (a) $\mu hU/T, w/h, \ell/h$
     (b) $U_m/U_p = \mu_p/\mu_m$
2.91 (a) $F/(\mu UR)$, r/R, ℓ/R
     (b) $U_m = \frac{1}{6}U_p$

3.1 $p = 433$ kPa < 500 kPa
     Blofeld will not suffer from the bends
3.7 $h = 1.0$ m
3.11 $h = 8.17$ m
3.15 450 m
3.19 $p = 0.133$ atm, $T = -55°$ C
3.23 $F = W/30 + \rho gAL/6$
3.27 $p_A = p_a + 4\rho gh$
3.31 (a) $p_1 - p_2 = (\rho_2 h_2 - \rho_1 h_1 - \rho_{Hg}\lambda)g$
     (b) $p_1 - p_2 = -8.9$ kPa

3.3 $\rho = 1252$ kg/m³
3.5 $\rho_u = 1.98$ slug/ft³
3.9 $p = p_a + \rho_a gz(1 + z/z_o)$
3.13 $p = 0.303$ atm, $T = -44°$ C
3.17 $\Delta p = -0.3$ psi, $\Delta T = -2°$ F
3.21 $\alpha = (\gamma - 1)g/(\gamma R)$
3.25 $W = 7865$ lb
3.29 $\Delta z = 1.14$ m
3.33 $p_A = p_a + 29.94\rho gh$

3.35 (a) $h = (p_1 - p_2)/(g\Delta\rho)$

(b) Kerosene: $h = 48.9$ mm

Glycerin: $h = 15.1$ mm

3.43 $U_j = 10.79$ m/sec, $z_{max} = 6.93$ m

3.47 0.054 N, -0.864 Pa

3.49 $d = D[1 + \pi^2\rho^2 gD^4(\ell - z)/(8\dot{m}^2)]^{-1/4}$

3.55 $I = 2.5\pi R^4$

3.57 $h_p = 0.3h$

3.61 $h = H/6$

3.65 $1.75\pi\rho g H^3$

3.69 $|\mathbf{F}| = 2.357\rho g H^3$

3.73 (a) $H = 3L\sin\phi$

(b) $F = 13.5\rho g L^3 \sin\phi$

3.79 $\mathbf{F} = 1.25\rho g h^3(2\mathbf{i} + \pi\mathbf{k})$, $z_{cp} = 2h/3$

3.83 $\mathbf{F} = -3.5\rho g H^3(\mathbf{i} + \mathbf{k})$

3.85 $\mathbf{F} = 3\rho g H^3(\mathbf{i} + \mathbf{k})$

3.89 $F_{buoy} = 0.25\rho g V$

3.91 $V = 16.0$ ft^3, $\rho = 19.4$ slug/ft^3

3.93 29% gold

3.97 $F_{buoy} = \rho_a R T h^2 e^{-gz/(RT)}[1 - e^{-gh/(RT)}]$

$z \approx (RT/g)\ell n(\rho_a/\rho_{He})$

3.103 $N = 4/3$

3.107 $\rho_c = 4\rho$

3.111 $V = \ell^3/5$

3.37 $\alpha = 39.9°$

3.39 $p - p_a = 37.57$ kPa

3.41 $p_2 - p_1 = 2.16$ kPa

3.45 (a) $U = \sqrt{2gh_1}$, $p_{min} = p_a - \rho g(h_1 + h_2)$

(b) $U = 8.86$ m/sec, $p_{min} = 0.225$ atm

3.51 (a) $p_b = p_a + 1.25\rho g h$

(b) $p_b = 2.21$ atm, $U = 21.0$ m/sec

3.59 $H = h/3$

3.63 $F \approx 20$ tons, $h_{cp} = 3.33$ ft

3.67 $H = 16.8$ m

3.71 $F = 7\rho g h^3/36$, $F_{tip} = \rho g h^3/18$

3.75 $\mathbf{F} = \rho g h^3(3\mathbf{i} + 4\mathbf{k})$, $z_{cp} = 2h/3$

3.77 $F = 2\rho g H^3$

3.81 (a) $\rho_2/\rho_1 = 2$

(b) $F_x = -\rho_1 g H^3/6$

3.87 (a) $\mathbf{F} = -\rho g \ell^3[3\mathbf{i} + (4 + \pi)\mathbf{k}]$

(b) $z_{cp} = 14\ell/9$

(c) $x_{cp} = 10\ell/[3(4 + \pi)]$

3.95 $\ell = \frac{11}{7}L$

3.99 $\mathbf{F} = 2\rho g H^3\mathbf{i}$

3.101 $\mathbf{F} = (3/16)\rho g D^2 w(2\mathbf{i} - \mathbf{k})$

3.105 $\rho_b = 0.42\rho$

3.109 $\rho_b = 0.625\rho$

3.113 $\mathbf{F} = \rho_a g H^3[\frac{3}{2}(1 + \frac{7}{9}\alpha)\mathbf{i} + \frac{5}{4}(1 + \frac{1}{2}\alpha)\mathbf{k}]$

4.1 (a) $\mathbf{a} = -U[U\cos(x + at) + a]\sin(x + at)\mathbf{i}$

(b) $\mathbf{a} = \mathbf{0}$

(c) $\mathbf{a} = g(x)df/dy\,\mathbf{i} + f(y)dg/dx\,\mathbf{j}$

4.7 $\mathbf{a} = 0$

4.9 (a) $\partial u/\partial t = -x/t^2$, $u\partial u/\partial x = x/t^2$

(b) $\partial u/\partial t = x/t^2$, $u\partial u/\partial x = x/t^2$

4.15 $\mathbf{r} = [x_o + (Ut/\ell)(y_o + \frac{1}{2}Ut)]\mathbf{i} + [y_o + Ut]\mathbf{j}$

4.17 $\mathbf{r} = x_o e^{2Ut/R}\mathbf{i} + y_o e^{-2Ut/R}\mathbf{j}$

4.21 $r = \sqrt{r_o^2 + Qt/\pi}$, $\theta = \theta_o$

4.25 $\boldsymbol{\omega} = (-z\sin\theta\mathbf{e}_r + 3z\cos\theta\mathbf{e}_\theta + r\sin\theta\mathbf{k})/r^4$

4.27 (a), (b) and (c) $\boldsymbol{\omega} = \mathbf{0}$

4.29 $\boldsymbol{\omega} = \mathbf{0}$ for all values of A

4.33 $\boldsymbol{\omega} = (U/\sqrt{\pi\nu t})\exp[-y/(2\sqrt{\nu t})]\mathbf{k}$

4.41 $x^2 y = $ constant

4.45 $y^2 - x^2 = $ constant

4.49 $r/\cos^2\theta = $ constant

4.53 $\dot{m} = 10\rho U h^2$, $\dot{\mathbf{P}} = 6.67\rho U^2 h^2\mathbf{i}$

$\dot{K} = 2.5\rho U^3 h^2$

4.57 $\dot{m} = 2\rho U A$

4.61 $dN_R/dt = -0.1$ people/sec

4.65 $dN_c/dt = -dN_g/dt - (U_g A_g + U_c A_c)/V$

4.3 (a) $\mathbf{a} = (\dot{A} + A^2)(x\mathbf{i} + y\mathbf{j})$

(b) $\mathbf{a} = (\dot{A} + 2A^2 x)x^2\mathbf{i} - (\dot{A} - 2A^2 y)y^2\mathbf{j}$

4.5 (a) $\mathbf{a} = (2U^2/x_o)(1 - x/x_o)^{-5}\mathbf{i}$

(b) $u_2/u_1 = 0.693$

4.11 $\mathbf{a} = (U/H)^2(x\mathbf{i} + y\mathbf{j})$

4.13 (a) $\mathbf{a} = -\Gamma^2/(4\pi^2 r^3)\,\mathbf{e}_r$

(b) $\mathbf{a} = -\mathcal{D}^2/(2\pi^2 r^5)\,\mathbf{e}_r$

4.19 $r = r_o$, $\theta = \theta_o + \Omega t$

4.23 (a) $\boldsymbol{\omega} = \mathbf{0}$

(b) $\boldsymbol{\omega} = (df_1/dx - df_2/dy)\mathbf{k}$

(c) $\boldsymbol{\omega} = -(U/\delta)e^{-y/\delta}\mathbf{k}$

4.31 (a) and (b) $\nabla \times \mathbf{u} \neq \mathbf{0} \implies$ rotational

4.39 $\Gamma = 0$

4.43 $y/x = $ constant

4.47 $(r - R^2/r)\sin\theta = $ constant

4.51 (a) $y - k^{-1}\sin[k(x - Ut)] = $ constant

(b) $y = (x - x_o)\cos(kx_o)$

4.55 $\dot{m} = 0$

4.59 $\dot{P}_z = -0.25\rho U^2 h^2$

4.63 $dN/dt = -\dot{N}_{pop} - nUA$

5.1 Easy: $\mathbf{i} \cos \phi - \mathbf{j} \sin \phi,\ U,\ \rho U^2 A \cos \phi,$
$-\rho U^2 A \sin \phi$
Hard: $\mathbf{i},\ U \cos \phi,\ \rho U^2 A \cos \phi,$
$-\rho U^2 A \sin \phi$

5.7 $\sum \mathbf{F}_i = (p_1 - p_2) H \mathbf{i}$

5.9 Inlet: $\frac{\pi}{4} D^2,\ -\mathbf{i},\ -U,\ -\frac{\pi}{4} D^2 U$
Outlet: $\frac{\pi}{4} D^2 \csc \alpha,\ \mathbf{i} \sin \alpha + \mathbf{j} \cos \alpha,$
$U \sin \alpha,\ \frac{\pi}{4} D^2 U$

5.17 $dh/dt = -0.06 U (d/D)^2$, tank emptying

5.21 $U_2 = \frac{8}{3} U_1$

5.25 $D = 0.268 H$

5.29 $d = D/5$

5.33 (b) $p_c = 5.05$ MPa

5.37 $u_{max} = 1.5 U_o$

5.41 $dM/dt = -3 \rho_a V_o r_o h$

5.45 (a) $V = 3(U_2 - U_1)$
(b) $\rho(U_2^2 - U_1^2 - \frac{1}{9} V^2 \cos \phi) = p_1 - p_2$
(c) $U_2^2 - U_1^2 = \frac{2}{9} V^2 \cos \phi$
(d) $\phi = \cos^{-1}(7/10) = 45.6°$

5.51 (a) $\mathbf{R} = -\frac{9}{64} \pi \rho U_j^2 d^2 \mathbf{i}$
(b) $\hat{U}_j = 1.25 U$

5.55 $\mathbf{F} = 0.32 \pi \rho U_i^2 d^2 \mathbf{i}$

5.59 $\Delta p = 1.4 \rho U^2$

5.63 $U_b = U/3$

5.67 $U_o = \frac{1}{2} U_i$
$R_y/R_x = 5(1 + 10 \sin \phi)/(2 + 5 \cos \phi)$

5.71 (a) $M(t) = M_o + \dot{m}t$
(b) $T = \dot{m} U \cos \alpha$
(c) $F_{buoy} = M_o g + \dot{m}(gt + U \sin \alpha)$

5.79 $\mathbf{F} = \rho U^2 A[(2 + \frac{5}{2} \cos \phi)\mathbf{i} - (2 - \frac{5}{2} \sin \phi)\mathbf{j}]$

5.83 $\mathbf{F} = \frac{1}{2} \pi D^2 (\Delta p + \rho U_i^2)\mathbf{i}$, Yes

5.87 $\mathbf{F} = -\frac{1}{4} \pi \rho U^2 (D^2 - d^2)(1 + \sin \phi)\mathbf{i}$

5.91 (a) $Mg = \frac{1}{4} \pi \rho W^2 d_o^2 F$, where
$F = 1 + \sqrt{1 - 2gh/W^2}$
(b) $Mg = 6.18$ kN

5.93 (a) $Mg = \frac{1}{4} \pi \rho W^2 d_o^2 F$, where
$F = 1 - \frac{4}{3} gh/W^2 + \sqrt{1 - 2gh/W^2} \cos \phi$
(b) $Mg = 0.464$ kN

5.99 $K = \frac{8}{7} U$

5.101 $V = \frac{1}{4} U,\ F_y = -\frac{1}{72}(40 - 3 \sin \alpha)\rho U^2 h$

5.105 $v_{min} = \frac{2}{3} U,\ C_p = -\frac{5}{18}$

5.111 (c) $U = 3.38$ knots

5.3 Inlet: $-\mathbf{i},\ -U,\ -9\rho U^2 H^2,\ 0$
Upper Outlet: $\mathbf{j},\ \frac{4}{3} U,\ 0,\ 8\rho U^2 H^2$
Lower Outlet: $-\mathbf{j},\ \frac{13}{15} U,\ 0,\ -\frac{507}{125} \rho U^2 H^2$

5.5 Inlet: $2h^2,\ -\mathbf{i} \cos \beta - \mathbf{j} \sin \beta,\ -U,\ -2h^2 U$
Outlet: $2h^2 \csc \beta,\ \mathbf{j},\ U \sin \beta,\ 2h^2 U$

5.11 $U = 3.1$ ft/sec, $u = 50.1$ ft/sec

5.13 $U_v = \frac{3}{16} U$

5.15 $\beta = \frac{4}{25}$

5.19 $V = -6U$

5.23 $u_{avg} = 1.45 U$ for observer on sphere

5.27 $u = U/6$

5.31 $z = 76.4d$

5.35 (a) $u(r) = -Vr/h$

5.39 $B = \frac{1}{2} A$

5.43 (a) $(U_2 - U_1)H = Vh$
(b) $\rho(U_2^2 - U_1^2 - \frac{h}{H} V^2 \cos \phi) = p_1 - p_2$
(c) $U_2 = U_1(H + h)/(H - h)$

5.47 $\cos \alpha = (h_1 - h_2)/(h_1 + h_2)$

5.49 $\phi = \sin^{-1}(1/3) = 19.5°$
$\Delta p = 0.707 \rho U^2$

5.53 $\tau = \frac{1}{16} D,\ \mathbf{F} = \frac{1}{2} \pi \rho U^2 D^2 \mathbf{i}$

5.57 $V_1 = 4.17 U,\ V_2 = 8.33 U,\ \phi = 77.5°$

5.61 $C_p = 0.80$

5.65 $\phi = 18.4°$

5.69 $\mathbf{F}_{top} = \dot{m} U[(\cos \alpha - 1)\mathbf{i} + \sin \alpha \mathbf{j}]$
$\mathbf{F}_{total} = 0$

5.73 $\Delta p = 0.43 \rho U^2$

5.75 $\Delta p = -0.44 \rho U^2$

5.77 $\mathbf{F} = 1.1 \pi \rho U^2 d^2 \mathbf{i}$

5.81 $\phi = 32.8°,\ F_x = -0.862 \rho U^2 A$

5.85 $\mathbf{F} = \frac{1}{32} \pi \rho V^2 D^2 (3\mathbf{i} + 2\mathbf{j})$

5.89 (a) $\mathbf{F}_{wall} = -\rho U^2 H \sin \phi \mathbf{j}$
(b) $U_1 = \frac{1}{2}(H/h - \cos \phi)U$
$U_2 = \frac{1}{2}(H/h + \cos \phi)U$
(c) $U_1 = U_2$ as required by symmetry

5.95 (a) $v = 2U$
(b) $U = 2\sqrt{gh/3}$

5.97 (a) $\mathbf{F} = -0.688 \pi \rho U^2 D^2 \mathbf{i}$
(b) $\mathbf{F} = -0.708 \pi \rho U^2 D^2 \mathbf{i}$, 2.9% larger

5.103 $p_1 - p_2 = \frac{17}{15} \rho U_\infty^2$

5.109 $M(t) = M_o \exp(-\sqrt{\rho_e A_e T} t)$

5.113 (b) $U(t) = u_e \ell n(M_o/M) - \mu_s gt$

6.1 $W_c = -1.36 \cdot 10^6$ ft·lb

6.5 $s = -R\ell np +$ constant

6.11 $P = \dot{m}[p/\rho + c_v \Delta T + 640 \dot{m}^2/(\pi^2 \rho^2 d^4)]$

6.3 (a) $n = 1.20$
(b) $W = -2.716 \cdot 10^6$ ft·lb
(c) $Q = -1745$ Btu

6.13 $\dot{Q} = 3925$ Btu/sec

6.17 $\dot{Q} = 240$ Btu/sec, No

6.21 $\alpha_1 - \alpha_2 = 0.03$

6.25 $\alpha = 1.106, 1.058, 1.037$

6.27 (a) Laminar, $f = 0.032$

    (b) Turbulent, $f = 0.026$

    (c) $k_s/D \approx 0.006$

6.31 $\bar{u} = 4.48$ ft/sec

6.33 $D = 19.5$ cm, $Re_D = 816 < 2300$

6.37 (a) $U_{in} = U/16$

    (b) $2gh_L/U^2 = g\Delta z/U^2 - 1$

6.41 $\dot{W}_t = 0.815\dot{m}gH$

6.45 $h_p = \Delta z + 1600\dot{m}^2/(\pi^2\rho^2gD^4)$

6.47 (a) $U_{ideal} = 2\sqrt{gH}$

    (b) $h_p = \frac{3}{2}H$

6.53 (a) $D = 0.01056L/(2g\Delta z/U^2 - 1)$

    (b) $D = 9.0$ cm

    (c) $D = 9.1$ cm

6.57 $gh_L/U^2 = 8975, 222, 28, 9$

6.59 $U = \frac{2}{3}\sqrt{gL}$

6.63 $p_1 - p_2 = 1018$ kPa

6.67 $\mathbf{R} = \pi\rho U^2 D^2 [\frac{13}{40}\mathbf{i} - \frac{9}{16}\mathbf{j}]$

6.69 Chézy-Manning: $\bar{u} = 1.70$ m/sec

    Colebrook: $\bar{u} = 1.61$ m/sec

6.71 $n = 0.033$

6.75 (b) $\alpha = 45°$

6.77 $Fr = 0.138$

6.79 $\bar{u}_{min} = 9.9$ m/sec, $\bar{u}_{down} = 19.8$ m/sec

6.81 $\bar{u}_{up} = 0.96$ m/sec; ripple travels 96 cm

6.85 Supercritical, $E/E_{min} = 1.52$

6.87 $Fr_1 = 0.042$

6.91 $Fr_1 = $ 2: $h_L/E_1 = 0.0908$

    $Fr_1 = $ 5: $h_L/E_1 = 0.4906$

    $Fr_1 = $ 10: $h_L/E_1 = 0.7271$

    $Fr_1 = $ 20: $h_L/E_1 = 0.8605$

6.15 $u_2 = 83.3$ m/sec

6.19 $\dot{Q} = \dot{m}^3/(2\pi^2\rho^2 d^4)$

6.23 (a) $U_{max} = 3U$

    (b) $\alpha_1 = 1$, $\alpha_2 = 2.7$

    (c) $h_L = 0.35U^2/g$

6.29 Glass: $f = 0.0090$, $h_L = 22.0$ m

    Wrt. iron: $f = 0.0137$, $h_L = 33.5$ m

    Cast iron: $f = 0.0198$, $h_L = 48.5$ m

6.35 (a) $2gh_L/U^2 = 2.2g\Delta z/U^2 - 1.05$

    (b) $2gh_L/U^2 = 0.52$

6.39 $K = 0.516$

6.43 (a) $\dot{W}_t = 0.90\dot{m}gH - 0.53\dot{m}U^2$

    (b) $\dot{W}_t = \dot{m}gH - 0.53\dot{m}U^2$

6.49 $h = 1.36d$

6.51 (a) $f = 1.25gD/U^2 - 0.012$

    (b) $k_s = 0.0002D$

6.55 (a) $U = \sqrt{2g\Delta z/(1 + 0.0263L/D)}$

    (b) $U = 2.424$ m/sec

    (c) $U = 2.417$ m/sec

6.61 $Re_D = 2035$

6.65 (b) $\bar{u}_1 = 7.45$ ft/sec, $\bar{u}_2 = 6.72$ ft/sec

    $\bar{u}_3 = 5.64$ ft/sec

    (c) $Re_{D_1} = 1.15 \cdot 10^5$, $Re_{D_2} = 1.56 \cdot 10^5$

    $Re_{D_3} = 1.74 \cdot 10^5$

6.73 Steel: $Q = 1391$ ft^3/sec, $Fr = 0.61$

    $Re_{D_h} = 1.55 \cdot 10^7$

    Weedy: $Q = 556$ ft^3/sec, $Fr = 0.24$

    $Re_{D_h} = 6.21 \cdot 10^6$

6.83 (b) $Fr = 1$

    (c) $U = 5.91$ m/sec, $Fr = 2.67$

6.89 $y_2 = 4.40$ ft, $h_L = 11.1$ in

6.93 $y_1 = 79.8$ cm

    $Fr_1 = 1.086$, $Fr_2 = 0.914$

    $\bar{u}_1 = 3.01$ m/sec, $\bar{u}_2 = 2.70$ m/sec

7.1 (a) $P = \dot{m}\Omega\ell(w_e \cos\phi - \Omega\ell)$

    (b) $w_e = 24.3$ m/sec

    (c) $\phi = 61.7°$

7.11 $Q = 4070$ gal/min, $\dot{W}_p = 612$ hp

7.15 (a) $Q = 151.9$ ft^3/sec, $\dot{W}_p = 2.41$ hp

    (b) $\alpha_2 = 67.7°$

7.21 Water: $h_p = 0.833$ ft

    Air: $h_p = $ 691 ft

    Helium: $h_p = $ 5005 ft

7.27 $\Omega_{min} = 1549$ rpm

7.31 $C_{Q^*} = 0.200$, $C_{H^*} = 0.0788$

    $C_{P^*} = 0.0208$, $\eta_p = 0.76$, $N_s = 3.0$

7.3 $\beta_2 = \tan^{-1}[(r_1/r_2)^2 \tan\beta_1]$

7.5 (b) $p_2 - p_1 = 306$ kPa

7.9 $Q = 20.26$ ft^3/sec, $\dot{W}_p = 588$ hp

7.13 $Q = 48.4$ ft^3/sec, $\dot{W}_p = 18.3$ hp

7.17 $\beta_1 = 6.9°, \beta_2 = 3.4°$

7.19 $\tilde{N}_s = 2733N_s$, $[\tilde{N}_s] = L^{3/4}/T^{3/2}$

7.23 (a) $Q^* = 1690$ gal/min, $\eta_p = 0.84$

    (b) 400 gal/min $< Q < 2400$ gal/min

7.25 $Q^* = 1960$ gal/min, $\eta_p = 0.81$

7.29 $C_{Q^*} = 0.0148$, $C_{H^*} = 0.151$

    $C_{P^*} = 0.00254$, $\eta_p = 0.88$, $N_s = 0.50$

7.33 $D = 1.32$ m, $\Omega = 350$ rpm

7.35 $\eta_p = 0.76$

7.39 $N_s = 0.084$, Single-jet Pelton wheel

7.43 (a) $\tau_{max} = \frac{9}{8}\rho QV_j r$ at $\phi = 120°$

(b) $\tau_{max} = \rho QV_j r$ at $\phi = 180°$

7.37 $N_s = 1.49$, Francis type

7.41 $\Omega = 135$ rpm

7.45 (a) $\Omega r = \frac{1}{2}\mathcal{C}_v \sqrt{2gh_p}$, $(\eta_t)_{max} = \mathcal{C}_v^2$

(b) $\mathcal{C}_v = 0.94$

(c) $\mathcal{C}_v = 0.89$

8.1 $U_{He} = 670$ mph

$U_{Hg} = 973$ mph

8.5 $z = 25$ m: $U = 168.6$ m/sec

$z = 2$ km: $U = 164.8$ m/sec

8.7 $T_o = 18°$ F

8.11 (b) $T = 365°$ C

8.15 $M = 0.029$ for $U = 100$ ft/sec

$M = 0.148$ for $U = 500$ ft/sec

$M = 0.298$ for $U = 1000$ ft/sec

8.19 $M = 3$

8.23 (a) $M_\infty = 0.288$, $\Delta T/T_\infty = -1.6\%$

(b) $M_\infty = 0.576$, $\Delta T/T_\infty = -6.4\%$

8.25 (a) $M = 0.527$

(b) $M = 0.893$

(c) $M = 0.236$

8.29 $M_1 = \sqrt{13} = 3.6$

8.33 Air: $p_t = 4.56$ atm, He: $p_t = 3.45$ atm

8.37 $p_2 = 79.0$ kPa

8.39 $M_1 = 4.08$

8.41 $M_1 = 3.53$, $T_t = 27°$ C, $p_{t1} = 221$ kPa

8.43 $M_1^2 = [1+\frac{1}{2}(\gamma-1)M_2^2]/[\gamma M_2^2 - \frac{1}{2}(\gamma-1)]$

8.47 $A_{throat} = 20$ cm^2

8.51 $p^*/p_t = 0.544$

8.55 $p_2 = 308$ kPa

8.59 $M_e = 0.271$

8.65 $A/A^* = \frac{1}{2}[2(2\gamma-1)/(\gamma+1)]^{(\gamma+1)/[2(\gamma-1)]}$

8.71 $M = 0.685$ or $M = 1.55$

8.73 $M_3 = 0.22$

8.75 (b) $p_t = 128$ kPa, $M_e = 0.595$

(c) $F = 298$ N

8.79 $L_{max} = 1.89$ m, $T_e = 17°$ C

8.83 (a) $L_i^* = 125$ cm

(b) Yes

8.93 Air: $q = 2.45 \cdot 10^4$ J/kg, $T_e = 415$ K

H$_2$: $q = 6.50 \cdot 10^6$ J/kg, $T_e = 681$ K

8.97 $q = -3543$ Btu/slug, $M_e = 2.56$

8.3 (a) $M = 1.16$, transonic

(b) $M = 0.24$, incompressible

(c) $M = 0.95$, transonic

(d) $M = 5.24$, hypersonic

8.9 $a_{sea} = 1477$ m/sec, $a_{sea}/a_{fresh} = 0.997$

8.13 Air: $p_t = 55.4$ psi, $T_t = 281°$ F

H$_2$: $p_t = 21.7$ psi, $T_t = 107°$ F

8.17 $T_{air} = -251°$ C

$T_{He} = -259°$ C

8.21 $M_A = 0.3$: $p_A/p_B = 1.198$

$M_A = 0.6$: $p_A/p_B = 1.053$

$T_{tA}/T_{tB} = 1$

8.27 (a) $M = 0.571$

(b) $M = 0.958$

(c) $M = 0.257$

8.31 $T_2 = 279$ K

8.35 Air: $M_2 = 0.435$, $T_{t2} = 1470$ K

$p_2 = 148$ kPa

CO$_2$: $M_2 = 0.401$, $T_{t2} = 1150$ K

$p_2 = 143$ kPa

8.45 $(M_2)_{min} = 0.378$, $(\rho_2/\rho_1)_{max} = 6$

8.49 $A/A^* = 2.64$, $p_t = 34.2$ psi, $T_t = 563°$ R

8.53 $p^*/p_t = 0.487$

8.57 $M_e = 0.112$

8.63 $p_t = 15.6$ atm

8.69 (a) $p_0 = 25.59$ psi, $p_1 = 13.51$ psi

$p_2 = 3.27$ psi, $M_3 = 0.58$

(b) $p_4 = 18.46$ psi, $T_4 = T_0$

8.77 $M_2^2 = 1/M_1^2$, $\rho_2/\rho_1 = M_1^2$

$p_2/p_1 = M_1^2$, $T_2/T_1 = 1$

8.81 $\overline{f} = 0.016$, $M_e = 1.18$

8.85 $M_e = 1.23$, $p_e = 42.7$ psi, $T_e = 600°$ F

8.91 $q = 13265$ Btu/slug, $p_e = 13.4$ psi

8.95 $M = 0.5$: $q = 2.92 \cdot 10^5$ J/kg

$M = 0.7$: $q = 4.76 \cdot 10^5$ J/kg

$M = 0.9$: $q = 5.47 \cdot 10^5$ J/kg

9.1 $\rho(x) = \rho_a(1 + x/x_o)^{-1}$, $x = \frac{2}{3}x_o$

9.5 No, Yes

9.9 Yes, Yes

9.13 $u(x,y) = Ux/h + \text{constant}$

9.3 $u(x,t) = u_o - (\rho_a/\rho - 1)x/\tau$

9.7 Yes, No

9.11 Yes, Yes

9.15 $v(x,y) = \frac{1}{2}C(x^2 - y^2) + \text{constant}$

9.17 $u_\theta(r, \theta) = (S/r^2)\sin\theta + \text{constant}/r$

9.21 $p(x, y, z, t) = -\rho gz + f(t)$

9.25 (a) Yes

    (b) Yes

    (c) $p(x, y) = p_t - \frac{1}{2}\rho A^2(x^2 + y^2)$

9.31 VW Bug: $h_o = 0.90h$

    Corvette: $h_o = 0.75h$

9.39 $U = 60$ m/sec, $p_t = 104.8$ kPa

9.43 $Q_{min} = 0.046$ m³/sec

9.49 $p_B - p_C = 0.81$ psi

9.19 $C = -1$, τ is arbitrary

9.23 $dp/dx = -42$ kPa/m

9.27 $\ell_1 = L_2 - 0.4\Omega^2 L_3^2/g$

    $\ell_2 = L_2 + 0.1\Omega^2 L_3^2/g$

9.29 $\Omega = 2\sqrt{gh}/R$

9.33 $h_o = h - a\ell/(2g)$

9.35 $p_{max} = p_a + \rho gh\cos\alpha$ at $z = 0$

9.41 $p_t = 1739$ psf, $U = 468$ ft/sec

9.47 $U = 8$ m/sec

9.51 $U_4 = 5.12$ m/sec

10.3 $\dot{m} = 0$

10.9 (a) $[\tilde{\psi}] = ML^{-1}T^{-1}$

10.13 $\phi = Ar^n\cos n\theta$, $F(z) = Az^n$

10.15 $\phi = U(x\cos\alpha + y\sin\alpha)$

    $\psi = U(y\cos\alpha - x\sin\alpha)$

    Uniform flow at angle α

10.19 $x = y = 0$, Yes

10.23 $\phi = U(x\cos\alpha + y\sin\alpha)$

    $\psi = U(y\cos\alpha - x\sin\alpha)$

10.29 $f_1(x) = -\frac{1}{2}Ax^2 - Bx + F$

    $f_2(y) = \frac{1}{2}Ay^2 + Cy + D$

    where A, B, C, D and F are constants

10.35 $p = p_\infty - \frac{1}{2}\rho Q^2/(2\pi r)^2$

    $r_{min} = 1.0$ cm

10.39 (a) $C_p = -(s/L)^2[1 + (t/\tau)^2]$

    (b) $|\mathbf{u}| = U\sqrt{1 + (t/\tau)^2}$

10.45 (a) $\psi = Ur\sin\theta + Uh(\theta - \pi)/(2\pi)$

    (b) $x = -h/(2\pi)$, $y = 0$

10.49 (a) $\psi = Ur\sin\theta - Q\theta/(2\pi)$

    $\phi = Ur\cos\theta - Q\ell nr/(2\pi)$

    (b) $p = p_\infty + \frac{1}{2}\rho[UQ/(2\pi) - Q^2/(2\pi x)^2]$

    (c) $x = Q/(2\pi U)$

10.53 $p_i = 14.46$ psi

10.57 (b) $D = \frac{8}{3}\rho U^2 R$

10.61 (a) $U_{min} = \sqrt{W/(2\pi\rho RH)}$

    $\Omega_{min} = \sqrt{W/(2\pi\rho R^3 H)}$

    (b) $U_{min} = 3.0$ ft/sec, $\Omega_{min} = 11.5$ Hz

10.63 (a) $\alpha = 30°$

    (b) $p_1 - p_2 = \frac{1}{2}\rho U^2 + 0.268\rho R\dot{U}(t)$

10.67 $dW/dt = g[1 + 2m_h/(\rho V)]^{-1}$

10.71 $\mathbf{F}_w = -\rho D^2/(8\pi h^3)\mathbf{j}$

10.5 $\Gamma = 0$

10.11 (a) $\phi = \mathcal{D}_p\cos\theta/(2\pi r)$

    (b) $\psi = -\mathcal{D}_p\sin\theta/(2\pi r)$

    (c) Identical to doublet with $\mathcal{D} = \mathcal{D}_p$

10.17 $y = \pm 2A\sqrt{x + A^2}$, $A = \text{constant}$

    Flow through a 360° turn

10.21 $x = y = 0$, No

10.25 $\phi = U(x^2 + y^2 - 2z^2)/(2h)$, Yes

10.27 $\phi = \psi_o(x^2 - y^2) + \text{constant}$

10.31 $\phi = r^n\cos n\theta + \text{constant}$

10.33 (a) $u_r = 2Ar\sin 2\theta$, $u_\theta = 2Ar\cos 2\theta$

    (b) $\phi = Ar^2\sin 2\theta$

    (c) $r = 0$

    (d) Plane at 45° to x axis

10.41 $r = r_o e^{Q\theta/\Gamma}$, $r_o = e^{-2\pi\psi/\Gamma}$

10.43 $\Gamma = \sqrt{4\pi - 1}Q$, $\alpha = 74°$

10.47 (a) $\psi = Vx$, $\phi = -Vy$

    (b) $\psi = Vr\cos\theta$, $\phi = -Vr\sin\theta$

    (c) $\psi = Vr\cos\theta + Q\theta/(2\pi)$

    $\phi = -Vr\sin\theta + Q\ell nr/(2\pi)$

    (d) $x = 0$, $y = Q/(2\pi V)$

10.51 (c) $\mathbf{F} = -\rho QU\mathbf{i}$ (a thrust)

10.55 $h = \frac{3}{2}R$, $u_{max} = \frac{13}{9}U$

10.59 (a) $\mathbf{F} = \pi\rho\Omega D^2 H(\tilde{u}\mathbf{i} - \tilde{v}\mathbf{j})$

    where $\tilde{u} = V\sin\alpha$, $\tilde{v} = U + V\cos\alpha$

    (b) $F_x = \pi\rho\Omega D^2 H\sqrt{V^2 - U^2}$

    (c) $F_x = 27$ tons

10.65 (a) $C_p = [1 - 4\sin^2\theta]\sin^2\omega t$

    $- 4St\cos\theta\cos\omega t$

    where $St \equiv \omega R/U_o$

    (b) $\overline{C}_p = \frac{1}{2}[1 - 4\sin^2\theta]$

11.1 (c) $d\tilde{\omega}/dt = \tilde{\omega}\cdot\nabla\mathbf{u}$

    where $\tilde{\omega} \equiv \omega/\rho$

11.5 $\Gamma = 1.11\cdot 10^5$ ft²/sec, $\Omega = 11.7$ rpm

11.9 $\Gamma_a = U\Delta x$

11.3 (a) $\mathbf{f}_{vortex} = \mathbf{0}$

    (b) $p + \frac{1}{2}\rho\mathbf{u}\cdot\mathbf{u} + \rho\mathcal{V} = \text{constant}$

11.7 $U = 143$ mph

11.11 (a) $\Gamma = -UL$

11.13 (a) $Re_L = 6.23 \cdot 10^6$
  (b) $Re_L = 6.25 \cdot 10^6$

11.17 $\tau_w = -\mu\omega_w$

11.21 $\delta = 3.2$ mm

11.25 $\ell_m = 0.01 H Re_H = \ell_e/6$

11.27 $u(y) = U(1 - e^{-v_w y/\nu})$, $v_w > 0$

11.31 $Re_L = 9.78 \cdot 10^6$, $\delta \approx 3.2$ in

11.33 (a) $(c_f)_t = 0.00047$, $(c_f)_f = 0.00305$
  (b) $(c_f)_t = 0.00094$, $(c_f)_f = 0.00395$

11.37 $\delta \approx 3.2$ in, $\tan^{-1}(\delta/x) = 0.6°$

11.39 $C_D = 0.144 Re_L^{-1/5}$

11.41 $\omega = -\tau_w/(30\kappa\mu)$, $\omega/\omega_{y=0} = 0.08$

11.43 $\delta/x = 0.37 Re_x^{-1/5}$
  $c_f = 0.0577 Re_x^{-1/5}$
  $D = 0.072\rho U^2 b L Re_L^{-1/5}$

11.47 (a) $c_f = 3\nu/(U\delta)$
  $\delta^* = \frac{3}{8}\delta$, $\theta = \frac{39}{280}\delta$
  $H = 2.69$
  (b) $\delta = 4.64 x Re_x^{-1/2}$

11.51 $c_f = 2C_Q$
  $\delta^* = \nu/(C_Q U)$, $\theta = \nu/(2C_Q U)$
  $H = 2$

11.55 $\delta = 10L\sqrt{\nu/(15U_o x)}$
  $c_f = 3.87 Re_x^{-1/2}$
  6% smaller than Falkner-Skan

11.65 (a) $w_f = \sqrt{8mg/(\pi\rho D^2 C_D)}$
  (b) $w = w_f \tanh(gt/w_f)$
  (d) 0.02 sec

11.15 Air: $U = 16.20$ ft/sec
  H$_2$O: $U = 1.08$ ft/sec

11.19 $\delta = 1.05$ in

11.23 $x = 1$ ft: $\delta = 0.21$ in, $c_f = 2.37 \cdot 10^{-3}$
  $x = 5$ ft: $\delta = 0.48$ in, $c_f = 1.06 \cdot 10^{-3}$

11.29 (a) $\partial u/\partial x \sim \frac{1}{6}(u_o/L)(x/L)^{-5/6}$
  (b) $\partial u/\partial y \sim \frac{1}{6}u_o^{3/2}/(\nu L)^{1/2}(x/L)^{-1/4}$
  (c) $(\partial u/\partial y)/(\partial u/\partial x)|_{x=L} \sim \sqrt{Re_L}$

11.35 (a) 11%
  (b) 4.2%
  (c) 4.7%
  (d) 0.9%

11.45 (a) $c_f = 2\nu/(U\delta)$
  $\delta^* = \frac{1}{2}\delta$, $\theta = \frac{1}{6}\delta$
  $H = 3.00$
  (b) $\delta = 3.46 x Re_x^{-1/2}$

11.49 (a) $c_f = 4\nu/(U\delta)$
  $\delta^* = \frac{3}{10}\delta$, $\theta = \frac{37}{315}\delta$
  $H = 2.55$
  (b) $\delta = 5.84 x Re_x^{-1/2}$

11.53 $\theta_E = \delta/4$, $\delta = 4\sqrt{\nu x/u_e}$
  20% smaller than Blasius

11.59 (c) $\theta = 0.671\sqrt{\nu x/U}$
  1.1% larger than Blasius

11.61 200 km

Bibliography

Abbott, I. H. and von Doenhoff, A. E. (1959), *Theory of Wing Sections, Including a Summary of Airfoil Data*, Dover Publications, Inc., New York, NY.

Adair, R. K. (1990), *The Physics of Baseball*, Harper Perennial, Harper & Row Publishers, Inc., New York, NY.

Adamson, A. W. (1960), *Physical Chemistry of Surfaces*, Interscience, New York, NY.

Anderson, J. D. (1989), *Introduction to Flight*, Third Ed., McGraw-Hill, New York, NY.

Anderson, J. D. (1990), *Modern Compressible Flow With Historical Perspective*, Second Ed., McGraw-Hill, New York, NY.

Anderson, J. D. (1995), *Computational Fluid Dynamics: The Basics with Applications*, McGraw-Hill, New York, NY.

ASHRAE (1981), *ASHRAE Handbook—Fundamentals*, American Society of Heating, Refrigerating, and Air Conditioning Engineers, Inc., Atlanta, GA.

Bathe, W. W. (1984), *Fundamentals of Gas Turbines*, John Wiley & Sons, Inc., New York, NY.

Betz, A. (1966), *Introduction to the Theory of Flow Machines*, Pergamon Press, Oxford, England.

Blevins, R. D. (1984), *Applied Fluid Dynamics Handbook*, Van Nostrand Reinhold, New York, NY.

Bradshaw, P. (1972), "The Understanding and Prediction of Turbulent Flow," *The Aeronautical Journal*, Vol. 76, No. 739, pp. 403-418.

Buckingham, E. (1915), "Model Experiments and the Forms of Empirical Equations," *Transactions of the ASME*, Vol. 37, p. 263.

Carrier, G. F., Krook, M. and Pearson, C. E. (1966), *Functions of a Complex Variable*, McGraw-Hill, Inc., New York, NY.

Cengel, T. A. and Boles, M. A. (1994), *Thermodynamics*, McGraw-Hill, Inc., New York, NY.

Chow, V.-T. (1959), *Open Channel Hydraulics*, McGraw-Hill, Inc., New York, NY.

Churchill, R. V. and Brown, J. W. (1990), *Complex Variables and Applications*, Fifth Ed., McGraw-Hill, Inc., New York, NY.

Cleveland Hydraulic Institute (1979), *Engineering Data Book*, Cleveland Hydraulic Institute, Cleveland, OH.

Colebrook, C. F. (1939), "Turbulent Flow in Pipes, with Particular Reference to the Transition between the Smooth and Rough Pipe Laws," *Journal of the Institution of Civil Engineers, London*, Vol. 11, pp. 133-156.

Coles, D. E. and Hirst, E. A. (1969), *Computation of Turbulent Boundary Layers-1968 AFOSR-IFP-Stanford Conference*, Vol. II, Stanford University, CA.

Crane Company (1982), "Flow of Fluids through Valves, Fittings and Pipes," New York: Crane Company, Technical Paper No. 410.

555

Csanady, G. T. (1964), *Theory of Turbomachines*, McGraw-Hill, New York, NY.

Dixon, S. L. (1978), *Fluid Mechanics of Turbomachinery*, Third Ed., Pergamon Press, Oxford, England.

Emanuel, G. (1994), *Analytical Fluid Dynamics*, CRC Press, Boca Raton, FL.

Eringen, A. C. (1980), *Mechanics of Continua*, Robert E. Krieger Publishing Co., Huntington, NY.

Falkner, V. M. and Skan, S. W. (1931), "Some Approximate Solutions of the Boundary Layer Equations," *Phil. Mag.*, Vol. 12, p. 865.

Forsythe, W. E. (1964), *Smithsonian Physical Tables*, 9th Rev. Ed., Smithsonian Institution, Washington, DC.

Glauert, H. (1948), *The Elements of Airfoil and Airscrew Theory*, Second Ed., Cambridge University Press, Cambridge, England.

Granger, R. A. (1988), *Experiments in Fluid Mechanics*, Holt, Rinehart and Winston, Inc., New York, NY.

Hartree, D. R. (1937), "On an Equation Occurring in Falkner and Skan's Approximate Treatment of the Equations of the Boundary Layer," *Proc. Cambr. Phil. Soc.*, Vol. 33, Part II, p. 223.

Hess, J. L. and Smith, A. M. O. (1966), "Calculation of Potential Flow about Arbitrary Bodies," *Progress in Aeronautical Sciences*, Vol. 8, pp. 1-138, D. Kuchemann (ed.), Pergamon Press, Oxford, England.

Hess, J. L. (1975), "Review of Integral-Equation Techniques for Solving Potential Flow Problems, with Emphasis on the Surface-Source Method," *Comput. Methods Appl. Mech. Eng.*, Vol. 5, pp. 145-196.

Hildebrand, F. B. (1976), *Advanced Calculus for Applications*, Prentice-Hall, Inc., Englewood Cliffs, NJ.

Hoerner, S. F. (1965), *Fluid-Dynamic Drag*, Hoerner Fluid Dynamics, Bricktown, NJ.

Hoerner, S. F. and Borst, H. V. (1975), *Fluid-Dynamic Lift*, Hoerner Fluid Dynamics, Bricktown, NJ.

Hopkins, E. J. and Inouye, M. (1971), "An Evaluation of Theories for Predicting Turbulent Skin Friction and Heat Transfer on Flat Plates at Supersonic and Hypersonic Mach Numbers," *AIAA Journal*, Vol. 9, No. 6, pp. 993-1003.

Jeans, J. (1962), *An Introduction to the Kinetic Theory of Gases*, Cambridge University Press, Cambridge, England.

Jepson, R. W. (1976), *Analysis of Flow in Pipe Networks*, Ann Arbor Science Publishers, Ann Arbor, MI.

Kellogg, O. D. (1953), *Foundations of Potential Theory*, Dover Publications, Inc., New York, NY.

Kevorkian, J. and Cole, J. D. (1981), *Perturbation Methods in Applied Mathematics*, Springer-Verlag, New York, NY.

Kline, S. J. (1999), *The Low-Down on Entropy and Interpretive Thermodynamics*, DCW Industries, La Cañada, CA.

Knupp, P. and Steinberg, S. (1993), *The Fundamentals of Grid Generation*, CRC Press, Boca Raton, FL.

Kolmogorov, A. N. (1941), "Local Structures of Turbulence in Incompressible Viscous Fluid for Very Large Reynolds Number," *Doklady AN. SSSR*, Vol. 30, pp. 299-303 [translated in *Proc. Roy. Soc.*, Vol. A434, pp. 9-13 (1991)].

Lakshminarayana, B. (1996), *Fluid Dynamics and Heat Transfer of Turbomachinery*, John Wiley & Sons, Inc., New York, NY.

Landahl, M. T. and Mollo-Christensen, E. (1992), *Turbulence and Random Processes in Fluid Mechanics*, Second Ed., Cambridge University Press, New York, NY.

Landau, L. D. and Lifshitz, E. M. (1966), *Fluid Mechanics*, Addison-Wesley, Reading, MA.

Lee, J. F. and Sears, F. W. (1963), *Thermodynamics*, Second Ed., Addison-Wesley, Reading, MA.

Liepmann, H. W. and Roshko, A. (1963), *Elements of Gasdynamics*, John Wiley & Sons, Inc., New York, NY.

Lighthill, M. J. and Whitham, G. B. (1955), "On Kinematic Waves: II. A Theory of Traffic Flow on Long Crowded Roads," *Proc. Roy. Soc. London*, Vol. A229, pp. 317{345.

Logan, E. S. (1981), *Turbomachinery: Basic Theory and Applications*, Dekker, New York, NY.

Maxworthy, T. (1968), "A Storm in a Teacup," *Journal of Applied Mechanics*, Vol. 35, No. 4.

Milne-Thompson, L. M. (1960), *Theoretical Hydrodynamics*, Fourth Ed., Macmillan Publishing Company, New York, NY.

Moody, L. F. (1944), "Friction Factors for Pipe Flow," *Transactions of the ASME*, Vol. 66, pp. 671-684.

Morrison, R. B. (1962), *Design Data for Aeronautics and Astronautics*, John Wiley and Sons, Inc., New York, NY.

Nayfeh, A. H. (1981), *Introduction to Perturbation Techniques*, John Wiley and Sons, Inc., New York, NY.

Panton, R. L. (1996), *Incompressible Flow*, Second Ed., John Wiley & Sons, Inc., New York, NY.

Patel, V. C. (1968), "The Effects of Curvature on the Turbulent Boundary Layer," Reports and Memoranda No. 3599, Engineering Dept., Cambridge University, Cambridge, England.

Reynolds, O. (1883), "An Experimental Investigation of the Circumstances which Determine whether the Motion of Water Shall Be Direct or Sinuous and of the Law of Resistance in Parallel Channels," *Philosophical Transactions of the Royal Society of London*, Vol. 174, pp. 935-982.

Reynolds, O. (1895), "On the Dynamical Theory of Incompressible Viscous Fluids and the Determination of the Criterion," *Philosophical Transactions of the Royal Society of London, Series A*, Vol. 186, pp. 123-164 [reprinted in *Proc. Roy. Soc.*, London, Vol. A451, pp. 5-47 (1995)].

Reynolds, W. C. and Perkins, H. C. (1977), *Engineering Thermodynamics*, McGraw-Hill, New York, NY.

Robertson, J. M. (1965), *Hydrodynamics in Theory and Application*, Prentice-Hall, Inc., Englewood Cliffs, NJ.

Rouse, H. (1946), *Elementary Mechanics of Fluids*, John Wiley & Sons, Inc., New York, NY.

Sabersky, R. H., Acosta, A. J. and Hauptmann, E. G. (1989), *Fluid Flow*, Macmillan Publishing Company, New York, NY.

Saffman, P. G. (1993), *Vortex Dynamics*, Cambridge University Press, Cambridge, England.

Schlichting, H. and Gersten, K. (2000), *Boundary Layer Theory*, Eighth Ed., Springer-Verlag, Berlin, Germany.

Shames, I. H. (1992), *Mechanics of Fluids*, Third Ed., McGraw-Hill, New York, NY.

Shapiro, A. H. (1953), *The Dynamics and Thermodynamics of Compressible Fluid Flow*, Volume 1, Ronald Press, New York, NY.

Shepherd, D. G. (1956), *Principles of Turbomachinery*, Macmillan Publishing Company, New York, NY.

Stewartson, K. (1954), "Further Solutions of the Falkner-Skan Equation," *Proc. Camb. Phil. Soc.*, Vol. 50, pp. 454-465.

Streeter, V. L. (1961), *Handbook of Fluid Dynamics*, McGraw Hill, New York, NY.

Taylor, E. S. (1974), *Dimensional Analysis for Engineers*, Clarendon Press, Oxford, England.

Tennekes, H. and Lumley, J. L. (1983), *A First Course in Turbulence*, MIT Press, Cambridge, MA.

Thompson, J. F., Soni, B. K. and Weatherill, N. P. (1998), *Handbook of Grid Generation*, CRC Press, Boca Raton, FL.

U. S. Government Printing Office (1974), *The U. S. Standard Atmosphere, 1974*, U. S. Government Printing Office, Washington, DC.

Valentine, H. R. (1967), *Applied Hydrodynamics*, Second Ed., Plenum, New York, NY.

Van Dyke, M. D. (1975), *Perturbation Methods in Fluid Mechanics*, Parabolic Press, Stanford, CA.

Van Dyke, M. D. (1982), *An Album of Fluid Motion*, Parabolic Press, Stanford, CA.

Van Wylen, G. J. and Sonntag, R. E. (1986), *Fundamentals of Classical Thermodynamics*, Third Ed., John Wiley & Sons, Inc., New York, NY.

Vincenti, W.G. (1990), *What Engineers Know and How They Know It*, The Johns Hopkins University Press, Baltimore, MD.

Wark, K. (1966), *Thermodynamics*, McGraw-Hill, New York, NY.

White, F. M. (1994), *Fluid Mechanics*, Third Ed., McGraw-Hill, New York, NY.

Whittaker, E. T. and Watson, G. N. (1963), *A Course of Modern Analysis*, Fourth Ed., Cambridge University Press, Cambridge, England.

Wilcox, D. C. (1995), *Perturbation Methods in the Computer Age*, DCW Industries, La Cañada, CA.

Wilcox, D. C. (1998), *Turbulence Modeling for CFD*, Second Edition, DCW Industries, La Cañada, CA.

Wilcox, D. C. (2000), *Basic Fluid Mechanics*, Second Edition, DCW Industries, La Cañada, CA.

Index

A

Absolute angular momentum, 304-305, 309
Absolute momentum, 173,200
Absolute pressure (*defined*), 18
Absolute temperature (*defined*), 6
Absolute velocity, 201, 308-309, 311-313, 323, 543
Absolute vorticity, 544
Acceleration:
    advective (*defined*), 137
    centrifugal (*defined*), 543
    convective (*defined*), 137
    Coriolis (*defined*), 543
    gravitational (*defined*), 10
    in Cartesian coordinates, 540
    in cylindrical coordinates, 541
    in rotating coordinates, 543-544
    in spherical coordinates, 542-543
    in streamline coordinates, 544-545
    instantaneous (*see Acceleration: unsteady*)
    unsteady (*defined*), 137
Acoustic wave, 336, 338-339, 346
Added mass (*see Virtual mass*)
Adiabatic flow, 340, 343, 355, 360, 388-389, 398, 401
    reversible (*see Isentropic flow*)
    with friction, 355-363
Adiabatic process (*defined*), 235
Adverse pressure gradient, 491, 493-495
Airfoil:
    drag, 494, 500-503
    flow past, 470-472, 494
    lift, 439, 494, 500-503
    pressure surface, 494, 501
    stall, 314, 494-495, 503
    suction surface, 494, 501
    thickness, 494
Alternate depths (*defined*), 280
Andrade's equation, 28, 40
Anemometer:
    hot-wire, 498
    laser-Doppler, 498
Angle of attack, 494, 502
Angular-momentum conservation, 303-305
    differential form, 304
    integral form, 304
Archimedes' Principle, 81, 112
Area:
    centroid of, 101
    moment of inertia of, 101
Area ratio for isentropic flow, 350-351
    tables of, 525-531

Aspect ratio, 296
Atmosphere:
    adiabatic, 115
    exponential, 88
    lapse rate, 87
    U. S. Standard, 86-88
Atmospheric pressure (*defined*), 18
Attached flow (*defined*), 480
Average velocity:
    in control-volume analysis, 181, 198-199
    in pipe flow, 245-248
    in turbulent flow, 496
Axial-flow turbomachines, 302, 317, 320-322
Axisymmetric flow:
    equations of motion for, 541

B

Barotropic fluid, 410, 506
Beltrami flow, 506
Bends:
    diving hazard, 86, 114
    in pipes, 254, 260
Bernoulli's equation:
    compared to energy equation, 401-403
    conditions for, 398-400, 403
    derivation from Euler's equation, 398-400
    for turbomachines, 310
    for unsteady potential flow, 420-421
    introduced, 91
BG units, 8
Blasius:
    boundary-layer solution, 486-489
    equation, 487
    formula for pipe friction factor, 293
    formula for skin friction, 488
Blast wave, 69
Blower, 301, 321
Bluff body, 495
Body force (*defined*), 173
    conservative (*defined*), 399
Boiling, 18-19
Boiling point (*defined*), 19
Bottom slope (open channel), 267, 278
Bound vortex, 472
Boundary conditions:
    free-surface, 93, 266, 268
    inviscid-flow, 469-470
    no-slip, 29, 469-470

Boundary layer:
    displacement thickness, 483-484, 487, 495, 499
    energy thickness, 515
    equations, 475-481
    laminar, 486-495
    momentum-integral equation:
        constant pressure, 483
        general incompressible, 483
        suction, 515
    momentum thickness, 483, 485, 487, 499
    on a flat plate, 486-489, 499
    on a porous surface, 510
    pressure within, 478
    separation, 257-260, 314, 434-435, 439, 471,
        480, 484, 491-495, 516
    shape factor, 483, 487-488, 499
    thickness, 487, 499
    transitional, 496
    turbulent, 495, 498-499
    with pressure gradient, 490-492
Brake horsepower, 316
British Gravitational (BG) units, 8
British thermal unit (Btu), 8
Brook, 268
Buckingham Π theorem, 49-59
Bulk modulus, 22
Buoyancy, 112-113

C

Calorically-perfect gas (*defined*), 21
Capillary action, 25
Capillary wave, 54
Cavitation, 18-19, 37
Celerity, 274
Center of pressure (*defined*), 99
Centimeter-Gram-Second (CGS) units, 7
Centrifugal pump, 306-321
    performance curves, 316-320
    velocity triangles, 310-314
Centroid of common areas, 101
CGS units, 7
Channel-roughness table, 272
Chézy coefficient, 271
Chézy equation, 271
Chézy-Manning equation, 68, 271
Choked flow, 352, 363, 367-368, 376
Circulation (*defined*), 142
    aerodynamic (*defined*), 142
Clausius inequality, 238
Colebrook formula, 252
Combined first and second laws (*see Gibbs' equation*)
Complex velocity potential, 455
Compressibility (*defined*), 22
Compressible flow:
    classification, 334-335
    tables, 525-536
    with friction, 355-363
    with heat transfer, 363-369
    with varying area, 340-344, 350-354

Compressor, 302, 309, 316, 320-322
Conformal mapping, 430
Conservation form of a differential equation, 386
Conservation laws/differential form:
    angular momentum, 304
    energy, 244-248, 388-398
    mass, 386-387
    momentum (linear), 387-388
Conservation laws/integral form:
    angular momentum, 304
    energy, 240-244
    mass, 171-172
    momentum (linear), 172-173
Conservation of difficulty, 139
Conservative force (*defined*), 91, 399
Contact angle, 25
Continuity equation (*defined*), 386
Continuum approximation, 13-15
Control volume (*defined*), 151
    accelerating, 200-205
    deforming, 186-188
    guidelines for selection, 174-176
    moving, 182-185
Convective acceleration (*defined*), 137
Conversion factors, 10
Coriolis acceleration (*defined*), 543
Couette flow, 29-30, 198-199
Critical depth (*defined*), 279
Critical (open-channel) flow, 276, 280
Critical point, 236
Critical slope, 298
Curl operator, 537, 539-542
Curveball, 440
Curved-surface hydrostatic force, 106-109
Cylinder:
    drag coefficient for, 503
    inviscid flow past, 33, 94, 430-444
    large Reynolds number flow past, 148, 435
    pressure variation on, 434
    rotating, 435-442
Cylindrical coordinates:
    continuity equation, 541
    Navier-Stokes equation, 541
    potential-flow relations, 417

D

d'Alembert's paradox, 33-34, 465, 469, 474
Darcy friction factor (*defined*), 250
Darcy-Weisbach formula, 250, 269
Dean number, 67
Density (*defined*), 14
    of air, 15, 520
    of common liquids, 520
    of water, 15, 520
Dependent dimensions, 5
Derivative:
    Eulerian, 137
    following a fluid particle, 137
    material, 139
    substantial, 139

Differential manometer, 117
Diffuser, 259
Diffusion, 238, 471, 473-495, 497
Dimensional homogeneity (*defined*), 48
Dimensionless groupings, 49-61, 317-320
    list of, 60-61
Dimensions and units, 5-10, 46-49
Dipole, 455
Directional derivative, 538
Displacement thickness, 483-484, 487, 495, 499
Dissipation, 238, 338
Divergence operator, 537, 539-542
Dividing streamline (*defined*), 258
Doublet (*defined*), 426
Drag:
    aerodynamic, 4, 202-203
    friction (*defined*), 502
    pressure (*defined*), 501
Drag coefficient:
    of common objects, 504
    of flat plates, 489
    of rotating cylinder, 439
    of spheres and cylinders, 503
Dynamic head (*defined*), 91, 399
Dynamic pressure (*defined*), 91, 399
Dynamic similitude, 34, 61-65
Dyne, 7

E

Eckert number, 66
EE units, 8
Efficiency (*defined*), 316
Ekman number, 66
Energy conservation:
    differential form, 244-248, 388-398
    integral form, 240-244
English Engineering (EE) units, 8
Enthalpy (*defined*), 21
Entrance length, 249
Entrance loss, 256, 263
Entropy (*defined*), 238
    change across a shock, 349
    for a perfect gas, 239
Equation of state:
    general, 236
    perfect gas (*defined*), 20
Equipotential lines (*defined*), 419
Equivalent length (*defined*), 255
Euler turbomachine equations, 306-310
Euler's equation, 388
Eulerian description, 133-139
Evaporation, 18-19
Expansion loss, 258
Extensive property (*defined*), 149

F

Falkner-Skan solution, 490-492
Fan, 301, 321

Fanno flow, 355-363
    choking length, 363
    equations, 359
    Mollier diagram, 362
    normal shock in, 363
    tables, 531-533
Fastball, 440
Favorable pressure gradient, 491, 493
Finite-difference grid (*defined*), 170
First law of thermodynamics, 237
Flat-plate flow:
    incompressible, 486-489, 498-499
    laminar, 486-490
    turbulent, 498-499
Flettner-Rotor ship, 441, 462
Fluid (*defined*), 12
    frictionless (*defined*), 32
    ideal (*defined*), 32
    Newtonian (*defined*), 27
    Non-Newtonian, 27, 41
    perfect (*defined*), 32
    real (*defined*), 32
    viscous (*defined*), 32
Fluid particle (*defined*), 14
Fluid statics (*defined*), 81
Flume, 296, 298
Force potential (*defined*), 91, 399
Forces, aerodynamic, 4, 465
Francis turbine, 322, 325
Free surface (*defined*), 93
Friction drag, 502
Friction factor (*defined*), 250
    compressibility effect on, 359
    laminar, 250
    roughness effect on, 252
    turbulent, 250
Frictionless fluid (*defined*), 32
Froude number (*defined*), 60
Fully-developed flow:
    pipes and channels (*defined*), 248

G

Gage pressure (*defined*), 18
Galilean invariance, 202, 383, 389, 395-397
Galilean transformation, 182-184, 186, 201-202, 274,
    336-337, 346, 383, 395-396
Gas-property tables, 519, 521-523
Gibbs' equation, 238
Gradient operator, 537, 539-542
Gradually-varied flow, 273-274, 277-278
Granny B, xxii, 135
Grashof number, 66
Gravitational acceleration (*defined*), 10
Gravity waves, 274
Guide vanes, 306

H

Hagen-Poiseuille flow, 31
Hazen-Williams formula, 68

Head:
   given up to a turbine (*defined*), 246
   supplied by a pump (*defined*), 246
Head loss (*defined*), 246
   in bends, 260
   in contractions, 257
   in expansions, 258
   in hydraulic jumps, 281
   in inlets, 256, 261
   major (*defined*), 254
   minor (*defined*), 254
Head loss coefficient (*defined*), 255
Helmholtz's theorem, 465, 467-470, 472
Home experiments:
   Bernoulli's equation, 92
   Buoyancy, 113
   Magnus force, 442
   Surface tension, 27
   Turbulence, 500
   Vorticity, 486
Hoover Dam, 105
Horsepower (*defined*), 8
Hot-wire anemometer, 498
Hull speed, theoretical, 46
Hurricane, 141, 201, 408, 458, 465, 506
Hydraulic diameter (*defined*), 255
Hydraulic jack, 116
Hydraulic jump, 281-282
Hydraulic power (*defined*), 316
Hydraulic radius (*defined*), 267
   for common geometries, 268
Hydrometer, 112
Hydrostatic force:
   on curved surfaces, 106-109
   on planar surfaces, 97-105
Hydrostatic relation for pressure, 81
Hypersonic flow, 335, 379

I

Ideal fluid (*defined*), 32
Images, method of, 450-453
Impeller, 306
Impulse turbine, 322
Incipient separation, 491
Inclined manometer, 118
Incompressible (*defined*), 22-23
   kinematical interpretation, 390-391
Independent dimensions, 5, 46
Indicial equations (*defined*), 51
Indicial method (*defined*), 50
Inertial frame, 200
Intensive property (*defined*), 149
Internal energy (*defined*), 21
Inverse boundary-layer methods, 495
Inversion layer, 88
Inviscid fluid (*see Fluid: frictionless*)
Irreversible process (*defined*), 235
Irrotational flow (*defined*), 140
Isentropic flow (*defined*), 239
   tables, 525-531
Isentropic process, 398

J

Jet pump, 209, 218
Joule (*defined*), 7

K

Kaplan turbine, 325
Kármán-Schoenherr correlation, 499-500
Kelvin's circulation theorem, 472
Killing worms, 440
Kinematic viscosity (*defined*), 28
   for air, 28, 522
   for common gases, 522-523
   for common liquids, 522
   for water, 28, 522
Kinematics (*defined*), 133
Kinetic-energy correction factor (*defined*), 245
Kutta condition, 472
Kutta-Joukowski lift theorem, 438, 440, 470

L

Lagrangian description, 133-139
Laminar, transitional and turbulent flow, 496-498
Laplace's equation, 416
Laplacian operator, 540-542
Lapse rate, 87
Laser-Doppler anemometer, 498
Laval nozzle, 350-354
   area ratio, 351
   choked flow, 352
   design case, 353
   fictitious area, 350
   overexpanded, 353
   sonic case, 353
   subsonic case, 352-353
   supersonic case, 353
   underexpanded, 353
Lawn sprinkler, 304-305
Leadfoot D, xxii, 135
Lift (*defined*), 470
Lift coefficient:
   of airfoils, 502
   of rotating cylinder, 439
Loss:
   major (*defined*), 254
   minor (*defined*), 254
Lubrication theory, 30

M

Mach angle (*defined*), 340
Mach cone (*defined*), 340
Mach number (*defined*), 60
Mach wave (*defined*), 340
Magnus force, 438, 440-441
Major loss (*defined*), 254
Manning correlation, 271
Manning roughness coefficient, 68, 271
   table of, 272

Manometer:
    differential, 117
    inclined, 118
    U-tube, 89, 95
Mass conservation,
    differential form, 386-387
    integral form, 171-172
Material derivative, 139
Mean free path, 35, 346
Mechanical power (*defined*), 316
Meniscus, 25
Meridional plane, 307
Method of images, 450-453
Micromanometer, 118
Minor loss (*defined*), 254
Molecular weight, 20, 36
Mollier diagram, 362, 368
Moment of inertia, 100
    of common geometries, 101
Momentum, absolute, 173, 200
Momentum conservation,
    differential form, 387-388
    integral form, 172-173
Momentum-integral equation:
    constant pressure, 483
    general incompressible, 483
    suction, 515
Momentum thickness, 483, 485, 487, 499
Moody diagram, 253
Multiple-pipe systems, 262-265
    network, 265
    parallel, 264
    series, 263
Multistage turbomachines, 303

N

Navier-Stokes equation, 473
    in Cartesian coordinates, 540
    in cylindrical coordinates, 541
    in rotating coordinates, 543
    in spherical coordinates, 542
Newtonian fluid (*defined*), 27
Noninertial frame, 201
Non-Newtonian fluid, 27, 41
Normal shock:
    relations, 347
    tables, 525-531
No-slip boundary condition, 29, 469-470
Nozzle (*see Laval nozzle*)
Number density (*defined*), 13
Nusselt number, 66

O

Open-channel flow:
    critical depth (*defined*), 279
    critical flow, 276, 280
    critical slope, 298
    gradually-varied flow, 273-274, 277-278
    hydraulic jump, 281-282
Open-channel flow (*continued*):
    rapidly-varied flow, 273-274
    specific energy (*defined*), 278
    subcritical flow (*defined*), 276
    supercritical flow (*defined*), 276
    surface-waves, 273-278
    uniform flow, 269-270
Overexpanded nozzle, 353

P

Parallel-axis theorem, 100
Particle isenthalpic (*defined*), 401
Pascal (*defined*), 18
Pathline (*defined*), 145
Péclet number, 66
Pelton wheel, 322-325
Perfect fluid, 32
Perfect gas (*defined*), 20
    constant (*defined*), 20, 36, 519
    equation of state, 20, 519
Pipe flow:
    fully developed (*defined*), 248
    laminar, 31-32, 249-250
    noncircular, 254-256
    turbulent, 249-262
    with roughness, 252-254
Pipe systems, 262-265
    network, 265
    parallel, 264
    series, 263
Pitot-static tube, 96-97
Pitot tube, 94-95
    direction finding, 463
    in supersonic flow, 374, 377
Potential flow:
    complex-variable methods, 429, 455
    cylindrical coordinates, 415-418
    fundamental solutions:
        doublet, 426-428
        potential vortex, 423-424
        source/sink, 424-426
        uniform flow, 422-423
    past accelerating cylinder, 442-444
    past cylinder, 430-444
    past Rankine oval, 444-446
    past rotating cylinder, 435-442
    past wedge, 446-448
    singular points, 429
    superposition, 418
Power-law velocity, 498-499
Power-law viscosity, 28, 40
Power-specific speed, 331
Prandtl number (*defined*), 60
Pressure:
    absolute (*defined*), 18, 519
    atmospheric (*defined*), 18, 519
    center of (*defined*), 99
    dynamic (*defined*), 91, 399
    gage (*defined*), 18, 519

Pressure (*continued*):
    gradient (*defined*), 81
      adverse, 491, 493-495
      favorable, 491, 493
    stagnation (*defined*), 345
    static (*defined*), 96
    total (*defined*), 345
    vapor, 18, 89, 523
Pressure coefficient (*defined*), 434
Pressure drag (*defined*), 501
Primitive-variable form of a differential equation, 387
Propeller turbine, 322, 325
Pump:
    axial, 317, 320-321
    centrifugal, 310-321
    efficiency (*defined*), 316
    mixed-flow, 320-322
    performance curves, 316-320
      dimensional, 317
      dimensionless, 319
    specific speed (*defined*), 320

Q

Q, 327, 509, 517
Quantum-mechanical effects, 71, 371
Quasi-linear (*defined*), 139
Quasi-one-dimensional, 266, 341
Quasi-static (*defined*), 235
Quasi-steady (*defined*), 235

R

Rankine oval, 444-446
Rankine-Hugoniot relations, 374
Rapidly-varied flow, 273-274
Rayleigh flow, 363-369
    choking, 367
    equations, 365
    Mollier diagram, 368
    normal shock in, 368
    tables, 534-536
Reaction turbine, 322
Reservoir temperature (*defined*), 344-345
Reversible process (*defined*), 235
Reynolds number (*defined*), 60
Reynolds Transport Theorem, 151-154
    for an infinitesimal element, 384-387
Richardson number, 66
Rigid-body rotation (*defined*), 141
Rossby number, 66
Rotating-cylinder viscometer, 41
Rotational flow (*defined*), 140
Rotor, 303
Roughness:
    for completely-rough surface, 270-271
    Manning coefficient, 68, 271-272
    sand-grain, 252

S

Schlieren photograph, 354
Second law of thermodynamics, 237-238
Secondary dimensions, 5
Secondary motion, 180, 260
Separation, 148, 257-260, 314, 434-435, 439, 471,
    480, 484, 491-495, 516
    bubble (*defined*), 257
    point (*defined*), 435, 493
Separation of variables, 431
Shaft work (*defined*), 242
Shape factor, 483, 487-488, 499
Shock strength (*defined*), 347
Shock wave:
    detached, 374
    normal, 346-350
SI units, 6-7
    prefixes, 7
Similarity solution, 490
Similitude, 34, 61-65
Sink, 424-426
Sinker, 440
Skin-friction coefficient (*defined*), 29
Slope of channel bottom, 267, 278
Sluice gate, 280
Sonic point (*defined*), 344
Sound, speed of (*see Speed of sound*)
Source, 424-426
SpaceShipOne, 2, 4, 115
Specific energy (open channel), 278-280
Specific gravity (*defined*), 16
Specific heat (*defined*), 21
    of common gases, 519
    ratio (*defined*), 21
      of common gases, 519
Specific speed, 320
Specific volume (*defined*), 16
Specific weight (*defined*), 16
Speed of sound, 336-339
    of air, 521
    of common fluids, 521
    of perfect gas, 338
    of water, 521
Sphere, drag coefficient for, 503
Spherical coordinates, 542
    continuity equation, 542
    Navier-Stokes equation, 542
Spiral vortex, 458
Stagnation:
    point (*defined*), 94
    pressure (*defined*), 345
    temperature (*defined*), 345
Stagnation-point flow, 448-450
Stall, 314, 494-495, 503
Stanton number, 61
Static pressure (*defined*), 96
Stator, 303
Steady flow (*defined*), 139
Stokes' First Problem, 162
Stokes flow, 55
Stokes' Second Problem, 160, 162

STP (Standard Temperature and Pressure), 13
Stratified flow, 22, 66
Stratosphere, 87
Streakline (*defined*), 145
Streamfunction (*defined*), 417
    relation to streamlines, 418-419
Streamline (*defined*), 94
    coordinates, 544-545
    differential equation of, 145
Streamtube, 340-344
    equations, 343
Strouhal number (*defined*), 60
Subcritical flow:
    open channel (*defined*), 276
    rotating cylinder, 437
Subsonic flow (*defined*), 335
Sudden contraction, 257
Sudden expansion, 258
Supercritical flow:
    open channel (*defined*), 276
    rotating cylinder, 437
Superposition:
    fluid statics, 110-111
    potential flow, 418
Supersonic flow (*defined*), 335
Surface force (*defined*), 173
Surface tension, 23-27
    of common liquids, 521
Surface wave, 273-278
Surge, 316
Sutherland's law, 28, 522
System (*defined*), 148
Système Internationale (SI) units, 6-7

T

Temperature:
    absolute (*defined*), 17
    reservoir (*defined*), 344
    stagnation (*defined*), 345
    total (*defined*), 344
Theoretical hull speed, 46
Thermal conductivity, 60
Thermally-perfect gas, 21
Thermodynamic equilibrium, 235
Thermodynamics:
    first law of, 237
    second law of, 237-238
Throat of a nozzle (*defined*), 344
Thwaites' method, 516
Total density (*defined*), 345
Total pressure (*defined*), 345
Total temperature (*defined*), 344
Towing tank, 64
Transition to turbulence, 496-498
Transonic flow (*defined*), 335
Troposphere, 86
Turbine:
    efficiency (*defined*), 316
    impulse, 322
    Pelton, 322-325
    hydraulic jump, 281-282

Turbine (*continued*):
    reaction, 322
    power-specific speed (*defined*), 331
    wind, 302
Turbomachinery:
    capacity coefficient, 318
    classification, 301-303
    head coefficient, 318
    power coefficient, 318
    specific speed, 320
Turbulent boundary layer, 495, 498-499
Turbulent flow, 495-500

U

Underexpanded nozzle, 353
Uniform open-channel flow, 269-270
Units (*see Dimensions and units*)
Universal gas constant, 20
Unsteady Bernoulli equation, 420
Unsteady flow (*defined*), 139
USCS units, 6
U. S. Customary System (USCS) of units, 6
U. S. Standard atmosphere, 86-88
U-tube manometer, 89, 95

V

Van der Waal's equation, 37
Vapor, 236
Vapor pressure, 18, 89, 523
    of common liquids, 523
Vector calculus, 156-157
    ∇ operator, 537-538
Velocity potential (*defined*), 416
Velocity triangles, 310-314
Vena contracta, 257
Venturi meter, 409
Virtual mass (*defined*), 443
Viscometer, 41
Viscosity, 27-28
    Andrade's equation, 28, 522
    kinematic (*defined*), 28, 522
    of common fluids, 522
    power-law, 28
    Sutherland's law, 28, 522-523
Volume-flow rate (*defined*), 266
Volute, 306
Vortex:
    bound, 472
    potential, 423-424
    spiral, 458
    stretching, 468, 473
    wingtip, 148
Vortex shedding, 71
Vorticity (*defined*), 140
    in Cartesian coordinates, 540
    in cylindrical coordinates, 541
    in rotating coordinates, 544
    in spherical coordinates, 542
    in streamline coordinates, 545

Vorticity equation:
    inviscid, 468
    viscous, 473

W

Wake, 33, 435, 494, 500
Wave:
    acoustic, 336, 338-339, 346
    capillary, 69
    gravity, 274
    Mach (*defined*), 340
    shock (*see Shock wave*)
Wave speed:
    acoustic, 336-339
    free-surface, 273-277
Weber number (*defined*), 61-66
Weir, 129
Wetted area (*defined*), 255
Wetting angle, 25
Wind tunnel, 62
Wind turbine, 302
Windmill, 301
Wingtip vortices, 148

Made in the USA

Other Publications by DCW Industries

Basic Fluid Mechanics, Second Edition, D. C. Wilcox (2000): This book is appropriate for a two-term, junior or senior level undergraduate series of courses, or as an introductory text for graduate students with minimal prior knowledge of fluid mechanics. The first part of the book provides sufficient material for an introductory course, focusing primarily on the control-volume approach. With a combination of dimensional analysis and the control-volume method, the text discusses pipe flow, open-channel flow, elements of turbomachine theory and one-dimensional compressible flows. The balance of the text can be presented in a subsequent course, focusing on the differential equations of fluid mechanics. Topics covered include a rigorous development of the Navier-Stokes equation, potential-flow, exact Navier-Stokes solutions, boundary layers, simple viscous compressible flows, centered expansions and oblique shocks. The book is accompanied by a CD with a variety of useful programs, fluid mechanics photos and movies.

Study Guide for Basic Fluid Mechanics, Second Edition, C. P. Landry and D. C. Wilcox (1999): A companion for **Basic Fluid Mechanics** that contains 129 examples worked in complete detail to help readers benefit from step-by-step solutions. All problems are explicitly worked start to finish with comments indicating why steps are taken to achieve solutions.

The Low-Down on Entropy and Interpretive Thermodynamics, S. J. Kline (1999): This is the final book written by Professor Stephen Kline of Stanford University. It is a delightful treatise on the subtleties of entropy and the second law of thermodynamics.

Turbulence Modeling for CFD, Second Edition, D. C. Wilcox (1998): A first- or second-year graduate text on modern methods for formulating and analyzing engineering models of turbulence. Presents a comprehensive discussion of algebraic, one-equation, two-equation and stress-transport models. Emphasizes an integrated balance of similarity solutions, perturbation methods and numerical integration schemes to test and formulate rational models. Includes a brief introduction to DNS, LES and Chaos theory. Accompanied by a compact disk with a variety of useful programs, including an industrial-strength boundary-layer program with a menu driven input-data preparation utility.

Perturbation Methods in the Computer Age, D. C. Wilcox (1995): Advanced undergraduate or first-year graduate text on asymptotic and perturbation methods. Discusses asymptotic expansion of integrals, including Laplace's method, stationary phase and steepest descent. Introduces the general principles of singular perturbation theory, including examples for both ODE's and PDE's. Covers multiple-scale analysis, including the method of averaging and the WKB method. Shows, through a collection of practical examples, how useful asymptotics can be when used in conjunction with computational methods.

Visit our World Wide Web Home Page (http://www.dcwindustries.com) for complete details about our books and special sales that we conduct from time to time.